Generalized Additive Models

An Introduction with R

SECOND EDITION

CHAPMAN & HALL/CRC
Texts in Statistical Science Series

Statistical Theory: A Concise Introduction
F. Abramovich and Y. Ritov

Practical Multivariate Analysis, Fifth Edition
A. Afifi, S. May, and V.A. Clark

Practical Statistics for Medical Research
D.G. Altman

Interpreting Data: A First Course in Statistics
A.J.B. Anderson

Introduction to Probability with R
K. Baclawski

Linear Algebra and Matrix Analysis for Statistics
S. Banerjee and A. Roy

Modern Data Science with R
B. S. Baumer, D. T. Kaplan, and N. J. Horton

Mathematical Statistics: Basic Ideas and Selected Topics, Volume I, Second Edition
P. J. Bickel and K. A. Doksum

Mathematical Statistics: Basic Ideas and Selected Topics, Volume II
P. J. Bickel and K. A. Doksum

Analysis of Categorical Data with R
C. R. Bilder and T. M. Loughin

Statistical Methods for SPC and TQM
D. Bissell

Introduction to Probability
J. K. Blitzstein and J. Hwang

Bayesian Methods for Data Analysis, Third Edition
B.P. Carlin and T.A. Louis

Second Edition
R. Caulcutt

The Analysis of Time Series: An Introduction, Sixth Edition
C. Chatfield

Introduction to Multivariate Analysis
C. Chatfield and A.J. Collins

Problem Solving: A Statistician's Guide, Second Edition
C. Chatfield

Statistics for Technology: A Course in Applied Statistics, Third Edition
C. Chatfield

Analysis of Variance, Design, and Regression : Linear Modeling for Unbalanced Data, Second Edition
R. Christensen

Bayesian Ideas and Data Analysis: An Introduction for Scientists and Statisticians
R. Christensen, W. Johnson, A. Branscum, and T.E. Hanson

Modelling Binary Data, Second Edition
D. Collett

Modelling Survival Data in Medical Research, Third Edition
D. Collett

Introduction to Statistical Methods for Clinical Trials
T.D. Cook and D.L. DeMets

Applied Statistics: Principles and Examples
D.R. Cox and E.J. Snell

Multivariate Survival Analysis and Competing Risks
M. Crowder

Statistical Analysis of Reliability Data
M.J. Crowder, A.C. Kimber, T.J. Sweeting, and R.L. Smith

An Introduction to Generalized Linear Models, Third Edition
A.J. Dobson and A.G. Barnett

Nonlinear Time Series: Theory, Methods, and Applications with R Examples
R. Douc, E. Moulines, and D.S. Stoffer

Introduction to Optimization Methods and Their Applications in Statistics
B.S. Everitt

Extending the Linear Model with R: Generalized Linear, Mixed Effects and Nonparametric Regression Models, Second Edition
J.J. Faraway

Linear Models with R, Second Edition
J.J. Faraway

A Course in Large Sample Theory
T.S. Ferguson

Multivariate Statistics: A Practical Approach
B. Flury and H. Riedwyl

Readings in Decision Analysis
S. French

Discrete Data Analysis with R: Visualization and Modeling Techniques for Categorical and Count Data
M. Friendly and D. Meyer

Markov Chain Monte Carlo: Stochastic Simulation for Bayesian Inference, Second Edition
D. Gamerman and H.F. Lopes

Bayesian Data Analysis, Third Edition
A. Gelman, J.B. Carlin, H.S. Stern, D.B. Dunson, A. Vehtari, and D.B. Rubin

Multivariate Analysis of Variance and Repeated Measures: A Practical Approach for Behavioural Scientists
D.J. Hand and C.C. Taylor

Practical Longitudinal Data Analysis
D.J. Hand and M. Crowder

Logistic Regression Models
J.M. Hilbe

Richly Parameterized Linear Models: Additive, Time Series, and Spatial Models Using Random Effects
J.S. Hodges

Statistics for Epidemiology
N.P. Jewell

Stochastic Processes: An Introduction, Second Edition
P.W. Jones and P. Smith

The Theory of Linear Models
B. Jørgensen

Pragmatics of Uncertainty
J.B. Kadane

Principles of Uncertainty
J.B. Kadane

Graphics for Statistics and Data Analysis with R
K.J. Keen

Mathematical Statistics
K. Knight

Introduction to Functional Data Analysis
P. Kokoszka and M. Reimherr

Introduction to Multivariate Analysis: Linear and Nonlinear Modeling
S. Konishi

Nonparametric Methods in Statistics with SAS Applications
O. Korosteleva

Modeling and Analysis of Stochastic Systems, Second Edition
V.G. Kulkarni

Exercises and Solutions in Biostatistical Theory
L.L. Kupper, B.H. Neelon, and S.M. O'Brien

Exercises and Solutions in Statistical Theory
L.L. Kupper, B.H. Neelon, and S.M. O'Brien

Design and Analysis of Experiments with R
J. Lawson

Design and Analysis of Experiments with SAS
J. Lawson

A Course in Categorical Data Analysis
T. Leonard

Statistics for Accountants
S. Letchford

Introduction to the Theory of Statistical Inference
H. Liero and S. Zwanzig

Statistical Theory, Fourth Edition
B.W. Lindgren

Stationary Stochastic Processes: Theory and Applications
G. Lindgren

Statistics for Finance
E. Lindström, H. Madsen, and J. N. Nielsen

The BUGS Book: A Practical Introduction to Bayesian Analysis
D. Lunn, C. Jackson, N. Best, A. Thomas, and D. Spiegelhalter

Introduction to General and Generalized Linear Models
H. Madsen and P. Thyregod

Time Series Analysis
H. Madsen

Pólya Urn Models
H. Mahmoud

Randomization, Bootstrap and Monte Carlo Methods in Biology, Third Edition
B.F.J. Manly

Statistical Regression and Classification: From Linear Models to Machine Learning
N. Matloff

Introduction to Randomized Controlled Clinical Trials, Second Edition
J.N.S. Matthews

Statistical Rethinking: A Bayesian Course with Examples in R and Stan
R. McElreath

Statistical Methods in Agriculture and Experimental Biology, Second Edition
R. Mead, R.N. Curnow, and A.M. Hasted

Statistics in Engineering: A Practical Approach
A.V. Metcalfe

Statistical Inference: An Integrated Approach, Second Edition
H. S. Migon, D. Gamerman, and F. Louzada

Beyond ANOVA: Basics of Applied Statistics
R.G. Miller, Jr.

A Primer on Linear Models
J.F. Monahan

Stochastic Processes: From Applications to Theory
P.D Moral and S. Penev

Applied Stochastic Modelling, Second Edition
B.J.T. Morgan

Elements of Simulation
B.J.T. Morgan

Probability: Methods and Measurement
A. O'Hagan

Introduction to Statistical Limit Theory
A.M. Polansky

Applied Bayesian Forecasting and Time Series Analysis
A. Pole, M. West, and J. Harrison

Statistics in Research and Development, Time Series: Modeling, Computation, and Inference
R. Prado and M. West

Essentials of Probability Theory for Statisticians
M.A. Proschan and P.A. Shaw

Introduction to Statistical Process Control
P. Qiu

Sampling Methodologies with Applications
P.S.R.S. Rao

A First Course in Linear Model Theory
N. Ravishanker and D.K. Dey

Essential Statistics, Fourth Edition
D.A.G. Rees

Stochastic Modeling and Mathematical Statistics: A Text for Statisticians and Quantitative Scientists
F.J. Samaniego

Statistical Methods for Spatial Data Analysis
O. Schabenberger and C.A. Gotway

Bayesian Networks: With Examples in R
M. Scutari and J.-B. Denis

Large Sample Methods in Statistics
P.K. Sen and J. da Motta Singer

Spatio-Temporal Methods in Environmental Epidemiology
G. Shaddick and J.V. Zidek

Decision Analysis: A Bayesian Approach
J.Q. Smith

Analysis of Failure and Survival Data
P. J. Smith

Applied Statistics: Handbook of GENSTAT Analyses
E.J. Snell and H. Simpson

Applied Nonparametric Statistical Methods, Fourth Edition
P. Sprent and N.C. Smeeton

Data Driven Statistical Methods
P. Sprent

Generalized Linear Mixed Models: Modern Concepts, Methods and Applications
W. W. Stroup

Survival Analysis Using S: Analysis of Time-to-Event Data
M. Tableman and J.S. Kim

Applied Categorical and Count Data Analysis
W. Tang, H. He, and X.M. Tu

Elementary Applications of Probability Theory, Second Edition
H.C. Tuckwell

Introduction to Statistical Inference and Its Applications with R
M.W. Trosset

Understanding Advanced Statistical Methods
P.H. Westfall and K.S.S. Henning

Statistical Process Control: Theory and Practice, Third Edition
G.B. Wetherill and D.W. Brown

Generalized Additive Models: An Introduction with R, Second Edition
S. Wood

Epidemiology: Study Design and Data Analysis, Third Edition
M. Woodward

Practical Data Analysis for Designed Experiments
B.S. Yandell

Texts in Statistical Science

Generalized Additive Models

An Introduction with R

SECOND EDITION

Simon N. Wood

University of Bristol, UK

CRC Press is an imprint of the
Taylor & Francis Group, an **informa** business
A CHAPMAN & HALL BOOK

CRC Press
Taylor & Francis Group
6000 Broken Sound Parkway NW, Suite 300
Boca Raton, FL 33487-2742

ISBN 13: 978-1-4987-2833-1 (hbk)

Visit the Taylor & Francis Web site at
http://www.taylorandfrancis.com

and the CRC Press Web site at
http://www.crcpress.com

Contents

Preface **xvii**

1 Linear Models **1**

1.1 A simple linear model 2

Simple least squares estimation 2

1.1.1 Sampling properties of $\hat{\beta}$ 3

1.1.2 So how old is the universe? 4

1.1.3 Adding a distributional assumption 7

Testing hypotheses about β 7

Confidence intervals 8

1.2 Linear models in general 9

1.3 The theory of linear models 11

1.3.1 Least squares estimation of $\boldsymbol{\beta}$ 12

1.3.2 The distribution of $\hat{\boldsymbol{\beta}}$ 13

1.3.3 $(\hat{\beta}_i - \beta_i)/\hat{\sigma}_{\hat{\beta}_i} \sim t_{n-p}$ 13

1.3.4 F-ratio results I 14

1.3.5 F-ratio results II 14

1.3.6 The influence matrix 16

1.3.7 The residuals, $\hat{\boldsymbol{\epsilon}}$, and fitted values, $\hat{\boldsymbol{\mu}}$ 16

1.3.8 Results in terms of $\mathbf{X}$ 17

1.3.9 The Gauss Markov Theorem: What's special about least squares? 18

1.4 The geometry of linear modelling 19

1.4.1 Least squares 19

1.4.2 Fitting by orthogonal decompositions 20

1.4.3 Comparison of nested models 21

1.5 Practical linear modelling 22

1.5.1 Model fitting and model checking 23

1.5.2 Model `summary` 28

1.5.3 Model selection 30

1.5.4 Another model selection example 31

A follow-up 34

1.5.5 Confidence intervals 35

1.5.6 Prediction 36

1.5.7 Co-linearity, confounding and causation 36

1.6 Practical modelling with factors 38
1.6.1 Identifiability 39
1.6.2 Multiple factors 40
1.6.3 'Interactions' of factors 41
1.6.4 Using factor variables in R 43
1.7 General linear model specification in R 46
1.8 Further linear modelling theory 47
1.8.1 Constraints I: General linear constraints 47
1.8.2 Constraints II: 'Contrasts' and factor variables 48
1.8.3 Likelihood 49
1.8.4 Non-independent data with variable variance 50
1.8.5 Simple AR correlation models 52
1.8.6 AIC and Mallows' statistic 52
1.8.7 The wrong model 54
1.8.8 Non-linear least squares 54
1.8.9 Further reading 56
1.9 Exercises 56

2 Linear Mixed Models 61
2.1 Mixed models for balanced data 61
2.1.1 A motivating example 61
The wrong approach: A fixed effects linear model 62
The right approach: A mixed effects model 64
2.1.2 General principles 65
2.1.3 A single random factor 66
2.1.4 A model with two factors 69
2.1.5 Discussion 74
2.2 Maximum likelihood estimation 74
2.2.1 Numerical likelihood maximization 76
2.3 Linear mixed models in general 77
2.4 Linear mixed model maximum likelihood estimation 78
2.4.1 The distribution of $\mathbf{b}|\mathbf{y}, \hat{\boldsymbol{\beta}}$ given $\boldsymbol{\theta}$ 79
2.4.2 The distribution of $\hat{\boldsymbol{\beta}}$ given $\boldsymbol{\theta}$ 80
2.4.3 The distribution of $\hat{\boldsymbol{\theta}}$ 81
2.4.4 Maximizing the profile likelihood 81
2.4.5 REML 83
2.4.6 Effective degrees of freedom 83
2.4.7 The EM algorithm 84
2.4.8 Model selection 85
2.5 Linear mixed models in R 86
2.5.1 Package `nlme` 86
2.5.2 Tree growth: An example using `lme` 87
2.5.3 Several levels of nesting 91
2.5.4 Package `lme4` 93
2.5.5 Package `mgcv` 94

2.6 Exercises 95

3 Generalized Linear Models 101

3.1 GLM theory 102

3.1.1 The exponential family of distributions 103

3.1.2 Fitting generalized linear models 105

3.1.3 Large sample distribution of $\hat{\beta}$ 107

3.1.4 Comparing models 108

Deviance 108

Model comparison with unknown ϕ 109

AIC 110

3.1.5 Estimating ϕ, Pearson's statistic and Fletcher's estimator 110

3.1.6 Canonical link functions 111

3.1.7 Residuals 112

Pearson residuals 112

Deviance residuals 113

3.1.8 Quasi-likelihood 113

3.1.9 Tweedie and negative binomial distributions 115

3.1.10 The Cox proportional hazards model for survival data 116

Cumulative hazard and survival functions 118

3.2 Geometry of GLMs 119

3.2.1 The geometry of IRLS 121

3.2.2 Geometry and IRLS convergence 122

3.3 GLMs with R 124

3.3.1 Binomial models and heart disease 125

3.3.2 A Poisson regression epidemic model 131

3.3.3 Cox proportional hazards modelling of survival data 136

3.3.4 Log-linear models for categorical data 138

3.3.5 Sole eggs in the Bristol channel 142

3.4 Generalized linear mixed models 147

3.4.1 Penalized IRLS 148

3.4.2 The PQL method 149

3.4.3 Distributional results 151

3.5 GLMMs with R 151

3.5.1 `glmmPQL` 151

3.5.2 `gam` 154

3.5.3 `glmer` 155

3.6 Exercises 156

4 Introducing GAMs 161

4.1 Introduction 161

4.2 Univariate smoothing 162

4.2.1 Representing a function with basis expansions 162

A very simple basis: Polynomials 162

The problem with polynomials 162

The piecewise linear basis 164
Using the piecewise linear basis 165
4.2.2 Controlling smoothness by penalizing wiggliness 166
4.2.3 Choosing the smoothing parameter, λ, by cross validation 169
4.2.4 The Bayesian/mixed model alternative 172
4.3 Additive models 174
4.3.1 Penalized piecewise regression representation of an additive model 175
4.3.2 Fitting additive models by penalized least squares 177
4.4 Generalized additive models 180
4.5 Summary 182
4.6 Introducing package `mgcv` 182
4.6.1 Finer control of `gam` 184
4.6.2 Smooths of several variables 187
4.6.3 Parametric model terms 189
4.6.4 The `mgcv` help pages 191
4.7 Exercises 191

5 Smoothers 195
5.1 Smoothing splines 195
5.1.1 Natural cubic splines are smoothest interpolators 196
5.1.2 Cubic smoothing splines 198
5.2 Penalized regression splines 199
5.3 Some one-dimensional smoothers 201
5.3.1 Cubic regression splines 201
5.3.2 A cyclic cubic regression spline 202
5.3.3 P-splines 204
5.3.4 P-splines with derivative based penalties 206
5.3.5 Adaptive smoothing 207
5.3.6 SCOP-splines 208
5.4 Some useful smoother theory 210
5.4.1 Identifiability constraints 211
5.4.2 ‘Natural’ parameterization, effective degrees of freedom and smoothing bias 211
5.4.3 Null space penalties 214
5.5 Isotropic smoothing 214
5.5.1 Thin plate regression splines 215
Thin plate splines 215
Thin plate regression splines 217
Properties of thin plate regression splines 218
Knot-based approximation 219
5.5.2 Duchon splines 221
5.5.3 Splines on the sphere 222
5.5.4 Soap film smoothing over finite domains 223
5.6 Tensor product smooth interactions 227

5.6.1 Tensor product bases 227
5.6.2 Tensor product penalties 229
5.6.3 ANOVA decompositions of smooths 232
Numerical identifiability constraints for nested terms 233
5.6.4 Tensor product smooths under shape constraints 234
5.6.5 An alternative tensor product construction 235
What is being penalized? 236
5.7 Isotropy versus scale invariance 237
5.8 Smooths, random fields and random effects 239
5.8.1 Gaussian Markov random fields 240
5.8.2 Gaussian process regression smoothers 241
5.9 Choosing the basis dimension 242
5.10 Generalized smoothing splines 243
5.11 Exercises 245

6 GAM theory **249**
6.1 Setting up the model 249
6.1.1 Estimating β given $\boldsymbol{\lambda}$ 251
6.1.2 Degrees of freedom and scale parameter estimation 251
6.1.3 Stable least squares with negative weights 252
6.2 Smoothness selection criteria 255
6.2.1 Known scale parameter: UBRE 255
6.2.2 Unknown scale parameter: Cross validation 256
Leave-several-out cross validation 257
Problems with ordinary cross validation 257
6.2.3 Generalized cross validation 258
6.2.4 Double cross validation 260
6.2.5 Prediction error criteria for the generalized case 261
6.2.6 Marginal likelihood and REML 262
6.2.7 The problem with $\log|\mathbf{S}_\lambda|_+$ 264
6.2.8 Prediction error criteria versus marginal likelihood 266
Unpenalized coefficient bias 267
6.2.9 The 'one standard error rule' and smoother models 268
6.3 Computing the smoothing parameter estimates 269
6.4 The generalized Fellner-Schall method 269
6.4.1 General regular likelihoods 271
6.5 Direct Gaussian case and performance iteration (PQL) 272
6.5.1 Newton optimization of the GCV score 273
6.5.2 REML 276
$\log|\mathbf{S}_\lambda|_+$ and its derivatives 277
The remaining derivative components 278
6.5.3 Some Newton method details 280
6.6 Direct nested iteration methods 280
6.6.1 Prediction error criteria 282
6.6.2 Example: Cox proportional hazards model 283

Derivatives with respect to smoothing parameters 285
Prediction and the baseline hazard 286
6.7 Initial smoothing parameter guesses 287
6.8 GAMM methods 288
6.8.1 GAMM inference with mixed model estimation 289
6.9 Bigger data methods 290
6.9.1 Bigger still 291
6.10 Posterior distribution and confidence intervals 293
6.10.1 Nychka's coverage probability argument 294
Interval limitations and simulations 296
6.10.2 Whole function intervals 300
6.10.3 Posterior simulation in general 300
6.11 AIC and smoothing parameter uncertainty 301
6.11.1 Smoothing parameter uncertainty 302
6.11.2 A corrected AIC 303
6.12 Hypothesis testing and p-values 304
6.12.1 Approximate p-values for smooth terms 305
Computing T_r 307
Simulation performance 308
6.12.2 Approximate p-values for random effect terms 309
6.12.3 Testing a parametric term against a smooth alternative 312
6.12.4 Approximate generalized likelihood ratio tests 313
6.13 Other model selection approaches 315
6.14 Further GAM theory 316
6.14.1 The geometry of penalized regression 316
6.14.2 Backfitting GAMs 318
6.15 Exercises 320

7 GAMs in Practice: `mgcv` **325**
7.1 Specifying smooths 325
7.1.1 How smooth specification works 327
7.2 Brain imaging example 328
7.2.1 Preliminary modelling 329
7.2.2 Would an additive structure be better? 333
7.2.3 Isotropic or tensor product smooths? 334
7.2.4 Detecting symmetry (with `by` variables) 336
7.2.5 Comparing two surfaces 337
7.2.6 Prediction with `predict.gam` 339
Prediction with `lpmatrix` 341
7.2.7 Variances of non-linear functions of the fitted model 342
7.3 A smooth ANOVA model for diabetic retinopathy 343
7.4 Air pollution in Chicago 346
7.4.1 A single index model for pollution related deaths 349
7.4.2 A distributed lag model for pollution related deaths 351
7.5 Mackerel egg survey example 353

7.5.1 Model development 354
7.5.2 Model predictions 357
7.5.3 Alternative spatial smooths and geographic regression 359
7.6 Spatial smoothing of Portuguese larks data 361
7.7 Generalized additive mixed models with R 365
7.7.1 A space-time GAMM for sole eggs 365
Soap film improvement of boundary behaviour 368
7.7.2 The temperature in Cairo 371
7.7.3 Fully Bayesian stochastic simulation: `jagam` 374
7.7.4 Random wiggly curves 376
7.8 Primary biliary cirrhosis survival analysis 377
7.8.1 Time dependent covariates 380
7.9 Location-scale modelling 383
7.9.1 Extreme rainfall in Switzerland 384
7.10 Fuel efficiency of cars: Multivariate additive models 387
7.11 Functional data analysis 390
7.11.1 Scalar on function regression 390
Prostate cancer screening 391
A multinomial prostate screening model 393
7.11.2 Function on scalar regression: Canadian weather 395
7.12 Other packages 397
7.13 Exercises 398

A Maximum Likelihood Estimation 405
A.1 Invariance 405
A.2 Properties of the expected log-likelihood 406
A.3 Consistency 409
A.4 Large sample distribution of $\hat{\boldsymbol{\theta}}$ 410
A.5 The generalized likelihood ratio test (GLRT) 411
A.6 Derivation of $2\lambda \sim \chi^2_r$ under H_0 411
A.7 AIC in general 414
A.8 Quasi-likelihood results 416

B Some Matrix Algebra 419
B.1 Basic computational efficiency 419
B.2 Covariance matrices 420
B.3 Differentiating a matrix inverse 420
B.4 Kronecker product 421
B.5 Orthogonal matrices and Householder matrices 421
B.6 QR decomposition 422
B.7 Cholesky decomposition 422
B.8 Pivoting 423
B.9 Eigen-decomposition 424
B.10 Singular value decomposition 425
B.11 Lanczos iteration 426

C Solutions to Exercises **429**
C.1 Chapter 1 429
C.2 Chapter 2 432
C.3 Chapter 3 438
C.4 Chapter 4 440
C.5 Chapter 5 441
C.6 Chapter 6 443
C.7 Chapter 7 447

Bibliography **455**

Index **467**

Preface

This book is designed for readers wanting a compact introduction to linear models, generalized linear models, generalized additive models, and the mixed model extensions of these, with particular emphasis on generalized additive models. The aim is to provide a practically oriented theoretical treatment of how the models and methods work, underpinning material on application of the models using R.

Linear models are statistical models in which a univariate response is modelled as the sum of a 'linear predictor' and a zero mean random error term. The linear predictor depends on some predictor variables, measured with the response variable, and some unknown parameters, which must be estimated. A key feature of the models is that the linear predictor depends *linearly* on the parameters. Statistical inference is usually based on the assumption that the response variable is normally distributed. Linear models are used widely in most branches of science, in the analysis of designed experiments, and for other modelling tasks such as polynomial regression. The linearity of the models endows them with some rather elegant theory which is explored in chapter 1, alongside practical examples of their use. Linear mixed models allow the model for the random component of the response data to be as richly structured as the model for the mean, by introducing a linear predictor for this random component which depends on Gaussian random variables known as *random effects*, rather than directly on parameters. These models are the subject of chapter 2.

Generalized linear models (GLMs) somewhat relax the strict linearity assumption of linear models, by allowing the expected value of the response to depend on a smooth monotonic function of the linear predictor. Similarly the assumption that the response is normally distributed is relaxed by allowing it to follow any distribution from the exponential family (for example, normal, Poisson, binomial, gamma, etc.). Inference for GLMs is based on likelihood theory, as is explained in chapter 3, where the practical use of these models is also covered, along with their extension to generalized linear mixed models.

A generalized additive model (GAM) is a GLM in which the linear predictor depends linearly on smooth functions of predictor variables. The exact parametric form of these functions is unknown, as is the degree of smoothness appropriate for each of them. To use GAMs in practice requires some extensions to GLM methods:

1. The smooth functions must be represented somehow.
2. The degree of smoothness of the functions must be made controllable, so that models with varying degrees of smoothness can be explored.

3. Some means for selecting the most appropriate degree of smoothness from data is required, if the models are to be useful for more than purely exploratory work.

This book provides an introduction to the framework for Generalized Additive Modelling in which (1) is addressed using basis expansions of the smooth functions, (2) is addressed by estimating models by penalized likelihood maximization, in which wiggly models are penalized more heavily than smooth models in a controllable manner, and (3) is performed using methods based on marginal likelihood maximization, cross validation or sometimes AIC or Mallows' statistic. Chapter 4 introduces this framework, chapter 5 provides details of the wide range of smooth model components that fall within it, while chapter 6 covers details of the theory. Chapter 7 illustrates the practical use of GAMs using the R package `mgcv`. Chapters 5–7 exploit the duality between smooths and random effects, especially from a Bayesian perspective, and include generalized additive mixed models (GAMMs).

I assume that most people are interested in statistical models in order to use them, rather than to gaze upon the mathematical beauty of their structure, and for this reason I have tried to keep this book practically focused. However, I think that practical work tends to progress much more smoothly if it is based on solid understanding of how the models and methods used actually work. For this reason, the book includes fairly full explanations of the theory underlying the methods, including the underlying geometry, where this is helpful. Given that the applied modelling involves using computer programs, the book includes a good deal of material on statistical modelling in R. This approach is standard when writing about practical statistical analysis, but in addition chapter 4 attempts to introduce GAMs by having the reader 'build their own' GAM using R: I hope that this provides a useful way of quickly gaining a rather solid familiarity with the fundamentals of the GAM framework presented in this book. Once the basic framework is mastered from chapter 4, the theory in chapters 5 and 6 is really just filling in details, albeit practically important ones.

The book includes a moderately high proportion of practical examples which reflect the reality that statistical modelling problems are usually quite involved, and rarely require only straightforward brain-free application of some standard model. This means that some of the examples are fairly lengthy, but do provide illustration of the process of producing practical models of some scientific utility, and of checking those models. They also provide much greater scope for the reader to decide that what I've done is utter rubbish. Working through this book from linear models to GAMMs, it is striking that as model flexibility increases, so that models become better able to describe the reality that we believe generated a set of data, so the methods for inference become more approximate. This book is offered in the belief that it is usually better to be able to say something approximate about the right model, rather than something more precise about the wrong model.

Life is too short to spend too much of it reading statistics texts. This book is, of course, an exception to this rule and should be read from cover to cover. However, if you don't feel inclined to follow this suggestion, here are some alternatives.

- For those who are already very familiar with linear models and GLMs, but want to use GAMs with a reasonable degree of understanding: work through chapter

4 and read chapters 5 and 7, trying some exercises from both. Use chapter 6 for reference. Perhaps skim the other chapters.

- For those who want to focus only on practical modelling in R, rather than theory, work through the following: 1.5, 1.6.4, 1.7, 2.5, 3.3, 3.5 and chapter 7.
- For those familiar with the basic idea of setting up a GAM using basis expansions and penalties, but wanting to know more about the underlying theory and practical application: work through chapters 5 and 6, and 7.
- For those not interested in GAMs, but wanting to know about linear models, GLMs and mixed models, work through chapters 1 to 3.

The book is written to be accessible to numerate researchers and students from the last two years of an undergraduate mathematics/statistics programme upwards. It is designed to be used either for self-study, or as the text for the 'regression modelling' strand of mathematics and/or statistics degree programme. Some familiarity with statistical concepts is assumed, particularly with notions such as random variable, expectation, variance and distribution. Some knowledge of matrix algebra is also needed, although Appendix B is intended to provide all that is needed beyond the fundamentals. Note that R is available at `cran.r-project.org` as is this book's data package, `gamair`.

What is new in the second edition? Firstly, mixed models are introduced much earlier with a new chapter 2 and alongside the GLMs of chapter 3. Consequently the equivalence of smooths and Gaussian random fields is now more visible, including in the rewritten introduction to GAMs in Chapter 4. The range of smoothers covered is substantially enlarged to include adaptive smoothing, P-splines with derivative penalties, Duchon splines, splines on the sphere, soap film smoothers, Gaussian Markov random fields, Gaussian process smoothers and more: these now occupy the whole of chapter 5. Chapter 6 covers estimation and inference theory for GAMs from a much updated perspective, with prediction error and marginal likelihood smoothness estimation criteria as alternatives, and the modern range of estimation strategies brought to bear on these, including methods for large data sets and models, and for smooth models beyond the exponential family. AIC, p-value and penalty based model selection are covered, based on the much improved methods available since the first edition. Chapter 7 looks at GAMs in practice, with new examples covering topics such as smooth ANOVA models, survival modelling, location scale modelling of extremes, signal regression and other functional data analysis, spatio-temporal modelling, geographic regression, single index models, distributed lag models, multivariate additive models and Bayesian simulation. The second edition of this book was made possible in large part by an EPSRC fellowship grant (EP/K005251/1).

I thank the people who have, in various ways, helped me out in the writing of this book, or in the work that led to it. Among these, are Thilo Augustin, Lucy Augustin, Nicole Augustin, Miguel Bernal, Steve Blythe, David Borchers, Mark Bravington, Steve Buckland, Chris Chatfield, Richard Cormack, Anthony Davison, Yousra El Bachir, Julian Faraway, Matteo Fasiolo, José Pedro Granadeiro, Sonja Greven, Chong Gu, Yannig Goude, Bill Gurney, John Halley, Joe Horwood, Sharon Hedley, Thomas Kneib, Peter Jupp, Alain Le Tetre, Stephan Lang, Zheyuan Li, Finn Lindgren, Mike Lonergan, Giampiero Marra, Brian Marx, Dave Miller, Raphael Nedellec, Henric Nilsson, Roger Nisbet, Roger Peng, Charles Paxton, Natalya Pya, Phil Reiss, Fabian Scheipl, Björn Stollenwerk, Yorgos Stratoudaki, the R core team, in particular Kurt Hornik and Brian Ripley, Rob Calver and the CRC team, the Little Italians, and the RiederAlpinists. The `mgcv` package would not be what it is without the large number of people who sent me bug reports and suggestions and without Bravington (2013) — thank you. Before the dawn of England's post-factual age, the `mgcv` project was started with European Union funding.

Chapter 1

Linear Models

How old is the universe? The standard Big Bang model of the origin of the universe says that it expands uniformly, and locally, according to Hubble's law,

$$y = \beta x,$$

where y is the relative velocity of any two galaxies separated by distance x, and β is 'Hubble's constant' (in standard astrophysical notation $y \equiv v$, $x \equiv d$ and $\beta \equiv H_0$). β^{-1} gives the approximate age of the universe, but β is unknown and must somehow be estimated from observations of y and x, made for a variety of galaxies at different distances from us.

Figure 1.1 plots velocity against distance for 24 galaxies, according to measurements made using the Hubble Space Telescope. Velocities are assessed by measuring the Doppler effect red shift in the spectrum of light observed from the galaxies concerned, although some correction for 'local' velocity components is required. Distance measurement is much less direct, and is based on the 1912 discovery, by

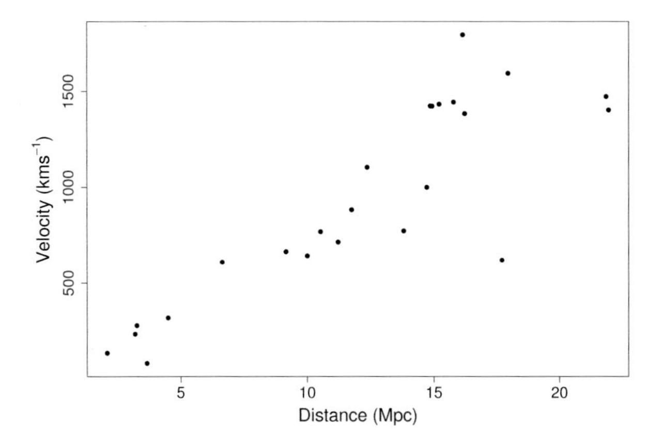

Figure 1.1 *A Hubble diagram showing the relationship between distance, x, and velocity, y, for 24 galaxies containing Cepheid stars. The data are from the Hubble Space Telescope key project to measure the Hubble constant as reported in Freedman et al. (2001).*

Henrietta Leavitt, of a relationship between the period of a certain class of variable stars, known as the Cepheids, and their luminosity. The intensity of Cepheids varies regularly with a period of between 1.5 and something over 50 days, and the mean intensity increases predictably with period. This means that, if you can find a Cepheid, you can tell how far away it is, by comparing its apparent brightness to its period predicted intensity.

It is clear, from the figure, that the observed data do not follow Hubble's law exactly, but given the measurement process, it would be surprising if they did. Given the apparent variability, what can be inferred from these data? In particular: (i) what value of β is most consistent with the data? (ii) what range of β values is consistent with the data and (iii) are some particular, theoretically derived, values of β consistent with the data? Statistics is about trying to answer these three sorts of questions.

One way to proceed is to formulate a linear statistical model of the way that the data were generated, and to use this as the basis for inference. Specifically, suppose that, rather than being governed directly by Hubble's law, the observed velocity is given by Hubble's constant multiplied by the observed distance plus a 'random variability' term. That is

$$y_i = \beta x_i + \epsilon_i, \quad i = 1 \ldots 24, \tag{1.1}$$

where the ϵ_i terms are independent random variables such that $\mathbb{E}(\epsilon_i) = 0$ and $\mathbb{E}(\epsilon_i^2) = \sigma^2$. The random component of the model is intended to capture the fact that if we gathered a replicate set of data, for a new set of galaxies, Hubble's law would not change, but the apparent random variation from it would be different, as a result of different measurement errors. Notice that it is not implied that these errors are completely unpredictable: their mean and variance are assumed to be fixed; it is only their particular values, for any particular galaxy, that are not known.

1.1 A simple linear model

This section develops statistical methods for a simple linear model of the form (1.1), allowing the key concepts of linear modelling to be introduced without the distraction of any mathematical difficulty.

Formally, consider n observations, x_i, y_i, where y_i is an observation on random variable, Y_i, with expectation, $\mu_i \equiv \mathbb{E}(Y_i)$. Suppose that an appropriate model for the relationship between x and y is:

$$Y_i = \mu_i + \epsilon_i \text{ where } \mu_i = x_i \beta. \tag{1.2}$$

Here β is an unknown parameter and the ϵ_i are mutually independent zero mean random variables, each with the same variance σ^2. So the model says that Y is given by x multiplied by a constant plus a random term. Y is an example of a *response variable*, while x is an example of a *predictor variable*. Figure 1.2 illustrates this model for a case where $n = 8$.

Simple least squares estimation

How can β, in model (1.2), be estimated from the x_i, y_i data? A sensible approach is to choose a value of β that makes the model fit closely to the data. To do this we

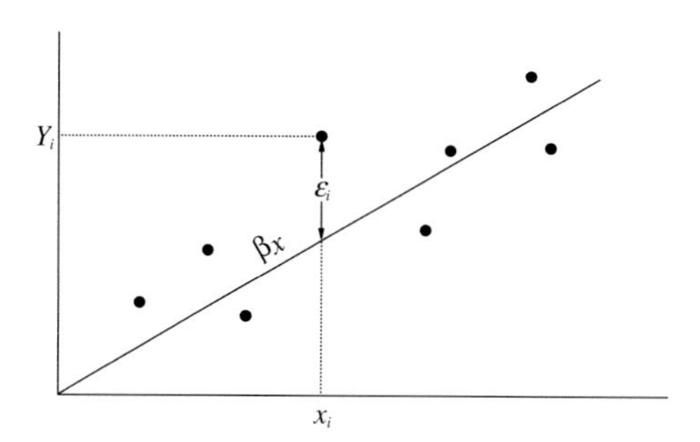

Figure 1.2 *Schematic illustration of a simple linear model with one explanatory variable.*

need to define a measure of how well, or how badly, a model with a particular β fits the data. One possible measure is the residual sum of squares of the model:

$$\mathcal{S} = \sum_{i=1}^{n}(y_i - \mu_i)^2 = \sum_{i=1}^{n}(y_i - x_i\beta)^2.$$

If we have chosen a good value of β, close to the true value, then the model predicted μ_i should be relatively close to the y_i, so that $\mathcal{S}$ should be small, whereas poor choices will lead to μ_i far from their corresponding y_i, and high values of $\mathcal{S}$. Hence β can be estimated by minimizing $\mathcal{S}$ with respect to (w.r.t.) β and this is known as the method of *least squares*.

To minimize $\mathcal{S}$, differentiate w.r.t. β,

$$\frac{\partial \mathcal{S}}{\partial \beta} = -\sum_{i=1}^{n} 2x_i(y_i - x_i\beta)$$

and set the result to zero to find $\hat{\beta}$, the least squares estimate of β:

$$-\sum_{i=1}^{n} 2x_i(y_i - x_i\hat{\beta}) = 0 \Rightarrow \sum_{i=1}^{n} x_i y_i - \hat{\beta}\sum_{i=1}^{n} x_i^2 = 0 \Rightarrow \hat{\beta} = \sum_{i=1}^{n} x_i y_i \Big/ \sum_{i=1}^{n} x_i^2.^*$$

1.1.1 *Sampling properties of* $\hat{\beta}$

To evaluate the reliability of the least squares estimate, $\hat{\beta}$, it is useful to consider the sampling properties of $\hat{\beta}$. That is, we should consider some properties of the distribution of $\hat{\beta}$ values that would be obtained from repeated independent replication of the x_i, y_i data used for estimation. To do this, it is helpful to introduce the concept of an *estimator*, obtained by replacing the observations, y_i, in the estimate of $\hat{\beta}$ by the random variables, Y_i:

$$\hat{\beta} = \sum_{i=1}^{n} x_i Y_i \Big/ \sum_{i=1}^{n} x_i^2.$$

* $\partial^2 \mathcal{S}/\partial \beta^2 = 2\sum x_i^2$ which is clearly positive, so a minimum of $\mathcal{S}$ has been found.

Clearly the *estimator*, $\hat{\beta}$, is a random variable and we can therefore discuss its distribution. For now, consider only the first two moments of that distribution.

The expected value of $\hat{\beta}$ is obtained as follows:

$$\mathbb{E}(\hat{\beta}) = \mathbb{E}\left(\sum_{i=1}^n x_i Y_i / \sum_{i=1}^n x_i^2\right) = \sum_{i=1}^n x_i \mathbb{E}(Y_i) / \sum_{i=1}^n x_i^2 = \sum_{i=1}^n x_i^2 \beta / \sum_{i=1}^n x_i^2 = \beta.$$

So $\hat{\beta}$ is an unbiased estimator — its expected value is equal to the true value of the parameter that it is supposed to estimate.

Unbiasedness is a reassuring property, but knowing that an estimator gets it right on average does not tell us much about how good any one particular estimate is likely to be. For this we also need to know how much estimates would vary from one replicate data set to the next — we need to know the estimator variance.

From general probability theory we know that if $Y_1, Y_2, \ldots, Y_n$ are *independent* random variables and $a_1, a_2, \ldots a_n$ are real constants then

$$\text{var}\left(\sum_i a_i Y_i\right) = \sum_i a_i^2 \text{var}(Y_i).$$

But we can write

$$\hat{\beta} = \sum_i a_i Y_i \text{ where } a_i = x_i / \sum_i x_i^2,$$

and from the original model specification we have that $\text{var}(Y_i) = \sigma^2$ for all i. Hence,

$$\text{var}(\hat{\beta}) = \sum_i x_i^2 / \left(\sum_i x_i^2\right)^2 \sigma^2 = \left(\sum_i x_i^2\right)^{-1} \sigma^2. \tag{1.3}$$

In most circumstances σ^2 is an unknown parameter and must also be estimated. Since σ^2 is the variance of the ϵ_i, it makes sense to estimate it using the variance of the 'estimated' ϵ_i, the model *residuals*, $\hat{\epsilon}_i = y_i - x_i\hat{\beta}$. An unbiased estimator of σ^2 is:

$$\hat{\sigma^2} = \frac{1}{n-1} \sum_i (y_i - x_i\hat{\beta})^2$$

(proof of unbiasedness is given later for the general case). Plugging this into (1.3) obviously gives an unbiased estimator of the variance of $\hat{\beta}$.

1.1.2 So how old is the universe?

The least squares calculations derived above are available as part of the statistical package and environment R. The function `lm` fits linear models to data, including the simple example currently under consideration. The Cepheid distance–velocity data shown in figure 1.1 are stored in a data frame[†] `hubble`. The following R code fits the model and produces the (edited) output shown.

[†] A data frame is just a two-dimensional array of data in which the values of different variables (which may have different types) are stored in different named columns.

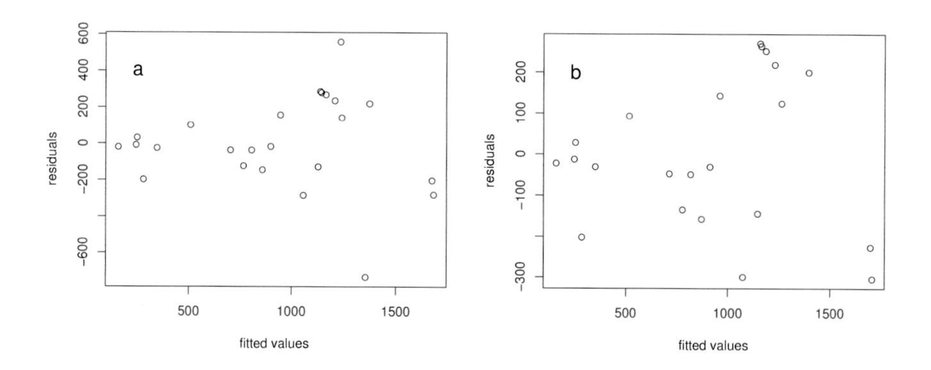

Figure 1.3 *Residuals against fitted values for* (a) *the model (1.1) fitted to all the data in figure 1.1 and* (b) *the same model fitted to data with two substantial outliers omitted.*

```
> library(gamair) ## contains 'hubble'
> data(hubble)
> hub.mod <- lm(y ~ x - 1, data=hubble)
> summary(hub.mod)

Call:
lm(formula = y ~ x - 1, data = hubble)

Coefficients:
  Estimate Std. Error
x   76.581      3.965
```

The call to `lm` passed two arguments to the function. The first is a *model formula*, `y ~ x - 1`, specifying the model to be fitted: the name of the response variable is to the left of '`~`' while the predictor variable is specified on the right; the '`-1`' term indicates that the model has no 'intercept' term, i.e., that the model is a straight line through the origin. The second (optional) argument gives the name of the data frame in which the variables are to be found. `lm` takes this information and uses it to fit the model by least squares: the results are returned in a 'fitted model object', which in this case has been assigned to an object called `hub.mod` for later examination. '`<-`' is the assignment operator, and `hub.mod` is created by this assignment (overwriting any previously existing object of this name).

The `summary` function is then used to examine the fitted model object. Only part of its output is shown here: $\hat{\beta}$ and the estimate of the standard error of $\hat{\beta}$ (the square root of the estimated variance of $\hat{\beta}$, derived above). Before using these quantities it is important to check the model assumptions. In particular we should check the plausibility of the assumptions that the ϵ_i are independent and all have the same variance. The way to do this is to examine residual plots.

The 'fitted values' of the model are defined as $\hat{\mu}_i = \hat{\beta}x_i$, while the residuals are simply $\hat{\epsilon}_i = y_i - \hat{\mu}_i$. A plot of residuals against fitted values is often useful and the following produces the plot in figure 1.3(a).

```
plot(fitted(hub.mod),residuals(hub.mod),xlab="fitted values",
     ylab="residuals")
```

What we would like to see, in such a plot, is an apparently random scatter of residuals around zero, with no trend in either the mean of the residuals, or their variability, as the fitted values increase. A trend in the mean violates the independence assumption, and is usually indicative of something missing in the model structure, while a trend in the variability violates the constant variance assumption. The main problematic feature of figure 1.3(a) is the presence of two points with very large magnitude residuals, suggesting a problem with the constant variance assumption. It is probably prudent to repeat the model fit, with and without these points, to check that they are not having undue influence on our conclusions.[‡] The following code omits the offending points and produces a new residual plot shown in figure 1.3(b).

```
> hub.mod1 <- lm(y ~ x - 1,data=hubble[-c(3,15),])
> summary(hub.mod1)

Call:
lm(formula = y ~ x - 1, data = hubble[-c(3, 15), ])

Coefficients:
  Estimate Std. Error
x    77.67       2.97

> plot(fitted(hub.mod1),residuals(hub.mod1),
+      xlab="fitted values",ylab="residuals")
```

The omission of the two large outliers has improved the residuals and changed $\hat{\beta}$ somewhat, but not drastically.

The Hubble constant estimates have units of $(\mathrm{km})\mathrm{s}^{-1}\ (\mathrm{Mpc})^{-1}$. A Mega-parsec is 3.09×10^{19}km, so we need to divide $\hat{\beta}$ by this amount, in order to obtain Hubble's constant with units of s^{-1}. The approximate age of the universe, in seconds, is then given by the reciprocal of $\hat{\beta}$. Here are the two possible estimates expressed in years:

```
> hubble.const <- c(coef(hub.mod),coef(hub.mod1))/3.09e19
> age <- 1/hubble.const
> age/(60^2*24*365)
12794692825 12614854757
```

Both fits give an age of around 13 billion years. So we now have an idea of the best estimate of the age of the universe, but, given the measurement uncertainties, what age range would be consistent with the data?

[‡]The most common mistake made by students in first courses on regression is simply to drop data with large residuals, without further investigation. Beware of this.

1.1.3 Adding a distributional assumption

So far everything done with the simple model has been based only on the model equations and the two assumptions of independence and equal variance for the response variable. To go further and find confidence intervals for β, or test hypotheses related to the model, a further distributional assumption will be necessary.

Specifically, assume that $\epsilon_i \sim N(0, \sigma^2)$ for all i, which is equivalent to assuming $Y_i \sim N(x_i\beta, \sigma^2)$. We have already seen that $\hat{\beta}$ is just a weighted sum of the Y_i, but the Y_i are now assumed to be normal random variables, and a weighted sum of normal random variables is itself a normal random variable. Hence the estimator, $\hat{\beta}$, must be a normal random variable. Since we have already established the mean and variance of $\hat{\beta}$, we have that

$$\hat{\beta} \sim N\left(\beta, \left(\sum x_i^2\right)^{-1} \sigma^2\right). \tag{1.4}$$

Testing hypotheses about β

One thing we might want to do, is to try and evaluate the consistency of some hypothesized value of β with the data. For example, some Creation Scientists estimate the age of the universe to be 6000 years, based on a reading of the Bible. This would imply that $\beta = 163 \times 10^6$.[§] The consistency with data of such a hypothesized value for β can be based on the probability that we would have observed the $\hat{\beta}$ actually obtained, if the true value of β was the hypothetical one.

Specifically, we can test the null hypothesis, $\text{H}_0 : \beta = \beta_0$, versus the alternative hypothesis, $\text{H}_1 : \beta \neq \beta_0$, for some specified value β_0, by examining the probability of getting the observed $\hat{\beta}$, or one further from β_0, assuming H_0 to be true. If σ^2 were known then we could work directly from (1.4), as follows.

The probability required is known as the **p-value** of the test. It is the *probability of getting a value of $\hat{\beta}$ at least as favourable to* H_1 *as the one actually observed, if* H_0 *is actually true.*[¶] In this case it helps to distinguish notationally between the estimate, $\hat{\beta}_{\text{obs}}$, and estimator $\hat{\beta}$. The p-value is then

$$\begin{aligned} p &= \Pr\left(|\hat{\beta} - \beta_0| \geq |\hat{\beta}_{\text{obs}} - \beta_0| \,\middle|\, \text{H}_0 \right) \\ &= \Pr\left(|\hat{\beta} - \beta_0|/\sigma_{\hat{\beta}} \geq |\hat{\beta}_{\text{obs}} - \beta_0|/\sigma_{\hat{\beta}} \,\middle|\, \text{H}_0 \right) \\ &= \Pr(|Z| > |z|) \end{aligned}$$

where $Z \sim N(0, 1)$, $z = (\hat{\beta}_{\text{obs}} - \beta_0)/\sigma_{\hat{\beta}}$ and $\sigma^2_{\hat{\beta}} = (\sum x_i^2)^{-1}\sigma^2$. Hence, having formed z, the p-value is easily worked out, using the cumulative distribution function for the standard normal built into any statistics package. Small p-values suggest that the data are inconsistent with H_0, while large values indicate consistency. 0.05 is often used as the arbitrary boundary between 'small' and 'large' in this context.

[§]This isn't really valid, of course, since the Creation Scientists are not postulating a Big Bang theory.

[¶]This definition holds for any hypothesis test, if the specific '$\hat{\beta}$' is replaced by the general '*a test statistic*'.

In reality σ^2 is usually unknown. Broadly the same testing procedure can still be adopted, by replacing σ with $\hat{\sigma}$, but we need to somehow allow for the extra uncertainty that this introduces (unless the sample size is very large). It turns out that if $\text{H}_0 : \beta = \beta_0$ is true then

$$T \equiv \frac{\hat{\beta} - \beta_0}{\hat{\sigma}_{\hat{\beta}}} \sim t_{n-1}$$

where n is the sample size, $\hat{\sigma}^2_{\hat{\beta}} = (\sum x_i^2)^{-1}\hat{\sigma}^2$, and t_{n-1} is the t-distribution with $n-1$ degrees of freedom. This result is proven in section 1.3. It is clear that large magnitude values of T favour H_1, so using T as the test statistic, in place of $\hat{\beta}$, we can calculate a p-value by evaluating

$$p = \Pr(|T| > |t|)$$

where $T \sim t_{n-1}$ and $t = (\hat{\beta}_{\text{obs}} - \beta_0)/\hat{\sigma}_{\hat{\beta}}$. This is easily evaluated using the c.d.f. of the t distributions, built into any decent statistics package. Here is some code to evaluate the p-value for H_0 : 'the Hubble constant is 163000000'.

```
> cs.hubble <- 163000000
> t.stat <- (coef(hub.mod1)-cs.hubble)/vcov(hub.mod1)[1,1]^0.5
> pt(t.stat,df=21)*2 # 2 because of |T| in p-value definition
3.906388e-150
```

i.e., as judged using t, the data would be hugely improbable if $\beta = 1.63 \times 10^8$. It would seem that the hypothesized value can be rejected rather firmly (in this case, using the data with the outliers increases the p-value by a factor of 1000 or so).

Hypothesis testing is useful when there are good reasons to stick with some null hypothesis until there are compelling grounds to reject it. This is often the case when comparing models of differing complexity: it is often a good idea to retain the simpler model until there is quite firm evidence that it is inadequate. Note one interesting property of hypothesis testing. If we choose to reject a null hypothesis whenever the p-value is less than some fixed level, α (often termed the *significance level* of a test), then we will inevitably reject a proportion, α, of correct null hypotheses. We could try and reduce the probability of such mistakes by making α very small, but in that case we pay the price of reducing the probability of rejecting H_0 when it is false.

Confidence intervals

Having seen how to test whether a *particular* hypothesized value of β is consistent with the data, the question naturally arises of what *range* of values of β would be consistent with the data? To answer this, we need to select a definition of 'consistent': a common choice is to say that any parameter value is consistent with the data if it results in a p-value of ≥ 0.05, when used as the null value in a hypothesis test.

Sticking with the Hubble constant example, and working at a significance level of 0.05, we would have rejected any hypothesized value for the constant that resulted in a t value outside the range $(-2.08, 2.08)$, since these values would result in p-values of less than 0.05. The R function `qt` can be used to find such ranges:

e.g., `qt(c(0.025,0.975),df=21)` returns the range of the middle 95% of t_{21} random variables. So we would accept any β_0 fulfilling:

$$-2.08 \leq \frac{\hat{\beta} - \beta_0}{\hat{\sigma}_{\hat{\beta}}} \leq 2.08$$

which rearranges to give the interval

$$\hat{\beta} - 2.08\hat{\sigma}_{\hat{\beta}} \leq \beta_0 \leq \hat{\beta} + 2.08\hat{\sigma}_{\hat{\beta}}.$$

Such an interval is known as a '95% confidence interval' for β.

The defining property of a 95% confidence interval is this: if we were to gather an infinite sequence of independent replicate data sets, and calculate 95% confidence intervals for β from each, then 95% of these intervals would include the true β, and 5% would not. It is easy to see how this comes about. By construction, a hypothesis test with a significance level of 5% rejects the correct null hypothesis for 5% of replicate data sets, and accepts it for the other 95% of replicates. Hence 5% of 95% confidence intervals must exclude the true parameter, while 95% include it.

A 95% CI for the Hubble constant (in the usual astrophysicists' units) is given by:

```
> sigb <- summary(hub.mod1)$coefficients[2]
> h.ci <- coef(hub.mod1)+qt(c(0.025,0.975),df=21)*sigb
> h.ci
[1] 71.49588 83.84995
```

This can be converted to a confidence interval for the age of the universe, in years:

```
> h.ci <- h.ci*60^2*24*365.25/3.09e19 # convert to 1/years
> sort(1/h.ci)
[1] 11677548698 13695361072
```

i.e., the 95% CI is (11.7,13.7) billion years. Actually this 'Hubble age' is the age of the universe if it has been expanding freely, essentially unfettered by gravitation and other effects since the Big Bang. In reality some corrections are needed to get a better estimate, and at time of writing this is about 13.8 billion years.[‖]

1.2 Linear models in general

The simple linear model, introduced above, can be generalized by allowing the response variable to depend on multiple predictor variables (plus an additive constant). These extra predictor variables can themselves be transformations of the original predictors. Here are some examples, for each of which a response variable datum, y_i, is treated as an observation on a random variable, Y_i, where $\mathbb{E}(Y_i) \equiv \mu_i$, the ϵ_i are zero mean random variables, and the β_j are model parameters, the values of which are unknown and will need to be estimated using data.

[‖] If that makes you feel young, recall that the stuff you are made of is also that old. Feeling small is better justified: there are estimated to be something like 10^{24} stars in the universe, which is approximately the number of full stops it would take to cover the surface of the earth (if it was a smooth sphere).

1. $\mu_i = \beta_0 + x_i\beta_1, \;\; Y_i = \mu_i + \epsilon_i$, is a straight line relationship between y and predictor variable, x.
2. $\mu_i = \beta_0 + x_i\beta_1 + x_i^2\beta_2 + x_i^3\beta_3, \;\; Y_i = \mu_i + \epsilon_i$, is a cubic model of the relationship between y and x.
3. $\mu_i = \beta_0 + x_i\beta_1 + z_i\beta_2 + \log(x_i z_i)\beta_3, \;\; Y_i = \mu_i + \epsilon_i$, is a model in which y depends on predictor variables x and z and on the log of their product.

Each of these is a linear model because the ϵ_i terms and the model parameters, β_j, enter the model in a linear way. Notice that the predictor variables can enter the model non-linearly. Exactly as for the simple model, the parameters of these models can be estimated by finding the β_j values which make the models best fit the observed data, in the sense of minimizing $\sum_i (y_i - \mu_i)^2$. The theory for doing this will be developed in section 1.3, and that development is based entirely on re-writing the linear model using matrices and vectors.

To see how this re-writing is done, consider the straight line model given above. Writing out the μ_i equations for all n pairs, (x_i, y_i), results in a large system of linear equations:

$$\begin{aligned} \mu_1 &= \beta_0 + x_1\beta_1 \\ \mu_2 &= \beta_0 + x_2\beta_1 \\ \mu_3 &= \beta_0 + x_3\beta_1 \\ . &\quad . \\ . &\quad . \\ \mu_n &= \beta_0 + x_n\beta_1 \end{aligned}$$

which can be re-written in matrix-vector form as

$$\begin{bmatrix} \mu_1 \\ \mu_2 \\ \mu_3 \\ . \\ . \\ \mu_n \end{bmatrix} = \begin{bmatrix} 1 & x_1 \\ 1 & x_2 \\ 1 & x_3 \\ . & . \\ . & . \\ 1 & x_n \end{bmatrix} \begin{bmatrix} \beta_0 \\ \beta_1 \end{bmatrix}.$$

So the model has the general form $\boldsymbol{\mu} = \mathbf{X}\boldsymbol{\beta}$, i.e., the expected value vector $\boldsymbol{\mu}$ is given by a **model matrix** (also known as a design matrix), $\mathbf{X}$, multiplied by a parameter vector, $\boldsymbol{\beta}$. All linear models can be written in this general form.

As a second illustration, the cubic example, given above, can be written in matrix vector form as

$$\begin{bmatrix} \mu_1 \\ \mu_2 \\ \mu_3 \\ . \\ . \\ \mu_n \end{bmatrix} = \begin{bmatrix} 1 & x_1 & x_1^2 & x_1^3 \\ 1 & x_2 & x_2^2 & x_2^3 \\ 1 & x_3 & x_3^2 & x_3^3 \\ . & . & . & . \\ . & . & . & . \\ 1 & x_n & x_n^2 & x_n^3 \end{bmatrix} \begin{bmatrix} \beta_0 \\ \beta_1 \\ \beta_2 \\ \beta_3 \end{bmatrix}.$$

Models in which data are divided into different groups, each of which are assumed to have a different mean, are less obviously of the form $\boldsymbol{\mu} = \mathbf{X}\boldsymbol{\beta}$, but in fact they can be written in this way, by use of dummy indicator variables. Again, this is most easily seen by example. Consider the model

$$\mu_i = \beta_j \text{ if observation } i \text{ is in group } j,$$

and suppose that there are three groups, each with 2 data. Then the model can be re-written

$$\begin{bmatrix} \mu_1 \\ \mu_2 \\ \mu_3 \\ \mu_4 \\ \mu_5 \\ \mu_6 \end{bmatrix} = \begin{bmatrix} 1 & 0 & 0 \\ 1 & 0 & 0 \\ 0 & 1 & 0 \\ 0 & 1 & 0 \\ 0 & 0 & 1 \\ 0 & 0 & 1 \end{bmatrix} \begin{bmatrix} \beta_0 \\ \beta_1 \\ \beta_2 \end{bmatrix}.$$

Variables indicating the group to which a response observation belongs are known as *factor* variables. Somewhat confusingly, the groups themselves are known as *levels* of a factor. So the above model involves one factor, 'group', with three levels. Models of this type, involving factors, are commonly used for the analysis of designed experiments. In this case the model matrix depends on the design of the experiment (i.e., on which units belong to which groups), and for this reason the terms 'design matrix' and 'model matrix' are often used interchangeably. Whatever it is called, $\mathbf{X}$ is absolutely central to understanding the theory of linear models, generalized linear models and generalized additive models.

1.3 The theory of linear models

This section shows how the parameters, $\boldsymbol{\beta}$, of the linear model

$$\boldsymbol{\mu} = \mathbf{X}\boldsymbol{\beta}, \quad \mathbf{y} \sim N(\boldsymbol{\mu}, \mathbf{I}_n\sigma^2)$$

can be estimated by least squares. It is assumed that $\mathbf{X}$ is a full rank matrix, with n rows and p columns. It will be shown that the resulting estimator, $\hat{\boldsymbol{\beta}}$, is unbiased, has the lowest variance of any possible linear estimator of $\boldsymbol{\beta}$, and that, given the normality of the data, $\hat{\boldsymbol{\beta}} \sim N(\boldsymbol{\beta}, (\mathbf{X}^\mathsf{T}\mathbf{X})^{-1}\sigma^2)$. Results are also derived for setting confidence limits on parameters and for testing hypotheses about parameters — in particular the hypothesis that several elements of $\boldsymbol{\beta}$ are simultaneously zero.

In this section it is important not to confuse the *length* of a vector with its *dimension*. For example $(1, 1, 1)^\mathsf{T}$ has dimension 3 and length $\sqrt{3}$. Also note that no distinction has been made notationally between random variables and particular observations of those random variables: it is usually clear from the context which is meant.

1.3.1 *Least squares estimation of* $\boldsymbol{\beta}$

Point estimates of the linear model parameters, $\boldsymbol{\beta}$, can be obtained by the method of least squares, that is by minimizing

$$\mathcal{S} = \sum_{i=1}^{n} (y_i - \mu_i)^2,$$

with respect to $\boldsymbol{\beta}$. To use least squares with a linear model written in general matrix-vector form, first recall the link between the Euclidean length of a vector and the sum of squares of its elements. If $\mathbf{v}$ is any vector of dimension, n, then

$$\|\mathbf{v}\|^2 \equiv \mathbf{v}^\mathsf{T}\mathbf{v} \equiv \sum_{i=1}^{n} v_i^2.$$

Hence

$$\mathcal{S} = \|\mathbf{y} - \boldsymbol{\mu}\|^2 = \|\mathbf{y} - \mathbf{X}\boldsymbol{\beta}\|^2.$$

Since this expression is simply the squared (Euclidian) length of the vector $\mathbf{y} - \mathbf{X}\boldsymbol{\beta}$, its value will be unchanged if $\mathbf{y} - \mathbf{X}\boldsymbol{\beta}$ is rotated. This observation is the basis for a practical method for finding $\hat{\boldsymbol{\beta}}$, and for developing the distributional results required to use linear models.

Specifically, like any real matrix, $\mathbf{X}$ can always be decomposed

$$\mathbf{X} = \mathbf{Q}\begin{bmatrix} \mathbf{R} \\ \mathbf{0} \end{bmatrix} = \mathbf{Q}_\mathrm{f}\mathbf{R} \tag{1.5}$$

where $\mathbf{R}$ is a $p \times p$ upper triangular matrix,[†] and $\mathbf{Q}$ is an $n \times n$ orthogonal matrix, the first p columns of which form $\mathbf{Q}_\mathrm{f}$ (see B.6). Recall that orthogonal matrices rotate or reflect vectors, but do not change their length. Orthogonality also means that $\mathbf{Q}\mathbf{Q}^\mathsf{T} = \mathbf{Q}^\mathsf{T}\mathbf{Q} = \mathbf{I}_n$. Applying $\mathbf{Q}^\mathsf{T}$ to $\mathbf{y} - \mathbf{X}\boldsymbol{\beta}$ implies that

$$\|\mathbf{y} - \mathbf{X}\boldsymbol{\beta}\|^2 = \|\mathbf{Q}^\mathsf{T}\mathbf{y} - \mathbf{Q}^\mathsf{T}\mathbf{X}\boldsymbol{\beta}\|^2 = \left\|\mathbf{Q}^\mathsf{T}\mathbf{y} - \begin{bmatrix} \mathbf{R} \\ \mathbf{0} \end{bmatrix}\boldsymbol{\beta}\right\|^2.$$

Writing $\mathbf{Q}^\mathsf{T}\mathbf{y} = \begin{bmatrix} \mathbf{f} \\ \mathbf{r} \end{bmatrix}$, where $\mathbf{f}$ is vector of dimension p, and hence $\mathbf{r}$ is a vector of dimension $n - p$, yields

$$\|\mathbf{y} - \mathbf{X}\boldsymbol{\beta}\|^2 = \left\|\begin{bmatrix} \mathbf{f} \\ \mathbf{r} \end{bmatrix} - \begin{bmatrix} \mathbf{R} \\ \mathbf{0} \end{bmatrix}\boldsymbol{\beta}\right\|^2 = \|\mathbf{f} - \mathbf{R}\boldsymbol{\beta}\|^2 + \|\mathbf{r}\|^2.^\ddagger$$

The length of $\mathbf{r}$ does not depend on $\boldsymbol{\beta}$, while $\|\mathbf{f} - \mathbf{R}\boldsymbol{\beta}\|^2$ can be reduced to zero by choosing $\boldsymbol{\beta}$ so that $\mathbf{R}\boldsymbol{\beta}$ equals $\mathbf{f}$. Hence

$$\hat{\boldsymbol{\beta}} = \mathbf{R}^{-1}\mathbf{f} \tag{1.6}$$

[†] i.e., $R_{i,j} = 0$ if $i > j$.

[‡] If the last equality isn't obvious recall that $\|\mathbf{x}\|^2 = \sum_i x_i^2$, so if $\mathbf{x} = \begin{bmatrix} \mathbf{v} \\ \mathbf{w} \end{bmatrix}$, $\|\mathbf{x}\|^2 = \sum_i v_i^2 + \sum_i w_i^2 = \|\mathbf{v}\|^2 + \|\mathbf{w}\|^2$.

is the least squares estimator of $\boldsymbol{\beta}$. Notice that $\|\mathbf{r}\|^2 = \|\mathbf{y} - \mathbf{X}\hat{\boldsymbol{\beta}}\|^2$, the *residual sum of squares* for the model fit.

1.3.2 *The distribution of* $\hat{\boldsymbol{\beta}}$

The distribution of the estimator, $\hat{\boldsymbol{\beta}}$, follows from that of $\mathbf{Q}^\mathsf{T}\mathbf{y}$. Multivariate normality of $\mathbf{Q}^\mathsf{T}\mathbf{y}$ follows from that of $\mathbf{y}$, and since the covariance matrix of $\mathbf{y}$ is $\mathbf{I}_n\sigma^2$, the covariance matrix of $\mathbf{Q}^\mathsf{T}\mathbf{y}$ is

$$\mathbf{V}_{\mathbf{Q}^\mathsf{T}\mathbf{y}} = \mathbf{Q}^\mathsf{T}\mathbf{I}_n\mathbf{Q}\sigma^2 = \mathbf{I}_n\sigma^2.$$

Furthermore,

$$\mathbb{E}\begin{bmatrix} \mathbf{f} \\ \mathbf{r} \end{bmatrix} = \mathbb{E}(\mathbf{Q}^\mathsf{T}\mathbf{y}) = \mathbf{Q}^\mathsf{T}\mathbf{X}\boldsymbol{\beta} = \begin{bmatrix} \mathbf{R} \\ \mathbf{0} \end{bmatrix}\boldsymbol{\beta}$$
$$\Rightarrow \mathbb{E}(\mathbf{f}) = \mathbf{R}\boldsymbol{\beta} \text{ and } \mathbb{E}(\mathbf{r}) = \mathbf{0},$$

i.e., we have that

$$\mathbf{f} \sim N(\mathbf{R}\boldsymbol{\beta}, \mathbf{I}_p\sigma^2) \text{ and } \mathbf{r} \sim N(\mathbf{0}, \mathbf{I}_{n-p}\sigma^2)$$

with both vectors independent of each other.

Turning to the properties of $\hat{\boldsymbol{\beta}}$ itself, unbiasedness follows immediately:

$$\mathbb{E}(\hat{\boldsymbol{\beta}}) = \mathbf{R}^{-1}\mathbb{E}(\mathbf{f}) = \mathbf{R}^{-1}\mathbf{R}\boldsymbol{\beta} = \boldsymbol{\beta}.$$

Since the covariance matrix of $\mathbf{f}$ is $\mathbf{I}_p\sigma^2$, it follows that the covariance matrix of $\hat{\boldsymbol{\beta}}$ is

$$\mathbf{V}_{\hat{\beta}} = \mathbf{R}^{-1}\mathbf{I}_p\mathbf{R}^{-\mathsf{T}}\sigma^2 = \mathbf{R}^{-1}\mathbf{R}^{-\mathsf{T}}\sigma^2. \tag{1.7}$$

Furthermore, since $\hat{\boldsymbol{\beta}}$ is just a linear transformation of the normal random variables $\mathbf{f}$, it must have a multivariate normal distribution,

$$\hat{\boldsymbol{\beta}} \sim N(\boldsymbol{\beta}, \mathbf{V}_{\hat{\beta}}).$$

The foregoing distributional result is not usually directly useful for making inferences about $\boldsymbol{\beta}$, since σ^2 is generally unknown and must be estimated, thereby introducing an extra component of variability that should be accounted for.

1.3.3 $(\hat{\beta}_i - \beta_i)/\hat{\sigma}_{\hat{\beta}_i} \sim t_{n-p}$

Since the elements of $\mathbf{r}$ are i.i.d. $N(0, \sigma^2)$, the r_i/σ are i.i.d. $N(0, 1)$ random variables, and hence

$$\frac{1}{\sigma^2}\|\mathbf{r}\|^2 = \frac{1}{\sigma^2}\sum_{i=1}^{n-p} r_i^2 \sim \chi^2_{n-p}.$$

The mean of a χ^2_{n-p} r.v. is $n - p$, so this result is sufficient (but not necessary: see exercise 7) to imply that

$$\hat{\sigma}^2 = \|\mathbf{r}\|^2/(n-p) \tag{1.8}$$

is an unbiased estimator of σ^2.[§] The independence of the elements of $\mathbf{r}$ and $\mathbf{f}$ also implies that $\hat{\boldsymbol{\beta}}$ and $\hat{\sigma}^2$ are independent.

Now consider a single parameter estimator, $\hat{\beta}_i$, with standard deviation, $\sigma_{\hat{\beta}_i}$, given by the square root of element i, i of $\mathbf{V}_{\hat{\beta}}$. An unbiased estimator of $\mathbf{V}_{\hat{\beta}}$ is $\hat{\mathbf{V}}_{\hat{\beta}} = \mathbf{V}_{\hat{\beta}}\hat{\sigma}^2/\sigma^2 = \mathbf{R}^{-1}\mathbf{R}^{-\mathsf{T}}\hat{\sigma}^2$, so an estimator, $\hat{\sigma}_{\hat{\beta}_i}$, is given by the square root of element i, i of $\hat{\mathbf{V}}_{\hat{\beta}}$, and it is clear that $\hat{\sigma}_{\hat{\beta}_i} = \sigma_{\hat{\beta}_i}\hat{\sigma}/\sigma$. Hence

$$\frac{\hat{\beta}_i - \beta_i}{\hat{\sigma}_{\hat{\beta}_i}} = \frac{\hat{\beta}_i - \beta_i}{\sigma_{\hat{\beta}_i}\hat{\sigma}/\sigma} = \frac{(\hat{\beta}_i - \beta_i)/\sigma_{\hat{\beta}_i}}{\sqrt{\frac{1}{\sigma^2}\|\mathbf{r}\|^2/(n-p)}} \sim \frac{N(0,1)}{\sqrt{\chi^2_{n-p}/(n-p)}} \sim t_{n-p}$$

(where the independence of $\hat{\beta}_i$ and $\hat{\sigma}^2$ has been used). This result enables confidence intervals for β_i to be found, and is the basis for hypothesis tests about individual β_i's (for example, $\text{H}_0 : \beta_i = 0$).

1.3.4 F-ratio results I

Sometimes it is useful to be able to test $\text{H}_0 : \mathbf{C}\boldsymbol{\beta} = \mathbf{d}$, where $\mathbf{C}$ is $q \times p$ and rank q $(< p)$. Under H_0 we have $\mathbf{C}\hat{\boldsymbol{\beta}} - \mathbf{d} \sim N(\mathbf{0}, \mathbf{C}\mathbf{V}_{\hat{\beta}}\mathbf{C}^{\mathsf{T}},)$, from basic properties of the transformation of normal random vectors. Forming a Cholesky decomposition $\mathbf{L}^{\mathsf{T}}\mathbf{L} = \mathbf{C}\mathbf{V}_{\hat{\beta}}\mathbf{C}^{\mathsf{T}}$ (see B.7), it is then easy to show that $\mathbf{L}^{-\mathsf{T}}(\mathbf{C}\hat{\boldsymbol{\beta}} - \mathbf{d}) \sim N(\mathbf{0}, \mathbf{I})$, so,

$$(\mathbf{C}\hat{\boldsymbol{\beta}} - \mathbf{d})^{\mathsf{T}}(\mathbf{C}\mathbf{V}_{\hat{\beta}}\mathbf{C}^{\mathsf{T}})^{-1}(\mathbf{C}\hat{\boldsymbol{\beta}} - \mathbf{d}) = \\ (\mathbf{C}\hat{\boldsymbol{\beta}} - \mathbf{d})^{\mathsf{T}}\mathbf{L}^{-1}\mathbf{L}^{-\mathsf{T}}(\mathbf{C}\hat{\boldsymbol{\beta}} - \mathbf{d}) \sim \sum_{i=1}^{q} N(0,1)^2 \sim \chi^2_q.$$

As in the previous section, plugging in $\hat{\mathbf{V}}_{\hat{\beta}} = \mathbf{V}_{\hat{\beta}}\hat{\sigma}^2/\sigma^2$ gives the computable test statistic and its distribution under H_0:

$$\frac{1}{q}(\mathbf{C}\hat{\boldsymbol{\beta}} - \mathbf{d})^{\mathsf{T}}(\mathbf{C}\hat{V}_{\hat{\beta}}\mathbf{C}^{\mathsf{T}})^{-1}(\mathbf{C}\hat{\boldsymbol{\beta}} - \mathbf{d}) = \frac{\sigma^2}{q\hat{\sigma}^2}(\mathbf{C}\hat{\boldsymbol{\beta}} - \mathbf{d})^{\mathsf{T}}(\mathbf{C}V_{\hat{\beta}}\mathbf{C}^{\mathsf{T}})^{-1}(\mathbf{C}\hat{\boldsymbol{\beta}} - \mathbf{d}) \\ = \frac{(\mathbf{C}\hat{\boldsymbol{\beta}} - \mathbf{d})^{\mathsf{T}}(\mathbf{C}V_{\hat{\beta}}\mathbf{C}^{\mathsf{T}})^{-1}(\mathbf{C}\hat{\boldsymbol{\beta}} - \mathbf{d})/q}{\frac{1}{\sigma^2}\|\mathbf{r}\|^2/(n-p)} \sim \frac{\chi^2_q/q}{\chi^2_{n-p}/(n-p)} \sim F_{q,n-p}. \quad (1.9)$$

This result can be used to compute a p-value for the test.

1.3.5 F-ratio results II

An alternative F-ratio test derivation is also useful. Consider testing the simultaneous equality to zero of several model parameters. Such tests are useful for making

[§]Don't forget that $\|\mathbf{r}\|^2 = \|\mathbf{y} - \mathbf{X}\hat{\boldsymbol{\beta}}\|^2$.

inferences about factor variables and their interactions, since each factor or interaction is typically represented by several elements of $\boldsymbol{\beta}$. More specifically suppose that the model matrix is partitioned $\mathbf{X} = [\mathbf{X}_0 : \mathbf{X}_1]$, where $\mathbf{X}_0$ and $\mathbf{X}_1$ have $p - q$ and q columns, respectively. Let $\boldsymbol{\beta}_0$ and $\boldsymbol{\beta}_1$ be the corresponding sub-vectors of $\boldsymbol{\beta}$, and consider testing

$$\text{H}_0 : \boldsymbol{\beta}_1 = \mathbf{0} \quad \text{versus} \quad \text{H}_1 : \boldsymbol{\beta}_1 \neq \mathbf{0}.$$

Any test involving the comparison of a linear model with a simplified null version of the model can be written in this form, by re-ordering of the columns of $\mathbf{X}$ or by re-parameterization. Now

$$\mathbf{Q}^\mathsf{T}\mathbf{X}_0 = \begin{bmatrix} \mathbf{R}_0 \\ \mathbf{0} \end{bmatrix}$$

where $\mathbf{R}_0$ is the first $p - q$ rows and columns of $\mathbf{R}$ ($\mathbf{Q}$ and $\mathbf{R}$ are from (1.5)). So rotating $\mathbf{y} - \mathbf{X}_0\boldsymbol{\beta}_0$ using $\mathbf{Q}^\mathsf{T}$ implies that

$$\|\mathbf{y} - \mathbf{X}_0\boldsymbol{\beta}_0\|^2 = \|\mathbf{Q}^\mathsf{T}\mathbf{y} - \mathbf{Q}^\mathsf{T}\mathbf{X}_0\boldsymbol{\beta}_0\|^2 = \|\mathbf{f}_0 - \mathbf{R}_0\boldsymbol{\beta}_0\|^2 + \|\mathbf{f}_1\|^2 + \|\mathbf{r}\|^2$$

where $\mathbf{f}$ has been partitioned so that $\mathbf{f} = \begin{bmatrix} \mathbf{f}_0 \\ \mathbf{f}_1 \end{bmatrix}$ ($\mathbf{f}_1$ being of dimension q). Hence $\|\mathbf{f}_1\|^2$ is the increase in residual sum of squares that results from dropping $\mathbf{X}_1$ from the model (i.e. setting $\boldsymbol{\beta}_1 = \mathbf{0}$).

Now, since $\mathbb{E}(\mathbf{f}) = \mathbf{R}\boldsymbol{\beta}$ and $\mathbf{R}$ is upper triangular, then $\mathbb{E}(\mathbf{f}_1) = \mathbf{0}$ if $\boldsymbol{\beta}_1 = \mathbf{0}$ (i.e. if H_0 is true). Also, we already know that the elements of $\mathbf{f}_1$ are independent normal r.v.s with variance σ^2. Hence, if H_0 is true, $\mathbf{f}_1 \sim N(\mathbf{0}, \mathbf{I}_q\sigma^2)$ and consequently

$$\frac{1}{\sigma^2}\|\mathbf{f}_1\|^2 \sim \chi^2_q.$$

So, forming an 'F-ratio statistic', F, assuming H_0, and recalling the independence of $\mathbf{f}$ and $\mathbf{r}$ we have

$$F = \frac{\|\mathbf{f}_1\|^2/q}{\hat{\sigma}^2} = \frac{\frac{1}{\sigma^2}\|\mathbf{f}_1\|^2/q}{\frac{1}{\sigma^2}\|\mathbf{r}\|^2/(n-p)} \sim \frac{\chi^2_q/q}{\chi^2_{n-p}/(n-p)} \sim F_{q,n-p}$$

which can be used to compute a p-value for H_0, thereby testing the significance of model terms. Notice that for comparing any null model with residual sum of squares, RSS_0, to a full model with residual sum of squares, RSS_1, the preceding derivation holds, but the test statistic can also be computed (without any initial re-ordering or re-parameterization) as

$$F = \frac{(\text{RSS}_0 - \text{RSS}_1)/q}{\text{RSS}_1/(n-p)}.$$

In slightly more generality, if $\boldsymbol{\beta}$ is partitioned into sub-vectors $\boldsymbol{\beta}_0, \boldsymbol{\beta}_1 \ldots, \boldsymbol{\beta}_m$ (each usually relating to a different effect), of dimensions $q_0, q_1, \ldots, q_m$, then $\mathbf{f}$ can also be so partitioned, $\mathbf{f}^\mathsf{T} = [\mathbf{f}_0^\mathsf{T}, \mathbf{f}_1^\mathsf{T}, \ldots, \mathbf{f}_m^\mathsf{T}]$, and tests of

$$\text{H}_0 : \boldsymbol{\beta}_j = \mathbf{0} \text{ versus } \text{H}_1 : \boldsymbol{\beta}_j \neq \mathbf{0}$$

are conducted using the result that under H_0

$$F = \frac{\|\mathbf{f}_j\|^2/q_j}{\hat{\sigma}^2} \sim F_{q_j,n-p},$$

with F larger than this suggests, if the alternative is true. This is the result used to draw up sequential ANOVA tables for a fitted model, of the sort produced by a single argument call to `anova` in R. Note, however, that the hypothesis test about $\boldsymbol{\beta}_j$ is only valid in general if $\boldsymbol{\beta}_k = \mathbf{0}$ for all k such that $j < k \leq m$: this follows from the way that the test was derived, and is the reason that the ANOVA tables resulting from such procedures are referred to as 'sequential' tables. The practical upshot is that, if models are reduced in a different order, the p-values obtained will be different. The exception to this is if the $\hat{\boldsymbol{\beta}}_j$'s are mutually independent, in which case all the tests are simultaneously valid, and the ANOVA table for a model is not dependent on the order of model terms: such independent $\hat{\boldsymbol{\beta}}_j$'s usually arise only in the context of balanced data, from designed experiments.

Notice that sequential ANOVA tables are very easy to calculate: once a model has been fitted by the QR method, all the relevant 'sums of squares' are easily calculated directly from the elements of $\mathbf{f}$, with $\|\mathbf{r}\|^2$ providing the residual sum of squares.

1.3.6 *The influence matrix*

One matrix that will feature extensively in the discussion of GAMs is the *influence matrix* (or *hat matrix*) of a linear model. This is the matrix which yields the fitted value vector, $\hat{\boldsymbol{\mu}}$, when post-multiplied by the data vector, $\mathbf{y}$. Recalling the definition of $\mathbf{Q}_\text{f}$, as being the first p columns of $\mathbf{Q}$, $\mathbf{f} = \mathbf{Q}_\text{f}^\mathsf{T}\mathbf{y}$ and so

$$\hat{\boldsymbol{\beta}} = \mathbf{R}^{-1}\mathbf{Q}_\text{f}^\mathsf{T}\mathbf{y}.$$

Furthermore $\hat{\boldsymbol{\mu}} = \mathbf{X}\hat{\boldsymbol{\beta}}$ and $\mathbf{X} = \mathbf{Q}_\text{f}\mathbf{R}$ so

$$\hat{\boldsymbol{\mu}} = \mathbf{Q}_\text{f}\mathbf{R}\mathbf{R}^{-1}\mathbf{Q}_\text{f}^\mathsf{T}\mathbf{y} = \mathbf{Q}_\text{f}\mathbf{Q}_\text{f}^\mathsf{T}\mathbf{y}$$

i.e., the matrix $\mathbf{A} \equiv \mathbf{Q}_\text{f}\mathbf{Q}_\text{f}^\mathsf{T}$ is the influence (hat) matrix such that $\hat{\boldsymbol{\mu}} = \mathbf{A}\mathbf{y}$.

The influence matrix has a couple of interesting properties. Firstly, the trace of the influence matrix is the number of (identifiable) parameters in the model, since

$$\text{tr}(\mathbf{A}) = \text{tr}\left(\mathbf{Q}_\text{f}\mathbf{Q}_\text{f}^\mathsf{T}\right) = \text{tr}\left(\mathbf{Q}_\text{f}^\mathsf{T}\mathbf{Q}_\text{f}\right) = \text{tr}(\mathbf{I}_p) = p.$$

Secondly, $\mathbf{AA} = \mathbf{A}$, a property known as *idempotency*. Again the proof is simple:

$$\mathbf{AA} = \mathbf{Q}_\text{f}\mathbf{Q}_\text{f}^\mathsf{T}\mathbf{Q}_\text{f}\mathbf{Q}_\text{f}^\mathsf{T} = \mathbf{Q}_\text{f}\mathbf{I}_p\mathbf{Q}_\text{f}^\mathsf{T} = \mathbf{Q}_\text{f}\mathbf{Q}_\text{f}^\mathsf{T} = \mathbf{A}.$$

1.3.7 *The residuals, $\hat{\boldsymbol{\epsilon}}$, and fitted values, $\hat{\boldsymbol{\mu}}$*

The influence matrix is helpful in deriving properties of the fitted values, $\hat{\boldsymbol{\mu}}$, and residuals, $\hat{\boldsymbol{\epsilon}}$. $\hat{\boldsymbol{\mu}}$ is unbiased, since $\mathbb{E}(\hat{\boldsymbol{\mu}}) = \mathbb{E}(\mathbf{X}\hat{\boldsymbol{\beta}}) = \mathbf{X}\mathbb{E}(\hat{\boldsymbol{\beta}}) = \mathbf{X}\boldsymbol{\beta} = \boldsymbol{\mu}$. The covariance matrix of the fitted values is obtained from the fact that $\hat{\boldsymbol{\mu}}$ is a linear transformation of the random vector $\mathbf{y}$, which has covariance matrix $\mathbf{I}_n\sigma^2$, so that

$$\mathbf{V}_{\hat{\mu}} = \mathbf{A}\mathbf{I}_n\mathbf{A}^\mathsf{T}\sigma^2 = \mathbf{A}\sigma^2,$$

by the idempotence (and symmetry) of $\mathbf{A}$. The distribution of $\hat{\boldsymbol{\mu}}$ is degenerate multivariate normal.

Similar arguments apply to the residuals.

$$\hat{\boldsymbol{\epsilon}} = (\mathbf{I} - \mathbf{A})\mathbf{y},$$

so

$$\mathbb{E}(\hat{\boldsymbol{\epsilon}}) = \mathbb{E}(\mathbf{y}) - \mathbb{E}(\hat{\boldsymbol{\mu}}) = \boldsymbol{\mu} - \boldsymbol{\mu} = \mathbf{0}.$$

Proceeding as in the fitted value case we have

$$\mathbf{V}_{\hat{\epsilon}} = (\mathbf{I}_n - \mathbf{A})\mathbf{I}_n(\mathbf{I}_n - \mathbf{A})^\mathsf{T}\sigma^2 = (\mathbf{I}_n - 2\mathbf{A} + \mathbf{A}\mathbf{A})\,\sigma^2 = (\mathbf{I}_n - \mathbf{A})\,\sigma^2.$$

Again, the distribution of the residuals will be degenerate normal. The results for the residuals are useful for model checking, since they allow the residuals to be standardized so that they should have constant variance if the model is correct.

1.3.8 Results in terms of $\mathbf{X}$

The presentation so far has been in terms of the method actually used to fit linear models in practice (the QR decomposition approach¶), which also greatly facilitates the derivation of the distributional results required for practical modelling. However, for historical reasons, these results are more usually presented in terms of the model matrix, $\mathbf{X}$, rather than the components of its QR decomposition. For completeness some of the results are restated here, in terms of $\mathbf{X}$.

Firstly consider the covariance matrix of $\boldsymbol{\beta}$. This turns out to be $(\mathbf{X}^\mathsf{T}\mathbf{X})^{-1}\sigma^2$, which is easily seen to be equivalent to (1.7) as follows:

$$\mathbf{V}_{\hat{\beta}} = (\mathbf{X}^\mathsf{T}\mathbf{X})^{-1}\sigma^2 = \left(\mathbf{R}^\mathsf{T}\mathbf{Q}_\mathrm{f}^\mathsf{T}\mathbf{Q}_\mathrm{f}\mathbf{R}\right)^{-1}\sigma^2 = \left(\mathbf{R}^\mathsf{T}\mathbf{R}\right)^{-1}\sigma^2 = \mathbf{R}^{-1}\mathbf{R}^{-\mathsf{T}}\sigma^2.$$

The expression for the least squares estimates is $\hat{\boldsymbol{\beta}} = (\mathbf{X}^\mathsf{T}\mathbf{X})^{-1}\mathbf{X}^\mathsf{T}\mathbf{y}$, which is equivalent to (1.6):

$$\hat{\boldsymbol{\beta}} = (\mathbf{X}^\mathsf{T}\mathbf{X})^{-1}\mathbf{X}^\mathsf{T}\mathbf{y} = \mathbf{R}^{-1}\mathbf{R}^{-\mathsf{T}}\mathbf{R}^\mathsf{T}\mathbf{Q}_\mathrm{f}^\mathsf{T}\mathbf{y} = \mathbf{R}^{-1}\mathbf{Q}_\mathrm{f}^\mathsf{T}\mathbf{y} = \mathbf{R}^{-1}\mathbf{f}.$$

Given this last result, it is easy to see that the influence matrix can be written:

$$\mathbf{A} = \mathbf{X}(\mathbf{X}^\mathsf{T}\mathbf{X})^{-1}\mathbf{X}^\mathsf{T}.$$

These results are of largely historical and theoretical interest: they should not generally be used for computational purposes, and derivation of the distributional results is much more difficult if one starts from these formulae.

¶Some software still fits models by solution of $\mathbf{X}^\mathsf{T}\mathbf{X}\hat{\boldsymbol{\beta}} = \mathbf{X}^\mathsf{T}\mathbf{y}$, but this is less computationally stable than the orthogonal decomposition method described here, although it is a bit faster.

1.3.9 The Gauss Markov Theorem: What's special about least squares?

How good are least squares estimators? In particular, might it be possible to find better estimators, in the sense of having lower variance while still being unbiased? The Gauss Markov theorem shows that least squares estimators have the lowest variance of all unbiased estimators that are linear (meaning that the data only enter the estimation process in a linear way).

Theorem 1. *Suppose that* $\boldsymbol{\mu} \equiv \mathbb{E}(\mathbf{Y}) = \mathbf{X}\boldsymbol{\beta}$ *and* $\mathbf{V}_y = \sigma^2\mathbf{I}$*, and let* $\tilde{\phi} = \mathbf{c}^\mathsf{T}\mathbf{Y}$ *be any* unbiased *linear estimator of* $\phi = \mathbf{t}^\mathsf{T}\boldsymbol{\beta}$*, where* $\mathbf{t}$ *is an arbitrary vector. Then:*

$$\text{var}(\tilde{\phi}) \geq \text{var}(\hat{\phi})$$

where $\hat{\phi} = \mathbf{t}^\mathsf{T}\hat{\boldsymbol{\beta}}$*, and* $\hat{\boldsymbol{\beta}} = (\mathbf{X}^\mathsf{T}\mathbf{X})^{-1}\mathbf{X}^\mathsf{T}\mathbf{Y}$ *is the least squares estimator of* $\boldsymbol{\beta}$*. Notice that, since* $\mathbf{t}$ *is arbitrary, this theorem implies that each element of* $\hat{\boldsymbol{\beta}}$ *is a minimum variance unbiased estimator.*

Proof. Since $\tilde{\phi}$ is a linear transformation of $\mathbf{Y}$, $\text{var}(\tilde{\phi}) = \mathbf{c}^\mathsf{T}\mathbf{c}\sigma^2$. To compare variances of $\hat{\phi}$ and $\tilde{\phi}$ it is also useful to express $\text{var}(\hat{\phi})$ in terms of $\mathbf{c}$. To do this, note that unbiasedness of $\tilde{\phi}$ implies that

$$\mathbb{E}(\mathbf{c}^\mathsf{T}\mathbf{Y}) = \mathbf{t}^\mathsf{T}\boldsymbol{\beta} \Rightarrow \mathbf{c}^\mathsf{T}\mathbb{E}(\mathbf{Y}) = \mathbf{t}^\mathsf{T}\boldsymbol{\beta} \Rightarrow \mathbf{c}^\mathsf{T}\mathbf{X}\boldsymbol{\beta} = \mathbf{t}^\mathsf{T}\boldsymbol{\beta} \Rightarrow \mathbf{c}^\mathsf{T}\mathbf{X} = \mathbf{t}^\mathsf{T}.$$

So the variance of $\hat{\phi}$ can be written as

$$\text{var}(\hat{\phi}) = \text{var}(\mathbf{t}^\mathsf{T}\hat{\boldsymbol{\beta}}) = \text{var}(\mathbf{c}^\mathsf{T}\mathbf{X}\hat{\boldsymbol{\beta}}).$$

This is the variance of a linear transformation of $\hat{\boldsymbol{\beta}}$, and the covariance matrix of $\hat{\boldsymbol{\beta}}$ is $(\mathbf{X}^\mathsf{T}\mathbf{X})^{-1}\sigma^2$, so

$$\text{var}(\hat{\phi}) = \text{var}(\mathbf{c}^\mathsf{T}\mathbf{X}\hat{\boldsymbol{\beta}}) = \mathbf{c}^\mathsf{T}\mathbf{X}(\mathbf{X}^\mathsf{T}\mathbf{X})^{-1}\mathbf{X}^\mathsf{T}\mathbf{c}\sigma^2 = \mathbf{c}^\mathsf{T}\mathbf{A}\mathbf{c}\sigma^2$$

(where $\mathbf{A}$ is the influence or hat matrix). Now the variances of the two estimators can be directly compared, and it can be seen that $\text{var}(\tilde{\phi}) \geq \text{var}(\hat{\phi})$ iff

$$\mathbf{c}^\mathsf{T}(\mathbf{I} - \mathbf{A})\mathbf{c} \geq 0.$$

This condition will always be met, because it is equivalent to:

$$\{(\mathbf{I} - \mathbf{A})\mathbf{c}\}^\mathsf{T}(\mathbf{I} - \mathbf{A})\mathbf{c} \geq 0$$

by the idempotency and symmetry of $\mathbf{A}$ and hence of $(\mathbf{I} - \mathbf{A})$, but this last condition is saying that a sum of squares can not be less than 0, which is clearly true. □

Notice that this theorem uses independence and equal variance assumptions, but does not assume normality. Of course there is a sense in which the theorem is intuitively rather unsurprising, since it says that the minimum variance estimators are those obtained by seeking to minimize the residual variance.

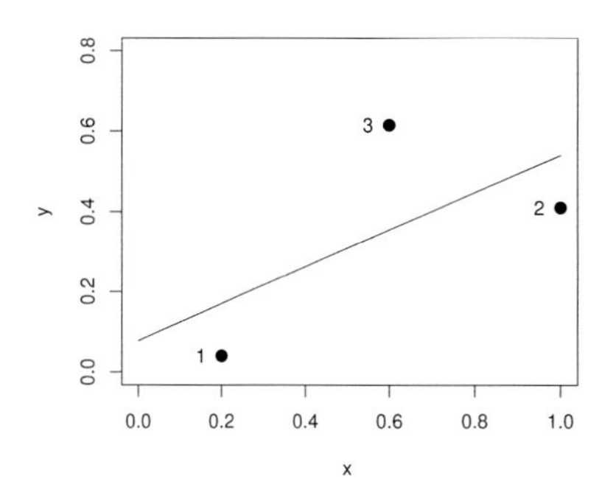

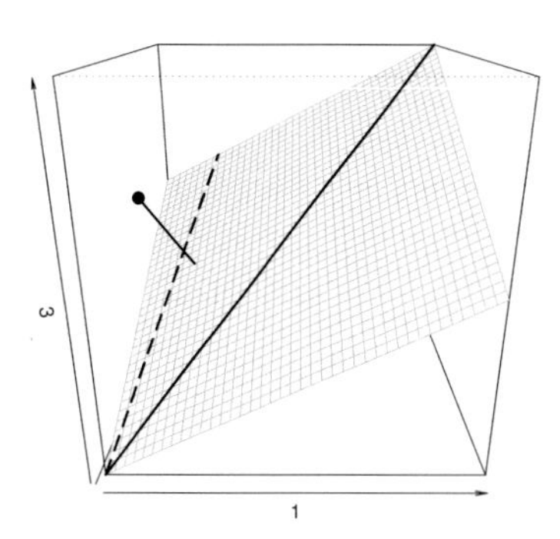

Figure 1.4 *The geometry of least squares. The left panel shows a straight line model fitted to 3 data by least squares. The right panel gives a geometric interpretation of the fitting process. The 3-dimensional space shown is spanned by 3 orthogonal axes: one for each response variable. The observed response vector,* $\mathbf{y}$, *is shown as a point (•) within this space. The columns of the model matrix define two directions within the space: the thick and dashed lines from the origin. The model states that* $\mathbb{E}(\mathbf{y})$ *could be any linear combination of these vectors, i.e., anywhere in the 'model subspace' indicated by the grey plane. Least squares fitting finds the closest point in the model subspace to the response data (•): the 'fitted values'. The short thick line joins the response data to the fitted values: it is the 'residual vector'.*

1.4 The geometry of linear modelling

A full understanding of what is happening when models are fitted by least squares is facilitated by taking a geometric view of the fitting process. Some of the results derived in the last few sections become rather obvious when viewed in this way.

1.4.1 Least squares

Again consider the linear model,

$$\boldsymbol{\mu} = \mathbf{X}\boldsymbol{\beta}, \quad \mathbf{y} \sim N(\boldsymbol{\mu}, \mathbf{I}_n\sigma^2),$$

where $\mathbf{X}$ is an $n \times p$ model matrix. But now consider an n-dimensional Euclidean space, $\Re^n$, in which $\mathbf{y}$ defines the location of a single point. The space of all possible linear combinations of the columns of $\mathbf{X}$ defines a subspace of $\Re^n$, the elements of this space being given by $\mathbf{X}\boldsymbol{\beta}$, where $\boldsymbol{\beta}$ can take any value in $\Re^p$: this space will be referred to as the *space of* $\mathbf{X}$ (strictly the *column* space). So, a linear model states that $\boldsymbol{\mu}$, the expected value of $\mathbf{Y}$, lies in the space of $\mathbf{X}$. Estimating a linear model by least squares, amounts to finding the point, $\hat{\boldsymbol{\mu}} \equiv \mathbf{X}\hat{\boldsymbol{\beta}}$, in the space of $\mathbf{X}$, that is closest to the observed data $\mathbf{y}$. Equivalently, $\hat{\boldsymbol{\mu}}$ is the orthogonal projection of $\mathbf{y}$ on to the space of $\mathbf{X}$. An obvious, but important, consequence of this is that the residual vector, $\hat{\boldsymbol{\epsilon}} = \mathbf{y} - \hat{\boldsymbol{\mu}}$, is orthogonal to all vectors in the space of $\mathbf{X}$.

Figure 1.4 illustrates these ideas for the simple case of a straight line regression model for 3 data (shown conventionally in the left hand panel). The response data

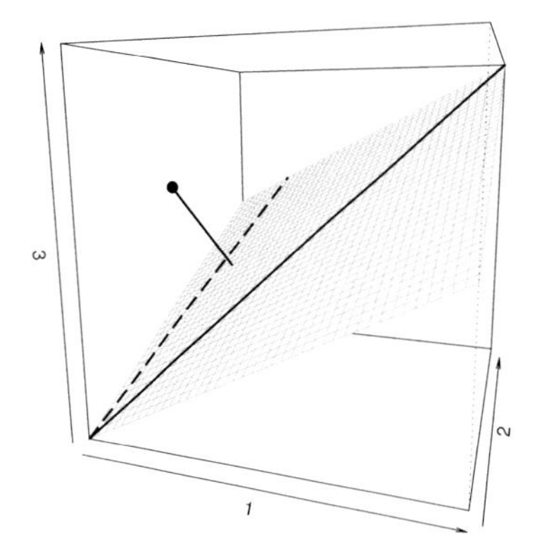

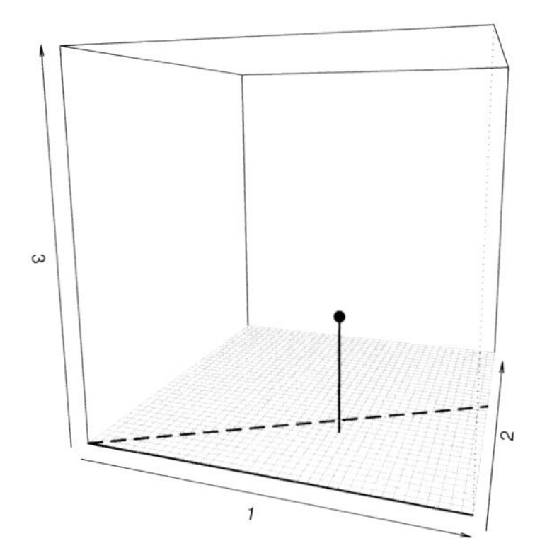

Figure 1.5 *The geometry of fitting via orthogonal decompositions. The left panel illustrates the geometry of the simple straight line model of 3 data introduced in figure 1.4. The right hand panel shows how this original problem appears after rotation by* $\mathbf{Q}^\mathsf{T}$*, the transpose of the orthogonal factor in a QR decomposition of* $\mathbf{X}$*. Notice that in the rotated problem the model subspace only has non-zero components relative to* p *axes (2 axes for this example), while the residual vector has only zero components relative to those same axes.*

and model are

$$\mathbf{y} = \begin{bmatrix} .04 \\ .41 \\ .62 \end{bmatrix} \text{ and } \boldsymbol{\mu} = \begin{bmatrix} 1 & 0.2 \\ 1 & 1.0 \\ 1 & 0.6 \end{bmatrix} \begin{bmatrix} \beta_1 \\ \beta_2 \end{bmatrix}.$$

Since $\boldsymbol{\beta}$ is unknown, the model simply says that $\boldsymbol{\mu}$ could be any linear combination of the vectors $[1, 1, 1]^\mathsf{T}$ and $[.2, 1, .6]^\mathsf{T}$. As the right hand panel of figure 1.4 illustrates, fitting the model by least squares amounts to finding the particular linear combination of the columns of these vectors that is as close to $\mathbf{y}$ as possible (in terms of Euclidean distance).

1.4.2 Fitting by orthogonal decompositions

Recall that the actual calculation of least squares estimates involves first forming the QR decomposition of the model matrix, so that

$$\mathbf{X} = \mathbf{Q} \begin{bmatrix} \mathbf{R} \\ \mathbf{0} \end{bmatrix},$$

where $\mathbf{Q}$ is an $n \times n$ orthogonal matrix and $\mathbf{R}$ is a $p \times p$ upper triangular matrix. Orthogonal matrices rotate vectors (without changing their length) and the first step in least squares estimation is to rotate both the response vector, $\mathbf{y}$, and the columns of the model matrix, $\mathbf{X}$, in exactly the same way, by pre-multiplication with $\mathbf{Q}^\mathsf{T}$.[‖]

[‖] In fact the QR decomposition is not uniquely defined, in that the sign of rows of $\mathbf{Q}$, and corresponding columns of $\mathbf{R}$, can be switched, without changing $\mathbf{X}$ — these sign changes are equivalent to reflections

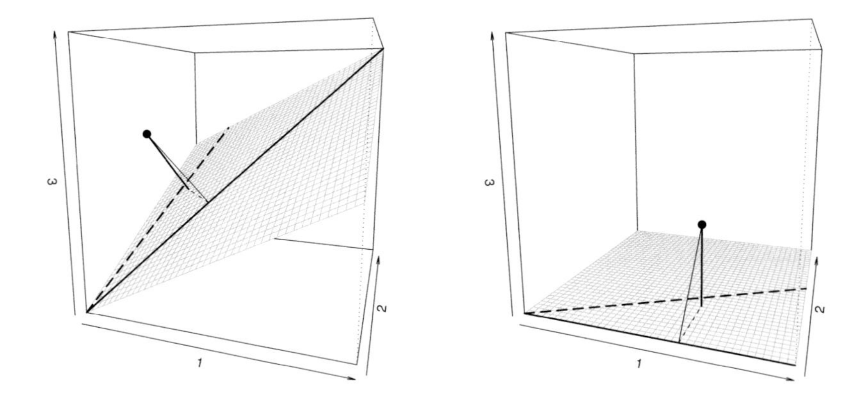

Figure 1.6 *The geometry of nested models.*

Figure 1.5 illustrates this rotation for the example shown in figure 1.4. The left panel shows the response data and model space, for the original problem, while the right hand panel shows the data and model space after rotation by $\mathbf{Q}^\mathsf{T}$. Notice that, since the problem has simply been rotated, the relative position of the data and basis vectors (columns of $\mathbf{X}$) has not changed. What has changed is that the problem now has a particularly convenient orientation relative to the axes. The first two components of the fitted value vector can now be read directly from axes 1 and 2, while the third component is simply zero. By contrast, the residual vector has zero components relative to axes 1 and 2, and its non-zero component can be read directly from axis 3. In terms of section 1.3.1, these vectors are $[\mathbf{f}^\mathsf{T}, \mathbf{0}^\mathsf{T}]^\mathsf{T}$ and $[\mathbf{0}^\mathsf{T}, \mathbf{r}^\mathsf{T}]^\mathsf{T}$, respectively.

The $\hat{\beta}$ corresponding to the fitted values is now easily obtained. Of course we usually require fitted values and residuals to be expressed in terms of the un-rotated problem, but this is simply a matter of reversing the rotation using $\mathbf{Q}$, i.e.,

$$\hat{\mu} = \mathbf{Q}\begin{bmatrix} \mathbf{f} \\ \mathbf{0} \end{bmatrix}, \text{ and } \hat{\epsilon} = \mathbf{Q}\begin{bmatrix} \mathbf{0} \\ \mathbf{r} \end{bmatrix}.$$

1.4.3 Comparison of nested models

A linear model with model matrix $\mathbf{X}_0$ is nested within a linear model with model matrix $\mathbf{X}_1$ if they are models for the same response data, and the columns of $\mathbf{X}_0$ span a subspace of the space spanned by the columns of $\mathbf{X}_1$. Usually this simply means that $\mathbf{X}_1$ is $\mathbf{X}_0$ with some extra columns added.

The vector of the difference between the fitted values of two nested linear models is entirely within the subspace of the larger model, and is therefore orthogonal to the residual vector for the larger model. This fact is geometrically obvious, as figure

of vectors, and the sign leading to maximum numerical stability is usually selected in practice. These reflections don't introduce any extra conceptual difficulty, but can make plots less easy to understand, so I have suppressed them in this example.

1.6 illustrates, but it is a key reason why F-ratio statistics have a relatively simple distribution (under the simpler model).

Figure 1.6 is again based on the same simple straight line model that forms the basis for figures 1.4 and 1.5, but this time also illustrates the least squares fit of the simplified model

$$y_i = \beta_0 + \epsilon_i,$$

which is nested within the original straight line model. Again, both the original and rotated versions of the model and data are shown. This time the fine continuous line shows the projection of the response data onto the space of the simpler model, while the fine dashed line shows the vector of the difference in fitted values between the two models. Notice how this vector is orthogonal both to the reduced model subspace and the full model residual vector.

The right panel of figure 1.6 illustrates that the rotation, using the transpose of the orthogonal factor, $\mathbf{Q}$, of the full model matrix, has also lined up the problem very conveniently for estimation of the reduced model. The fitted value vector for the reduced model now has only one non-zero component, which is the component of the rotated response data (•) relative to axis 1. The residual vector has gained the component that the fitted value vector has lost, so it has zero component relative to axis 1, while its other components are the positions of the rotated response data relative to axes 2 and 3.

So, much of the work required for estimating the simplified model has already been done, when estimating the full model. Note, however, that if our interest had been in comparing the full model to the model

$$y_i = \beta_1 x_i + \epsilon_i,$$

then it would have been necessary to reorder the columns of the full model matrix, in order to avoid extra work in this way.

1.5 Practical linear modelling

This section covers practical linear modelling, via an extended example: the analysis of data reported by Baker and Bellis (1993), which they used to support a theory of 'sperm competition' in humans. The basic idea is that it is evolutionarily advantageous for males to (subconciously) increase their sperm count in proportion to the opportunities that their mate may have had for infidelity. Such behaviour has been demonstrated in a wide variety of other animals, and using a sample of student and staff volunteers from Manchester University, Baker and Bellis set out to see if there is evidence for similar behaviour in humans. Two sets of data will be examined: `sperm.comp1` contains data on sperm count, time since last copulation and proportion of that time spent together, for single copulations, from 15 heterosexual couples; `sperm.comp2` contains data on median sperm count, over multiple copulations, for 24 heterosexual couples, along with the weight, height and age of the male and female of each couple, and the volume of one teste of the male. From these data, Baker and Bellis concluded that sperm count increases with the proportion of time,

`lm`	Estimates a linear model by least squares. Returns a fitted model object of class `lm` containing parameter estimates plus other auxiliary results for use by other functions.
`plot`	Produces model checking plots from a fitted model object.
`summary`	Produces summary information about a fitted model, including parameter estimates, associated standard errors, p-values, r^2, etc.
`anova`	Used for model comparison based on F-ratio testing.
`AIC`	Extract Akaike's information criterion for a model fit.
`residuals`	Extract an array of model residuals from a fitted model.
`fitted`	Extract an array of fitted values from a fitted model object.
`predict`	Obtain predicted values from a fitted model, either for new values of the predictor variables, or for the original values. Standard errors of the predictions can also be returned.

Table 1.1 *Some standard linear modelling functions. Strictly all of these functions except* `lm` *itself end* `.lm`*, but when calling them with an object of class* `lm` *this may be omitted.*

since last copulation, that a couple have spent apart, and that sperm count increases with female weight.

In general, practical linear modelling is concerned with finding an appropriate model to explain the relationship of a response (random) variable to some predictor variables. Typically, the first step is to decide on a linear model that can reasonably be supposed capable of describing the relationship, in terms of the predictors included and the functional form of their relationship to the response. In the interests of ensuring that the model is not too restrictive this 'full' model is often more complicated than is necessary, in that the most appropriate value for a number of its parameters may, in fact, be zero. Part of the modelling process is usually concerned with 'model selection': that is deciding which parameter values ought to be zero. At each stage of model selection it is necessary to estimate model parameters by least squares fitting, and it is equally important to check the model assumptions (particularly equal variance and independence) by examining diagnostic plots. Once a model has been selected and estimated, its parameter estimates can be interpreted, in part with the aid of confidence intervals for the parameters, and possibly with other follow-up analyses. In R these practical modelling tasks are facilitated by a large number of functions for linear modelling, some of which are listed in table 1.1.

1.5.1 Model fitting and model checking

The first thing to do with the sperm competition data is to have a look at them.

```
library(gamair)
pairs(sperm.comp1[,-1])
```

produces the plot shown in figure 1.7. The columns of the data frame are plotted against each other pairwise (with each pairing transposed between lower left and upper right of the plot); the first column has been excluded from the plot as it sim-

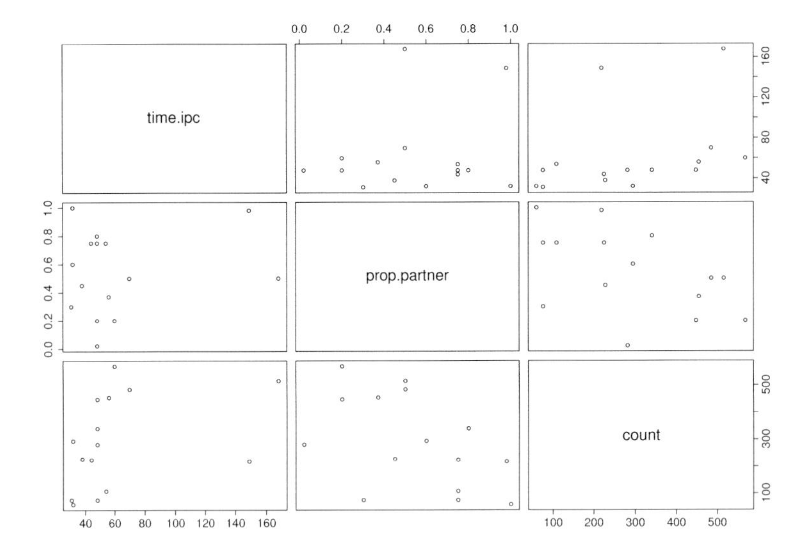

Figure 1.7 *Pairs plot of the sperm competition data from Baker and Bellis (1993). 'count' is sperm count (millions) from one copulation, 'time.ipc' is time (hours) since the previous copulation and 'prop.partner' is the proportion of the time since the previous copulation that the couple have spent together.*

ply contains subject identification labels. The clearest pattern seems to be of some decrease in sperm count as the proportion of time spent together increases.

Following Baker and Bellis, a reasonable initial model might be

$$y_i = \beta_0 + t_i\beta_1 + p_i\beta_2 + \epsilon_i, \tag{1.10}$$

where y_i is sperm count (`count`), t_i is the time since last inter-pair copulation (`time.ipc`) and p_i is the proportion of time, since last copulation, that the pair have spent together (`prop.partner`). As usual, the β_j are unknown parameters and the ϵ_i are i.i.d. $N(0, \sigma^2)$ random variables. Really this model defines the *class* of models thought to be appropriate: it is not immediately clear whether either of β_1 or β_2 are non-zero.

The following fits the model (1.10) and stores the results in an object called `sc.mod1`.

```
sc.mod1 <- lm(count ~ time.ipc + prop.partner, sperm.comp1)
```

The first argument to `lm` is a model formula, specifying the structure of the model to be fitted. In this case, the response (to the left of ~) is `count`, and this is to depend on variables `time.ipc` and `prop.partner`. By default, the model will include an intercept term, unless it is suppressed by a '`-1`' in the formula. The second argument to `lm` supplies a data frame within which the variables in the formula can be found.

The terms on the right hand side of the model formula specify how the model matrix, **X**, is to be specified. In fact, in this example, the terms give the model matrix columns directly. It is possible to check the model matrix of a linear model:

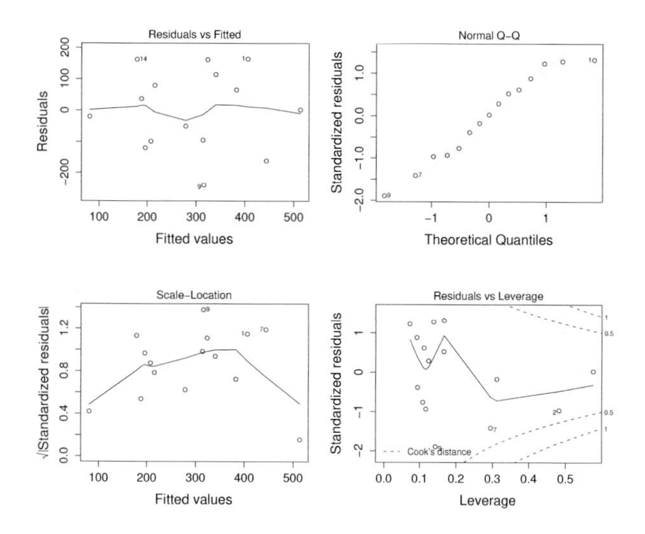

Figure 1.8 *Model checking plots for the linear model* `sc.mod1`.

```
> model.matrix(sc.mod1)
   (Intercept) time.ipc prop.partner
1            1       60         0.20
2            1      149         0.98
3            1       70         0.50
4            1      168         0.50
5            1       48         0.20
6            1       32         1.00
7            1       48         0.02
8            1       56         0.37
9            1       31         0.30
10           1       38         0.45
11           1       48         0.75
12           1       54         0.75
13           1       32         0.60
14           1       48         0.80
15           1       44         0.75
```

Having fitted the model, it is important to check the plausibility of the assumptions, graphically.

```
par(mfrow=c(2,2))  # split the graphics device into 4 panels
plot(sc.mod1)      # (uses plot.lm as sc.mod1 is class 'lm')
```

The resulting plots, shown in figure 1.8, require some explanation. In two of the plots the residuals have been scaled, by dividing them by their estimated standard deviation (see section 1.3.7). If the model assumptions are met, then this standardization should result in residuals that look like $N(0,1)$ random deviates.

- The upper left plot shows the model residuals, $\hat{\epsilon}_i$, against the model fitted values,

$\hat{\mu}_i$, where $\hat{\boldsymbol{\mu}} = \mathbf{X}\hat{\boldsymbol{\beta}}$ and $\hat{\boldsymbol{\epsilon}} = \mathbf{y} - \hat{\boldsymbol{\mu}}$. The residuals should be evenly scattered above and below zero (the distribution of fitted values is not of interest). A trend in the mean of the residuals would violate the assumption of independent response variables, and usually results from an erroneous model structure: e.g., assuming a linear relationship with a predictor, when a quadratic is required, or omitting an important predictor variable. A trend in the variability of the residuals suggests that the variance of the response is related to its mean, violating the constant variance assumption: transformation of the response, or use of a GLM, may help in such cases. The plot shown does not indicate any problem.

- The lower left plot is a scale-location plot. The raw residuals are standardized by dividing by their estimated standard deviation, $\hat{\sigma}\sqrt{1 - A_{ii}}$ ($\mathbf{A}$ is the influence matrix). The square root of the absolute value of each standardized residual is then plotted against the equivalent fitted value. It can be easier to judge the constant variance assumption from such a plot, and the square root transformation reduces the skew in the distribution, which would otherwise be likely to occur. Again, the plot shown gives no reason to doubt the constant variance assumption.
- The upper right panel is a normal QQ (quantile-quantile) plot. The standardized residuals are sorted and plotted against the quantiles of a standard normal distribution. If the residuals are normally distributed then the resulting plot should look like a straight line relationship, perturbed by some correlated random scatter. The current plot fits this description, so the normality assumption seems plausible.
- The lower right panel plots the standardized residuals against the *leverage* of each datum. The leverage is simply A_{ii}, which measures the *potential* for the ith datum to influence the overall model fit. A large residual combined with high leverage implies that the corresponding datum has a substantial influence on the overall fit. A quantitative summary of how much influence each datum actually has is provided by its *Cook's distance*. If $\hat{\mu}_i^{[k]}$ is the ith fitted value when the kth datum is omitted from the fit, then Cook's distance is

$$d_k = \frac{1}{(p+1)\hat{\sigma}^2} \sum_{i=1}^{n} (\hat{\mu}_i^{[k]} - \hat{\mu}_i)^2, \tag{1.11}$$

  where p and n are the numbers of parameters and data, respectively. A large value of d_k indicates a point that has a substantial influence on the model results. If the Cook's distance values indicate that model estimates may be very sensitive to just a few data, then it usually prudent to repeat any analysis without the offending points, in order to check the robustness of the modelling conclusions. d_k can be shown to be a function of leverage and standardized residual, so contours of Cook's distance are shown on the plot, from which the Cook's distance for any datum can be read. In this case none of the points look wildly out of line.

The QQ-plot shows the reference line that a 'perfect' set of residuals would follow, while the other plots have a simple smooth overlaid, in order to guide the eye when looking for patterns in the residuals (these can be very useful for large datasets). By default the 'most extreme' three points in each plot are labelled with their row labels from the original data frame, so that the corresponding data can readily be checked. The 9th datum is flagged in all four plots in figure 1.8. It should be checked:

```
> sperm.comp1[9,]
  subject time.ipc prop.partner count
9       P       31          0.3    76
```

This subject has quite a low count, but not the lowest in the frame. Examination of the plot of `count` against `prop.partner` indicates that the point adds substantially to the uncertainty surrounding the relationship, but it is hard to see a good reason to remove it, particularly since, if anything, it is obscuring the relationship, rather than exaggerating it.

Since the assumptions of model (1.10) appear reasonable, we can proceed to examine the fitted model object. Typing the name of an object in R causes the default print method for the object to be invoked (`print.lm` in this case).

```
> sc.mod1

Call:
lm(formula= count ~ time.ipc + prop.partner, data=sperm.comp1)

Coefficients:
 (Intercept)      time.ipc  prop.partner
     357.418         1.942      -339.560
```

The intercept parameter (β_0) is estimated to be 357.4. Notionally, this would be the count expected if `time.ipc` and `prop.partner` were zero, but the value is biologically implausible if interpreted in this way. Given that the smallest observed `time.ipc` was 31 hours we cannot really expect to predict the count near zero. The remaining two parameter estimates are $\hat{\beta}_1$ and $\hat{\beta}_2$, and are labelled by the name of the variable to which they relate. In both cases they give the expected increase in count for a unit increase in their respective predictor variable. Note the important point that the absolute values of the parameter estimates are only interpretable *relative* to the variable which they multiply. For example, we are not entitled to conclude that the effect of `prop.partner` is much greater than that of `time.ipc`, on the basis of the relative magnitudes of the respective parameters: they are measured in completely different units.

One point to consider is whether `prop.partner` is the most appropriate predictor variable. Perhaps the total time spent together (in hours) would be a better predictor.

```
sc.mod2 <- lm(count ~ time.ipc + I(prop.partner*time.ipc),
              sperm.comp1)
```

would fit such a model. The term `I(prop.partner*time.ipc)` indicates that, rather than use proportion of time together as a predictor, total time should be used. The `I()` function is used to 'protect' the product `prop.partner*time.ipc` within the model formula. This is necessary because symbols like `+` and `*` have special meanings within model formulae (see section 1.7): by protecting terms using `I()`, the usual arithmetic meanings are restored. Examination of diagnostic plots for `sc.mod2` shows that two points have much greater influence on the fit than the others, so for the purposes of this section it seems sensible to stick with the biologists' preferred model structure and use `prop.partner`.

1.5.2 *Model* `summary`

The `summary`** function provides more information about the fitted model.

```
> summary(sc.mod1)

Call:
lm(formula= count ~ time.ipc + prop.partner, data=sperm.comp1)

Residuals:
     Min       1Q   Median       3Q      Max
-239.740  -96.772    2.171   96.837  163.997

Coefficients:
             Estimate Std. Error t value Pr(>|t|)
(Intercept)  357.4184    88.0822   4.058  0.00159 **
time.ipc       1.9416     0.9067   2.141  0.05346 .
prop.partner -339.5602  126.2535  -2.690  0.01969 *
---
Signif. codes:  0 `***' 0.001 `**' 0.01 `*' 0.05 `.' 0.1 ` ' 1

Residual standard error: 136.6 on 12 degrees of freedom
Multiple R-Squared: 0.4573,     Adjusted R-squared: 0.3669
F-statistic: 5.056 on 2 and 12 DF,  p-value: 0.02554
```

The explanations of the parts of this output are as follows:

`Call` simply reminds you of the call that generated the object being summarized.

`Residuals` gives a five figure summary of the residuals: this should indicate any gross departure from normality. For example, very skewed residuals might lead to very different magnitudes for `Q1` and `Q2`, or to the median residual being far from 0 (the mean residual is always zero if the model includes an intercept: see exercise 6).

`Coefficients` gives a table relating to the estimated parameters of the model. The first two columns are the least squares estimates ($\hat{\beta}_j$) and the estimated standard errors associated with those estimates ($\hat{\sigma}_{\hat{\beta}_j}$), respectively. The standard error calculations follow sections 1.3.2 and 1.3.3. The third column gives the parameter estimates divided by their estimated standard errors: $T_j \equiv \hat{\beta}_j/\hat{\sigma}_{\hat{\beta}_j}$, which is a standardized measure of how far each parameter estimate is from zero. It was shown in section 1.3.3 (p. 13) that under $\text{H}_0 : \beta_j = 0$,

$$T_j \sim t_{n-p}, \tag{1.12}$$

**Note that calling the `summary` function with a fitted linear model object, `x`, actually results in the following: `summary` looks at the class of `x`, finds that it is `"lm"` and passes it to `summary.lm`; `summary.lm` calculates a number of interesting quantities from `x` which it returns in a list, `y`, of class `lm.summary`; unless `y` is assigned to an object, R prints it, using the print method `print.lm.summary`.

where n is the number of data and p the number of model parameters estimated. i.e., if the null hypothesis is true, then the observed T_j should be consistent with having been drawn from a t_{n-p} distribution. The final `Pr(>|t|)` column provides the measure of that consistency, namely the probability that the magnitude of a t_{n-p} random variable would be at least as large as the observed T_j. This quantity is known as the **p-value** of the test of $\text{H}_0 : \beta_j = 0$. A large p-value indicates that the data are consistent with the hypothesis, in that the observed T_j is quite a probable value for a t_{n-p} deviate, so that there is no reason to doubt the hypothesis underpinning (1.12). Conversely a small p-value suggests that the hypothesis is wrong, since the observed T_j is a rather improbable observation from the t_{n-p} distribution implied by $\beta_j = 0$. Various arbitrary **significance levels** are often used as the boundary p-values for deciding whether to accept or reject hypotheses. Some are listed at the foot of the table, and the p-values are flagged according to which, if any, they fall below.

`Residual standard error` gives $\hat{\sigma}$ where $\hat{\sigma}^2 = \sum \hat{\epsilon}_i^2/(n-p)$ (see section 1.3.3). $n-p$ is the 'residual degrees of freedom'.

`Multiple R-squared` is an estimate of the proportion of the variance in the data explained by the regression:

$$r^2 = 1 - \frac{\sum \hat{\epsilon}_i^2/n}{\sum (y_i - \bar{y})^2/n}$$

where $\bar{y}$ is the mean of the y_i. The fraction in this expression is basically an estimate of the proportion variance not explained by the regression.

`Adjusted R-squared`. The problem with r^2 is that it always increases when a new predictor variable is added to the model, no matter how useless that variable is for prediction. Part of the reason for this is that the variance estimates used to calculate r^2 are biased in a way that tends to inflate r^2. If unbiased estimators are used we get the adjusted r^2

$$r^2_{\text{adj}} = 1 - \frac{\sum \hat{\epsilon}_i^2/(n-p)}{\sum (y_i - \bar{y})^2/(n-1)}.$$

A high value of r^2_{adj} indicates that the model is doing well at explaining the variability in the response variable.

`F-statistic`. The final line, giving an F-statistic and p-value, is testing the null hypothesis that the data were generated from a model with only an intercept term, against the alternative that the fitted model generated the data. This line is really about asking if the whole model is of any use. The theory of such tests is covered in section 1.3.5.

Note that if the model formula contains '`-1`' to suppress the intercept, then `summary.lm` has the feature that r^2 is computed with the mean of the response data replaced by zero. This avoids r^2 being negative, but generally causes the r^2 to increase massively (and meaninglessly) if you drop the intercept from a model. The same feature applies to the F-ratio statistic — instead of comparing the fitted model to the model with just an intercept, it is compared to the model in which the mean is zero. Personally, when the model contains no intercept, I always re-compute the r^2 values using the observed mean of the data in place of R's default zero.

The summary of `sc.mod1` suggests that there is evidence that the model is better than one including just a constant (p-value = 0.02554). There is quite clear evidence that `prop.partner` is important in predicting sperm count (p-value = 0.019), but less evidence that `time.ipc` matters (p-value = 0.053). Indeed, using the conventional significance level of 0.05, we might be tempted to conclude that `time.ipc` does not affect count at all. Finally note that the model leaves most of the variability in count unexplained, since $r^2_{\rm adj}$ is only 37%.

1.5.3 Model selection

From the model summary it appears that `time.ipc` may not be necessary: the associated p-value of 0.053 does not provide strong evidence that the true value of β_1 is non-zero. By the 'true value' is meant the value of the parameter in the model imagined to have actually generated the data; or equivalently, the value of the parameter applying to the whole population of couples from which, at least conceptually, our particular sample has been randomly drawn. The question then arises of whether a simpler model, without any dependence on `time.ipc`, might be appropriate. This is a question of *model selection*. Usually it is a good idea to avoid overcomplicated models, dependent on irrelevant predictor variables, for reasons of interpretability and efficiency. Interpretations about causality will be made more difficult if a model contains spurious predictors, but estimates using such a model will also be less precise, as more parameters than necessary have been estimated from the finite amount of uncertain data available.

Several approaches to model selection are based on hypothesis tests about model terms, and can be thought of as attempting to find the simplest model consistent with a set of data, where consistency is judged relative to some threshold p-value. For the sperm competition model the p-value for `time.ipc` is greater than 0.05, so this predictor might be a candidate for dropping.

```
> sc.mod3 <- lm(count ~ prop.partner, sperm.comp1)
> summary(sc.mod3)
(edited)
Coefficients:
             Estimate Std. Error t value Pr(>|t|)
(Intercept)    451.50      86.23   5.236 0.000161 ***
prop.partner  -292.23     140.40  -2.081 0.057727 .
---

Residual standard error: 154.3 on 13 degrees of freedom
Multiple R-Squared:  0.25,      Adjusted R-squared: 0.1923
F-statistic: 4.332 on 1 and 13 DF,  p-value: 0.05773
```

These results provide a good example of why it is dangerous to apply automatic model selection procedures unthinkingly. In this case dropping `time.ipc` has made the estimate of the parameter multiplying `prop.partner` less precise: indeed this term also has a p-value greater than 0.05 according to this new fit. Furthermore, the new model has a much reduced r^2, while the model's overall p-value does not give strong evidence that it is better than a model containing only an intercept. The only sensible choice here is to revert to `sc.mod1`. The statistical evidence indicates that it is better than the intercept only model, and dropping its possibly 'non-significant' term has led to a much worse model.

Hypothesis testing is not the only approach to model selection. One alternative is to try and find the model that gets as close as possible to the true model, rather than to find the simplest model consistent with data. In this case we can attempt to find the model which does the best job of predicting the $\mathbb{E}(y_i)$. Selecting models in order to minimize Akaike's Information Criterion (AIC) is one way of trying to do this (see section 1.8.6). In R, the `AIC` function can be used to calculate the AIC statistic for different models.

```
> sc.mod4 <- lm(count ~ 1, sperm.comp1) # null model
> AIC(sc.mod1,sc.mod3,sc.mod4)
        df      AIC
sc.mod1  4 194.7346
sc.mod3  3 197.5889
sc.mod4  2 199.9031
```

This alternative model selection approach also suggests that the model with both `time.ipc` and `prop.partner` is best.

So, on the basis of `sperm.comp1`, there seems to be reasonable evidence that sperm count increases with `time.ipc` but decreases with `prop.partner`: exactly as Baker and Bellis concluded.

1.5.4 Another model selection example

The second data set from Baker and Bellis (1993) is `sperm.comp2`. This gives median sperm count for 24 couples, along with ages (years), heights (cm) and weights (kg) for the male and female of each couple and volume (cm^3) of one teste for the male of the couple (`m.vol`). There are quite a number of missing values for the predictors, particularly for `m.vol`, but, for the 15 couples for which there is an `m.vol` measurement, the other predictors are also available. The number of copulations over which the median count has been taken varies widely from couple to couple. Ideally one should probably allow within couple and between couple components to the random variability component of the data to allow for this, but this will not be done here. Following Baker and Bellis it seems reasonable to start from a model including linear effects of all predictors, i.e.,

$$\texttt{count}_i = \beta_0 + \beta_1\texttt{f.age}_i + \beta_2\texttt{f.weight}_i + \beta_3\texttt{f.height}_i + \beta_4\texttt{m.age}_i \\ + \beta_5\texttt{m.weight}_i + \beta_6\texttt{m.height}_i + \beta_7\texttt{m.vol} + \epsilon_i$$

The following estimates and summarizes the model, and plots diagnostics.

```
> sc2.mod1 <- lm(count ~ f.age + f.height + f.weight + m.age +
+                 m.height + m.weight + m.vol, sperm.comp2)
> plot(sc2.mod1)
> summary(sc2.mod1)
```

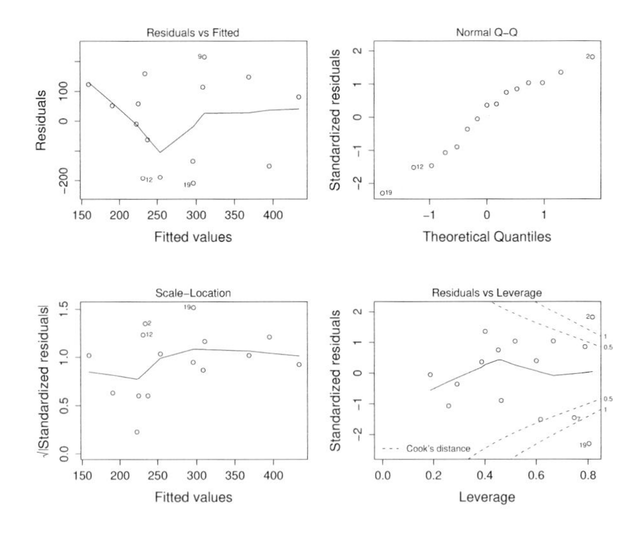

Figure 1.9 *Model checking plots for the* `sc2.mod1` *model.*

```
[edited]
Coefficients:
              Estimate Std. Error t value Pr(>|t|)
(Intercept) -1098.518   1997.984  -0.550    0.600
f.age          10.798     22.755   0.475    0.650
f.height       -4.639     10.910  -0.425    0.683
f.weight       19.716     35.709   0.552    0.598
m.age          -1.722     10.219  -0.168    0.871
m.height        6.009     10.378   0.579    0.581
m.weight       -4.619     12.655  -0.365    0.726
m.vol           5.035     17.652   0.285    0.784

Residual standard error: 205.1 on 7 degrees of freedom
Multiple R-Squared: 0.2192,     Adjusted R-squared: -0.5616
F-statistic: 0.2807 on 7 and 7 DF,  p-value: 0.9422
```

The resulting figure 1.9 looks reasonable, but datum 19 appears to produce the most extreme point on all 4 plots. Checking row 19 of the data frame shows that the male of this couple is rather heavy (particularly for his height), and has a large `m.vol` measurement, but a count right near the bottom of the distribution (actually down at the level that might be expected to cause fertility problems, if this is typical). Clearly, whatever we conclude from these data will need to be double-checked without this observation. Notice, from the summary, how poorly this model does at explaining the count variability: the adjusted r^2 is actually negative, an indication that we have a large number of irrelevant predictors in the model.

There are only 15 data from which to estimate the 8 parameters of the full model: it would be better to come up with something more parsimonious. One possibility would be to use the `step` function in R to perform model selection automatically. `step` takes a fitted model and repeatedly drops the term that leads to the largest decrease in AIC. By default it also tries adding back in each single term already dropped, to see if that leads to a reduction in AIC. See `?step` for more details. In this case using `AIC` suggests a rather complicated model with only `m.weight` dropped. It seems sensible to switch to hypothesis testing based model selection, and ask whether there is really good evidence that all these terms are necessary. One approach is to perform 'backwards model selection', by repeatedly removing the *single* term with highest p-value, above some threshold (e.g., 0.05), and then refitting the resulting reduced model, until all terms have significant p-values. For example the first step in this process would remove `m.age`:

```
> sc2.mod2 <- lm(count ~ f.age + f.height + f.weight +
+                m.height + m.weight + m.vol,sperm.comp2)
> summary(sc2.mod2)
[edited]
Coefficients:
            Estimate Std. Error t value Pr(>|t|)
(Intercept) -1054.770   1856.843  -0.568    0.586
f.age           8.847     18.359   0.482    0.643
f.height       -5.119      9.871  -0.519    0.618
f.weight       20.259     33.334   0.608    0.560
m.height        6.033      9.727   0.620    0.552
m.weight       -4.473     11.834  -0.378    0.715
m.vol           4.506     16.281   0.277    0.789

Residual standard error: 192.3 on 8 degrees of freedom
Multiple R-Squared: 0.216,      Adjusted R-squared: -0.372
F-statistic: 0.3674 on 6 and 8 DF,  p-value: 0.8805
```

Relative to `sc2.mod1`, the reduced model has different estimates for each of the remaining parameter, as well as smaller standard error estimates for each parameter, and (consequently) different p-values. This is part of the reason for only dropping one term at a time: when we drop one term from a model, it is quite possible for some remaining terms to have their p-values massively reduced. For example, if two terms are highly correlated it is possible for both to be significant individually, but both to have very high p-values if present together. This occurs because the terms are to some extent interchangeable predictors: the information provided by one is much the same as the information provided by the other, so that one must be present in the model but both are not needed. If we were to drop several terms from a model at once, we might miss such effects.

Proceeding with backwards selection, we would drop `m.vol` next. This allows rather more of the couples to be used in the analysis. Continuing in the same way leads to the dropping of `m.weight`, `f.height`, `m.height` and finally `f.age` before arriving at a final model which includes only `f.weight`.

```
> sc2.mod7 <- lm(count ~ f.weight,sperm.comp2)
> summary(sc2.mod7)
[edited]
```

```
Coefficients:
              Estimate Std. Error t value Pr(>|t|)
(Intercept) -1002.281    489.352  -2.048   0.0539 .
f.weight       22.397      8.629   2.595   0.0173 *

Residual standard error: 147.3 on 20 degrees of freedom
Multiple R-Squared: 0.252,      Adjusted R-squared: 0.2146
F-statistic: 6.736 on 1 and 20 DF,  p-value: 0.01730
```

This model does appear to be better than a model containing only a constant, according both to a hypothesis test at the 5% level and AIC.

Apparently then, only female weight influences sperm count. This concurs with the conclusion of Baker and Bellis (1993), who interpreted the findings to suggest that males might 'invest' more in females with a higher reproductive potential. However, in the light of the residual plots we need to re-analyze the data without observation 19, before having too much confidence in the conclusions. This is easily done:

```
> sc <- sperm.comp2[-19,]
> sc3.mod1 <- lm(count ~ f.age + f.height + f.weight + m.age +
+                m.height + m.weight + m.vol, sc)
> summary(sc3.mod1)
[edited]
Coefficients:
            Estimate Std. Error t value Pr(>|t|)
(Intercept) 1687.406   1251.338   1.348   0.2262
f.age         55.248     15.991   3.455   0.0136 *
f.height      21.381      8.419   2.540   0.0441 *
f.weight     -88.992     31.737  -2.804   0.0310 *
m.age        -17.210      6.555  -2.626   0.0393 *
m.height     -11.321      6.869  -1.648   0.1504
m.weight       6.885      7.287   0.945   0.3812
m.vol         48.996     13.938   3.515   0.0126 *
--- [edited]
```

`m.vol` now has the lowest p-value. Repeating the whole backwards selection process, every term now drops out except for `m.vol`, leading to the much less interesting conclusion that the data only really supply evidence that size of testes influences sperm count. Given the rather tedious plausibility of this conclusion, it probably makes sense to prefer it to the conclusion based on the full data set.

A follow-up

Given the biological conclusions from the analysis of `sperm.comp2`, it would make sense to revisit the analysis of `sperm.comp1`. Baker and Bellis do not report `m.vol` values for these data, but the same couples feature in both datasets and are identified by label, so the required values can be obtained:

```
sperm.comp1$m.vol <-
  sperm.comp2$m.vol[sperm.comp2$pair %in% sperm.comp1$subject]
```

Repeating the same sort of backwards selection we end up selecting a 1 term model:

```
> sc1.mod1 <- lm(count ~ m.vol, sperm.comp1)
> summary(sc1.mod1)

Call:
lm(formula = count ~ m.vol, data = sperm.comp1)

Residuals:
     Min       1Q   Median       3Q      Max
-187.236  -55.028   -8.606   75.928  156.257

Coefficients:
            Estimate Std. Error t value Pr(>|t|)
(Intercept)  -58.694    121.619  -0.483   0.6465
m.vol         23.247      7.117   3.266   0.0171 *
---
Signif. codes:  0 `***' 0.001 `**' 0.01 `*' 0.05 `.' 0.1 ` ' 1

Residual standard error: 120.8 on 6 degrees of freedom
Multiple R-Squared:  0.64,      Adjusted R-squared:  0.58
F-statistic: 10.67 on 1 and 6 DF,  p-value: 0.01711
```

Although based on only 8 couples, this must call into question the original analysis, which concluded that time since last copulation and proportion of time spent together controlled sperm count. There is at least a suggestion that the explanation for sperm count variability may be rather more prosaic than the explanation suggested by sperm competition theory.

1.5.5 Confidence intervals

Exactly as in section 1.1.3 the results from section 1.3.3 can be used to obtain confidence intervals for the parameters. In general, for a p parameter model of n data, a $(1-2\alpha)100\%$ confidence interval for the jth parameter is

$$\hat{\beta}_j \pm t_{n-p}(\alpha)\hat{\sigma}_{\hat{\beta}_j},$$

where $t_{n-p}(\alpha)$ is the value below which a t_{n-p} random variable would lie with probability α.

As an example of its use, the following calculates a 95% confidence interval for the mean increase in `count` per cm^3 increase in `m.vol`.

```
> sc.c <- summary(sc1.mod1)$coefficients
> sc.c   # check info extracted from summary
             Estimate Std. Error    t value   Pr(>|t|)
(Intercept) -58.69444 121.619433 -0.4826075 0.64647664
m.vol        23.24653   7.117239  3.2662284 0.01711481
> sc.c[2,1]+qt(c(.025,.975),6)*sc.c[2,2]
[1]  5.831271 40.661784   # 95% CI
```

1.5.6 Prediction

It is possible to predict the expected value of the response at new values of the predictor variables using the `predict` function. For example: what are the model predicted counts for `m.vol` values of 10, 15, 20 and 25? The following obtains the answer along with associated standard errors (see section 1.3.7).

```
> df <- data.frame(m.vol=c(10,15,20,25))
> predict(sc1.mod1,df,se=TRUE)
$fit
       1        2        3        4
173.7708 290.0035 406.2361 522.4688


$se.fit
       1        2        3        4
60.39178 43.29247 51.32314 76.98471
```

The first line creates a data frame containing the predictor variable values at which predictions are required. The second line calls `predict` with the fitted model object and new data from which to predict. `se=TRUE` tells the function to return standard errors along with the predictions.

1.5.7 Co-linearity, confounding and causation

Consider the case in which a predictor variable x is the variable which really controls response variable y, but at the same time, it is also highly correlated with a variable z, which plays no role at all in setting the level of y. The situation is represented by the following simulation:

```
set.seed(1); n <- 100; x <- runif(n)
z <- x + rnorm(n)*.05
y <- 2 + 3 * x + rnorm(n)
```

Despite the fact that z played no role in generating y, the correlation between x and z leads to the following

```
> summary(lm(y~z))

Coefficients:
            Estimate Std. Error t value Pr(>|t|)
(Intercept)   2.1628     0.2245   9.632 7.60e-16 ***
z             2.7342     0.3836   7.127 1.75e-10 ***
```

i.e., there seems to be strong evidence that z is predictive of y. Clearly there must be something wrong with interpreting this as implying that z is actually controlling y, since in this case we know that it isn't. Now the problem here is that both z and y are being controlled by a *confounding* variable that is absent from the model. Since the data were simulated we are in the happy position of knowing what the confounder is. What happens if we include it?

```
> summary(lm(y ~ x + z))

Call:
lm(formula = y ~ x + z)
```

```
Residuals:
    Min      1Q  Median      3Q     Max
-2.8311 -0.7273 -0.0537  0.6338  2.3359

Coefficients:
            Estimate Std. Error t value Pr(>|t|)
(Intercept)   2.1305     0.2319   9.188  7.6e-15 ***
x             1.3750     2.3368   0.588    0.558
z             1.4193     2.2674   0.626    0.533
---
Signif. codes:  0 '***' 0.001 '**' 0.01 '*' 0.05 '.' 0.1 ' ' 1

Residual standard error: 1.056 on 97 degrees of freedom
Multiple R-squared: 0.3437,     Adjusted R-squared: 0.3302
F-statistic:  25.4 on 2 and 97 DF,  p-value: 1.345e-09
```

In this case the linear model fitting has 'shared out' the real dependence on x between x and z (look at the estimated slope parameters), and has given x, the true predictor, a higher p-value than z, suggesting that if anything we should drop x! Notice also that both slope parameter estimates have very high standard errors. Because they are so highly correlated it is not possible to distinguish which is actually driving y, and this is reflected in the high standard errors for both parameters. Notice also the importance of dropping one of x and z here: if we don't then our inferences will be very imprecise.

In this simulated example we know what the truth is, but in most real data analyses we only have the data to tell us. We can not usually know whether there are hidden confounder variables or not. Similarly, even if we do have a complete set of possible predictors, we do not usually have special knowledge of which ones really control the response: we only have the statistical analysis to go on.

These issues are ubiquitous when trying to interpret the results of analysing observational data. For example, a majority of men find a low ratio of waist to hip size attractive in women. A recent study (Brosnan and Walker, 2009) found that men who did not share the usual preference were more likely to have autistic children. Now, if you view this association as causative, then you might be tempted to postulate all sorts of theories about the psychology of autistic brains and the inheritance of autism. However, it is known that being exposed to higher levels of testosterone in the womb is also associated with autism, and women with higher levels of testosterone are less likely to have a low waist to hip ratio. Since these women are more likely to select men who are attracted to them, the original association may have a rather obvious explanation.

So, correlation between known predictor variables, and with un-observed predictors, causes interpretational problems and tends to inflate variances. Does this mean that linear modelling is useless? No. Discovering *associations* can be a very useful part of the scientific process of uncovering *causation*, but it is often the case that we don't care about causation. If all I want to do is to be able to predict y, I don't much care whether I do so by using the variable that actually controlled y, or by using a variable that is a proxy for the controlling variable.

To really establish causation, it is usually necessary to do an experiment in which the putative causative variable is manipulated to see what effect it has on the response. In doing this, it is necessary to be very careful not to introduce correlations between the values of the manipulated variable and other variables (known or unknown) that might influence the response. In experiments carried out on individual units (e.g., patients in a clinical trial) the usual way that this is achieved is to allocate units randomly to the different values of the manipulated variable(s). This *randomization* removes any possibility that characteristics of the units that may effect the response can be correlated with the manipulated variable. Analysis of such randomized experiments often involves factor variables, and these will be covered next.

1.6 Practical modelling with factors

Most of the models covered so far have been for situations in which the predictor variables are variables that measure some quantity (*metric variables*), but there are many situations in which the predictor variables are more qualitative in nature, and serve to divide the responses into groups. Examples might be eye colour of subjects in a psychology experiment, which of three alternative hospitals were attended by patients in a drug trial, manufacturers of cars used in crash tests, etc. Variables like these, which serve to classify the units on which the response variable has been measured into distinct categories, are known as *factor variables*, or simply *factors*. The different categories of the factor are known as *levels* of the factor. For example, levels of the factor 'eye colour' might be 'blue', 'brown', 'grey' and 'green', so that we would refer to eye colour as a factor with four levels. Note that 'levels' is quite confusing terminology: there is not necessarily any natural ordering of the levels of a factor. Hence 'levels' of a factor might best be thought of as 'categories' of a factor, or 'groups' of a factor.

Factor variables are handled using dummy variables. Each factor variable can be replaced by as many dummy variables as there are levels of the factor — one for each level For each response datum, only one of these dummy variables will be non-zero: the dummy variable for the single level that applies to that response. Consider an example to see how this works in practice: 9 laboratory rats are fed too much, so that they divide into 3 groups of 3: 'fat', 'very fat' and 'enormous'. Their blood insulin levels are then measured 10 minutes after being fed a standard amount of sugar. The investigators are interested in the relationship between insulin levels and the factor 'rat size'. Hence a model could be set up in which the predictor variable is the factor 'rat size', with the three levels 'fat', 'very fat' and 'enormous'. Writing y_i for the ith insulin level measurement, a suitable model might be:

$$\mathbb{E}(Y_i) \equiv \mu_i = \begin{cases} \beta_0 & \text{if rat is fat} \\ \beta_1 & \text{if rat is very fat} \\ \beta_2 & \text{if rat is enormous} \end{cases}$$

and this is easily written in linear model form, using a dummy predictor variable for each level of the factor:

$$\begin{bmatrix} \mu_1 \\ \mu_2 \\ \mu_3 \\ \mu_4 \\ \mu_5 \\ \mu_6 \\ \mu_7 \\ \mu_8 \\ \mu_9 \end{bmatrix} = \begin{bmatrix} 1 & 0 & 0 \\ 1 & 0 & 0 \\ 1 & 0 & 0 \\ 0 & 1 & 0 \\ 0 & 1 & 0 \\ 0 & 1 & 0 \\ 0 & 0 & 1 \\ 0 & 0 & 1 \\ 0 & 0 & 1 \end{bmatrix} \begin{bmatrix} \beta_0 \\ \beta_1 \\ \beta_2 \end{bmatrix}.$$

A key difference between dummy variables and directly measured predictor variables is that the dummy variables, and parameters associated with a factor, are almost always treated as a group during model selection — it does not usually make sense for a subset of the dummy variables associated with a factor to be dropped or included on their own: either all are included or none. The F-ratio tests derived in section 1.3.5 are designed for hypothesis testing in this situation.

1.6.1 Identifiability

When modelling with factor variables, model 'identifiability' becomes an important issue. It is quite easy to set up models, involving factors, in which it is impossible to estimate the parameters uniquely, because an infinite set of alternative parameter vectors would give rise to exactly the same expected value vector. A simple example illustrates the problem. Consider again the fat rat example, but suppose that we wanted to formulate the model in terms of an overall mean insulin level, α, and deviations from that level, β_j, associated with each level of the factor. The model would be something like:

$$\mu_i = \alpha + \beta_j \text{ if rat } i \text{ is rat size level } j$$

(where j is 0, 1 or 2, corresponding to 'fat', 'very fat' or 'enormous'). The problem with this model is that there is not a one-to-one correspondence between the parameters and the fitted values, so that the parameters can not be uniquely estimated from the data. This is easy to see. Consider any particular set of α and $\boldsymbol{\beta}$ values, giving rise to a particular $\boldsymbol{\mu}$ value: any constant c could be added to α and simultaneously subtracted from each element of $\boldsymbol{\beta}$ without changing the value of $\boldsymbol{\mu}$. Hence there is an infinite set of parameters giving rise to each $\boldsymbol{\mu}$ value, and therefore the parameters can not be estimated uniquely. The model is not 'identifiable'.

This situation can be diagnosed directly from the model matrix. Written out in full, the example model is

$$\begin{bmatrix} \mu_1 \\ \mu_2 \\ \mu_3 \\ \mu_4 \\ \mu_5 \\ \mu_6 \\ \mu_7 \\ \mu_8 \\ \mu_9 \end{bmatrix} = \begin{bmatrix} 1 & 1 & 0 & 0 \\ 1 & 1 & 0 & 0 \\ 1 & 1 & 0 & 0 \\ 1 & 0 & 1 & 0 \\ 1 & 0 & 1 & 0 \\ 1 & 0 & 1 & 0 \\ 1 & 0 & 0 & 1 \\ 1 & 0 & 0 & 1 \\ 1 & 0 & 0 & 1 \end{bmatrix} \begin{bmatrix} \alpha \\ \beta_0 \\ \beta_1 \\ \beta_2 \end{bmatrix}.$$

But the columns of the model matrix are not independent, and this lack of full column rank means that the formulae for finding the least squares parameter estimates break down.[††] Identifiable models have model matrices of full column rank; unidentifiable models are column rank deficient.

The solution to the identifiability problem is to impose just enough linear constraints on the model parameters that the model becomes identifiable. For example, in the fat rat model we could impose the constraint that

$$\sum_{j=0}^{2} \beta_j = 0.$$

This would immediately remove the identifiability problem, but does require use of a linearly constrained least squares method (see sections 1.8.1 and 1.8.2). A simpler constraint is to set one of the unidentifiable parameters to zero, which requires only that the model is re-written without the zeroed parameter, rather than a modified fitting method. For example, in the fat rat case we could set α to zero, and recover the original identifiable model. This is perfectly legitimate, since the reduced model is capable of reproducing any expected values that the original model could produce.

In the one factor case, this discussion of identifiability may seem to be unnecessarily complicated, since it is so easy to write down the model directly in an identifiable form. However, when models involve more than one factor variable, the issue can not be avoided. For more on imposing constraints see sections 1.8.1 and 1.8.2.

1.6.2 *Multiple factors*

It is frequently the case that more than one factor variable should be included in a model, and this is straightforward to do. Continuing the fat rat example, it might be that the sex of the rats is also a factor in insulin production, and that

$$\mu_i = \alpha + \beta_j + \gamma_k \text{ if rat } i \text{ is rat size level } j \text{ and sex } k$$

[††]In terms of section 1.3.1, $\mathbf{R}$ will not be full rank, and will hence not be invertible; in terms of section 1.3.8, $\mathbf{X}^\mathsf{T}\mathbf{X}$ will not be invertible.

is an appropriate model, where k is 0 or 1 for male or female. Written out in full (assuming the rats are MMFFFMFMM) the model is

$$\begin{bmatrix} \mu_1 \\ \mu_2 \\ \mu_3 \\ \mu_4 \\ \mu_5 \\ \mu_6 \\ \mu_7 \\ \mu_8 \\ \mu_9 \end{bmatrix} = \begin{bmatrix} 1 & 1 & 0 & 0 & 1 & 0 \\ 1 & 1 & 0 & 0 & 1 & 0 \\ 1 & 1 & 0 & 0 & 0 & 1 \\ 1 & 0 & 1 & 0 & 0 & 1 \\ 1 & 0 & 1 & 0 & 0 & 1 \\ 1 & 0 & 1 & 0 & 1 & 0 \\ 1 & 0 & 0 & 1 & 0 & 1 \\ 1 & 0 & 0 & 1 & 1 & 0 \\ 1 & 0 & 0 & 1 & 1 & 0 \end{bmatrix} \begin{bmatrix} \alpha \\ \beta_0 \\ \beta_1 \\ \beta_2 \\ \gamma_0 \\ \gamma_1 \end{bmatrix}.$$

It is immediately obvious that the model matrix is of column rank 4, implying that two constraints are required to make the model identifiable. You can see the lack of column independence by noting that column 5 is column 1 minus column 6, while column 2 is column 1 minus columns 3 and 4. An obvious pair of constraints would be to set $\beta_0 = \gamma_0 = 0$, so that the full model is

$$\begin{bmatrix} \mu_1 \\ \mu_2 \\ \mu_3 \\ \mu_4 \\ \mu_5 \\ \mu_6 \\ \mu_7 \\ \mu_8 \\ \mu_9 \end{bmatrix} = \begin{bmatrix} 1 & 0 & 0 & 0 \\ 1 & 0 & 0 & 0 \\ 1 & 0 & 0 & 1 \\ 1 & 1 & 0 & 1 \\ 1 & 1 & 0 & 1 \\ 1 & 1 & 0 & 0 \\ 1 & 0 & 1 & 1 \\ 1 & 0 & 1 & 0 \\ 1 & 0 & 1 & 0 \end{bmatrix} \begin{bmatrix} \alpha \\ \beta_1 \\ \beta_2 \\ \gamma_1 \end{bmatrix}.$$

When you specify models involving factors in R, it will automatically impose identifiability constraints for you, and by default these constraints will be that the parameter for the 'first' level of each factor is zero. Note that 'first' is essentially arbitrary here — the order of levels of a factor is not important. However, if you need to change which level is 'first', in order to make parameters more interpretable, see the `relevel` function.

1.6.3 *'Interactions' of factors*

In the examples considered so far, the effect of factor variables has been purely additive, but it is possible that a response variable may react differently to the combination of two factors relative to what would be predicted by simply adding the effect of the two factors separately. For example, if examining patient blood cholesterol levels, we might consider the factors 'hypertensive' (yes/no) and 'diabetic' (yes/no). Being hypertensive or diabetic would be expected to raise cholesterol levels, but being both is likely to raise cholesterol levels much more than would be predicted from just adding up the apparent effects when only one condition is present. When the effect of two factor variables together differs from the sum of their separate effects, then they are said to *interact*, and an adequate model in such situations requires *interaction terms*. Put another way, if the effects of one factor change in response to another factor, then the factors are said to interact.

Let us continue with the fat rat example, but now suppose that how insulin level depends on size varies with sex. An appropriate model is then

$$\mu_i = \alpha + \beta_j + \gamma_k + \delta_{jk} \text{ if rat } i \text{ is rat size level } j \text{ and sex } k,$$

where the δ_{jk} terms are the parameters for the interaction of rat size and sex. Writing this model out in full it is clear that it is spectacularly unidentifiable:

$$\begin{bmatrix} \mu_1 \\ \mu_2 \\ \mu_3 \\ \mu_4 \\ \mu_5 \\ \mu_6 \\ \mu_7 \\ \mu_8 \\ \mu_9 \end{bmatrix} = \begin{bmatrix} 1 & 1 & 0 & 0 & 1 & 0 & 1 & 0 & 0 & 0 & 0 & 0 \\ 1 & 1 & 0 & 0 & 1 & 0 & 1 & 0 & 0 & 0 & 0 & 0 \\ 1 & 1 & 0 & 0 & 0 & 1 & 0 & 1 & 0 & 0 & 0 & 0 \\ 1 & 0 & 1 & 0 & 0 & 1 & 0 & 0 & 0 & 1 & 0 & 0 \\ 1 & 0 & 1 & 0 & 0 & 1 & 0 & 0 & 0 & 1 & 0 & 0 \\ 1 & 0 & 1 & 0 & 1 & 0 & 0 & 0 & 1 & 0 & 0 & 0 \\ 1 & 0 & 0 & 1 & 0 & 1 & 0 & 0 & 0 & 0 & 0 & 1 \\ 1 & 0 & 0 & 1 & 1 & 0 & 0 & 0 & 0 & 0 & 1 & 0 \\ 1 & 0 & 0 & 1 & 1 & 0 & 0 & 0 & 0 & 0 & 1 & 0 \end{bmatrix} \begin{bmatrix} \alpha \\ \beta_0 \\ \beta_1 \\ \beta_2 \\ \gamma_0 \\ \gamma_1 \\ \delta_{00} \\ \delta_{01} \\ \delta_{10} \\ \delta_{11} \\ \delta_{20} \\ \delta_{21} \end{bmatrix}.$$

In fact, for this simple example, with rather few rats, we now have more parameters than data. Of course there are many ways of constraining this model to achieve identifiability. One possibility (the default in R) is to set $\beta_0 = \gamma_0 = \delta_{00} = \delta_{01} = \delta_{10} = \delta_{20} = 0$. The resulting model can still produce any fitted value vector that the full model can produce, but all the columns of its model matrix are independent, so that the model is identifiable:

$$\begin{bmatrix} \mu_1 \\ \mu_2 \\ \mu_3 \\ \mu_4 \\ \mu_5 \\ \mu_6 \\ \mu_7 \\ \mu_8 \\ \mu_9 \end{bmatrix} = \begin{bmatrix} 1 & 0 & 0 & 0 & 0 & 0 \\ 1 & 0 & 0 & 0 & 0 & 0 \\ 1 & 0 & 0 & 1 & 0 & 0 \\ 1 & 1 & 0 & 1 & 1 & 0 \\ 1 & 1 & 0 & 1 & 1 & 0 \\ 1 & 1 & 0 & 0 & 0 & 0 \\ 1 & 0 & 1 & 1 & 0 & 1 \\ 1 & 0 & 1 & 0 & 0 & 0 \\ 1 & 0 & 1 & 0 & 0 & 0 \end{bmatrix} \begin{bmatrix} \alpha \\ \beta_1 \\ \beta_2 \\ \gamma_1 \\ \delta_{11} \\ \delta_{21} \end{bmatrix}.$$

Of course, the more factor variables are present, the more interactions are possible, and the higher the order of the possible interactions: for example if three factors are present then each factor could interact with each factor, giving three possible 'two-way' interactions, while all the factors could also interact together, giving a three-way interaction (e.g., the way in which insulin level's dependence on rat size is influenced by sex is itself influenced by blood group — perhaps with interactions beyond two-way, equations are clearer than words).

Notice one very convenient fact about interactions. The model matrix columns corresponding to the interactions of two or more factors consist of the element-wise product of all possible combinations of the model matrix columns for the main effects of the factors. This applies with or without identifiability constraints, which makes the imposition of identifiability constraints on interactions especially convenient. In fact this 'column product' way of defining interactions is so convenient that in most software it is used as the definition of an interaction between factors and metric variables and, less naturally, even between different metric variables.

1.6.4 Using factor variables in R

It is very easy to work with factor variables in R. All that is required is that you let R know that a particular variable is a factor variable. For example, suppose z is a variable declared as follows:

```
> z <- c(1,1,1,2,2,1,3,3,3,3,4)
> z
 [1] 1 1 1 2 2 1 3 3 3 3 4
```

and it is to be treated as a factor with four levels. The functions `as.factor` or `factor` will ensure that `z` is treated as a factor:

```
> z <- as.factor(z)
> z
 [1] 1 1 1 2 2 1 3 3 3 3 4
Levels: 1 2 3 4
```

Notice that, when a factor variable is printed, a list of its levels is also printed — this provides an easy way to tell if a variable is a factor variable. Note also that the digits of `z` are treated purely as labels: the numbers 1 to 4 could have been any labels. For example, `x` could be a factor with 3 levels, declared as follows:

```
> x <- c("A","A","C","C","C","er","er")
> x
[1] "A"  "A"  "C"  "C"  "C"  "er" "er"
> x <- factor(x)
> x
[1] A  A  C  C  C  er er
Levels: A C er
```

Once a variable is declared as a factor variable, then R can process it automatically within model formulae, by replacing it with the appropriate number of binary dummy variables (and imposing any necessary identifiability constraints).

As an example of the use of factor variables, consider the `PlantGrowth` data frame supplied with R. These are data on the growth of plants under control conditions and two different treatment conditions. The factor `group` has three levels `cntrl`, `trt1` and `trt2`, and it is believed that the growth of the plants depends on this factor. First check that `group` is already a factor variable:

```
> PlantGrowth$group
 [1] ctrl ctrl ctrl ctrl ctrl ctrl ctrl ctrl ctrl ctrl trt1
```

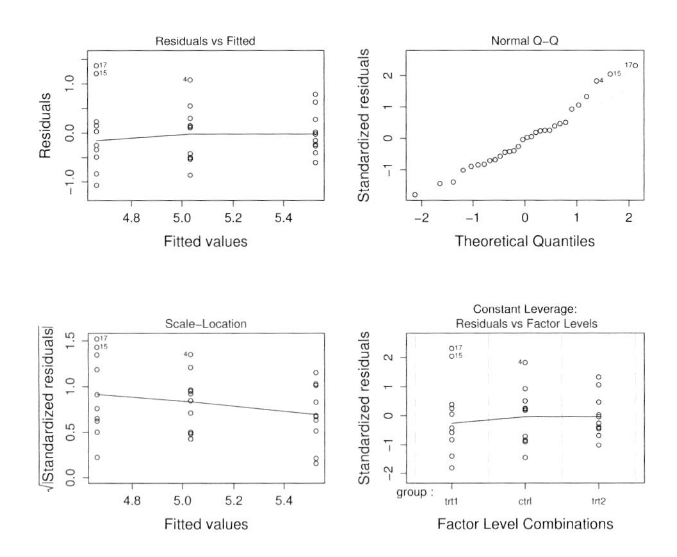

Figure 1.10 *Model checking plots for the plant growth example. Note that, since the leverages are all the same in this case, the lower right plot is now simplified.*

```
[12] trt1 trt1 trt1 trt1 trt1 trt1 trt1 trt1 trt1 trt2 trt2
[23] trt2 trt2 trt2 trt2 trt2 trt2 trt2 trt2
Levels: ctrl trt1 trt2
```

...since a list of levels is reported, it must be. If it had not been then

```
PlantGrowth$group <- as.factor(PlantGrowth$group)
```

would have converted it. The response variable for these data is `weight` of the plants at some set time after planting, and the aim is to investigate whether the `group` factor controls this, and if so to what extent.

```
> pgm.1 <- lm(weight ~ group, data=PlantGrowth)
> plot(pgm.1)
```

As usual, the first thing to do, after fitting a model, is to check the residual plots shown in figure 1.10. In this case there is some suggestion of decreasing variance with increasing mean, but the effect does not look very pronounced, so it is probably safe to proceed.

```
> summary(pgm.1)
[edited]
Coefficients:
            Estimate Std. Error t value Pr(>|t|)
(Intercept)   5.0320     0.1971  25.527   <2e-16 ***
grouptrt1    -0.3710     0.2788  -1.331   0.1944
grouptrt2     0.4940     0.2788   1.772   0.0877 .
---
[edited]
```

Notice how R reports an intercept parameter and parameters for the two treatment levels, but, in order to obtain an identifiable model, it has not included a parameter for the control level of the group factor. So the estimated overall mean weight (in the population that these plants represent, given control conditions) is 5.032, while treatment 1 is estimated to lower this weight by 0.37, and treatment 2 to increase it by 0.49. However, neither parameter individually appears to be significantly different from zero. (Don't forget that `model.matrix(pgm.1)` can be used to check up on the form of the model matrix used in the fit.)

Model selection based on the summary output is very difficult for models containing factors. It makes little sense to drop the dummy variable for just one level of a factor from a model, and if we did, what would we then do about the model identifiability constraints? Usually, it is only of interest to test whether the whole factor variable should be in the model or not, and this amounts to testing whether all its associated parameters are simultaneously zero or not. The F-ratio tests derived in section 1.3.5 are designed for just this purpose, and in R, such tests can be invoked using the `anova` function. For example, we would compare `pgm.1` to a model in which the expected response is given by a single parameter that does not depend on `group`:

```
> pgm.0 <- lm(weight ~ 1, data=PlantGrowth)
> anova(pgm.0,pgm.1)
Analysis of Variance Table

Model 1: weight ~ 1
Model 2: weight ~ group
  Res.Df     RSS Df Sum of Sq      F  Pr(>F)
1     29 14.2584
2     27 10.4921  2    3.7663 4.8461 0.01591 *
---
Signif. codes:  0 '***' 0.001 '**' 0.01 '*' 0.05 '.' 0.1 ' ' 1
```

The output gives the F-ratio statistic used to test the null hypothesis that the simpler model is correct, against the alternative that `pgm.1` is correct. If the null is true then the probability of obtaining so large an F value is only 0.016, suggesting that the null hypothesis is not very plausible.

So we see that the data provide evidence for an effect of the `group` factor variable on `weight`, which appeared marginal or absent when we examined p-values for the individual model parameters. This comes about because we have, in effect, considered all the parameters associated with the factor simultaneously, thereby obtaining a more powerful test than any of the single parameter tests could be. In the light of this analysis, the most promising treatment to look at is clearly treatment 2, since this gives the largest and 'most significant' effect, and it is a positive effect.

Another useful function, particularly with models containing many terms, is `drop1` which drops each term singly from the full model, and then computes the p-value or AICs for comparing each reduced-by-one model and the full model.

1.7 General linear model specification in R

Having seen several examples of the use of `lm`, it is worth briefly reviewing the way in which models are specified using model formulae in R. The main components of a formula are all present in the following example

```
y ~ a*b + x:z + offset(v) -1
```

Note the following:

- `~` separates the response variable, on the left, from the 'linear predictor', on the right. So in the example `y` is the response and `a`, `b`, `x`, `z` and `v` are the predictors.
- $+$ indicates that the response depends on what is to the left of $+$ *and* what is to the right of it. Hence within formulae '+' should be thought of as 'and' rather than 'the sum of'.
- `:` indicates that the response depends on the *interaction* of the variables to the left and right of `:`. Interactions are obtained by forming the element-wise products of all model matrix columns corresponding to the two terms that are interacting and appending the resulting columns to the model matrix.
- `*` means that the response depends on whatever is to the left of `*` and whatever is to the right of it *and* their interaction, i.e. `a*b` is just a shorter way of writing `a + b + a:b`.
- `offset(v)` is equivalent to including `v` as a model matrix column with the corresponding parameter fixed at 1.
- `-1` means that the default intercept term should not be included in the model. Note that, for models involving factor variables, this often has no real impact on the model structure, but simply reduces the number of identifiability constraints by one, while changing the interpretation of some parameters.

Because of the way that some symbols change their usual meaning in model formulae, it is necessary to take special measures if the usual meaning is to be restored to arithmetic operations within a formula. This is accomplished by using the identity function `I()` which simply evaluates its argument and returns it. For example, if we wanted to fit the model:

$$y_i = \beta_0 + \beta_1(x_i + z_i) + \beta_2 v_i + \epsilon_i$$

then we could do so using the model formula

```
y ~ I(x+z) + v
```

Note that there is no need to 'protect' arithmetic operations within arguments to other functions in this way. For example

$$y_i = \beta_0 + \beta_1 \log(x_i + z_i) + \beta_2 v_i + \epsilon_i$$

would be fitted correctly by

```
y ~ log(x+z) + v
```

1.8 Further linear modelling theory

This section covers linear models with constraints on the parameters (including a discussion of contrasts), the connection with maximum likelihood estimation, AIC and Mallows' C_p, linear models for non-independent data with non-constant variance and non-linear models.

1.8.1 Constraints I: General linear constraints

It is often necessary, particularly when working with factor variables, to impose constraints on the linear model parameters of the general form

$$\mathbf{C}\boldsymbol{\beta} = \mathbf{0},$$

where $\mathbf{C}$ is an $m \times p$ matrix of known coefficients. The general approach to imposing such constraints is to rewrite the model in terms of $p - m$ unconstrained parameters. There are a number of ways of doing this, but a simple general approach uses the QR decomposition of $\mathbf{C}^\mathsf{T}$. Let

$$\mathbf{C}^\mathsf{T} = \mathbf{U} \begin{bmatrix} \mathbf{P} \\ \mathbf{0} \end{bmatrix}$$

where $\mathbf{U}$ is a $p \times p$ orthogonal matrix and $\mathbf{P}$ is an $m \times m$ upper triangular matrix. $\mathbf{U}$ can be partitioned $\mathbf{U} \equiv (\mathbf{D} : \mathbf{Z})$ where $\mathbf{Z}$ is a $p \times (p - m)$ matrix.

It turns out that

$$\boldsymbol{\beta} = \mathbf{Z}\boldsymbol{\beta}_z$$

will meet the constraints for any value of the $p - m$ dimensional vector $\boldsymbol{\beta}_z$. This is easy to see:

$$\mathbf{C}\boldsymbol{\beta} = \begin{bmatrix} \mathbf{P}^\mathsf{T} & \mathbf{0} \end{bmatrix} \begin{bmatrix} \mathbf{D}^\mathsf{T} \\ \mathbf{Z}^\mathsf{T} \end{bmatrix} \mathbf{Z}\boldsymbol{\beta}_z = \begin{bmatrix} \mathbf{P}^\mathsf{T} & \mathbf{0} \end{bmatrix} \begin{bmatrix} \mathbf{0} \\ \mathbf{I}_{p-m} \end{bmatrix} \boldsymbol{\beta}_z = \mathbf{0}.$$

Hence to minimize $\|\mathbf{y} - \mathbf{X}\boldsymbol{\beta}\|^2$ w.r.t. $\boldsymbol{\beta}$, subject to $\mathbf{C}\boldsymbol{\beta} = \mathbf{0}$, the following algorithm can be used.

1. Find the QR decomposition of $\mathbf{C}^\mathsf{T}$: the final $p - m$ columns of the orthogonal factor define $\mathbf{Z}$.
2. Minimize the unconstrained sum of squares $\|\mathbf{y} - \mathbf{X}\mathbf{Z}\boldsymbol{\beta}_z\|^2$ w.r.t. $\boldsymbol{\beta}_z$ to obtain $\hat{\boldsymbol{\beta}}_z$.
3. $\hat{\boldsymbol{\beta}} = \mathbf{Z}\hat{\boldsymbol{\beta}}_z$.

Note that, in practice, it is computationally inefficient to form $\mathbf{Z}$ explicitly, when we only need to be able to post-multiply $\mathbf{X}$ by it, and pre-multiply $\boldsymbol{\beta}_z$ by it. The reason for this is that $\mathbf{Z}$ is completely defined as the product of m 'Householder rotations': simple matrix operations that can be applied very rapidly to any vector or matrix (see section 5.4.1, p. 211). R includes routines for multiplication by the orthogonal factor of a QR decomposition which makes use of these efficiencies. See B.5 and B.6 for further details on QR decomposition.

1.8.2 *Constraints II: 'Contrasts' and factor variables*

There is another approach to imposing identifiability constraints on models involving factor variables. To explain it, it is worth revisiting the basic identifiability problem with a simple example. Consider the model

$$y_i = \mu + \alpha_j + \epsilon_i \text{ if } y_i \text{ is from group } j$$

and suppose that there are three groups with two observations in each. The model matrix in this case is

$$\mathbf{X} = \begin{bmatrix} 1 & 1 & 0 & 0 \\ 1 & 1 & 0 & 0 \\ 1 & 0 & 1 & 0 \\ 1 & 0 & 1 & 0 \\ 1 & 0 & 0 & 1 \\ 1 & 0 & 0 & 1 \end{bmatrix},$$

but this is not of full column rank: any of its columns could be made up from a linear combination of the other 3. In geometric terms the model space is of dimension 3 and not 4, and this means that we can not estimate all 4 model parameters, but at most 3 parameters. Numerically this problem would manifest itself in the rank deficiency of $\mathbf{R}$ in equation (1.6), which implies that $\mathbf{R}^{-1}$ does not exist.

One approach to this issue is to remove one of the model matrix columns, implicitly treating the corresponding parameter as zero. This gives the model matrix full column rank, so that the remaining parameters are estimable, but since the model space is unaltered, we have not fundamentally changed the model. It can still predict every set of fitted values that the original model could predict. By default, R drops the model matrix column corresponding to the first level of each factor, in order to impose identifiability on models with factors. For the simple example, this results in

$$\mathbf{X}' = \begin{bmatrix} 1 & 0 & 0 \\ 1 & 0 & 0 \\ 1 & 1 & 0 \\ 1 & 1 & 0 \\ 1 & 0 & 1 \\ 1 & 0 & 1 \end{bmatrix}.$$

To generalize this approach, it helps to write out this deletion in a rather general way. For example, if we re-write the original model matrix in partitioned form, $\mathbf{X} = [\mathbf{1} : \mathbf{X}_1]$, where $\mathbf{1}$ is a column of 1s, then

$$\mathbf{X}' = [\mathbf{1} : \mathbf{X}_1\mathbf{C}_1] \text{ where } \mathbf{C}_1 = \begin{bmatrix} 0 & 0 \\ 1 & 0 \\ 0 & 1 \end{bmatrix}.$$

Now all that $\mathbf{C}_1$ has done is to replace the 3 columns of $\mathbf{X}_1$ by a 2 column linear combination of them, which cannot be combined to give $\mathbf{1}$. On reflection, *any* matrix which did these two things would have served as well as the particular $\mathbf{C}_1$ actually given, in terms of making the model identifiable. This observation leads to the idea of choosing alternative matrix elements for $\mathbf{C}_1$, in order to enhance the interpretability of the parameters actually estimated, for some models.

Matrices like $\mathbf{C}_1$ are known as *contrast matrices*[†] and several different types are available in R (see `?contrasts` and `options("contrasts")`). The degree of interpretability of some of the contrasts is open to debate. For the $\mathbf{C}_1$ given, suppose that parameters μ, α'_2 and α'_3 are estimated: μ would now be interpretable as the mean for group 1, while α'_2 and α'_3 would be the differences between the means for groups 2 and 3, and the mean for group 1.

With all contrast matrices, the contrast matrix itself can be used to transform back from the parameters actually estimated, to (non-unique) estimates in the original redundant parameterization, e.g.

$$\hat{\boldsymbol{\alpha}} = \mathbf{C}_1 \hat{\boldsymbol{\alpha}}'.$$

The contrast approach generalizes easily to models with multiple factors and their interactions. Again a simple example makes things clear. Consider the model:

$$y_i = \mu + \alpha_j + \beta_k + \gamma_{jk} + \epsilon_i \text{ if } y_i \text{ is from groups } j \text{ and } k.$$

The unconstrained model matrix for this might have the form $\mathbf{X} = [\mathbf{1} : \mathbf{X}_\alpha : \mathbf{X}_\beta : \mathbf{X}_\alpha \odot \mathbf{X}_\beta]$, where $\mathbf{X}_\alpha$ and $\mathbf{X}_\beta$ are the columns generated by $\boldsymbol{\alpha}$ and $\boldsymbol{\beta}$ and $\mathbf{X}_\alpha \odot \mathbf{X}_\beta$ are the columns corresponding to $\boldsymbol{\gamma}$, which are in fact generated by element-wise multiplication of all possible pairings of the column from $\mathbf{X}_\alpha$ and $\mathbf{X}_\beta$.

To make this model identifiable we would choose contrast matrices $\mathbf{C}_\alpha$ and $\mathbf{C}_\beta$ for $\boldsymbol{\alpha}$ and $\boldsymbol{\beta}$, respectively, and then form the following identifiable model matrix:

$$\mathbf{X}' = [\mathbf{1} : \mathbf{X}_\alpha \mathbf{C}_\alpha : \mathbf{X}_\beta \mathbf{C}_\beta : (\mathbf{X}_\alpha \mathbf{C}_\alpha) \odot (\mathbf{X}_\beta \mathbf{C}_\beta)]$$

(in this case $\boldsymbol{\gamma} = \mathbf{C}_\alpha \otimes \mathbf{C}_\beta \gamma'$, where $\otimes$ is the Kronecker product, see B.4). Some further information can be found in Venables and Ripley (2002).

1.8.3 Likelihood

The theory developed in section 1.3 is quite sufficient to justify the approach of estimating linear models by least squares, but although it doesn't directly strengthen the case, it is worth understanding the link between the method of least squares and the method of maximum likelihood for normally distributed data.

The basic idea of likelihood is that, given some parameter values, a statistical model allows us to write down the probability, or probability density, of any set of data, and in particular of the set actually observed. In some sense, parameter values which cause the model to suggest that the observed data are probable, are more 'likely' than parameter values that suggest that what was observed was improbable. In fact it seems reasonable to use as estimates of the parameters those values which

[†]Actually, in much of the literature on linear models, the given $\mathbf{C}_1$ would not be called a 'contrast', as its columns are not orthogonal to $\mathbf{1}$, but R makes no terminological distinction.

maximize the probability of the data according to the model: these are the 'maximum likelihood estimates' of the parameters.

Appendix A covers the properties of maximum likelihood estimation in greater detail. For the moment consider the likelihood for the parameters of a linear model. According to the model, the joint p.d.f. of the response data is

$$f_{\boldsymbol{\beta},\sigma^2}(\mathbf{y}) = (2\pi\sigma^2)^{-n/2} e^{-\|\mathbf{y}-\mathbf{X}\boldsymbol{\beta}\|^2/(2\sigma^2)}.$$

Now suppose that the observed data are plugged into this expression and it is treated as a function of its parameters $\boldsymbol{\beta}$ and σ^2. This is known as the likelihood function

$$L(\boldsymbol{\beta}, \sigma^2) = (2\pi\sigma^2)^{-n/2} e^{-\|\mathbf{y}-\mathbf{X}\boldsymbol{\beta}\|^2/(2\sigma^2)},$$

and it is important to note that $\mathbf{y}$ is now representing the actual observed data, rather than arguments of a p.d.f. To estimate the parameters, L should be maximized w.r.t. them, and it is immediately apparent that the value of $\boldsymbol{\beta}$ maximizing L will be the value minimizing

$$\mathcal{S} = \|\mathbf{y} - \mathbf{X}\boldsymbol{\beta}\|^2$$

(irrespective of the value of σ^2 or its MLE).

In itself this connection is of little interest, but it suggests how to estimate linear models when data do not meet the constant variance assumption, and may not even be independent. To this end consider the linear model

$$\boldsymbol{\mu} = \mathbf{X}\boldsymbol{\beta}, \quad \mathbf{y} \sim N(\boldsymbol{\mu}, \mathbf{V}\sigma^2),$$

where $\mathbf{V}$ is any positive definite[‡] matrix. In this case the likelihood for $\boldsymbol{\beta}$ is

$$L(\boldsymbol{\beta}) = \frac{1}{\sqrt{(2\pi\sigma^2)^n|\mathbf{V}|}} e^{-(\mathbf{y}-\mathbf{X}\boldsymbol{\beta})^\mathsf{T}\mathbf{V}^{-1}(\mathbf{y}-\mathbf{X}\boldsymbol{\beta})/(2\sigma^2)}$$

and if $\mathbf{V}$ is known then maximum likelihood estimation of $\boldsymbol{\beta}$ is achieved by minimizing

$$\mathcal{S}_v = (\mathbf{y} - \mathbf{X}\boldsymbol{\beta})^\mathsf{T}\mathbf{V}^{-1}(\mathbf{y} - \mathbf{X}\boldsymbol{\beta}).$$

In fact the likelihood approach can be taken further, since if $\mathbf{V}$ depends on unknown parameters then these too can be estimated by maximum likelihood estimation: this is what is done in linear mixed modelling, which is discussed in Chapter 2.

1.8.4 *Non-independent data with variable variance*

In the previous section a modified least squares criterion was developed for linear model parameter estimation, when data follow a general multivariate normal distribution, with unknown mean and covariance matrix known to within a constant of

[‡]A matrix $\mathbf{A}$ is positive definite iff $\mathbf{x}^\mathsf{T}\mathbf{A}\mathbf{x} > 0$ for any non-zero vector $\mathbf{x}$. Equivalent to this condition is the condition that all the eigenvalues of $\mathbf{A}$ must be strictly positive. Practical tests for positive definiteness are examination of the eigenvalues of the matrix or (more efficiently) seeing if a Cholesky decomposition of the matrix is possible (this must be performed without pivoting, otherwise only positive semi-definiteness is tested): see B.7.

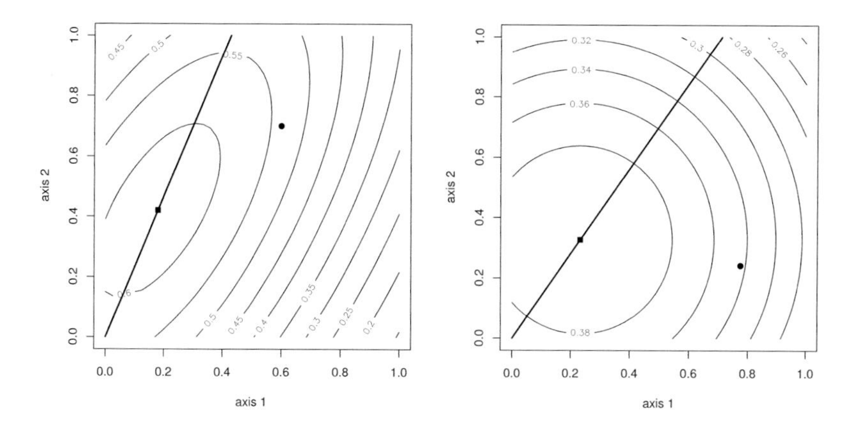

Figure 1.11 *Fitting a linear model to data that are not independent with constant variance. The example is a straight line through the origin, fit to two data. In both panels the response data values give the co-ordinates of the point • and the straight line is the vector defining the 'model subspace' (the line along which the fitted values could lie). It is assumed that the data arise from the multivariate normal distribution contoured in the left panel. The right panel shows the fitting problem after transformation in the manner described in section 1.8.4: the response data and model subspace have been transformed and this implies a transformation of the distribution of the response. The transformed data are an observation from a radially symmetric multivariate normal density.*

proportionality. It turns out to be possible to transform this fitting problem so that it has exactly the form of the fitting problem for independent data with constant variance. Having done this, all inference about the model parameters can proceed using the methods of section 1.3.

First let $\mathbf{L}$ be any matrix such that $\mathbf{L}^\mathsf{T}\mathbf{L} = \mathbf{V}$: a Cholesky decomposition is usually the easiest way to obtain this (see section B.7). Then

$$\begin{aligned} \mathcal{S}_v &= (\mathbf{y} - \mathbf{X}\boldsymbol{\beta})^\mathsf{T}\mathbf{V}^{-1}(\mathbf{y} - \mathbf{X}\boldsymbol{\beta}) \\ &= (\mathbf{y} - \mathbf{X}\boldsymbol{\beta})^\mathsf{T}\mathbf{L}^{-1}\mathbf{L}^{-\mathsf{T}}(\mathbf{y} - \mathbf{X}\boldsymbol{\beta}) \\ &= \|\mathbf{L}^{-\mathsf{T}}\mathbf{y} - \mathbf{L}^{-\mathsf{T}}\mathbf{X}\boldsymbol{\beta}\|^2, \end{aligned}$$

and this least squares objective can be minimized by the methods already met (i.e., form a QR decomposition of $\mathbf{L}^{-\mathsf{T}}\mathbf{X}$, etc.)

It is not just the model fitting that carries over from the theory of section 1.3. Since $\mathbf{L}^{-\mathsf{T}}\mathbf{y}$ is a linear transformation of a normal random vector, it must have a multivariate normal distribution, and it is easily seen that $\mathbb{E}(\mathbf{L}^{-\mathsf{T}}\mathbf{y}) = \mathbf{L}^{-\mathsf{T}}\mathbf{X}\boldsymbol{\beta}$ while the covariance matrix of $\mathbf{L}^{-\mathsf{T}}\mathbf{y}$ is

$$\mathbf{V}_{\mathbf{L}^{-\mathsf{T}}\mathbf{y}} = \mathbf{L}^{-\mathsf{T}}\mathbf{V}\mathbf{L}^{-1}\sigma^2 = \mathbf{L}^{-\mathsf{T}}\mathbf{L}^\mathsf{T}\mathbf{L}\mathbf{L}^{-1}\sigma^2 = \mathbf{I}\sigma^2,$$

i.e., $\mathbf{L}^{-\mathsf{T}}\mathbf{y} \sim N(\mathbf{L}^{-\mathsf{T}}\mathbf{X}\boldsymbol{\beta}, \mathbf{I}\sigma^2)$. In other words, the transformation has resulted in a new linear modelling problem, in which the response data are independent normal random variables with constant variance: exactly the situation which allows all the results from section 1.3 to be used for inference about $\boldsymbol{\beta}$.

Figure 1.11 illustrates the geometry of the transformation for a simple linear model,

$$y_i = \beta x_i + \epsilon_i, \quad \epsilon_i \sim N(\mathbf{0}, \mathbf{V}),$$

where $\mathbf{y}^\mathsf{T} = (.6, .7)$, $\mathbf{x}^\mathsf{T} = (.3, .7)$, $\beta = .6$ and $\mathbf{V} = \begin{bmatrix} .6 & .5 \\ .5 & 1.1 \end{bmatrix}$. The left panel shows the geometry of the original fitting problem, while the right panel shows the geometry of the transformed fitting problem.

1.8.5 *Simple AR correlation models*

Considerable computational simplification occurs when the correlation is given by a simple auto-regression (AR) model. For example residuals with a simple AR1 structure are generated as $\epsilon_i = \epsilon_{i-1} + \rho e_i$ where the e_i are independent $N(0, \sigma^2)$. Such a model is sometimes appropriate when the data are observed at regular time intervals. The covariance matrix for such a model is

$$\mathbf{V} = \sigma^2 \begin{bmatrix} 1 & \rho & \rho^2 & \rho^3 & . \\ \rho & 1 & \rho & \rho^2 & . \\ \rho^2 & \rho & 1 & \rho & . \\ . & . & . & . & . \end{bmatrix}$$

and the inverse of the Cholesky factor of $\mathbf{V}$ is

$$\mathbf{L}^{-1} = \begin{bmatrix} 1 & -\rho/\sqrt{1-\rho^2} & 0 & 0 & . \\ 0 & 1/\sqrt{1-\rho^2} & -\rho/\sqrt{1-\rho^2} & 0 & . \\ 0 & 0 & 1/\sqrt{1-\rho^2} & -\rho/\sqrt{1-\rho^2} & . \\ . & . & . & . & . \end{bmatrix} \sigma^{-1}.$$

Now because $\mathbf{L}^{-1}$ has only 2 non-zero diagonals the computation of $\mathbf{L}^{-\mathsf{T}}\mathbf{y}$ amounts to a simple differencing operation on the elements of $\mathbf{y}$ computable with $O(n)$ operations, while $\mathbf{L}^{-\mathsf{T}}\mathbf{X}$ simply differences the columns of $\mathbf{X}$ at $O(np)$ cost. Hence computing with such an AR1 model is very efficient. If the likelihood of the model is required note that $|\mathbf{V}|^{-1/2} = |\mathbf{L}^{-1}|$: the latter is just the product of the leading diagonal elements of $\mathbf{L}^{-1}$, which again has $O(n)$ cost. This computational convenience carries over to higher order AR models, which also have banded inverse Cholesky factors.

1.8.6 *AIC and Mallows' statistic*

Consider again the problem of selecting between nested models, which is usually equivalent to deciding whether some terms from the model should simply be set to zero. The most natural way to do this would be to select the model with the smallest residual sum of squares or largest likelihood, but this is always the largest model under consideration. Model selection methods based on F-ratio or t-tests address this problem by selecting the simplest model consistent with the data, where consistency is judged using some significance level (threshold p-value) that has to be chosen more or less arbitrarily.

In this section an alternative approach is developed, based on the idea of trying to select the model that should do the best job of predicting $\boldsymbol{\mu} \equiv \mathbb{E}(\mathbf{y})$, rather than the model that gets as close as possible to $\mathbf{y}$. Specifically, we select the model that minimizes an estimate of

$$K = \|\boldsymbol{\mu} - \mathbf{X}\hat{\boldsymbol{\beta}}\|^2.$$

We have

$$\begin{aligned} K = \|\boldsymbol{\mu} - \mathbf{X}\hat{\boldsymbol{\beta}}\|^2 &= \|\boldsymbol{\mu} - \mathbf{A}\mathbf{y}\|^2 = \|\mathbf{y} - \mathbf{A}\mathbf{y} - \boldsymbol{\epsilon}\|^2 \\ &= \|\mathbf{y} - \mathbf{A}\mathbf{y}\|^2 + \boldsymbol{\epsilon}^\mathsf{T}\boldsymbol{\epsilon} - 2\boldsymbol{\epsilon}^\mathsf{T}(\mathbf{y} - \mathbf{A}\mathbf{y}) \\ &= \|\mathbf{y} - \mathbf{A}\mathbf{y}\|^2 + \boldsymbol{\epsilon}^\mathsf{T}\boldsymbol{\epsilon} - 2\boldsymbol{\epsilon}^\mathsf{T}(\boldsymbol{\mu} + \boldsymbol{\epsilon}) + 2\boldsymbol{\epsilon}^\mathsf{T}\mathbf{A}(\boldsymbol{\mu} + \boldsymbol{\epsilon}) \\ &= \|\mathbf{y} - \mathbf{A}\mathbf{y}\|^2 - \boldsymbol{\epsilon}^\mathsf{T}\boldsymbol{\epsilon} - 2\boldsymbol{\epsilon}^\mathsf{T}\boldsymbol{\mu} + 2\boldsymbol{\epsilon}^\mathsf{T}\mathbf{A}\boldsymbol{\mu} + 2\boldsymbol{\epsilon}^\mathsf{T}\mathbf{A}\boldsymbol{\epsilon}. \end{aligned}$$

Now, $\mathbb{E}(\boldsymbol{\epsilon}^\mathsf{T}\boldsymbol{\epsilon}) = \mathbb{E}(\sum_i \epsilon_i^2) = n\sigma^2$, $\mathbb{E}(\boldsymbol{\epsilon}^\mathsf{T}\boldsymbol{\mu}) = \mathbb{E}(\boldsymbol{\epsilon}^\mathsf{T})\boldsymbol{\mu} = 0$ and $\mathbb{E}(\boldsymbol{\epsilon}^\mathsf{T}\mathbf{A}\boldsymbol{\mu}) = \mathbb{E}(\boldsymbol{\epsilon}^\mathsf{T})\mathbf{A}\boldsymbol{\mu} = 0$. Finally, using the fact that a scalar is its own trace:

$$\mathbb{E}[\mathrm{tr}\left(\boldsymbol{\epsilon}^\mathsf{T}\mathbf{A}\boldsymbol{\epsilon}\right)] = \mathbb{E}[\mathrm{tr}\left(\mathbf{A}\boldsymbol{\epsilon}\boldsymbol{\epsilon}^\mathsf{T}\right)] = \mathrm{tr}\left(\mathbf{A}\mathbb{E}[\boldsymbol{\epsilon}\boldsymbol{\epsilon}^\mathsf{T}]\right) = \mathrm{tr}\left(\mathbf{A}\mathbf{I}\right)\sigma^2 = \mathrm{tr}\left(\mathbf{A}\right)\sigma^2.$$

Hence, replacing terms involving $\boldsymbol{\epsilon}$ by their expectation we can estimate K using

$$\hat{K} = \|\mathbf{y} - \mathbf{A}\mathbf{y}\|^2 - n\sigma^2 + 2\mathrm{tr}\left(\mathbf{A}\right)\sigma^2. \tag{1.13}$$

Using the fact that $\mathrm{tr}\left(\mathbf{A}\right) = p$ and dividing through by σ^2 we get 'Mallows' C_p' (Mallows, 1973)

$$C_p = \|\mathbf{y} - \mathbf{X}\hat{\boldsymbol{\beta}}\|^2/\sigma^2 + 2p - n.$$

Model selection by C_p minimization works well if σ^2 is known, but for most models σ^2 must be estimated, and using the estimate derived from the model fit has the unfortunate consequence that C_p ceases to depend on which model has been fitted, in any meaningful way. To avoid this, σ^2 is usually fixed at the estimate given by the fit of the largest candidate model (unless it really is known, of course).

An alternative that handles σ more satisfactorily is to replace K by the likelihood based equivalent

$$K' = \|\boldsymbol{\mu} - \mathbf{X}\hat{\boldsymbol{\beta}}\|^2/\sigma^2 + n\log(2\pi\sigma).$$

Writing $l(\boldsymbol{\beta}, \sigma)$ as the model log likelihood, and re-using the calculation that led to Mallows' C_p we have the estimate

$$\hat{K}' = -2l(\hat{\boldsymbol{\beta}}, \sigma^2) + 2p - n,$$

which is minimized by whichever model minimises Akaike's information criteria (Akaike, 1973)

$$\mathrm{AIC} = -2l(\hat{\boldsymbol{\beta}}, \sigma^2) + 2p.$$

If we use the MLE , $\hat{\sigma}^2 = \|\mathbf{y} - \mathbf{X}\hat{\boldsymbol{\beta}}\|^2/n$, then

$$\text{AIC} = -2l(\hat{\boldsymbol{\beta}}, \hat{\sigma}^2) + 2(p+1). \tag{1.14}$$

See section A.7 for justification in the context of general likelihoods.

Notice how the above derivation does not involve any assumption that directly implies that the models to be compared must be nested: however, a more detailed examination of the comparison of models by AIC would suggest that the comparison will be more reliable for nested models, since in that case some of the neglected terms in the approximation of K cancel. Notice also that if the true model is not in the set of models under consideration then the properties of (1.14) will be less ideal. Indeed in this case (1.14) would itself tend to favour more complex models, since it would be impossible to match $\boldsymbol{\mu}$ exactly: as sample size increases and estimates become more precise, this tendency starts to overcome the negative effects of overfitting and leads to more complex models being selected. The tendency for AIC to favour more complex models with increasing sample size is often seen in practice: presumably because the true model is rarely in the set of candidate models.

1.8.7 The wrong model

The derivation that led to Mallows' C_p implies that $\|\mathbf{y} - \hat{\boldsymbol{\mu}}\|^2 \to \|\boldsymbol{\mu} - \hat{\boldsymbol{\mu}}\|^2 - 2\text{tr}(\mathbf{A})\sigma^2 + n\sigma^2$ as the sample size $n \to \infty$, and this applies whether or not $\mathbb{E}(\hat{\boldsymbol{\mu}}) = \boldsymbol{\mu}$. Hence in the large sample limit $\hat{\boldsymbol{\mu}}$ will minimize $\|\boldsymbol{\mu} - \hat{\boldsymbol{\mu}}\|^2$, whether or not $\hat{\boldsymbol{\mu}}$ is biased. This result is useful when we consider the properties of regression splines.

1.8.8 Non-linear least squares

Some non-linear models can be estimated by iterative approximation by a linear model. At each iterate the fitted approximating linear model suggests improved parameter estimates, and at convergence the parameter estimates are least squares estimates. In addition, the approximating linear model at convergence can be used as the basis for approximate inference about the parameters.

Formally, consider fitting the model:

$$\mathbb{E}(\mathbf{y}) \equiv \boldsymbol{\mu} = \mathbf{f}(\boldsymbol{\beta})$$

to response data $\mathbf{y}$, when $\mathbf{f}$ is a non-linear vector valued function of $\boldsymbol{\beta}$. An obvious fitting objective is:

$$\mathcal{S} = \sum_{i=1}^{n} \{y_i - f_i(\boldsymbol{\beta})\}^2 = \|\mathbf{y} - \mathbf{f}(\boldsymbol{\beta})\|^2$$

and, if the functions, f_i, are sufficiently well behaved, then this non-linear least squares problem can be solved by iterative linear least squares. To do this, we start from a guess at the best fit parameters, $\hat{\boldsymbol{\beta}}^{[k]}$, and then take a first order Taylor expansion of f_i around $\hat{\boldsymbol{\beta}}^{[k]}$ so that the fitting objective becomes

$$\mathcal{S} \approx \mathcal{S}^{[k]} = \|\mathbf{y} - \mathbf{f}(\hat{\boldsymbol{\beta}}^{[k]}) + \mathbf{J}^{[k]}\hat{\boldsymbol{\beta}}^{[k]} - \mathbf{J}^{[k]}\boldsymbol{\beta}\|^2$$

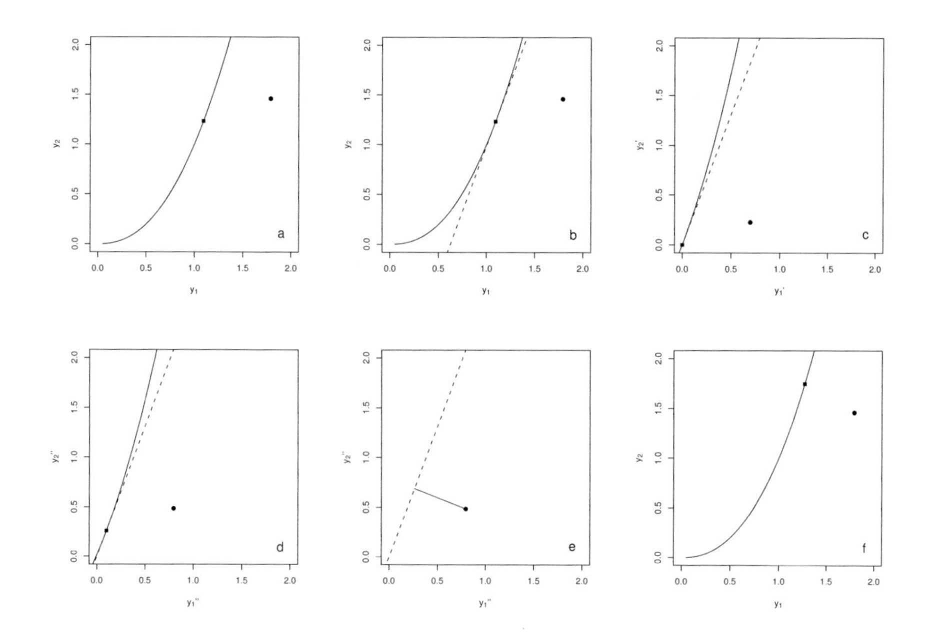

Figure 1.12 *Geometry of a single iteration of the Gauss-Newton approach to non-linear model fitting. The example is fitting the model $\mathbb{E}(y_i) = \exp(\beta x_i)$ to x_i, y_i data ($i = 1, 2$).* (a) *plots y_2 against y_1 (•) with the curve illustrating the possible values for the expected values of the response variable: as the value of β changes $\mathbb{E}(\mathbf{y})$ follows this curve – the 'model manifold'. An initial guess at the parameter gives the fitted values plotted as ■.* (b) *The tangent space to the model manifold is found and illustrated by the dashed line.* (c) *The model, tangent and data are linearly translated so that the current estimate of the fitted values is at the origin.* (d) *The model, tangent and data are linearly translated so that $\mathbf{J}\hat{\boldsymbol{\beta}}^{[k]}$ gives the location of the current estimate of the fitted values.* (e) *$\hat{\boldsymbol{\beta}}^{[k+1]}$ is estimated by finding the closest point in the tangent space to the response data, by least squares.* (f) *shows the original model and data, with the next estimate of the fitted values, obtained using $\hat{\boldsymbol{\beta}}^{[k+1]}$. The steps can now be repeated until the estimates converge.*

where $\mathbf{J}^{[k]}$ is the 'Jacobian' matrix such that $J_{ij}^{[k]} = \partial f_i / \partial \beta_j$ (derivatives evaluated at $\hat{\boldsymbol{\beta}}^{[k]}$, of course). Defining a vector of *pseudodata*,

$$\mathbf{z}^{[k]} = \mathbf{y} - \mathbf{f}(\hat{\boldsymbol{\beta}}^{[k]}) + \mathbf{J}^{[k]}\hat{\boldsymbol{\beta}}^{[k]},$$

the objective can be re-written

$$\mathcal{S}^{[k]} = \|\mathbf{z}^{[k]} - \mathbf{J}^{[k]}\boldsymbol{\beta}\|^2,$$

and, since this is a linear least squares problem, it can be minimized with respect to $\boldsymbol{\beta}$ to obtain an improved estimated parameter vector $\hat{\boldsymbol{\beta}}^{[k+1]}$. If $\mathbf{f}(\boldsymbol{\beta})$ is not too non-linear then this process can be iterated until the $\hat{\boldsymbol{\beta}}^{[k]}$ sequence converges, to the final least squares estimate $\hat{\boldsymbol{\beta}}$.

Figure 1.12 illustrates the geometrical interpretation of this method. Because the non-linear model is being approximated by a linear model, large parts of the theory of linear models carry over as approximations in the current context. For example, under the assumption of equal variance, σ^2, and independence of the response variable, the covariance matrix of the parameters is simply: $(\mathbf{J}^\mathsf{T}\mathbf{J})^{-1}\sigma^2$, where $\mathbf{J}$ is evaluated at convergence.

If $\mathbf{f}$ is sufficiently non-linear that convergence does not occur, then a simple 'step reduction' approach will stabilize the fitting. The vector $\Delta = \hat{\boldsymbol{\beta}}^{[k+1]} - \hat{\boldsymbol{\beta}}^{[k]}$ is treated as a trial step. If $\hat{\boldsymbol{\beta}}^{[k+1]}$ does not decrease $\mathcal{S}$, then trial steps $\hat{\boldsymbol{\beta}}^{[k]} + \alpha\Delta$ are taken, with ever decreasing α, until $\mathcal{S}$ does decrease (of course, $0 < \alpha < 1$). Geometrically, this is equivalent to performing an updating step by fitting to the average $\mathbf{y}\alpha + (1 - \alpha)\mathbf{f}(\boldsymbol{\beta}^{[k]})$, rather than original data $\mathbf{y}$: viewed in this way it is clear that for small enough α, each iteration must decrease $\mathcal{S}$, until a minimum is reached. It is usual to halve α each time that a step is unsuccessful, and to start each iteration with twice the α value which finally succeeded at the previous step (or $\alpha = 1$ if this is less). If it is *necessary* to set $\alpha < 1$ at the final step of the iteration then it is likely that inference based on the final approximating linear model will be somewhat unreliable.

1.8.9 Further reading

The literature on linear models is rather large, and there are many book length treatments. For a good introduction to linear models with R, see Faraway (2014). Other sources on the linear modelling functionality in R are Chambers (1993) and Venables and Ripley (2002). Numerical estimation of linear models is covered in Golub and Van Loan (2013, chapter 5). Dobson and Barnett (2008), McCullagh and Nelder (1989) and Davison (2003) also consider linear models as part of broader treatments. For R itself see R Core Team (2016).

1.9 Exercises

1. Four car journeys in London of length 1, 3, 4 and 5 kilometres took 0.1, 0.4, 0.5 and 0.6 hours, respectively. Find a least squares estimate of the mean journey speed.
2. Given n observations x_i, y_i, find the least squares estimate of β in the linear model: $y_i = \mu_i + \epsilon_i$, $\mu_i = \beta$.
3. Which, if any, of the following common linear model assumptions are required for $\hat{\boldsymbol{\beta}}$ to be unbiased: (i) The Y_i are independent, (ii) the Y_i all have the same variance, (iii) the Y_i are normally distributed?
4. Write out the following three models in the form $\mathbf{y} = \mathbf{X}\boldsymbol{\beta} + \boldsymbol{\epsilon}$, ensuring that all the parameters left in the written out model are identifiable. In all cases y is the response variable, ϵ the residual 'error' and other Greek letters indicate model parameters.

   (a) The 'balanced one-way ANOVA model':

$$y_{ij} = \alpha + \beta_i + \epsilon_{ij}$$

where $i = 1 \ldots 3$ and $j = 1 \ldots 2$.

(b) A model with two explanatory factor variables and only one observation per combination of factor variables:

$$y_{ij} = \alpha + \beta_i + \gamma_j + \epsilon_{ij}.$$

The first factor (β) has 3 levels and the second factor has 4 levels.

(c) A model with two explanatory variables: a factor variable and a continuous variable, x.

$$y_i = \alpha + \beta_j + \gamma x_i + \epsilon_i \quad \text{if obs. } i \text{ is from factor level } j.$$

Assume that $i = 1 \ldots 6$, that the first two observations are for factor level 1 and the remaining 4 for factor level 2 and that the x_i's are 0.1, 0.4, 0.5, 0.3, 0.4 and 0.7.

5. Consider some data for deformation (in mm), y_i, of 3 different types of alloy, under different loads (in kg), x_i. When there is no load, there is no deformation, and the deformation is expected to vary linearly with load, in exactly the same way for all three alloys. However, as the load increases the three alloys deviate from this ideal linear behaviour in slightly different ways, with the relationship becoming slightly curved (possibly suggesting quadratic terms). The loads are known very precisely, so errors in x_i's can by ignored, whereas the deformations, y_i, are subject to larger measurement errors that do need to be taken into account. Define a linear model suitable for describing these data, assuming that the same 6 loads are applied to each alloy, and write it out in the form $\mathbf{y} = \mathbf{X}\boldsymbol{\beta} + \boldsymbol{\epsilon}$.

6. Show that for a linear model with model matrix $\mathbf{X}$, fitted to data $\mathbf{y}$, with fitted values $\hat{\boldsymbol{\mu}}$,

$$\mathbf{X}^\mathsf{T}\hat{\boldsymbol{\mu}} = \mathbf{X}^\mathsf{T}\mathbf{y}.$$

What implication does this have for the residuals of a model which includes an intercept term?

7. Equation (1.8) in section 1.3.3 gives an unbiased estimator of σ^2, but in the text unbiasedness was only demonstrated assuming that the response data were normally distributed. By considering $\mathbb{E}(r_i^2)$, show that the estimator (1.8) is unbiased whatever the distribution of the response, provided that the response data are independent with constant variance.

8. The `MASS` library contains a data frame `Rubber` on wear of tyre rubber. The response, `loss`, measures rubber loss in grammes per hour. The predictors are `tens`, a measure of the tensile strength of the rubber with units of kgm^{-2}, and `hard`, the hardness of the rubber in Shore[§] units. Modelling interest focuses on predicting wear from hardness and tensile strength.

(a) Starting with a model in which `loss` is a polynomial function of `tens` and

[§]Measures hardness as the extent of the rebound of a diamond tipped hammer, dropped on the test object.

hard, with all terms up to third order present, perform backwards model selection, based on hypothesis testing, to select an appropriate model for these data.

(b) Note the AIC scores of the various models that you have considered. It would also be possible to do model selection based on AIC, and the `step` function in R provides a convenient way of doing this. After reading the `step` help file, use it to select an appropriate model for the `loss` data, by AIC.

(c) Use the `contour` and `predict` functions in R to produce a contour plot of model predicted `loss`, against `tens` and `hard`. You may find functions `seq` and `rep` helpful, as well.

9. The R data frame `warpbreaks` gives the number of `breaks` per fixed length of wool during weaving, for two different `wool` types, and three different weaving `tensions`. Using a linear model, establish whether there is evidence that the effect of tension on break rate is dependent on the type of wool. If there is, use `interaction.plot` to examine the nature of the dependence.

10. This question is about modelling the relationship between stopping distance of a car and its speed at the moment that the driver is signalled to stop. Data on this are provided in R data frame `cars`. It takes a more or less fixed 'reaction time' for a driver to apply the car's brakes, so that the car will travel a distance directly proportional to its speed before beginning to slow. A car's kinetic energy is proportional to the square of its speed, but the brakes can only dissipate that energy, and slow the car, at a roughly constant rate per unit distance travelled: so we expect that once braking starts, the car will travel a distance proportional to the square of its initial speed, before stopping.

(a) Given the information provided above, fit a model of the form

$$\texttt{dist}_i = \beta_0 + \beta_1 \texttt{speed}_i + \beta_2 \texttt{speed}_i^2 + \epsilon_i$$

to the data in `cars`, and from this starting model, select the most appropriate model for the data using both AIC and hypothesis testing methods.

(b) From your selected model, estimate the average time that it takes a driver to apply the brakes (there are 5280 feet in a mile).

(c) When selecting between different polynomial models for a set of data, it is often claimed that one should not leave in higher powers of some continuous predictor, while removing lower powers of that predictor. Is this sensible?

11. This question relates to the material in sections 1.3.1 to 1.3.5, and it may be useful to review sections B.5 and B.6 of Appendix B. R function `qr` computed the QR decomposition of a matrix, while function `qr.qry` provides a means of efficiently pre-multiplying a vector by the $\mathbf{Q}$ matrix of the decomposition and `qr.R` extracts the $\mathbf{R}$ matrix. See `?qr` for details.
The question concerns calculation of estimates and associated quantities for the linear model, $y_i = \mathbf{X}_i\boldsymbol{\beta} + \epsilon_i$, where the ϵ_i are i.i.d. $N(0, \sigma^2)$.

(a) Write an R function which will take a vector of response variables, `y`, and a model matrix, X, as arguments, and compute the least squares estimates of associated parameters, $\boldsymbol{\beta}$, based on QR decomposition of X.

(b) Test your function by using it to estimate the parameters of the model

$$\texttt{dist}_i = \beta_0 + \beta_1 \texttt{speed}_i + \beta_2 \texttt{speed}_i^2 + \epsilon_i$$

for the data found in R data frame cars. Note that:

```
X <- model.matrix(dist ~ speed + I(speed^2),cars)
```

will produce a suitable model matrix. Check your answers against those produced by the lm function.

(c) Extend your function to also return the estimated standard errors of the parameter estimators, and the estimated residual variance. Again, check your answers against what lm produces, using the cars model. Note that solve(R) or more efficiently backsolve(R,diag(ncol(R))) will produce the inverse of an upper triangular matrix R.

(d) Use R function pt to produce p-values for testing the null hypothesis that each β_i is zero (against a two sided alternative). Again check your answers against a summary of an equivalent lm fit.

(e) Extend your fitting function again, to produce the p-values associated with a sequential ANOVA table for your model (see section 1.3.5). Again test your results by comparison with the results of applying the anova function to an equivalent lm fit. Note that pf is the function giving the c.d.f. of F-distributions.

12. R data frame InsectSprays contains counts of insects in each of several plots. The plots had each been sprayed with one of six insecticides. A model for these data might be

$$y_i = \mu + \beta_j \text{ if } i^{\text{th}} \text{ observation is for spray } j.$$

A possible identifiability constraint for this model is that $\sum_j \beta_j = 0$. In R, construct the rank deficient model matrix, for this model, and the coefficient matrix for the sum to zero constraint on the parameters. Using the methods of section 1.8.1, impose the constraint via QR decomposition of the constraint matrix, fit the model (using lm with your constrained model matrix), and then obtain the estimates of the original parameters. You will need to use R functions qr, qr.qy and qr.qty as part of this. Confirm that your estimated parameters meet the constraint.

13. The R data frame trees contains data on Height, Girth and Volume of 31 felled cherry trees. A possible model for these data is

$$\texttt{Volume}_i = \beta_1 \texttt{Girth}_i^{\beta_2} \texttt{Height}_i^{\beta_3} + \epsilon_i,$$

which can be fitted by non-linear least squares using the method of section 1.8.8.

(a) Write an R function to evaluate (i) the vector of $\mathbb{E}(\texttt{Volume})$ estimates given a vector of β_j values and vectors of `Girth` and `Height` measurements and (ii) the 31×3 'Jacobian' matrix with $(i, j)^{\text{th}}$ element $\partial\mathbb{E}(\texttt{Volume}_i)/\partial\beta_j$, returning these in a two item list. (Recall that $\partial x^y/\partial y = x^y \log(x)$.)

(b) Write R code to fit the model, to the `trees` data, using the method of section 1.8.8. Starting values of .002, 2 and 1 are reasonable.

(c) Evaluate approximate standard error estimates for your estimated model parameters.

Chapter 2

Linear Mixed Models

In general, linear mixed models extend the linear model,

$$\mathbf{y} = \mathbf{X}\boldsymbol{\beta} + \boldsymbol{\epsilon}, \quad \boldsymbol{\epsilon} \sim N(\mathbf{0}, \mathbf{I}\sigma^2),$$

to

$$\mathbf{y} = \mathbf{X}\boldsymbol{\beta} + \mathbf{Z}\mathbf{b} + \boldsymbol{\epsilon}, \;\; \mathbf{b} \sim N(\mathbf{0}, \boldsymbol{\psi}_\theta), \;\; \boldsymbol{\epsilon} \sim N(\mathbf{0}, \boldsymbol{\Lambda}_\theta),$$

where random vector, $\mathbf{b}$, contains *random effects*, with zero expected value and covariance matrix $\boldsymbol{\psi}_\theta$, with unknown parameters in $\boldsymbol{\theta}$; $\mathbf{Z}$ is a model matrix for the random effects. $\boldsymbol{\Lambda}_\theta$ is a positive definite matrix which can be used to model residual autocorrelation: its elements are usually determined by some simple model, with few (or no) unknown parameters (here considered to be part of $\boldsymbol{\theta}$). Often $\boldsymbol{\Lambda}_\theta = \mathbf{I}\sigma^2$. $\mathbf{b}$ and $\boldsymbol{\epsilon}$ are independent.

The idea is to allow a linear model structure for the random component of the response data, $\mathbf{y}$, which is as rich as the linear model structure used to model the systematic component. Except in the special case of the analysis of balanced designed experiments, inference for such models relies on some general theory for maximum likelihood estimation. It is also worth noting the resemblance of the distributional assumption for $\mathbf{b}$ to the specification of a prior in Bayesian analysis: some of the computations performed with this model will also correspond to Bayesian computations, although the inferential framework is not really Bayesian.

2.1 Mixed models for balanced data

This section briefly covers why random effects models are useful, and the topic of linear mixed models for balanced data. The latter is not essential to the rest of this book, but does provide an understanding of the degrees of freedom used in classical mixed model ANOVA computations, for example. Readers who skip straight to section 2.3 will not miss anything essential to the rest of the book.

2.1.1 A motivating example

Plant leaves have tiny holes, called 'stomata', through which they take up air, but also lose water. Most non-tropical plants photosynthesize in such a way that, on sunny days, they are limited by how much carbon dioxide they can obtain through

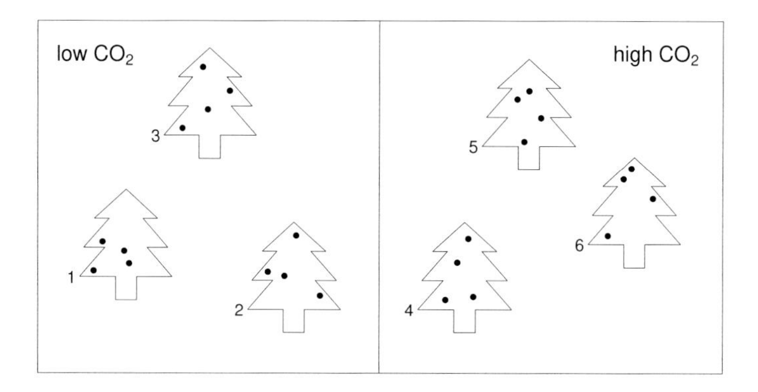

Figure 2.1 *Schematic diagram of the* CO_2 *experiment.*

these stomata. The 'problem' for a plant is that if its stomata are too small, it will not be able to get enough carbon dioxide, and if they are too large it will lose too much water on sunny days. Given the importance of this to such plants, it seems likely that stomatal size will depend on the concentration of carbon dioxide in the atmosphere. This may have climate change implications if increasing the amount of CO_2 in the atmosphere causes plants to release less water: water vapour is the most important greenhouse gas.

Consider an experiment* in which tree seedlings are grown under 2 levels of carbon dioxide concentration, with 3 trees assigned to each treatment, and suppose that after 6 months' growth stomatal area is measured at each of 4 random locations on each plant (the sample sizes here are artificially small). Figure 2.1 shows the experimental layout, schematically.

The wrong approach: A fixed effects linear model

A model of these data should include a (2 level) factor for CO_2 treatment, but also a (6 level) factor for individual tree, since we have multiple measurements on each tree and must expect some variability in stomatal area from tree to tree. So a suitable linear model is

$$y_i = \alpha_j + \beta_k + \epsilon_i \text{ if observation } i \text{ is for CO}_2 \text{ level } j, \text{ tree } k,$$

where y_i is the ith stomatal area measurement, α_j is the population mean stomatal area at CO_2 level j, β_k is the deviation of tree k from that mean and the ϵ_i are independent $N(0, \sigma^2)$ random variables. Now if this is a fixed effects model, we have two problems:

1. The α_j's and β_k's are completely confounded. Trees are 'nested' within treatment,

*One important part of the design of such an experiment would be to ensure that the trees are grown under natural, *variable* light levels. At constant *average* light levels the plants are not CO_2 limited.

with 3 trees in one treatment and 3 in the other: any number you like could be added to α_1 and simultaneously subtracted from β_1, β_2 and β_3, without changing the model predictions at all, and the same goes for α_2 and the remaining β_k's.

2. We really want to learn about trees in general, but this is not possible with a model in which there is a fixed effect for each particular tree: unless the tree effects happen to turn out to be negligible, we cannot use the model to predict what happens to a tree other than one of the six in the experiment.

The following R session illustrates problem 1. First compare models with and without the tree factor (β_k):

```
> m1 <- lm(area ~ CO2 + tree, stomata)
> m0 <- lm(area ~ CO2, stomata)
> anova(m0,m1)
Analysis of Variance Table

Model 1: area ~ CO2
Model 2: area ~ CO2 + tree
  Res.Df    RSS Df Sum of Sq      F   Pr(>F)
1     22 2.1348
2     18 0.8604  4    1.2744 6.6654 0.001788 **
```

Clearly, there is strong evidence for tree to tree differences, which means that with this model we can not tell whether CO_2 had an effect or not. To re-emphasize this point, here is what happens if we attempt to test for a CO_2 effect:

```
> m2 <- lm(area ~ tree, stomata)
> anova(m2,m1)
Analysis of Variance Table

Model 1: area ~ tree
Model 2: area ~ CO2 + tree
  Res.Df    RSS Df  Sum of Sq F Pr(>F)
1     18 0.8604
2     18 0.8604  0 -2.220e-16
```

The confounding of the CO_2 and tree factors means that the models being compared here are really the same model: as a result, they give the same residual sum of squares and have the same residual degrees of freedom — 'comparing' them tells us nothing about the effect of CO_2.

In many ways this problem comes about because our model is simply too flexible. Individual tree effects are allowed to take any value whatsoever, which amounts to saying that each individual tree is completely different to every other individual tree: having results for 6 trees will tell us nothing whatsoever about a 7th. This is not a sensible starting point for a model aimed at analysing data like these. We really expect trees of a particular species to behave in broadly similar ways so that a representative (preferably random) sample of trees, from the wider population of such trees, *will* allow us to make inferences about that wider population of trees. Treating the individual trees, not as completely unique individuals but as a random sample from the target population of trees, will allow us to estimate the CO_2 effect *and* to generalize beyond the 6 trees in the experiment.

The right approach: A mixed effects model

The key to establishing whether CO_2 has an effect is to recognise that the CO_2 factor and tree factors are different in kind. The CO_2 effects are fixed characteristics of the whole population of trees that we are trying to learn about. In contrast, the tree effect will vary randomly from tree to tree in the population. We are not primarily interested in the values of the tree effect for the particular trees used in the experiment: if we had used a different 6 trees these effects would have taken different values anyway. But we can not simply ignore the tree effect without inducing dependence between the response observations (area), and hence violating the independence assumption of the linear model. In this circumstance it makes sense to model the distribution of tree effects across the population of trees, and to suppose that the particular tree effects that occur in the experiment are just independent observations from this distribution. That is, the CO_2 effect will be modelled as a fixed effect, but the tree effect will be modelled as a random effect. Here is a model set up in this way:

$$y_i = \alpha_j + b_k + \epsilon_i \text{ if observation } i \text{ is for CO}_2 \text{ level } j \text{ and tree } k, \tag{2.1}$$

where $b_k \sim N(0, \sigma_b^2)$, $\epsilon_i \sim N(0, \sigma^2)$ and all the b_k and ϵ_i are mutually independent random variables. Now testing for tree effects can proceed exactly as it did in the fixed effects case, by comparing the least squares fits of models with and without the tree effects. But this mixed effects model also lets us test CO_2 effects, whether or not there is evidence for a tree effect.

All that is required is to average the data at each level of the random effect, i.e., at each tree. For balanced data, such as we have here, the key feature of a mixed effects model is that this 'averaging out' of a random effect automatically implies a simplified mixed effects model for the aggregated data: the random effect is absorbed into the independent residual error term. It is easy to see that the model for the average stomatal area per tree must be

$$\bar{y}_k = \alpha_j + e_k \text{ if tree } k \text{ is for CO}_2 \text{ level } j, \tag{2.2}$$

where the e_k are independent $N(0, \sigma_b^2 + \sigma^2/4)$ random variables.

Now it is a straightforward matter to test for a CO_2 effect in R. First aggregate the data for each tree:

```
> st <- aggregate(data.matrix(stomata),
+          by=list(tree=stomata$tree),mean)
> st$CO2 <- as.factor(st$CO2);st
  tree     area CO2 tree
1    1 1.623374   1    1
2    2 1.598643   1    2
3    3 1.162961   1    3
4    4 2.789238   2    4
5    5 2.903544   2    5
6    6 2.329761   2    6
```

and then fit the model implied by the aggregation.

```
> m3 <- lm(area ~ CO2, st)
```

```
> anova(m3)
Analysis of Variance Table

Response: area
          Df  Sum Sq Mean Sq F value   Pr(>F)
CO2        1 2.20531 2.20531  27.687 0.006247 **
Residuals  4 0.31861 0.07965
```

There is strong evidence for a CO_2 effect here, and we would now proceed to examine the estimate of this fixed effect (e.g., using `summary(m3)`). Usually with a mixed model the variances of the random effects are of more interest than the effects themselves, so in this example σ_b^2 should be estimated.

Let RSS_i stand for the residual sum of squares for model i. From the usual theory of linear models we have that:

$$\hat{\sigma}^2 = \text{RSS}_1/18$$

(RSS_1 is the residual sum of squares from fitting (2.1)) and

$$\widehat{\sigma_b^2 + \sigma^2/4} = \text{RSS}_3/4$$

(RSS_3 is the residual sum of squares from fitting (2.2)). Both estimators are unbiased. Hence, an unbiased estimator for σ_b^2 is

$$\hat{\sigma}_b^2 = \text{RSS}_3/4 - \text{RSS}_1/72.$$

This can easily be evaluated in R.

```
> summary(m3)$sigma^2 - summary(m1)$sigma^2/4
[1] 0.06770177
```

2.1.2 *General principles*

To see how the ideas from the previous section generalize, consider data from a designed experiment and an associated linear mixed model for the data, in which the response variable depends only on factor variables and their interactions (which may be random or fixed). Assume that the data are *balanced* with respect to the model, meaning that for each factor or interaction in the model, the same number of data have been collected at each of its levels. In this case:

- Aggregated data, obtained by averaging the response at each level of any factor or interaction, will be described by a mixed model, derived from the original mixed model by the averaging process.
- Models for different aggregations will enable inferences to be made about different fixed and random factors, using standard methods for ordinary linear models. Note that not all aggregations will be useful, and the random effects themselves can not be 'estimated' in this way.
- The variances of the random effects can be estimated from combinations of the the usual residual variance estimates from models for different aggregations.

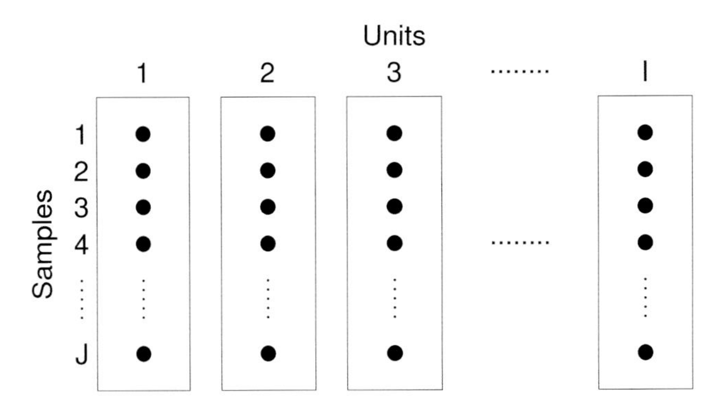

Figure 2.2 *Schematic illustration of the balanced one-way experimental layout discussed in section 2.1.3. Rectangles are experimental units and •'s indicate measurements.*

These principles are useful for two reasons. Firstly, the classical mixed model analyses for designed experiments can be derived using them. Secondly, they provide a straightforward explanation for the degrees of freedom of the reference distributions used in mixed model hypothesis testing: the degrees of freedom are always those that apply to the aggregated model appropriate for testing hypotheses about the effect concerned. For example, in the CO_2 analysis the hypothesis tests about the CO_2 effect were conducted with reference to an $F_{1,4}$ distribution, with these degrees of freedom being those appropriate to the aggregated model, used for the test.

To illustrate and reinforce these ideas two further simple examples of the analysis of 'standard designs' will be covered, before returning to the general mixed models that are of more direct relevance to GAMs.

2.1.3 A single random factor

Consider an experimental design in which you have J measurements from each of I units, illustrated schematically in figure 2.2. Suppose that we are interested in establishing whether there are differences between the units, but are not really interested in the individual unit effects: rather in quantifying how much variability can be ascribed to differences between units. This would suggest using a random effect term for units.

A concrete example comes from animal breeding. For a breeding program to be successful we need to know that variability in the targeted trait has a substantial enough genetic component that we can expect to alter it by selective breeding. Consider a pig breeding experiment in which I pregnant sows are fed a standard diet, and the fat content of J of each of their piglets is measured. The interesting questions here are: is there evidence for litter to litter variability in fat content (which would be consistent with genetic variation in this trait) and if so how large is this component, in relation to the piglet to piglet variability within a litter? Notice here that we are not interested in how piglet fat content varies from particular sow to particular sow in the experiment, but rather in the variability between sows in general. This suggests using a random effect for sow in a model for such data.

So a suitable model is

$$y_{ij} = \alpha + b_i + \epsilon_{ij}, \tag{2.3}$$

where α is the fixed parameter for the population mean, $i = 1 \ldots I$, $j = 1 \ldots J$, $b_i \sim N(0, \sigma_b^2)$, $\epsilon_{ij} \sim N(0, \sigma^2)$ and all the b_i and ϵ_{ij} terms are mutually independent.

The first question of interest is whether $\sigma_b^2 > 0$, i.e., whether there is evidence that the factor variable contributes to the variance of the response. Formally we would like to test $\text{H}_0 : \sigma_b^2 = 0$ against $\text{H}_1 : \sigma_b^2 > 0$. To do this, simply note that the null hypothesis is exactly equivalent to $\text{H}_0 : b_i = 0 \; \forall \; i$, since both formulations of H_0 imply that the data follow,

$$y_{ij} = \alpha + \epsilon_{ij}. \tag{2.4}$$

Hence we can test the null hypothesis by using standard linear modelling methods to compare (2.4) to (2.3) using an F-ratio test (ANOVA).

Fitting (2.3) to the data will also yield the usual estimate of σ^2,

$$\hat{\sigma}^2 = \text{RSS}_1/(n - I),$$

where RSS_1 is the residual sum of squares from fitting the model to data, $n = IJ$ is the number of data, and $n - I$ is the residual degrees of freedom from this model fit.

So far the analysis with the mixed model has been identical to what would have been done with a fixed effects model, but now consider estimating σ_b^2. The 'obvious' method of just using the sample variance of the $\hat{b}_i$'s, 'estimated' by least squares, is not to be recommended, as such estimators are biased. Instead we make use of the model that results from averaging at each level of the factor:

$$\bar{y}_{i\cdot} = \alpha + b_i + \frac{1}{J}\sum_{j=1}^{J} \epsilon_{ij}.$$

Now define a new set of I random variables,

$$e_i = b_i + \frac{1}{J}\sum_{j=1}^{J} \epsilon_{ij}.$$

The e_i's are clearly mutually independent, since their constituent random variables are independent and no two e_i's share a constituent random variable. They are also zero mean normal random variables, since each is a sum of zero mean normal random variables. It is also clear that

$$\text{var}(e_i) = \sigma_b^2 + \sigma^2/J.$$

Hence, the model for the aggregated data becomes

$$\bar{y}_{i\cdot} = \alpha + e_i, \tag{2.5}$$

where the e_i are i.i.d. $N(0, \sigma_b^2 + \sigma^2/J)$ random variables. If RSS_2 is the residual sum of squares when this model is fitted by least squares, then the usual unbiased residual variance estimate gives

$$\text{RSS}_2/(I-1) = \hat{\sigma}_b^2 + \hat{\sigma}^2/J.$$

Re-arrangement and substitution of the previous $\hat{\sigma}^2$ implies that

$$\hat{\sigma}_b^2 = \text{RSS}_2/(I-1) - \hat{\sigma}^2/J$$

is an unbiased estimator of σ_b^2.

Now consider a practical industrial example. An engineering test for longitudinal stress in rails involves measuring the time it takes certain ultrasonic waves to travel along the rail. To be a useful test, engineers need to know the average travel time for rails and the variability to expect between rails, as well as the variability in the measurement process. The `Rail` data frame available with R package `nlme` provides 3 measurements of travel time for each of 6 randomly chosen rails. This provides an obvious application for model (2.3). First examine the data.

```
> library(nlme)  # load nlme 'library', which contains data
> data(Rail)     # load data
> Rail
   Rail travel
1     1     55
2     1     53
3     1     54
4     2     26
5     2     37
.     .      .
.     .      .
17    6     85
18    6     83
```

Now fit model (2.3) as a fixed effects model, and use this model to test $\text{H}_0 : \sigma_b^2 = 0$, i.e., to test for evidence of differences between rails.

```
> m1 <- lm(travel ~ Rail,Rail)
> anova(m1)
Analysis of Variance Table

Response: travel
          Df Sum Sq Mean Sq F value    Pr(>F)
Rail       5 9310.5  1862.1  115.18 1.033e-09 ***
Residuals 12  194.0    16.2
---
Signif. codes:  0 '***' 0.001 '**' 0.01 '*' 0.05 '.' 0.1 ' ' 1
```

So there is strong evidence to reject the null hypothesis and accept rail to rail differences as real. As we saw theoretically, so far the analysis does not differ from that for a fixed effects model, but to estimate σ_b^2 involves averaging at each level of the random effect and fitting model (2.5) to the resulting averages. R function `aggregate` will achieve the required averaging.

```
> rt <-  # average over Rail effect
+ aggregate(data.matrix(Rail),by=list(Rail$Rail),mean)
> rt
  Group.1 Rail    travel
1       2    1 31.66667
2       5    2 50.00000
3       1    3 54.00000
4       6    4 82.66667
5       3    5 84.66667
6       4    6 96.00000
```

It is now possible to fit (2.5) and calculate $\hat{\sigma}_b$ and $\hat{\sigma}$, as described above:

```
> m0 <- lm(travel ~ 1,rt)    # fit model to aggregated data
> sig <- summary(m1)$sigma # sig^2 is resid. var. component
> sigb <- (summary(m0)$sigma^2 - sig^2/3)^0.5
>                  # sigb^2 is the variance component for rail
> sigb
[1] 24.80547
> sig
[1] 4.020779
```

So, there is a fairly large amount of rail to rail variability, whereas the measurement error is relatively small. In this case the model intercept, α, is confounded with the random effects, b_j, so α must be estimated from the fit of model (2.5).

```
> summary(m0)
Coefficients:
            Estimate Std. Error t value Pr(>|t|)
(Intercept)    66.50      10.17   6.538  0.00125 **
```

Model checking proceeds by looking at residual plots, from the fits to both the original and the aggregated data, since, approximately, these should look like samples of i.i.d. normal random variables. However, there would have to be a really grotesque violation of the normality assumption for the b_j before you could hope to pick it up from examination of 6 residuals.

2.1.4 *A model with two factors*

Now consider an experiment in which each observation is grouped according to two factors. A schematic diagram of such a design is shown in figure 2.3. Suppose that one factor is to be modelled as a fixed effect and one as a random effect. A typical example is a randomized block design for an agricultural field trial, testing different fertilizer formulations. The response variable would be yield of the crop concerned, assessed by harvesting at the end of the experiment. Because crop yields depend on many uncontrolled soil related factors, it is usual to arrange the experiment in blocks, within which it is hoped that the soil will be fairly homogeneous. Treatments are randomly arranged within the blocks. For example, a field site might be split into 4 adjacent blocks with 15 plots in each block, each plot being randomly assigned one of five replicates of each of 3 fertilizer treatments. The idea is that differences within blocks should be smaller than differences between blocks — i.e., variability

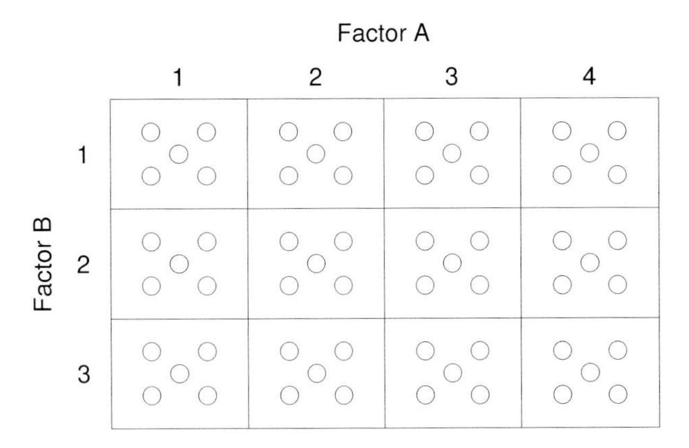

Figure 2.3 *A schematic diagram of a two factor design of the sort discussed in section 2.1.4, with 3 levels of one factor, 4 levels of another and 5 observations for each combination of factor levels. Note that this diagram is not intended to represent the actual physical layout of any experiment.*

in conditions within a block will be smaller than variability across the whole field. A suitable model for the data would include a block effect, to account for this block to block variability, and it makes sense to treat it as a random effect since we are not in the least interested in the particular values of the block effects, but view them as representing variability in environment with location. The treatments, on the other hand, would be modelled as fixed effects, since the values of the treatment effects are fixed properties of the crop population in general. (If we repeated the experiment in a different location, the fertilizer effects should be very similar, whereas the particular values of the block effects would be unrelated to the block effects in the first experiment, apart from having a similar distribution.)

So, a model for the kth observation at level i of fixed effect A and level j of random effect B is

$$y_{ijk} = \mu + \alpha_i + b_j + (\alpha b)_{ij} + \epsilon_{ijk}, \tag{2.6}$$

where $b_j \sim N(0, \sigma_b^2)$, $(\alpha b)_{ij} \sim N(0, \sigma_{\alpha b}^2)$ and $\epsilon_{ijk} \sim N(0, \sigma^2)$, and all these random variables are mutually independent. μ is the overall population mean, the α_i are the I fixed effects for factor A, the b_j are the J random effects for factor B, and the $(\alpha b)_{ij}$ are the IJ interaction terms for the interaction between the factors (an interaction term involving a random effect must also be a random term).

Testing $\text{H}_0 : \sigma_{\alpha b}^2 = 0$ is logically equivalent to testing $\text{H}_0 : (\alpha b)_{ij} = 0 \,\forall\, ij$, in a fixed effects framework. Hence this hypothesis can be tested by the usual ANOVA/F-ratio test comparison of models with and without the interaction terms. If RSS_1 now denotes the residual sum of squares from fitting (2.6) by least squares then

$$\hat{\sigma}^2 = \text{RSS}_1/(n - IJ).$$

In a purely fixed effects context it only makes sense to test for main effects if the interaction terms are not significant, and can hence be treated as zero. In the mixed effects case, because the interaction is a random effect, it is possible to make inferences about the main effects whether or not the interaction terms are significant. This can be done by averaging the K data at each level of the interaction. The averaging, together with model (2.6), implies the following model for the averages:

$$\bar{y}_{ij\cdot} = \mu + \alpha_i + b_j + (\alpha b)_{ij} + \frac{1}{K}\sum_{k=1}^{K} \epsilon_{ijk}.$$

Defining

$$e_{ij} = (\alpha b)_{ij} + \frac{1}{K}\sum_{k=1}^{K} \epsilon_{ijk},$$

it is clear that, since the e_{ij} are each sums of zero mean normal random variables, they are also zero mean normal random variables. Also, since the $(\alpha b)_{ij}$ and ϵ_{ijk} are mutually independent random variables, and no $(\alpha b)_{ij}$ or ϵ_{ijk} is a component of more than one e_{ij}, the e_{ij} are mutually independent. Furthermore

$$\text{var}(e_{ij}) = \sigma^2_{\alpha b} + \sigma^2/K.$$

Hence the simplified model is

$$\bar{y}_{ij\cdot} = \mu + \alpha_i + b_j + e_{ij}, \tag{2.7}$$

where the e_{ij} are i.i.d. $N(0, \sigma^2_{\alpha b} + \sigma^2/K)$ random variables. The null hypothesis, $\text{H}_0 : \alpha_i = 0 \; \forall \; i$, is tested by comparing the least squares fits of (2.7) and $\bar{y}_{ij\cdot} = \mu + b_j + e_{ij}$, in the usual way, by F-ratio testing. Similarly $\text{H}_0 : \sigma^2_b = 0$ is logically equivalent to $\text{H}_0 : b_j = 0 \; \forall \; j$, and is hence tested by F-ratio test comparison of (2.7) and $\bar{y}_{ij\cdot} = \mu + \alpha_i + e_{ij}$. The residual sum of squares for model (2.7), RSS_2, say, is useful for unbiased estimation of the interaction variance:

$$\hat{\sigma}^2_{\alpha b} + \hat{\sigma}^2/K = \text{RSS}_2/(IJ - I - J + 1)$$

and hence,

$$\hat{\sigma}^2_{\alpha b} = \text{RSS}_2/(IJ - I - J + 1) - \hat{\sigma}^2/K.$$

Averaging the data once more, over the levels of factor B, induces the model

$$\bar{y}_{\cdot j\cdot} = \mu + \frac{1}{I}\sum_{i=1}^{I} \alpha_i + b_j + \frac{1}{I}\sum_{i=1}^{I} e_{ij}.$$

Defining $\mu' = \mu + \frac{1}{I}\sum_i \alpha_i$ and $e_j = b_j + \frac{1}{I}\sum_i e_{ij}$ this model becomes

$$\bar{y}_{\cdot j\cdot} = \mu' + e_j, \text{ where } e_j \sim N(0, \sigma^2_b + \sigma^2_{\alpha b}/I + \sigma^2/(IK)). \tag{2.8}$$

Hence, if RSS_3 is the residual sum of squares of model (2.8), an unbiased estimator of σ^2_b is given by

$$\hat{\sigma}^2_b = \text{RSS}_3/(J - 1) - \hat{\sigma}^2_{\alpha b}/I - \hat{\sigma}^2/(IK).$$

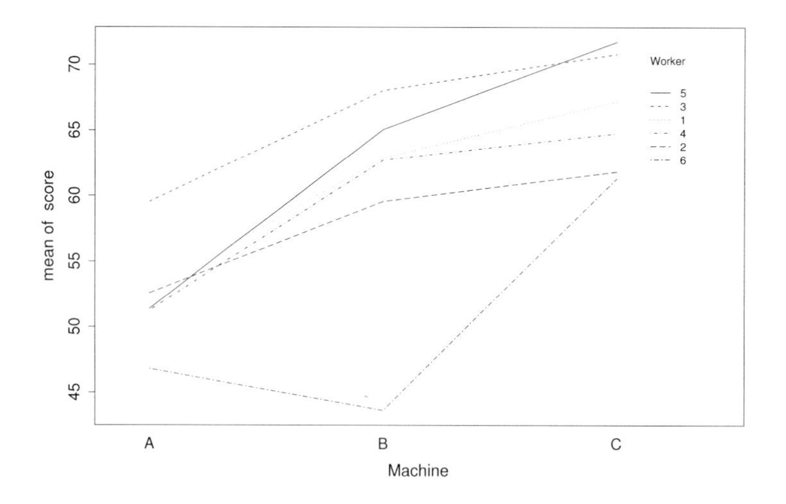

Figure 2.4 *Plot of the* `Machines` *data discussed in section 2.1.4.*

Now consider a practical example. The `Machines` data frame, from the `nlme` package, contains data from an industrial experiment comparing 3 different machine types. The aim of the experiment was to determine which machine type resulted in highest worker productivity. 6 workers were randomly selected to take part in the trial, with each worker operating each machine 3 times (presumably after an appropriate period of training designed to eliminate any 'learning effect'). The following produces the plot shown in figure 2.4

```
> library(nlme)
> data(Machines)
> names(Machines)
[1] "Worker"   "Machine" "score"
> attach(Machines)   # make data available without 'Machines$'
> interaction.plot(Machine,Worker,score)
```

From the experimental aims, it is clear that fixed machine effects and random worker effects are appropriate. We are interested in the effects of these particular machine types, but are only interested in the worker effects in as much as they reflect variability between workers in the population of workers using this type of machine. Put another way, if the experiment were repeated somewhere else (with different workers) we would expect the estimates of the machine effects to be quite close to the results obtained from the current experiment, while the individual worker effects would be quite different (although with similar variability, we hope). So model (2.6) is appropriate, with the α_i representing the fixed machine effects, b_j representing the random worker effects, and $(\alpha b)_{ij}$ representing the worker machine interaction (i.e., the fact that different workers may work better on different machines).

Fitting the full model, we can immediately test $\mathrm{H}_0 : \sigma^2_{\alpha b} = 0$.

```
> m1 <- lm(score ~ Worker*Machine,Machines)
> m0 <- lm(score ~ Worker + Machine,Machines)
> anova(m0,m1)
Analysis of Variance Table
Model 1: score ~ Worker + Machine
Model 2: score ~ Worker + Machine + Worker:Machine
  Res.Df    RSS Df Sum of Sq     F    Pr(>F)
1     46 459.82
2     36  33.29 10    426.53 46.13 < 2.2e-16 ***
```

We must accept $\text{H}_1 : \sigma^2_{\alpha b} \neq 0$. There is very strong evidence for an interaction between machine and worker. σ^2 can now be estimated:

```
> summary(m1)$sigma^2
[1] 0.9246296
```

To examine the main effects we can aggregate at each level of the interaction,

```
Mach <- aggregate(data.matrix(Machines),by=
        list(Machines$Worker,Machines$Machine),mean)
Mach$Worker <- as.factor(Mach$Worker)
Mach$Machine <- as.factor(Mach$Machine)
```

and fit model (2.7) to the resulting data.

```
> m0 <- lm(score ~ Worker + Machine,Mach)
> anova(m0)
Analysis of Variance Table
Response: score
          Df Sum Sq Mean Sq F value    Pr(>F)
Worker     5 413.96   82.79  5.8232 0.0089495 **
Machine    2 585.09  292.54 20.5761 0.0002855 ***
Residuals 10 142.18   14.22
```

The very low p-values again indicate that $\text{H}_0 : \sigma_b^2 = 0$ and $\text{H}_0 : \alpha_1 = \alpha_2 = \alpha_3 = 0$ should be rejected in favour of the obvious alternatives. There is strong evidence for differences between machine types and for variability between workers. Going on to examine the fixed effect estimates, using standard fixed effects methods, indicates that machine C leads to substantially increased productivity.

Estimation of $\sigma^2_{\alpha b}$, the interaction variance, is straightforward.

```
> summary(m0)$sigma^2 - summary(m1)$sigma^2/3
[1] 13.90946
```

Aggregating once more and fitting (2.8), we can estimate the worker variance component, σ_b^2.

```
> M <- aggregate(data.matrix(Mach),by=list(Mach$Worker),mean)
> m00 <- lm(score ~ 1, M)
> summary(m00)$sigma^2 - (summary(m0)$sigma^2)/3
[1] 22.96118
```

Residual plots should be checked for `m1`, `m0` and `m00`. If this is done, then it is tempting to try and see how robust the results are to the omission of worker 6 on machine B (see figure 2.4), but this requires methods that can cope with unbalanced data, which will be considered shortly.

2.1.5 Discussion

Although practical examples were presented above, this theory for mixed models of balanced experimental data is primarily of theoretical interest, for understanding the results used in classical mixed model ANOVA tables, and for motivating the use of particular reference distributions when conducting hypothesis tests for mixed models. In practice the somewhat cumbersome analysis based on aggregating data would usually be eschewed in favour of using specialist mixed modelling software, such as that accessible via R function `lme` from the `nlme` library.

Before leaving the topic of balanced data altogether, it is worth noting the reason that ordinary linear model theory can be used for inference with balanced models. The first reason relates to the estimator of the residual variance, $\hat{\sigma}^2$. In ordinary linear modelling, $\hat{\sigma}^2$ does not depend in any way on the values of the model parameters $\boldsymbol{\beta}$ and is independent of $\hat{\boldsymbol{\beta}}$ (see section 1.3.3). This fact is not altered if some elements of $\boldsymbol{\beta}$ are themselves random variables. Hence the usual estimator of $\hat{\sigma}^2$, based on the least squares estimate of a linear model, remains valid, and unbiased, for a linear mixed model.

The second reason relates to the estimators of the fixed effect parameters. In a fixed effects setting, consider two subsets $\boldsymbol{\beta}_1$ and $\boldsymbol{\beta}_2$ of the parameter vector, $\boldsymbol{\beta}$, with corresponding model matrix columns $\mathbf{X}_1$ and $\mathbf{X}_2$. If $\mathbf{X}_1$ and $\mathbf{X}_2$ are orthogonal, meaning that $\mathbf{X}_1^\mathsf{T}\mathbf{X}_2 = \mathbf{0}$, then the least squares estimators $\hat{\boldsymbol{\beta}}_1$ and $\hat{\boldsymbol{\beta}}_2$ will be independent: so inferences about $\boldsymbol{\beta}_1$ do not depend in any way on the value of $\boldsymbol{\beta}_2$. This situation is unaltered if we now move to a mixed effects model, and suppose that $\boldsymbol{\beta}_2$ is actually a random vector. Hence, in a mixed model context, we can still use fixed effects least squares methods for inferences about any fixed effects whose estimators are independent of all the random effects in the model. So, when successively aggregating data (and models), we can use least squares methods to make inferences about a fixed effect as soon as the least squares estimator of that fixed effect becomes independent of all random effects in the aggregated model. Generally such independence only occurs for balanced data from designed experiments.

Finally, note that least squares methods are not useful for 'estimating' the random effects. This is, in part, because identifiability constraints are generally required in order to estimate effects, but imposing such constraints on random effects fundamentally modifies the model, by changing the random effect distributions.

2.2 Maximum likelihood estimation

We have come as far as we can relying only on least squares. A more general approach to mixed models, as well as the generalized linear models of the next chapter, will require use of general large sample results from the theory of maximum likelihood estimation. Appendix A derives these results. This section simply summarises what we need in order to proceed.

A statistical model for a data vector, $\mathbf{y}$, defines a probability density (or mass) function for the random vector of which $\mathbf{y}$ is an observation, $f_\theta(\mathbf{y})$. $\boldsymbol{\theta}$ denotes the unknown parameters of the model. The aim is to make inferences about $\boldsymbol{\theta}$ based on $\mathbf{y}$. Of course, f_θ may depend on other observed variables (covariates) and known parameters, but there is no need to make this explicit in the notation. The key idea behind maximum likelihood estimation is:

> Values of $\boldsymbol{\theta}$ that make $f_{\theta}(\mathbf{y})$ larger for the observed $\mathbf{y}$ are more *likely* to be correct than values that make $f_{\theta}(\mathbf{y})$ smaller.

So we judge the goodness of fit of a value for $\boldsymbol{\theta}$ using the log likelihood function†

$$l(\boldsymbol{\theta}) = \log f_{\theta}(\mathbf{y}),$$

that is the log of the p.d.f. (or p.m.f.) of $\mathbf{y}$ evaluated at the observed $\mathbf{y}$, and considered as a function of $\boldsymbol{\theta}$. The maximum likelihood estimate of $\boldsymbol{\theta}$ is then simply

$$\hat{\boldsymbol{\theta}} = \underset{\theta}{\text{argmax}}\, l(\boldsymbol{\theta}).$$

Subject to some regularity conditions and in the large sample limit

$$\hat{\boldsymbol{\theta}} \sim N(\boldsymbol{\theta}, \boldsymbol{\mathcal{I}}^{-1}), \tag{2.9}$$

where $\boldsymbol{\theta}$ denotes the true parameter value, and $\boldsymbol{\mathcal{I}}$ is the negative expected Hessian matrix of the log likelihood, so that $\mathcal{I}_{ij} = -\mathbb{E}(\partial^2 l/\partial\theta_i\partial\theta_j)$. $\boldsymbol{\mathcal{I}}$ is known as the (Fisher) information matrix, and actually the same result holds if we replace it by the 'observed information matrix', $\hat{\boldsymbol{\mathcal{I}}}$, where $\hat{\mathcal{I}}_{ij} = -\left.\partial^2 l/\partial\theta_i\partial\theta_j\right|_{\hat{\theta}}$. This result can be used to compute approximate confidence intervals for elements of $\boldsymbol{\theta}$, or to obtain p-values for tests about elements of $\boldsymbol{\theta}$.

There are also large sample results useful for model selection. Consider two models, where model 0 is a reduced version of model 1 (i.e., the models are 'nested'), and suppose we want to test the null hypothesis that model 0 is correct. Let p_j be the number of identifiable parameters for model j. If both models are estimated by MLE then in the large sample limit, assuming a regular likelihood,

$$2\{l(\hat{\boldsymbol{\theta}}_1) - l(\hat{\boldsymbol{\theta}}_0)\} \sim \chi^2_{p_1-p_0} \text{ if model 0 is correct,}$$

i.e., under repeated sampling of $\mathbf{y}$ the generalized log likelihood ratio statistic, $2\{l(\hat{\boldsymbol{\theta}}_1) - l(\hat{\boldsymbol{\theta}}_0)\}$, should follow a $\chi^2_{p_1-p_0}$ distribution, if model 0 is correct. Otherwise $l(\hat{\boldsymbol{\theta}}_0)$ should be sufficiently much lower than $l(\hat{\boldsymbol{\theta}}_1)$ that the statistic will be too large for consistency with $\chi^2_{p_1-p_0}$. As with any hypothesis test, consistency of the data with the null model is judged using a p-value, here $\Pr[2\{l(\hat{\boldsymbol{\theta}}_1) - l(\hat{\boldsymbol{\theta}}_0)\} \le \chi^2_{p_1-p_0}]$ (the l.h.s. of the inequality is the *observed* test statistic).

A popular alternative model selection approach is to select between models on the basis of an estimate of their closeness to $f_0(\mathbf{y})$, the true density (or mass) function of $\mathbf{y}$. The idea is to try to estimate the expected Kullback-Leibler divergence

$$\mathbb{E}_{\hat{\theta}} \int \{\log f_0(\mathbf{y}) - \log f_{\hat{\theta}}(\mathbf{y})\} f_0(\mathbf{y}) d\mathbf{y}.$$

†The *log* likelihood is used for reasons of computational convenience (the likelihood itself tends to underflow to zero), and because the key distributional results involve the log likelihood.

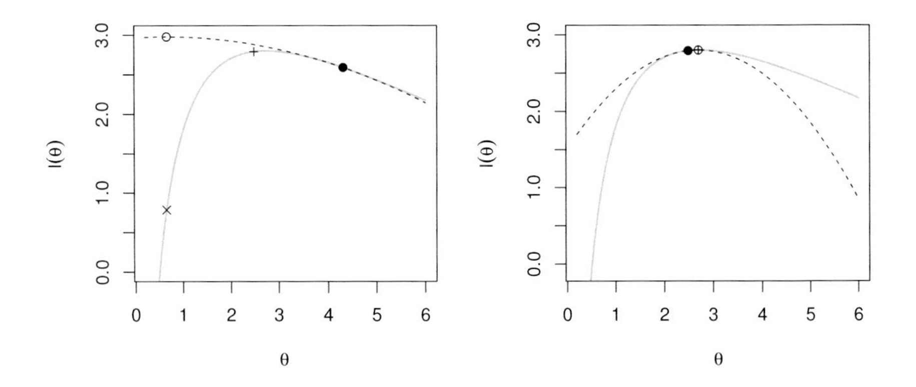

Figure 2.5 *Illustration of two steps of Newton's method as described in section 2.2.1. In both panels the grey curve is the log likelihood to be maximised, and the dashed curve is the quadratic Taylor approximation about the point indicated by •. The maximum of the quadratic approximation is indicated by ○. Left: the first Newton step is initially too large, resulting in decreased likelihood (×) so that step halving is needed to obtain the improved likelihood (+). Right: for the next step the quadratic approximation is excellent, with a maximum corresponding to that of the log-likelihood.*

A particular estimate of this quantity is minimized by whichever model minimizes

$$\text{AIC} = -2l(\hat{\boldsymbol{\theta}}) + 2p,$$

where p is the number of identifiable model parameters (usually the dimension of $\boldsymbol{\theta}$). From a set of not necessarily nested models we select the one with the lowest AIC.

2.2.1 Numerical likelihood maximization

It is unusual to be able maximize the log likelihood directly, and numerical optimization methods are usually required. Some variant of Newton's method is often the method of choice, for reasons of speed and reliability, and because it is based on the same Hessian of the log likelihood that appears in the large sample result (2.9). Note that optimization software often minimizes by default, but maximizing $l(\boldsymbol{\theta})$ is exactly equivalent to minimizing $-l(\boldsymbol{\theta})$.

The basic principle of Newton's method is to approximate $l(\boldsymbol{\theta})$ by a second order Taylor expansion about the current parameter, guess, $\boldsymbol{\theta}_0$. The maximizer of the approximating quadratic is taken as the next trial parameter vector $\boldsymbol{\theta}'$. If $l(\boldsymbol{\theta}') < l(\boldsymbol{\theta}_0)$ then we repeatedly set $\boldsymbol{\theta}' \leftarrow (\boldsymbol{\theta}' + \boldsymbol{\theta}_0)/2$ until the new log likelihood is not worse than the old one. The whole processes is repeated until $\partial l/\partial\boldsymbol{\theta} \simeq \mathbf{0}$. See figure 2.5.

Formally we are using the approximation

$$l(\boldsymbol{\theta}_0 + \boldsymbol{\Delta}) \simeq l(\boldsymbol{\theta}_0) + \nabla l^{\mathsf{T}}\boldsymbol{\Delta} + \boldsymbol{\Delta}^{\mathsf{T}}\nabla^2 l\boldsymbol{\Delta}/2,$$

where $\nabla l = \partial l/\partial\boldsymbol{\theta}|_{\theta_0}$ and $\nabla^2 l = \partial^2 l/\partial\boldsymbol{\theta}\partial\boldsymbol{\theta}^{\mathsf{T}}|_{\theta_0}$. Assuming $-\nabla^2 l$ is positive definite, the maximum of the right hand side is found by differentiating and setting to zero, to obtain

$$\boldsymbol{\Delta} = -(\nabla^2 l)^{-1}\nabla l,$$

so $\boldsymbol{\theta}' = \boldsymbol{\theta}_0 + \boldsymbol{\Delta}$. An important detail is that $\boldsymbol{\Delta}$ will be an ascent direction if *any* positive definite matrix, $\mathbf{H}$, is used in place of $-(\nabla^2 l)^{-1}$. This is because, for small enough (positive) step size α, a first order Taylor expansion then implies that $l(\boldsymbol{\theta}_0 + \alpha\boldsymbol{\Delta}) \to l(\boldsymbol{\theta}_0) + \alpha \nabla l^{\mathsf{T}} \mathbf{H} \nabla l$, and the second term on the right hand side is positive if $\mathbf{H}$ is positive definite.

This means that if the Hessian of the negative log likelihood is indefinite, then it should be perturbed to make it positive definite, in order to guarantee that the Newton step will actually increase the log likelihood. This is sometimes achieved by adding a small multiple of the identity matrix to the Hessian, or by eigen-decomposing the Hessian and re-setting any eigenvalues ≤ 0 to positive values. Other consequences of not requiring the exact Hessian matrix to get an ascent direction are that it is possible to substitute the information matrix for the Hessian when this is convenient ('Fisher scoring'), and also that finite difference approximations to the Hessian often work well. Quasi-Newton methods are a variant on Newton's method in which an approximation to the Hessian is updated at each step, using only the gradients of the log likelihood at the start and end of the step. See (Wood, 2015, §5.1) for more.

2.3 Linear mixed models in general

Recall that the general linear mixed model can conveniently be written as

$$\mathbf{y} = \mathbf{X}\boldsymbol{\beta} + \mathbf{Z}\mathbf{b} + \boldsymbol{\epsilon}, \quad \mathbf{b} \sim N(\mathbf{0}, \boldsymbol{\psi}_\theta), \quad \boldsymbol{\epsilon} \sim N(\mathbf{0}, \boldsymbol{\Lambda}_\theta) \tag{2.10}$$

where $\boldsymbol{\psi}_\theta$ is a positive definite covariance matrix for the random effects $\mathbf{b}$, and $\mathbf{Z}$ is a matrix of fixed coefficients describing how the response variable, $\mathbf{y}$, depends on the random effects (it is a model matrix for the random effects). $\boldsymbol{\psi}_\theta$ depends on some parameters, $\boldsymbol{\theta}$, which will be the prime target of statistical inference about the random effects (the exact nature of the dependence is model specific). Finally, $\boldsymbol{\Lambda}_\theta$ is a positive definite matrix which usually has a simple structure depending on few or no unknown parameters: often it is simply $\mathbf{I}\sigma^2$, or sometimes the covariance matrix of a simple auto-regressive residual model, yielding a banded $\boldsymbol{\Lambda}_\theta^{-1}$ and hence efficient computation. Notice that the model states that $\mathbf{y}$ is a linear combination of normal random variables, implying that it has a multivariate normal distribution: $\mathbf{y} \sim N(\mathbf{X}\boldsymbol{\beta}, \mathbf{Z}\boldsymbol{\psi}_\theta\mathbf{Z}^{\mathsf{T}} + \boldsymbol{\Lambda}_\theta)$.

As a simple example of this general formulation, recall the rails example from section 2.1.3. The model for the jth response on the ith rail is

$$y_{ij} = \alpha + b_i + \epsilon_{ij}, \quad b_i \sim N(0, \sigma_b^2), \quad \epsilon_{ij} \sim N(0, \sigma^2), \tag{2.11}$$

with all the b_i and ϵ_{ij} mutually independent. There were 6 rails with 3 measurements on each. In the general linear mixed model form the model is therefore

$$\begin{bmatrix} y_{11} \\ y_{12} \\ y_{13} \\ y_{21} \\ y_{22} \\ y_{23} \\ y_{31} \\ y_{32} \\ y_{33} \\ y_{41} \\ y_{42} \\ y_{43} \\ y_{51} \\ y_{52} \\ y_{53} \\ y_{61} \\ y_{62} \\ y_{63} \end{bmatrix} = \begin{bmatrix} 1 \\ 1 \\ 1 \\ 1 \\ 1 \\ 1 \\ 1 \\ 1 \\ 1 \\ 1 \\ 1 \\ 1 \\ 1 \\ 1 \\ 1 \\ 1 \\ 1 \\ 1 \end{bmatrix} \begin{bmatrix} \alpha \end{bmatrix} + \begin{bmatrix} 1 & 0 & 0 & 0 & 0 & 0 \\ 1 & 0 & 0 & 0 & 0 & 0 \\ 1 & 0 & 0 & 0 & 0 & 0 \\ 0 & 1 & 0 & 0 & 0 & 0 \\ 0 & 1 & 0 & 0 & 0 & 0 \\ 0 & 1 & 0 & 0 & 0 & 0 \\ 0 & 0 & 1 & 0 & 0 & 0 \\ 0 & 0 & 1 & 0 & 0 & 0 \\ 0 & 0 & 1 & 0 & 0 & 0 \\ 0 & 0 & 0 & 1 & 0 & 0 \\ 0 & 0 & 0 & 1 & 0 & 0 \\ 0 & 0 & 0 & 1 & 0 & 0 \\ 0 & 0 & 0 & 0 & 1 & 0 \\ 0 & 0 & 0 & 0 & 1 & 0 \\ 0 & 0 & 0 & 0 & 1 & 0 \\ 0 & 0 & 0 & 0 & 0 & 1 \\ 0 & 0 & 0 & 0 & 0 & 1 \\ 0 & 0 & 0 & 0 & 0 & 1 \end{bmatrix} \begin{bmatrix} b_1 \\ b_2 \\ b_3 \\ b_4 \\ b_5 \\ b_6 \end{bmatrix} + \begin{bmatrix} \epsilon_{11} \\ \epsilon_{12} \\ \epsilon_{13} \\ \epsilon_{21} \\ \epsilon_{22} \\ \epsilon_{23} \\ \epsilon_{31} \\ \epsilon_{32} \\ \epsilon_{33} \\ \epsilon_{41} \\ \epsilon_{42} \\ \epsilon_{43} \\ \epsilon_{51} \\ \epsilon_{52} \\ \epsilon_{53} \\ \epsilon_{61} \\ \epsilon_{62} \\ \epsilon_{63} \end{bmatrix}$$

where $\mathbf{b} \sim N(\mathbf{0}, \mathbf{I}_6\sigma_b^2)$ and $\boldsymbol{\epsilon} \sim N(\mathbf{0}, \mathbf{I}_{18}\sigma^2)$. In this case the parameter vector, $\boldsymbol{\theta}$, contains σ_b^2 and σ^2.

2.4 Maximum likelihood estimation for the linear mixed model

The likelihood for the parameters $\boldsymbol{\beta}$ and $\boldsymbol{\theta}$ of the linear mixed model (2.10) could in principle be based on the p.d.f. implied by the fact that $\mathbf{y} \sim N(\mathbf{X}\boldsymbol{\beta}, \mathbf{Z}\boldsymbol{\psi}_\theta\mathbf{Z}^\mathsf{T} + \boldsymbol{\Lambda}_\theta)$, but this involves the inverse of the $n \times n$ matrix $\mathbf{Z}\boldsymbol{\psi}_\theta\mathbf{Z}^\mathsf{T} + \boldsymbol{\Lambda}_\theta$ which is computationally unattractive. A more convenient expression results by obtaining the marginal distribution, $f(\mathbf{y}|\boldsymbol{\beta})$ (which is the likelihood for $\boldsymbol{\beta}$), by integrating out $\mathbf{b}$ from the joint density of $\mathbf{y}$ and $\mathbf{b}$, $f(\mathbf{y}, \mathbf{b}|\boldsymbol{\beta})$.

From standard probability theory, $f(\mathbf{y}, \mathbf{b}|\boldsymbol{\beta}) = f(\mathbf{y}|\mathbf{b}, \boldsymbol{\beta})f(\mathbf{b})$, and from (2.10),

$$f(\mathbf{y}|\mathbf{b}, \boldsymbol{\beta}) = (2\pi)^{-n/2}|\boldsymbol{\Lambda}_\theta|^{-1/2}\exp\{-\|\mathbf{y} - \mathbf{X}\boldsymbol{\beta} - \mathbf{Z}\mathbf{b}\|^2_{\Lambda_\theta^{-1}}/2\},$$

where $\|\mathbf{x}\|^2_{\Lambda^{-1}} = \mathbf{x}^\mathsf{T}\boldsymbol{\Lambda}^{-1}\mathbf{x}$, while, if p is the dimension of $\mathbf{b}$,

$$f(\mathbf{b}) = (2\pi)^{-p/2}|\boldsymbol{\psi}_\theta|^{-1/2}\exp\{-\mathbf{b}^\mathsf{T}\boldsymbol{\psi}_\theta^{-1}\mathbf{b}/2\}.$$

Now let $\hat{\mathbf{b}}$ be the maximizer of $\log f(\mathbf{y}, \mathbf{b}|\boldsymbol{\beta})$ (and hence $f(\mathbf{y}, \mathbf{b}|\boldsymbol{\beta})$) for a given $\boldsymbol{\beta}$, and consider evaluating the likelihood $f(\mathbf{y}|\boldsymbol{\beta})$ by integration:

$$\begin{aligned} f(\mathbf{y}|\boldsymbol{\beta}) &= \int f(\mathbf{y}, \mathbf{b}|\boldsymbol{\beta})d\mathbf{b} = \int \exp\{\log f(\mathbf{y}, \mathbf{b}|\boldsymbol{\beta})\}d\mathbf{b} \\ &= \int \exp\left\{\log f(\mathbf{y}, \hat{\mathbf{b}}|\boldsymbol{\beta}) + \frac{1}{2}(\mathbf{b} - \hat{\mathbf{b}})^\mathsf{T}\frac{\partial^2 \log f(\mathbf{y}, \mathbf{b}|\boldsymbol{\beta})}{\partial\mathbf{b}\partial\mathbf{b}^\mathsf{T}}(\mathbf{b} - \hat{\mathbf{b}})\right\} d\mathbf{b} \end{aligned}$$

where the second line is obtained by Taylor expansion about $\hat{\mathbf{b}}$, and there is no remainder term because the higher order derivatives of $\log f(\mathbf{y}, \mathbf{b}|\boldsymbol{\beta})$ w.r.t. $\mathbf{b}$ are identically zero (since $f(\mathbf{b})$ and $f(\mathbf{y}|\mathbf{b}, \boldsymbol{\beta})$ are Gaussian). Hence

$$f(\mathbf{y}|\boldsymbol{\beta}) = f(\mathbf{y}, \hat{\mathbf{b}}|\boldsymbol{\beta}) \int \exp\{-(\mathbf{b} - \hat{\mathbf{b}})^{\mathsf{T}}(\mathbf{Z}^{\mathsf{T}}\boldsymbol{\Lambda}_\theta^{-1}\mathbf{Z} + \boldsymbol{\psi}_\theta^{-1})(\mathbf{b} - \hat{\mathbf{b}})/2\} d\mathbf{b}. \quad (2.12)$$

Like any p.d.f. a multivariate normal p.d.f. must integrate to 1, i.e.,

$$\int \frac{|\boldsymbol{\Sigma}^{-1}|^{1/2}}{(2\pi)^{p/2}} \exp\left\{-\frac{1}{2}(\mathbf{b} - \hat{\mathbf{b}})^{\mathsf{T}}\boldsymbol{\Sigma}^{-1}(\mathbf{b} - \hat{\mathbf{b}})\right\} d\mathbf{b} = 1$$
$$\Rightarrow \int \exp\left\{-\frac{1}{2}(\mathbf{b} - \hat{\mathbf{b}})^{\mathsf{T}}\boldsymbol{\Sigma}^{-1}(\mathbf{b} - \hat{\mathbf{b}})\right\} d\mathbf{b} = \frac{(2\pi)^{p/2}}{|\boldsymbol{\Sigma}^{-1}|^{1/2}}.$$

Applying this result to the integral in (2.12) we obtain the likelihood for $\boldsymbol{\beta}$ and $\boldsymbol{\theta}$,

$$f(\mathbf{y}|\boldsymbol{\beta}) = f(\mathbf{y}, \hat{\mathbf{b}}|\boldsymbol{\beta}) \frac{(2\pi)^{p/2}}{|\mathbf{Z}^{\mathsf{T}}\boldsymbol{\Lambda}_\theta^{-1}\mathbf{Z} + \boldsymbol{\psi}_\theta^{-1}|^{1/2}}.$$

Explicitly, twice the log likelihood ($l = \log f(\mathbf{y}|\boldsymbol{\beta})$) is therefore

$$2l(\boldsymbol{\beta}, \boldsymbol{\theta}) = -\|\mathbf{y} - \mathbf{X}\boldsymbol{\beta} - \mathbf{Z}\hat{\mathbf{b}}\|^2_{\Lambda_\theta^{-1}} - \hat{\mathbf{b}}^{\mathsf{T}}\boldsymbol{\psi}_\theta^{-1}\hat{\mathbf{b}}$$
$$- \log|\boldsymbol{\Lambda}_\theta| - \log|\boldsymbol{\psi}_\theta| - \log|\mathbf{Z}^{\mathsf{T}}\boldsymbol{\Lambda}_\theta^{-1}\mathbf{Z} + \boldsymbol{\psi}_\theta^{-1}| - n\log(2\pi), \quad (2.13)$$

where $\hat{\mathbf{b}}$ is dependent on $\boldsymbol{\beta}$ and $\boldsymbol{\theta}$.

Since only the first two terms on the r.h.s. of (2.13) depend on $\boldsymbol{\beta}$ (don't forget that $\hat{\mathbf{b}}$ depends on $\boldsymbol{\beta}$) it follows that for any $\boldsymbol{\theta}$, the maximum likelihood estimator $\hat{\boldsymbol{\beta}}$ can be found by simply minimizing

$$\|\mathbf{y} - \mathbf{X}\boldsymbol{\beta} - \mathbf{Z}\mathbf{b}\|^2_{\Lambda_\theta^{-1}} + \mathbf{b}^{\mathsf{T}}\boldsymbol{\psi}_\theta^{-1}\mathbf{b}, \quad (2.14)$$

jointly w.r.t. $\boldsymbol{\beta}$ and $\mathbf{b}$ (also yielding $\hat{\mathbf{b}}$). So, given $\boldsymbol{\theta}$ values, $\hat{\boldsymbol{\beta}}$ can be found by a quadratic optimization which has an explicit solution. This fact can be exploited by basing inference about $\boldsymbol{\theta}$ on the *profile likelihood* $l_p(\boldsymbol{\theta}) = l(\boldsymbol{\theta}, \hat{\boldsymbol{\beta}}_\theta)$, where $\hat{\boldsymbol{\beta}}_\theta$ is the minimizer of (2.14) given $\boldsymbol{\theta}$. Exploiting the direct computation of $\hat{\boldsymbol{\beta}}$ in this way reduces the computational burden of iteratively seeking the MLE of $\boldsymbol{\theta}$ by numerical methods.

2.4.1 *The distribution of* $\mathbf{b}|\mathbf{y}, \hat{\boldsymbol{\beta}}$ *given* $\boldsymbol{\theta}$

Now consider the distribution of $\mathbf{b}|\mathbf{y}, \hat{\boldsymbol{\beta}}$. We know that $f(\mathbf{b}|\mathbf{y}, \hat{\boldsymbol{\beta}}) \propto f(\mathbf{y}, \mathbf{b}|\hat{\boldsymbol{\beta}})$, so, defining $\tilde{\mathbf{y}} = \mathbf{y} - \mathbf{X}\hat{\boldsymbol{\beta}}$,

$$\begin{aligned}\log f(\mathbf{b}|\mathbf{y}, \hat{\boldsymbol{\beta}}) &= -(\tilde{\mathbf{y}} - \mathbf{Z}\mathbf{b})^{\mathsf{T}}\boldsymbol{\Lambda}_\theta^{-1}(\tilde{\mathbf{y}} - \mathbf{Z}\mathbf{b})/2 - \mathbf{b}^{\mathsf{T}}\boldsymbol{\psi}_\theta^{-1}\mathbf{b}/2 + k_1 \\
&= -(\tilde{\mathbf{y}}^{\mathsf{T}}\boldsymbol{\Lambda}_\theta^{-1}\tilde{\mathbf{y}} - 2\mathbf{b}^{\mathsf{T}}\mathbf{Z}^{\mathsf{T}}\boldsymbol{\Lambda}_\theta^{-1}\tilde{\mathbf{y}} + \mathbf{b}^{\mathsf{T}}\mathbf{Z}^{\mathsf{T}}\boldsymbol{\Lambda}_\theta^{-1}\mathbf{Z}\mathbf{b} + \mathbf{b}^{\mathsf{T}}\boldsymbol{\psi}_\theta^{-1}\mathbf{b})/2 + k_1 \\
&= -\{\mathbf{b} - (\mathbf{Z}^{\mathsf{T}}\boldsymbol{\Lambda}_\theta^{-1}\mathbf{Z} + \boldsymbol{\psi}_\theta^{-1})^{-1}\mathbf{Z}^{\mathsf{T}}\boldsymbol{\Lambda}_\theta^{-1}\tilde{\mathbf{y}}\}^{\mathsf{T}}(\mathbf{Z}^{\mathsf{T}}\boldsymbol{\Lambda}_\theta^{-1}\mathbf{Z} + \boldsymbol{\psi}_\theta^{-1}) \\
&\qquad \{\mathbf{b} - (\mathbf{Z}^{\mathsf{T}}\boldsymbol{\Lambda}_\theta^{-1}\mathbf{Z} + \boldsymbol{\psi}_\theta^{-1})^{-1}\mathbf{Z}^{\mathsf{T}}\boldsymbol{\Lambda}_\theta^{-1}\tilde{\mathbf{y}}\}/2 + k_2,\end{aligned}$$

where k_1 and k_2 are constants not involving $\mathbf{b}$. The final expression is recognisable as the kernel of a multivariate normal p.d.f. so

$$\mathbf{b}|\mathbf{y}, \hat{\boldsymbol{\beta}} \sim N(\hat{\mathbf{b}}, (\mathbf{Z}^\mathsf{T}\boldsymbol{\Lambda}_\theta^{-1}\mathbf{Z} + \boldsymbol{\psi}_\theta^{-1})^{-1}), \tag{2.15}$$

where $\hat{\mathbf{b}} = (\mathbf{Z}^\mathsf{T}\boldsymbol{\Lambda}_\theta^{-1}\mathbf{Z} + \boldsymbol{\psi}_\theta^{-1})^{-1}\mathbf{Z}^\mathsf{T}\boldsymbol{\Lambda}_\theta^{-1}\tilde{\mathbf{y}}$, which is readily seen to be the minimiser of (2.14). $\hat{\mathbf{b}}$ is sometimes referred to as the *maximum a posteriori*, or MAP, estimate of the random effects or alternatively as the *predicted* random effects vector.

2.4.2 *The distribution of $\hat{\boldsymbol{\beta}}$ given $\boldsymbol{\theta}$*

Finally, consider the distribution of $\hat{\boldsymbol{\beta}}$ for a given $\boldsymbol{\theta}$. An obvious way to approach this is to use the log likelihood based on $\mathbf{y} \sim N(\mathbf{X}\boldsymbol{\beta}, \mathbf{Z}\boldsymbol{\psi}\mathbf{Z}^\mathsf{T} + \boldsymbol{\Lambda})$ (where $\boldsymbol{\theta}$ subscripts have been dropped to avoid clutter). Then for a given $\boldsymbol{\theta}$ the MLE is the weighted least squares estimate $\hat{\boldsymbol{\beta}} = \{\mathbf{X}^\mathsf{T}(\mathbf{Z}\boldsymbol{\psi}\mathbf{Z}^\mathsf{T} + \boldsymbol{\Lambda})^{-1}\mathbf{X}\}^{-1}\mathbf{X}^\mathsf{T}(\mathbf{Z}\boldsymbol{\psi}\mathbf{Z}^\mathsf{T} + \boldsymbol{\Lambda})^{-1}\mathbf{y}$. Since $\mathbb{E}(\mathbf{y}) = \mathbf{X}\boldsymbol{\beta}$, $\mathbb{E}(\hat{\boldsymbol{\beta}}) = \boldsymbol{\beta}$, and because $\hat{\boldsymbol{\beta}}$ is a linear transformation of $\mathbf{y}$, which has covariance matrix $\mathbf{Z}\boldsymbol{\psi}\mathbf{Z}^\mathsf{T} + \boldsymbol{\Lambda}$, the covariance matrix of $\hat{\boldsymbol{\beta}}$ is

$$\{\mathbf{X}^\mathsf{T}(\mathbf{Z}\boldsymbol{\psi}\mathbf{Z}^\mathsf{T} + \boldsymbol{\Lambda})^{-1}\mathbf{X}\}^{-1}\mathbf{X}^\mathsf{T}(\mathbf{Z}\boldsymbol{\psi}\mathbf{Z}^\mathsf{T} + \boldsymbol{\Lambda})^{-1}(\mathbf{Z}\boldsymbol{\psi}\mathbf{Z}^\mathsf{T} + \boldsymbol{\Lambda})(\mathbf{Z}\boldsymbol{\psi}\mathbf{Z}^\mathsf{T} + \boldsymbol{\Lambda})^{-1}\mathbf{X} \\ \{\mathbf{X}^\mathsf{T}(\mathbf{Z}\boldsymbol{\psi}\mathbf{Z}^\mathsf{T} + \boldsymbol{\Lambda})^{-1}\mathbf{X}\}^{-1} = \{\mathbf{X}^\mathsf{T}(\mathbf{Z}\boldsymbol{\psi}\mathbf{Z}^\mathsf{T} + \boldsymbol{\Lambda})^{-1}\mathbf{X}\}^{-1}.$$

So

$$\hat{\boldsymbol{\beta}} \sim N(\boldsymbol{\beta}, \{\mathbf{X}^\mathsf{T}(\mathbf{Z}\boldsymbol{\psi}\mathbf{Z}^\mathsf{T} + \boldsymbol{\Lambda})^{-1}\mathbf{X}\}^{-1}). \tag{2.16}$$

Again, the inverse of $\mathbf{Z}\boldsymbol{\psi}\mathbf{Z}^\mathsf{T} + \boldsymbol{\Lambda}$ is potentially computationally costly, with $O(n^3)$ floating point computational cost for n data. However, it turns out that an identical covariance matrix can be obtained by treating $\boldsymbol{\beta}$ as a vector of random effects with improper uniform (prior) distributions, reducing the cost to $O(np^2)$, where p is the number of model coefficients. In this case $\boldsymbol{\beta}$ is effectively part of $\mathbf{b}$ so that (2.15) applies and can be re-written as

$$\begin{bmatrix} \mathbf{b}|\mathbf{y} \\ \boldsymbol{\beta}|\mathbf{y} \end{bmatrix} \sim N\left(\begin{bmatrix} \hat{\mathbf{b}} \\ \hat{\boldsymbol{\beta}} \end{bmatrix}, \begin{bmatrix} \mathbf{Z}^\mathsf{T}\boldsymbol{\Lambda}^{-1}\mathbf{Z} + \boldsymbol{\psi}^{-1} & \mathbf{Z}^\mathsf{T}\boldsymbol{\Lambda}^{-1}\mathbf{X} \\ \mathbf{X}^\mathsf{T}\boldsymbol{\Lambda}^{-1}\mathbf{Z} & \mathbf{X}^\mathsf{T}\boldsymbol{\Lambda}^{-1}\mathbf{X} \end{bmatrix}^{-1} \right). \tag{2.17}$$

It turns out that the block of the above covariance matrix relating to $\boldsymbol{\beta}$ is identical to the frequentist covariance matrix for $\hat{\boldsymbol{\beta}}$ in (2.16). Using a standard result on the inverse of a symmetric partitioned matrix[‡] the covariance matrix block of (2.17) corresponding to $\boldsymbol{\beta}$ is $[\mathbf{X}^\mathsf{T}\{\boldsymbol{\Lambda}^{-1} - \boldsymbol{\Lambda}^{-1}\mathbf{Z}(\mathbf{Z}^\mathsf{T}\boldsymbol{\Lambda}^{-1}\mathbf{Z} + \boldsymbol{\psi}^{-1})^{-1}\mathbf{Z}^\mathsf{T}\boldsymbol{\Lambda}^{-1}\}\mathbf{X}]^{-1}$. This turns out to be identical to $\{\mathbf{X}^\mathsf{T}(\mathbf{Z}\boldsymbol{\psi}\mathbf{Z}^\mathsf{T} + \boldsymbol{\Lambda})^{-1}\mathbf{X}\}^{-1}$, because $\boldsymbol{\Lambda}^{-1} - \boldsymbol{\Lambda}^{-1}\mathbf{Z}(\mathbf{Z}^\mathsf{T}\boldsymbol{\Lambda}^{-1}\mathbf{Z} + \boldsymbol{\psi}^{-1})^{-1}\mathbf{Z}^\mathsf{T}\boldsymbol{\Lambda}^{-1}$ is equal to the inverse of $\mathbf{Z}\boldsymbol{\psi}\mathbf{Z}^\mathsf{T} + \boldsymbol{\Lambda}$, as the following shows,

[‡]Defining $\mathbf{D} = \mathbf{B} - \mathbf{C}^\mathsf{T}\mathbf{A}^{-1}\mathbf{C}$, it is easy, but tedious, to check that

$$\begin{bmatrix} \mathbf{A} & \mathbf{C} \\ \mathbf{C}^\mathsf{T} & \mathbf{B} \end{bmatrix}^{-1} = \begin{bmatrix} \mathbf{A}^{-1} + \mathbf{A}^{-1}\mathbf{C}\mathbf{D}^{-1}\mathbf{C}^\mathsf{T}\mathbf{A}^{-1} & -\mathbf{A}^{-1}\mathbf{C}\mathbf{D}^{-1} \\ -\mathbf{D}^{-1}\mathbf{C}^\mathsf{T}\mathbf{A}^{-1} & \mathbf{D}^{-1} \end{bmatrix}.$$

$$\begin{aligned}
\{\mathbf{\Lambda}^{-1} - \mathbf{\Lambda}^{-1}\mathbf{Z}(\mathbf{Z}^{\mathsf{T}}\mathbf{\Lambda}^{-1}\mathbf{Z} + \boldsymbol{\psi}^{-1})^{-1}\mathbf{Z}^{\mathsf{T}}\mathbf{\Lambda}^{-1}\}(\mathbf{Z}\boldsymbol{\psi}\mathbf{Z}^{\mathsf{T}} + \mathbf{\Lambda}) &= \\
\mathbf{\Lambda}^{-1}\mathbf{Z}\boldsymbol{\psi}\mathbf{Z}^{\mathsf{T}} + \mathbf{I} - \mathbf{\Lambda}^{-1}\mathbf{Z}(\mathbf{Z}^{\mathsf{T}}\mathbf{\Lambda}^{-1}\mathbf{Z} + \boldsymbol{\psi}^{-1})^{-1}\mathbf{Z}^{\mathsf{T}}\mathbf{\Lambda}^{-1}\mathbf{Z}\boldsymbol{\psi}\mathbf{Z}^{\mathsf{T}} & \\
-\mathbf{\Lambda}^{-1}\mathbf{Z}(\mathbf{Z}^{\mathsf{T}}\mathbf{\Lambda}^{-1}\mathbf{Z} + \boldsymbol{\psi}^{-1})^{-1}\mathbf{Z}^{\mathsf{T}} &= \\
\mathbf{I} + \mathbf{\Lambda}^{-1}\mathbf{Z}\{\boldsymbol{\psi} - (\mathbf{Z}^{\mathsf{T}}\mathbf{\Lambda}^{-1}\mathbf{Z} + \boldsymbol{\psi}^{-1})^{-1}(\mathbf{Z}^{\mathsf{T}}\mathbf{\Lambda}^{-1}\mathbf{Z}\boldsymbol{\psi} + \mathbf{I})\}\mathbf{Z}^{\mathsf{T}} &= \\
\mathbf{I} + \mathbf{\Lambda}^{-1}\mathbf{Z}\{\boldsymbol{\psi} - \boldsymbol{\psi}(\mathbf{Z}^{\mathsf{T}}\mathbf{\Lambda}^{-1}\mathbf{Z}\boldsymbol{\psi} + \mathbf{I})^{-1}(\mathbf{Z}^{\mathsf{T}}\mathbf{\Lambda}^{-1}\mathbf{Z}\boldsymbol{\psi} + \mathbf{I})\}\mathbf{Z}^{\mathsf{T}} &= \mathbf{I}.
\end{aligned}$$

2.4.3 *The distribution of* $\hat{\boldsymbol{\theta}}$

Inference about $\boldsymbol{\theta}$ is reliant on the large sample result that

$$\hat{\boldsymbol{\theta}} \sim N(\boldsymbol{\theta}, \hat{\boldsymbol{\mathcal{I}}}_p^{-1}) \tag{2.18}$$

where $\hat{\boldsymbol{\mathcal{I}}}_p$ is the negative Hessian of the log profile likelihood, with i, j^{th} element $-\partial^2 l_p/\partial\theta_i\partial\theta_j|_{\hat{\theta}}$. l_r from section (2.4.5) can also be used in place of l_p.

2.4.4 *Maximizing the profile likelihood*

In practice $l_p(\boldsymbol{\theta})$ is maximized numerically, with each trial value for $\boldsymbol{\theta}$ requiring (2.14) to be minimized for $\hat{\mathbf{b}}, \hat{\boldsymbol{\beta}}$. The $\hat{\boldsymbol{\beta}}$ computed at convergence is the MLE, of course, while the $\hat{\mathbf{b}}$ are the predicted random effects. To make this concrete, let us repeat the simple rail example from section 2.1.3 and 2.3 using a maximum likelihood approach. Here is a function to evaluate the negative log profile likelihood, l_r. In this case $\mathbf{\Lambda}_\theta = \mathbf{I}\sigma^2$ and $\boldsymbol{\psi}_\theta = \mathbf{I}\sigma_b^2$, so $\boldsymbol{\theta} = (\log\sigma_b, \log\sigma)^{\mathsf{T}}$. The log parameterization ensures that the variance components remain positive.

```
llm <- function(theta,X,Z,y) {
  ## untransform parameters...
  sigma.b <- exp(theta[1])
  sigma <- exp(theta[2])
  ## extract dimensions...
  n <- length(y); pr <- ncol(Z); pf <- ncol(X)
  ## obtain \hat \beta, \hat b...
  X1 <- cbind(X,Z)
  ipsi <- c(rep(0,pf),rep(1/sigma.b^2,pr))
  b1 <- solve(crossprod(X1)/sigma^2+diag(ipsi),
              t(X1)%*%y/sigma^2)
  ## compute log|Z'Z/sigma^2 + I/sigma.b^2|...
  ldet <- sum(log(diag(chol(crossprod(Z)/sigma^2 +
              diag(ipsi[-(1:pf)])))))
  ## compute log profile likelihood...
  l <- (-sum((y-X1%*%b1)^2)/sigma^2 - sum(b1^2*ipsi) -
  n*log(sigma^2) - pr*log(sigma.b^2) - 2*ldet - n*log(2*pi))/2
  attr(l,"b") <- as.numeric(b1) ## return \hat beta and \hat b
  -l
}
```

Notice how (2.14) is minimized to find `b1` which contains $\hat{\boldsymbol{\beta}}$ followed by $\hat{\mathbf{b}}$. The determinant of a positive definite matrix is the square of the product of the terms on the leading diagonal of its Cholesky factor (see appendix B.7), and this is used to compute $\log|\mathbf{Z}^\mathsf{T}\boldsymbol{\Lambda}_\theta^{-1}\mathbf{Z} + \boldsymbol{\psi}_\theta^{-1}|$. We then have all the ingredients to evaluate $l(\boldsymbol{\theta}, \hat{\boldsymbol{\beta}})$ using (2.13). Before returning, the estimates/predictions $\hat{\boldsymbol{\beta}}, \hat{\mathbf{b}}$ are attached to the log profile likelihood value as an attribute. Finally the negative of the log likelihood is returned, since maximization of l_p is equivalent to minimizing $-l_p$ and most optimization routines minimize by default.

The following fits the rail model by maximizing l_p using R function `optim`.[§]

```
> library(nlme) ## for Rail data
> options(contrasts=c("contr.treatment","contr.treatment"))
> Z <- model.matrix(~ Rail$Rail - 1) ## r.e. model matrix
> X <- matrix(1,18,1)              ## fixed model matrix
> ## fit the model...
> rail.mod <- optim(c(0,0),llm,hessian=TRUE,
                          X=X,Z=Z,y=Rail$travel)
> exp(rail.mod$par) ## variance components
[1] 22.629166  4.024072
> solve(rail.mod$hessian) ## approx cov matrix for theta
             [,1]          [,2]
[1,]  0.0851408546 -0.0004397245
[2,] -0.0004397245  0.0417347933
> attr(llm(rail.mod$par,X,Z,Rail$travel),"b")
[1]  66.50000 -34.46999 -16.32789 -12.36961  15.99803
[7]  17.97717  29.19229
```

The estimated variance components and intercept term should be compared to those obtained in section 2.1.3. `optim` can return the Hessian of its objective function, which is $\hat{\mathcal{I}}_p$, and can therefore be inverted to estimate the covariance matrix of $\hat{\boldsymbol{\theta}}$. The final line of output is $\hat{\beta}$ followed by $\hat{\mathbf{b}}$.

Of course, for practical analysis we would usually use specialist software to estimate a linear mixed model. A major reason for this is computational efficiency. In many applications $\mathbf{Z}$ has a very large number of columns, but also has a sparse structure with many zero elements. Similarly, $\boldsymbol{\psi}$ or its inverse are often sparse (block diagonal, for example). Exploiting this sparse structure (i.e., avoiding multiplications and additions between number pairs containing a zero) is essential for efficient computation. The major packages for linear mixed modelling in R are `nlme` and `lme4`. The former is designed to exploit the sparsity that arises when models have a nested structure, while the latter uses sparse direct matrix methods (e.g., Davis, 2006) to exploit any sparsity pattern. Before looking at these packages there are some more theoretical issues to deal with.

[§]The setting of contrast options ensures that we get the desired form for $\mathbf{Z}$ since `Rail$Rail` is originally declared as an ordered factor.

2.4.5 *REML*

A problem with maximum likelihood estimation of variance components is that it tends to underestimate them. The most obvious example of this is the MLE of σ^2 for the linear model, which is $\hat{\sigma}^2 = \|\mathbf{y} - \mathbf{X}\boldsymbol{\beta}\|^2/n$. This is clearly biased as shown by comparison with the unbiased estimator (1.8) derived in section 1.3.3 (p. 13). This tendency is not limited to the residual variance and gets worse as the number of fixed effects increase. Patterson and Thompson (1971) proposed REML (restricted maximum likelihood) as a bias reducing alternative to maximum likelihood. The original approach is motivated by considering the estimation of particular contrasts, but here it is more helpful to follow Laird and Ware (1982). They observed that the restricted likelihood can be obtained by integrating $\mathbf{b}$ *and* $\boldsymbol{\beta}$ out of $f(\mathbf{y}, \mathbf{b}|\boldsymbol{\beta})$.

The integral can be performed using exactly the same approach taken to arrive at (2.13) in section 2.4, resulting in

$$2l_r(\boldsymbol{\theta}) = -\|\mathbf{y} - \mathbf{X}\hat{\boldsymbol{\beta}} - \mathbf{Z}\hat{\mathbf{b}}\|^2_{\boldsymbol{\Lambda}_\theta^{-1}} - \hat{\mathbf{b}}^{\mathsf{T}}\boldsymbol{\psi}_\theta^{-1}\hat{\mathbf{b}} - \log|\boldsymbol{\Lambda}_\theta| - \log|\boldsymbol{\psi}_\theta|$$
$$- \log\begin{vmatrix} \mathbf{Z}^{\mathsf{T}}\boldsymbol{\Lambda}_\theta^{-1}\mathbf{Z} + \boldsymbol{\psi}_\theta^{-1} & \mathbf{Z}^{\mathsf{T}}\boldsymbol{\Lambda}_\theta^{-1}\mathbf{X} \\ \mathbf{X}^{\mathsf{T}}\boldsymbol{\Lambda}_\theta^{-1}\mathbf{Z} & \mathbf{X}^{\mathsf{T}}\boldsymbol{\Lambda}_\theta^{-1}\mathbf{X} \end{vmatrix} - (n - M)\log(2\pi), \quad (2.19)$$

where M is the dimension of $\boldsymbol{\beta}$ and, as before, $\hat{\boldsymbol{\beta}}$ and $\hat{\mathbf{b}}$ are dependent on $\boldsymbol{\theta}$ and must be recomputed afresh for each value of $\boldsymbol{\theta}$ for which l_r is evaluated. $l_r(\boldsymbol{\theta})$ can be used exactly as $l_p(\boldsymbol{\theta})$ is used with one exception. l_r can not be used to compare models with differing fixed effect structures (e.g., in a generalized likelihood ratio test statistic or AIC comparison): l_r is simply not comparable between models with different fixed effect structures. $\hat{\boldsymbol{\beta}}$ and $\hat{\mathbf{b}}$ are also used exactly as before (although some people object to referring to $\hat{\boldsymbol{\beta}}$ as 'REML estimates' of $\boldsymbol{\beta}$, since l_r is not a function of $\boldsymbol{\beta}$).

2.4.6 *Effective degrees of freedom*

A notion that will prove helpful in the context of smoothing is the *effective degrees of freedom* of a model. To see why the idea might be useful, consider the example of a p dimensional random effect $\mathbf{b} \sim (\mathbf{0}, \mathbf{I}\sigma_b^2)$. How many degrees of freedom are associated with $\mathbf{b}$? Clearly if $\sigma_b = 0$ then $\mathbf{b}$ makes no contribution to the model and the answer must be 0. On the other hand if $\sigma_b \to \infty$, then $\mathbf{b}$ will behave like a fixed effect parameter, and the answer is presumably p. This suggests that the effective degrees of freedom for $\mathbf{b}$ should increase with σ_b, from 0 up to p.

One approach to arriving at a quantitative definition of the effective degrees of freedom is to consider REML estimation of σ^2 when $\boldsymbol{\Lambda} = \mathbf{I}\sigma^2$. To save ink write $\mathbf{b}$ and $\boldsymbol{\beta}$ in a single vector $\mathcal{B}^{\mathsf{T}} = (\mathbf{b}^{\mathsf{T}}, \boldsymbol{\beta}^{\mathsf{T}})$, and define corresponding model matrix and precision matrix

$$\mathcal{X} = (\mathbf{Z}, \mathbf{X}) \text{ and } \mathbf{S} = \begin{bmatrix} \boldsymbol{\psi}_\theta^{-1} & \mathbf{0} \\ \mathbf{0} & \mathbf{0} \end{bmatrix}.$$

Then the parts of (2.19) dependent on σ^2 can be written

$$-\|\mathbf{y} - \mathbf{X}\hat{\boldsymbol{\beta}} - \mathbf{Z}\hat{\mathbf{b}}\|^2/\sigma^2 - \hat{\mathbf{b}}^{\mathsf{T}}\boldsymbol{\psi}_\theta^{-1}\hat{\mathbf{b}} - n\log\sigma^2 - \log|\mathcal{X}^{\mathsf{T}}\mathcal{X}/\sigma^2 + \mathbf{S}|.$$

Differentiating with respect to σ^2 and equating to zero yields¶

$$\|\mathbf{y} - \mathbf{X}\hat{\beta} - \mathbf{Z}\hat{\mathbf{b}}\|^2/\sigma^4 - n/\sigma^2 + \text{tr}\{(\mathcal{X}^\mathsf{T}\mathcal{X}/\sigma^2 + \mathbf{S})^{-1}\mathcal{X}^\mathsf{T}\mathcal{X}/\sigma^4\} = 0,$$

which implies that

$$\hat{\sigma}^2 = \frac{\|\mathbf{y} - \mathbf{X}\hat{\beta} - \mathbf{Z}\hat{\mathbf{b}}\|^2}{n - \tau}$$

where $\tau = \text{tr}\{(\mathcal{X}^\mathsf{T}\mathcal{X}/\hat{\sigma}^2 + \mathbf{S})^{-1}\mathcal{X}^\mathsf{T}\mathcal{X}/\hat{\sigma}^2\}$. By comparison with the usual unbiased estimator of σ^2, (1.8) p. 13, this suggests considering τ as the *effective* degrees of freedom of the mixed model. It is relatively straightforward to show that $M \le \tau \le M + p$ where M and p are the number of fixed and random effects, respectively.

2.4.7 The EM algorithm

The profile log likelihood, l_p, or restricted log likelihood, l_r, can be numerically optimized using Newton's method. However Newton's method can sometimes be slow if the starting values for the parameters are poor, and it can therefore be useful to start fitting using another approach that converges more rapidly when far from the optimum values. The *EM algorithm* (Dempster et al., 1977) is such a method (see Davison, 2003; Wood, 2015, for introductions).

Starting from parameter guesses, $\hat{\boldsymbol{\theta}}, \hat{\boldsymbol{\beta}}$, the following steps are iterated:

1. Find the distribution of $\mathbf{b}|\mathbf{y}$ according to the current parameter estimates.
2. Treating the distribution from 1 as fixed (rather than depending on $\boldsymbol{\theta}, \boldsymbol{\beta}$), find an expression for $Q(\boldsymbol{\theta}, \boldsymbol{\beta}) = \mathbb{E}_{\mathbf{b}|\mathbf{y}}\{\log f(\mathbf{y}, \mathbf{b}|\boldsymbol{\beta})\}$ as a function of $\boldsymbol{\theta}, \boldsymbol{\beta}$, using the distribution from 1. The $\mathbf{y}$ are treated as fixed, here. (This is the E-step.)
3. Maximize the expression for $Q(\boldsymbol{\theta}, \boldsymbol{\beta})$ w.r.t. the parameters to obtain updated estimates $\hat{\boldsymbol{\theta}}, \hat{\boldsymbol{\beta}}$. (This is the M-step.)

Note that the expectation in step 2 is taken with respect to the *fixed* distribution from step 1, which depends on the current parameter *estimates*. When evaluating $Q(\boldsymbol{\theta}, \boldsymbol{\beta})$, we view $\log f(\mathbf{y}, \mathbf{b}|\boldsymbol{\beta})$ as a function of $\boldsymbol{\theta}, \boldsymbol{\beta}$, but do not treat the distribution of $\mathbf{b}|\mathbf{y}$ as depending on these parameters.

There are two key points about this algorithm.

1. Each step of the algorithm can be shown to increase the log-likelihood $l(\boldsymbol{\theta}, \boldsymbol{\beta})$ (until a turning point of the likelihood is reached, hopefully the MLE).
2. $Q(\boldsymbol{\theta}, \boldsymbol{\beta})$ is often much easier to evaluate and maximize than $l(\boldsymbol{\theta}, \boldsymbol{\beta})$ itself.

To appreciate exactly what the E-step involves, it helps to derive Q. First note that if $k(\boldsymbol{\theta}) = -\log|\boldsymbol{\Lambda}_\theta|/2 - \log|\boldsymbol{\psi}_\theta|/2 - (n-p)\log(2\pi)/2$, then

$$\begin{aligned}\log f(\mathbf{y}, \mathbf{b}|\boldsymbol{\beta}) &= -\|\mathbf{y} - \mathbf{X}\boldsymbol{\beta} - \mathbf{Z}\mathbf{b}\|^2_{\Lambda_\theta^{-1}}/2 - \mathbf{b}^\mathsf{T}\boldsymbol{\psi}_\theta^{-1}\mathbf{b}/2 + k(\boldsymbol{\theta}) \\ &= -\|\mathbf{y} - \mathbf{X}\boldsymbol{\beta}\|^2_{\Lambda_\theta^{-1}}/2 - \mathbf{b}^\mathsf{T}\mathbf{Z}^\mathsf{T}\boldsymbol{\Lambda}_\theta^{-1}\mathbf{Z}\mathbf{b}/2 + (\mathbf{y} - \mathbf{X}\boldsymbol{\beta})^\mathsf{T}\boldsymbol{\Lambda}_\theta^{-1}\mathbf{Z}\mathbf{b} \\ &\quad -\mathbf{b}^\mathsf{T}\boldsymbol{\psi}_\theta^{-1}\mathbf{b}/2 + k(\boldsymbol{\theta}).\end{aligned}$$

¶The derivatives of $-\|\mathbf{y} - \mathbf{X}\boldsymbol{\beta} - \mathbf{Z}\mathbf{b}\|^2/\sigma^2 - \mathbf{b}^\mathsf{T}\boldsymbol{\psi}_\theta^{-1}\mathbf{b}$ w.r.t. $\mathbf{b}$ are all 0 at $\hat{\mathbf{b}}$, which is why $\hat{\mathbf{b}}$'s dependence on σ^2 adds nothing to the derivative.

To find the expectation of $\log f(\mathbf{y}, \mathbf{b}|\boldsymbol{\beta})$ w.r.t. $f(\mathbf{b}|\mathbf{y})$ requires use of the standard results $\mathbb{E}(\mathbf{b}^{\mathsf{T}}\mathbf{A}\mathbf{b}) = \text{tr}\{\mathbf{A}\mathbb{E}(\mathbf{b}\mathbf{b}^{\mathsf{T}})\} = \text{tr}\{\mathbf{A}\mathbf{V}_b\} + \mathbb{E}(\mathbf{b})^{\mathsf{T}}\mathbf{A}\mathbb{E}(\mathbf{b})$, where $\mathbf{V}_b$ is the covariance matrix of $\mathbf{b}$. From (2.15) in section 2.4.1, $\mathbb{E}_{b|y}(\mathbf{b}) = \hat{\mathbf{b}}$ and $\mathbf{V}_b = (\mathbf{Z}^{\mathsf{T}}\boldsymbol{\Lambda}_\theta^{-1}\mathbf{Z} + \boldsymbol{\psi}_\theta^{-1})^{-1}$, so taking the required expectations we have

$$\begin{aligned} Q(\boldsymbol{\theta}, \boldsymbol{\beta}) &= -\|\mathbf{y} - \mathbf{X}\boldsymbol{\beta}\|^2_{\Lambda_\theta^{-1}}/2 - \text{tr}\{\mathbf{Z}^{\mathsf{T}}\boldsymbol{\Lambda}_\theta^{-1}\mathbf{Z}(\mathbf{Z}^{\mathsf{T}}\boldsymbol{\Lambda}_\theta^{-1}\mathbf{Z} + \boldsymbol{\psi}_\theta^{-1})^{-1}\}/2 \\ &\quad -\hat{\mathbf{b}}^{\mathsf{T}}\mathbf{Z}^{\mathsf{T}}\boldsymbol{\Lambda}_\theta^{-1}\mathbf{Z}\hat{\mathbf{b}}/2 + (\mathbf{y} - \mathbf{X}\boldsymbol{\beta})^{\mathsf{T}}\boldsymbol{\Lambda}_\theta^{-1}\mathbf{Z}\hat{\mathbf{b}} - \hat{\mathbf{b}}^{\mathsf{T}}\boldsymbol{\psi}_\theta^{-1}\hat{\mathbf{b}}/2 \\ &\quad -\text{tr}\{\boldsymbol{\psi}_\theta^{-1}(\mathbf{Z}^{\mathsf{T}}\boldsymbol{\Lambda}_\theta^{-1}\mathbf{Z} + \boldsymbol{\psi}_\theta^{-1})^{-1}\}/2 + k(\boldsymbol{\theta}) \\ &= -\|\mathbf{y} - \mathbf{X}\boldsymbol{\beta}\|^2_{\Lambda_\theta^{-1}}/2 - \hat{\mathbf{b}}^{\mathsf{T}}\mathbf{Z}^{\mathsf{T}}\boldsymbol{\Lambda}_\theta^{-1}\mathbf{Z}\hat{\mathbf{b}}/2 + (\mathbf{y} - \mathbf{X}\boldsymbol{\beta})^{\mathsf{T}}\boldsymbol{\Lambda}_\theta^{-1}\mathbf{Z}\hat{\mathbf{b}} \\ &\quad -\hat{\mathbf{b}}^{\mathsf{T}}\boldsymbol{\psi}_\theta^{-1}\hat{\mathbf{b}}/2 - p/2 + k(\boldsymbol{\theta}). \end{aligned}$$

Remember that when optimizing $Q(\boldsymbol{\theta}, \boldsymbol{\beta})$ w.r.t. $\boldsymbol{\beta}$ and $\boldsymbol{\theta}$, $\hat{\mathbf{b}}$ remains fixed at its value from step 1 of the iteration. $\hat{\mathbf{b}}$ only changes at the next step 1. This contrasts to the situation when optimizing l_p or l_r, when $\hat{\mathbf{b}}$ and $\hat{\boldsymbol{\beta}}$ must be re-computed for each new trial value for $\boldsymbol{\theta}$.

The algorithm is a very reliable way of maximizing the likelihood and can also be used with the REML criterion, but is rather slow to converge near the MLE. Hence it is often best to start optimization off using EM steps, before switching to Newton's method (Pinheiro and Bates, 2000).

2.4.8 Model selection

The results of section 2.4.2 provide the basis for obtaining approximate confidence intervals for $\boldsymbol{\beta}$, albeit by fixing $\boldsymbol{\theta}$ at its estimated value. It is also possible to use the results to construct simple Wald tests for fixed effects.

The random effect parameters, $\boldsymbol{\theta}$, are more awkward. The fundamental difficulty is that many tests of interest restrict some parameters to the edge of the feasible parameter space, invalidating the usual large sample generalized likelihood ratio tests and other simple testing procedures.

AIC is also problematic for model comparison (Greven and Kneib, 2010). The difficulty is that we can only base AIC on the the log restricted likelihood, l_r, if the fixed effect structure of all models is identical, but if we use the log likelihood, l_p, instead then variance parameters are biased downwards, biasing AIC model selection towards models with a simpler random effects structure. See section 6.11 (p. 301).

Despite the difficulties matters are far from hopeless. Confidence intervals for elements of $\boldsymbol{\theta}$ can be computed using (3.4.3), and if those intervals suggest that variance components are bounded comfortably away from zero then it is clear that they are needed in the model. Similarly if a variance component is estimated as being effectively zero, then we can safely drop that term. In cases of doubt we may also use a GLRT as a rough guide for model comparison: very large or small p-values still suggest a clear-cut result and it is only p-values of similar size to our decision threshold for inclusion/exclusion that are problematic. Note also that some interesting tests do not involve restricting parameters to the edge of the feasible space, and in that case there is no problem.

There are also several reliable tests available for testing variance components for equality to zero. Crainiceanu and Ruppert (2004) produced an exact test for a linear mixed model with one variance component and Greven et al. (2008) proposed an approximate method to extend this to models with multiple variance components. Wood (2013b) proposed an alternative test exploiting the link between mixed models and penalized regression which is also applicable beyond the Gaussian setting (see section 6.12.2, p. 309). However, at time of writing, none of these are directly available in the major linear mixed modelling packages in R.

2.5 Linear mixed models in R

There are several packages for linear mixed modelling in R, of which `nlme` and the `lme4` are particularly noteworthy. Package `mgcv` can also fit linear mixed models with relatively simple random effects structures. `nlme` also provides nonlinear mixed models (see Pinheiro and Bates, 2000), while `lme4` and `mgcv` also provide generalized linear mixed models.

2.5.1 *Package* `nlme`

The main model fitting function of interest is called `lme`. A call to the `lme` function is similar to a call to `lm`, except that an extra argument specifying the random effects structure must also be supplied to the model. By default, `lme` works with a slightly more restricted structure for linear mixed models than the very general form (2.10), given in section 2.3. Specifically `lme` assumes that your data are grouped according to the levels of some factor(s), and that the same random effects structure is required for each group, with random effects independent between groups. Assuming just one level of grouping, the model for the data in the ith group is then

$$\mathbf{y}_i = \mathbf{X}_i\boldsymbol{\beta} + \mathbf{Z}_i\mathbf{b}_i + \boldsymbol{\epsilon}_i, \quad \mathbf{b}_i \sim N(\mathbf{0}, \boldsymbol{\psi}_\theta), \quad \boldsymbol{\epsilon}_i \sim N(\mathbf{0}, \boldsymbol{\Lambda}_\theta\sigma^2). \tag{2.20}$$

Careful attention should be paid to which terms have an i subscript, and hence depend on group, and which are common to all groups. Note, in particular, that $\boldsymbol{\beta}$, $\boldsymbol{\psi}_\theta$ and $\boldsymbol{\Lambda}_\theta$, the unknowns for the fixed effects and random effects, respectively, are assumed to be the same for all groups, as is the residual variance, σ^2. This form of mixed effects model, which was introduced by Laird and Ware (1982), is a sensible default because it is very common in practical applications, and because model fitting for this structure is more efficient than for the general form (2.10). However, it is important to realize that (2.20) is only a special case of (2.10). Indeed if we treat all the data as belonging to a single group then (2.20) is exactly (2.10), with no special structure imposed.

Because of `lme`'s default behaviour, you need to provide two parts to the random effects specification: a part that specifies $\mathbf{Z}_i$ and a part specifying the grouping factor(s). By default, $\boldsymbol{\psi}_\theta$ is assumed to be a general positive definite matrix to be estimated, but it is also possible to specify that it should have a more restricted form (for example $\mathbf{I}\sigma_b^2$). The simplest way to specify the random effects structure is with a one sided formula. For example `~x|g` would set up $\mathbf{Z}_i$ according to the `~x` part of the formula while the levels of the factor variable, `g`, would be used to split the data into groups (i.e., the levels of `g` are effectively the group index, i). The random effects formula is one sided, because there is no choice about the response variable — it must be whatever was specified in the fixed effects formula. So an example call to `lme` looks something like this:

```
lme(y ~ x + z, dat, ~ x|g)
```

where the response is `y`, the fixed effects depend on `x` and `z`, the random effects depend only on `x`, the data are grouped according to factor `g`, and all data are in data frame `dat`. An alternative way of specifying the same model is:

```
lme(y ~ x + z, dat, list(g = ~x))
```

and in fact this latter form is the one we will eventually use with GAMMs.

As an example, model (2.11), from sections 2.1.3 and 2.3, can easily be fitted.

```
> library(nlme)
> lme(travel ~ 1, Rail, list(Rail = ~ 1))
Linear mixed-effects model fit by REML
  Data: Rail
  Log-restricted-likelihood: -61.0885
  Fixed: travel ~ 1
(Intercept)
       66.5

Random effects:
 Formula: ~ 1 | Rail
        (Intercept) Residual
StdDev:    24.80547 4.020779

Number of Observations: 18
Number of Groups: 6
```

Because REML has been used for estimation the results are identical to those obtained in section 2.1.3. If we had used MLE, by specifying `method="ML"` in the call to `lme`, then the results would have corresponded to those obtained in section 2.4.4.

2.5.2 *Tree growth: An example using* `lme`

The `nlme` package includes a data frame called `Loblolly`, containing growth data on Loblolly pine trees. `height`, in feet (data are from the US), and `age`, in years, are recorded for 14 individual trees. A factor variable `Seed`, with 14 levels, indicates the identity of individual trees. Interest lies in characterising the population level mean growth trajectory of Loblolly pines, but it is clear that we would expect a good deal of tree to tree variation, and probably also some degree of autocorrelation in the random component of height.

From examination of data plots, the following initial model might be appropriate

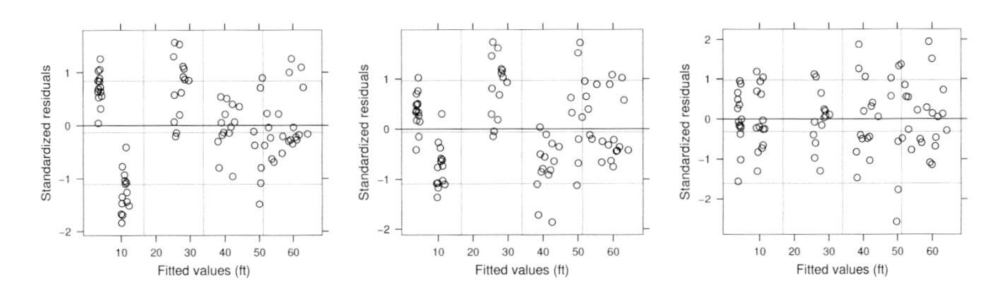

Figure 2.6 *Default residual plots for models* m0, m1 *and* m2 *(left to right). There is a clear trend in the mean of the residuals for the first two models, which model* m2 *eliminates.*

for the ith measurement on the jth tree:

$$\texttt{height}_{ji} = \beta_0 + \beta_1 \texttt{age}_{ji} + \beta_2 \texttt{age}_{ji}^2 + \beta_3 \texttt{age}_{ji}^3 \\ + b_0 + b_{j1} \texttt{age}_{ji} + b_{j2} \texttt{age}_{ji}^2 + b_{j3} \texttt{age}_{ji}^3 + \epsilon_{j,i}$$

where the $\epsilon_{j,i}$ are zero mean normal random variables, with correlation given by $\rho(\epsilon_{j,i}, \epsilon_{j,i-k}) = \phi^k$, and ϕ is an unknown parameter: this $\epsilon_{j,i}$ model is an autoregressive model of order 1 (if the ages had been unevenly spaced then a continuous generalization of this is available, which would then be more appropriate). The ϵ terms are independent between different trees. As usual $\boldsymbol{\beta}$ denotes the fixed effects and $\mathbf{b}_j \sim N(0, \boldsymbol{\psi})$ denotes the random effects.

This model can be estimated using `lme`, but to avoid convergence difficulties in the following analysis, two preparatory steps are useful. Firstly, it is worth centring the `age` variable as follows:

```
Loblolly$age <- Loblolly$age - mean(Loblolly$age)
```

without such centring, polynomial terms can become highly correlated which can cause numerical difficulties. An alternative to centering would be to use the `poly` function to set up an orthogonal polynomial basis.

Secondly, for this analysis the default fitting method fails without some adjustment. `lme` fits start by using the EM algorithm to get reasonably close to the optimal parameter estimates, and then switch to Newton's method, which converges more quickly. The number of EM steps to take, and the maximum number of Newton steps to allow, are both controllable via the `control` argument of `lme`. The `lmeControl` function offers a convenient way of producing a `control` list, with some elements modified from their default. For example

```
lmc <- lmeControl(niterEM=500,msMaxIter=100)
```

produces a `control` list in which the number of EM iterations is set to 500, and the maximum number of Newton iterations is set to 100. For future reference, note that `niterEM` should rarely be increased from its default 0 when calling `gamm`.

The model can now be estimated.

```
m0 <- lme(height ~ age + I(age^2) + I(age^3),Loblolly,
          random = list(Seed = ~ age + I(age^2) + I(age^3)),
          correlation = corAR1(form = ~ age|Seed),control=lmc)
```

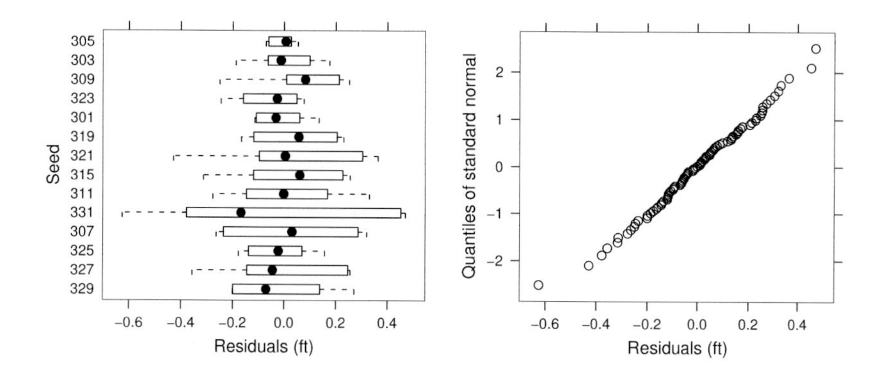

Figure 2.7 *Further residual plots for model* m2. *The left panel shows boxplots of the residuals for each tree, while the right plot is a normal QQ-plot for the residuals.*

The random argument specifies that there should be a different cubic term for each tree, while the correlation argument specifies an autoregressive model for the residuals for each tree. form=~age|Seed indicates that age is the variable determining the ordering of residuals, and that the correlation applies within measurements made on one tree, but not between measurements on different trees.

The command plot(m0) produces the default residual plot shown in the left panel of figure 2.6. The plot shows a clear trend in the mean of the residuals: the model seems to underestimate the first group of measurements, made at age 5, and then overestimate the next group, made at age 10, before somewhat underestimating the next group, which correspond to year 15. This suggests a need for a more flexible model, so fourth and fifth order polynomials were also tried.

```
m1 <- lme(height ~ age + I(age^2) + I(age^3) + I(age^4),
          Loblolly, list(Seed = ~ age + I(age^2) + I(age^3)),
          cor = corAR1(form = ~age|Seed), control=lmc)
plot(m1)
m2 <- lme(height ~ age +I(age^2) +I(age^3) +I(age^4) +I(age^5),
          Loblolly,list(Seed = ~ age + I(age^2) + I(age^3)),
          cor = corAR1(form = ~ age|Seed), control=lmc)
plot(m2)
```

The resulting residuals plots are shown in the middle and right panels of figure 2.6. m1 does lead to a slight improvement, but only m2 is really satisfactory. Further model checking plots can now be produced for m2.

```
plot(m2,Seed~resid(.))
qqnorm(m2,~resid(.))
qqnorm(m2,~ranef(.))
```

The resulting plots are shown in figures 2.7 and 2.8, and suggest that the model is reasonable.

An obvious question is whether the elaborate model structure, with random cubic and autocorrelated within-tree errors, is really required. First try dropping the autocorrelation component.

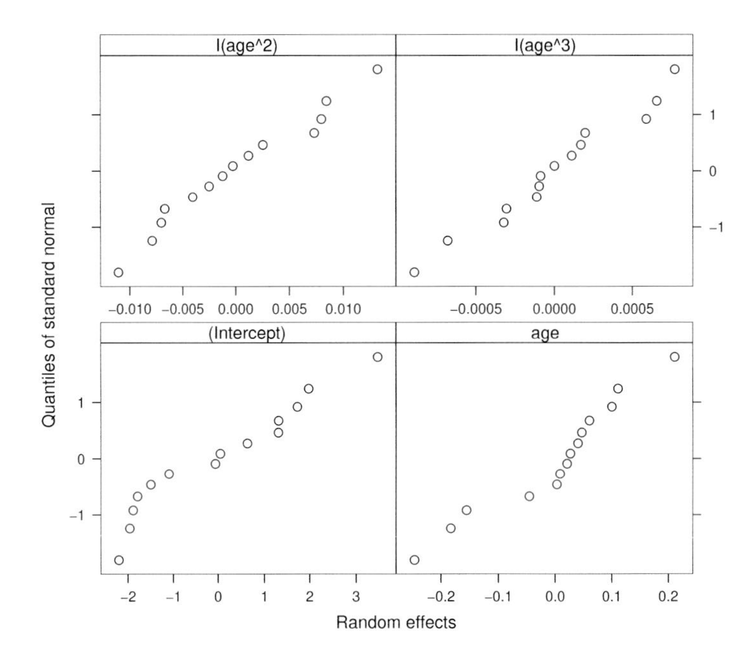

Figure 2.8 *Normal QQ-plots for the predicted random effects from model* m2. *The plots should look like correlated random scatters around straight lines, if the normality assumptions for the random effects are reasonable: only* $\hat{b}_1$ *shows any suggestion of any problem, but it is not enough to cause serious concern.*

```
> m3 <- lme(height ~ age+I(age^2)+I(age^3)+I(age^4)+I(age^5),
+     Loblolly,list(Seed = ~ age+I(age^2)+I(age^3)),control=lmc)
> anova(m3,m2)
   Model df    AIC    BIC  logLik   Test L.Ratio p-value
m3     1 17 250.46 290.53 -108.23
m2     2 18 239.36 281.78 -101.68 1 vs 2 13.1041   3e-04
```

The anova command is actually conducting a generalized likelihood ratio test here, which rejects m3 in favour of m2. Note that in this case the GLRT assumptions are met: m3 is effectively setting the autocorrelation parameter ϕ to zero, which is in the middle of its possible range, not on a boundary. anova also reports the AIC for the models, which also suggest that m2 is preferable. There seems to be strong evidence for auto-correlation in the within-tree residuals.

Perhaps the random effects model could be simplified, by dropping the dependence of tree-specific growth on the cube of age.

```
> m4 <- lme(height~age+I(age^2)+I(age^3)+I(age^4)+I(age^5),
+            Loblolly,list(Seed=~age+I(age^2)),
+            correlation=corAR1(form=~age|Seed),control=lmc)
```

```
> anova(m4,m2)
   Model df    AIC    BIC  logLik   Test L.Ratio p-value
m4     1 14 253.76 286.75 -112.88
m2     2 18 239.36 281.78 -101.68 1 vs 2 22.4004   2e-04
```

Recall that the GLRT test is somewhat problematic here, since m4 is m2 with some variance parameters set to the edge of the feasible parameter space; however, a likelihood ratio statistic so large that it would have given rise to a p-value of .0002, for a standard GLRT, is strong grounds for rejecting m4 in favour of m2 in the current case. Comparison of AIC scores (which could also have been obtained using AIC(m4,m2)) suggests quite emphatically that m2 is the better model.

Another obvious model to try is one with a less general random effects structure. The models so far have allowed the random effects for any tree to be correlated in a very general way: it has simply been assumed that $\mathbf{b}_j \sim N(0, \psi_\theta)$, where the only restriction on the matrix ψ_θ, is that it should be positive definite. Perhaps a less flexible model would suffice: for example, ψ_θ might be a diagonal matrix (with positive diagonal elements). Such a structure (and indeed many other structures) can be specified in the call to lme.

```
> m5 <- lme(height~age+I(age^2)+I(age^3)+I(age^4)+I(age^5),
+          Loblolly,list(Seed=pdDiag(~age+I(age^2)+I(age^3))),
+          correlation=corAR1(form=~age|Seed),control=lmc)
```

Here the pdDiag function indicates that the covariance matrix for the random effects at each level of Seed should have a (positive definite) diagonal structure. m5 can be compared to m2.

```
> anova(m2,m5)
   Model df      AIC      BIC    logLik   Test L.Ratio p-value
m2     1 18 239.3576 281.7783 -101.6788
m5     2 12 293.7081 321.9886 -134.8540 1 vs 2 66.3505  <.0001
```

Again, both the GLRT test and AIC comparison favour the more general model m2. In this case the GLRT assumptions are met: m5 amounts to setting the random effects covariances in m2 to zero, but since covariances can be positive or negative this is not on the boundary of the parameter space and the GLRT assumptions hold. The nlme package includes very many useful utilities for examining and plotting grouped data, one of which is the following, for plotting data and model predictions together on a unit by unit basis. See figure 2.9.

```
plot(augPred(m2))
```

2.5.3 Several levels of nesting

When using mixed models it is quite common to have several levels of nesting present in a model. For example, in the machine type and worker productivity model (2.6), of section 2.1.4, there are random effects for worker and each worker-machine combination. lme can accommodate such structures as follows

```
> lme(score ~ Machine,Machines,list(Worker = ~1, Machine = ~1))
Linear mixed-effects model fit by REML
  Data: Machines
  Log-restricted-likelihood: -107.8438
```

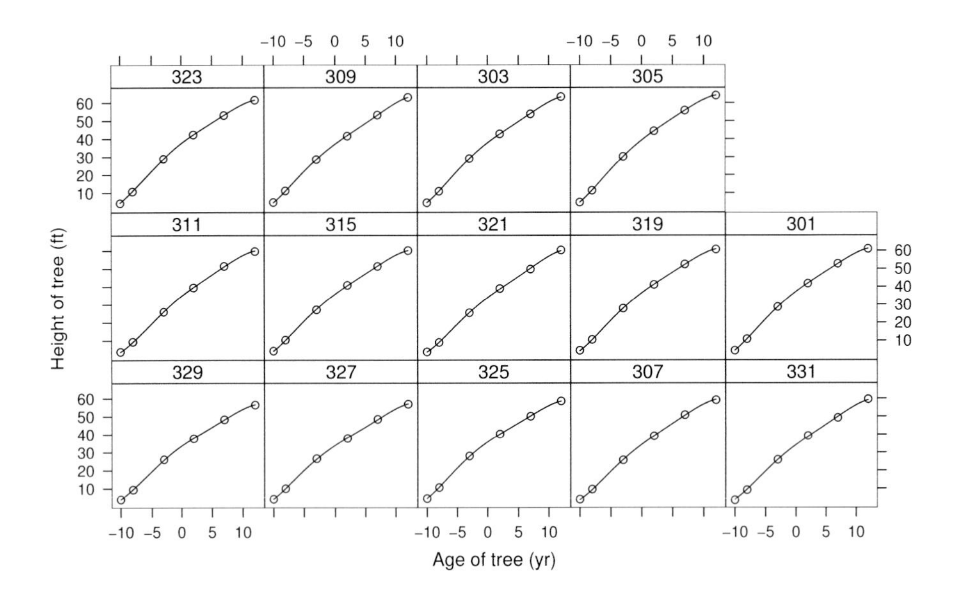

Figure 2.9 *Model predictions from* `m4` *at the individual tree level, overlaid on individual Loblolly pine growth data. The panel titles are the value of the* `Seed` *tree identifier.*

```
  Fixed: score ~ Machine
(Intercept)    MachineB    MachineC
  52.355556    7.966667   13.916667

Random effects:
 Formula: ~1 | Worker
        (Intercept)
StdDev:    4.781049

 Formula: ~1 | Machine %in% Worker
        (Intercept)  Residual
StdDev:    3.729536 0.9615768

Number of Observations: 54
Number of Groups:
             Worker Machine %in% Worker
                  6                  18
```

Notice how any grouping factor in the random effects list is assumed to be nested within the grouping factors to its left.

This section can only hope to scratch the surface of what is possible with `lme`: for a much fuller account, see Pinheiro and Bates (2000).

2.5.4 *Package* lme4

The nlme package is designed to efficiently exploit the nested structure of linear models such as

$$y_{ijk} = \alpha + b_i + c_{ij} + \epsilon_{ijk}$$

where b_i and c_{ij} are Gaussian random effects. In consequence nlme is less computationally efficient with non-nested models such as

$$y_{ijk} = \alpha + b_i + c_j + \epsilon_{ijk},$$

even when the random effects model matrices for both models have the same proportion of zero entries (the same 'sparsity').

The lme4 package is designed to be equally efficient in both nested and non nested cases, by using direct sparse matrix methods (e.g., Davis, 2006) for the model estimation. Sparse matrix methods are designed to be memory efficient by only storing the non-zero elements of matrices, and floating point efficient, by only computing the elements of matrix results that are not structurally zero. Use of sparse matrix methods is complicated by the problem of *infill*: that the result of a matrix operation on one or more sparse matrices need not be sparse. However, the linear mixed model likelihood can be efficiently computed using sparse methods, by using sparse Cholesky decomposition as the basis for the computations (see, e.g., the Matrix package in R). For example, the solve and determinant calculations in the simple code in section 2.4.4 would both be replaced with code based on sparse Cholesky decomposition.

The lmer function from package lme4 is designed for linear mixed modelling, and its use is similar to other R modelling functions, such as lm. Model specification uses a single model formula combining fixed and random effects. Fixed effects are specified exactly as for lm, whereas random effects are of the form (x|g) where g is a grouping factor, and x is interpreted as the right hand side of a model formula specifying the random effect model matrix nested in each level of g. In contrast to lme, the ordering of the random effects is unimportant. At time of writing there are no facilities for the sort of correlation structure that lme supports.

Consider repeating the Machines model from the previous section.

```
> library(lme4)
> a1 <- lmer(score ~ Machine + (1|Worker) + (1|Worker:Machine),
+            data=Machines)
> a1
Linear mixed model fit by REML ['lmerMod']
Formula: score ~ Machine + (1 | Worker) + (1 | Worker:Machine)
   Data: Machines
REML criterion at convergence: 215.6876
Random effects:
 Groups         Name        Std.Dev.
 Worker:Machine (Intercept) 3.7295
 Worker         (Intercept) 4.7811
 Residual                   0.9616
Number of obs: 54, groups:  Worker:Machine, 18; Worker, 6
```

```
Fixed Effects:
(Intercept)      MachineB      MachineC
     52.356         7.967        13.917
```

The estimates are the same as those from `lme`, of course. Now try an alternative model in which the random effects for machine are correlated between machines used by the same worker, but independent between workers, and compare the models using a generalized likelihood ratio test, and AIC.

```
> a2 <- lmer(score ~ Machine + (1|Worker) + (Machine-1|Worker),
+            data=Machines)
> AIC(a1,a2)
   df      AIC
a1  6 227.6876
a2 11 230.3112
> anova(a1,a2)
refitting model(s) with ML (instead of REML)
Data: Machines
Models:
a1: score ~ Machine + (1 | Worker) + (1 | Worker:Machine)
a2: score ~ Machine + (1 | Worker) + (Machine - 1 | Worker)
   Df    AIC   BIC  logLik deviance  Chisq Chi Df Pr(>Chisq)
a1  6 237.27 249.2 -112.64   225.27
a2 11 238.42 260.3 -108.21   216.42 8.8516      5     0.1151
```

AIC suggests that this new model is overcomplicated. This is confirmed by the generalized likelihood ratio test. The latter is valid because the null model, `a1`, is restricting the variances of the `Machine` effects to be equal (not zero), and the correlations between machines to be zero (which is not at the edge of the feasible parameter space). See the `lme4` help files and package vignettes for more.

2.5.5 *Package* `mgcv`

Function `gam` from package `mgcv` can also fit mixed models provided they have simple independent Gaussian random effects. `mgcv` does not exploit sparsity of the random effects model matrix at all, but its relatively efficient optimizers mean that it can still be computationally competitive for modest numbers of random effects. Random effects are specified as special cases of 'smooths' using terms like `s(z,x,v,bs="re")` in the model formula arguments of functions `gam` or `bam`. The model matrix specified by such a term is whatever results from `model.matrix(~z:x:v-1)` in R, and the corresponding random effects covariance matrix is simply $\mathbf{I}\sigma_b^2$.

Again refit the simple `Machines` model.

```
> library(mgcv)
> b1 <- gam(score ~ Machine + s(Worker,bs="re") +
+        s(Machine,Worker,bs="re"),data=Machines,method="REML")

> gam.vcomp(b1)
```

```
Standard deviations and 0.95 confidence intervals:

                       std.dev     lower      upper
s(Worker)            4.7810626 2.2498659 10.159965
s(Machine,Worker) 3.7295240 2.3828104  5.837371
scale                0.9615766 0.7632535  1.211432
```

Within the confines of simple i.i.d. random effects an alternative model might allow different worker variances for each machine.

```
> b2 <- gam(score ~ Machine + s(Worker,bs="re") +
+      s(Worker,bs="re",by=Machine),data=Machines,method="REML")
> gam.vcomp(b2)

Standard deviations and 0.95 confidence intervals:

                       std.dev     lower      upper
s(Worker)            3.7859468 1.7987315  7.968612
s(Worker):MachineA 1.9403242 0.2531895 14.869726
s(Worker):MachineB 5.8740228 2.9883339 11.546281
s(Worker):MachineC 2.8454688 0.8299327  9.755842
scale                0.9615766 0.7632536  1.211432

Rank: 5/5
> AIC(b1,b2)
         df      AIC
b1 18.85995 165.1905
b2 18.98557 165.6204
```

The AIC comparison here still suggests that the simpler model is marginally preferable, but note that it is a 'conditional' AIC as opposed to the 'marginal' AIC produced by `lme4`. The distinction is covered in section 6.11 (p. 301).

2.6 Exercises

1. A pig breeding company was interested in investigating litter to litter variability in piglet weight (after a fixed growth period). Six sows were selected randomly from the company's breeding stock, impregnated and 5 (randomly selected) piglets from each resulting litter were then weighed at the end of a growth period. The data were entered into an R data frame, `pig`, with weights recorded in column `w` and a column, `sow`, containing a factor variable indicating which litter the piglet came from. The following R session is part of the analysis of these data using a simple mixed model for piglet weight.

```
> pig$w
 [1]  9.6 10.1 11.2 11.1 10.5  9.5  9.6  9.4  9.5  9.5 11.5
 [12] 10.9 10.8 10.7 11.7 10.7 11.2 11.2 10.9 10.5 12.3 12.1
 [23] 11.2 12.3 11.7 11.2 10.3  9.9 11.1 10.5
> pig$sow
 [1]  1 1 1 1 1 2 2 2 2 2 3 3 3 3 3 4 4 4 4 4 5 5 5 5 5 6 6
 [28] 6 6 6
Levels: 1 2 3 4 5 6
```

```
> m1<-lm(w~sow,data=pig)
> anova(m1)
Analysis of Variance Table

Response: w
          Df  Sum Sq Mean Sq F value    Pr(>F)
sow        5 15.8777  3.1755  14.897 1.086e-06 ***
Residuals 24  5.1160  0.2132

> piggy<-aggregate(data.matrix(pig),
+                 by=list(sow=pig$sow),mean)
> m0<-lm(w~1,data=piggy)
> summary(m1)$sigma^2
[1] 0.2131667
> summary(m0)$sigma^2
[1] 0.6351067
```

(a) The full mixed model being used in the R session has a random effect for litter/sow and a fixed mean. Write down a full mathematical specification of the model.

(b) Specify the hypothesis being tested by the `anova` function, both in terms of the parameters of the mixed model, and in words.

(c) What conclusion would you draw from the printed ANOVA table? Again state your conclusions both in terms of the model parameters and in terms of what this tells you about pigs.

(d) Using the given output, obtain an (unbiased) estimate of the between litter variance in weight, in the wider population of pigs.

2. Consider a model with two random effects of the form:

$$y_{ij} = \alpha + b_i + c_j + \epsilon_{ij}$$

where $i = 1, \ldots, I$, $j = 1, \ldots, J$, $b_i \sim N(0, \sigma_b^2)$, $c_j \sim N(0, \sigma_c^2)$ and $\epsilon_{ij} \sim N(0, \sigma^2)$ and all these r.v.'s are mutually independent. If the model is fitted by least squares then

$$\hat{\sigma}^2 = RSS/(IJ - I - J + 1)$$

is an unbiased estimator of σ^2, where RSS is the residual sum of squares from the model fit.

(a) Show that, if the above model is correct, the averages $\bar{y}_{i\cdot} = \sum_j y_{ij}/J$ are governed by the model:

$$\bar{y}_{i\cdot} = a + e_i$$

where the e_i are i.i.d. $N(0, \sigma_b^2 + \sigma^2/J)$ and a is a random intercept term. Hence suggest how to estimate σ_b^2.

(b) Show that the averages $\bar{y}_{\cdot j} = \sum_i y_{ij}/I$ are governed by the model:

$$\bar{y}_{\cdot j} = a' + e'_j$$

where the e'_j are i.i.d. $N(0, \sigma_c^2 + \sigma^2/I)$ and a' is a random intercept parameter. Suggest an estimator for σ_c^2.

3. Data were collected on blood cholesterol levels and blood pressure for a group of patients regularly attending an outpatient clinic for a non-heart-disease related condition. Measurements were taken each time the patient attended the clinic. A possible model for the resulting data is

$$y_{ij} = \mu + a_i + \beta x_{ij} + \epsilon_{ij}, \quad a_i \sim N(0, \sigma_a^2) \text{ and } \epsilon_{ij} \sim N(0, \sigma^2),$$

where y_{ij} is the jth blood pressure measurement for the ith patient and x_{ij} is the corresponding cholesterol measurement. β is a fixed parameter relating blood pressure to cholesterol concentration and a_i is a random coefficient for the ith patient. Assume (somewhat improbably) that the same number of measurements are available for each patient.

(a) Explain how you would test $\text{H}_0 : \sigma_a^2 = 0$ vs. $\text{H}_1 : \sigma_a^2 > 0$ and test $\text{H}_0 : \beta = 0$ vs. $\text{H}_1 : \beta \neq 0$, using standard software for ordinary linear modelling.

(b) Explain how β and σ_a^2 could be estimated. You should write down the models involved, but should assume that these would be fitted using standard linear modelling software.

4. Write out the following three models in the general form,

$$\mathbf{y} = \mathbf{X}\boldsymbol{\beta} + \mathbf{Z}\mathbf{b} + \boldsymbol{\epsilon}, \quad \mathbf{b} \sim N(\mathbf{0}, \boldsymbol{\psi}_\theta) \text{ and } \boldsymbol{\epsilon} \sim N(0, \mathbf{I}\sigma^2),$$

where $\mathbf{Z}$ is a matrix containing known coefficients which determine how the response, $\mathbf{y}$, depends on the random effects $\mathbf{b}$ (i.e. it is a 'model matrix' for the random effects). $\boldsymbol{\psi}_\theta$ is the covariance matrix of the random effects $\mathbf{b}$. You should ensure that $\mathbf{X}$ is specified so that the fixed effects are identifiable (you don't need to do this for $\mathbf{Z}$) and don't forget to specify $\boldsymbol{\psi}_\theta$.

(a) The model from question 3, assuming four patients and two measurements per patient.

(b) The mixed effects model from section 2.1.1, assuming only two measurements per tree.

(c) Model (2.6) from section 2.1.4, assuming that $I = 2$, $J = 3$ and $K = 3$.

5.(a) Show that if $\mathbf{X}$ and $\mathbf{Z}$ are independent random vectors, both of the same dimension, and with covariance matrices $\boldsymbol{\Sigma}_x$ and $\boldsymbol{\Sigma}_z$, then the covariance matrix of $\mathbf{X} + \mathbf{Z}$ is $\boldsymbol{\Sigma}_x + \boldsymbol{\Sigma}_z$.

(b) Consider a study examining patients' blood insulin levels 30 minutes after eating, y, in relation to sugar content, x, of the meal eaten. Suppose that each of 3 patients had their insulin levels measured for each of 3 sugar levels, and that an appropriate linear mixed model for the jth measurement on the ith patient is

$$y_{ij} = \alpha + \beta x_{ij} + b_i + \epsilon_{ij}, \quad b_i \sim N(0, \sigma^2), \text{ and } \epsilon_{ij} \sim N(0, \sigma^2),$$

where all the random effects and residuals are mutually independent.

i. Write this model out in matrix vector form.
ii. Find the covariance matrix for the response vector $\mathbf{y}$.

6. The R data frame `Oxide` from the `nlme` library contains data from a quality control exercise in the semiconductor industry. The object of the exercise was to investigate sources of variability in the thickness of oxide layers in silicon wafers. The dataframe contains the following columns:

   `Thickness` is the thickness of the oxide layer (in nanometres, as far as I can tell).

   `Source` is a two level factor indicating which of two possible suppliers the sample came from.

   `Site` is a 3 level factor, indicating which of three sites on the silicon wafer the thickness was measured.

   `Lot` is a factor variable with levels indicating which particular batch of Silicon wafers this measurement comes from.

   `Wafer` is a factor variable with levels labelling the individual wafers examined.

   The investigators are interested in finding out if there are systematic differences between the two sources, and expect that thickness may vary systematically across the three sites; they are only interested in the lots and wafers in as much as they are representative of a wider population of lots and wafers.

   (a) Identify which factors you would treat as random and which as fixed, in a linear mixed model analysis of these data.
   (b) Write down a model that might form a suitable basis for beginning to analyse the `Oxide` data.
   (c) Perform a complete analysis of the data, including model checking. Your aim should be to identify the sources of thickness variability in the data and any fixed effects causing thickness variability.

7. Starting from model (2.6) in section 2.1.4, re-analyse the `Machines` data using `lme`. Try to find the most appropriate model, taking care to examine appropriate model checking plots. Make sure that you test whether the interaction in (2.6) is appropriate. Similarly test whether a more complex random effects structure would be appropriate: specifically one in which the machine-worker interaction is correlated within worker. If any data appear particularly problematic in the checking plots, repeat the analysis, and see if the conclusions change.

8. This question follows on from question 7. *Follow-up multiple comparisons* are a desirable part of some analyses. This question is about how to do this in practice. In the analysis of the Machines data the ANOVA table for the fixed effects indicates that there are significant differences between machine types, so an obvious follow-up analysis would attempt to assess exactly where these differences lie. Obtaining Bonferroni corrected intervals for each of the 3 machine to machine differences would be one way to proceed, and this is easy to do.
   First note that provided you have set up the default contrasts using

```
options(contrasts=c("contr.treatment","contr.treatment"))
```

(*before calling* `lme`, *of course*) then `lme` will set your model up in such a way that the coefficients associated with the `Machine` effect correspond to the *difference* between the second and first machines, and between the third and first machines. Hence the `intervals` function can produce two of the required comparisons automatically. However, by default the `intervals` function uses the 95% confidence level, which needs to be modified if you wish to Bonferroni correct for the fact that 3 comparisons are being made. If your model object is `m1` then

```
intervals(m1,level=1-0.05/3,which="fixed")
```

will produce 2 of the required intervals. Note the Bonferroni correction '3'. The option `which="fixed"` indicates that only fixed effect intervals are required. The third comparison, between machines B and C, can easily be obtained by changing the way that the factor variable `Machine` is treated, so that machine type B or C count as the 'first machine' when setting up the model. The `relevel` function can be used to do this.

```
levels(Machines$Machine) # check the level names
## reset levels so that 'first level' is "B" ...
Machines$Machine<-relevel(Machines$Machine,"B")
```

Now re-fit the model and re-run the `intervals` function for the new fit. This will yield the interval for the remaining comparison (plus one of the intervals you already have, of course). What are the Bonferroni corrected 95% intervals for the 3 possible comparisons? How would you interpret them?

9. The data frame `Gun` (library `nlme`) is from a trial examining methods for firing naval guns. Two firing methods were compared, with each of a number of teams of 3 gunners; the gunners in each team were matched to have similar physique (Slight, Average or Heavy). The response variable `rounds` is rounds fired per minute, and there are 3 explanatory factor variables, `Physique` (levels `Slight`, `Medium` and `Heavy`); `Method` (levels `M1` and `M2`) and `Team` with 9 levels. The main interest is in determining which method and/or physique results in the highest firing rate and in quantifying team-to-team variability in firing rate.

   (a) Identify which factors should be treated as random and which as fixed, in the analysis of these data.

   (b) Write out a suitable mixed model as a starting point for the analysis of these data.

   (c) Analyse the data using `lme` in order to answer the main questions of interest. Include any necessary follow-up multiple comparisons (as in the previous question) and report your conclusions.

Chapter 3

Generalized Linear Models

Generalized linear models* (Nelder and Wedderburn, 1972) allow for response distributions other than normal, and for a degree of non-linearity in the model structure. A GLM has the basic structure

$$g(\mu_i) = \mathbf{X}_i\boldsymbol{\beta},$$

where $\mu_i \equiv \mathbb{E}(Y_i)$, g is a smooth monotonic 'link function', $\mathbf{X}_i$ is the ith row of a model matrix, $\mathbf{X}$, and $\boldsymbol{\beta}$ is a vector of unknown parameters. In addition, a GLM usually makes the distributional assumptions that the Y_i are independent and

$$Y_i \sim \text{some exponential family distribution.}$$

The *exponential family* of distributions includes many distributions that are useful for practical modelling, such as the Poisson, binomial, gamma and normal distributions. The comprehensive reference for GLMs is McCullagh and Nelder (1989), while Dobson and Barnett (2008) provides a thorough introduction.

A generalized linear mixed model (GLMM) follows from a linear mixed model in the same way as a GLM follows from a linear model. We now have

$$g(\mu_i) = \mathbf{X}_i\boldsymbol{\beta} + \mathbf{Z}_i\mathbf{b}, \quad \mathbf{b} \sim N(\mathbf{0}, \boldsymbol{\psi})$$

but otherwise the model for Y_i is as above. For now, concentrate on the GLM case.

Because generalized linear models are specified in terms of the 'linear predictor', $\boldsymbol{\eta} \equiv \mathbf{X}\boldsymbol{\beta}$, many of the general ideas and concepts of linear modelling carry over, with a little modification, to generalized linear modelling. Basic model formulation is much the same as for linear models, except that a link function and distribution must be chosen. Of course, if the identity function is chosen as the link, along with the normal distribution, then ordinary linear models are recovered as a special case.

The generalization comes at some cost: model fitting now has to be done iteratively, and distributional results, used for inference, are now approximate and justified by large sample limiting results, rather than being exact. But before going further into these issues, consider a couple of simple examples.

Example 1: In the early stages of a disease epidemic, the rate at which new cases

*Note that there is a distinction between 'generali**zed**' and 'general' linear models — the latter term being sometimes used to refer to all linear models other than simple straight lines.

occur can often increase exponentially through time. Hence, if μ_i is the expected number of new cases on day t_i, a model of the form

$$\mu_i = \gamma \exp(\delta t_i)$$

might be appropriate, where γ and δ are unknown parameters. Such a model can be turned into GLM form by using a log link so that

$$\log(\mu_i) = \log(\gamma) + \delta t_i = \beta_0 + t_i \beta_1$$

(by definition of $\beta_0 = \log(\gamma)$ and $\beta_1 = \delta$). Note that the right hand side of the model is now linear in the parameters. The response variable is the number of new cases per day and, since this is a count, the Poisson distribution is probably a reasonable distribution to try. So the GLM for this situation uses a Poisson response distribution, log link, and linear predictor $\beta_0 + t_i \beta_1$.

Example 2: The rate of capture of prey items, y_i, by a hunting animal, tends to increase with increasing density of prey, x_i, but to eventually level off, when the predator is catching as much as it can cope with. A suitable model for this situation might be

$$\mu_i = \frac{\alpha x_i}{h + x_i},$$

where α is an unknown parameter, representing the maximum capture rate, and h is an unknown parameter, representing the prey density at which the capture rate is half the maximum rate. Obviously this model is non-linear in its parameters, but, by using a reciprocal link, the right hand side can be made linear in the parameters:

$$\frac{1}{\mu_i} = \frac{1}{\alpha} + \frac{h}{\alpha}\frac{1}{x_i} = \beta_0 + \frac{1}{x_i}\beta_1$$

(here $\beta_0 \equiv 1/\alpha$ and $\beta_1 \equiv h/\alpha$). In this case the standard deviation of prey capture rate might be approximately proportional to the mean rate, suggesting the use of a gamma distribution for the response, and completing the model specification.

Of course we are not restricted to the simple straight line forms of the examples, but can have any structure for the linear predictor that was possible for linear models.

3.1 The theory of GLMs

Estimation and inference with GLMs is based on the theory of maximum likelihood estimation, although the maximization of the likelihood turns out to require an iterative least squares approach, related to the method of section 1.8.8 (p. 54). This section begins by introducing the exponential family of distributions, which allows a general method to be developed for maximizing the likelihood of a GLM. Inference for GLMs is then discussed, based on general results of likelihood theory (see section 2.2, p.74, and appendix A for derivations). In this section it is sometimes useful to distinguish between the response data, y, and the random variable of which it is an observation, Y, so they are distinguished notationally: this has not been done for estimates and estimators.

3.1.1 *The exponential family of distributions*

The response variable in a GLM can have any distribution from the *exponential family*. A distribution belongs to the exponential family of distributions if its probability density function, or probability mass function, can be written as

$$f_\theta(y) = \exp\left[\{y\theta - b(\theta)\}/a(\phi) + c(y,\phi)\right],$$

where b, a and c are arbitrary functions, ϕ an arbitrary 'scale' parameter, and θ is known as the 'canonical parameter' of the distribution (in the GLM context, θ will completely depend on the model parameters $\boldsymbol{\beta}$, but it is not necessary to make this explicit yet).

For example, the normal distribution is a member of the exponential family since

$$\begin{aligned} f_\mu(y) &= \frac{1}{\sigma\sqrt{2\pi}}\exp\left[-\frac{(y-\mu)^2}{2\sigma^2}\right] \\ &= \exp\left[\frac{-y^2+2y\mu-\mu^2}{2\sigma^2} - \log(\sigma\sqrt{2\pi})\right] \\ &= \exp\left[\frac{y\mu-\mu^2/2}{\sigma^2} - \frac{y^2}{2\sigma^2} - \log(\sigma\sqrt{2\pi})\right], \end{aligned}$$

which is of exponential form, with $\theta = \mu$, $b(\theta) = \theta^2/2 \equiv \mu^2/2$, $a(\phi) = \phi = \sigma^2$ and $c(\phi, y) = -y^2/(2\phi) - \log(\sqrt{\phi 2\pi}) \equiv -y^2/(2\sigma^2) - \log(\sigma\sqrt{2\pi})$. Table 3.1 gives a similar breakdown for the distributions implemented for GLMs in R.

It is possible to obtain general expressions for the mean and variance of exponential family distributions, in terms of a, b and ϕ. The log likelihood of θ, given a particular y, is simply $\log\{f_\theta(y)\}$ considered as a function of θ. That is

$$l(\theta) = \{y\theta - b(\theta)\}/a(\phi) + c(y,\phi),$$

and so

$$\frac{\partial l}{\partial \theta} = \{y - b'(\theta)\}/a(\phi).$$

Treating l as a random variable, by replacing the particular observation y by the random variable Y, enables the expected value of $\partial l/\partial\theta$ to be evaluated:

$$\mathbb{E}\left(\frac{\partial l}{\partial \theta}\right) = \{\mathbb{E}(Y) - b'(\theta)\}/a(\phi).$$

Using the general result that $\mathbb{E}(\partial l/\partial\theta) = 0$ (see (A.1) in section A.2) and re-arranging implies that

$$\mathbb{E}(Y) = b'(\theta), \tag{3.1}$$

i.e., the mean of any exponential family random variable is given by the first derivative of b w.r.t. θ, where the form of b depends on the particular distribution. This equation is the key to linking the GLM model parameters, $\boldsymbol{\beta}$, to the canonical parameters of the exponential family. In a GLM, $\boldsymbol{\beta}$ determines the mean of the response

	Normal	Poisson	Binomial	Gamma	Inverse Gaussian
$f(y)$	$\frac{1}{\sigma\sqrt{2\pi}}\exp\left\{\frac{-(y-\mu)^2}{2\sigma^2}\right\}$	$\frac{\mu^y\exp(-\mu)}{y!}$	$\binom{n}{y}\left(\frac{\mu}{n}\right)^y\left(1-\frac{\mu}{n}\right)^{n-y}$	$\frac{1}{\Gamma(\nu)}\left(\frac{\nu}{\mu}\right)^{\nu} y^{\nu-1}\exp\left(-\frac{\nu y}{\mu}\right)$	$\sqrt{\frac{\gamma}{2\pi y^3}}\exp\left\{\frac{-\gamma(y-\mu)^2}{2\mu^2 y}\right\}$
Range	$-\infty < y < \infty$	$y = 0, 1, 2, \ldots$	$y = 0, 1, \ldots, n$	$y > 0$	$y > 0$
θ	μ	$\log(\mu)$	$\log\left(\frac{\mu}{n-\mu}\right)$	$-\frac{1}{\mu}$	$-\frac{1}{2\mu^2}$
ϕ	σ^2	1	1	$\frac{1}{\nu}$	$\frac{1}{\gamma}$
$a(\phi)$	$\phi(=\sigma^2)$	$\phi(=1)$	$\phi(=1)$	$\phi\left(=\frac{1}{\nu}\right)$	$\phi\left(=\frac{1}{\gamma}\right)$
$b(\theta)$	$\frac{\theta^2}{2}$	$\exp(\theta)$	$n\log\left(1+e^{\theta}\right)$	$-\log(-\theta)$	$-\sqrt{-2\theta}$
$c(y,\phi)$	$-\frac{1}{2}\left\{\frac{y^2}{\phi}+\log(2\pi\phi)\right\}$	$-\log(y!)$	$\log\binom{n}{y}$	$\nu\log(\nu y)-\log\{y\Gamma(\nu)\}$	$-\frac{1}{2}\left\{\log(2\pi y^3\phi)+\frac{1}{\phi y}\right\}$
$V(\mu)$	1	μ	$\mu(1-\mu/n)$	μ^2	μ^3
$g_c(\mu)$	μ	$\log(\mu)$	$\log\left(\frac{\mu}{n-\mu}\right)$	$\frac{1}{\mu}$	$\frac{1}{\mu^2}$
$D(y,\hat{\mu})$	$(y-\hat{\mu})^2$	$2y\log\left(\frac{y}{\hat{\mu}}\right) - 2(y-\hat{\mu})$	$2\left\{y\log\left(\frac{y}{\hat{\mu}}\right) + (n-y)\log\left(\frac{n-y}{n-\hat{\mu}}\right)\right\}$	$2\left\{\frac{y-\hat{\mu}}{\hat{\mu}}-\log\left(\frac{y}{\hat{\mu}}\right)\right\}$	$\frac{(y-\hat{\mu})^2}{\hat{\mu}^2 y}$

Table 3.1 *Some exponential family distributions. Note that when $y = 0$, $y\log(y/\hat{\mu})$ is taken to be zero (its limit as $y \to 0$).*

variable and, via (3.1), thereby determines the canonical parameter for each response observation.

Differentiating the likelihood once more yields

$$\frac{\partial^2 l}{\partial \theta^2} = -b''(\theta)/a(\phi),$$

and plugging this into the general result, $\mathbb{E}(\partial^2 l/\partial\theta^2) = -\mathbb{E}\{(\partial l/\partial\theta)^2\}$ (the derivatives are evaluated at the true θ value, see result (A.3), section A.2), gives

$$b''(\theta)/a(\phi) = \mathbb{E}\left[\{Y - b'(\theta)\}^2\right]/a(\phi)^2,$$

which re-arranges to the second useful general result:

$$\text{var}(Y) = b''(\theta)a(\phi).$$

a could in principle be any function of ϕ, and when working with GLMs there is no difficulty in handling any form of a, if ϕ is known. However, when ϕ is unknown matters become awkward, unless we can write $a(\phi) = \phi/\omega$, where ω is a known constant. This restricted form in fact covers all the cases of practical interest here (see, e.g., table 3.1). For example, $a(\phi) = \phi/\omega$ allows the possibility of unequal variances in models based on the normal distribution, but in most cases ω is simply 1. Hence we now have

$$\text{var}(Y) = b''(\theta)\phi/\omega. \tag{3.2}$$

In subsequent sections it is convenient to write $\text{var}(Y)$ as a function of $\mu \equiv \mathbb{E}(Y)$, and, since μ and θ are linked via (3.1), we can always define a function $V(\mu) = b''(\theta)/\omega$, such that $\text{var}(Y) = V(\mu)\phi$. Several such functions are listed in table 3.1.

3.1.2 Fitting generalized linear models

Recall that a GLM models an n-vector of independent response variables, $\mathbf{Y}$, where $\boldsymbol{\mu} \equiv \mathbb{E}(\mathbf{Y})$, via

$$g(\mu_i) = \mathbf{X}_i\boldsymbol{\beta} \text{ and } Y_i \sim f_{\theta_i}(y_i),$$

where $f_{\theta_i}(y_i)$ indicates an exponential family distribution, with canonical parameter θ_i, which is determined by μ_i (via equation 3.1) and hence ultimately by $\boldsymbol{\beta}$. Given vector $\mathbf{y}$, an observation of $\mathbf{Y}$, maximum likelihood estimation of $\boldsymbol{\beta}$ is possible. Since the Y_i are mutually independent, the likelihood of $\boldsymbol{\beta}$ is

$$L(\boldsymbol{\beta}) = \prod_{i=1}^{n} f_{\theta_i}(y_i),$$

and hence the log-likelihood of $\boldsymbol{\beta}$ is

$$l(\boldsymbol{\beta}) = \sum_{i=1}^{n} \log\{f_{\theta_i}(y_i)\} = \sum_{i=1}^{n} \{y_i\theta_i - b_i(\theta_i)\}/a_i(\phi) + c_i(\phi, y_i),$$

where the dependence of the right hand side on $\boldsymbol{\beta}$ is through the dependence of the θ_i on $\boldsymbol{\beta}$. Notice that the functions a, b and c may vary with i. For example, this allows different binomial denominators, n_i, for each observation of a binomial response, or different (but known to within a constant) variances for normal responses. On the other hand, ϕ is assumed to be the same for all i. As discussed in the previous section, for practical work it suffices to consider only cases where we can write $a_i(\phi) = \phi/\omega_i$, where ω_i is a known constant (usually 1), in which case

$$l(\boldsymbol{\beta}) = \sum_{i=1}^{n} \omega_i\{y_i\theta_i - b_i(\theta_i)\}/\phi + c_i(\phi, y_i).$$

To maximize the log likelihood via Newton's method (see section 2.2.1) requires the gradient vector and Hessian matrix of l.

$$\frac{\partial l}{\partial \beta_j} = \frac{1}{\phi}\sum_{i=1}^{n} \omega_i \left(y_i \frac{\partial \theta_i}{\partial \beta_j} - b_i'(\theta_i)\frac{\partial \theta_i}{\partial \beta_j} \right),$$

and by the chain rule

$$\frac{\partial \theta_i}{\partial \beta_j} = \frac{\mathrm{d}\theta_i}{\mathrm{d}\mu_i}\frac{\partial \mu_i}{\partial \beta_j} = \frac{\mathrm{d}\theta_i}{\mathrm{d}\mu_i}\frac{\partial \eta_i}{\partial \beta_j}\frac{\mathrm{d}\mu_i}{\mathrm{d}\eta_i} = \frac{X_{ij}}{g'(\mu_i)b''(\theta_i)}.$$

The final equality follows from the facts that $\mathrm{d}\eta_i/\mathrm{d}\mu_i = g'(\mu_i)$ (so that $\mathrm{d}\mu_i/\mathrm{d}\eta_i = 1/g'(\mu_i)$), $\partial\eta_i/\partial\beta_j = X_{ij}$, and by differentiating (3.1), $\mathrm{d}\theta_i/\mathrm{d}\mu_i = 1/b_i''(\theta_i)$. Hence after some re-arrangement, followed by substitution from (3.1) and (3.2),

$$\frac{\partial l}{\partial \beta_j} = \frac{1}{\phi}\sum_{i=1}^{n} \frac{y_i - b_i'(\theta_i)}{g'(\mu_i)b_i''(\theta_i)/\omega_i} X_{ij} = \frac{1}{\phi}\sum_{i=1}^{n} \frac{y_i - \mu_i}{g'(\mu_i)V(\mu_i)} X_{ij}. \tag{3.3}$$

Differentiating again we get

$$\frac{\partial^2 l}{\partial \beta_j \partial \beta_k} = -\frac{1}{\phi}\sum_{i=1}^{n} \left\{ \frac{X_{ik}X_{ij}}{g'(\mu_i)^2 V(\mu_i)} + \frac{(y_i - \mu_i)V'(\mu_i)X_{ik}X_{ij}}{g'(\mu_i)^2 V(\mu_i)^2} \right.$$
$$\left. + \frac{(y_i - \mu_i)X_{ij}g''(\mu_i)X_{ik}}{g'(\mu_i)^3 V(\mu_i)} \right\} = -\frac{1}{\phi}\sum_{i=1}^{n} \frac{X_{ik}X_{ij}\alpha(\mu_i)}{g'(\mu_i)^2 V(\mu_i)},$$

where $\alpha(\mu_i) = 1 + (y_i - \mu_i)\{V'(\mu_i)/V(\mu_i) + g''(\mu_i)/g'(\mu_i)\}$. Notice that the expression for $\mathbb{E}(\partial^2 l/\partial\beta_j\partial\beta_k)$ is the same, but with $\alpha(\mu_i) = 1$. Hence defining $\mathbf{W} = \text{diag}(w_i)$ where $w_i = \alpha(\mu_i)/\{g'(\mu_i)^2 V(\mu_i)\}$, the Hessian of the log likelihood is $-\mathbf{XWX}/\phi$, while the expected Hessian is obtained by setting $\alpha(\mu_i) = 1$. The weights computed with $\alpha(\mu_i) = 1$ will be referred to as *Fisher weights.*

Defining $\mathbf{G} = \text{diag}\{g'(\mu_i)/\alpha(\mu_i)\}$, the log likelihood gradient vector can we written as $\mathbf{X}^\mathsf{T}\mathbf{WG}(\mathbf{y} - \boldsymbol{\mu})/\phi$. Then a single Newton update takes the form

$$\begin{aligned} \boldsymbol{\beta}^{[k+1]} &= \boldsymbol{\beta}^{[k]} + (\mathbf{XWX})^{-1}\mathbf{X}^\mathsf{T}\mathbf{WG}(\mathbf{y} - \boldsymbol{\mu}) \\ &= (\mathbf{XWX})^{-1}\mathbf{X}^\mathsf{T}\mathbf{W}\{\mathbf{G}(\mathbf{y} - \boldsymbol{\mu}) + \mathbf{X}\boldsymbol{\beta}^{[k]}\} \\ &= (\mathbf{XWX})^{-1}\mathbf{X}^\mathsf{T}\mathbf{Wz} \end{aligned}$$

where $z_i = g'(\mu_i)(y_i - \mu_i)/\alpha(\mu_i) + \eta_i$ (remember $\boldsymbol{\eta} \equiv \mathbf{X}\boldsymbol{\beta}$). The update has the same form whether we use a full Newton update, or a 'Fisher scoring' update in which the expected Hessian replaces the Hessian, so that we set $\alpha(\mu_i) = 1$. In either case the thing to notice is that the update equations are the least squares estimates of $\boldsymbol{\beta}$ resulting from minimising the weighted least squares objective

$$\sum_{i=1}^{n} w_i(z_i - \mathbf{X}_i\boldsymbol{\beta})^2.$$

In consequence GLMs can be estimated by the *iteratively re-weighted least square* (IRLS) algorithm, which is as follows.

1. Initialize $\hat{\mu}_i = y_i + \delta_i$ and $\hat{\eta}_i = g(\hat{\mu}_i)$, where δ_i is usually zero, but may be a small constant ensuring that $\hat{\eta}_i$ is finite. Iterate the following two steps to convergence.
2. Compute pseudodata $z_i = g'(\hat{\mu}_i)(y_i - \hat{\mu}_i)/\alpha(\hat{\mu}_i) + \hat{\eta}_i$, and iterative weights $w_i = \alpha(\hat{\mu}_i)/\{g'(\hat{\mu}_i)^2 V(\hat{\mu}_i)\}$.
3. Find $\hat{\boldsymbol{\beta}}$, the minimizer of the weighted least squares objective

$$\sum_{i=1}^{n} w_i(z_i - \mathbf{X}_i\boldsymbol{\beta})^2,$$

then update $\hat{\boldsymbol{\eta}} = \mathbf{X}\hat{\boldsymbol{\beta}}$ and $\hat{\mu}_i = g^{-1}(\hat{\eta}_i)$.

Convergence can be based on monitoring the change in deviance (a scaled form of the negative log likelihood; see below) from iterate to iterate, stopping when it is near zero, or by testing whether the log likelihood derivatives are close enough to zero. If divergence occurs at any step then $\boldsymbol{\beta}^{[k+1]}$ can be successively moved back towards $\boldsymbol{\beta}^{[k]}$ (usually by 'step halving') until the deviance decreases.

Notice that the algorithm is the same whether we choose to use the full Newton or Fisher scoring variants: all that changes is $\alpha(\mu_i)$. However, the full Newton method can occasionally generate negative weights, w_i, which complicates fitting. If all weights are non negative, then the working model fitting problem is solved by defining $\sqrt{\mathbf{W}} = \text{diag}(\sqrt{w_i})$, and applying the orthogonal fitting methods discussed in section 1.3.1 (p. 12) to the objective $\|\sqrt{\mathbf{W}}\mathbf{z} - \sqrt{\mathbf{W}}\mathbf{X}\|^2$, i.e., to the objective of an ordinary linear model fitting problem with response data $\sqrt{\mathbf{W}}\mathbf{z}$ and model matrix $\sqrt{\mathbf{W}}\mathbf{X}$. Obviously this can not work if any w_i are negative. In that case we are forced either to use the less stable approach of solving $\mathbf{X}^\mathsf{T}\mathbf{W}\mathbf{X}\boldsymbol{\beta} = \mathbf{X}^\mathsf{T}\mathbf{W}\mathbf{z}$ directly, or to use the orthogonal scheme with correction covered in section 6.1.3 (p. 252).

3.1.3 Large sample distribution of $\hat{\boldsymbol{\beta}}$

Distributional results for GLMs are not exact, but are based instead on large sample approximations, making use of general properties of maximum likelihood estimators (see appendix A or section 2.2, p. 74). From the general properties of maximum likelihood estimators we have that, in the large sample limit,

$$\hat{\boldsymbol{\beta}} \sim N(\boldsymbol{\beta}, \boldsymbol{\mathcal{I}}^{-1}) \text{ or } \hat{\boldsymbol{\beta}} \sim N(\boldsymbol{\beta}, \hat{\boldsymbol{\mathcal{I}}}^{-1})$$

where the *information* matrix $\mathcal{I} = \mathbb{E}(\hat{\mathcal{I}})$ and $\hat{\mathcal{I}}$ is the Hessian of the negative log likelihood (see section A.2). We saw in the previous section that $\hat{\mathcal{I}} = \mathbf{X}^\mathsf{T}\mathbf{W}\mathbf{X}/\phi$ (or alternatively $\mathcal{I}$, if $\mathbf{W}$ contains the Fisher weights). Hence in the large sample limit

$$\hat{\boldsymbol{\beta}} \sim N(\boldsymbol{\beta}, (\mathbf{X}^\mathsf{T}\mathbf{W}\mathbf{X})^{-1}\phi).$$

For distributions with known scale parameter, ϕ, this result can be used directly to find confidence intervals for the parameters, but if the scale parameter is unknown (e.g., for the normal distribution), then it must be estimated (see section 3.1.5), and intervals must be based on an appropriate t distribution.

3.1.4 Comparing models

Consider testing

$$\text{H}_0 : \mathbf{g}(\boldsymbol{\mu}) = \mathbf{X}_0\boldsymbol{\beta}_0$$

against

$$\text{H}_1 : \mathbf{g}(\boldsymbol{\mu}) = \mathbf{X}_1\boldsymbol{\beta}_1,$$

where $\boldsymbol{\mu}$ is the expectation of a response vector, $\mathbf{Y}$, whose elements are independent random variables from the same member of the exponential family of distributions, and where $\mathbf{X}_0 \subset \mathbf{X}_1$. If we have an observation, $\mathbf{y}$, of the response vector, then a generalized likelihood ratio test can be performed. Let $l(\hat{\boldsymbol{\beta}}_0)$ and $l(\hat{\boldsymbol{\beta}}_1)$ be the maximized log likelihoods of the two models. If H_0 is true then in the large sample limit,

$$2\{l(\hat{\boldsymbol{\beta}}_1) - l(\hat{\boldsymbol{\beta}}_0)\} \sim \chi^2_{p_1 - p_0}, \tag{3.4}$$

where p_i is the number of (identifiable) parameters, $\boldsymbol{\beta}_i$, in model i (see sections A.5 and A.6 for derivation of this result). If the null hypothesis is false then model 1 will tend to have a substantially higher likelihood than model 0, so that twice the difference in log likelihoods would be too large for consistency with the relevant χ^2 distribution.

The approximate result (3.4) is only directly useful if the log likelihoods of the models concerned can be calculated. In the case of GLMs estimated by IRLS, this is only the case if the scale parameter, ϕ, is known. Hence the result can be used directly with Poisson and binomial models, but not with the normal,† gamma or inverse Gaussian distributions, where the scale parameter is not known. What to do in these latter cases will be discussed shortly.

Deviance

When working with GLMs in practice, it is useful to have a quantity that can be interpreted in a similar way to the residual sum of squares, in ordinary linear modelling. This quantity is the *deviance* of the model and is defined as

$$D = 2\{l(\hat{\boldsymbol{\beta}}_{\max}) - l(\hat{\boldsymbol{\beta}})\}\phi \tag{3.5}$$

$$= \sum_{i=1}^{n} 2\omega_i \left\{ y_i(\tilde{\theta}_i - \hat{\theta}_i) - b(\tilde{\theta}_i) + b(\hat{\theta}_i) \right\}, \tag{3.6}$$

†Of course, for the normal distribution with the identity link we use the results of chapter 1.

where $l(\hat{\boldsymbol{\beta}}_{\max})$ indicates the maximized likelihood of the saturated model: the model with one parameter per data point. $l(\hat{\boldsymbol{\beta}}_{\max})$ is the highest value that the log likelihood could possibly have, given the data. For exponential family distributions it is computed by simply setting $\hat{\boldsymbol{\mu}} = \mathbf{y}$ and evaluating the likelihood. $\tilde{\boldsymbol{\theta}}$ and $\hat{\boldsymbol{\theta}}$ denote the maximum likelihood estimates of the canonical parameters for the saturated model and model of interest, respectively. Notice how the deviance is defined to be independent of ϕ. Table 3.1 lists the contributions of a single datum to the deviance, for several distributions — these are the terms inside the summation in the definition of the deviance.

Related to the deviance is the *scaled deviance*,

$$D^* = D/\phi,$$

which does depend on the scale parameter. For the binomial and Poisson distributions, where $\phi = 1$, the deviance and scaled deviance are the same.

By the generalized likelihood ratio test result (3.4) we might expect that, if the model is correct, then approximately

$$D^* \sim \chi^2_{n-p}, \tag{3.7}$$

in the large sample limit. Actually such an argument is bogus, since the limiting argument justifying (3.4) relies on the number of parameters in the model staying fixed, while the sample size tends to infinity, but the saturated model has as many parameters as data. Asymptotic results *are* available for some of the distributions in table 3.1, to justify (3.7) as a large sample approximation under many circumstances (see McCullagh and Nelder, 1989), and it is exact for the normal case. However, it breaks down entirely for the binomial with binary data.

Given the definition of deviance, it is easy to see that the likelihood ratio test, with which this section started, can be performed by re-expressing the twice log-likelihood ratio statistic as $D_0^* - D_1^*$. Then under H_0, in the large sample limit,

$$D_0^* - D_1^* \sim \chi^2_{p_1 - p_0}, \tag{3.8}$$

where D_i^* is the deviance of model i with p_i identifiable parameters. But again, this is only useful if the scale parameter is known so that D^* can be calculated.

Model comparison with unknown ϕ

Under H_0 we have the approximate results

$$D_0^* - D_1^* \sim \chi^2_{p_1 - p_0} \text{ and } D_1^* \sim \chi^2_{n-p},$$

and, if $D_0^* - D_1^*$ and D_1^* are treated as asymptotically independent, this implies that

$$F = \frac{(D_0^* - D_1^*)/(p_1 - p_0)}{D_1^*/(n - p_1)} \sim F_{p_1 - p_0, n - p_1},$$

in the large sample limit (a result which is exactly true in the ordinary linear model special case, of course). The useful property of F is that it can be calculated without knowing ϕ, which can be cancelled from top and bottom of the ratio yielding, under H_0, the approximate result that

$$F = \frac{(D_0 - D_1)/(p_1 - p_0)}{D_1/(n - p_1)} \sim F_{p_1 - p_0, n - p_1}. \tag{3.9}$$

This can be used for hypothesis testing based model comparison when ϕ is unknown. The disadvantage is the dubious distributional assumption for D_1^* on which it is based.

An obvious alternative would be to use an estimate, $\hat{\phi}$, to obtain an estimate, $\hat{D}_i^* = D_i/\hat{\phi}$, for each model, and then to use (3.8) for hypothesis testing. However if we use the estimate (3.10) for this purpose then it is readily seen that $\hat{D}_0^* - \hat{D}_1^*$ is simply $(p_1 - p_0) \times F$, so our test would be exactly equivalent to using the F-ratio result (3.9), but with $F_{p_1 - p_0, \infty}$ as the reference distribution. Clearly direct use of (3.9) is a more conservative approach, and hence usually to be preferred: it at least makes some allowance for the uncertainty in estimating the scale parameter.

AIC

Akaike's information criterion (AIC) is an approach to model selection in which models are selected to minimize an estimate of the expected Kullback-Leibler divergence between the fitted model and the 'true model'. The criterion is

$$\text{AIC} = -2l + 2p,$$

where l is the maximized log likelihood of the model and p the number of model parameters that have had to be estimated. The model with the lowest AIC is selected. Models under comparison need not be nested, although some neglected terms in the approximation of the expected K-L divergence will cancel if they are. A simple derivation is provided in section A.7.

In some quarters AIC has acquired cult status as the one true way to model selection, in a similar way to p-values, somewhat earlier. This is unfortunate nonsense. Like p-values, AIC is useful as a rough and ready quantitative aid to selecting between models, nothing more.

3.1.5 Estimating ϕ, Pearson's statistic and Fletcher's estimator

As we have seen, the MLEs of the parameters, $\boldsymbol{\beta}$, can be obtained without knowing the scale parameter, ϕ, but, in those cases in which this parameter is unknown, it must usually be estimated. Approximate result (3.7) provides one obvious estimator. The expected value of a χ^2_{n-p} random variable is $n - p$, so equating the observed $D^* = D/\phi$ to its approximate expected value we have

$$\hat{\phi}_D = \hat{D}/(n - p). \tag{3.10}$$

A second estimator is based on the *Pearson statistic*, which is defined as

$$X^2 = \sum_{i=1}^{n} \frac{(y_i - \hat{\mu}_i)^2}{V(\hat{\mu}_i)}.$$

Clearly X^2/ϕ would be the sum of squares of a set of zero mean, unit variance, random variables, having $n-p$ degrees of freedom, suggesting ‡ that if the model is adequate then approximately $X^2/\phi \sim \chi^2_{n-p}$: this approximation turns out to be well founded. Setting the observed Pearson statistic to its expected value we get

$$\hat{\phi}_P = \hat{X}^2/(n-p).$$

Note that it is straightforward to show that

$$X^2 = \|\sqrt{\mathbf{W}}(\mathbf{z} - \mathbf{X}\hat{\boldsymbol{\beta}})\|^2,$$

where $\mathbf{W}$ and $\mathbf{z}$ are the Fisher weights and pseudodata, evaluated at convergence.

Unfortunately $\hat{\phi}_D$ tends to be negatively biased for count data with a low mean. The Pearson estimator, $\hat{\phi}_P$, is less biased, but can be disastrously unstable for low count data. For example, with Poisson data, where $V(\mu_i) = \mu_i$, then a non zero count when μ_i is very small makes an enormous contribution to $\hat{\phi}_P$. Fletcher (2012) proposes an alternative estimator which largely overcomes these problems:

$$\hat{\phi} = \frac{\hat{\phi}_P}{1+\bar{s}}, \tag{3.11}$$

where $\bar{s} = n^{-1}\sum_{i=1}^n V'(\hat{\mu}_i)(y_i - \hat{\mu}_i)/V(\hat{\mu}_i)$. This is especially useful for quasi-likelihood models for count data (see section 3.1.8), which are often used for over-dispersed data and where the problems with the $\hat{\phi}_P$ and $\hat{\phi}_D$ can become extreme. At the time of writing the `glm` function in R uses $\hat{\phi}_P$, while the `gam` function from package `mgcv` uses $\hat{\phi}$.

3.1.6 Canonical link functions

The canonical link, g_c, for a distribution, is the link function such that $g_c(\mu_i) = \theta_i$, where θ_i is the canonical parameter of the distribution. For example, for the Poisson distribution the canonical link is the log function (see table 3.1 for other examples). Use of the canonical link means that $\theta_i = \mathbf{X}_i\boldsymbol{\beta}$ (where $\mathbf{X}_i$ is the ith row of $\mathbf{X}$). Using the canonical link also causes Fisher scoring and the full Newton methods of fitting to coincide, since it causes $\alpha(\mu_i) = 1$. Obviously this also means that the Hessian of the log likelihood is equal to its expected value.

Canonical links tend to have some nice properties, such as ensuring that μ stays within the range of the response variable, but they also have a more subtle advantage which is derived here. Recall that likelihood maximization involves differentiating the log likelihood with respect to each β_j, and setting the results to zero, to obtain the system of equations

$$\frac{\partial l}{\partial \beta_j} = \sum_{i=1}^n \omega_i \left(y_i \frac{\partial \theta_i}{\partial \beta_j} - \mu_i \frac{\partial \theta_i}{\partial \beta_j} \right) = 0 \;\forall\; j.$$

‡Recall that if $\{Z_i : i = 1 \ldots n\}$ are a set of i.i.d. $N(0,1)$ r.v.'s then $\sum Z_i^2 \sim \chi^2_n$.

But if the canonical link is being used then $\partial\theta_i/\partial\beta_j = X_{ij}$, and if, as is often the case, $\omega_i = 1 \; \forall \; i$ this system of equations reduces to

$$\mathbf{X}^\mathsf{T}\mathbf{y} - \mathbf{X}^\mathsf{T}\hat{\boldsymbol{\mu}} = \mathbf{0},$$

i.e., to $\mathbf{X}^\mathsf{T}\mathbf{y} = \mathbf{X}^\mathsf{T}\hat{\boldsymbol{\mu}}$. Now consider the case in which $\mathbf{X}$ contains a column of 1's: this implies that one of the equations in this system is simply $\sum_i y_i = \sum_i \hat{\mu}_i$. Similarly any other weighted summation, where the weights are given by model matrix columns (or a linear combination of these), is conserved between the raw data and the fitted values.

One practical upshot of this is that the residuals will sum to zero, for any GLM with an intercept term and canonical link: this 'observed unbiasedness' is a reassuring property. Another practical use of the result is in categorical data analysis using log-linear models, where it provides a means of ensuring, via specification of the model, that totals which were built into the design of a study can be preserved in any model: see section 3.3.4.

3.1.7 Residuals

Model checking is perhaps the most important part of applied statistical modelling. In the case of ordinary linear models it is based on examination of the model residuals, which contain all the information in the data not explained by the systematic part of the model. Examination of residuals is also the chief means for model checking in the case of GLMs, but in this case the standardization of residuals is both necessary and a little more difficult.

For GLMs the main reason for not simply examining the raw residuals, $\hat{\epsilon}_i = y_i - \hat{\mu}_i$, is the difficulty of checking the validity of the assumed mean variance relationship from the raw residuals. For example, if a Poisson model is employed, then the variance of the residuals should increase in direct proportion to the size of the fitted values, $\hat{\mu}_i$. However, if raw residuals are plotted against fitted values it takes an extraordinary ability to judge whether the residual variability is increasing in proportion to the mean, as opposed to, say, the square root or square of the mean. For this reason it is usual to standardize GLM residuals so that, if the model assumptions are correct, the standardized residuals should have approximately equal variance and behave, as far as possible, like residuals from an ordinary linear model.

Pearson residuals

The most obvious way to standardize the residuals is to divide them by a quantity proportional to their standard deviation according to the fitted model. This gives rise to the *Pearson residuals*

$$\hat{\epsilon}_i^p = \frac{y_i - \hat{\mu}_i}{\sqrt{V(\hat{\mu}_i)}},$$

which should approximately have zero mean and variance ϕ, if the model is correct. These residuals should not display any trend in mean or variance when plotted against the fitted values, or any covariates (whether included in the model or not). The name

'Pearson residuals' relates to the fact that the sum of squares of the Pearson residuals gives the Pearson statistic discussed in section 3.1.5.

Note that the Pearson residuals are the residuals of the working linear model from the converged IRLS method, divided by the square roots of the converged IRLS weights.

Deviance residuals

In practice the distribution of the Pearson residuals can be quite asymmetric around zero, so that their behaviour is not as close to ordinary linear model residuals as might be hoped for. The *deviance residuals* are often preferable in this respect. The deviance residuals are arrived at by noting that the deviance plays much the same role for GLMs that the residual sum of squares plays for ordinary linear models: indeed for an ordinary linear model the deviance *is* the residual sum of squares. In the ordinary linear model case, the deviance is made up of the sum of the squared residuals. That is, the residuals are the square roots of the components of the deviance with the appropriate sign attached.

So, writing d_i as the component of the deviance contributed by the ith datum (i.e. the ith term in the summation in (3.6)) we have

$$D = \sum_{i=1}^{n} d_i$$

and, by analogy with the ordinary linear model, we can define

$$\hat{\epsilon}_i^d = \text{sign}(y_i - \hat{\mu}_i)\sqrt{d_i}.$$

As required the sum of squares of these 'deviance residuals' gives the deviance itself.

Now if the deviance were calculated for a model where all the parameters were known, then (3.7) would become $D^* \sim \chi^2_n$, and this might suggest that for a single datum $d_i/\phi \sim \chi^2_1$, implying that $\epsilon_i^d \sim N(0, \phi)$. Of course (3.7) can not reasonably be applied to a single datum, but nonetheless it suggests that we might expect the deviance residuals to behave something like $N(0, \phi)$ random variables, for a well fitting model, especially in cases for which (3.7) is expected to be a reasonable approximation.

3.1.8 Quasi-likelihood

The treatment of GLMs has so far assumed that the distribution of the response variable is a known member of the exponential family. If there is a good reason to suppose that the response follows a particular distribution then it is appealing to base models on that distribution, but in many cases the nature of the response distribution is not known so precisely, and it is only possible to specify what the relationship between the variance of the response and its mean should be. That is, the function $V(\mu)$ can be specified, but little more. The question then arises of whether it is possible to develop theory for fitting and inference with GLMs, starting from the position of specifying only the mean-variance relationship.

It turns out that it is possible to develop satisfactory methods, based on the notion of **quasi-likelihood**. The fact that it is possible implies that getting the distribution exactly right in generalized linear modelling is rather unimportant, provided that the mean-variance relationship is appropriate.

Consider an observation, y_i, of a random variable with mean μ_i and known variance function, $V(\mu_i)$. Then the log quasi-likelihood for μ_i, given y_i, is defined to be

$$q_i(\mu_i) = \int_{y_i}^{\mu_i} \frac{y_i - z}{\phi V(z)} dz. \tag{3.12}$$

As we will see, the key feature of this function is that it shares many useful properties of l_i, the log-likelihood corresponding to a single observation, but only requires knowledge of V rather than the full distribution of Y_i. Provided that the data are observations of independent random variables, we can define a log quasi-likelihood for the mean vector, $\boldsymbol{\mu}$, of all the response data, or any parameter vector defining $\boldsymbol{\mu}$ as

$$q(\boldsymbol{\mu}) = \sum_{i=1}^{n} q_i(\mu_i).$$

The key property of q is that, for the purposes of inference with GLMs, it behaves in a very similar manner to the log-likelihood, but only requires knowledge of the variance function in order to define it.

Consider, for example, obtaining maximum quasi-likelihood parameter estimates of the GLM parameters $\boldsymbol{\beta}$. Differentiating q w.r.t. β_j yields

$$\frac{\partial q}{\partial \beta_j} = \sum_{i=1}^{n} \frac{y_i - \mu_i}{\phi V(\mu_i)} \frac{\partial \mu_i}{\partial \beta_j},$$

so that the parameter estimates are solutions to the equations

$$\sum_{i=1}^{n} \frac{(y_i - \mu_i)}{V(\mu_i) g'(\mu_i)} X_{ij} = 0 \;\; \forall \;\; j, \tag{3.13}$$

but this is exactly the system that results when (3.3) is equated to zero: that is the system that must be solved to find the MLE for a GLM. Hence the maximum quasi-likelihood parameter estimates can be found by the usual GLM IRLS method, which in any case only requires knowledge of $V(\mu)$.

Furthermore, the log quasi-likelihood shares just enough properties with the log-likelihood that the results on the large sample distribution of the parameter estimators, given in section 3.1.3, also hold for the maximum quasi-likelihood estimators of the parameters. Similarly, the large sample distributional results of section 3.1.4, underpinning hypothesis testing with GLMs, hold when the log-likelihood, l, is replaced by the log quasi-likelihood, q. The theoretical basis for these assertions is provided in section A.8.

Note that the log quasi-likelihood of the saturated model is always zero, so the quasi-deviance of a GLM is simply

$$D_q = -2q(\hat{\boldsymbol{\mu}})\phi.$$

Obviously the discussion of residuals and scale parameter estimation also carries over from the likelihood to the quasi-likelihood case, again with no more than the replacement of l by q.

The practical use of the quasi-likelihood approach requires that the integral in (3.12) be evaluated, but this is possible for most practically useful forms of V: see McCullagh and Nelder (1989). `quasi(variance="mu^3")$dev.resids` or similar can be typed in R, to access the form of q_i for any particular mean-variance relationship there implemented. For mean-variance relationships corresponding to an exponential family distribution from table 3.1, the form of the quasi-deviance corresponds exactly to the form of the deviance for that family.

One major practical use of quasi-likelihood is to provide a means of modelling count data that are more variable than the Poisson or binomial distributions (with their fixed scale parameters) predict: the quasi-likelihood approach assumes that ϕ is unknown. Such 'over-dispersed' data are common in practice. Another practical use is to provide a means of modelling data with a mean-variance relationship for which there is no convenient exponential family distribution: for example continuous data for which the variance is expected to be proportional to the mean cubed.

3.1.9 Tweedie and negative binomial distributions

The Tweedie and negative binomial distributions are practically useful and are each exponential family distributions given a single extra parameter. The Tweedie distribution (Tweedie, 1984) has a scale parameter, ϕ, and a variance function $V(\mu) = \mu^p$, where $\mu = \mathbb{E}(y)$ and p is a parameter outside the interval $(0, 1)$. For $p > 1$, $\mu > 0$. Here consider only the most tractable and practically interesting case where $1 < p < 2$. In this case a Tweedie random variable, Y, can be characterized as a sum of $N \sim \text{Poi}(\lambda)$ gamma random variables where $\lambda = \mu^{2-p}/\{\phi(2-p)\}$. Y is non-negative and $\Pr(Y = 0) = e^{-\lambda}$. Otherwise the density of Y is

$$f(y) = a(y, \phi, p) \exp[\mu^{1-p}\{y/(1-p) - \mu/(2-p)\}/\phi],$$

where the normalizing constant is given by

$$a(y, \phi, p) = \frac{1}{y} \sum_{j=1}^{\infty} W_j$$

and, defining $\alpha = (2-p)/(1-p)$,

$$\log W_j = j\{\alpha \log(p-1) - \log(\phi)/(p-1) - \log(2-p)\} \\ - \log\Gamma(j+1) - \log\Gamma(-j\alpha) - j\alpha \log y.$$

The sum is interesting in that the early terms are near zero, as are the later terms, so that it has to be summed from the middle. Dunn and Smyth (2005) give the details, showing that the series maximum is around $j_{\max} = y^{2-p}/\{\phi(2-p)\}$. The density tends to a Poisson probability function as $p \to 1$ and $\phi \to 1$. As $p \to 2$ it tends to a gamma density. Since the normalizing constant cancels from the deviance, we can fit

Tweedie models with known p using standard GLM methods. The constant only has to be evaluated if we need to calculate AIC, or estimate p itself by MLE.

Tweedie densities are often a good option in circumstances where we might otherwise have used quasi-likelihood. R package `mgcv` allows p to be estimated routinely as part of model fitting, using the `tw` family with `gam`.

Negative binomial random variables are discrete and non-negative, and are often used to model count data that are over-dispersed relative to Poisson. Various parameterizations are used, but for GLMs and GAMs the usual one involves extra parameter θ such that $V(\mu) = \mu + \mu^2/\theta$. The scale parameter $\phi = 1$. Likelihood based estimation of θ is possible. An alternative is to choose θ so that $\hat{\phi} \simeq 1$. The name arises because the number of failures before a target number of successes in a sequence of binary trials has a negative binomial distribution. A more useful characterization, in the GLM context of modelling over-dispersion, is that a negative binomial random variable arises from a mixture of Poisson distributions, having gamma distributed means. Again `gam` in R package `mgcv` can accept family `nb` and estimate θ as part of model estimation.

3.1.10 *The Cox proportional hazards model for survival data*

Suppose that we observe event time data along with covariates which may help to predict the event times. Examples are times of death or relapse of patients undergoing treatment, times to failure of machinery or time to see the next whale on a marine survey. A popular model for such data is the Cox (1972) proportional hazards model. The hazard function, $h(t)$, is defined as the probability per unit time of the event occurring at t, given that it occurs at t or later (i.e., given that it has not yet occurred). It is sometimes known as the instantaneous per capita death rate.

The Cox model is concerned with modelling the effects of the covariates on the hazard, rather than in the hazard itself. For example, does a drug treatment increase or decrease the hazard for patients with a particular disease? The model for the hazard function for the ith subject is

$$h_i(t) = h_0(t) \exp\{\mathbf{X}_i(t)\boldsymbol{\beta}\}$$

where $\mathbf{X}_i(t)$ is a vector of covariates, which may be time dependent, $\boldsymbol{\beta}$ is a parameter vector, and $h_0(t)$ is a common *baseline hazard*. So the idea is that the covariates for each subject modify the baseline hazard, to give the subject-specific hazard.

Cox (1972) treats $h_0(t)$ as a nuisance parameter, and rather than estimate it, he considers how to estimate $\boldsymbol{\beta}$ when $h_0(t)$ is an arbitrary non-negative function. If $h_0(t)$ is really arbitrary then it could be zero between the times at which events actually occurred, in which case these intervals could contain no information about $\boldsymbol{\beta}$. So any procedure which successfully used information between the event times for inference about $\boldsymbol{\beta}$ would necessarily be making the assumption that h_0 was non-zero over these intervals, and would therefore contradict the requirement to allow $h_0(t)$ to be arbitrary. Hence we only have information about $\boldsymbol{\beta}$ at the times at which events actually occurred. In that case our data simply consist of observations of which subject experienced the event at each time, out of those 'at risk' of experiencing it at that time.

Suppose that event times are all distinct, and that the event for the ith subject occurred at time t_i. Let $R(i)$ denote the set of all subjects at risk at time t_i. The model probability that the ith subject was the one with the event at t_i is simply

$$\exp\{\mathbf{X}_i(t_i)\boldsymbol{\beta}\}/\sum_{j\in R(i)}\exp\{\mathbf{X}_j(t_i)\boldsymbol{\beta}\}$$

(h_0 terms having cancelled). Hence the likelihood is

$$\prod_i \exp\{\mathbf{X}_i(t_i)\boldsymbol{\beta}\}/\sum_{j\in R(i)}\exp\{\mathbf{X}_j(t_i)\boldsymbol{\beta}\},$$

where the product is taken over the events. Note that there is nothing to prevent subjects leaving the study without their event occurring. Such subjects are known as *right censored*. They contribute to the summation over $R(i)$ in the denominator of the likelihood terms, but not to the numerator. *Left censoring*, in which we only know that a subject's event occurred before some specified time, can not be dealt with so easily (in that case we really have to consider the full likelihood of $\boldsymbol{\beta}$ and $h_0(t)$). Note that $\mathbf{X}_j(t)\boldsymbol{\beta}$ does not include an intercept term, as this would be completely confounded with h_0 and would cancel from the likelihood. The likelihood is referred to as a *partial likelihood*, because of the decision not to estimate h_0.

The Cox proportional hazard model can be estimated as a Poisson GLM, following Whitehead (1980). At each event time (not censoring times), t_i, an artificial response variable, y_{ij}, is produced for every subject still at risk (i.e., who has not yet dropped out, nor experienced their event). The response is set to 1 if the subject's event occurs at this time, and zero otherwise (there will be several 1's if several subjects have the same event time). Corresponding to each y_{ij} value is the corresponding $\mathbf{X}_j(t_i)$ for the subject. The generalized linear model

$$y_{ij} \sim \text{Poi}(\mu_{ij}), \quad \log(\mu_{ij}) = \alpha_i + \mathbf{X}_j(t_i)\boldsymbol{\beta}$$

is then fitted to the data, where α_i is an intercept term for the ith time. Notice that if the original data consists of a single event time and a fixed $\mathbf{X}_j$ for each subject then the production of artificial data multiplies the size of the model fitting problem by a factor of $O(n)$. This is not an efficient way to proceed in this fixed covariate setting.

The fits are equivalent because the Poisson likelihood is identical to the partial likelihood up to a constant of proportionality. To see this, again assume that events occur at distinct times, and suppose that $\mathbf{X}_i(t_i)$ denotes the model matrix row corresponding to the event at time t_i while $R(i)$ denotes the set of (indices of) subjects at risk at t_i. Since the Poisson probability function is $\mu^y\exp(-\mu)/y!$ and y is either 0 or 1 for the artifical data, the contribution to the Poisson likelihood from the artificial data for t_i is

$$L_i = \frac{\exp\{\alpha_i + \mathbf{X}_i(t_i)\boldsymbol{\beta}\}}{\exp[\sum_{j\in R(i)}\exp\{\alpha_i + \mathbf{X}_j(t_i)\boldsymbol{\beta}\}]}.$$

Now the log link is canonical for the Poisson distribution, so because of the inclusion of the α_i coefficients, the results of section 3.1.6 imply that the sum of the model expected values for the artificial data at t_i must equal the sum of the artificial response data. That is,

$$\sum_{j \in R(i)} \hat{\mu}_{ij} = \sum_{j \in R(i)} \exp\{\hat{\alpha}_i + \mathbf{X}_j(t_i)\boldsymbol{\beta}\} = 1.$$

This immediately implies that the denominator of L_i is e, while

$$\exp(\hat{\alpha}_i) = \left[\sum_{j \in R(i)} \exp\{\mathbf{X}_j(t_i)\boldsymbol{\beta}\} \right]^{-1}. \tag{3.14}$$

Hence the $\hat{\alpha}_i$ can be profiled out of the likelihood, to obtain the profile likelihood

$$L(\boldsymbol{\beta}) = \prod_i \frac{\exp\{\mathbf{X}_i(t_i)\boldsymbol{\beta}\}}{e \sum_{j \in R(i)} \exp\{\mathbf{X}_j(t_i)\boldsymbol{\beta}\}}.$$

Since this is the same as the partial likelihood to within a multiplicative constant, we will make identical inferences about $\boldsymbol{\beta}$ whether we use the Poisson model on the artificial data or the original partial likelihood. Whitehead (1980) also shows that if the artificial data are generated as described above when several events occur at the same time, then this is equivalent to using the Peto (1972) correction for ties with the partial likelihood.

Cumulative hazard and survival functions

A subject's probability of surviving until time t without its event occurring is given by the *survival function* (also know as the 'survivor function')

$$S(t) = \exp\left\{ -\int_0^t h(x)dx \right\}$$

and $H(t) = \int_0^t h(x)dx$ is known as the *cumulative hazard function* (so $S(t) = \exp\{-H(t)\}$). If t_j are the event times, then an estimator of the baseline cumulative hazard is given by

$$\hat{H}_0(t) = \sum_{t_j \le t} \exp(\hat{\alpha}_j),$$

which given (3.14), and its equivalent for ties, is the Breslow (1972) estimate of the baseline cumulative hazard. Hence we have the interpretation

$$\exp(\alpha_j) = \int_{t_{j-1}^+}^{t_j} h_0(x)dx,$$

which might also be arrived at from first principles. Now consider the cumulative hazard function for a particular subject, with model matrix row $\mathbf{X}_i(t)$. If we make the approximation that $\mathbf{X}_i(t)$ is constant over $(t_{j-1}, t_j]$ then

$$\int_{t_{j-1}^+}^{t_j} h(x)dx = \int_{t_{j-1}^+}^{t_j} h_0(x)\exp\{\mathbf{X}_i(t_j)\boldsymbol{\beta}\}dx = \exp\{\alpha_j + \mathbf{X}_i(t_j)\boldsymbol{\beta}\}$$

and the estimated cumulative hazard for this subject is

$$\hat{H}(t) = \sum_{t_j \le t} \exp\{\hat{\alpha}_j + \mathbf{X}_i(t_j)\hat{\boldsymbol{\beta}}\}.$$

Notice how this is easily obtained by prediction from the fitted Poisson GLM.

For the fixed covariate case in which $\mathbf{X}_i(t)$ is independent of t, then standard expressions for computing the uncertainty of $\hat{H}$ and $\hat{S}$ are given in section 6.6.2, but if we have the Poisson fit available then we can use that instead, thereby also dealing with the time varying covariate case. Let $\boldsymbol{\Sigma}$ denote the matrix with all unit elements apart from above the leading diagonal, where they are all zero. Let $\tilde{\mathbf{X}}$ denote the matrix mapping $\boldsymbol{\alpha}$ and $\boldsymbol{\beta}$ to the Poisson model linear predictor at each event time for subject i: $\hat{\boldsymbol{\eta}} = \tilde{\mathbf{X}} \begin{bmatrix} \hat{\boldsymbol{\alpha}} \\ \hat{\boldsymbol{\beta}} \end{bmatrix}$. Then $\hat{\mathbf{H}} = \boldsymbol{\Sigma} \exp(\hat{\boldsymbol{\eta}})$ is the vector of $\hat{H}(t)$ values for each inter-event interval, and the Jacobian of $\hat{\mathbf{H}}$ w.r.t. the model coefficients is $\mathbf{J} = \boldsymbol{\Sigma}\text{diag}\{\exp(\hat{\boldsymbol{\eta}})\}\tilde{\mathbf{X}}$. So the approximate covariance matrix for $\hat{\mathbf{H}}$ is $\mathbf{J}\mathbf{V}\mathbf{J}^\mathsf{T}$ where $\mathbf{V}$ is the covariance matrix of the coefficients of the Poisson model. Hence we can produce approximate confidence intervals for $H(t)$ and $S(t)$.

The residuals of the Poisson model are very difficult to interpret, but more useful residuals can readily be computed, as described in Klein and Moeschberger (2003). If $\hat{H}_i$ is the estimated cumulative hazard for the ith subject at its event or censoring time, and d_i is one for an event and zero for censoring, then the Martingale residual for that subject is defined as

$$M_i = d_i - \hat{H}_i$$

where $\hat{H}_i$ is readily computed from the Poisson model by summing up the fitted values corresponding to the ith subject. Deviance residuals are then defined as

$$r_i = \text{sign}(M_i)\sqrt{-2\{M_i + d_i \log(\hat{H}_i)\}}.$$

Section 3.3.3 provides a simple example.

3.2 Geometry of GLMs

The geometry of GLMs and GLM fitting is less straightforward than the geometry of ordinary linear models, since the likelihood used to judge model fit does not generally mean that the fit can be judged by Euclidian distance between model and data. Figure 3.1 illustrates the geometry of GLMs, using the example of the fit to 3 data of a 2 parameter GLM with a gamma distribution and a log link. The flat model subspace of section 1.4 (p. 19) is now replaced by a curved 'model manifold', consisting of all the possible fitted value vectors predictable by the model. Since Euclidean distance between model manifold and data is no longer the measure of fit being used, then different means must be employed to illustrate the geometry of estimation. The black lines, in the right panel of figure 3.1, show all the combinations of the response variables, which give rise to the same estimated model. Notice how these lines are not generally parallel, and are not generally orthogonal to the model manifold.

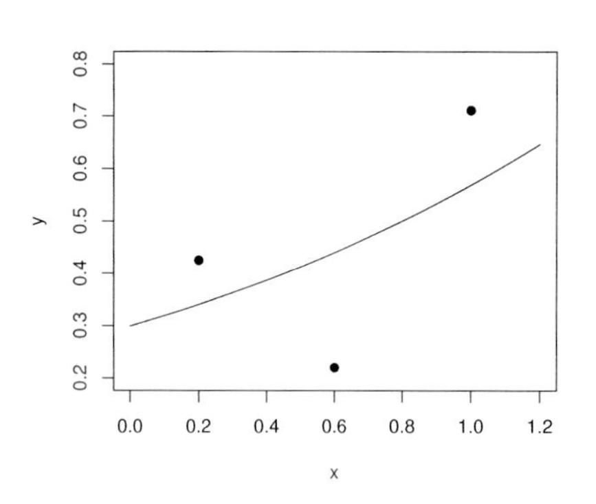

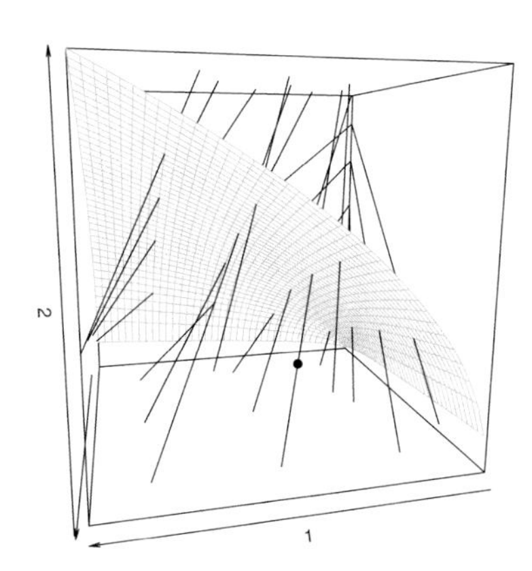

Figure 3.1 *The geometry of GLMs. The left panel illustrates the best fit of the generalized linear model* $\mathbb{E}(y) \equiv \mu = \exp(\beta_0 + \beta_1 x)$ *to the three* x, y *data shown, assuming that each* y_i *is an observation of a gamma distributed random variable with mean given by the model. The right panel illustrates the geometry of GLM fitting using this model as an example. The unit cube shown, represents a space within which the vector* $(y_1, y_2, y_3)^\mathsf{T}$ *defines a single point,* $\bullet$. *The grey surface shows all possible predicted values (within the unit cube) according to the model, i.e., it represents all possible* $(\mu_1, \mu_2, \mu_3)^\mathsf{T}$ *values. As the parameters* β_0 *and* β_1 *are allowed to vary, over all their possible values, this is the surface that the corresponding model 'fitted values' trace out: the 'model manifold'. The continuous lines, which each start at one face of the cube and leave at another, are lines of equivalent fit: the values of the response data* $(y_1, y_2, y_3)^\mathsf{T}$ *lying on such a line each result in the same maximum likelihood estimates of* β_0, β_1 *and hence the same* $(\mu_1, \mu_2, \mu_3)^\mathsf{T}$. *Notice how the equivalent fit lines are neither parallel to each other nor orthogonal to the model manifold.*

To understand figure 3.1, it may help to consider what the figure would look like for some different 2 parameter models.

1. For an ordinary linear model, the model manifold would be a flat plane, to which all the lines of equal fit would be orthogonal (and hence parallel to each other).
2. For a GLM assuming a normal distribution (but non-identity link) the lines of equal fit would be orthogonal to the (tangent space of the) model manifold where they meet it.
3. For a 2 parameter fit to 4 data, lines of equal fit would become planes of equal fit.

In general, the geometric picture presented in figure 3.1 applies to any GLM. With more data the lines of equal fit become $n - p$ dimensional planes of equal fit, where n and p are the number of data and parameters, respectively: for any fixed $\boldsymbol{\beta}$, equation (3.13) gives the restrictions on $\mathbf{y}$ defining such a plane. Note that these planes can intersect — a point which will be returned to later. For discrete response

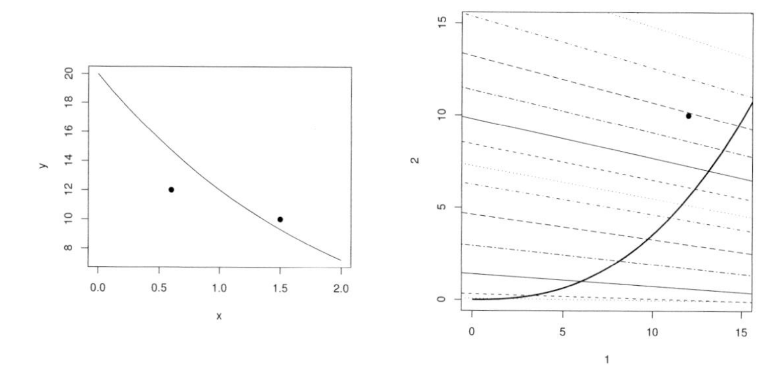

Figure 3.2 *Geometry of the GLM,* $\mathbb{E}(y_i) \equiv \mu_i = 20\exp(-\beta x_i)$ *where* $y_i \sim$ *gamma and* $i = 1, 2$. *The left panel illustrates the maximum likelihood estimate of the model (continuous curve) fitted to the 2* x, y *data shown as* •. *The right panel illustrates the fitting geometry. The* 15×15 *square is part of the space* $\Re^2$ *in which* (y_1, y_2) *defines a single point,* •. *The bold curve is the 'model manifold': it consists of all possible points* (μ_1, μ_2) *according to the model (i.e., as* β *varies* (μ_1, μ_2) *traces out this curve). The fine lines are examples of lines of equal fit. All points* (y_1, y_2) *lying on one of these lines share the same MLE of* β *and hence* (μ_1, μ_2)*: this MLE is where the equal fit line cuts the model manifold. The lines of equal fit are plotted for* $\beta =$ *.1, .2, .3, .4, .5, .6, .7, .8, .9, 1, 1.2, 1.5, 2, 3, 4. (*$\beta =$ *.1, .7 and 2 are represented by unbroken lines, with the* $\beta = 2$ *line being near the bottom of the plot. The* $\beta = .1$ *line is outside the plotting region in this plot, but appears in subsequent plots.)*

data the pictures are no different, although the lines of equal fit strictly make sense only under continuous generalizations of the likelihoods (generally obtainable by replacing factorials by appropriate gamma functions in the probability functions). Only for the normal distribution are the lines/planes of equal fit orthogonal to the model manifold wherever they meet it. For other distributions the lines/planes of equal fit may sometimes be parallel to each other, but are never all orthogonal to the model manifold.

3.2.1 *The geometry of IRLS*

The geometry of the IRLS estimation algorithm is most easily appreciated by considering the fit of a one parameter model to two response data. Figure 3.2 illustrates the geometry of such a model: in this case a GLM with a log link and gamma errors, but similar pictures can be constructed for a GLM with any combination of link and distributional assumption.

Now the key problems in fitting a GLM are that the model manifold is not flat, and that the lines of equal fit are not orthogonal to the model manifold where they meet it. The IRLS method linearly translates and re-scales the fitting problem, so that at the current estimate of $\boldsymbol{\mu}$, the model manifold and intersecting line of equal fit are orthogonal, and, in the re-scaled space, the location of the current estimate of $\boldsymbol{\mu}$ is given by $\mathbf{X}$ multiplied by the current $\boldsymbol{\beta}$ estimate. This re-scaling results in a fitting problem that can be treated as locally linear, so that the $\boldsymbol{\beta}$ estimate can be updated by least squares.

Figure 3.3 illustrates how the IRLS steps involved in forming pseudodata and weighting it, effectively transform the fitting problem into one that can be approximately solved by linear least squares. The figure illustrates the transformations involved in one IRLS step, which are redone repeatedly as the IRLS method is iterated to convergence.

3.2.2 *Geometry and IRLS convergence*

Figure 3.4 illustrates the geometry of fitting a model, $\mathbb{E}(y_i) \equiv \mu_i = \exp(-\beta x_i)$, where the y_i are normally distributed and there are two data, y_i, to fit, for which $x_1 = .6$ and $x_2 = 1.5$. As in the previous two sections, lines of equal fit are shown on a plot in which a response vector $(y_1, y_2)^\mathsf{T}$ would define a single point and the set of all possible fitted values $(\mu_1, \mu_2)^\mathsf{T}$, according to the model, is shown as a thick curve. In this example, the lines of equal fit intersect and cross in the top left hand corner of the plot (corresponding to very poor model fit). This crossing is problematic: in particular, the results of IRLS fitting to data lying in the upper left corner will depend on the initial parameter estimate from which the IRLS process is started, since each such data point lies on the intersection of two equal fit lines. If the IRLS iteration is started from fitted values in the top right of the plot then fitted values nearer the top right will be estimated, while starting the iteration with fitted values at the bottom left of the plot will result in estimated fitted values that are different, and closer to the bottom left of the plot.

That this happens in practice is easily demonstrated in R, by fitting to the data $y_1 = .02$, $y_2 = .9$, illustrated as • in figure 3.4.

```
> ms <- exp(-x*4)    # set initial values at lower left
> glm(y ~ I(-x)-1,family=gaussian(link=log),mustart=ms)
Coefficients:
5.618
Residual Deviance: 0.8098        AIC: 7.868
> ms <- exp(-x*0.1)  # set initial values at upper right
> glm(y ~ I(-x)-1,family=gaussian(link=log),mustart=ms)
Coefficients:
0.544
Residual Deviance: 0.7017        AIC: 7.581
```

Notice that the second fit actually has higher likelihood (lower deviance) — the fits are not equivalent in terms of likelihood. The type of fitting geometry that gives rise to these ambiguities does not always occur: for example some models have parallel lines/planes of equal fit. But for any model with intersecting lines/planes of equal fit there is some scope for ambiguity. Fortunately, if the model is a good model, it is often the case that data lying in the region of ambiguity are rather improbable. In the example in figure 3.4, the problematic region consists entirely of data that the model

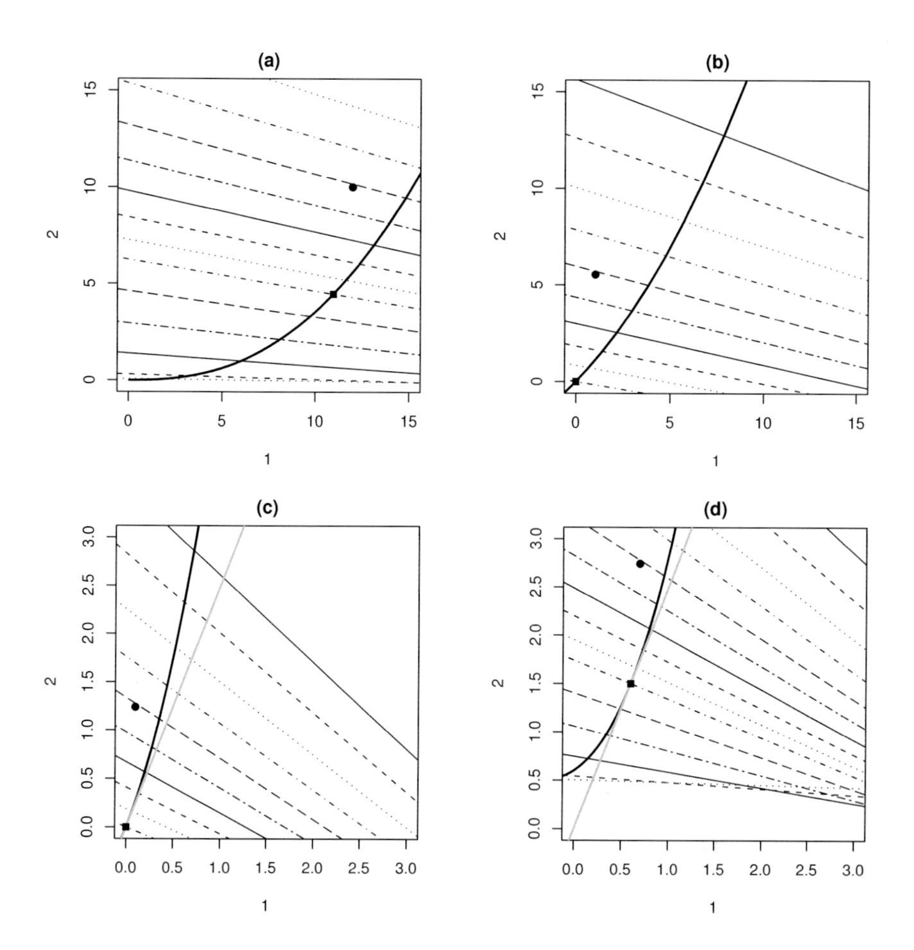

Figure 3.3 *Geometry of the IRLS estimation of a GLM, based on the example shown in figure 3.2. (a) shows the geometry of the fitting problem — the model manifold is the thick black curve, the equal fit lines are the thin lines (as figure 3.2), the data are at • and the current estimates of the fitted values, $\boldsymbol{\mu}^{[k]}$, are at ■. (b) The problem is re-centred around the current fitted values (y_i is replaced by $y_i - \mu_i^{[k]}$). (c) The problem is linearly re-scaled so that the columns of $\mathbf{X}$ now span the tangent space to the model manifold at ■. The tangent space is illustrated by the grey line (this step replaces $y_i - \mu_i^{[k]}$ by $g'(\mu_i^{[k]})(y_i - \mu_i^{[k]})$). (d) The problem is linearly translated so that the location of ■ is now given by $\mathbf{X}\boldsymbol{\beta}^{[k]}$. For most GLMs the problem would now have to be rescaled again by multiplying the components relative to each axis by $\sqrt{W_i}$, where the W_i are the iterative weights: this would ensure that the equal estimate line through ■ is orthogonal to the tangent space. In the current example these weights are all 1, so that the required orthogonality already holds. Now for the transformed problem, in the vicinity of ■, the model manifold can be approximated by the tangent space, to which the equal fit lines are approximately orthogonal: hence an updated estimate of $\boldsymbol{\mu}$ and $\boldsymbol{\beta}$ can be obtained by finding the least squares projection of the transformed data, •, onto the tangent space (grey line).*

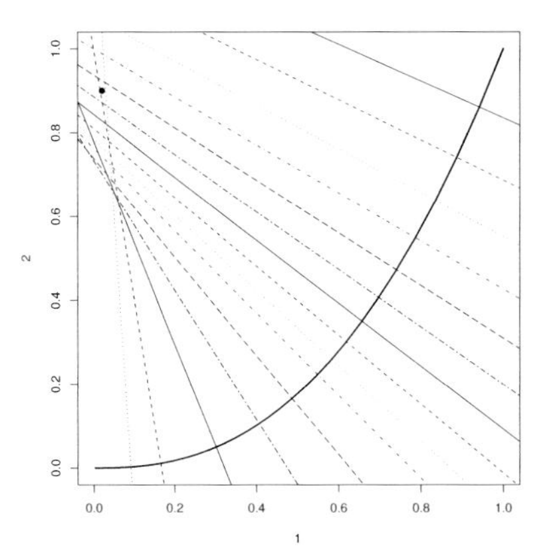

Figure 3.4 *Geometry of fitting and convergence problems. The geometry of a one parameter GLM with a log link and normal errors is illustrated. The thick curve is the model manifold — within the unit square, it contains all the possible fitted values of the data, according to the model. The thin lines are the equal fit lines (levels as in figure 3.2). Notice how the lines of equal fit meet and cross each other at the top left of the plot. Data in this overlap region will yield model likelihoods having local minima at more than one parameter value. Consideration of the operation of the IRLS fitting method reveals that, in this situation, it may converge to different estimates depending on the initial values used to start the fitting process.* • *illustrates the location of a problematic response vector, used to illustrate non-unique convergence in the text.*

can only fit very poorly. It follows that very poor models of data may yield estimation problems of this sort, but it is not uncommon for very poor models to be a feature of early attempts to model any complex set of data. If such problems are encountered then it can be better to proceed by linear modelling of transformed response data, until good enough candidate models have been identified to switch back to GLMs.

If reasonable starting values are chosen, then the ambiguity in the fitting process is unlikely to cause major problems: the algorithm will converge to one of the local minima of the likelihood, after all. However, the ambiguity can cause more serious convergence problems for GAM estimation by 'performance iteration', when it becomes possible to cycle between alternative minima without ever converging.

3.3 GLMs with R

The `glm` function provides the means for using GLMs in R. Its use is similar to that of the `lm` function but with two differences. The right hand side of the model formula, specifying the form for the linear predictor, now gives the link function of the mean of the response, rather than the mean of the response directly. Also `glm` takes a `family` argument, which is used to specify the distribution from the exponential family to use, and the link function that is to go with it. In this section the use of the

CK value	Patients with heart attack	Patients without heart attack
20	2	88
60	13	26
100	30	8
140	30	5
180	21	0
220	19	1
260	18	1
300	13	1
340	19	1
380	15	0
420	7	0
460	8	0

Table 3.2 *Data (from Hand et al., 1994) on heart attack probability as a function of CK level.*

`glm` function with a variety of simple GLMs will be presented, to illustrate the wide variety of model structures that the GLM encompasses.

3.3.1 Binomial models and heart disease

Early diagnosis of heart attack is important if the best care is to be given to patients. One suggested diagnostic aid is the level of the enzyme creatinine kinase (CK) in the blood stream. A study was conducted (Smith, 1967) in which the level of CK was measured for 360 patients suspected of suffering from a heart attack. Whether or not each patient had really suffered a heart attack was established later, after more prolonged medical investigation. The data are given in table 3.2. The original paper classified patients according to ranges of CK level, but in the table only midpoints of the range have been given.

It would be good to be able to base diagnostic criteria on data like these, so that CK level can be used to estimate the probability that a patient has had a heart attack. We can go some way towards such a goal, by constructing a model which tries to explain the proportion of patients suffering a heart attack, from the CK levels. In the following the data were read into a data.frame called `heart`. It contains variables `ha`, `ok` and `ck`, giving numbers of patients who subsequently turned out to have had, or not to have had, heart attacks, at each CK level. It makes sense to plot the observed proportions against CK level first.

```
p <- heart$ha/(heart$ha+heart$ok)
plot(heart$ck,p,xlab="Creatinine kinase level",
     lab="Proportion Heart Attack")
```

The resulting plot is figure 3.5.

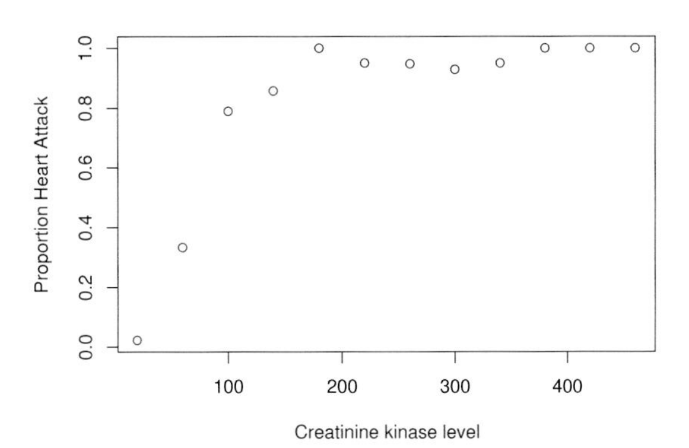

Figure 3.5 *Observed proportion of patients subsequently diagnosed as having had a heart attack, against CK level at admittance.*

A particularly convenient model for describing these proportions is

$$\mathbb{E}(p_i) = \frac{e^{\beta_0+\beta_1 x_i}}{1+e^{\beta_0+\beta_1 x_i}},$$

where p_i is the probability of heart attack at CK level x_i. This curve is sigmoid in shape, and bounded between 0 and 1. (Obviously the `heart` data do not show the lower tail of this proposed sigmoid curve.) This means that the expected number of heart attack sufferers is given by

$$\mu_i \equiv \mathbb{E}(p_i N_i) = \frac{e^{\beta_0+\beta_1 x_i}}{1+e^{\beta_0+\beta_1 x_i}} N_i,$$

where N_i is the known total number of patients at each CK level. This model is somewhat non-linear in its parameters, but if the 'logit' link,

$$g(\mu_i) = \log\left(\frac{\mu_i}{N_i - \mu_i}\right),$$

is applied to it we obtain

$$g(\mu_i) = \beta_0 + \beta_1 x_i,$$

the r.h.s. of which is linear in the model parameters. The logit link is the canonical link for binomial models, and hence the default in R.

In R there are two ways of specifying binomial models with `glm`.

1. The response variable can be the observed proportion of successful binomial trials, in which case a vector giving the number of trials must be supplied as the `weights` argument to `glm`. For binary data, no weights vector need be supplied, as the default weights of 1 suffice.

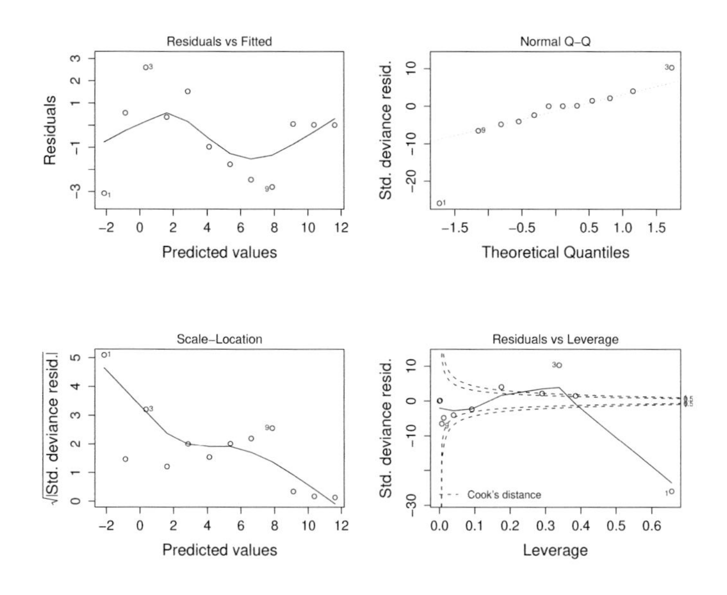

Figure 3.6 *Model checking plots for the first attempt to fit the CK data.*

2. The response variable can be supplied as a two column array, in which the first column gives the number of binomial 'successes', and the second column is the number of binomial 'failures'.

For the current example the second method will be used. Supplying two arrays on the l.h.s. of the model formula involves using `cbind`. Here is a `glm` call which will fit the heart attack model:

```
mod.0 <- glm(cbind(ha,ok) ~ ck, family=binomial(link=logit),
             data=heart)
```

or we could have used

```
mod.0 <- glm(cbind(ha,ok) ~ ck, family=binomial, data=heart)
```

since the logit link is canonical for the binomial and hence the R default. Here is the default information printed about the model:

```
> mod.0

Call: glm(formula=cbind(ha,ok)~ck,family=binomial,data=heart)

Coefficients:
(Intercept)           ck
   -2.75834      0.03124

Degrees of Freedom: 11 Total (i.e. Null);  10 Residual
Null Deviance:        271.7
Residual Deviance: 36.93        AIC: 62.33
```

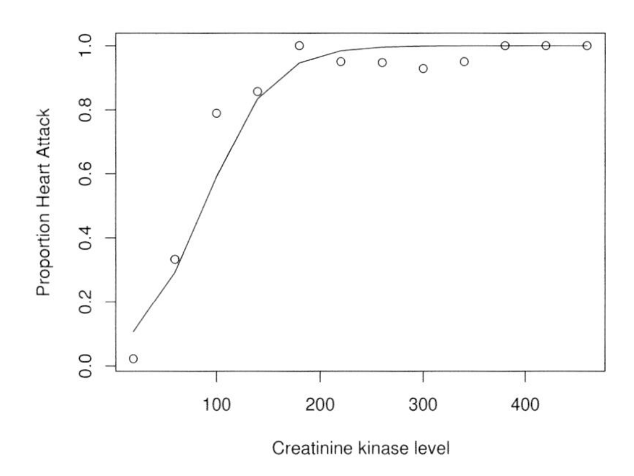

Figure 3.7 *Predicted and observed probability of heart attack, against CK level* (`mod.0`).

The `Null deviance` is the deviance for a model with just a constant term, while the `Residual deviance` is the deviance of the fitted model (and also the scaled deviance in the case of a binomial model). These can be combined to give the *proportion deviance explained*, a generalization of r^2, as follows:

```
> (271.7-36.93)/271.7
[1] 0.864078
```

AIC is the Akaike Information Criteria for the model, discussed in sections 3.1.4 and A.7 (it could also have been extracted using `AIC(mod.0)`).

The deviance is quite high for the χ^2_{10} random variable that it should approximate if the model is fitting well. In fact

```
> 1-pchisq(36.93,10)
[1] 5.819325e-05
```

shows that there is a very small probability of a χ^2_{10} random variable being as large as 36.93. The residual plots (shown in figure 3.6) also suggest a poor fit.

```
par(mfrow=c(2,2))
plot(mod.0)
```

The plots have the same interpretation as the model checking plots for an ordinary linear model, discussed in detail in section 1.5.1 (p. 23), except that it is now the deviance residuals that are plotted, the `Predicted values` are on the scale of the linear predictor, rather than the response, and some departure from a straight line relationship in the normal QQ-plot is often to be expected. The plots are not easy to interpret when there are so few data, but there appears to be a trend in the mean of the residuals plotted against fitted value, which would cause concern. Furthermore, the first point has very high influence. Note that the interpretation of the residuals would be much more difficult for binary data: exercise 2 explores simple approaches that can be taken in the binary case.

Notice how the problems do not stand out so clearly from a plot of the fitted values overlaid on the raw estimated probabilities (see figure 3.7):

```
plot(heart$ck, p, xlab="Creatinine kinase level",
     ylab="Proportion Heart Attack")
lines(heart$ck, fitted(mod.0))
```

Note also that the fitted values provided by `glm` for binomial models are the estimated p_i, rather than the estimated μ_i.

The residual plots suggest trying a cubic linear predictor, rather than the initial straight line.

```
> mod.2 <- glm(cbind(ha,ok) ~ ck + I(ck^2) + I(ck^3),
+              family=binomial, data=heart)
> mod.2

Call: glm(formula=cbind(ha,ok)~ck+I(ck^2)+I(ck^3),
          family=binomial,data=heart)

Coefficients:
(Intercept)           ck      I(ck^2)      I(ck^3)
 -5.786e+00    1.102e-01   -4.648e-04    6.448e-07

Degrees of Freedom: 11 Total (i.e. Null);  8 Residual
Null Deviance:      271.7
Residual Deviance: 4.252        AIC: 33.66
> par(mfrow=c(2,2))
> plot(mod.2)
```

Clearly 4.252 is not too large for consistency with a χ^2_8 distribution (it is less than the expected value, in fact) and the AIC has improved substantially. The residual plots (figure 3.8) now show less clear patterns than for the previous model, although if we had more data then such a departure from constant variance would be a cause for concern. Furthermore the fit is clearly closer to the data now (see figure 3.9):

```
par(mfrow=c(1,1))
plot(heart$ck,p,xlab="Creatinine kinase level",
     ylab="Proportion Heart Attack")
lines(heart$ck,fitted(mod.2))
```

We can also get R to test the null hypothesis that `mod.0` is correct against the alternative that `mod.2` is required. Somewhat confusingly the `anova` function is used to do this, although it is an analysis of **deviance** (i.e., a generalized likelihood ratio test) that is being performed, and not an analysis of variance.

```
> anova(mod.0,mod.2,test="Chisq")
Analysis of Deviance Table

Model 1: cbind(ha, ok) ~ ck
Model 2: cbind(ha, ok) ~ ck + I(ck^2) + I(ck^3)
  Resid. Df Resid. Dev Df Deviance P(>|Chi|)
1        10     36.929
2         8      4.252  2   32.676 8.025e-08
```

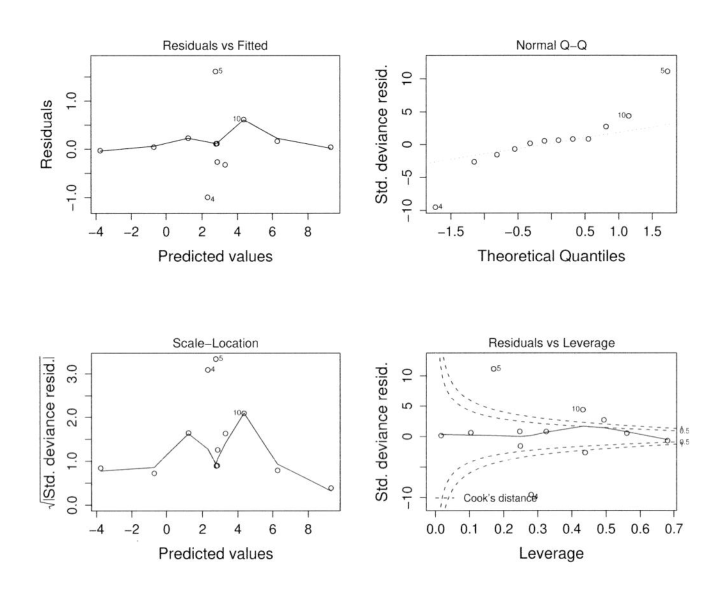

Figure 3.8 *Model checking plots for the second attempt to fit the CK data.*

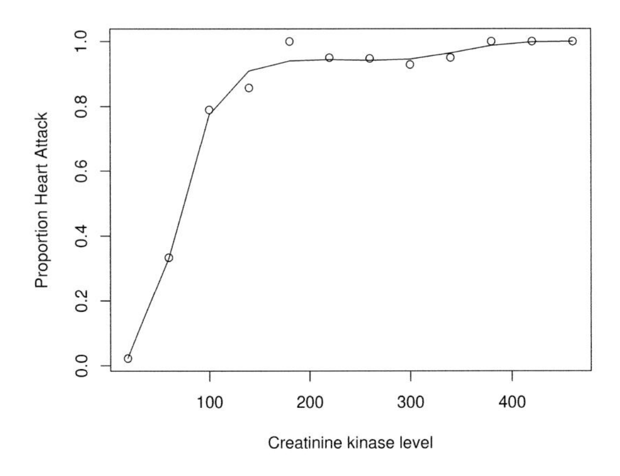

Figure 3.9 *Predicted and observed probability of heart attack, against CK level (*`mod.2`*).*

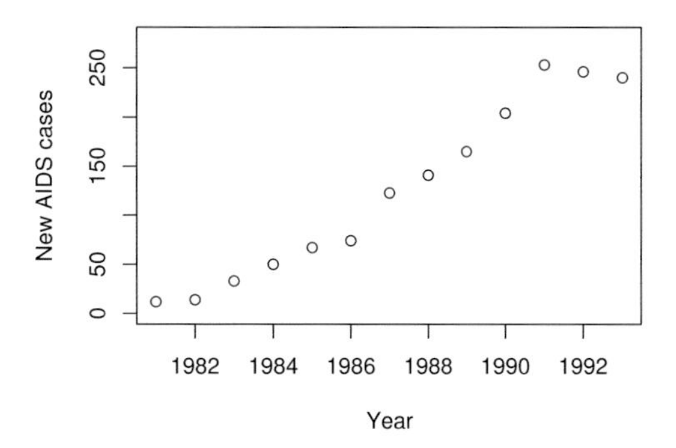

Figure 3.10 *AIDS cases per year in Belgium.*

A p-value this low indicates very strong evidence against the null hypothesis: we really do need model 2. Recall that this comparison of models has a much firmer theoretical basis than the examination of the individual deviances had.

3.3.2 A Poisson regression epidemic model

The introduction to this chapter included a simple model for the early stages of an epidemic. Venables and Ripley (2002) provide some data on the number of new AIDS cases each year, in Belgium, from 1981 onwards. The data can be entered into R and plotted as follows.

```
y <- c(12,14,33,50,67,74,123,141,165,204,253,246,240)
t <- 1:13
plot(t+1980,y,xlab="Year",ylab="New AIDS cases",ylim=c(0,280))
```

Figure 3.10 shows the resulting plot. The scientifically interesting question is whether the data provide any evidence that the increase in the underlying rate of new case generation is slowing. The simple model from the introduction provides a plausible model from which to start investigating this question. The model assumes that the underlying expected number of cases per year, μ_i, increases according to:

$$\mu_i = \gamma \exp(\delta t_i)$$

where γ and δ are unknown parameters, and t_i is time in years since the start of the data. A log link turns this into a GLM,

$$\log(\mu_i) = \log(\gamma) + \delta t_i = \beta_0 + t_i\beta_1,$$

and we assume that $y_i \sim \text{Poi}(\mu_i)$ where y_i is the observed number of new cases in year t_i. The y_i are assumed independent. This is essentially a model of unchecked spread of the disease.

The following fits the model (the log link is canonical for the Poisson distribution, and hence the R default) and checks it.

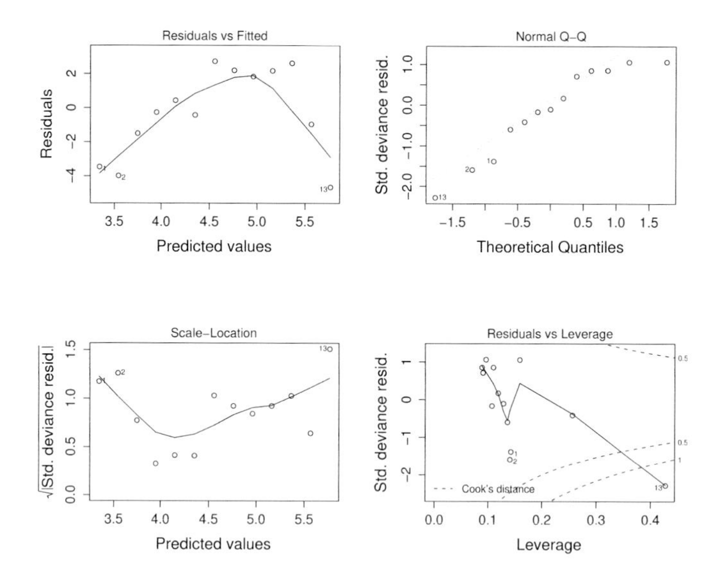

Figure 3.11 *Residual plots for* m0 *fitted to the AIDS data.*

```
> m0 <- glm(y ~ t, poisson)
> m0

Call:  glm(formula = y ~ t, family = poisson)

Coefficients:
(Intercept)            t
     3.1406       0.2021

Degrees of Freedom: 12 Total (i.e. Null);  11 Residual
Null Deviance:      872.2
Residual Deviance: 80.69       AIC: 166.4
> par(mfrow=c(2,2))
> plot(m0)
```

The deviance is very high for the observation of a χ^2_{11} random variable that it ought to approximate if the model is a good fit. The residual plots shown in figure 3.11 are also worrying. In particular the clear pattern in the mean of the residuals, plotted against the fitted values, shows violation of the independence assumption and probably results from omission of something important from the model. Since, for this model, the fitted values increase monotonically with time, we would get the same sort of pattern if residuals were plotted against time — i.e., it appears that a quadratic term in time could usefully be added to the model. Also worrying is the very high influence of the final year's data, evident in the residuals versus leverage plot. Note that the interpretation of residual plots can become difficult if the Poisson mean is low

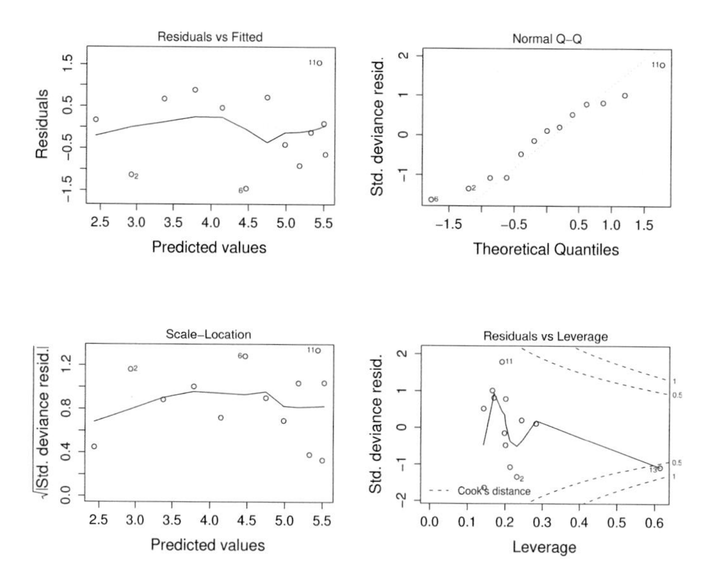

Figure 3.12 *Residual plots for* `m1` *fitted to the AIDS data.*

so that the data are mostly zeroes and ones. In such cases the simulation approaches covered in exercise 2 can prove useful, if adapted to the Poisson case.

It seems sensible to amend the model by adding a quadratic term to obtain:

$$\mu_i = \exp(\beta_0 + \beta_1 t_i + \beta_2 t_i^2).$$

This model allows situations other than unrestricted spread of the disease to be represented. The following fits and checks it:

```
> m1 <- glm(y ~ t + I(t^2), poisson)
> plot(m1)
> summary(m1)
Call:
glm(formula = y ~ t + I(t^2), family = poisson)

Deviance Residuals:
     Min        1Q    Median        3Q       Max
-1.45903  -0.64491   0.08927   0.67117   1.54596

Coefficients:
             Estimate Std. Error z value Pr(>|z|)
(Intercept)  1.901459   0.186877  10.175  < 2e-16 ***
t            0.556003   0.045780  12.145  < 2e-16 ***
I(t^2)      -0.021346   0.002659  -8.029 9.82e-16 ***
---
Signif. codes: 0 '***' 0.001 '**' 0.01 '*' 0.05 '.' 0.1

(Dispersion parameter for poisson family taken to be 1)
```

```
    Null deviance: 872.2058  on 12  degrees of freedom
Residual deviance:   9.2402  on 10  degrees of freedom
AIC: 96.924     Number of Fisher Scoring iterations: 4
```

The residual plots shown in figure 3.12 are now much improved: the clear trend in the mean has gone, the (vertical) spread of the residuals is reasonably even, the influence of point 13 is much reduced and the QQ-plot is straighter. The fuller model summary shown for this model also indicates improvement: the deviance is now acceptable, i.e., close to what is expected for a χ^2_{10} r.v., and the AIC has dropped massively. All in all this model appears to be quite reasonable.

Notice how the structure of the `glm summary` is similar to an `lm summary`. The standard errors and p-values in the table of coefficient estimates are now based on the large sample distribution of the parameter estimators given in section 3.1.3. The `z value` column simply reports the parameter estimates divided by their estimated standard deviations. Since no dispersion parameter estimate is required for the Poisson, these z-values should be observations of N(0,1) r.v.s, if the true value of the corresponding parameter is zero (at least in the large sample limit), and the reported p-value is based on this distributional approximation. For `mod.1` the reported p-values are very low, i.e., for each parameter there is clear evidence that it is not zero.

Examination of the coefficient summary table indicates that the hypothesis that $\beta_2 = 0$ can be firmly rejected, providing clear evidence that `mod.1` is preferable to `mod.0`. The same question can also be addressed using a generalized likelihood ratio test:

```
> anova(m0,m1,test="Chisq")
Analysis of Deviance Table

Model 1: y ~ t
Model 2: y ~ t + I(t^2)
  Resid. Df Resid. Dev Df Deviance P(>|Chi|)
1        11     80.686
2        10      9.240  1   71.446 2.849e-17
```

The conclusion is the same as before: the tiny p-value indicates that `mod.0` should be firmly rejected in favour of `mod.1`. The `test="Chisq"` argument to `anova` is justified because the scale parameter is known for this model: had it been estimated it would be preferable to set `test` to `"F"`.

The hypothesis testing approach to model selection is appropriate here, as the main question of interest is whether there is evidence, from these data, that the epidemic is spreading un-checked, or not. It would be prudent not to declare that things are improving if the evidence is not quite firm that this is true. If we had been more interested in simply finding the best model for predicting the data then comparison of AIC would be more appropriate, but leads to the same conclusion for these data.

The parameter β_1 can be interpreted as the rate of spread of the disease at the epidemic start: that is as a sort of intrinsic rate of increase of the disease in a new population where no control measures are in place. Notice how the estimate of this parameter has actually increased substantially between the first and second models:

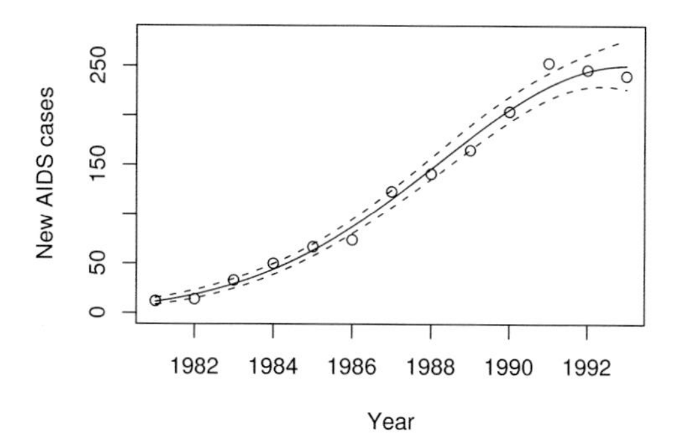

Figure 3.13 *Underlying AIDS case rate according to model* `m1` *shown as a continuous curve, with 95% confidence limits shown as dashed curves.*

it would have been possible to be quite badly misled if we had stuck with the first poorly fitting model. An approximate confidence interval for β_1 can be obtained in the usual manner, based on the large sample results from section 3.1.3. The required estimate and standard error are easily extracted using the `summary` function, as the following illustrates:

```
> beta.1 <- summary(m1)$coefficients[2,]
> ci <- c(beta.1[1]-1.96*beta.1[2],beta.1[1]+1.96*beta.1[2])
> ci # print 95% CI for beta_1
0.4662750 0.6457316
```

The use of the critical points of the standard normal distribution is appropriate, because the scale parameter is known for this model. Had it been estimated, then we would have had to use critical points from a t distribution, with degrees of freedom set to the residual degrees of freedom of the model (i.e., number of data less number of estimated β parameters).

Another obvious thing to want to do is to use the model to find a confidence interval for the underlying rate of case generation at any time. The following R code illustrates how to use the `predict.glm` function to find CI's for the underlying rate over the whole period of the data, and plot these.

```
new.t <- seq(1,13,length=100)
fv <- predict(m1,data.frame(t=new.t),se=TRUE)
plot(t+1980,y,xlab="Year",ylab="New AIDS cases",ylim=c(0,280))
lines(new.t+1980,exp(fv$fit))
lines(new.t+1980,exp(fv$fit+2*fv$se.fit),lty=2)
lines(new.t+1980,exp(fv$fit-2*fv$se.fit),lty=2)
```

The plot is shown in figure 3.13. Notice that by default the `predict.glm` function predicts on the scale of the linear predictor: we have to apply the inverse of the link function to get back onto the original response scale.

So the data provide quite firm evidence to suggest that the unfettered exponential increase model is overly pessimistic: by the end of the data there is good evidence that the rate of increase is slowing. Of course, this model contains no mechanistic content — it says nothing about how or why the slowing might be occurring: as such it is entirely inappropriate for prediction beyond the range of the data. The model allows us to be reasonably confident that the apparent slowing in the rate of increase in new cases is real, and not just the result of chance variation, but it says little or nothing about what may happen later.

3.3.3 Cox proportional hazards modelling of survival data

The following data from Klein and Moeschberger (2003) are times to death, relapse or last follow-up of patients with non-Hodgkins lymphoma, after receiving one of two alternative bone marrow treatments at the Ohio state university bone marrow transplant unit. The 'Allo' treatment is a bone marrow transplant from a matched sibling donor, while the 'Auto' treatment involves removal of the patient's bone marrow and then replacing it after chemotherapy.

	Time (Days)											
Allo	28	32	49	84	357	$\overline{933}$	$\overline{1078}$	$\overline{1183}$	$\overline{1560}$	$\overline{2114}$	$\overline{2144}$	
Auto	42	53	57	63	81	140	176	$\overline{210}$	252	$\overline{476}$	524	$\overline{1037}$

Times with an over-bar are censored: they record the last time that the patient was recorded, by which time they had neither died, nor relapsed. This is clearly a small sample from a statistical point of view, but not from a human point of view, and it is important to try to determine whether there is really evidence for a difference between the treatments.

A possible model for these data is the Cox proportional hazards model discussed in section 3.1.10. To fit the model as described in that section we first need to produce appropriate artificial data for the equivalent Poisson model. The data frame `bone` contains the data. It has three columns: `t` is the time, `d` is a binary indicator: 1 for death or relapse and 0 for censoring, and `trt` is a factor with levels `allo` and `auto`. The following simple routine will take such a data frame and automatically produce the equivalent data frame for the Poisson model. Its first argument is the data frame and its next two arguments are the names of the time and censoring columns and the final argument is the name to give the artificial Poisson response.

```
psurv <- function(surv,time="t",censor="d",event="z") {
## create data frame to fit Cox PH as Poisson model.
## surv[[censor]] should be 1 for event or zero for censored.
  if (event %in% names(surv)) warning("event name clashes")
  surv <- as.data.frame(surv)[order(surv[[time]]),] # t order
  et <- unique(surv[[time]][surv[[censor]]==1]) # unique times
  es <- match(et,surv[[time]]) # starts of risk sets in surv
  n <- nrow(surv); t <- rep(et,1+n-es) # times for risk sets
  st <- cbind(0,
     surv[unlist(apply(matrix(es),1,function(x,n) x:n,n=n)),])
  st[st[[time]]==t&st[[censor]]!=0,1] <- 1 # signal events
  st[[time]] <- t ## reset event time to risk set time
  names(st)[1] <- event
  st
} ## psurv
```

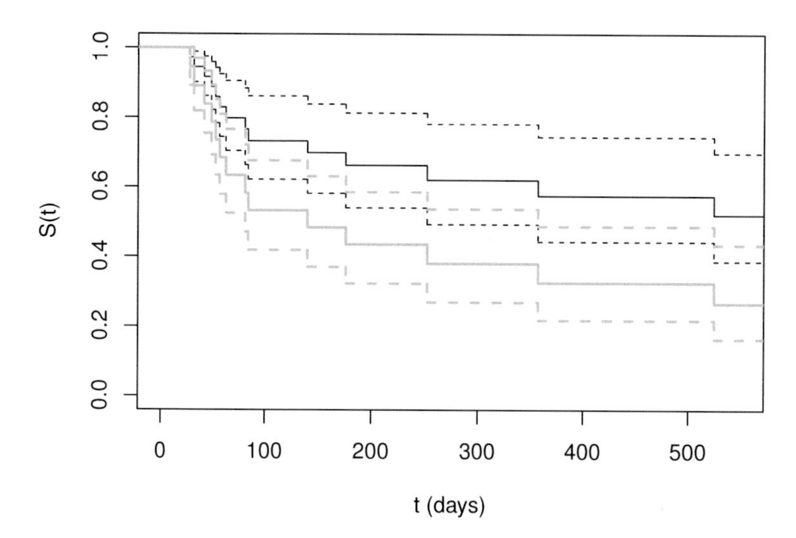

Figure 3.14 *Estimated survival curves and one standard error limits for the allogenic treatment (black) and autogenic treatment (grey), discussed in section 3.3.3.*

It is now straightforward to fit the model, first transforming the `bone` data frame, adding an event time factor variable, `tf`, and then calling `glm` for the actual fitting.

```
require(gamair); data(bone); bone$id <- 1:nrow(bone)
pb <- psurv(bone); pb$tf <- factor(pb$t)
b <- glm(z ~ tf + trt - 1,poisson,pb)
```

The `glm` formula includes a '`-1`', to suppress the intercept, while `tf` is the first term on the right hand side, to ensure that it is the contrast free term, and we really do get a separate α_j for each event time.

The `pb` data frame has 5 columns and 227 rows. Its first 23 rows consist of a copy of the `bone` data frame, but with the extra `z` column set to zero for all rows except that originally corresponding to time 28, and the time and `tf` column set to 28 for all rows. The next 22 rows contain the rows of `bone` except for the row for time 28. The `z` column is zero except for the row originally corresponding to time 32, with the time and `tf` column set to 32 for all rows. The next 21 rows leave out rows for times 28 and 32 from `bone`, and so it goes on. Note that censoring times do not produce a block, and that all subjects with a censoring time between two events are dropped from the block for the second event.

Residuals are now easily computed. For example

```
chaz <- tapply(fitted(b),pb$id,sum) ## by subject cum. hazard
mrsd <- bone$d - chaz ## Martingale residuals
```

It is also straightforward to test the significance of the treatment, either by looking at the `summary`, or by

```
> drop1(b,test="Chisq") ## test for effect - no evidence
Model:
z ~ tf + trt - 1
```

```
       Df Deviance    AIC     LRT Pr(>Chi)
<none>       75.198 133.2
tf     14   261.702 291.7 186.504   <2e-16 ***
trt     1    76.799 132.8   1.601   0.2057
```

So although the effect of `auto` relative to `allo` is estimated to be 0.70, the effect is not significant, and its inclusion is only barely supported by AIC. The positive effect means that the hazard is estimated to be higher for the `auto` group. To visualize what is happening it can help to plot the survival curve estimates for the two groups. The following code does this by following section 3.1.10.

```
te <- sort(unique(bone$t[bone$d==1])) ## event times
## predict survivor function for "allo"...
pd <- data.frame(tf=factor(te),trt=bone$trt[1])
fv <- predict(b,pd)
H <- cumsum(exp(fv)) ## cumulative hazard
plot(stepfun(te,c(1,exp(-H))),do.points=FALSE,ylim=c(0,1),
     xlim=c(0,550),main="",ylab="S(t)",xlab="t (days)")
## add s.e. bands...
X <- model.matrix(~tf+trt-1,pd)
J <- apply(exp(fv)*X,2,cumsum)
se <- diag(J%*%vcov(b)%*%t(J))^.5
lines(stepfun(te,c(1,exp(-H+se))),do.points=FALSE,lty=2)
lines(stepfun(te,c(1,exp(-H-se))),do.points=FALSE,lty=2)
```

Adding similar code for the `auto` group produces figure 3.14. For much more on survival analysis see Collett (2015), Klein and Moeschberger (2003) and Therneau and Grambsch (2000).

3.3.4 Log-linear models for categorical data

The following table classifies a random sample of women and men according to their belief in the afterlife:

	Believer	Non-Believer
Female	435	147
Male	375	134

The data (reported in Agresti, 1996) come from the US General Social Survey (1991), and the 'non-believer' category includes 'undecideds'. Are there differences between males and females in the holding of this belief? We can address this question by using analysis of deviance to compare the fit of two competing models of these data: one in which belief is modelled as independent of gender, and a second in which there is some interaction between belief and gender. First consider the model of independence. If y_i is an observation of the counts in one of the cells of the table, then we could model the expected number of counts as

$$\mu_i \equiv \mathbb{E}(Y_i) = n\gamma_k\alpha_j \text{ if } y_i \text{ is data for gender } k, \text{ and faith } j,$$

where n is the total number of people surveyed, α_1 the proportion of believers, α_2 the proportion of non-believers and γ_1 and γ_2 the proportions of women and men, respectively. Taking logs of this model yields

$$\eta_i \equiv \log(\mu_i) = \log(n) + \log(\gamma_k) + \log(\alpha_j).$$

So defining $\tilde{n} = \log(n)$, $\tilde{\gamma}_k = \log(\gamma_k)$ and $\tilde{\alpha}_j = \log(\alpha_j)$ the model can be written as

$$\begin{bmatrix} \eta_1 \\ \eta_2 \\ \eta_3 \\ \eta_4 \end{bmatrix} = \begin{bmatrix} 1 & 1 & 0 & 0 & 1 \\ 1 & 1 & 0 & 1 & 0 \\ 1 & 0 & 1 & 0 & 1 \\ 1 & 0 & 1 & 1 & 0 \end{bmatrix} \begin{bmatrix} \tilde{n} \\ \tilde{\gamma}_1 \\ \tilde{\gamma}_2 \\ \tilde{\alpha}_1 \\ \tilde{\alpha}_2 \end{bmatrix}.$$

This is clearly a GLM structure, but is obviously not identifiable. Dropping $\tilde{\gamma}_1$ and $\tilde{\alpha}_1$ solves the identifiability problem yielding

$$\begin{bmatrix} \eta_1 \\ \eta_2 \\ \eta_3 \\ \eta_4 \end{bmatrix} = \begin{bmatrix} 1 & 0 & 1 \\ 1 & 0 & 0 \\ 1 & 1 & 1 \\ 1 & 1 & 0 \end{bmatrix} \begin{bmatrix} \tilde{n} \\ \tilde{\gamma}_2 \\ \tilde{\alpha}_2 \end{bmatrix}.$$

Note how gender and faith are both factor variables with two levels in this model.

If the counts in the contingency table occurred independently at random, then the obvious distribution to use would be Poisson. In fact even when the total number of subjects in the table, or even some other marginal totals, are fixed, then it can be shown that the correct likelihood can be written as a product of Poisson p.m.f.s, conditional on the various fixed quantities. Hence provided that the fitted model is forced to match the fixed total, and any fixed marginal totals, the Poisson is still the distribution to use. As was shown in section 3.1.6, forcing the model to match certain fixed totals in the data is simply a matter of insisting on certain terms being retained in the model.

The simple 'independence' model is easily estimated in R. First enter the data and check it:

```
> al <- data.frame(y=c(435,147,375,134),gender=
+    as.factor(c("F","F","M","M")),faith=as.factor(c(1,0,1,0)))
> al
    y gender faith
1 435      F     1
2 147      F     0
3 375      M     1
4 134      M     0
```

Since gender and faith are both factor variables, model specification is very easy. The following fits the model and checks that the model matrix is as expected:

```
> mod.0 <- glm(y ~ gender + faith, data=al, family=poisson)
> model.matrix(mod.0)
  (Intercept) genderM faith1
1           1       0      1
2           1       0      0
3           1       1      1
4           1       1      0
```

Now look at the fitted model object `mod.0`

```
> mod.0

Call:  glm(formula=y~gender+faith,family=poisson,data=al)

Coefficients:
(Intercept)      genderM       faith1
     5.0100      -0.1340       1.0587

Degrees of Freedom: 3 Total (i.e. Null);  1 Residual
Null Deviance:       272.7
Residual Deviance: 0.162        AIC: 35.41

> fitted(mod.0)
      1       2       3       4
432.099 149.901 377.901 131.099
```

The fit appears to be quite close, and it would be somewhat surprising if a model with interactions between faith and gender did significantly better. Nevertheless such a model could be:

$$\eta_i \equiv \log(\mu_i) = \tilde{n} + \tilde{\gamma}_k + \tilde{\alpha}_j + \tilde{\zeta}_{kj} \text{ if } y_i \text{ is data for gender } k \text{ and faith } j,$$

where $\tilde{\zeta}_{kj}$ is an 'interaction parameter'. This model allows each combination of faith and gender to vary independently. As written, the model has rather a large number of un-identifiable terms.

$$\begin{bmatrix} \eta_1 \\ \eta_2 \\ \eta_3 \\ \eta_4 \end{bmatrix} = \begin{bmatrix} 1 & 1 & 0 & 0 & 1 & 0 & 1 & 0 & 0 \\ 1 & 1 & 0 & 1 & 0 & 1 & 0 & 0 & 0 \\ 1 & 0 & 1 & 0 & 1 & 0 & 0 & 0 & 1 \\ 1 & 0 & 1 & 1 & 0 & 0 & 0 & 1 & 0 \end{bmatrix} \begin{bmatrix} \tilde{n} \\ \tilde{\gamma}_1 \\ \tilde{\gamma}_2 \\ \tilde{\alpha}_1 \\ \tilde{\alpha}_2 \\ \tilde{\zeta}_{11} \\ \tilde{\zeta}_{12} \\ \tilde{\zeta}_{21} \\ \tilde{\zeta}_{22} \end{bmatrix}.$$

But this is easily reduced to something identifiable:

$$\begin{bmatrix} \eta_1 \\ \eta_2 \\ \eta_3 \\ \eta_4 \end{bmatrix} = \begin{bmatrix} 1 & 0 & 1 & 0 \\ 1 & 0 & 0 & 0 \\ 1 & 1 & 1 & 1 \\ 1 & 1 & 0 & 0 \end{bmatrix} \begin{bmatrix} \tilde{n} \\ \tilde{\gamma}_2 \\ \tilde{\alpha}_2 \\ \tilde{\zeta}_{22} \end{bmatrix}.$$

The following fits the model, checks the model matrix and prints the fitted model object:

```
> mod.1 <- glm(y ~ gender*faith, data=al, family=poisson)
> model.matrix(mod.1)
```

```
  (Intercept) genderM faith1 genderM:faith1
1           1       0      1              0
2           1       0      0              0
3           1       1      1              1
4           1       1      0              0
> mod.1

Call:  glm(formula=y ~ gender*faith,family=poisson,data=al)

Coefficients:
  (Intercept)       genderM        faith1  genderM:faith1
      4.99043      -0.09259       1.08491        -0.05583

Degrees of Freedom: 3 Total (i.e. Null);  0 Residual
Null Deviance:       272.7
Residual Deviance: 9.659e-14    AIC: 37.25
```

To test whether there is evidence for an interaction between gender and faith the null hypothesis that `mod.0` is correct is tested against the more general alternative that `mod.1` is correct, using analysis of deviance.

```
> anova(mod.0,mod.1,test="Chisq")
Analysis of Deviance Table

Model 1: y ~ gender + faith
Model 2: y ~ gender * faith
  Resid. Df Resid. Dev Df Deviance P(>|Chi|)
1         1    0.16200
2         0  9.659e-14  1  0.16200   0.68733
```

A p-value of 0.69 suggests that there is no evidence to reject model 0 and the hypothesis of no association between gender and belief in the afterlife.

Notice that, in fact, the model with the interaction is the saturated model, which is why its deviance is numerically zero, and there was not really any need to fit it and compare it with the independence model explicitly — in this case we could just as well have examined the deviance of the independence model. However, the general approach taken for this simple 2-way contingency table can easily be generalized to multi-way tables and to arbitrary number of groups. In other words, the approach outlined here can be extended to produce a rather general approach for analyzing categorical data using log-linear GLMs.

Finally, note that the fitted values for `mod.0` had the odd property that although the fitted values and original data are different, the total number of men and women is conserved between data and fitted values, as is the total number of believers and non-believers. This results from the fact that the log link is canonical for the Poisson distribution, so by the results of section 3.1.6, $\mathbf{X}^\mathsf{T}\mathbf{y} = \mathbf{X}^\mathsf{T}\hat{\boldsymbol{\mu}}$. The summations equated on the two sides of this last equation are the total number of subjects, the total number of males and the total number of believers: this explains the match between fitted values and data in respect of these totals.

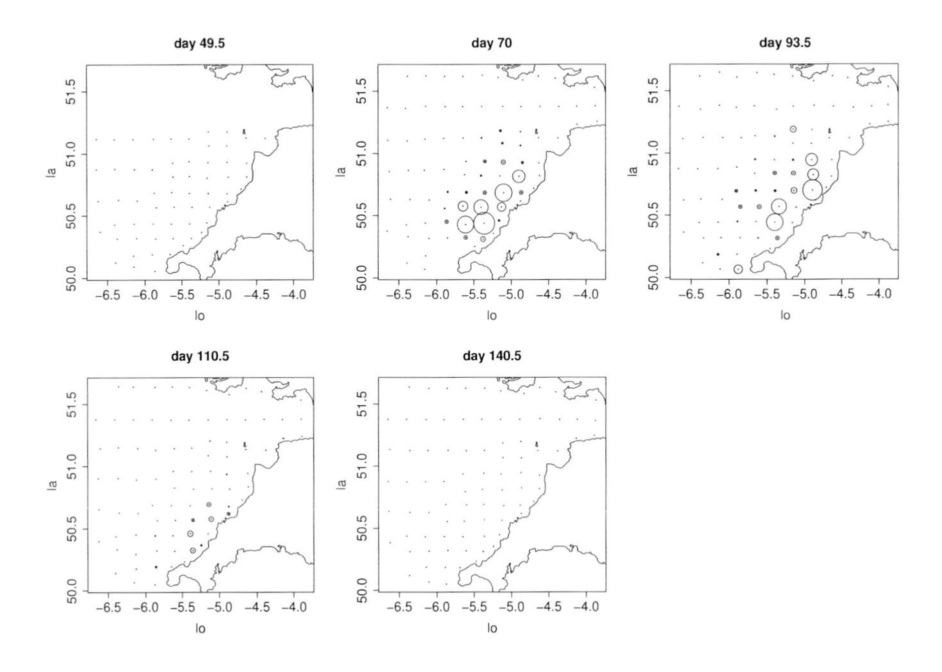

Figure 3.15 *Density per* m^2 *sea surface of stage I sole eggs in the Bristol channel. The days given are the Julian day of the survey midpoint (day 1 is January 1). The symbol sizes are proportional to egg density and a simple dot indicates a station where no eggs were found.*

3.3.5 Sole eggs in the Bristol channel

Fish stock assessment is difficult because adult fish are not easy to survey: they tend to actively avoid fishing gear, so that turning number caught into an assessment of the number in the sea is rather difficult. To get around this problem, fisheries biologists sometimes try and count fish eggs, and work back to the number of adult fish required to produce the estimated egg population. These 'egg production methods' are appealing because eggs are straightforward to sample. This section concerns a simple attempt to model data on sole eggs in the Bristol channel. The data (available in Dixon, 2003) are measurements of density of eggs per square metre of sea surface in each of 4 identifiable egg developmental stages, at each of a number of sampling stations in the Bristol channel on the west coast of England. The samples were taken during 5 cruises spaced out over the spawning season. Figure 3.15 shows the survey locations and egg densities for stage I eggs for each of the 5 surveys. Similar plots could be produced for stages II–IV. For further information on this stock, see Horwood (1993) and Horwood and Greer Walker (1990).

The biologists' chief interest is in estimating the *rate* at which eggs are spawned at any time and place within the survey arena, so this is the quantity that needs to be estimated from the data. To this end it helps that the durations of the egg stages are known (they vary somewhat with temperature, but temperature is known for each sample). Basic demography suggests that a reasonable model for the density of eggs (per day per square metre of sea surface), at any age, a, and location-time with covariates $\mathbf{x}$, would be

$$d(a, \mathbf{x}) = S(\mathbf{x})e^{-\delta(\mathbf{x})a}.$$

That is, the density of eggs of age a is given by the product of the local spawning rate S and the local survival rate. δ is the per capita mortality rate, and, given this rate, we expect a proportion $\exp(-\delta a)$ of eggs to reach age a. Both S and δ are assumed to be functions of some covariates.

What we actually observe are not egg densities per unit age, per m^2 sea surface, but egg densities *in particular developmental stages* per m^2 sea surface: y_i, say. To relate the model to the data we need to integrate the model egg density over the age range of the developmental stage to which any particular datum relates. That is, if a_i^- and a_i^+ are the lower and upper age limits for the egg stage to which y_i relates, then the model should be

$$\mathbb{E}(y_i) \equiv \mu_i = \int_{a_i^-}^{a_i^+} d(z, \mathbf{x}_i) dz.$$

Evaluation of the integral would be straightforward, but does not enable the model to be expressed in the form of a GLM. However, if the integral is approximated so that the model becomes

$$\mu_i = \Delta_i d(\bar{a}_i, \mathbf{x}_i),$$

where $\Delta_i = a_i^+ - a_i^-$ and $\bar{a}_i = (a_i^+ + a_i^-)/2$, then progress can be made, since in that case

$$\log(\mu_i) = \log(\Delta_i) + \log\{S(\mathbf{x}_i)\} - \delta(\mathbf{x}_i)\bar{a}_i. \tag{3.15}$$

The right hand side of this model can be expressed as the linear predictor of a GLM, with terms representing $\log(S)$ and δ as functions of covariates and with $\log(\Delta)$ treated as an 'offset' term — essentially a column of the model matrix with associated parameter fixed at 1.

For the sole eggs, a reasonable starting model might represent $\log(S)$ as a cubic function of longitude, `lo`, latitude, `la`, and time, `t`. Mortality might be modelled by a simpler function — say a quadratic in `t`. It remains only to decide on a distributional assumption. The eggs are sampled by hauling a net vertically through the water and counting the number of eggs caught in it. This might suggest a Poisson model, but most such data display overdispersion relative to Poisson, and additionally, the data are not available as raw counts but rather as densities per m^2 sea surface. These considerations suggest using quasi-likelihood, with the variance proportional to the mean.

The following R code takes the `sole` data frame, calculates the mean ages and offset terms required and fits the suggested model. Since polynomial models can lead to numerical stability problems if not handled carefully, the covariates are all translated and scaled before fitting.

```
> sole$off <- log(sole$a.1-sole$a.0)# model offset term
> sole$a<-(sole$a.1+sole$a.0)/2     # mean stage age
```

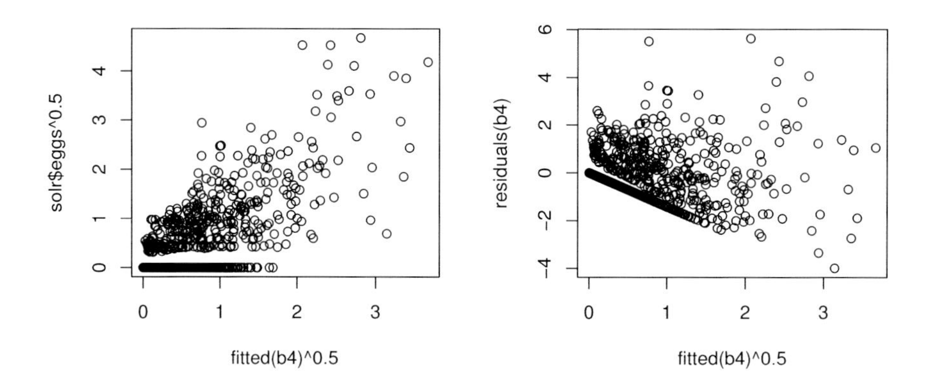

Figure 3.16 *Residual plots for the final sole egg model.*

```
> solr<-sole                              # make copy for rescaling
> solr$t<-solr$t-mean(sole$t)
> solr$t<-solr$t/var(sole$t)^0.5
> solr$la<-solr$la-mean(sole$la)
> solr$lo<-solr$lo-mean(sole$lo)
> b <- glm(eggs ~ offset(off)+lo+la+t+I(lo*la)+I(lo^2)+I(la^2)
+          +I(t^2)+I(lo*t)+I(la*t)+I(lo^3)+I(la^3)+I(t^3)+
+          I(lo*la*t)+I(lo^2*la)+I(lo*la^2)+I(lo^2*t)+
+          I(la^2*t)+I(la*t^2)+I(lo*t^2)+ a +I(a*t)+I(t^2*a),
+          family=quasi(link=log,variance="mu"),data=solr)
> summary(b)

Call:
glm(formula = eggs~offset(off)+lo+la+t+I(lo*la)+
    I(lo^2)+I(la^2)+I(t^2)+I(lo*t)+I(la*t)+I(lo^3)+
    I(la^3)+I(t^3)+I(lo*la*t)+I(lo^2*la)+I(lo*la^2)
    +I(lo^2*t)+I(la^2*t)+I(la*t^2)+I(lo*t^2)+
    a+I(a*t)+I(t^2*a),family = quasi(link=log,
    variance = "mu"), data = solr)

Deviance Residuals:
     Min        1Q    Median        3Q       Max
-4.10474  -0.35127  -0.10418  -0.01289   5.66956

Coefficients:
                Estimate Std. Error t value Pr(>|t|)
(Intercept)     -0.03836    0.14560  -0.263 0.792202
lo               5.22548    0.39436  13.251  < 2e-16 ***
```

```
la                 -5.94345    0.50135 -11.855  < 2e-16 ***
t                  -2.43222    0.25761  -9.442  < 2e-16 ***
I(lo * la)          3.38576    0.61797   5.479 4.99e-08 ***
I(lo^2)            -3.98406    0.36744 -10.843  < 2e-16 ***
I(la^2)            -4.21517    0.56228  -7.497 1.10e-13 ***
I(t^2)             -1.77607    0.26279  -6.758 1.97e-11 ***
I(lo * t)           0.20029    0.35117   0.570 0.568518
I(la * t)           1.82637    0.47332   3.859 0.000119 ***
I(lo^3)            -3.46452    0.49554  -6.991 4.03e-12 ***
I(la^3)             8.53152    1.28587   6.635 4.48e-11 ***
I(t^3)              0.70085    0.12397   5.653 1.87e-08 ***
I(lo * la * t)     -1.10150    0.90738  -1.214 0.224959
I(lo^2 * la)        5.20779    0.88873   5.860 5.65e-09 ***
I(lo * la^2)      -12.87497    1.24298 -10.358  < 2e-16 ***
I(lo^2 * t)         0.79928    0.54238   1.474 0.140774
I(la^2 * t)         5.42159    1.08911   4.978 7.14e-07 ***
I(la * t^2)        -1.14220    0.46440  -2.459 0.014021 *
I(lo * t^2)         0.65862    0.36929   1.783 0.074705 .
a                  -0.12285    0.02184  -5.624 2.21e-08 ***
I(a * t)            0.09456    0.04615   2.049 0.040635 *
I(t^2 * a)         -0.18310    0.05998  -3.053 0.002306 **
---
Sig. codes:  0 `***' 0.001 `**' 0.01 `*' 0.05 `.' 0.1 ` ' 1

(Dispersion parameter for quasi family taken to be 1.051635)

    Null deviance: 3108.86  on 1574  degrees of freedom
Residual deviance:  913.75  on 1552  degrees of freedom
AIC: NA

Number of Fisher Scoring iterations: 7
```

The summary information suggests dropping the `lo*t` term (it seems unreasonable to drop the constant altogether). Rather than re-type the whole `glm` command again, it is easier to use:

```
b1 <- update(b, ~ . - I(lo*t))
```

which re-fits the model, dropping the term specified. Repeating the process suggests dropping `lo*la*t`, `lo*t^2` and finally `lo^2*t`, after which all the remaining terms are significant at the 5% level. If `b4` is the final reduced model, then it can be tested against the full model:

```
> anova(b,b4,test="F")
Analysis of Deviance Table
[edited]
  Resid. Df Resid. Dev   Df Deviance      F Pr(>F)
1      1552     913.75
2      1556     919.28   -4    -5.54 1.3161 0.2618
```

which gives no reason not to accept the simplified model.

The default residual plots are unhelpful for this model, because of the large number of zeroes in the data, corresponding to areas where there really are no eggs. This

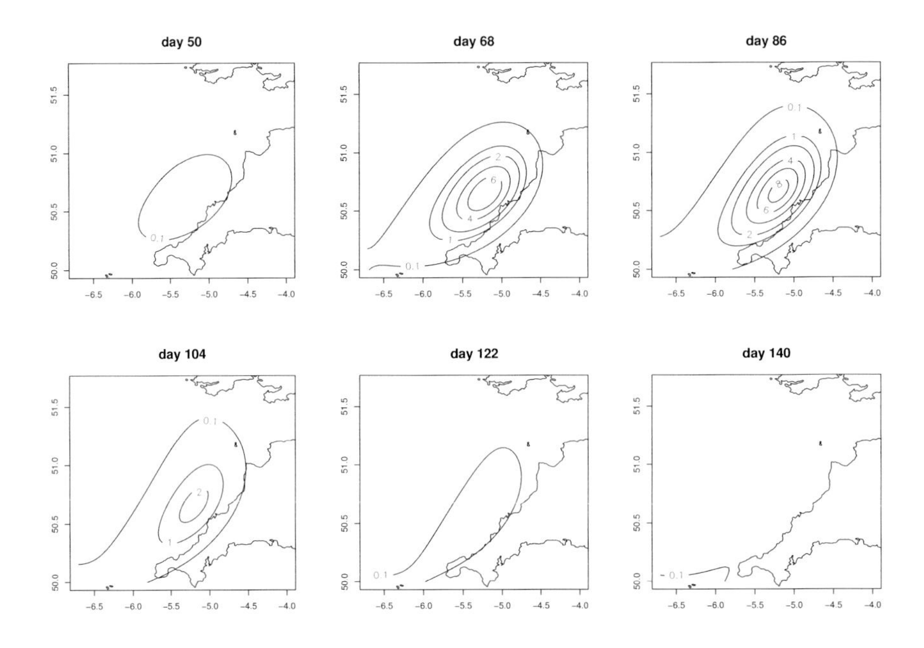

Figure 3.17 *Model predicted sole spawning rates in the Bristol channel at various times in the spawning season.*

tends to lead to some very small values for the linear predictor, corresponding to zero predictions, which in turn lead to rather distorted plots. The following residual plots are perhaps more useful.

```
par(mfrow=c(1,2)) # split graph window into 2 panels
plot(fitted(b4)^0.5,solr$eggs^0.5) # fitted vs. data plot
plot(fitted(b4)^0.5,residuals(b4)) # resids vs. sqrt(fitted)
```

The plots are shown in figure 3.16. The most noticeable features of both plots relate to the large number of zeros in the data, with the lower boundary line in the right hand plot corresponding entirely to zeros, for which the raw residual is simply the negative of the fitted value. The plots are clearly far from perfect, but it is unlikely that great improvements can be made to them with models of this general type.

The fitted model can be used for prediction of the spawning rate over the Bristol channel, by setting up a data frame containing the times and locations at which predictions are required, the age at which prediction is required — always zero — and the offset required — zero if spawning rate per square metre is the desired output. The time and location co-ordinates must be scaled in the same way as was done for fitting, of course. Figure 3.17 shows model predicted spawning rates produced in this way from model b4. It has been possible to get surprisingly far with the analysis of these data using a simple GLM approach, but the fitting did become somewhat unwieldy when it came to specifying that spawning rate should be a smooth function of location and time. For a less convenient spatial spawning distribution, it is doubtful that a satisfactory model could have been produced in this manner. This is part of the motivation for seeking to extend the way in which GLMs are specified, to allow a more compact and flexible way of specifying smooth functional relationships within the models, i.e., part of the motivation for developing GAMs.

3.4 Generalized linear mixed models

Recall that a GLMM models an exponential family random variable, Y_i, with expected value μ_i using

$$g(\mu_i) = \mathbf{X}_i\boldsymbol{\beta} + \mathbf{Z}_i\mathbf{b}, \quad \mathbf{b} \sim N(\mathbf{0}, \boldsymbol{\psi}_\theta).$$

That is, a GLMM is a GLM in which the linear predictor depends on some Gaussian random effects, $\mathbf{b}$, multiplied by a random effects model matrix $\mathbf{Z}$. The main difficulty in moving from linear mixed models to GLMMs is that it is no longer possible to evaluate the log likelihood exactly, since it is not generally analytically tractable to integrate $\mathbf{b}$ out of the joint density of $\mathbf{y}$ and $\mathbf{b}$ to obtain the likelihood.

One effective way to proceed is to follow the approach of section 2.4, Taylor expanding around, $\hat{\mathbf{b}}$, the mode of $f(\mathbf{y}, \mathbf{b}|\boldsymbol{\beta})$, to get

$$f(\mathbf{y}|\boldsymbol{\beta}) \simeq \int \exp\left\{\log f(\mathbf{y}, \hat{\mathbf{b}}|\boldsymbol{\beta}) + \frac{1}{2}(\mathbf{b}-\hat{\mathbf{b}})^\mathsf{T}\frac{\partial^2 \log f(\mathbf{y}, \mathbf{b}|\boldsymbol{\beta})}{\partial\mathbf{b}\partial\mathbf{b}^\mathsf{T}}(\mathbf{b}-\hat{\mathbf{b}})\right\} d\mathbf{b}.$$

This differs from section 2.4 in being approximate, since we have neglected the higher order terms in the Taylor expansion, rather than those terms being identically zero. However, having accepted this approximation the rest of the evaluation of the integral follows section 2.4 exactly. In the current case the required Hessian is $-\mathbf{Z}^\mathsf{T}\mathbf{W}\mathbf{Z}/\phi - \boldsymbol{\psi}^{-1}$, where $\mathbf{W}$ is the IRLS weight vector based on the $\boldsymbol{\mu}$ implied by $\hat{\mathbf{b}}$ and $\boldsymbol{\beta}$. So

$$f(\mathbf{y}|\boldsymbol{\beta}) \simeq f(\mathbf{y}, \hat{\mathbf{b}}|\boldsymbol{\beta})\frac{(2\pi)^{p/2}}{|\mathbf{Z}^\mathsf{T}\mathbf{W}\mathbf{Z}/\phi + \boldsymbol{\psi}_\theta^{-1}|^{1/2}}.$$

This sort of integral approximation is known as 'Laplace approximation'. Substituting the explicit form for the random effects density and taking logs gives

$$l(\boldsymbol{\theta}, \boldsymbol{\beta}) \simeq \log f(\mathbf{y}|\hat{\mathbf{b}}, \boldsymbol{\beta}) - \hat{\mathbf{b}}^\mathsf{T}\boldsymbol{\psi}_\theta^{-1}\hat{\mathbf{b}}/2 - \log|\boldsymbol{\psi}_\theta|/2 - \log|\mathbf{Z}^\mathsf{T}\mathbf{W}\mathbf{Z}/\phi + \boldsymbol{\psi}_\theta^{-1}|/2. \tag{3.16}$$

Notice that as well as direct dependencies, the right hand side depends on $\boldsymbol{\beta}$ and $\boldsymbol{\theta}$ via $\hat{\mathbf{b}}$ and $\mathbf{W}$. The dependence of $\mathbf{W}$ on $\boldsymbol{\beta}$ means that the MAP estimate and MLE of $\boldsymbol{\beta}$ no longer correspond exactly, in contrast to the LMM case. Nonetheless it is convenient to use the MAP estimates, since they can be computed easily along with the corresponding $\hat{\mathbf{b}}$, using a penalized version of the IRLS method used for GLMs. If we do this then we can also define a Laplace approximate profile likelihood $l_p(\boldsymbol{\theta}) = l(\boldsymbol{\theta}, \hat{\boldsymbol{\beta}})$, where $\hat{\boldsymbol{\beta}}$ is the MAP estimate given $\boldsymbol{\theta}$.

Laplace approximate REML also follows in a similar way, from the linear model case in section 2.4.5. If $\hat{\boldsymbol{\beta}}$ is the MAP estimate given $\boldsymbol{\theta}$,

$$l_r(\boldsymbol{\theta}) \simeq \log f(\mathbf{y}|\hat{\mathbf{b}}, \hat{\boldsymbol{\beta}}) - \hat{\mathbf{b}}^\mathsf{T}\psi_\theta^{-1}\hat{\mathbf{b}}/2 - \log|\psi_\theta|/2$$
$$- \log\begin{vmatrix} \mathbf{Z}^\mathsf{T}\mathbf{W}\mathbf{Z}/\phi + \psi_\theta^{-1} & \mathbf{Z}^\mathsf{T}\mathbf{W}\mathbf{X}/\phi \\ \mathbf{X}^\mathsf{T}\mathbf{W}\mathbf{Z}/\phi & \mathbf{X}^\mathsf{T}\mathbf{W}\mathbf{X}/\phi \end{vmatrix}/2 + M\log(2\pi)/2. \quad (3.17)$$

As in the LMM case l_p or l_r can be optimized numerically to find $\hat{\boldsymbol{\theta}}$, with each trial value for $\boldsymbol{\theta}$ requiring computation of the corresponding $\hat{\boldsymbol{\beta}}$, $\hat{\mathbf{b}}$. In contrast to the linear case, these computations are no longer direct, but also require iteration, but the penalized IRLS scheme used is generally fast and reliable.

Laplace approximation is discussed in more detail in Davison (2003) or Wood (2015). Shun and McCullagh (1995) show that it generally provides well founded inference provided that the number of random effects is increasing at no higher rate than $n^{1/3}$. For example, consider data with two or three non-Gaussian observations per subject, and a corresponding model with one random effect per subject: clearly Laplace approximation is unlikely to give valid results. In such situations it may be possible to use numerical integration (sometimes known as 'quadrature') to evaluate the (restricted) likelihood, provided that the required integral decomposes into a set of numerically tractable low dimensional integrals.[§]

3.4.1 *Penalized IRLS*

A penalized version of the IRLS algorithm of section 3.1.2 is used to compute $\hat{\boldsymbol{\beta}}$ and $\hat{\mathbf{b}}$. We seek

$$\hat{\boldsymbol{\beta}}, \hat{\mathbf{b}} = \underset{\beta,b}{\operatorname{argmax}} \log f(\mathbf{y}, \mathbf{b}|\boldsymbol{\beta}) = \underset{\beta,b}{\operatorname{argmax}} \left\{\log f(\mathbf{y}|\mathbf{b}, \boldsymbol{\beta}) - \mathbf{b}^\mathsf{T}\psi_\theta^{-1}\mathbf{b}/2\right\}, \quad (3.18)$$

where terms not dependent on $\mathbf{b}$ and $\boldsymbol{\beta}$ have been dropped from the final objective (which is often referred to as the *penalized likelihood*). To de-clutter the notation, write $\mathbf{b}$ and $\boldsymbol{\beta}$ in a single vector $\mathcal{B}^\mathsf{T} = (\mathbf{b}^\mathsf{T}, \boldsymbol{\beta}^\mathsf{T})$, and define corresponding model matrix and precision matrix

$$\mathcal{X} = (\mathbf{Z}, \mathbf{X}) \text{ and } \mathbf{S} = \begin{bmatrix} \psi_\theta^{-1} & \mathbf{0} \\ \mathbf{0} & \mathbf{0} \end{bmatrix}.$$

Following section 3.1.2, but now including the quadratic penalty term, the Hessian of the penalized likelihood is $-\mathcal{X}^\mathsf{T}\mathbf{W}\mathcal{X}/\phi - \mathbf{S}$, where $\mathbf{W}$ is the diagonal weight matrix (Fisher or full Newton) implied by the current $\hat{\boldsymbol{\mu}}$, which in turn depends on the current estimate $\mathcal{B}^{[k]}$. The corresponding gradient vector can be written as $\mathcal{X}^\mathsf{T}\mathbf{W}\mathbf{G}(\mathbf{y} - \hat{\boldsymbol{\mu}})/\phi - \mathbf{S}\mathcal{B}^{[k]}$, again following section 3.1.2.

A single Newton update step now has the form

$$\mathcal{B}^{[k+1]} = \mathcal{B}^{[k]} + (\mathcal{X}^\mathsf{T}\mathbf{W}\mathcal{X} + \phi\mathbf{S})^{-1}\{\mathcal{X}^\mathsf{T}\mathbf{W}\mathbf{G}(\mathbf{y} - \hat{\boldsymbol{\mu}}) - \phi\mathbf{S}\mathcal{B}^{[k]}\},$$

[§] Direct numerical integration of high dimensional integrals without special structure is generally computationally infeasible, since if we need m function evaluations to perform a one-dimensional integral, we will generally need of the order of m^d evaluations for a d dimensional integral.

but substituting $\mathcal{B}^{[k]} = (\mathcal{X}^\mathsf{T}\mathbf{W}\mathcal{X} + \phi\mathbf{S})^{-1}(\mathcal{X}^\mathsf{T}\mathbf{W}\mathcal{X} + \phi\mathbf{S})\mathcal{B}^{[k]}$ into this expression and re-arranging we have

$$\mathcal{B}^{[k+1]} = (\mathcal{X}^\mathsf{T}\mathbf{W}\mathcal{X} + \phi\mathbf{S})^{-1}\mathcal{X}^\mathsf{T}\mathbf{W}\{\mathbf{G}(\mathbf{y} - \hat{\boldsymbol{\mu}}) + \mathcal{X}\mathcal{B}^{[k]}\},$$

which is immediately recognisable as the minimizer of the penalized weighted least squares objective¶

$$\|\mathbf{z} - \mathcal{X}\mathcal{B}\|_W^2 + \phi\mathcal{B}^\mathsf{T}\mathbf{S}\mathcal{B} = \|\mathbf{z} - \mathbf{X}\boldsymbol{\beta} - \mathbf{Z}\mathbf{b}\|_W^2 + \phi\mathbf{b}^\mathsf{T}\psi_\theta^{-1}\mathbf{b}$$

where $z_i = g'(\hat{\mu}_i)(y_i - \hat{\mu}_i) + \hat{\eta}_i$ exactly as in section 3.1.2. The penalized IRLS (PIRLS) algorithm is then...

1. Initialize $\hat{\mu}_i = y_i + \delta_i$ and $\hat{\eta}_i = g(\hat{\mu}_i)$, where δ_i is usually zero, but may be a small constant ensuring that $\hat{\eta}_i$ is finite. Iterate the following two steps to convergence.
2. Compute pseudodata $z_i = g'(\hat{\mu}_i)(y_i - \hat{\mu}_i)/\alpha(\hat{\mu}_i) + \hat{\eta}_i$, and iterative weights $w_i = \alpha(\hat{\mu}_i)/\{g'(\hat{\mu}_i)^2 V(\hat{\mu}_i)\}$.
3. Find $\hat{\boldsymbol{\beta}}, \hat{\mathbf{b}}$, to minimize the weighted least squares objective

$$\|\mathbf{z} - \mathbf{X}\boldsymbol{\beta} - \mathbf{Z}\mathbf{b}\|_W^2 + \phi\mathbf{b}^\mathsf{T}\phi_\theta^{-1}\mathbf{b}$$

   and then update $\hat{\boldsymbol{\eta}} = \mathbf{X}\hat{\boldsymbol{\beta}} + \mathbf{Z}\hat{\mathbf{b}}$ and $\hat{\mu}_i = g^{-1}(\hat{\eta}_i)$.

Convergence is now judged according to the penalized likelihood/deviance or its gradient.

3.4.2 *The PQL method*

The nested optimization required to optimize the Laplace approximate profile or restricted log likelihoods is somewhat involved and can be computationally costly. It is therefore tempting to instead perform a PIRLS iteration, estimating $\boldsymbol{\theta}, \phi$ at each step based on the working mixed model:

$$\mathbf{z}|\mathbf{b}, \boldsymbol{\beta} \sim N(\mathbf{X}\boldsymbol{\beta} + \mathbf{Z}\mathbf{b}, \mathbf{W}^{-1}\phi), \quad \mathbf{b} \sim N(\mathbf{0}, \psi_\theta),$$

where Fisher weights are used. That is, step 3 of the PIRLS algorithm finds $\hat{\boldsymbol{\theta}}, \hat{\boldsymbol{\beta}}, \hat{\mathbf{b}}, \hat{\phi}$ by fitting the above working linear mixed model using the methods of section 2.4 (p. 78). This approach is known as the PQL (penalized quasi-likelihood‖) method in the context of mixed models (Breslow and Clayton, 1993). In the spline smoothing context the approach is known as 'performance oriented iteration' (Gu, 1992).

The obvious question is why it should be valid to use the Gaussian linear mixed model profile or restricted likelihood, when $\mathbf{z}|\mathbf{b}, \boldsymbol{\beta}$ clearly does not have a Gaussian distribution. Consider the restricted likelihood for the working model under a

¶Recall $\|\mathbf{x}\|_W^2 = \mathbf{x}^\mathsf{T}\mathbf{W}\mathbf{x}$.

‖The names arises from the way that the method was originally justified, which was somewhat different to the approach taken here.

Gaussian assumption for $\mathbf{z}$. Following sections 2.4.5 and 2.4.6 we have

$$2l_r = -\|\mathbf{z} - \mathcal{X}\hat{\mathcal{B}}\|_W^2/\phi - \hat{\mathcal{B}}^\mathsf{T}\mathbf{S}\hat{\mathcal{B}} - n\log\phi + \log|\mathbf{S}|_+ \\ - \log|\mathcal{X}^\mathsf{T}\mathbf{W}\mathcal{X}/\phi + \mathbf{S}| - (n-M)\log(2\pi), \quad (3.19)$$

where $\log|\mathbf{S}|_+$ ($= -\log|\psi|_+$) is the product of the non zero eigenvalues of $\mathbf{S}$ (a generalized determinant). From section 2.4.6 (p. 83) the REML estimate of ϕ is then

$$\hat{\phi} = \frac{\|\mathbf{z} - \mathcal{X}\hat{\mathcal{B}}\|_W^2}{n-\tau} \text{ where } \tau = \text{tr}\{(\mathcal{X}^\mathsf{T}\mathbf{W}\mathcal{X}/\phi + \mathbf{S})^{-1}\mathcal{X}^\mathsf{T}\mathbf{W}\mathcal{X}/\phi\}.$$

This estimate is also just the Pearson estimate of section 3.1.5 with effective degrees of freedom (see section 2.4.6) in place of the number of model coefficients: i.e., it is a reasonable estimator without any REML justification.

Now consider the QR decomposition $\mathbf{QR} = \sqrt{\mathbf{W}}\mathcal{X}$, where $\mathbf{Q}$ is column orthogonal and $\mathbf{R}$ is upper triangular (remember that Fisher weights are being used here), and let $\mathbf{f} = \mathbf{Q}^\mathsf{T}\sqrt{\mathbf{W}}\mathbf{z}$. Then $\|\mathbf{f} - \mathbf{R}\mathcal{B}\|^2 + c = \|\mathbf{z} - \mathcal{X}\mathcal{B}\|_W^2$ where $c = \|\mathbf{z}\|_W^2 - \|\mathbf{f}\|^2$, and following section 1.3.2 (p. 13) we have $\mathbb{E}(\mathbf{f}) = \mathbf{R}\mathcal{B}$ with the covariance matrix of $\mathbf{f}$ being $\mathbf{I}\phi$. Now if $n/(M+p) \to \infty$ the multivariate central limit theorem implies that $\mathbf{f}$ has a multivariate normal distribution. So without assuming normality of $\mathbf{z}$, we can legitimately base inference about $\mathcal{B}$ and $\boldsymbol{\theta}$ on the Gaussian linear mixed model

$$\mathbf{f}|\mathcal{B} \sim N(\mathbf{R}\mathcal{B}, \mathbf{I}\phi) \text{ where } \mathbf{b} \sim N(\mathbf{0}, \psi_\theta),$$

for which the restricted likelihood is given by

$$2l_r = -\|\mathbf{f} - \mathbf{R}\hat{\mathcal{B}}\|^2/\phi - \hat{\mathcal{B}}^\mathsf{T}\mathbf{S}\hat{\mathcal{B}} - p\log\phi + \log|\mathbf{S}|_+ \\ - \log|\mathbf{R}^\mathsf{T}\mathbf{R}/\phi + \mathbf{S}| - p\log(2\pi). \quad (3.20)$$

For fixed ϕ, (3.20) and (3.19) are identical to within an additive constant and hence will be maximized by the same $\hat{\boldsymbol{\theta}}$ and $\hat{\mathcal{B}}$. It would be foolish to base inference about ϕ on (3.20), because that would ignore the information about ϕ contained in c. However, if we plug the above Pearson estimate, $\hat{\phi}$, into (3.20) and optimize for $\boldsymbol{\theta}$ and $\mathcal{B}$ then we obtain exactly the $\hat{\boldsymbol{\theta}}$, $\hat{\mathcal{B}}$ and $\hat{\phi}$ that optimize (3.19). This then, is the justification for the PQL approach. Note that we could justify using the profile likelihood of the working model in the same way.

In many cases PQL is highly effective and produces estimates very close to the full Laplace approximation based estimates, but for some types of data it is problematic. Firstly, performance is likely to be poor whenever the Pearson scale estimator is poor, such as when modelling over-dispersed low-mean count data (see section 3.1.5). Secondly, some implementations of PQL use standard linear mixed modelling software for which there is no option to fix ϕ. In many cases estimating a supposedly known ϕ is not a problem, and can be viewed as a useful model check, but it leads to poor performance in the case of binary data, where there is almost no information with which to estimate ϕ. Thirdly, situations in which there are only a few observations per random effect undermine the central limit theorem justification for

PQL. A final general disadvantage is that we do not have access to the restricted or profile likelihood for the model itself, so that GLR testing or AIC based on those is not possible.

3.4.3 *Distributional results*

The argument that justifies PQL also justifies using (2.16) and (2.17) from section 2.4.2 (p. 80), with $\mathbf{\Lambda}_\theta^{-1} = \mathbf{W}^{-1}\phi$, as large sample distributional results for approximate inference about $\boldsymbol{\beta}$, and the predictive distribution of $\mathbf{b}$. That is, (2.16) becomes the large sample result

$$\hat{\boldsymbol{\beta}} \sim N(\boldsymbol{\beta}, \{\mathbf{X}^\mathsf{T}(\mathbf{Z}\boldsymbol{\psi}\mathbf{Z}^\mathsf{T} + \mathbf{W}^{-1}\phi)^{-1}\mathbf{X}\}^{-1}),$$

although it is often preferable to compute using the large sample result

$$\mathcal{B}|\mathbf{y} \sim N(\hat{\mathcal{B}}, (\mathcal{X}^\mathsf{T}\mathbf{W}\mathcal{X}/\phi + \mathbf{S})^{-1}), \tag{3.21}$$

based on (2.17), where the covariance matrix for $\boldsymbol{\beta}$ implied by the second result again corresponds to the covariance matrix for $\hat{\boldsymbol{\beta}}$ implied by the first.

Exactly as in the linear model case (section 2.4.3) we can use the large sample result

$$\hat{\boldsymbol{\theta}} \sim N(\boldsymbol{\theta}, \hat{\mathcal{I}}^{-1})$$

where $\hat{\mathcal{I}}$ is the negative Hessian of the profile or restricted likelihood, based on either the PQL working linear mixed model, or on the full Laplace approximation.

3.5 GLMMs with R

GLMMs estimated by PQL are implemented by routine `glmmPQL` supplied with Venables and Ripley's `MASS` library, one of the recommended packages in R. The `glmer` function from the `lme4` package implements the full Laplace approximation and also has facilities for quadrature based evaluation of the likelihood. For simple i.i.d. random effects `gam` in package `mgcv` is also a possibility, again using full Laplace approximation.

3.5.1 `glmmPQL`

`glmmPQL` calls are much like `lme` calls except that it is now necessary to supply a `family`. Also `glmmPQL` will accept offsets, unlike `lme` (at time of writing). As you would expect from section 3.4.2, `glmmPQL` operates by iteratively calling `lme`, and it returns the fitted `lme` model object for the working model at convergence.

To illustrate its use, consider again the sole egg modelling undertaken in section 3.3.5. One issue with the data used in that exercise is that, at any sampling station, the counts for the four different egg stages are all taken from the same net sample. It is therefore very likely that there is a 'sampling station' component to the variance of the data, which means that the data for different stages at a station can not be treated as independent. This effect can easily be checked by examining the residuals from the final model in section 3.3.5 for evidence of a 'station effect'.

```
> rf <- residuals(b4,type="d") # extract deviance residuals
> ## create an identifier for each sampling station
> solr$station <- factor(with(solr,paste(-la,-lo,-t,sep="")))
>
> ## is there evidence of a station effect in the residuals?
> solr$rf <-rf
> rm <- lme(rf~1,solr,random=~1|station)
> rm0 <- lm(rf~1,solr)
> anova(rm,rm0)
    Model df     AIC     BIC   logLik   Test L.Ratio p-value
rm      1  3 3319.57 3335.66 -1656.79
rm0     2  2 3582.79 3593.52 -1789.40 1 vs 2 265.220  <.0001
```

The above output compares two models for the residuals. `rm0` models them as i.i.d. normal random variables, with some unknown mean, while `rm` models the residuals as having a mean depending on a station dependent random effect. The GLRT test, performed by the `anova` function, so clearly rejects the i.i.d. model that we probably don't need to worry that its assumptions are not strictly met. AIC also suggests that there is a real sampling station effect.

One way of modelling the sampling station effect would be to suppose that the mean of each stage count, at each station, is multiplied by a log-normal random variable, $\exp(b_i)$, where $b_i \sim N(0, \sigma_b^2)$ is a station specific random effect. The resulting GLMM can be estimated as follows.

```
> form <- eggs ~ offset(off)+lo+la+t+I(lo*la)+I(lo^2)+
            I(la^2)+I(t^2)+I(lo*t)+I(la*t)+I(lo^3)+I(la^3)+
            I(t^3)+I(lo*la*t)+I(lo^2*la)+I(lo*la^2)+I(lo^2*t)+
            I(la^2*t)+I(la*t^2)+I(lo*t^2)+ # end log spawn
            a +I(a*t)+I(t^2*a)
> b <- glmmPQL(form,random=list(station=~1),
            family=quasi(link=log,variance="mu"),data=solr)

> summary(b)

[edited]
                   Value Std.Error   DF    t-value p-value
(Intercept)     0.025506 0.1813214 1178   0.140665  0.8882
lo              4.643018 0.5179583  374   8.964077  0.0000
la             -4.878785 0.6313637  374  -7.727375  0.0000
t              -2.101037 0.3091183  374  -6.796872  0.0000
I(lo * la)      4.221226 0.8216752  374   5.137342  0.0000
I(lo^2)        -4.895147 0.4573844  374 -10.702480  0.0000
I(la^2)        -5.187457 0.7565845  374  -6.856415  0.0000
I(t^2)         -1.469416 0.1961255  374  -7.492220  0.0000
I(lo * t)      -0.246946 0.3646591  374  -0.677197  0.4987
I(la * t)       1.578309 0.4576964  374   3.448376  0.0006
I(lo^3)        -3.956541 0.6598010  374  -5.996567  0.0000
I(la^3)         5.524490 1.5175128  374   3.640490  0.0003
I(t^3)          0.633109 0.1359888  374   4.655593  0.0000
```

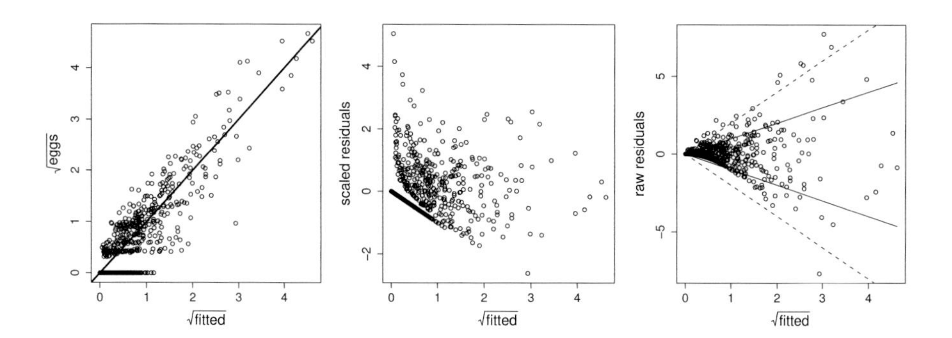

Figure 3.18 *Model checking plots for the GLMM of the Bristol channel sole data. The left panel shows the relationship between raw data and fitted values. The middle panel plots Pearson residuals against fitted values: there are a handful of rather high residuals at low predicted densities. The right panel shows raw residuals against fitted values, with reference lines illustrating where 1 residual standard deviation and 2 residual standard deviations from the residual mean should lie, for each fitted value.*

```
I(lo * la * t)   -0.040474 0.8643458  374  -0.046826  0.9627
I(lo^2 * la)      6.700204 1.0944157  374   6.122175  0.0000
I(lo * la^2)    -11.539919 1.5509149  374  -7.440717  0.0000
I(lo^2 * t)       0.517189 0.6008440  374   0.860770  0.3899
I(la^2 * t)       4.879013 1.0328137  374   4.724001  0.0000
I(la * t^2)      -0.548011 0.4099971  374  -1.336623  0.1822
I(lo * t^2)       0.822542 0.3576837  374   2.299635  0.0220
a                -0.121312 0.0125463 1178  -9.669168  0.0000
I(a * t)          0.092769 0.0270447 1178   3.430218  0.0006
I(t^2 * a)       -0.187472 0.0362511 1178  -5.171498  0.0000

[edited]
```

The summary** suggests dropping `I(lo * la * t)`. Refitting without this term, and then examining the significance of terms in the resulting model, suggests dropping `I(lo*t)`. Continuing in the same way we drop `I(lo^2*t)` and `I(la*t^2)`, before all terms register as significant at the 5% level. Note that GLR testing is not possible with models estimated in this way — we do not actually know the value of the maximized likelihood for the model. Supposing that the final fitted model is again called `b4`, some residual plots can now be produced.

```
fv <- exp(fitted(b4)+solr$off) # note need to add offset
resid <- solr$egg-fv           # raw residuals
plot(fv^.5,solr$eggs^.5); abline(0,1,lwd=2)
plot(fv^.5,resid/fv^.5); plot(fv^.5,resid)
```

** `summary(b)` could have been replaced by `anova(b,type="marginal")`, the latter being the more useful function for models with factors.

```
fl <- sort(fv^.5)
## add 1 s.d. and 2 s.d. reference lines
lines(fl,fl);lines(fl,-fl);lines(fl,2*fl,lty=2)
lines(fl,-2*fl,lty=2)
```

The results are shown in figure 3.18. Comparison of these plots with the equivalent plots in section 3.3.5 highlights how substantial the station effect appears to be, and this is emphasized by examining the estimated station effect.

```
> intervals(b4,which="var-cov")
Approximate 95% confidence intervals

 Random Effects:
  Level: station
                      lower      est.    upper
sd((Intercept)) 0.8398715 0.9599066 1.097097

 Within-group standard error:
    lower      est.     upper
0.5565463 0.5777919 0.5998485
```

The station effect is somewhat larger than the variability in the working residuals.

Figure 3.19 shows the predicted spawning rates over the Bristol channel, at various times of year. Relative to the predictions made using GLMs in section 3.3.5, the peak spawning densities are now slightly lower; however, given the clear evidence for a station effect, the current model is probably better supported than the GLM.

3.5.2 `gam`

Function `gam` from R package `mgcv` can fit models with simple i.i.d. Gaussian random effects, such as those in the sole egg model just estimated, using Laplace approximate restricted or profile likelihood. The station random effect would be specified by adding the terms `s(station,bs="re")` to the model formula, `form`, above. The following function call then fits exactly the same model as previously

```
b <- gam(form,family=quasi(link=log,variance="mu"),data=solr,
         method="REML")
```

Since quasi families do not have full likelihoods `gam` uses the notion of extended quasi-likelihood from McCullagh and Nelder (1989, section 9.6). A reasonable alternative here would also be to use the Tweedie distribution (see section 3.1.9): family `Tweedie` can be used for this with fixed p parameter, or `tw` can be used if p should be estimated as part of fitting. The variance component estimates can be extracted using `gam.vcomp(b)`, while `summary` and `anova` method functions can be used in the same way as with `lm`.

Prediction from the fitted model uses the `predict.gam` function. Using something like `predict(b,newd,exclude="s(station)")` would give predictions in which the random effects for `station` are set to zero. Prediction plots and residuals are very similar to those shown in figures 3.19 and 3.18.

Generically `s(...,bs="re")` terms can accept any number of covariates,

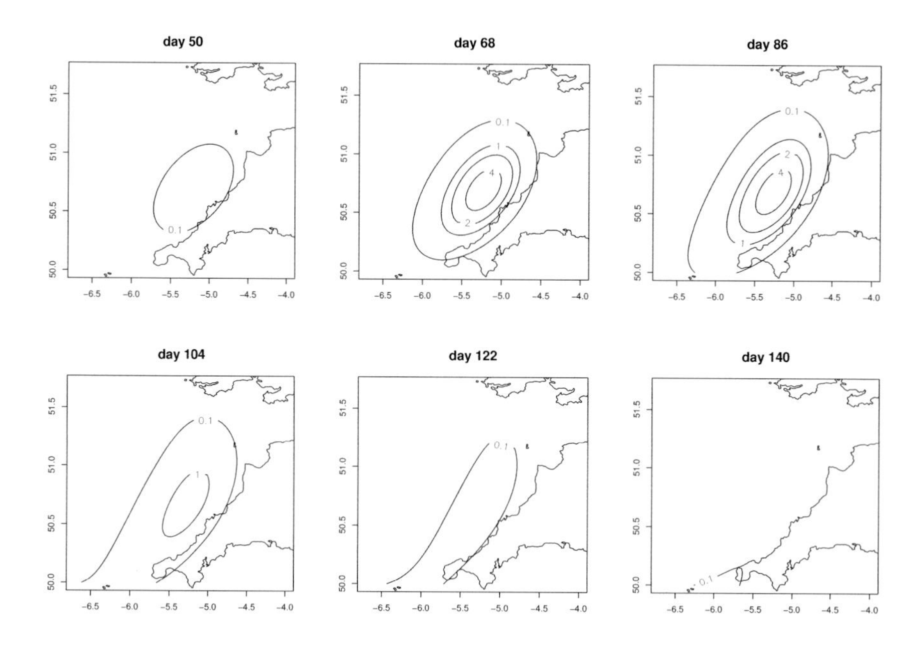

Figure 3.19 *Predicted spawning distributions of Bristol channel sole according to the GLMM of section 3.5. Notice how the spawning distributions are less peaked than those shown in figure 3.17.*

which can be a mixture of factor and/or metric variables. If more than one covariate is supplied then a random interaction of the covariates is produced. For example `s(a,b,c,bs="re")` contributes a random effect with model matrix generated by `model.matrix(~a:b:c-1)` and covariance matrix proportional to an identity matrix. So if `x` is a metric variable and `g` a grouping factor and we want to include a random intercept and random slope against `x` for each level of `g`, we would include

```
s(g,by="re") + s(x,g,bs="re")
```

All the slopes and intercepts would be independent. Note that `s(x,bs="re")` would produce a single random slope parameter applying to all observations — this is rarely sensible.

3.5.3 `glmer`

`glmer` operates just like `lmer`, but also accepts a `glm family` object specifying the distribution and link to use. At time of writing it is not possible to use `quasi` families with `glmer` and there appears to be an incompatibility between `glmer` and available Tweedie families. However, if we are willing to accept a quasi-Poisson model as appropriate, then a reasonable approximation is to round the egg densities, after multiplication by a multiplier, and treat the result as Poisson distributed. For example if we set `solr$egg1 <- round(solr$egg * 5)` then we can fit the model, by appending `(1|station)` to the original `form`, while replacing `egg` by `egg1` as the response:

```
glmer(form,family=poisson,data=solr)
```

The resulting residual plots are acceptable and the correlation coefficient between the fitted values and those produced by `gam`, above, is over 0.9999. Of course, when predicting we have to remember to correct for the fact that we are now modelling 5× egg density. Some experimentation was required to find the factor of 5 which gives no apparent over-dispersion, while not losing too much information to rounding.

3.6 Exercises

1. A Bernoulli random variable, Y, takes the value 1 with probability p and 0 with probability $1-p$, so that its probability function is:

$$f(y) = p^y(1-p)^{1-y}, \quad y = 0 \text{ or } 1.$$

   (a) Find $\mu \equiv E(Y)$.

   (b) Show that the Bernoulli distribution is a member of the exponential family of distributions, by showing that its probability function can be written as

$$f(y) = \exp\left[\{y\theta - b(\theta)\}/a(\phi) + c(y,\phi)\right],$$

   for appropriately defined θ, $b(\theta)$, ϕ, a and c. (See section 3.1.1.)

   (c) What will the canonical link be for the Bernoulli distribution?

2. Residual checking for non-Gaussian error models is not always as straightforward as it is in the Gaussian case, and the problems are particularly acute in the case of binary data. This question explores this issue.

   (a) The following code fits a GLM to data simulated from a simple binomial model and examines the default residual plots.

```
n <- 100; m <- 10
x <- runif(n)
lp <- 3*x-1
mu <- binomial()$linkinv(lp)
y <- rbinom(1:n,m,mu)
par(mfrow=c(2,2))
plot(glm(y/m ~ x,family=binomial,weights=rep(m,n)))
```

   Run the code several times to get a feel for the range of results that are possible even when the model is correct (as it clearly is in this case).

   (b) Explore how the plots change as m (the number of binomial trials) is reduced to 1. Also examine the effect of sample size n.

   (c) By repeatedly simulating data from a fitted model, and then refitting to the simulated data, you can build up a picture of how the residuals should behave when the distributional assumption is correct, and the data are really independent. Write code to take a `glm` object fitted to binary data, simulate binary data given the model fitted values, refit the model to the resulting data, and extract residuals from the resulting fits. Functions `fitted` and `rbinom` are useful here.

(d) If `rsd` contains your residual vector then

```
plot(sort(rsd),(1:length(rsd)-.5)/length(rsd))
```

produces a plot of the 'empirical CDF' of the residuals, which can be useful for characterizing their distribution. By repeatedly simulating residuals, as in the previous part, you can produce a 'simulation envelope' showing, e.g., where the middle 95% of these 'empirical CDFs' should lie, if the model assumptions are met: such envelopes provide a guide to whether an observed 'real' residual distribution is reasonable, or not. Based on your answer to the previous part, write code to do this.

(e) Plots of the residuals against fitted values, or predictors, are also hard to interpret for models of binary data. A simple check for lack of independence in the residuals can be obtained by ordering the residuals according to the values of the fitted values or a predictor, and checking whether these ordered residuals show fewer (or more) runs of values above and below zero than they should if independent. The command

`rsd <- rsd[sort(fv,return.index=TRUE)$ix]` will put `rsd` into the order corresponding to increasing `fv`. It is possible to simulate from the distribution of runs of residuals that should occur, under independence, as part of the simulation loop used in the previous part: modify your code to check whether the residuals appear non-independent with respect to fitted values.

3. This question looks at data covering guilty verdicts involving multiple murders in Florida between 1976 and 1987. The data are classified by skin 'colour' of victim and defendant and by whether or not the sentence was death.

Victim's Race	Defendant's Race	Death Penalty	No Death Penalty
White	White	53	414
	Black	11	37
Black	White	0	16
	Black	4	139

Data are from Radelet and Pierce (1991) *Florida Law Review:431-34*, as reported in Agresti (1996).

(a) What proportion of black defendants and what proportion of white defendants were sentenced to death?

(b) Find the log-linear model which best explains these data.

(c) How would you interpret your best fit model?

4. If a random variable, Y_i, has expected value μ_i, and variance $\phi V(\mu_i)$, where ϕ is a scale parameter and $V(\mu_i) = \mu_i$, then find the the quasi-likelihood of μ_i, given an observation y_i, by evaluation (3.12). Confirm that the corresponding deviance is equivalent to the deviance obtained if $Y_i \sim \text{Poi}(\mu_i)$.

5. If a linear model is to be estimated by minimizing $\sum_i w_i(y_i - \mathbf{X}_i\boldsymbol{\beta})^2$ w.r.t. $\boldsymbol{\beta}$,

show that the formal expression for the resulting parameter estimates is $\hat{\boldsymbol\beta} = (\mathbf{X}^\mathsf{T}\mathbf{W}\mathbf{X})^{-1}\mathbf{X}^\mathsf{T}\mathbf{W}\mathbf{y}$ where $\mathbf{W}$ is a diagonal matrix such that $W_{i,i} = w_i$. (Note that the expression is theoretically useful, but would not be used for practical computation).

6. This question relates to the IRLS method of section 3.1.2, and gives an alternative insight into the distribution of the parameter estimators $\hat{\boldsymbol\beta}$. Let y_i be independent random variables with mean μ_i, such that $g(\mu_i) = \eta_i \equiv \mathbf{X}_i\boldsymbol\beta$, where g is a link function, $\mathbf{X}$ a model matrix and $\boldsymbol\beta$ a parameter vector. Let the variance of y_i be $V(\mu_i)\phi$, where V is a known function, and ϕ a scale parameter. Define

$$z_i = g'(\mu_i)(y_i - \mu_i) + \eta_i \text{ and } w_i = \left\{V(\mu_i)g'(\mu_i)^2\right\}^{-1}.$$

   (a) Show that $\mathbb{E}(z_i) = \mathbf{X}_i\boldsymbol\beta$.
   (b) Show that the covariance matrix of $\mathbf{z}$ is $\mathbf{W}^{-1}\phi$, where $\mathbf{W}$ is a diagonal matrix with $W_{i,i} = w_i$.
   (c) If $\boldsymbol\beta$ is estimated by minimization of $\sum_i w_i(z_i - \mathbf{X}_i\boldsymbol\beta)^2$ show that the covariance matrix of the resulting estimates, $\hat{\boldsymbol\beta}$, is $(\mathbf{X}^\mathsf{T}\mathbf{W}\mathbf{X})^{-1}\phi$, and find $\mathbb{E}(\hat{\boldsymbol\beta})$.
   (d) The multivariate version of the central limit theorem implies that as the dimension of $\mathbf{z}$ tends to infinity, $\mathbf{X}^\mathsf{T}\mathbf{W}\mathbf{z}$ will tend to multivariate Gaussian. What does this imply about the large sample distribution of $\hat{\boldsymbol\beta}$?

7. Write R code to implement an IRLS scheme to fit the GLM defined in section 3.3.2 to the data given there. Use the `lm` function to fit the working linear model at each iteration.

8. This question is about GLMs for quite unusual models, handling non-linear parameters, and direct likelihood maximization as a flexible alternative to using GLMs. Data frame `harrier` has 2 columns: `Consumption.Rate` of Grouse by Hen Harriers (per day), and the corresponding `Grouse.Density` (per km^2). They have been digitized from figure 1 of Smout et al. (2010). Ecological theory suggests that the expected consumption rate, c, should be related to grouse density, d, by the model,

$$\mathbb{E}(c_i) = \frac{ad_i^m}{1 + atd_i^m},$$

   where a, t and m are unknown parameters. It is expected that the variance in consumption rate is proportional to the mean consumption rate.

   (a) Show that, for fixed m, a GLM relating c_i and d_i can be obtained by use of the reciprocal link.
   (b) For $m = 1$ estimate the model using `glm` with the `quasi` family.
   (c) Plot the model residuals against Grouse density, and interpret the plot.
   (d) Search for the approximate value of m which minimizes the model deviance, by repeatedly re-fitting the model with alternative trial values.
   (e) For your best fit model, with the optimal m, produce a plot showing the curve of predicted consumption against density overlaid on the raw data. Using the R function `predict`, with the `se` argument set to `TRUE`, add approximate 95% confidence limits to the plot.

(f) A more systematic way of fitting this model is to write a function, which takes the model parameters, and the consumption and density vectors as arguments, and evaluates the model likelihood (or quasi-likelihood in the current context). This likelihood can then be maximized using the R built in optimizer, `optim`. Write R code to do this (see question 4 for the quasi-likelihood).

Note that the `optim` will optionally return the approximate Hessian of a likelihood; so this general approach gives you easy access to everything you need for approximate inference about the parameters of the model, using the results covered in Appendix A. The approach is very general: it would be easy to estimate the full model discussed in Smout et al. (2010) in the same way.

9. R data frame `ldeaths` contains monthly death rates from 3 lung diseases in the UK over a period of several years (see `?ldeaths` for details and reference). One possible model for the data is that they can be treated as Poisson random variables with a seasonal component and a long term trend component, as follows:

$$\mathbb{E}(\texttt{deaths}_i) = \beta_0 + \beta_1 t_i + \alpha \sin(2\pi \texttt{toy}_i/12 + \phi),$$

where β_0, β_1, α and ϕ are parameters, t_i is time since the start of the data, and `toy` is time of year, in months (January being month 1).

(a) By making use of basic properties of sines and cosines, get this model into a form suitable for fitting using `glm`, and fit it. Use `as.numeric(ldeaths)` to treat `ldeaths` as a regular numeric vector rather than a time series object.

(b) Plot the raw data time series on a plot, with the predicted time-series overlaid.

(c) Is the model an adequate fit?

10. In 3.3.2 an approximate confidence interval for β_1 was found using the large sample distribution of the GLM parameter estimators. An alternative method of confidence interval calculation is sometimes useful, based on inversion of the generalized likelihood ratio test (see sections 3.1.4, A.5 and A.6). This works by finding the range of values of β_1 that would have been accepted as null hypotheses about β_1, using a GLRT. If the threshold for acceptance is 5% the resulting range gives a 95% confidence interval, if the threshold is 1% we get a 99% interval, and so on.

(a) Using `glm`, refit the AIDS model, with the quadratic time dependence, and save the resulting object.

(b) Write a loop which refits the same model for a sequence of *fixed* β_1 values centred on the MLE from part (a) and stores the resulting model log-likelihoods. In this step you are fitting the model under a sequence of null hypotheses about β_1. The way to fix β_1 in the model fit is to use an `offset` term in your model. e.g., if you want to fix β_1 at 0.5, you would replace the `t` term, in your model formula, with the term `offset(.5*t)`.

(c) Plot the log-likelihoods from the last part against the corresponding β_1 values, and add to your plot the line above which the log-likelihoods are high enough that the corresponding β_1 values would be included in a 95% confidence interval. From your plot, read off the 95% CI for β_1 and compare it to the interval obtained previously.

11. In an experiment comparing two varieties of soybeans, plants were grown on 48 plots and measurements of leaf weight were taken at regular intervals as the plants grew. The `nlme` data frame `Soybean` contains the resulting data and has the following columns:

    `Plot` is a factor variable with levels for each of the 48 plots.

    `weight` is the leaf weight in grammes.

    `Time` is the time in days since planting.

    `Variety` is either `F` or `P` indicating the variety of soybean.

    There is one observation for each variety in each plot at each time. Interest focuses on modelling the growth of soybeans over time and on establishing whether or not this differs between the varieties.
    In this question you may have to increase `niterEM` in the `control` list for `lme` and `glmmPQL`: see `?lmeControl`.

    (a) A possible model for the weights is

    $$w_{ijk} = \alpha_i + \beta_i t_k + \gamma_i t_k^2 + \delta_i t_k^3 + a_j + b_j t_k + \epsilon_{ijk}$$

    where w_{ijk} is the weight measurement for the ith variety in the jth plot at the kth time; $[a_j, b_j]^{\mathsf{T}} \sim N(\mathbf{0}, \boldsymbol{\psi})$ where $\boldsymbol{\psi}$ is a covariance matrix, and $\epsilon_{ijk} \sim N(0, \sigma^2)$. The random effects are independent of the residuals and independent of random effects with different j. The residuals are i.i.d.

    Fit this model using `lme` and confirm that the residual variance appears to increase with the (random effect conditional) mean.

    (b) To deal with the mean variance relationship, it might be appropriate to model the weights as being Gamma distributed, so that the model becomes a GLMM. e.g.

    $$\log(\mu_{ijk}) = \alpha_i + \beta_i t_k + \gamma_i t_k^2 + \delta_i t_k^3 + a_j + b_j t_k,$$

    where $\mu_{ijk} \equiv \mathbb{E}(w_{ijk})$ and $w_{ijk} \sim$ Gamma. Fit this GLMM, using `glmmPQL` from the `MASS` package, and confirm the improvement in residual plots that results.

    (c) Explore whether further improvements to the model could be made by modifications of the random or fixed effects model structures.

Chapter 4

Introducing GAMs

4.1 Introduction

A generalized additive model (Hastie and Tibshirani, 1986, 1990) is a generalized linear model with a linear predictor involving a sum of smooth functions of covariates. In general the model has a structure something like

$$g(\mu_i) = \mathbf{A}_i\boldsymbol{\theta} + f_1(x_{1i}) + f_2(x_{2i}) + f_3(x_{3i}, x_{4i}) + \ldots \tag{4.1}$$

where $\mu_i \equiv \mathbb{E}(Y_i)$ and $Y_i \sim \text{EF}(\mu_i, \phi)$. Y_i is a response variable, $\text{EF}(\mu_i, \phi)$ denotes an exponential family distribution with mean μ_i and scale parameter, ϕ, $\mathbf{A}_i$ is a row of the model matrix for any strictly parametric model components, $\boldsymbol{\theta}$ is the corresponding parameter vector, and the f_j are smooth functions of the covariates, x_k. The model allows for flexible specification of the dependence of the response on the covariates, but by specifying the model only in terms of 'smooth functions', rather than detailed parametric relationships, it is possible to avoid the sort of cumbersome and unwieldy models seen in section 3.3.5, for example. This flexibility and convenience comes at the cost of two new theoretical problems. It is necessary both to represent the smooth functions in some way and to choose how smooth they should be.

This chapter illustrates how GAMs can be represented using basis expansions for each smooth, each with an associated penalty controlling function smoothness. Estimation can then be carried out by penalized regression methods, and the appropriate degree of smoothness for the f_j can be estimated from data using cross validation or marginal likelihood maximization. To avoid obscuring the basic simplicity of the approach with a mass of technical detail, the most complicated model considered here will be a simple GAM with two univariate smooth components. Furthermore, the methods presented will not be those that are most suitable for general practical use, being rather the methods that enable the basic framework to be explained simply. The ideal way to read this chapter is sitting at a computer, working through the statistics, and its implementation in R, side by side. If adopting this approach recall that the help files for R functions can be accessed by typing `?` followed by the function name, at the command line (e.g., `?lm`, for help on the linear modelling function).

4.2 Univariate smoothing

The representation and estimation of component functions of a model is best introduced by considering a model containing one function of one covariate,

$$y_i = f(x_i) + \epsilon_i, \tag{4.2}$$

where y_i is a response variable, x_i a covariate, f a smooth function and the ϵ_i are independent $N(0, \sigma^2)$ random variables.

4.2.1 *Representing a function with basis expansions*

To estimate f, using the methods covered in chapters 1 to 3, requires that f be represented in such a way that (4.2) becomes a linear model. This can be done by choosing a *basis*, defining the space of functions of which f (or a close approximation to it) is an element. Choosing a basis amounts to choosing some *basis functions*, which will be treated as completely known: if $b_j(x)$ is the jth such basis function, then f is assumed to have a representation

$$f(x) = \sum_{j=1}^{k} b_j(x)\beta_j, \tag{4.3}$$

for some values of the unknown parameters, β_j. Substituting (4.3) into (4.2) clearly yields a linear model.

A very simple basis: Polynomials

As a simple example, suppose that f is believed to be a 4th order polynomial, so that the space of polynomials of order 4 and below contains f. A basis for this space is $b_1(x) = 1$, $b_2(x) = x$, $b_3(x) = x^2$, $b_4(x) = x^3$ and $b_5(x) = x^4$, so that (4.3) becomes

$$f(x) = \beta_1 + x\beta_2 + x^2\beta_3 + x^3\beta_4 + x^4\beta_5,$$

and (4.2) becomes the simple model

$$y_i = \beta_1 + x_i\beta_2 + x_i^2\beta_3 + x_i^3\beta_4 + x_i^4\beta_5 + \epsilon_i.$$

Figures 4.1 and 4.2 illustrate a basis function representation of a function, f, using a polynomial basis.

The problem with polynomials

Taylor's theorem implies that polynomial bases will be useful for situations in which interest focuses on properties of f in the vicinity of a single specified point. But when the questions of interest relate to f over its whole domain, polynomial bases are problematic (see exercise 1).

The difficulties are most easily illustrated in the context of interpolation. The middle panel of figure 4.3 illustrates an attempt to approximate the function shown in

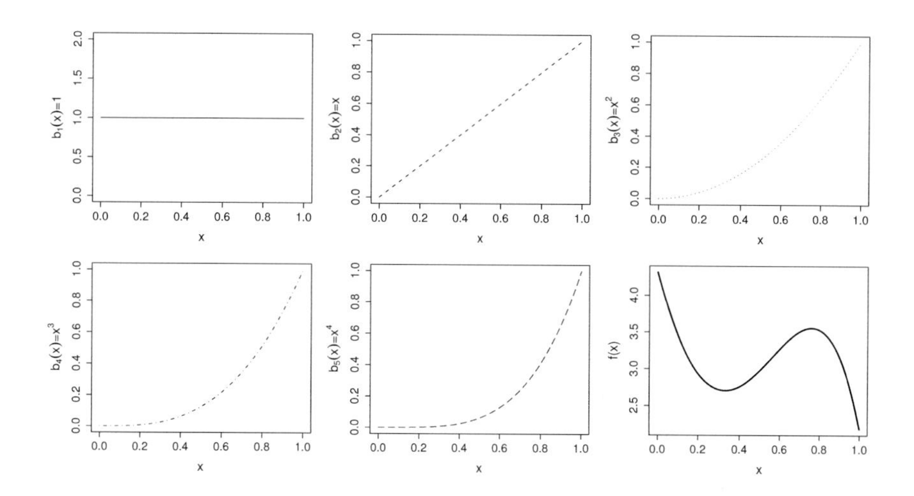

Figure 4.1 *Illustration of the idea of representing a function in terms of basis functions, using a polynomial basis. The first 5 panels (starting from top left) illustrate the 5 basis functions, $b_j(x)$, for a 4th order polynomial basis. The basis functions are each multiplied by a real valued parameter, β_j, and are then summed to give the final curve $f(x)$, an example of which is shown in the bottom right panel. By varying the β_j, we can vary the form of $f(x)$, to produce any polynomial function of order 4 or lower. See also figure 4.2*

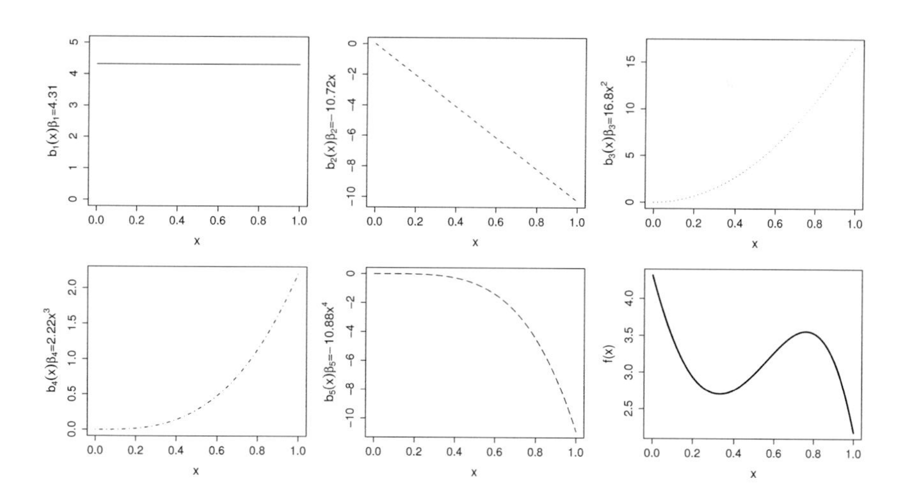

Figure 4.2 *An alternative illustration of how a function is represented in terms of basis functions. As in figure 4.1, a 4th order polynomial basis is illustrated. In this case the 5 basis functions, $b_j(x)$, each multiplied by its coefficient β_j, are shown in the first five figures (starting at top left). Simply summing these 5 curves yields the function, $f(x)$, shown at bottom right.*

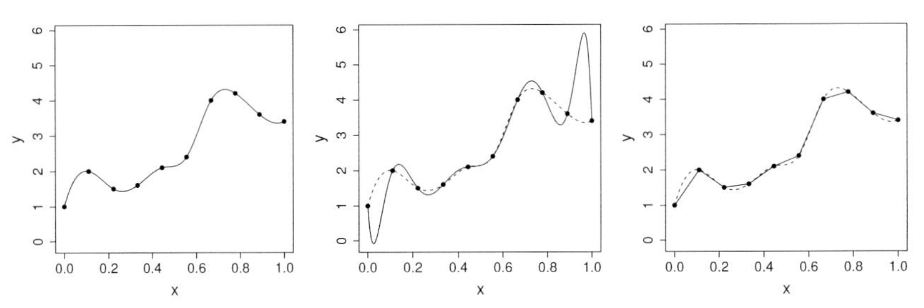

Figure 4.3 *The left panel shows a smooth function sampled at the points shown as black dots. The middle panel shows an attempt to reconstruct the function (dashed curve) by polynomial interpolation (black curve) of the black dots. The right panel shows the equivalent piecewise linear interpolant. The condition that the polynomial should interpolate the data and have all its derivatives continuous leads to quite wild oscillation.*

the left panel of figure 4.3, by polynomial interpolation of the points shown as black dots. The polynomial oscillates wildly in places, in order to accommodate the twin requirements to interpolate the data *and* to have all derivatives w.r.t. x continuous. If we relax the requirement for continuity of derivatives, and simply use a piecewise linear interpolant, then a much better approximation is obtained, as the right hand panel of figure 4.3 illustrates.

It clearly makes sense to use bases that are good at approximating known functions in order to represent unknown functions. Similarly, bases that perform well for interpolating exact observations of a function are also a good starting point for the closely related task of smoothing noisy observations of a function. In subsequent chapters we will see that piecewise linear bases can be improved upon by spline bases having continuity of just a few derivatives, but the piecewise linear case provides such a convenient illustration that we will stick with it for this chapter.

The piecewise linear basis

A basis for piecewise linear functions of a univariate variable x is determined entirely by the locations of the function's derivative discontinuities, that is by the locations at which the linear pieces join up. Let these *knots* be denoted $\{x_j^* : j = 1, \cdots k\}$, and suppose that $x_j^* > x_{j-1}^*$. Then for $j = 2, \ldots, k-1$

$$b_j(x) = \begin{cases} (x - x_{j-1}^*)/(x_j^* - x_{j-1}^*) & x_{j-1}^* < x \le x_j^* \\ (x_{i+j}^* - x)/(x_{j+1}^* - x_j^*) & x_j^* < x < x_{j+1}^* \\ 0 & \text{otherwise} \end{cases} \tag{4.4}$$

while

$$b_1(x) = \begin{cases} (x_2^* - x)/(x_2^* - x_1^*) & x < x_2^* \\ 0 & \text{otherwise} \end{cases}$$

and

$$b_k(x) = \begin{cases} (x - x_{k-1}^*)/(x_k^* - x_{k-1}^*) & x > x_{k-1}^* \\ 0 & \text{otherwise} \end{cases}$$

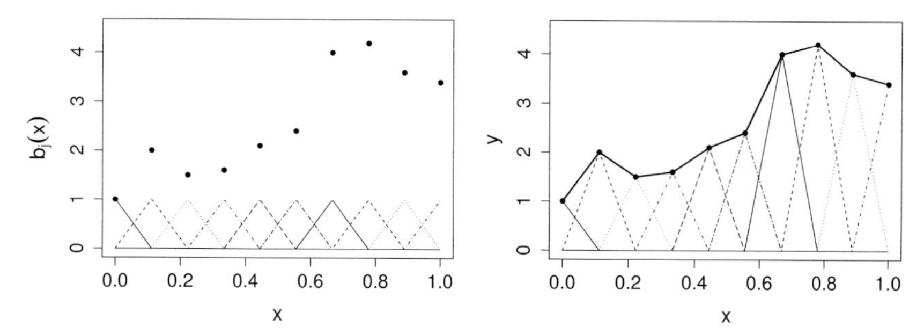

Figure 4.4 *The left panel shows an example tent function basis for interpolating the data shown as black dots. The continuous lines show the tent function basis functions, each of which peaks with value 1 at the x-axis value of one of the data points. The right panel illustrates how the basis functions are each multiplied by a coefficient, before being summed to give the interpolant, shown as the thick black line.*

So $b_j(x)$ is zero everywhere, except over the interval between the knots immediately to either side of x_j^*. $b_j(x)$ increases linearly from 0 at x_{j-1}^* to 1 at x_j^*, and then decreases linearly to 0 at x_{j+1}^*. Basis functions like this, that are non zero only over some finite intervals, are said to have *compact support*. Because of their shape the b_j are often known as *tent functions*. See figure 4.4.

Notice that an exactly equivalent way of defining $b_j(x)$ is as the linear interpolant of the data $\{x_i^*, \delta_i^j : i = 1, \ldots, k\}$ where $\delta_i^j = 1$ if $i = j$ and zero otherwise. This definition makes for very easy coding of the basis in R.

Using this basis to represent $f(x)$, (4.2) now becomes the linear model $\mathbf{y} = \mathbf{X}\boldsymbol{\beta} + \boldsymbol{\epsilon}$ where $X_{ij} = b_j(x_i)$.

Using the piecewise linear basis

Now consider an illustrative example. It is often claimed, at least by people with little actual knowledge of engines, that a car engine with a larger cylinder capacity will wear out less quickly than a smaller capacity engine. Figure 4.5 shows some data for 19 Volvo engines. The pattern of variation is not entirely clear, so (4.2) might be an appropriate model.

First read the data into R.

```
require(gamair); data(engine); attach(engine)
plot(size,wear,xlab="Engine capacity",ylab="Wear index")
```

Now write an R function defining $b_j(x)$

```
tf <- function(x,xj,j) {
## generate jth tent function from set defined by knots xj
  dj <- xj*0;dj[j] <- 1
  approx(xj,dj,x)$y
}
```

and use it to write an R function that will take a sequence of knots and an array of x values to produce a model matrix for the piecewise linear function.

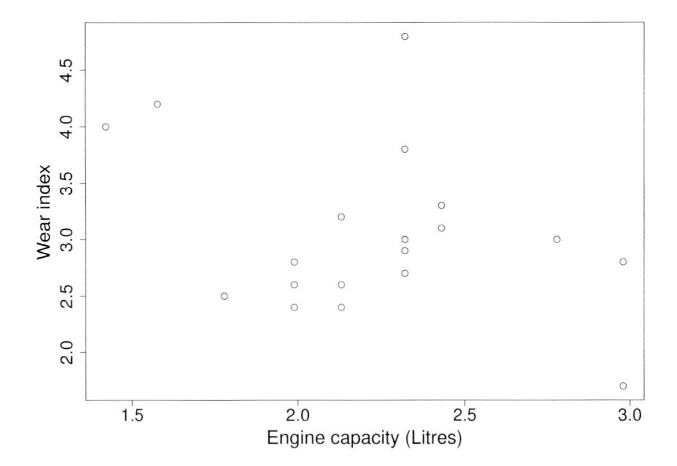

Figure 4.5 *Data on engine wear index versus engine capacity for 19 Volvo car engines, obtained from* `http://www3.bc.sympatico.ca/Volvo_Books/engine3.html`.

```
tf.X <- function(x,xj) {
## tent function basis matrix given data x
## and knot sequence xj
  nk <- length(xj); n <- length(x)
  X <- matrix(NA,n,nk)
  for (j in 1:nk) X[,j] <- tf(x,xj,j)
  X
}
```

All that is required now is to select a set of knots, x_j^*, and the model can be fitted. In the following a rank $k = 6$ basis is used, with the knots spread evenly over the range of the `size` data.

```
sj <- seq(min(size),max(size),length=6) ## generate knots
X <- tf.X(size,sj)                      ## get model matrix
b <- lm(wear ~ X - 1)                   ## fit model
s <- seq(min(size),max(size),length=200)## prediction data
Xp <- tf.X(s,sj)                        ## prediction matrix
plot(size,wear)                         ## plot data
lines(s,Xp %*% coef(b))                 ## overlay estimated f
```

The model fit looks quite plausible (figure 4.6), but the choice of degree of model smoothness, controlled here by the basis dimension, k, was essentially arbitrary. This issue must be addressed if a satisfactory theory for modelling with unknown functions is to be developed.

4.2.2 *Controlling smoothness by penalizing wiggliness*

One obvious possibility for choosing the degree of smoothing is to try to make use of the hypothesis testing methods from chapter 1, to select k by backwards selection.

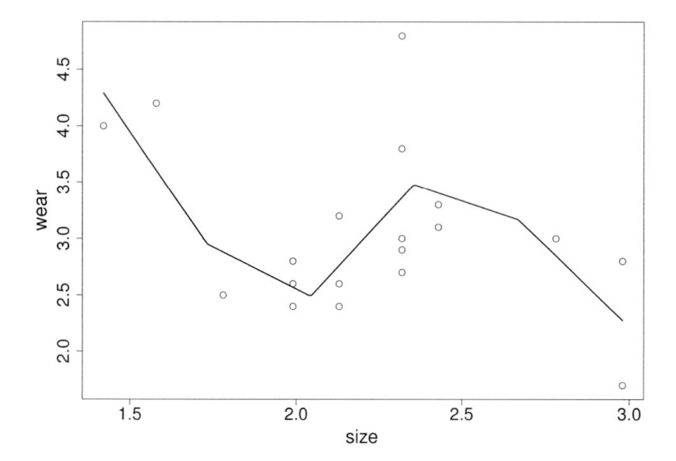

Figure 4.6 *Piecewise linear regression fit (continuous line) to data (○) on engine wear index versus engine capacity (size) for 19 Volvo car engines.*

However, such an approach is problematic, since a model based on $k-1$ evenly spaced knots will not generally be nested within a model based on k evenly spaced knots. It is possible to start with a fine grid of knots and simply drop knots sequentially, as part of backward selection, but the resulting uneven knot spacing can itself lead to poor model performance. Furthermore, the fit of such regression models tends to depend quite strongly on the locations chosen for the knots.

An alternative is to keep the basis dimension fixed at a size a little larger than it is believed could reasonably be necessary, but to control the model's smoothness by adding a 'wiggliness' penalty to the least squares fitting objective. For example, rather than fitting the model by minimizing

$$\|\mathbf{y} - \mathbf{X}\boldsymbol{\beta}\|^2,$$

it could be fitted by minimizing

$$\|\mathbf{y} - \mathbf{X}\boldsymbol{\beta}\|^2 + \lambda \sum_{j=2}^{k-1} \{f(x^*_{j-1}) - 2f(x^*_j) + f(x^*_{j+1})\}^2,$$

where the summation term measures wiggliness as a sum of squared second differences of the function at the knots (which crudely approximates the integrated squared second derivative penalty used in cubic spline smoothing: see section 5.1.2, p. 198). When f is very wiggly the penalty will take high values and when f is 'smooth' the penalty will be low.* If f is a straight line then the penalty is actually zero. So the penalty has a *null space* of functions that are un-penalized: the straight lines in this

*Note that even knot spacing has been assumed: uneven knot spacing would usually require some re-weighting of the penalty terms.

case. The dimension of the penalty null space is 2, since the basis for straight lines is 2-dimensional.

The *smoothing parameter*, λ, controls the trade-off between smoothness of the estimated f and fidelity to the data. $\lambda \to \infty$ leads to a straight line estimate for f, while $\lambda = 0$ results in an un-penalized piecewise linear regression estimate.

For the basis of tent functions, it is easy to see that the coefficients of f are simply the function values at the knots, i.e., $\beta_j = f(x_j^*)$. This makes it particularly straightforward to express the penalty as a quadratic form, $\boldsymbol{\beta}^\mathsf{T}\mathbf{S}\boldsymbol{\beta}$, in the basis coefficients (although in fact linearity of f in the basis coefficients is all that is required for this). Firstly note that

$$\begin{bmatrix} \beta_1 - 2\beta_2 + \beta_3 \\ \beta_2 - 2\beta_3 + \beta_4 \\ \beta_3 - 2\beta_4 + \beta_5 \\ . \\ . \end{bmatrix} = \begin{bmatrix} 1 & -2 & 1 & 0 & . & . & . \\ 0 & 1 & -2 & 1 & 0 & . & . \\ 0 & 0 & 1 & -2 & 1 & 0 & . \\ . & . & . & . & . & . & . \\ . & . & . & . & . & . & . \end{bmatrix} \begin{bmatrix} \beta_1 \\ \beta_2 \\ \beta_3 \\ . \\ . \end{bmatrix}$$

so that writing the right hand side as $\mathbf{D}\boldsymbol{\beta}$, by definition of $(k-2) \times k$ matrix $\mathbf{D}$, the penalty becomes

$$\sum_{j=2}^{k-1} (\beta_{j-1} - 2\beta_j + \beta_{j+1})^2 = \boldsymbol{\beta}^\mathsf{T}\mathbf{D}^\mathsf{T}\mathbf{D}\boldsymbol{\beta} = \boldsymbol{\beta}^\mathsf{T}\mathbf{S}\boldsymbol{\beta} \tag{4.5}$$

where $\mathbf{S} = \mathbf{D}^\mathsf{T}\mathbf{D}$ ($\mathbf{S}$ is obviously rank deficient by the dimension of the penalty null space). Hence the penalized regression fitting problem is to minimize

$$\|\mathbf{y} - \mathbf{X}\boldsymbol{\beta}\|^2 + \lambda\boldsymbol{\beta}^\mathsf{T}\mathbf{S}\boldsymbol{\beta} \tag{4.6}$$

w.r.t. $\boldsymbol{\beta}$. The problem of estimating the degree of smoothness for the model is now the problem of estimating the smoothing parameter λ. But before addressing λ estimation, consider $\boldsymbol{\beta}$ estimation given λ.

It is fairly straightforward to show (see exercise 3) that the formal expression for the minimizer of (4.6), the penalized least squares estimator of $\boldsymbol{\beta}$, is

$$\hat{\boldsymbol{\beta}} = \left(\mathbf{X}^\mathsf{T}\mathbf{X} + \lambda\mathbf{S}\right)^{-1}\mathbf{X}^\mathsf{T}\mathbf{y}. \tag{4.7}$$

Similarly the influence (hat) matrix, $\mathbf{A}$, for the model can be written

$$\mathbf{A} = \mathbf{X}\left(\mathbf{X}^\mathsf{T}\mathbf{X} + \lambda\mathbf{S}\right)^{-1}\mathbf{X}^\mathsf{T}.$$

Recall that $\hat{\boldsymbol{\mu}} = \mathbf{A}\mathbf{y}$. As with the un-penalized linear model, these expressions are not the ones to use for computation, for which the greater numerical stability offered by orthogonal matrix methods is to be preferred. For practical computation, therefore, note that

$$\left\| \begin{bmatrix} \mathbf{y} \\ \mathbf{0} \end{bmatrix} - \begin{bmatrix} \mathbf{X} \\ \sqrt{\lambda}\mathbf{D} \end{bmatrix} \boldsymbol{\beta} \right\|^2 = \|\mathbf{y} - \mathbf{X}\boldsymbol{\beta}\|^2 + \lambda\boldsymbol{\beta}^\mathsf{T}\mathbf{S}\boldsymbol{\beta}.$$

Obviously any matrix square root such that $\mathbf{D}^\mathsf{T}\mathbf{D} = \mathbf{S}$ could be substituted for the original $\mathbf{D}$ here, but at the moment there is no reason to use an alternative. The sum of squares term, on the left hand side, is just a least squares objective for a model in which the model matrix has been augmented by a square root of the penalty matrix, while the response data vector has been augmented with $k - 2$ zeros. Fitting the augmented linear model will therefore yield $\hat{\boldsymbol{\beta}}$.

To see a penalized regression spline in action, first note that $\mathbf{D}$ can be obtained in R using `diff(diag(k),differences=2)`, which applies second order differencing to each column of the rank k identity matrix. Now it is easy to write a simple function for fitting a penalized piecewise linear smoother.

```
prs.fit <- function(y,x,xj,sp) {
  X <- tf.X(x,xj)         ## model matrix
  D <- diff(diag(length(xj)),differences=2) ## sqrt penalty
  X <- rbind(X,sqrt(sp)*D) ## augmented model matrix
  y <- c(y,rep(0,nrow(D))) ## augmented data
  lm(y ~ X - 1)    ## penalized least squares fit
}
```

To use this function, we need to choose the basis dimension, k, the (evenly spaced) knot locations, x_j^*, and a value for the smoothing parameter, λ. Provided that k is large enough that the basis is more flexible than we expect to *need* to represent $f(x)$, then neither the exact choice of k, nor the precise selection of knot locations, has a great deal of influence on the model fit. Rather it is the choice of λ that now plays the crucial role in determining model flexibility, and ultimately the shape of $\hat{f}(x)$. In the following example $k = 20$ and the knots are evenly spread out over the range of observed engine sizes. It is the smoothing parameter, $\lambda = 2$, which really controls the behaviour of the fitted model.

```
sj <- seq(min(size),max(size),length=20) ## knots
b <- prs.fit(wear,size,sj,2) ## penalized fit
plot(size,wear)    ## plot data
Xp <- tf.X(s,sj)   ## prediction matrix
lines(s,Xp %*% coef(b)) ## plot the smooth
```

By changing the value of the smoothing parameter, λ, a variety of models of different smoothness can be obtained. Figure 4.7 illustrates this, but begs the question, which value of λ is 'best'?

4.2.3 Choosing the smoothing parameter, λ, by cross validation

If λ is too high then the data will be over-smoothed, and if it is too low then the data will be under-smoothed: in both cases this will mean that the estimate $\hat{f}$ will not be close to the true function f. Ideally, it would be good to choose λ so that $\hat{f}$ is as close as possible to f. A suitable criterion might be to choose λ to minimize

$$M = \frac{1}{n}\sum_{i=1}^{n}(\hat{f}_i - f_i)^2,$$

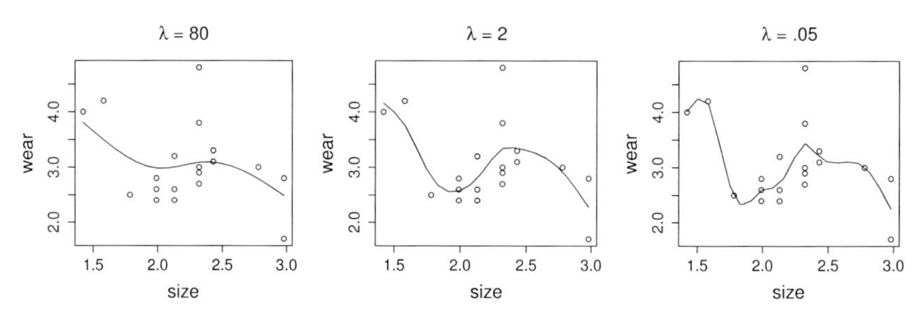

Figure 4.7 *Penalized piecewise linear fits to the engine wear versus capacity data, using three different values for the smoothing parameter, λ. Notice how penalization produces quite smooth estimates, despite the piecewise linear basis.*

where the notation $\hat{f}_i \equiv \hat{f}(x_i)$ and $f_i \equiv f(x_i)$ has been adopted for conciseness. Since f is unknown, M cannot be used directly, but it is possible to derive an estimate of $\mathbb{E}(M) + \sigma^2$, which is the expected squared error in predicting a new variable. Let $\hat{f}^{[-i]}$ be the model fitted to all data except y_i, and define the *ordinary cross validation* score

$$\mathcal{V}_o = \frac{1}{n}\sum_{i=1}^{n}(\hat{f}_i^{[-i]} - y_i)^2.$$

This score results from leaving out each datum in turn, fitting the model to the remaining data and calculating the squared difference between the missing datum and its predicted value: these squared differences are then averaged over all the data. Substituting $y_i = f_i + \epsilon_i$,

$$\begin{aligned}\mathcal{V}_o &= \frac{1}{n}\sum_{i=1}^{n}(\hat{f}_i^{[-i]} - f_i - \epsilon_i)^2 \\ &= \frac{1}{n}\sum_{i=1}^{n}(\hat{f}_i^{[-i]} - f_i)^2 - 2(\hat{f}_i^{[-i]} - f_i)\epsilon_i + \epsilon_i^2.\end{aligned}$$

Since $\mathbb{E}(\epsilon_i) = 0$, and ϵ_i and $\hat{f}_i^{[-i]}$ are independent, the second term in the summation vanishes if expectations are taken:

$$\mathbb{E}(\mathcal{V}_o) = \frac{1}{n}\mathbb{E}\left(\sum_{i=1}^{n}(\hat{f}_i^{[-i]} - f_i)^2\right) + \sigma^2.$$

Now, $\hat{f}^{[-i]} \approx \hat{f}$ with equality in the large sample limit, so $\mathbb{E}(\mathcal{V}_o) \approx \mathbb{E}(M) + \sigma^2$ also with equality in the large sample limit. Hence choosing λ in order to minimize $\mathcal{V}_o$ is a reasonable approach if the ideal would be to minimize M. Choosing λ to minimize $\mathcal{V}_o$ is known as ordinary cross validation.

Ordinary cross validation is a reasonable approach, in its own right, even without a mean square (prediction) error justification. If models are judged only by their ability to fit the data from which they were estimated, then complicated models are

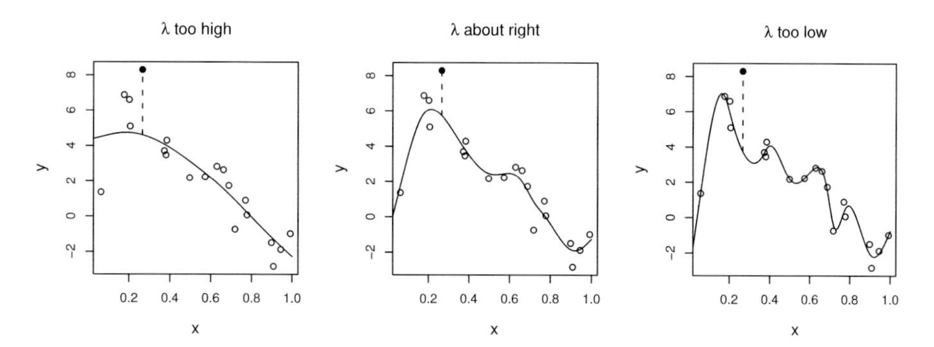

Figure 4.8 *Illustration of the principle behind cross validation. The fifth datum (•) has been omitted from fitting and the continuous curve shows a penalized regression spline fitted to the remaining data (∘). When the smoothing parameter is too high the spline fits many of the data poorly and does no better with the missing point. When λ is too low the spline fits the noise as well as the signal and the consequent extra variability causes it to predict the missing datum poorly. For intermediate λ the spline is fitting the underlying signal quite well, but smoothing through the noise: hence, the missing datum is reasonably well predicted. Cross validation leaves out each datum from the data in turn and considers the average ability of models fitted to the remaining data to predict the omitted data.*

always selected over simpler ones. Choosing a model in order to maximize the ability to predict data to which the model was not fitted, does not suffer from this problem, as figure 4.8 illustrates.

It is computationally costly to calculate $\mathcal{V}_o$ by leaving out one datum at a time, refitting the model to each of the n resulting data sets, but it can be shown that

$$\mathcal{V}_o = \frac{1}{n}\sum_{i=1}^{n}(y_i - \hat{f}_i)^2/(1 - A_{ii})^2,$$

where $\hat{f}$ is the estimate from fitting to all the data, and $\mathbf{A}$ is the corresponding influence matrix (see section 6.2.2, p. 256). In practice the A_{ii} are often replaced by their mean, $\text{tr}(\mathbf{A})/n$, resulting in the *generalized cross validation* score

$$\mathcal{V}_g = \frac{n\sum_{i=1}^{n}(y_i - \hat{f}_i)^2}{[n - \text{tr}(\mathbf{A})]^2}.$$

GCV has computational advantages over OCV, and it also has advantages in terms of invariance (see Wahba, 1990, p.53 or sections 6.2.2 and 6.2.3, p. 258). In any case, it can also be shown to minimize $\mathbb{E}(M)$ in the large sample limit.

Returning to the engine wear example, a simple direct search for the GCV optimal smoothing parameter can be made as follows.

```
rho = seq(-9,11,length=90)
n <- length(wear)
V <- rep(NA,90)
```

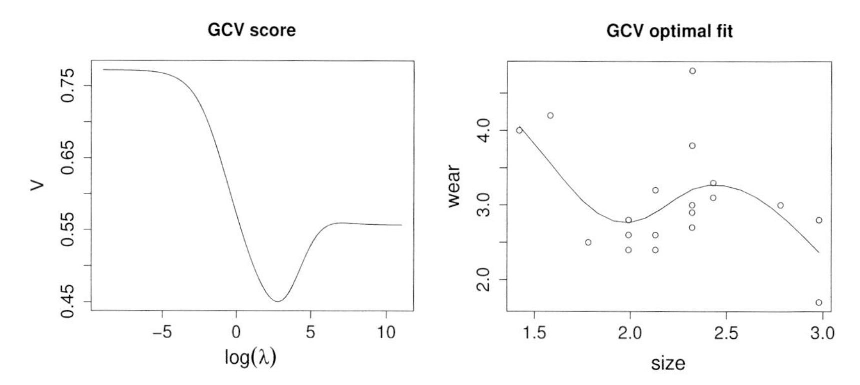

Figure 4.9 *Left panel: the GCV function for the engine wear example against log smoothing parameter. Right panel: the fitted model which minimizes the GCV score.*

```
for (i in 1:90) { ## loop through smoothing params
  b <- prs.fit(wear,size,sj,exp(rho[i])) ## fit model
  trF <- sum(influence(b)$hat[1:n])       ## extract EDF
  rss <- sum((wear-fitted(b)[1:n])^2)     ## residual SS
  V[i] <- n*rss/(n-trF)^2                 ## GCV score
}
```

Note that the `influence()` function returns a list of diagnostics including `hat`, an array of the elements on the leading diagonal of the influence/hat matrix of the augmented model. The first n of these are the leading diagonal of the influence matrix of the penalized model (see exercise 4).

For the example, `V[54]` is the lowest GCV score, so that the optimal smoothing parameter, from those tried, is $\hat{\lambda} \approx 18$. Plots of the GCV score and the optimal model are easily produced

```
plot(rho,V,type="l",xlab=expression(log(lambda)),
                    main="GCV score")
sp <- exp(rho[V==min(V)])      ## extract optimal sp
b <- prs.fit(wear,size,sj,sp) ## re-fit
plot(size,wear,main="GCV optimal fit")
lines(s,Xp %*% coef(b))
```

The results are shown in figure 4.9.

4.2.4 *The Bayesian/mixed model alternative*

At some level we introduce smoothing penalties because we believe that the truth is more likely to be smooth than wiggly. We might as well formalise this belief in a Bayesian way, and specify a prior distribution on function wiggliness. Perhaps the simplest choice is an exponential prior

$$\propto \exp(-\lambda \boldsymbol{\beta}^{\mathsf{T}} \mathbf{S} \boldsymbol{\beta}/\sigma^2)$$

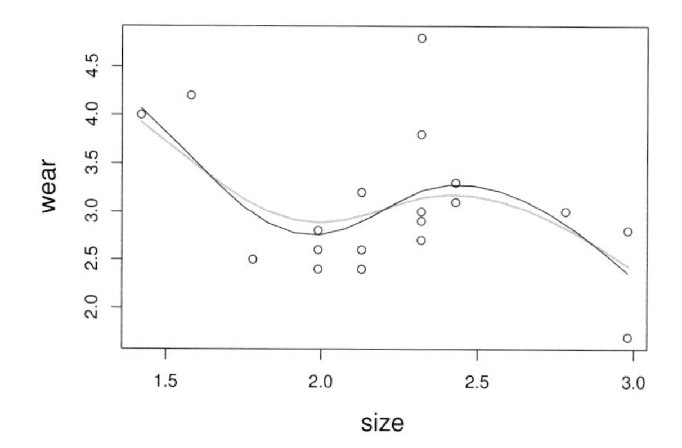

Figure 4.10 *Smooth model fits to the engine wear data with smoothing parameters estimated using marginal likelihood maximization (grey) or REML (black).*

(where scaling by σ^2 is introduced merely for later convenience), but this is immediately recognisable as being equivalent to an improper multivariate normal prior $\boldsymbol{\beta} \sim N(\mathbf{0}, \sigma^2 \mathbf{S}^- / \lambda)$. That is, the prior precision matrix is proportional to $\mathbf{S}$: because $\mathbf{S}$ is rank deficient by the dimension of the penalty null space, the prior covariance matrix is proportional to the pseudo-inverse† $\mathbf{S}^-$.

This Bayesian interpretation of the smoothing penalty gives the model the structure of a linear mixed model as discussed in chapter 2, and in consequence the MAP estimate of $\boldsymbol{\beta}$ is the solution (4.7) to (4.6), while

$$\boldsymbol{\beta}|\mathbf{y} \sim N(\hat{\boldsymbol{\beta}}, (\mathbf{X}^\mathsf{T}\mathbf{X} + \lambda\mathbf{S})^{-1}\sigma^2)$$

— the Bayesian posterior distribution of $\boldsymbol{\beta}$ (this is equivalent to (2.17), p. 80). Also, having given the model this extra structure opens up the possibility of estimating σ^2 and λ using marginal likelihood maximization or REML.

In this section we will re-parameterize slightly to get the smooth model into a form such that its marginal likelihood can be evaluated using the simple routine `llm` from section 2.4.4 (p. 81). R routine `optim` can be used to fit the model. The same re-parameterization allows the model to be easily estimated using `lme` (see section 2.5, p. 86). As we will see in chapter 6, this re-parameterization is not necessary: it just simplifies matters for the moment, and perhaps makes the relationship between fixed effects and the penalty null space clearer than might otherwise be the case.

The re-parameterization is to re-write the model in terms of parameters, $\boldsymbol{\beta}' = \mathbf{D}_+\boldsymbol{\beta}$ where

$$\mathbf{D}_+ = \begin{bmatrix} \mathbf{I}_2 & \mathbf{0} \\ & \mathbf{D} \end{bmatrix}.$$

So we now have $\mathbf{X}\boldsymbol{\beta} = \mathbf{X}\mathbf{D}_+^{-1}\boldsymbol{\beta}'$ and $\boldsymbol{\beta}^\mathsf{T}\mathbf{S}\boldsymbol{\beta} = \sum_{i=3}^k \beta_i'^2$. If we write the first 2

†Consider eigen-decomposition $\mathbf{S} = \mathbf{U}\boldsymbol{\Lambda}\mathbf{U}^\mathsf{T}$. Let $\boldsymbol{\Lambda}^-$ denote the diagonal matrix of the inverse of the non-zero eigenvalues, with zeroes in place of the inverse for any zero eigenvalues. Then $\mathbf{S}^- = \mathbf{U}\boldsymbol{\Lambda}^-\mathbf{U}^\mathsf{T}$.

elements of β' as β^* and the remainder as $\mathbf{b}$, the Bayesian smoothing prior becomes $\mathbf{b} \sim N(\mathbf{0}, \mathbf{I}\sigma^2/\lambda)$ (which is proper). β^* is completely unpenalized, so we treat this as a vector of fixed effects. To make the connection to a standard mixed model completely clear, let $\mathbf{X}^*$ now denote the first 2 columns of $\mathbf{X}\mathbf{D}_+^{-1}$, while $\mathbf{Z}$ is the matrix of the remaining columns. Then the smooth model has become

$$\mathbf{y} = \mathbf{X}^*\beta^* + \mathbf{Z}\mathbf{b} + \epsilon, \quad \mathbf{b} \sim N(\mathbf{0}, \mathbf{I}\sigma^2/\lambda), \quad \epsilon \sim N(\mathbf{0}, \mathbf{I}\sigma^2)$$

which is self-evidently in the standard form of a linear mixed model given in section 2.3, (p.77). Here is the code to re-parameterize the model and estimate it using `optim` and `llm` from section 2.4.4 (p. 81):

```
X0 <- tf.X(size,sj)          ## X in original parameterization
D <- rbind(0,0,diff(diag(20),difference=2))
diag(D) <- 1                   ## augmented D
X <- t(backsolve(t(D),t(X0)))  ## re-parameterized X
Z <- X[,-c(1,2)]; X <- X[,1:2] ## mixed model matrices
## estimate smoothing and variance parameters...
m <- optim(c(0,0),llm,method="BFGS",X=X,Z=Z,y=wear)
b <- attr(llm(m$par,X,Z,wear),"b") ## extract coefficients
## plot results...
plot(size,wear)
Xp1 <- t(backsolve(t(D),t(Xp))) ## re-parameterized pred. mat.
lines(s,Xp1 %*% as.numeric(b),col="grey",lwd=2)
```

The resulting plot is shown in figure 4.10.

Estimation using REML via `lme` is also easy. In `lme` terms all the data belong to a single group, so to use `lme` we must create a dummy grouping variable enforcing this. A covariance matrix proportional to the identity matrix is then specified via the `pdIdent` function.

```
library(nlme)
g <- factor(rep(1,nrow(X)))         ## dummy factor
m <- lme(wear ~ X - 1, random=list(g = pdIdent(~ Z-1)))
lines(s,Xp1 %*% as.numeric(coef(m))) ## and to plot
```

The curve of the estimated smooth is shown in black in figure 4.10. Notice how the REML based estimate (black) is more variable than the ML based estimate (grey), as expected from section 2.4.5 (p. 83).

4.3 Additive models

Now suppose that two explanatory variables, x and v, are available for a response variable, y, and that a simple additive model structure,

$$y_i = \alpha + f_1(x_i) + f_2(v_i) + \epsilon_i, \tag{4.8}$$

is appropriate. α is an intercept parameter, the f_j are smooth functions, and the ϵ_i are independent $N(0, \sigma^2)$ random variables.

There are two points to note about this model. Firstly, the assumption of additive effects is a fairly strong one: $f_1(x) + f_2(v)$ is a quite restrictive special case of the general smooth function of two variables $f(x, v)$. Secondly, the fact that the model now contains more than one function introduces an identifiability problem: f_1 and f_2 are each only estimable to within an additive constant. To see this, note that any constant could be simultaneously added to f_1 and subtracted from f_2, without changing the model predictions. Hence identifiability constraints have to be imposed on the model before fitting.

Provided that the identifiability issue is addressed, the additive model can be represented using penalized regression splines, estimated by penalized least squares and the degree of smoothing selected by cross validation or (RE)ML, in the same way as for the simple univariate model.

4.3.1 *Penalized piecewise regression representation of an additive model*

Each smooth function in (4.8) can be represented using a penalized piecewise linear basis. Specifically, let

$$f_1(x) = \sum_{j=1}^{k_1} b_j(x)\delta_j$$

where the δ_j are unknown coefficients, while the $b_j(x)$ are basis functions of the form (4.4), defined using a sequence of k_1 knots, x_j^*, evenly spaced over the range of x. Similarly

$$f_2(v) = \sum_{j=1}^{k_2} \mathcal{B}_j(v)\gamma_j$$

where the γ_j are the unknown coefficients and the $\mathcal{B}_j(v)$ are basis functions of the form (4.4), defined using a sequence of k_2 knots, v_j^*, evenly spaced over the range of v. Defining n-vector $\mathbf{f}_1 = [f_1(x_1), \ldots, f_1(x_n)]^\mathsf{T}$, we have $\mathbf{f}_1 = \mathbf{X}_1\boldsymbol{\delta}$ where $b_j(x_i)$ is element i, j of $\mathbf{X}_1$. Similarly, $\mathbf{f}_2 = \mathbf{X}_2\boldsymbol{\gamma}$, where $\mathcal{B}_j(v_i)$ is element i, j of $\mathbf{X}_2$.

A penalty of the form (4.5) is also associated with each function: $\boldsymbol{\delta}^\mathsf{T}\mathbf{D}_1^\mathsf{T}\mathbf{D}_1\boldsymbol{\delta} = \boldsymbol{\delta}^\mathsf{T}\bar{\mathbf{S}}_1\boldsymbol{\delta}$ for f_1 and $\boldsymbol{\gamma}^\mathsf{T}\mathbf{D}_2^\mathsf{T}\mathbf{D}_2\boldsymbol{\gamma} = \boldsymbol{\gamma}^\mathsf{T}\bar{\mathbf{S}}_2\boldsymbol{\gamma}$ for f_2.

Now it is necessary to deal with the identifiability problem. For estimation purposes, almost any linear constraint that removed the problem could be used, but most choices lead to uselessly wide confidence intervals for the constrained functions. The best constraints from this viewpoint are sum-to-zero constraints, such as

$$\sum_{i=1}^{n} f_1(x_i) = 0, \quad \text{or equivalently} \quad \mathbf{1}^\mathsf{T}\mathbf{f}_1 = 0,$$

where $\mathbf{1}$ is an n vector of 1's. Notice how this constraint still allows f_1 to have exactly the same shape as before constraint, with exactly the same penalty value. The constraint's only effect is to shift f_1, vertically, so that its mean value is zero.

To apply the constraint, note that we require $\mathbf{1}^\mathsf{T}\mathbf{X}_1\boldsymbol{\delta} = 0$ for all $\boldsymbol{\delta}$, which implies that $\mathbf{1}^\mathsf{T}\mathbf{X}_1 = \mathbf{0}$. To achieve this latter condition the column mean can be subtracted from each column of $\mathbf{X}_1$. That is, we define a column centred matrix

$$\tilde{\mathbf{X}}_1 = \mathbf{X}_1 - \mathbf{1}\mathbf{1}^\mathsf{T}\mathbf{X}_1/n$$

and set $\tilde{\mathbf{f}}_1 = \tilde{\mathbf{X}}_1\boldsymbol{\delta}$. It's easy to check that this constraint imposes no more than a shift in the level of $\mathbf{f}_1$:

$$\tilde{\mathbf{f}}_1 = \tilde{\mathbf{X}}_1\boldsymbol{\delta} = \mathbf{X}_1\boldsymbol{\delta} - \mathbf{1}\mathbf{1}^\mathsf{T}\mathbf{X}_1\boldsymbol{\delta}/n = \mathbf{X}_1\boldsymbol{\delta} - \mathbf{1}c = \mathbf{f}_1 - c$$

by definition of the scalar $c = \mathbf{1}^\mathsf{T}\mathbf{X}_1\boldsymbol{\delta}/n$. Finally note that the column centring reduces the rank of $\tilde{\mathbf{X}}_1$ to $k_1 - 1$, so that only $k_1 - 1$ elements of the k_1 vector $\boldsymbol{\delta}$ can be uniquely estimated. A simple identifiability constraint deals with this problem: a single element of $\boldsymbol{\delta}$ is set to zero, and the corresponding column of $\tilde{\mathbf{X}}_1$ and $\mathbf{D}$ is deleted.[‡] The column centred rank reduced basis will automatically satisfy the identifiability constraint. In what follows the tildes will be dropped, and it is assumed that the $\mathbf{X}_j$, $\mathbf{D}_j$, etc. are the constrained versions.

Here is an R function which produces constrained versions of $\mathbf{X}_j$ and $\mathbf{D}_j$.

```
tf.XD <- function(x,xk,cmx=NULL,m=2) {
## get X and D subject to constraint
  nk <- length(xk)
  X <- tf.X(x,xk)[,-nk]                     ## basis matrix
  D <- diff(diag(nk),differences=m)[,-nk] ## root penalty
  if (is.null(cmx)) cmx <- colMeans(X)
  X <- sweep(X,2,cmx)          ## subtract cmx from columns
  list(X=X,D=D,cmx=cmx)
}
```

`tf.XD` calls the functions producing the unconstrained basis and square root penalty matrices, given knot sequence `xk` and covariate values `x`. It drops a column of each resulting matrix and centres the remaining columns of the basis matrix. `cmx` is the vector of values to subtract from the columns of the `X`. For setting up a basis `cmx` should be `NULL`, in which case it is set to the column means of the basis matrix `X`. However, when using `tf.XD` to produce a basis matrix for *predicting* at new covariate values, it is essential that the basis matrix columns are centred using the same constants used for the *original* basis setup, so these must be supplied. Later code will clarify this.

Having set up constrained bases for the f_j it is now straightforward to re-express (4.8) as

$$\mathbf{y} = \mathbf{X}\boldsymbol{\beta} + \boldsymbol{\epsilon}$$

where $\mathbf{X} = (\mathbf{1}, \mathbf{X}_1, \mathbf{X}_2)$ and $\boldsymbol{\beta}^\mathsf{T} = (\alpha, \boldsymbol{\delta}^\mathsf{T}, \boldsymbol{\gamma}^\mathsf{T})$. Largely for later notational convenience it is useful to express the penalties as quadratic forms in the full coefficient vector $\boldsymbol{\beta}$, which is easily done by simply padding out $\bar{\mathbf{S}}_j$ with zeroes, as appropriate. For example,

$$\boldsymbol{\beta}^\mathsf{T}\mathbf{S}_1\boldsymbol{\beta} = (\alpha, \boldsymbol{\delta}^\mathsf{T}, \boldsymbol{\gamma}^\mathsf{T}) \begin{bmatrix} 0 & \mathbf{0} & \mathbf{0} \\ \mathbf{0} & \bar{\mathbf{S}}_1 & \mathbf{0} \\ \mathbf{0} & \mathbf{0} & \mathbf{0} \end{bmatrix} \begin{bmatrix} \alpha \\ \boldsymbol{\delta} \\ \boldsymbol{\gamma} \end{bmatrix} = \boldsymbol{\delta}^\mathsf{T}\bar{\mathbf{S}}_1\boldsymbol{\delta}.$$

[‡]The recipe given here is applicable to any basis which includes the constant function in its span, and has a penalty that is zero for constant functions. However, for bases that explicitly include a constant function, it is not sufficient to set any coefficient to zero: the coefficient for the constant is the one to constrain.

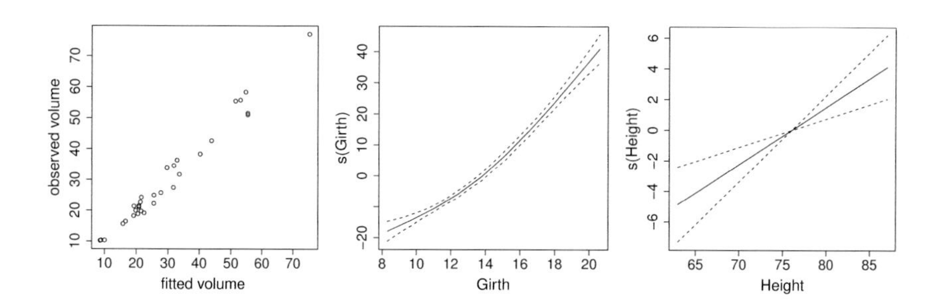

Figure 4.11 *The best fit two term additive model for the* `tree` *data. The left panel shows actual versus predicted tree volumes. The middle panel is the estimate of the smooth function of girth. The right panel is the estimate of the smooth function of height.*

4.3.2 *Fitting additive models by penalized least squares*

The coefficient estimates $\hat{\boldsymbol{\beta}}$ of the model (4.8) are obtained by minimization of the penalized least squares objective

$$\|\mathbf{y} - \mathbf{X}\boldsymbol{\beta}\|^2 + \lambda_1 \boldsymbol{\beta}^\mathsf{T}\mathbf{S}_1\boldsymbol{\beta} + \lambda_2\boldsymbol{\beta}^\mathsf{T}\mathbf{S}_2\boldsymbol{\beta},$$

where the smoothing parameters λ_1 and λ_2 control the weight to be given to the objective of making f_1 and f_2 smooth, relative to the objective of closely fitting the response data. For the moment, assume that these smoothing parameters are given.

Similarly to the single smooth case we have

$$\hat{\boldsymbol{\beta}} = \left(\mathbf{X}^\mathsf{T}\mathbf{X} + \lambda_1\mathbf{S}_1 + \lambda_2\mathbf{S}_2\right)^{-1}\mathbf{X}^\mathsf{T}\mathbf{y} \text{ and } \mathbf{A} = \mathbf{X}\left(\mathbf{X}^\mathsf{T}\mathbf{X} + \lambda_1\mathbf{S}_1 + \lambda_2\mathbf{S}_2\right)^{-1}\mathbf{X}^\mathsf{T},$$

but again these expressions are sub-optimal with regard to computational stability and it is better to re-write the objective as

$$\|\mathbf{y} - \mathbf{X}\boldsymbol{\beta}\|^2 + \lambda_1 \boldsymbol{\beta}^\mathsf{T}\mathbf{S}_1\boldsymbol{\beta} + \lambda_2\boldsymbol{\beta}^\mathsf{T}\mathbf{S}_2\boldsymbol{\beta} = \left\| \begin{bmatrix} \mathbf{y} \\ \mathbf{0} \end{bmatrix} - \begin{bmatrix} \mathbf{X} \\ \mathbf{B} \end{bmatrix} \boldsymbol{\beta} \right\|^2, \tag{4.9}$$

where

$$\mathbf{B} = \begin{bmatrix} \mathbf{0} & \sqrt{\lambda_1}\mathbf{D}_1 & \mathbf{0} \\ \mathbf{0} & \mathbf{0} & \sqrt{\lambda_2}\mathbf{D}_2 \end{bmatrix}$$

(or any other matrix such that $\mathbf{B}^\mathsf{T}\mathbf{B} = \lambda_1\mathbf{S}_1 + \lambda_2\mathbf{S}_2$).

As in the single smooth case, the right hand side of (4.9) is simply the unpenalized least squares objective for an augmented version of the model and corresponding response data. Hence, the model can be fitted by standard linear regression using stable orthogonal matrix based methods.

Here is a function to set up and fit a simple two term additive model, assuming the same number of knots for each smooth.

```
am.fit <- function(y,x,v,sp,k=10) {
  ## setup bases and penalties...
  xk <- seq(min(x),max(x),length=k)
  xdx <- tf.XD(x,xk)
  vk <- seq(min(v),max(v),length=k)
  xdv <- tf.XD(v,vk)
  ## create augmented model matrix and response...
  nD <- nrow(xdx$D)*2
  sp <- sqrt(sp)
  X <- cbind(c(rep(1,nrow(xdx$X)),rep(0,nD)),
             rbind(xdx$X,sp[1]*xdx$D,xdv$D*0),
             rbind(xdv$X,xdx$D*0,sp[2]*xdv$D))
  y1 <- c(y,rep(0,nD))
  ## fit model..
  b <- lm(y1 ~ X - 1)
  ## compute some useful quantities...
  n <- length(y)
  trA <- sum(influence(b)$hat[1:n]) ## EDF
  rsd <- y - fitted(b)[1:n] ## residuals
  rss <- sum(rsd^2)         ## residual SS
  sig.hat <- rss/(n-trA)       ## residual variance
  gcv <- sig.hat*n/(n-trA)     ## GCV score
  Vb <- vcov(b)*sig.hat/summary(b)$sigma^2 ## coeff cov matrix
  ## return fitted model...
  list(b=coef(b),Vb=Vb,edf=trA,gcv=gcv,fitted=fitted(b)[1:n],
       rsd=rsd,xk=list(xk,vk),cmx=list(xdx$cmx,xdv$cmx))
}
```

In addition to the quantities that we met in the single smooth case, `am.fit` also returns an estimate of the Bayesian covariance matrix for the model coefficients:

$$\hat{\mathbf{V}}_\beta = (\mathbf{X}^\mathsf{T}\mathbf{X} + \lambda_1\mathbf{S}_1 + \lambda_2\mathbf{S}_2)^{-1}\hat{\sigma}^2$$

where $\hat{\sigma}^2$ is taken as the residual sum of squares for the fitted model, divided by the effective residual degrees of freedom. Following section 4.2.4 the posterior distribution for $\boldsymbol{\beta}$ is

$$\boldsymbol{\beta}|\mathbf{y} \sim N(\hat{\boldsymbol{\beta}}, \mathbf{V}_\beta), \tag{4.10}$$

and this result can be used for further inference about $\boldsymbol{\beta}$ (see section 6.10, p. 293).

Let us use the routine to estimate an additive model for the data in R data frame `trees`. The data are `Volume`, `Girth` and `Height` for 31 felled cherry trees. Interest lies in predicting `Volume`, and we can try estimating the model

$$\texttt{Volume}_i = \alpha + f_1(\texttt{Girth}_i) + f_2(\texttt{Height}_i) + \epsilon_i.$$

Now that we have two smoothing parameters, grid searching for the GCV optimal values starts to become inefficient. Instead R function `optim` can be used to minimize the GCV score. The function to be optimised has to be in a particular form for use with `optim`: the optimization parameter vector must be the first argument, and the function must be real valued. A simple wrapper for `am.fit` suffices:

```
am.gcv <- function(lsp,y,x,v,k) {
## function suitable for GCV optimization by optim
  am.fit(y,x,v,exp(lsp),k)$gcv
}
```

Using log smoothing parameters for optimization ensures that the estimated smoothing parameters are non-negative. Fitting the model is now straightforward

```
## find GCV optimal smoothing parameters...
fit <- optim(c(0,0), am.gcv, y=trees$Volume, x=trees$Girth,
            v=trees$Height,k=10)
sp <- exp(fit$par) ## best fit smoothing parameters
## Get fit at GCV optimal smoothing parameters...
fit <- am.fit(trees$Volume,trees$Girth,trees$Height,sp,k=10)
```

Now let's plot the smooth effects. The following function will do this.

```
am.plot <- function(fit,xlab,ylab) {
## produces effect plots for simple 2 term
## additive model
  start <- 2 ## where smooth coeffs start in beta
  for (i in 1:2) {
    ## sequence of values at which to predict...
    x <- seq(min(fit$xk[[i]]), max(fit$xk[[i]]), length=200)
    ## get prediction matrix for this smooth...
    Xp <- tf.XD(x, fit$xk[[i]], fit$cmx[[i]])$X
    ## extract coefficients and cov matrix for this smooth
    stop <- start + ncol(Xp)-1; ind <- start:stop
    b <- fit$b[ind];Vb <- fit$Vb[ind,ind]
    ## values for smooth at x...
    fv <- Xp %*% b
    ## standard errors of smooth at x....
    se <- rowSums((Xp %*% Vb) * Xp)^.5
    ## 2 s.e. limits for smooth...
    ul <- fv + 2 * se; ll <- fv - 2 * se
    ## plot smooth and limits...
    plot(x, fv, type="l", ylim=range(c(ul,ll)), xlab=xlab[i],
         ylab=ylab[i])
    lines(x, ul, lty=2); lines(x, ll, lty=2)
    start <- stop + 1
  }
}
```

Calling it with the fitted tree model

```
par(mfrow=c(1,3))
plot(fit$fitted,trees$Vol,xlab="fitted volume ",
     ylab="observed volume")
am.plot(fit,xlab=c("Girth","Height"),
        ylab=c("s(Girth)","s(Height)"))
```

gives the result in figure 4.11. Notice that the smooth of `Height` is estimated to be a straight line, and as a result its confidence interval has zero width at some point.

The zero width point in the interval occurs because the sum to zero constraint exactly determines where the straight line must pass through zero.

As with the one dimensional smooth, the additive model could also be estimated as a linear mixed model, but let us move on.

4.4 Generalized additive models

Generalized additive models (GAMs) follow from additive models, as generalized linear models follow from linear models. That is, the linear predictor now predicts some known smooth monotonic function of the expected value of the response, and the response may follow any exponential family distribution, or simply have a known mean variance relationship, permitting the use of a quasi-likelihood approach. The resulting model has a general form something like (4.1) in section 4.1.

As an illustration, suppose that we would like to model the `trees` data using a GAM of the form:

$$\log\{\mathbb{E}(\texttt{Volume}_i)\} = f_1(\texttt{Girth}_i) + f_2(\texttt{Height}_i), \quad \texttt{Volume}_i \sim \text{gamma}.$$

This model is perhaps more natural than the additive model, as we might expect volume to be the product of some function of girth and some function of height, and it is reasonable to expect the variance in volume to increase with mean volume.

Whereas the additive model was estimated by penalized least squares, the GAM will be fitted by penalized likelihood maximization, and in practice this will be achieved by penalized iterative least squares (PIRLS).[§] For given smoothing parameters, the following steps are iterated to convergence.

1. Given the current linear predictor estimate, $\hat{\boldsymbol{\eta}}$, and corresponding estimated mean response vector, $\hat{\boldsymbol{\mu}}$, calculate:

   $$w_i = \frac{1}{V(\hat{\mu}_i)g'(\hat{\mu}_i)^2} \quad \text{and} \quad z_i = g'(\hat{\mu}_i)(y_i - \hat{\mu}_i) + \hat{\eta}_i$$

   where $\text{var}(Y_i) = V(\mu_i)\phi$, as in section 3.1.2, and g is the link function.

2. Defining $\mathbf{W}$ as the diagonal matrix such that $W_{ii} = w_i$, minimize

   $$\|\sqrt{\mathbf{W}}\mathbf{z} - \sqrt{\mathbf{W}}\mathbf{X}\boldsymbol{\beta})\|^2 + \lambda_1\boldsymbol{\beta}^{\mathsf{T}}\mathbf{S}_1\boldsymbol{\beta} + \lambda_2\boldsymbol{\beta}^{\mathsf{T}}\mathbf{S}_2\boldsymbol{\beta}$$

   w.r.t. $\boldsymbol{\beta}$ to obtain new estimate $\hat{\boldsymbol{\beta}}$, and hence updated estimates $\hat{\boldsymbol{\eta}} = \mathbf{X}\hat{\boldsymbol{\beta}}$ and $\hat{\mu}_i = g^{-1}(\hat{\eta}_i)$.

The penalized least squares problem at step 2 is exactly the problem already solved for the simple additive model. Note the link to section 3.4.1 (p. 148).

For the trees GAM, the link function, g, is the log function, so $g'(\mu_i) = \mu_i^{-1}$, while for the gamma distribution, $V(\mu_i) = \mu_i^2$ (see table 3.1, p. 104). Hence, for the log-link gamma model, we have:

$$w_i = 1 \quad \text{and} \quad z_i = (y_i - \hat{\mu}_i)/\hat{\mu}_i + \hat{\eta}_i.$$

[§]There is no simple trick to produce an unpenalized GLM whose likelihood is equivalent to the penalized likelihood of the GAM that we wish to fit.

So, given λ_1 and λ_2 it will be straightforward to obtain $\hat{\beta}$, but what should be used as the GCV score for this model? A natural choice is to use the GCV score for the final linear model in the PIRLS iteration (although this choice is poor for binary data: see section 6.2, p. 255 for better performing alternatives). It is easy to show that this GCV score is equivalent to the usual GCV score, but with the Pearson statistic replacing the residual sum of squares. Obviously we could also estimate the smoothing parameters by exploiting the Bayesian/mixed model connection of section 4.2.4, and estimating the model as a generalized linear mixed model using the methods of section 3.4 (p. 147).

The following function implements the PIRLS loop for the log-gamma model, and returns the required GCV score in its return list.

```
gam.fit <- function(y,x,v,sp,k=10) {
## gamma error log link 2 term gam fit...
  eta <- log(y) ## get initial eta
  not.converged <- TRUE
  old.gcv <- -100 ## don't converge immediately
  while (not.converged) {
    mu <- exp(eta)  ## current mu estimate
    z <- (y - mu)/mu + eta ## pseudodata
    fit <- am.fit(z,x,v,sp,k) ## penalized least squares
    if (abs(fit$gcv-old.gcv)<1e-5*fit$gcv) {
      not.converged <- FALSE
    }
    old.gcv <- fit$gcv
    eta <- fit$fitted ## updated eta
  }
  fit$fitted <- exp(fit$fitted) ## mu
  fit
}
```

Again a simple wrapper is needed in order to optimize the GCV score using `optim`

```
gam.gcv <- function(lsp,y,x,v,k=10) {
  gam.fit(y,x,v,exp(lsp),k=k)$gcv
}
```

Now fitting and plotting proceeds exactly as in the simple additive case.

```
fit <- optim(c(0,0),gam.gcv,y=trees$Volume,x=trees$Girth,
             v=trees$Height,k=10)
sp <- exp(fit$par)
fit <- gam.fit(trees$Volume,trees$Girth,trees$Height,sp)
par(mfrow=c(1,3))
plot(fit$fitted,trees$Vol,xlab="fitted volume ",
     ylab="observed volume")
am.plot(fit,xlab=c("Girth","Height"),
        ylab=c("s(Girth)","s(Height)"))
```

The resulting plots are shown in figure 4.12.

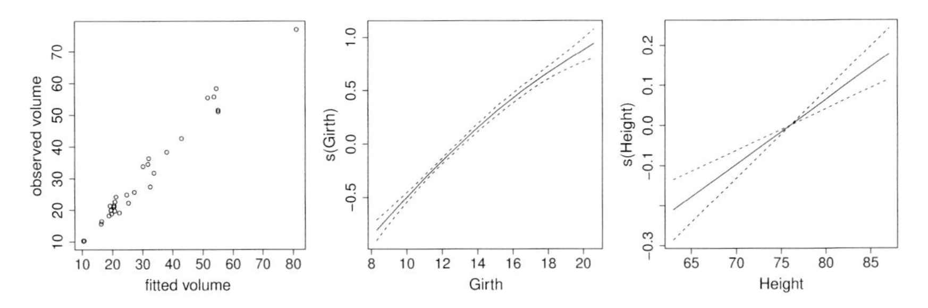

Figure 4.12 *The best fit two term generalized additive model for the* `tree` *data. The left panel shows actual versus predicted tree volumes. The middle panel is the estimate of the smooth function of girth. The right panel is the estimate of the smooth function of height.*

4.5 Summary

The preceding sections have illustrated how models based on smooth functions of predictor variables can be represented, and estimated, once a basis and wiggliness penalty have been chosen for each smooth in the model. Estimation is by penalized versions of the least squares and maximum-likelihood methods used for linear models and GLMs. Indeed technically GAMs are simply GLMs estimated subject to smoothing penalties. The most substantial difficulty introduced by this penalization is the need to select the degree of penalization, that is, to estimate the smoothing parameters. As we have seen, GCV provides one quite reasonable solution, and marginal likelihood provides an alternative.

The rest of this book will stick to the basic framework presented here, simply adding refinements to it. We will consider a variety of basis-penalty smoothers somewhat preferable to the piecewise linear basis given here, and some alternatives to GCV for smoothness estimation. More efficient and reliable computational methods will be developed, and the theoretical basis for inference will be more fully expounded. The link between smooths and random effects will also be developed, as will models based on linear functionals of smooths. However, throughout, functions are represented using penalized basis expansions, estimation of coefficients is by penalized likelihood maximisation and estimation of smoothing parameters uses a separate criterion, such as GCV or REML.

4.6 Introducing package `mgcv`

Before considering smoothers and GAM theory in more detail, it is worth briefly introducing the `mgcv` package. The `gam` function from `mgcv` is very much like the `glm` function covered in chapter 3. The main difference is that the `gam` model formula can include smooth terms, `s()` and `te()` (as well as the `te` variants `ti` and `t2`). Also there are a number of options available for controlling automatic smoothing parameter estimation, or for directly controlling model smoothness (summarized in table 4.1).

The cherry tree data provide a simple example with which to introduce the modelling functions available in R package mgcv.

```
library(mgcv)    ## load the package
data(trees)
ct1 <- gam(Volume ~ s(Height) + s(Girth),
           family=Gamma(link=log),data=trees)
```

This fits the generalized additive model

$$\log(\mathbb{E}[\texttt{Volume}_i]) = f_1(\texttt{Height}_i) + f_2(\texttt{Girth}_i) \quad \text{where } \texttt{Volume}_i \sim \text{gamma}$$

and the f_j are smooth functions. By default, the degree of smoothness of the f_j (within certain limits) is estimated by GCV. The results can be checked by typing the name of the fitted model object to invoke the print.gam print method, and by plotting the fitted model object. For example

```
> ct1

Family: Gamma
Link function: log

Formula:
Volume ~ s(Height) + s(Girth)

Estimated degrees of freedom:
1.00 2.42  total = 4.42

GCV score: 0.008082356
> plot(ct1,residuals=TRUE)
```

The resulting plot is displayed in the upper two panels of figure 4.13. Notice that the default print method reports the model distribution family, link function and formula, before displaying the effective degrees of freedom for each term (in the order that the terms appear in the model formula) and the whole model: in this case a nearly straight line, corresponding to about one degree of freedom, is estimated for the effect of height, while the effect of girth is estimated as a smooth curve with 2.4 degrees of freedom; the total degrees of freedom is the sum of these two, plus one degree of freedom for the model intercept. Finally, the GCV score for the fitted model is reported.

The plots show the estimated effects as solid lines/curves, with 95% confidence limits (strictly Bayesian credible intervals; see section 6.10, p. 293), based on (4.10), shown as dashed lines. The coincidence of the confidence limits and the estimated straight line, at the point where the line passes through zero on the vertical axis, is a result of the identifiability constraints applied to the smooth terms.[¶] The points shown on the plots are *partial residuals*. These are simply the Pearson residuals

[¶]The identifiability constraint is that the sum of the values of each curve, at the observed covariate values, must be zero: for a straight line, this condition exactly determines where the line must pass through zero, so there can be no uncertainty about this point.

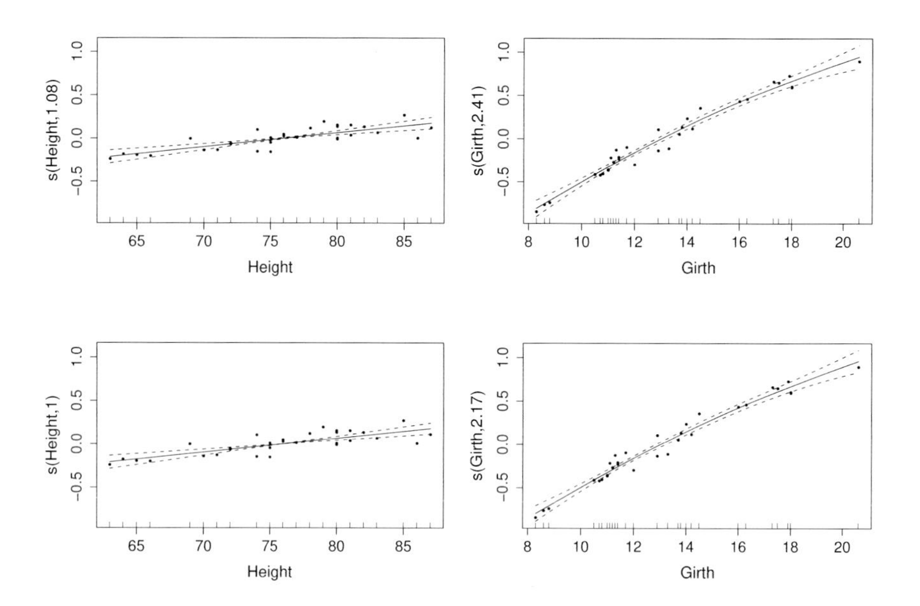

Figure 4.13 *Components of GAM model fits to the cherry tree data. The upper two panels are from* `ct1` *and the lower 2 from* `ct4`.

added to the smooth terms evaluated at the appropriate covariate values. For example, the residuals plotted in the top left panel of figure 4.13 are given by

$$\hat{\epsilon}_{1i}^{\text{partial}} = f_1(\texttt{Height}_\texttt{i}) + \hat{\epsilon}_i^p$$

plotted against $\texttt{Height}_i$. For a well fitting model the partial residuals should be evenly scattered around the curve to which they relate. The 'rug plots', along the bottom of each plot, show the values of the covariates of each smooth. The number in each y-axis caption is the effective degrees of freedom of the term being plotted.

4.6.1 *Finer control of* `gam`

The simple form of the `gam` call producing `ct1` hides a number of options that have been set to default values. The first of these is the choice of basis used to represent the smooth terms. The default is to use thin plate regression splines (section 5.5.1, p. 215), which have some appealing properties, but can be somewhat computationally costly for large data sets. The full range of available smoothers is covered in chapter 5. In the following, penalized cubic regression splines are selected using `s(...,bs="cr")`.

```
> ct2 <- gam(Volume ~ s(Height,bs="cr") + s(Girth,bs="cr"),
+            family=Gamma(link=log),data=trees)
> ct2
```

`scale`	The value of the scale parameter, or a negative value if it is to be estimated. For `method="GCV.Cp"` then `scale` > 0 implies Mallows' C_p/UBRE/AIC is used. `scale` $< 0 \Rightarrow$ implies GCV is used. `scale` $= 0 \Rightarrow$ UBRE/AIC for Poisson or binomial, otherwise GCV.
`gamma`	This multiplies the model degrees of freedom in the GCV or UBRE/AIC criteria. Hence as `gamma` is increased from 1 the 'penalty' per degree of freedom increases in the GCV or UBRE/AIC criterion and increasingly smooth models are produced. Increasing `gamma` to around 1.5 can usually reduce over-fitting, without much degradation in prediction error performance.
`sp`	An array of supplied smoothing parameters. When this array is non-null, a negative element signals that a smoothing parameter should be estimated, while a non-negative value is used as the smoothing parameter for the corresponding term. This is useful for directly controlling the smoothness of some terms.
`method`	Selects the smoothing parameter selection criterion: `GCV.Cp`, `GACV`, `ML` or `REML`.

Table 4.1 *Main arguments to* `gam` *for controlling the smoothness estimation process.*

```
Family: Gamma
Link function: log

Formula:
Volume ~ s(Height, bs = "cr") + s(Girth, bs = "cr")

Estimated degrees of freedom:
 1.000126 2.418591   total =  4.418718

GCV score:  0.008080546
```

As you can see, the change in basis has made very little difference to the fit. Plots are almost indistinguishable to those for `ct1`. This is re-assuring: it would be unfortunate if the model depended very strongly on details like the exact choice of basis. However, larger changes to the basis, such as using P-splines (section 5.3.3, p. 204), can make an appreciable difference.

Another choice, hidden in the previous two model fits, is the *dimension*, k, of the basis used to represent smooth terms. In the previous two fits, the (arbitrary) default, $k = 10$, was used. The choice of basis dimensions amounts to setting the *maximum* possible degrees of freedom allowed for each model term. The actual effective degrees of freedom for each term will usually be estimated from the data, by GCV or another smoothness selection criterion, but the upper limit on this estimate is $k - 1$: the basis dimension minus one degree of freedom due to the identifiability constraint on each smooth term. The following example sets k to 20 for the smooth of `Girth` (and illustrates, by the way, that there is no problem in mixing different bases).

```
> ct3 <- gam(Volume ~ s(Height) + s(Girth,bs="cr",k=20),
+            family=Gamma(link=log),data=trees)
> ct3

Family: Gamma
Link function: log

Formula:
Volume ~ s(Height) + s(Girth, bs = "cr", k = 20)

Estimated degrees of freedom:
 1.000003 2.424226   total =  4.424229

GCV score:  0.00808297
```

Again, this change makes boringly little difference in this case, and the plots (not shown) are indistinguishable from those for `ct1`. This insensitivity to basis dimension is not universal, of course, and checking of this choice is covered in section 5.9 (p. 242). One quite subtle point is worth being aware of. This is that a space of functions of dimension 20 will contain a larger subspace of functions with effective degrees of freedom 5, than will a function space of dimension 10 (the particular numbers being arbitrary here). Hence it is often the case that increasing k will change the effective degrees of freedom estimated for a term, even though both old and new estimated degrees of freedom are lower than the original $k - 1$.

Another choice is the parameter `gamma` which can be used to multiply the model effective degrees of freedom in the GCV or UBRE scores in order to (usually) increase the amount of smoothing selected. The default value is 1, but GCV is known to have some tendency to overfitting on occasion, and it has been suggested that using $\gamma \approx 1.5$ can somewhat correct this without compromising model fit (e.g., Kim and Gu, 2004). See section 6.2.4 for one justification. Applying this idea to the current model results in the bottom row of figure 4.13 and the following output.

```
> ct4 <- gam(Volume ~ s(Height) + s(Girth),
+            family=Gamma(link=log),data=trees,gamma=1.4)
> ct4

Family: Gamma
Link function: log

Formula:
Volume ~ s(Height) + s(Girth)

Estimated degrees of freedom:
 1.00011 2.169248   total =  4.169358

GCV score:  0.00922805

> plot(ct4,residuals=TRUE)
```

The heavier penalty on each degree of freedom in the GCV score has resulted in

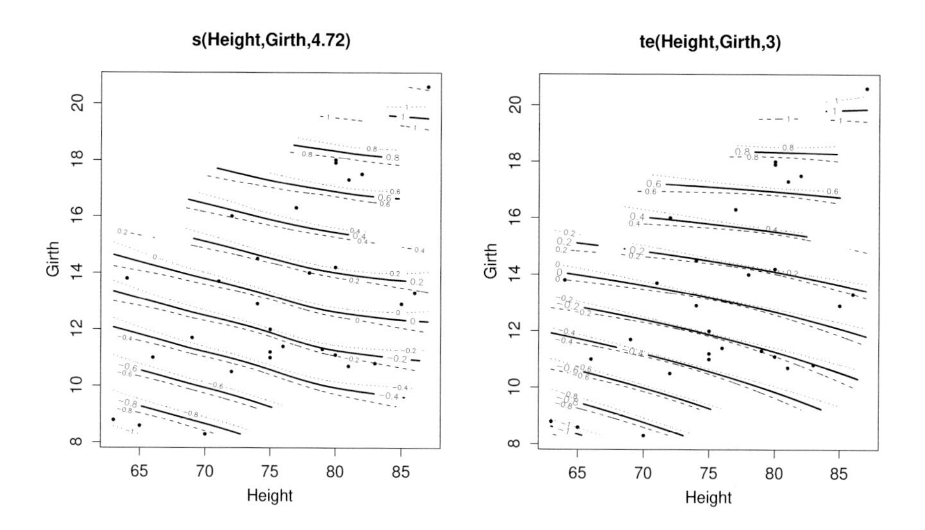

Figure 4.14 *Smooth functions of height and girth fitted to the cherry tree data, with degree of smoothing chosen by GCV. The left hand panel shows a thin plate regression spline fit (*`ct5`*), while the right panel shows a tensor product spline fit (*`ct6`*). For both plots the bold contours show the estimate of the smooth; the dashed contours show the smooth plus the standard error of the smooth and the dotted contours show the smooth less its standard error. The symbols show the locations of the covariate values on the height–girth plane. Parts of the smooths that are far away from covariate values have been excluded from the plots using the* `too.far` *argument to* `plot.gam`*.*

a model with fewer degrees of freedom, but the figure indicates that the change in estimates that this produces is barely perceptible.

4.6.2 *Smooths of several variables*

`gam` is not restricted to models containing only smooths of one predictor. In principle, smooths of any number of predictors are possible via two types of smooth. Within a model formula, `s()` terms, using the `"tp"`, `"ds"` or `"gp"` bases,‖ produce isotropic smooths of multiple predictors, while `te()` terms produce smooths of multiple predictors from tensor products of *any* singly penalized bases available for use with `s()` (including mixtures of different bases). The tensor product smooths are invariant to linear rescaling of covariates, and can be quite computationally efficient. Alternative versions `t2()` and `ti()` are available for different sorts of functional ANOVA decomposition. Section 5.7 (p. 237) compares isotropic and tensor product smoothers.

‖Or indeed `"sos"` or `"so"` bases.

By way of illustration, the following code fragments both fit the model

$$\log(\mathbb{E}[\texttt{Volume}_i]) = f(\texttt{Height}_i, \texttt{Girth}_i) \text{ where } \texttt{Volume}_i \sim \text{gamma},$$

and f is a smooth function. Firstly an isotropic thin plate regression spline is used:

```
> ct5 <- gam(Volume ~ s(Height,Girth,k=25),
+            family=Gamma(link=log),data=trees)
> ct5

Family: Gamma
Link function: log

Formula:
Volume ~ s(Height, Girth, k = 25)

Estimated degrees of freedom:
 4.668129   total =  5.668129

GCV score:  0.009358786

> plot(ct5,too.far=0.15)
```

yielding the left hand panel of figure 4.14. Secondly a tensor product smooth is used. Note that the `k` argument to `te` specifies the dimension for each marginal basis: if different dimensions are required for the marginal bases then `k` can also be supplied as an array. The basis dimension of the tensor product smooth is the product of the dimensions of the marginal bases.

```
> ct6 <- gam(Volume ~ te(Height,Girth,k=5),
+            family=Gamma(link=log),data=trees)
> ct6

Family: Gamma
Link function: log

Formula:
Volume ~ te(Height, Girth, k = 5)

Estimated degrees of freedom:
 3.000175   total =  4.000175

GCV score:  0.008197151

> plot(ct6,too.far=0.15)
```

Notice how the tensor product model has fewer degrees of freedom and a lower GCV score than the TPRS smooth. In fact, with just 3 degrees of freedom, the tensor product smooth model amounts to

$$\log(\mathbb{E}[\texttt{Volume}_i]) = \beta_0 + \beta_1\texttt{Height}_i + \beta_2\texttt{Girth}_i + \beta_3\texttt{Height}_i\texttt{Girth}_i,$$

the 'wiggly' components of the model having been penalized away altogether.

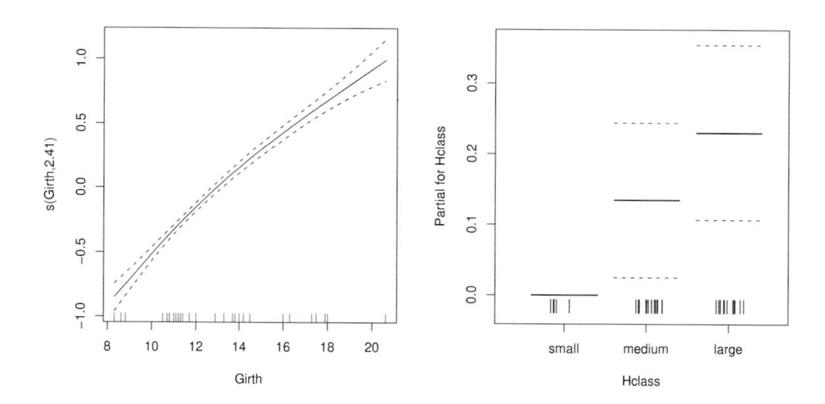

Figure 4.15 *Plot of model* `ct7`*, a* semi-parametric *model of cherry tree volume, with a factor for height and a smooth term for the dependence on girth. The left plot shows the smooth of girth, with 95% confidence interval, while the right panel shows the estimated effect, for each level of factor* `Hclass`*. The effect of being in the* `small` *height class is shown as zero, because the default contrasts have been used here, which set the parameter for the first level of each factor to zero.*

4.6.3 Parametric model terms

So far, only models consisting of smooth terms have been considered, but there is no difficulty in mixing smooth and parametric model components. For example, given that the model `ct1` smooth of height is estimated to be a straight line, we might as well fit the model:

```
gam(Volume ~ Height+s(Girth),family=Gamma(link=log),data=trees)
```

but to make the example more informative, let us instead suppose that the `Height` is actually only measured as a categorical variable. This can easily be arranged, by creating a factor variable which simply labels each tree as small, medium or large:

```
trees$Hclass <- factor(floor(trees$Height/10)-5,
                labels=c("small","medium","large"))
```

Now we can fit a generalized additive model to these data, using the `Hclass` variable as a factor variable, and plot the result (figure 4.15).

```
ct7 <- gam(Volume ~ Hclass + s(Girth),
           family=Gamma(link=log), data=trees)
par(mfrow=c(1,2)); plot(ct7,all.terms=TRUE)
```

Often, more information about a fitted model is required than is supplied by plots or the default print method, and various utility functions exist to provide this. For example the `anova` function can be used to investigate the approximate significance of model terms.

```
> anova(ct7)

Family: Gamma
```

```
Link function: log

Formula:
Volume ~ Hclass + s(Girth)

Parametric Terms:
       df     F p-value
Hclass  2 7.076 0.00358

Approximate significance of smooth terms:
           edf Est.rank     F  p-value
s(Girth) 2.414    9.000 54.43 1.98e-14
```

Clearly there is quite strong evidence that both height and girth matter (see section 6.12, for information on the p-value calculations for the smooth terms). Similarly, an approximate AIC value can be obtained for the model (see section 6.11, p. 301):

```
> AIC(ct7)
[1] 154.9411
```

The `summary` method provides considerable detail.

```
> summary(ct7)

Family: Gamma
Link function: log

Formula:
Volume ~ Hclass + s(Girth)

Parametric coefficients:
            Estimate Std. Error t value Pr(>|t|)
(Intercept)  3.12693    0.04814  64.949  < 2e-16 ***
Hclassmedium 0.13459    0.05428   2.479 0.020085 *
Hclasslarge  0.23024    0.06137   3.752 0.000908 ***

Approximate significance of smooth terms:
           edf Est.rank     F  p-value
s(Girth) 2.414    9.000 54.43 1.98e-14 ***

R-sq.(adj) =  0.967   Deviance explained = 96.9%
GCV score = 0.012076   Scale est. = 0.0099671  n = 31
```

Notice that, in this case, the significance of individual parameters of the parametric terms is given, rather than whole term significance. Other measures of fit are also reported, such as the adjusted r^2 and percentage deviance explained, along with the GCV score, an estimate of the scale parameter of the model, and the number of data fitted.

4.6.4 *The* mgcv *help pages*

mgcv has quite extensive help pages, both documenting functions and attempting to provide overviews of a topic. The easiest way to access the pages is via the HTML versions, by typing help.start() in R, then navigating to the mgcv pages and browsing. Several pages are well worth knowing about:

- mgcv-package offers an overview of the package and what it offers.
- family.mgcv gives an overview of the distributions available.
- smooth.terms gives an overview of the smooths types available.
- random.effects is an overview of random effects in mgcv.
- gam.models reviews some aspects of model specification; gam.selection covers model selection options.
- gam, bam, gamm and jagam cover the main modelling functions.

4.7 Exercises

1. This question is about illustrating the problems with polynomial bases. First run

```
set.seed(1)
x<-sort(runif(40)*10)^.5
y<-sort(runif(40))^0.1
```

   to simulate some apparently innocuous x, y data.

   (a) Fit 5th and 10th order polynomials to the simulated data using, e.g., lm(y~poly(x,5)).

   (b) Plot the x, y data, and overlay the fitted polynomials. (Use the predict function to obtain predictions on a fine grid over the range of the x data: only predicting at the data fails to illustrate the polynomial behaviour adequately).

   (c) One particularly simple basis for a cubic regression spline is $b_1(x) = 1$, $b_2(x) = x$ and $b_{j+2}(x) = |x - x_j^*|^3$ for $j = 1 \ldots q - 2$, where q is the basis dimension, and the x_j^* are knot locations. Use this basis to fit a rank 11 cubic regression spline to the x, y data (using lm and evenly spaced knots).

   (d) Overlay the predicted curve according to the spline model, onto the existing x, y plot, and consider which basis you would rather use.

2. Polynomial models of the data from question 1 can also provide an illustration of why orthogonal matrix methods are preferable to fitting models by solution of the 'normal equations' $\mathbf{X}^\mathsf{T}\mathbf{X}\boldsymbol{\beta} = \mathbf{X}^\mathsf{T}\mathbf{y}$. The bases produced by poly are actually orthogonal polynomial bases, which are a numerically stable way of representing polynomial models, but if a naïve basis is used then a numerically badly behaved model can be created.

```
form<-paste("I(x^",1:10,")",sep="",collapse="+")
form <- as.formula(paste("y~",form))
```

   produces the model formula for a suitably ill-behaved model. Fit this model using lm, extract the model matrix from the fitted model object using model.matrix, and re-estimate the model parameters by solving the 'normal equations' given above (see `?solve`). Compare the estimated coefficients in both cases, along with the fits. It is also instructive to increase the order of the polynomial by one or two and examine the results (and to decrease it to 5, say, in order to confirm that the QR and normal equations approaches agree if everything is 'well behaved'). Finally, note that the singular value decomposition (see B.10) provides a reliable way of diagnosing the linear dependencies that can cause problems when model fitting. `svd(X)$d` obtains the singular values of a matrix `X`. The largest divided by the smallest gives the 'condition number' of the matrix — a measure of how ill-conditioned computations with the matrix are likely to be.

3. Show that the $\boldsymbol{\beta}$ minimizing (4.6), in section 4.2.2, is given by (4.7).

4. Let $\mathbf{X}$ be an $n \times p$ model matrix, $\mathbf{S}$ a $p \times p$ penalty matrix, and $\mathbf{B}$ any matrix such that $\mathbf{B}^\mathsf{T}\mathbf{B} = \mathbf{S}$. If

$$\tilde{\mathbf{X}} = \begin{bmatrix} \mathbf{X} \\ \mathbf{B} \end{bmatrix}$$

is an augmented model matrix, show that the sum of the first n elements on the leading diagonal of $\tilde{\mathbf{X}}(\tilde{\mathbf{X}}^\mathsf{T}\tilde{\mathbf{X}})^{-1}\tilde{\mathbf{X}}^\mathsf{T}$ is $\text{tr}\left(\mathbf{X}(\mathbf{X}^\mathsf{T}\mathbf{X} + \mathbf{S})^{-1}\mathbf{X}^\mathsf{T}\right)$.

5. The 'obvious' way to estimate smoothing parameters is by treating smoothing parameters just like the other model parameters, $\boldsymbol{\beta}$, and to choose λ to minimize the residual sum of squares for the fitted model. What estimate of λ will such an approach always produce?

6. Show that for any function f, which has a basis expansion

$$f(x) = \sum_j \beta_j b_j(x),$$

it is possible to write

$$\int f''(x)^2 dx = \boldsymbol{\beta}^\mathsf{T}\mathbf{S}\boldsymbol{\beta},$$

where the coefficient matrix $\mathbf{S}$ can be expressed in terms of the known basis functions b_j (assuming that these possess at least two (integrable) derivatives). As usual $\boldsymbol{\beta}$ is a parameter vector with β_j in its jth element.

7. Show that for any function f which has a basis expansion

$$f(x, z) = \sum_j \beta_j b_j(x, z),$$

it is possible to write

$$\int \left(\frac{\partial^2 f}{\partial x^2}\right)^2 + 2\left(\frac{\partial^2 f}{\partial x \partial z}\right)^2 + \left(\frac{\partial f^2}{\partial z^2}\right)^2 dx dz = \boldsymbol{\beta}^\mathsf{T}\mathbf{S}\boldsymbol{\beta},$$

where the coefficient matrix $\mathbf{S}$ can be expressed in terms of the known basis functions b_j (assuming that these possess at least two (integrable) derivatives w.r.t. x and z). Again, $\boldsymbol{\beta}$ is a parameter vector with β_j in its jth element.

8. The additive model of section 4.3 can equally well be estimated as a mixed model.

(a) Write a function which converts the model matrix and penalty returned by `tf.XD` into mixed model form. Hint: because of the constraints the penalty null space is of dimension 1 now, leading to a slight modification of $\mathbf{D}_+$.

(b) Using your function from part (a) obtain the model matrices required to fit the two term additive tree model, and estimate it using `lme`. Because there are now two smooths, two `pdIdent` terms will be needed in the `random` list supplied to `lme`, which will involve two dummy grouping variables (which can just be differently named copies of the same variable).

(c) Produce residual versus fitted volume and raw volume against fitted volume plots.

(d) Produce plots of the two smooth effect estimates with partial residuals.

9. Following on from question 8, we can also estimate a GAM as a GLMM. This is particularly easy to implement using the PQL method of section 3.4.2 (p. 149).

(a) Modify the function `gam.fit` from section 4.4, so that in place of the call to `am.fit` there is an appropriate call to `lme` to estimate the coefficients and smoothing parameters of a working linear mixed model. The modified function should take a response vector and the model matrices from the previous question as inputs, and return the `lme` fitted model object for the working model at convergence.

(b) Use your function to fit the Gamma additive model of section 4.4 to the `trees` data.

(c) Produce plots of measured volume against predicted volume, and of residuals against the linear predictor of the model.

(d) Plot the smooth effects with partial residuals.

Chapter 5

Smoothers

The piecewise linear smoother of the previous chapter offers a perfectly reasonable way of representing the smooth functions in additive models, but substantial improvement is possible. In particular, if we represent the smooth model terms using *spline* bases, then it is possible to obtain substantially reduced function approximation error for a given dimension of smoothing basis. This chapter covers spline smoothers, and the equivalence between quadratically penalized smoothers and Gaussian random effects/fields.

The chapter starts by considering one dimensional spline smoothing, and then computationally efficient reduced rank penalized regression splines. Several one dimensional penalized regression smoothers are covered, including adaptive smoothers. Constraints, the 'natural' parameterization of a smooth and effective degrees of freedom are discussed next, followed by multidimensional smoothers. First isotropic smoothers: thin-plate and other Duchon splines, splines on the sphere and soap film smoothers. Isotropy is usually inappropriate when the arguments of a smooth have different units. Scale invariant 'tensor product' smoothers are therefore constructed next, and consideration given to the notion of a smooth interaction, and to decompositions involving smooth main effects and smooth interactions. There follows a discussion of the duality between smooths and Gaussian random effects/fields and of Gaussian Markov random field smoothers and Gaussian process smoothers. A slightly more advanced final section gives a brief introduction to the reproducing kernel Hilbert space approach to spline theory.

5.1 Smoothing splines

Almost all the smooths considered in this book are based in some way on splines, so it is worth spending a little time on the theoretical properties that make these functions so appealing for penalized regression. Rather than attempt full generality, the flavour of the theoretical ideas can be gleaned by considering some properties of cubic splines, first in the context of interpolation, and then of smoothing.

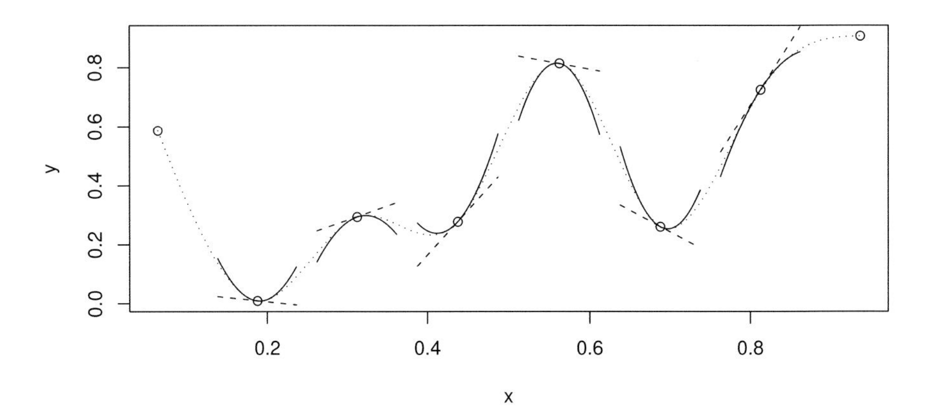

Figure 5.1 *A cubic spline is a curve constructed from sections of cubic polynomial joined together so that the curve is continuous up to second derivative. The spline shown (dotted curve) is made up of 7 sections of cubic. The points at which they are joined (○) (and the two end points) are known as the* knots *of the spline. Each section of cubic has different coefficients, but at the knots it will match its neighbouring sections in value and first two derivatives. Straight dashed lines show the gradients of the spline at the knots and the continuous curves are quadratics matching the first and second derivatives at the knots: these illustrate the continuity of first and second derivatives across the knots. This spline has zero second derivatives at the end knots: a* 'natural *spline'. Note that there are many alternative ways of representing such a cubic spline using basis functions: although all are equivalent, the link to the piecewise cubic characterization is not always transparent.*

5.1.1 *Natural cubic splines are smoothest interpolators*

Consider a set of points $\{x_i, y_i : i = 1, \ldots, n\}$ where $x_i < x_{i+1}$. The *natural cubic spline*, $g(x)$,* interpolating these points, is a function made up of sections of cubic polynomial, one for each $[x_i, x_{i+1}]$, which are joined together so that the whole spline is continuous to second derivative, while $g(x_i) = y_i$ and $g''(x_1) = g''(x_n) = 0$. Figure 5.1 illustrates such a cubic spline.

Of all functions that are continuous on $[x_1, x_n]$, have absolutely continuous first derivatives and interpolate $\{x_i, y_i\}$, $g(x)$ is the one that is smoothest in the sense of minimizing:

$$J(f) = \int_{x_1}^{x_n} f''(x)^2 dx.$$

Green and Silverman (1994) provide a neat proof of this, based on the original work of Schoenberg (1964). Let $f(x)$ be an interpolant of $\{x_i, y_i\}$, other than $g(x)$,

*In this chapter g denotes a generic smooth function, rather than the link function.

and let $h(x) = f(x) - g(x)$. We seek an expression for $J(f)$ in terms of $J(g)$.

$$\begin{aligned}\int_{x_1}^{x_n} f''(x)^2 dx &= \int_{x_1}^{x_n} \{g''(x) + h''(x)\}^2 dx \\ &= \int_{x_1}^{x_n} g''(x)^2 dx + 2\int_{x_1}^{x_n} g''(x)h''(x)dx + \int_{x_1}^{x_n} h''(x)^2 dx\end{aligned}$$

and integrating the second term on the second line, by parts, yields

$$\begin{aligned}\int_{x_1}^{x_n} g''(x)h''(x)dx &= g''(x_n)h'(x_n) - g''(x_1)h'(x_1) - \int_{x_1}^{x_n} g'''(x)h'(x)dx \\ &= -\int_{x_1}^{x_n} g'''(x)h'(x)dx = -\sum_{i=1}^{n-1} g'''(x_i^+)\int_{x_i}^{x_{i+1}} h'(x)dx \\ &= -\sum_{i=1}^{n-1} g'''(x_i^+)\{h(x_{i+1}) - h(x_i)\} = 0,\end{aligned}$$

where equality of lines 1 and 2 follows from the fact that $g''(x_1) = g''(x_n) = 0$. The equalities in line 2 result from the fact that $g(x)$ is made up of sections of cubic polynomial, so that $g'''(x)$ is constant over any interval (x_i, x_{i+1}); x_i^+ denotes an element of such an interval. The final equality to zero follows from the fact that both $f(x)$ and $g(x)$ are interpolants, and are hence equal at x_i, implying that $h(x_i) = 0$.

So we have shown that

$$\int_{x_1}^{x_n} f''(x)^2 dx = \int_{x_1}^{x_n} g''(x)^2 dx + \int_{x_1}^{x_n} h''(x)^2 dx \geq \int_{x_1}^{x_n} g''(x)^2 dx$$

with equality only if $h''(x) = 0$ for $x_1 < x < x_n$. However, $h(x_1) = h(x_n) = 0$, so in fact we have equality if and only if $h(x) = 0$ on $[x_1, x_n]$. In other words, any interpolant that is not identical to $g(x)$ will have a higher integrated squared second derivative. So there is a well defined sense in which the cubic spline is the smoothest possible interpolant through any set of data.

The smoothest interpolation property is not the only good property of cubic spline interpolants. In de Boor (1978, Chapter 5) a number of results are presented showing that cubic spline interpolation is optimal, or at least very good, in various respects. For example, if a 'complete' cubic spline, g, is used to approximate a function, $\tilde{f}$, by interpolating a set of points $\{x_i, \tilde{f}(x_i) : i = 1, \ldots, n\}$ and matching $\tilde{f}'(x_1)$ and $\tilde{f}'(x_n)$ then if $\tilde{f}(x)$ has 4 continuous derivatives:

$$\max|\tilde{f} - g| \leq \frac{5}{384}\max(x_{i+1} - x_i)^4 \max|\tilde{f}''''|. \tag{5.1}$$

The inequality is 'sharp' (meaning that 5/384 can not be improved upon).

These properties of spline interpolants suggest that splines ought to provide a good basis for representing smooth terms in statistical models. Whatever the true underlying smooth function is, a spline ought to be able to approximate it closely, and if we want to construct models from smooth functions of covariates, then representing those functions from smoothest approximations is intuitively appealing.

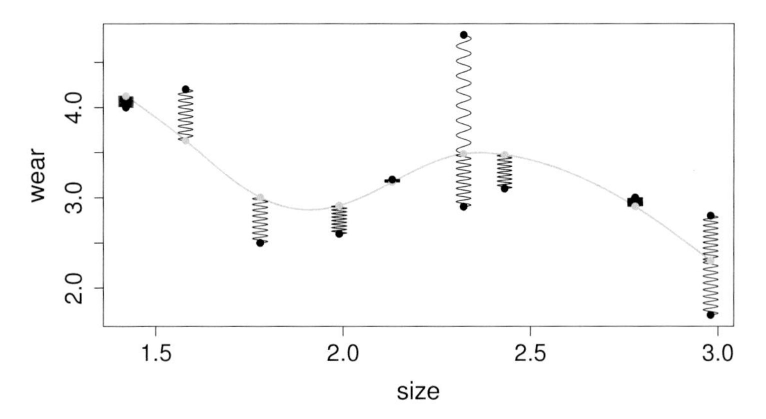

Figure 5.2 *Physical analogy for a cubic smoothing spline. The black points represent x, y data, while the grey curve is a thin flexible strip (e.g., of wood). If the strip is hooked up to the data by springs then the thin strip can be viewed as a* smoothing spline. *The springs are idealised here, having zero length under zero load, and only extending in the y direction. Also, strictly the analogy only holds if the vertical deflection of the grey curve is 'small'.*

5.1.2 Cubic smoothing splines

In statistical work y_i is usually measured with noise, and it is generally more useful to smooth x_i, y_i data, rather than interpolating them. To this end, rather than setting $g(x_i) = y_i$, it might be better to treat the $g(x_i)$ as n free parameters of the cubic spline, and to estimate them in order to minimize

$$\sum_{i=1}^{n}\{y_i - g(x_i)\}^2 + \lambda \int g''(x)^2 dx,$$

where λ is a tunable parameter, used to control the relative weight to be given to the conflicting goals of matching the data and producing a smooth g. The resulting $g(x)$ is a *smoothing spline* (Reinsch, 1967). Figure 5.2 illustrates the basic idea. In fact, of *all functions*, f, that are continuous on $[x_1, x_n]$, and have absolutely continuous first derivatives, $g(x)$ is the function minimizing:

$$\sum_{i=1}^{n}\{y_i - f(x_i)\}^2 + \lambda \int f''(x)^2 dx. \tag{5.2}$$

The proof is easy. Suppose that some other function, $f^*(x)$, minimized (5.2). In that case we could interpolate $\{x_i, f^*(x_i)\}$ using a cubic spline, $g(x)$. Now $g(x)$ and $f^*(x)$ have the same sum of squares term in (5.2), but by the properties of interpolating splines, $g(x)$ must have the lower integrated squared second derivative. Hence $g(x)$ yields a lower (5.2) than $f^*(x)$, and a contradiction, unless $f^* = g$.

Notice the obvious corollary that the same result holds substituting for the residual sum of squares any measure of fit dependent on f only via $f(x_1), \ldots f(x_n)$. In particular we might substitute a log likelihood.

So, rather than being chosen in advance, the cubic spline basis arises naturally from the specification of the smoothing objective (5.2), in which what is meant by model fit and by smoothness are precisely defined in a basis independent way.

Smoothing splines seem to be somewhat ideal smoothers. The only substantial problem is the fact that they have as many free parameters as there are data to be smoothed. This is wasteful, given that, in practice, λ will almost always be high enough that the resulting spline is much smoother than n degrees of freedom would suggest. Indeed, in section 5.4.2 we will see that many degrees of freedom of a spline are often suppressed completely by the penalty. For univariate smoothing with cubic splines the large number of parameters turns out not to be problematic (e.g. De Hoog and Hutchinson, 1987, give a stable $O(n)$ algorithm), but as soon as we try to deal with more covariates the computational expense becomes severe.

5.2 Penalized regression splines

An obvious compromise between retaining the good properties of splines and computational efficiency, is to use penalized regression splines, as introduced in Chapter 4. At its simplest, this involves constructing a spline basis (and associated penalties) for a much smaller data set than the one to be analysed, and then using that basis (plus penalties) to model the original data set. The covariate values in the smaller data set should be arranged to nicely cover the distribution of covariate values in the original data set. This penalized regression spline idea is presented in Wahba (1980) and Parker and Rice (1985), for example. An alternative approach is based on eigen-approximation and is covered in section 5.5.1.

Use of penalized regression splines raises the question of how many basis functions to use. No answer to this is generally possible without knowing the true functions that we are trying to estimate, but it is possible to say something about how the basis dimension should scale with sample size, n, as $n \to \infty$.

Consider the case of cubic interpolating spline approximation of a function $g(x)$, where the spline has k knots, spaced evenly along the x axis. We saw in section 5.1.1 that the approximation error of such a spline is $O(k^{-4})$. See figure 5.3(a). Now suppose that instead of interpolating g we have n/k noisy observations of g at each knot, as illustrated in figure 5.3(b). Without penalisation, the spline estimate of g is just the spline interpolant of the mean observation at each knot. Hence the standard error of the spline estimate is $O(\sqrt{k/n})$. For fixed k, in the $n/k \to \infty$ limit, the standard error vanishes, but the approximation error bias remains at $O(k^{-4})$. To allow the approximation error to also vanish in the limit we require k to grow with n. If we want neither the $O(\sqrt{k/n})$ standard error, nor the $O(k^{-4})$ bias to dominate in the $n \to \infty$ limit, then we should set $k = O(n^{1/9})$ to obtain a mean square error rate of $O(n^{-8/9})$.

Obviously the situation in which all the observations are at the knots is not the one that we usually face, so consider the regression spline with k evenly spaced knots, but where the observations are spread over the domain (in a non-pathological manner so that the standard error at any particular point within the range of the knots is still $O(\sqrt{k/n})$). Let $\hat{g}_\infty$ again denote the spline interpolating g at the knots. From

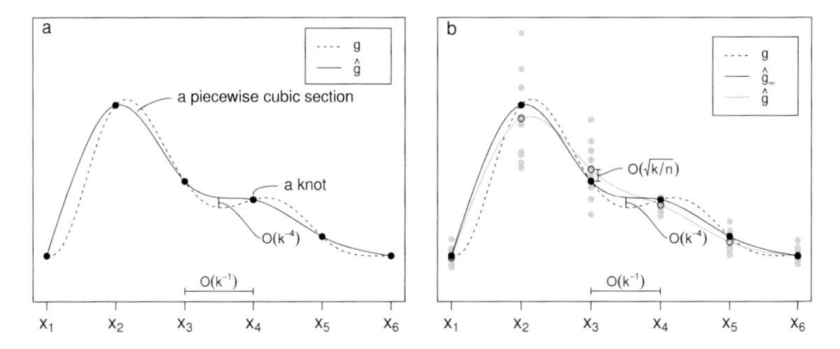

Figure 5.3 *(a) Cubic interpolating spline approximation $\hat{g}$ to function g observed at k knots x_j without error. The x-axis knot spacing is $O(k^{-1})$ so that the approximation error is $O(k^{-4})$. (b) Unpenalized cubic regression spline estimate, $\hat{g}$, to function g noisily observed n/k times at each x_j. The standard error of $\hat{g}$ is $O(\sqrt{k/n})$. In the $n/k \to \infty$ limit, the spline, $\hat{g}_\infty$, coincides with the interpolating spline in a, and has approximation error $O(k^{-4})$. To avoid approximation error or sampling error dominating in the $n \to \infty$ limit requires $k = O(n^{1/9})$.*

(5.1) the elements of $\mathbf{g} - \hat{\mathbf{g}}_\infty$ are bounded above by a constant $O(k^{-4})$, and so $n^{-1}\|\mathbf{g} - \hat{\mathbf{g}}_\infty\|$ must also be bounded above by a constant $O(k^{-4})$. Now in the large sample limit, the regression spline, $\hat{\mathbf{g}}$, minimises $n^{-1}\|\mathbf{g} - \hat{\mathbf{g}}\|$ (see section 1.8.7, p. 54). This must also be bounded above by a constant $0(k^{-4})$ since otherwise we could reduce $n^{-1}\|\mathbf{g} - \hat{\mathbf{g}}\|$ by substituting $\hat{g}_\infty$ for $\hat{g}$. $\hat{g}$ can not have a lower order bias term than this in general, since for some designs $\hat{g} = \hat{g}_\infty$ which has $O(k^{-4})$ bias. In the general regression spline case the average variance at the observation locations is $\text{tr}(\mathbf{A})\sigma^2/n = \sigma^2 k/n$, so again $k = O(n^{1/9})$ is required for optimality. See Agarwal and Studden (1980) for the original result.

Under penalization the situation is slightly more complicated. We could use $k = O(n^{1/9})$ and achieve an asymptotic mean square error rate of $O(n^{-8/9})$, but under smoothing parameter selection, this will often imply having no penalization in the large sample limit. In reality, when smoothing parameters are estimated, we would usually view a λ estimate near zero as a possible indication that k is too low, and therefore increase it. Hence asymptotics in which penalization persists in the large sample limit are more useful. For the cubic spline under REML smoothing parameter estimation this turns out to require that $k = O(n^{1/5})$, in which case the penalized cubic regression spline can achieve a mean square error of $O(n^{-4/5})$. This is somewhat below the optimal rate achievable by a non-parametric estimator of a 4th order continuous function which Cox (1983) shows is $O(n^{-8/9})$. This optimal rate can be achieved by cubic smoothing splines (Stone, 1982; Speckman, 1985) and penalized cubic regression splines (Hall and Opsomer, 2005), under particular assumptions about the rate at which the smoothing parameters change with n. Unfortunately there is little work on the rates actually achieved when smoothing parameters are estimated: Kauermann et al. (2009) is a notable exception, considering estimation under REML smoothing parameter selection, but assuming $k = O(n^{1/9})$ in the cu-

Basis functions for a cubic spline

$a_j^-(x) = (x_{j+1} - x)/h_j$	$c_j^-(x) = [(x_{j+1} - x)^3/h_j - h_j(x_{j+1} - x)]/6$
$a_j^+(x) = (x - x_j)/h_j$	$c_j^+(x) = [(x - x_j)^3/h_j - h_j(x - x_j)]/6$

Non-zero matrix elements — non-cyclic spline

$D_{i,i} = 1/h_i$ $B_{i,i} = (h_i + h_{i+1})/3$	$D_{i,i+1} = -1/h_i - 1/h_{i+1}$	$D_{i,i+2} = 1/h_{i+1}$ $i = 1 \ldots k-2$
$B_{i,i+1} = h_{i+1}/6$	$B_{i+1,i} = h_{i+1}/6$	$i = 1 \ldots k-3$

Non-zero matrix elements — cyclic spline

$\tilde{B}_{i-1,i} = \tilde{B}_{i,i-1} = h_{i-1}/6$ $\tilde{D}_{i-1,i} = \tilde{D}_{i,i-1} = 1/h_{i-1}$	$\tilde{B}_{i,i} = (h_{i-1} + h_i)/3$ $\tilde{D}_{i,i} = -1/h_{i-1} - 1/h_i$	$i = 2 \ldots k-1$
$\tilde{B}_{1,1} = (h_{k-1} + h_1)/3$ $\tilde{D}_{1,1} = -1/h_1 - 1/h_{k-1}$	$\tilde{B}_{1,k-1} = h_{k-1}/6$ $\tilde{D}_{1,k-1} = 1/h_{k-1}$	$\tilde{B}_{k-1,1} = h_{k-1}/6$ $\tilde{D}_{k-1,1} = 1/h_{k-1}$

Table 5.1 *Definitions of basis functions and matrices used to define a cubic regression spline.* $h_j = x_{j+1} - x_j$.

bic regression spline case, which corresponds to not penalizing in the large sample limit. Claeskens et al. (2009) explicitly deal with the asymptotically penalized and unpenalized regimes separately, but not under smoothing parameter estimation.

Despite the obvious deficiencies in the theory, the upshot of this sort of analysis is that we really only need the basis dimension to grow rather slowly with sample size in order to achieve statistical performance asymptotically indistinguishable from that of a full smoothing spline.

5.3 Some one-dimensional smoothers

This section covers the explicit representation of some basis-penalty smoothers useful for smoothing with respect to a single predictor variable. Cubic penalized regression splines are covered first, along with their cyclic version. P-splines are then discussed, and used to create an adaptive smoother. Note that the thin plate regression splines of section 5.5.1 can also be used for one-dimensional smoothing.

5.3.1 Cubic regression splines

There are many equivalent bases that can be used to represent cubic splines. One approach is to parameterize the spline in terms of its values at the knots. Consider defining a cubic spline function, $f(x)$, with k knots, $x_1 \ldots x_k$. Let $\beta_j = f(x_j)$ and $\delta_j = f''(x_j)$. Then the spline can be written as

$$f(x) = a_j^-(x)\beta_j + a_j^+(x)\beta_{j+1} + c_j^-(x)\delta_j + c_j^+(x)\delta_{j+1} \text{ if } x_j \le x \le x_{j+1}, \quad (5.3)$$

where the basis functions a_j^-, a_j^+, c_j^- and c_j^+ are defined in table 5.1. The conditions that the spline must be continuous to second derivative, at the x_j, and should have zero second derivative at x_1 and x_k, can be shown to imply (exercise 1) that

$$\mathbf{B}\boldsymbol{\delta}^- = \mathbf{D}\boldsymbol{\beta}, \quad (5.4)$$

where $\boldsymbol{\delta}^- = (\delta_2, \ldots, \delta_{k-1})^\mathsf{T}$, $\delta_1 = \delta_k = 0$ and $\mathbf{B}$ and $\mathbf{D}$ are defined in table 5.1.

Defining $\mathbf{F}^- = \mathbf{B}^{-1}\mathbf{D}$ and

$$\mathbf{F} = \begin{bmatrix} \mathbf{0} \\ \mathbf{F}^- \\ \mathbf{0} \end{bmatrix},$$

where $\mathbf{0}$ is a row of zeros, we have that $\boldsymbol{\delta} = \mathbf{F}\boldsymbol{\beta}$. Hence, the spline can be rewritten entirely in terms of $\boldsymbol{\beta}$ as

$$f(x) = a_j^-(x)\beta_j + a_j^+(x)\beta_{j+1} + c_j^-(x)\mathbf{F}_j\boldsymbol{\beta} + c_j^+(x)\mathbf{F}_{j+1}\boldsymbol{\beta} \text{ if } x_j \le x \le x_{j+1},$$

which can be re-written, once more, as

$$f(x) = \sum_{i=1}^{k} b_i(x)\beta_i$$

by implicit definition of new basis functions $b_i(x)$: figure 5.4 illustrates the basis. Hence, given a set of x values at which to evaluate the spline, it is easy to obtain a model matrix mapping $\boldsymbol{\beta}$ to the evaluated spline. It can further be shown (e.g., Lancaster and Šalkauskas, 1986, or exercise 2) that

$$\int_{x_1}^{x_k} f''(x)^2 dx = \boldsymbol{\beta}^\mathsf{T}\mathbf{D}^\mathsf{T}\mathbf{B}^{-1}\mathbf{D}\boldsymbol{\beta},$$

i.e. $\mathbf{S} \equiv \mathbf{D}^\mathsf{T}\mathbf{B}^{-1}\mathbf{D}$ is the penalty matrix for this basis.

In addition to having directly interpretable parameters, this basis does not require any re-scaling of the predictor variables before it can be used to construct a GAM, although, as with the chapter 3 basis, we do have to choose the locations of the knots, x_j. See Lancaster and Šalkauskas (1986) for more details. In `mgcv`, model terms like `s(x,bs="cr",k=15)` use this basis (basis dimension `k` defaults to 10 if not supplied).

5.3.2 *A cyclic cubic regression spline*

It is quite often appropriate for a model smooth function to be 'cyclic', meaning that the function has the same value and first few derivatives at its upper and lower

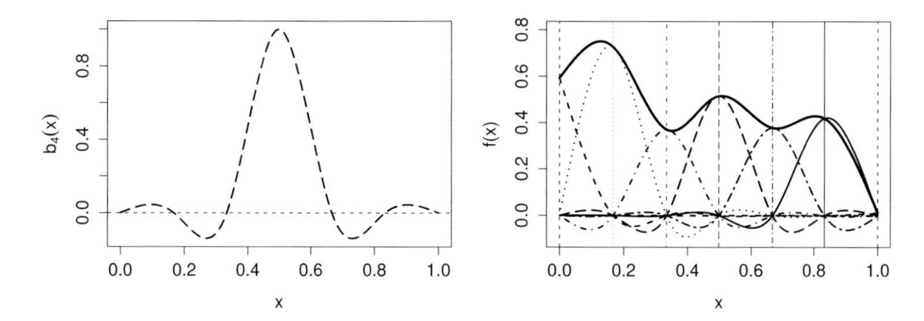

Figure 5.4 *The left hand panel illustrates one basis function, $b_4(x)$, for a cubic regression spline of the type discussed in section 5.3.1: this basis function takes the value one at one knot of the spline, and zero at all other knots (such basis functions are sometimes called 'cardinal basis functions'). The right hand panel shows how such basis functions are combined to represent a smooth curve. The various curves of medium thickness show the basis functions, $b_j(x)$, of a cubic regression spline, each multiplied by its associated coefficient β_j: these scaled basis functions are summed to get the smooth curve illustrated by the thick continuous curve. The vertical thin lines show the knot locations.*

boundaries. For example, in most applications, it would not be appropriate for a smooth function of time of year to change discontinuously at the year end. The penalized cubic regression spline of the previous section can be modified to produce such a smooth. The spline can still be written in the form (5.3), but we now have that $\beta_1 = \beta_k$ and $\delta_1 = \delta_k$. In this case then, we define vectors $\boldsymbol{\beta}^\mathsf{T} = (\beta_1, \ldots, \beta_{k-1})$ and $\boldsymbol{\delta}^\mathsf{T} = (\delta_1, \ldots, \delta_{k-1})$. The conditions that the spline must be continuous to second derivative at each knot, and that $f(x_1)$ must match $f(x_k)$ up to second derivative, are equivalent to

$$\tilde{\mathbf{B}}\boldsymbol{\delta} = \tilde{\mathbf{D}}\boldsymbol{\beta}$$

where $\tilde{\mathbf{B}}$ and $\tilde{\mathbf{D}}$ are defined in table 5.1. Similar reasoning to that employed in the previous section implies that the spline can be written as

$$f(x) = \sum_{i=1}^{k-1} \tilde{b}_i(x)\beta_i,$$

by appropriate definition of the basis functions $\tilde{b}_i(x)$: figure 5.5 illustrates this basis. A second derivative penalty also follows:

$$\int_{x_1}^{x_k} f''(x)^2 dx = \boldsymbol{\beta}^\mathsf{T}\tilde{\mathbf{D}}^\mathsf{T}\tilde{\mathbf{B}}^{-1}\tilde{\mathbf{D}}\boldsymbol{\beta}.$$

In `mgcv`, model terms like `s(x,bs="cc")` specify this basis. By default the spline will match at the smallest and largest x values, since these are the default locations of the outermost knots. Often this is inappropriate, and knots should be supplied by the user, via the `knots` argument to `gam`. If only two knots are supplied then

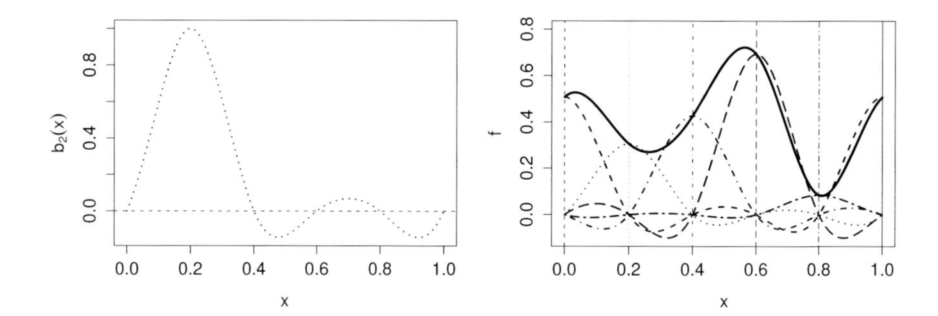

Figure 5.5 *The left hand panel illustrates one basis function, $b_2(x)$, for a cyclic cubic regression spline of the type discussed in section 5.3.2: this basis function takes the value one at one knot of the spline, and zero at all other knots — notice how the basis function values and first two derivatives match at $x = 0$ and $x = 1$. The right hand panel shows how such basis functions are combined to represent a smooth curve. The various curves of medium thickness show the basis functions, $b_j(x)$, of a cyclic cubic regression spline, each multiplied by its associated coefficient β_j: these scaled basis functions are summed to get the smooth curve illustrated by the thick continuous curve. The vertical thin lines show the knot locations.*

they are taken to be the outermost knots, and the remaining `k-2` knots are placed automatically between these.

5.3.3 P-splines

Yet another way to represent cubic splines (and splines of higher or lower order) is by use of the B-spline basis. The B-spline basis is appealing because the basis functions are strictly local — each basis function is only non-zero over the intervals between $m + 3$ adjacent knots, where $m + 1$ is the order of the basis (e.g., $m = 2$ for a cubic spline†). To define a k parameter B-spline basis, first define $k + m + 2$ knots, $x_1 < x_2 < \ldots < x_{k+m+2}$, where the interval over which the spline is to be evaluated lies within $[x_{m+2}, x_{k+1}]$ (so the first and last $m + 1$ knot locations are essentially arbitrary). An $(m + 1)^{\text{th}}$ order spline can then be represented as

$$f(x) = \sum_{i=1}^{k} B_i^m(x)\beta_i,$$

where the B-spline basis functions are most conveniently defined recursively as follows:

$$B_i^m(x) = \frac{x - x_i}{x_{i+m+1} - x_i} B_i^{m-1}(x) + \frac{x_{i+m+2} - x}{x_{i+m+2} - x_{i+1}} B_{i+1}^{m-1}(x) \quad i = 1, \ldots k$$

†The somewhat inconvenient definition of order is for compatibility with the notation usually used for smoothing splines.

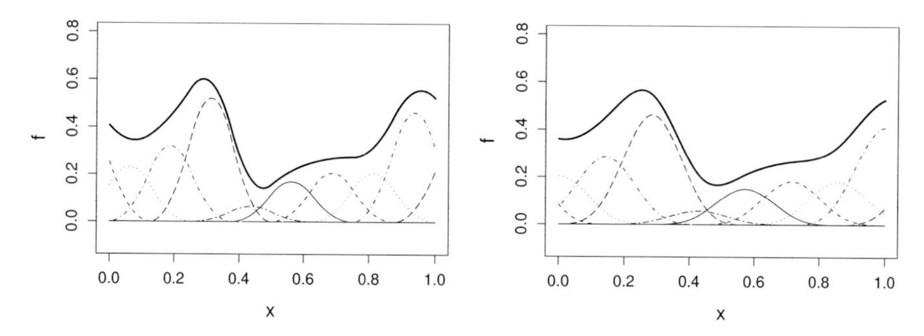

Figure 5.6 *Illustration of the representation of a smooth curve by rank 10 B-spline bases. The left plot shows a B-spline basis with $m = 1$. The thin curves show B-spline basis functions multiplied by their associated coefficients; each is non-zero over only 3 intervals. The sum of the coefficients multiplied by the basis functions gives the spline itself, represented by the thicker continuous curve. The right panel is the same, but for a basis for which $m = 2$: in this case each basis function is non-zero over 4 adjacent intervals. In both panels the knot locations are where each basis function peaks.*

and

$$B_i^{-1}(x) = \begin{cases} 1 & x_i \leq x < x_{i+1} \\ 0 & \text{otherwise} \end{cases}$$

(see, e.g., de Boor, 1978; Lancaster and Šalkauskas, 1986). For example, the following R code can be used to evaluate single B-spline basis functions at a series of x values:

```
bspline <- function(x,k,i,m=2)
# evaluate ith B-spline basis function of order m at the
# values in x, given knot locations in k
{ if (m==-1) { # base of recursion
    res <- as.numeric(x<k[i+1]&x>=k[i])
  } else {      # construct from call to lower order basis
    z0 <- (x-k[i])/(k[i+m+1]-k[i])
    z1 <- (k[i+m+2]-x)/(k[i+m+2]-k[i+1])
    res <- z0*bspline(x,k,i,m-1)+ z1*bspline(x,k,i+1,m-1)
  }
  res
}
```

Figure 5.6 illustrates the representation of functions using B-spline bases of two different orders. In R the function `splineDesign` from the `splines` package can be used to generate B-spline bases, while `cSplineDes` from package `mgcv` will generate cyclic B-spline bases.

B-splines were developed as a very stable basis for large scale spline interpolation (see de Boor, 1978, for further details), but for most statistical work with low rank penalized regression splines you would have to be using very poor numerical methods before the enhanced stability of the basis became noticeable. The real statistical interest in B-splines has resulted from the work of Eilers and Marx (1996) in using them to develop what they term *P-splines*.

P-splines are low rank smoothers using a B-spline basis, usually defined on evenly spaced knots, with a *difference penalty* applied directly to the parameters, β_i, to control function wiggliness. How this works is best seen by example. If we decide to penalize the squared difference between adjacent β_i values then the penalty is

$$\mathcal{P} = \sum_{i=1}^{k-1} (\beta_{i+1} - \beta_i)^2 = \boldsymbol{\beta}^\mathsf{T}\mathbf{P}^\mathsf{T}\mathbf{P}\boldsymbol{\beta}$$

where

$$\mathbf{P} = \begin{bmatrix} -1 & 1 & 0 & . & . \\ 0 & -1 & 1 & 0 & . \\ . & . & . & . & . \\ . & . & . & . & . \end{bmatrix} \text{ so that } \begin{bmatrix} \beta_2 - \beta_1 \\ \beta_3 - \beta_2 \\ . \\ . \end{bmatrix} = \mathbf{P}\boldsymbol{\beta},$$

and hence

$$\mathcal{P} = \boldsymbol{\beta}\mathbf{P}^\mathsf{T}\mathbf{P}\boldsymbol{\beta} = \boldsymbol{\beta}^\mathsf{T} \begin{bmatrix} 1 & -1 & 0 & . & . \\ -1 & 2 & -1 & . & . \\ 0 & -1 & 2 & . & . \\ . & . & . & . & . \\ . & . & . & . & . \end{bmatrix} \boldsymbol{\beta}.$$

Such penalties are very easily generated in R. For example,

```
k <- 6                               # example basis dimension
P <- diff(diag(k),differences=1)     # sqrt of penalty matrix
S <- t(P)%*%P                        # penalty matrix
```

Higher order penalties are produced by increasing the `differences` parameter. The only lower order penalty is the identity matrix.

P-splines are extremely easy to set up and use and allow a good deal of flexibility, in that any order of penalty can be combined with any order of B-spline basis, as the user sees fit. Their disadvantage is that the simplicity is somewhat diminished if uneven knot spacing is required, and that, relative to the more usual spline penalties, the discrete penalties are less easy to interpret in terms of the properties of the fitted smooth. They can be especially useful for MCMC based Bayesian inference, where the sparsity of the basis and the penalty can offer substantial computational savings.

In `mgcv`, terms like `s(x,bs="ps",m=c(2,3))` specify P-splines (`bs="cp"` for a cyclic version), with the `m` parameter specifying the order of the spline basis and penalty. Basis order 2 gives cubic P-splines, while penalty order 3 would specify a penalty built on third differences. If only a single `m` is specified then it is used for both. `m` defaults to 2.

5.3.4 P-splines with derivative based penalties

Actually it is not very difficult to use derivative based penalties with any B-spline basis, and the P-spline advantages of sparse basis and penalty and the ability to 'mix-and-match' basis and penalty orders carry over to this case as well. The penalty matrix setup requires a few lines of code, rather than two lines needed for P-splines, but it is still relatively straightforward (Wood, 2016b).

In this section let m_1 denote the order of basis with 3 corresponding to cubic, and let m_2 denote the order of differentiation required in the penalty

$$J = \int_a^b f^{[m_2]}(x)^2 dx = \boldsymbol{\beta}^\mathsf{T}\mathbf{S}\boldsymbol{\beta}.$$

We seek $\mathbf{S}$ and possibly its 'square root'. Let $x_1, x_2 \ldots x_{k-m+1}$ be the (ordered) 'interior knots' defining the B-spline basis: that is, the knots within whose range the spline and its penalty are to be evaluated (so $a = x_1$ and $b = x_{k-m+1}$). Let the inter-knot distances be $h_j = x_{j+1} - x_j$, for $0 < j \le k - m$.

1. For each interval $[x_j, x_{j+1}]$, generate $p + 1$ evenly spaced points within the interval. For $p = 0$ the point should be at the interval centre; otherwise the points always include the end points x_j and x_{j+1}. Let $\mathbf{x}'$ contain the unique x values so generated, in ascending order.
2. Obtain the matrix $\mathbf{G}$ mapping the spline coefficients to the m_2^{th} derivative of the spline at the points $\mathbf{x}'$.
3. If $p = 0$, $\mathbf{W} = \text{diag}(\mathbf{h})$.
4. It $p > 0$, let $p+1 \times p+1$ matrices $\mathbf{P}$ and $\mathbf{H}$ have elements $P_{ij} = (-1+2(i-1)/p)^j$ and $H_{ij} = (1 + (-1)^{i+j-2})/(i + j - 1)$ (i and j start at 1). Then compute matrix $\tilde{\mathbf{W}} = \mathbf{P}^{-\mathsf{T}}\mathbf{H}\mathbf{P}^{-1}$. Now compute $\mathbf{W} = \sum_q \mathbf{W}^q$ where each $\mathbf{W}^q$ is zero everywhere except at $W^q_{i+pq-p,j+pq-p} = h_q \tilde{W}_{ij}/2$, for $i = 1, \ldots, p+1$, $j = 1, \ldots, p+1$. $\mathbf{W}$ is banded with $2p + 1$ non-zero diagonals.
5. The diagonally banded penalty coefficient matrix is $\mathbf{S} = \mathbf{G}^\mathsf{T}\mathbf{W}\mathbf{G}$.
6. Optionally, compute the diagonally banded Cholesky decomposition $\mathbf{R}^\mathsf{T}\mathbf{R} = \mathbf{W}$, and form diagonally banded matrix $\mathbf{D} = \mathbf{R}\mathbf{G}$, such that $\mathbf{S} = \mathbf{D}^\mathsf{T}\mathbf{D}$.

`splines::splineDesign` in R can be used for step 2, while the single rank $p+1$ matrix inversion of $\mathbf{P}$ at step 4 can be pre-formed using `solve`. $\mathbf{P}$ is somewhat ill-conditioned for $p \ge 20$. However it is difficult to imagine a sensible application for which $p > 10$, and for $p \le 10$, $\mathbf{P}$'s condition number is $< 2 \times 10^4$. Note that $\mathbf{W}$ is formed without explicitly forming the $\mathbf{W}^q$ matrices. Step 6 can be accomplished by a banded Cholesky decomposition (for example `bandchol` in `mgcv`) but for applications with k less than 1000 or so, a dense Cholesky decomposition might be deemed efficient enough. Step 6 is preferable to construction of $\mathbf{D}$ by decomposition of $\mathbf{S}$, since $\mathbf{W}$ is positive definite by construction, while, for $m_2 > 0$, $\mathbf{S}$ is only positive semi-definite. In `mgcv`, terms like `s(x,bs="bs",m=c(3,2))` create such smooths (the example creates a cubic spline with second derivative penalty).

5.3.5 *Adaptive smoothing*

A nice illustration of the flexibility of P-splines is provided by adaptive smoothing. Suppose that we would like the amount of smoothing to vary with the smoothing

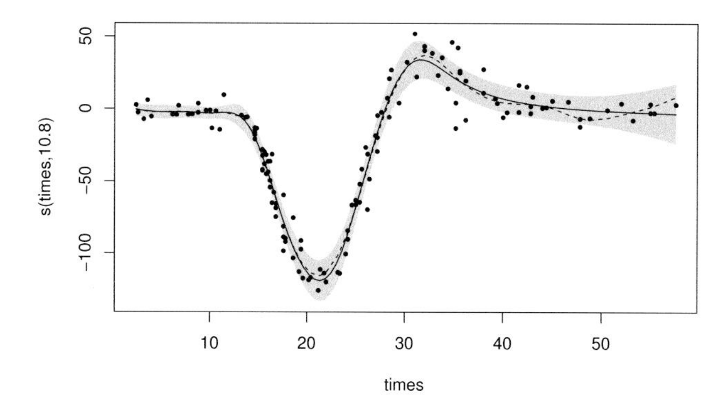

Figure 5.7 *An adaptive smooth of the acceleration of the head of a crash test dummy in a simulated motorcycle crash (*`mcycle` *data from* `MASS` *library). The continuous black curve is an adaptive smooth described in section 5.3.5, with 95% CI shown in grey. The dashed curve is a conventional cubic spline smoother. Smoothing parameter selected by GCV in both cases.*

covariate, x, say. For example consider a second order P-spline penalty in which the squared differences are now weighted

$$\mathcal{P}_a = \sum_{i=2}^{k-1} \omega_i(\beta_{i-1} - 2\beta_i + \beta_{i+1})^2 = \boldsymbol{\beta}^\mathsf{T}\mathbf{D}^\mathsf{T}\mathsf{diag}(\boldsymbol{\omega})\mathbf{D}\boldsymbol{\beta}$$

$$\text{where } \mathbf{D} = \begin{bmatrix} 1 & -2 & 1 & 0 & \cdot \\ 0 & 1 & -2 & 1 & \cdot \\ \cdot & \cdot & \cdot & \cdot & \cdot \end{bmatrix}.$$

Now let the weights ω_i vary smoothly with i and hence with x. An obvious way to do this is to use a B-spline basis expansion $\boldsymbol{\omega} = \mathbf{B}\boldsymbol{\lambda}$, where $\boldsymbol{\lambda}$ is the vector of basis coefficients. Then, writing $\mathbf{B}_{\cdot j}$ for the jth column of $\mathbf{B}$ we have

$$\boldsymbol{\beta}^\mathsf{T}\mathbf{D}^\mathsf{T}\mathsf{diag}(\boldsymbol{\omega})\mathbf{D}\boldsymbol{\beta} = \sum_j \lambda_j \boldsymbol{\beta}^\mathsf{T}\mathbf{D}^\mathsf{T}\mathsf{diag}(\mathbf{B}_{\cdot j})\mathbf{D}\boldsymbol{\beta} = \sum_j \lambda_j \boldsymbol{\beta}^\mathsf{T}\mathbf{S}_j\boldsymbol{\beta}.$$

Notice the B-spline basis property that the elements of $\mathbf{B}$ are non-negative, which ensures that the $\mathbf{S}_j$ are positive semi-definite. The point here is that the adaptive smoother can be represented using a multiply penalized B-spline basis. In `mgcv` a term such as `s(x,bs="ad",k=30,m=4)` would specify an adaptive smoother with a basis dimension of 30 and 4 smoothing parameters (the defaults are 40 and 5, respectively). `?adaptive.smooth` describes how cyclic and two dimensional versions are also possible. Figure 5.7 illustrates the application of such an adaptive smoother.

5.3.6 *SCOP-splines*

Another interesting application of the P-spline approach is shape constrained smoothing. Pya and Wood (2015) propose one method for constructing shape constrained splines in this way. Consider creating a monotonically increasing smoother,

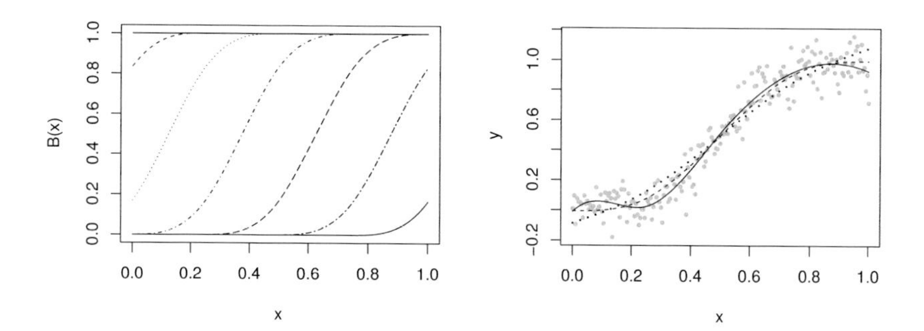

Figure 5.8 *Left. A 7-dimensional SCOP spline basis based on cubic B-splines; each curve shows one basis function. Right. Data sampled from a monotonic truth in grey, and some model fits shown as curves. The continuous curve is the un-penalized fit with a 7-dimensional B-spline basis. The dashed curve is SCOP-spline fit with negligible penalization. The dotted curve is a more heavily penalized SCOP-spline fit.*

represented using an order m B spline basis expansion (where $m = 3$ here denotes a cubic basis):

$$f(x) = \sum_{j=1}^{k} \beta_j B_{m,j}(x).$$

From (de Boor, 2001, p. 116), we have

$$f'(x) = \sum_j \beta_j B'_{m,j}(x) = (m-1)\sum_j \frac{\beta_j - \beta_{j-1}}{x_{j+m} - x_j} B_{m-1,j}(x),$$

where x_j are the (increasing) knot locations. Since the B-spline basis functions of any order are non-negative, a sufficient condition for $f'(x) > 0$ is that the β_j form an increasing sequence. Similarly a decreasing, concave or convex β_j sequence yields, respectively, a decreasing, concave or convex $f(x)$ (and combinations of monotonicity and convexity restrictions are possible).

Employing the convention $\sum_{i=2}^{1} x_i = 0$, suppose that we re-parameterize so that

$$\beta_j = \gamma_1 + \sum_{i=2}^{j} \exp(\gamma_i)$$

where the γ_j are unrestricted coefficients. Clearly the β_j sequence is increasing (in finite precision arithmetic, non-decreasing). Now suppose that we impose the quadratic penalty

$$J = \sum_{i=2}^{k-1} (\gamma_{j+1} - \gamma_j)^2$$

on γ. Clearly this tends to force the γ_i towards equality, which in turn means that we are smoothing towards a straight line. So we have a means for producing a monotonically increasing penalized spline. The price paid is that the spline is no longer linear in its coefficients, but the non-linearity is rather benign.

Further insight into the basis can be obtained as follows. Let $\tilde{\gamma}_1 = \gamma_1$ and $\tilde{\gamma}_i = \exp(\gamma_i)$ if $i > 2$, then $\boldsymbol{\beta} = \mathbf{B}\tilde{\boldsymbol{\gamma}}$ where $B_{ij} = 1$ if $i \ge j$ and 0 otherwise. So if $\mathbf{X}$ is the usual B-spline model matrix for $f(x)$ then we have $\mathbf{f} = \mathbf{XB}\tilde{\boldsymbol{\gamma}}$. Hence we could view the monotonic basis as being that evaluated in $\mathbf{XB}$, with basis functions

$$M_{m,j}(x) = \sum_{i=j}^{k} B_{m,i}(x).$$

Figure 5.8 illustrates such basis functions in its left panel, while the right panel shows some monotonic fits using the basis. If sum-to-zero constraints are required for a SCOP-spline then they should be implemented in the manner described in section 4.3.1 (p. 175). That is, the first (constant) column of $\mathbf{XB}$ should be dropped, and each remaining column should have its mean subtracted from each of its elements (i.e., the columns should be 'centred'). Pya and Wood (2015) provides the equivalent $\mathbf{B}$ matrices for monotonic decrease, and a variety of other constraints (as well as a complete framework for shape constrained additive modelling).

`mgcv`'s `"ps"` smooth constructor allows monotonic setup

```
ssp <- s(x,bs="ps",k=k); ssp$mono <- 1
sm <- smoothCon(ssp,data.frame(x))[[1]]
```

(`ssp$mono <- -1` would have set things up for monotonic decrease). The `sm` smooth object contains a model matrix `X` and penalty matrix `S` that we can then use in a Newton loop for fitting. The penalized least squares fits shown in the right hand panel of figure 5.8 were actually produced by the following very simple Newton loop.

```
X <- sm$X; XX <- crossprod(X); sp <- .5
gamma <- rep(0,k); S <- sm$S[[1]]
for (i in 1:20) {
  gt <- c(gamma[1],exp(gamma[2:k]))
  dg <- c(1,gt[2:k])
  g <- -dg*(t(X)%*%(y-X%*%gt)) + sp*S%*%gamma
  H <- dg*t(dg*XX)
  gamma <- gamma - solve(H+sp*S,g)
}
```

where `sp` is the smoothing parameter.[‡]

5.4 Some useful smoother theory

Before covering smoothing with respect to multiple covariates, it is worth considering two forms of re-parameterization of smooths. The first is the re-parameterization needed to absorb identifiability constraints into the smooth model matrix and penalty

[‡]Obviously in more serious applications a Newton loop should perform step length control, and should test for convergence, rather than just assuming that 20 iterations will do.

in practice, while the second is a reparameterization that helps understand the notion of the effective degrees of freedom of a smooth. It is also worth considering the construction of penalties for the otherwise un-penalized components of the smooth.

5.4.1 *Identifiability constraints*

Incorporating smooths into additive models involves the imposition of identifiability constraints, and it is convenient if these are dealt with by re-parameterization. The constraints are necessary because it is clearly not possible to estimate an intercept term for each smooth. The constraints that lead to sharpest inference about the constrained components are those that orthogonalise the smooth to the intercept term, i.e., $\sum_j f(x_j) = 0$ which states that the smooth sums to zero over the observed values of x_j.[§] If $\mathbf{X}$ is the model matrix for the smooth then the constraint becomes $\mathbf{1}^\mathsf{T}\mathbf{X}\boldsymbol{\beta} = 0$. Section 4.3.1 (p. 175) presents one method for incorporating such constraints, but an elegant alternative uses the general QR based approach of section 1.8.1 (p. 47).

The idea is to create a $k \times (k-1)$ matrix $\mathbf{Z}$ such that if $\boldsymbol{\beta} = \mathbf{Z}\boldsymbol{\beta}_z$, then $\mathbf{1}^\mathsf{T}\mathbf{X}\boldsymbol{\beta} = 0$ for any $\boldsymbol{\beta}_z$. Then we re-parameterize the smooth by setting its model matrix to $\mathbf{XZ}$ and its penalty matrix to $\mathbf{Z}^\mathsf{T}\mathbf{SZ}$. In section 1.8.1 $\mathbf{Z}$ is constructed using a QR decomposition of $\bar{\mathbf{x}} = \mathbf{X}^\mathsf{T}\mathbf{1}$, but actually the orthogonal factor in the QR decomposition consists of a single Householder reflection (see section B.5), making the formation of $\mathbf{XZ}$ and $\mathbf{Z}^\mathsf{T}\mathbf{SZ}$ particularly straightforward and efficient. In particular

$$\bar{\mathbf{x}} = \mathbf{H}\bar{\mathbf{x}}' = [\mathbf{D} : \mathbf{Z}]\bar{\mathbf{x}}' \text{ where } \bar{\mathbf{x}}' = \begin{bmatrix} \|\mathbf{x}\| \\ \mathbf{0} \end{bmatrix}, \ \mathbf{H} = (\mathbf{I} - 2\mathbf{u}\mathbf{u}^\mathsf{T}/\mathbf{u}^\mathsf{T}\mathbf{u}) \text{ and } \mathbf{u} = \bar{\mathbf{x}} - \bar{\mathbf{x}}'.$$

Multiplication by $\mathbf{H}$ is a factor of p more efficient in terms of floating point operations and storage costs if it is done as shown in section B.5, rather than by forming $\mathbf{H}$ explicitly. This means that rather than form $\mathbf{Z}$ explicitly, we should form $\mathbf{XZ}$ by computing $\mathbf{XH}$ and dropping its first column. The same approach can be used for forming $\mathbf{Z}^\mathsf{T}\mathbf{SZ}$ having first computed $\mathbf{Z}^\mathsf{T}\mathbf{S}$ by computing $\mathbf{HS}$ and then dropping its first row. Note also that if $\boldsymbol{\beta}_z' = (0, \boldsymbol{\beta}_z^\mathsf{T})^\mathsf{T}$ then $\mathbf{Z}\boldsymbol{\beta}_z = \mathbf{H}\boldsymbol{\beta}_z'$.

5.4.2 *'Natural' parameterization, effective degrees of freedom and smoothing bias*

Penalized smoothers, with a single penalty, can always be parameterized in such a way that the parameter estimators are independent with unit variance in the absence of the penalty, and the penalty matrix is diagonal. This 'natural' or Demmler and Reinsch (1975) parameterization is particularly helpful for understanding the way in which the penalty suppresses model degrees of freedom.

Consider a smooth with model matrix $\mathbf{X}$, parameter vector $\boldsymbol{\beta}$, wiggliness penalty coefficient matrix $\mathbf{S}$, and smoothing parameter λ. Suppose that the model is to be

[§] If f is estimated to be a straight line then the constraint determines exactly where it must pass through zero — in consequence confidence intervals for f will have zero width at this point

estimated by penalized least squares from data with variance σ^2. Forming the QR decomposition

$$\mathbf{X} = \mathbf{QR},$$

we can re-parameterize in terms of $\boldsymbol{\beta}'' = \mathbf{R}\boldsymbol{\beta}$, so that the model matrix is now $\mathbf{Q}$, and the penalty matrix becomes $\mathbf{R}^{-\mathsf{T}}\mathbf{S}\mathbf{R}^{-1}$. Eigen-decomposing the penalty matrix yields

$$\mathbf{R}^{-\mathsf{T}}\mathbf{S}\mathbf{R}^{-1} = \mathbf{U}\mathbf{D}\mathbf{U}^{\mathsf{T}},$$

where $\mathbf{U}$ is an orthogonal matrix, the columns of which are the eigenvectors of $\mathbf{R}^{-\mathsf{T}}\mathbf{S}\mathbf{R}^{-1}$, while $\mathbf{D}$ is a diagonal matrix of the corresponding eigenvalues, arranged in decreasing order. Reparameterization, via a rotation/reflection of the parameter space, now yields parameters $\boldsymbol{\beta}' = \mathbf{U}^{\mathsf{T}}\boldsymbol{\beta}''$, and correspondingly a model matrix $\mathbf{QU}$ and penalty matrix $\mathbf{D}$. If the penalty is not applied then the covariance matrix for these parameters is $\mathbf{I}\sigma^2$, since $\mathbf{U}$ is orthogonal, and the columns of $\mathbf{Q}$ are columns of an orthogonal matrix.

If the penalty is applied, then the Bayesian covariance matrix of the parameters (see sections 4.2.4 and 5.8) is simply the diagonal matrix $(\mathbf{I}+\lambda\mathbf{D})^{-1}\sigma^2$, from which the role of the penalty in limiting parameter variance is rather clear. The frequentist equivalent is equally simple: $(\mathbf{I}+\lambda\mathbf{D})^{-2}\sigma^2$.

Now in the natural parameterization, $\hat{\boldsymbol{\beta}}' = (\mathbf{I}+\lambda\mathbf{D})^{-1}\mathbf{U}^{\mathsf{T}}\mathbf{Q}^{\mathsf{T}}\mathbf{y}$, and hence the corresponding unpenalized estimate is $\tilde{\boldsymbol{\beta}}' = \mathbf{U}^{\mathsf{T}}\mathbf{Q}^{\mathsf{T}}\mathbf{y}$. That is the penalized parameter estimates are simply shrunken versions of the unpenalized coefficient estimates:

$$\hat{\beta}'_i = (1+\lambda D_{ii})^{-1}\tilde{\beta}'_i.$$

The shrinkage factors $(1+\lambda D_{ii})^{-1}$ lie in the interval $(0,1]$. Since the unpenalized coefficients have one degree of freedom each, the shrinkage factor, $(1+\lambda D_{ii})^{-1}$, can be viewed as the 'effective degrees of freedom' of $\hat{\beta}'_i$. This means that the total effective degrees of freedom of the smooth is

$$\sum_i (1+\lambda D_{ii})^{-1} = \text{tr}(\mathbf{F}) \text{ where } \mathbf{F} = (\mathbf{X}^{\mathsf{T}}\mathbf{X}+\lambda\mathbf{S})^{-1}\mathbf{X}^{\mathsf{T}}\mathbf{X}.$$

Notice how this is bounded between the number of zero eigenvalues of the penalty (when $\lambda \to \infty$), and the total number of coefficients (when $\lambda = 0$). Also notice how this notion of effective degrees of freedom corresponds to that given for mixed effects models in section 2.4.6 (p. 83).

The fact that the unpenalized estimators are unbiased also means that $\mathbb{E}(\hat{\beta}'_i) = (1+\lambda D_{ii})^{-1}\beta_i$: the shrinkage factors determine the relative *smoothing bias*. In the original parameterization this becomes $\mathbb{E}(\hat{\boldsymbol{\beta}}) = \mathbf{F}\boldsymbol{\beta}$.

So, in the natural parameterization, all parameters have the same degree of associated variability when un-penalized, and the penalty acts on each parameter independently (i.e., the degree of penalization of one parameter has no effect on the other parameters). In this parameterization the relative magnitudes of different D_{ii} terms directly indicate the relative degree of penalization of different components of the model space. $\mathbf{D}$ is uniquely defined — no matter what parameterization you start out with, the elements of $\mathbf{D}$ are always the same.

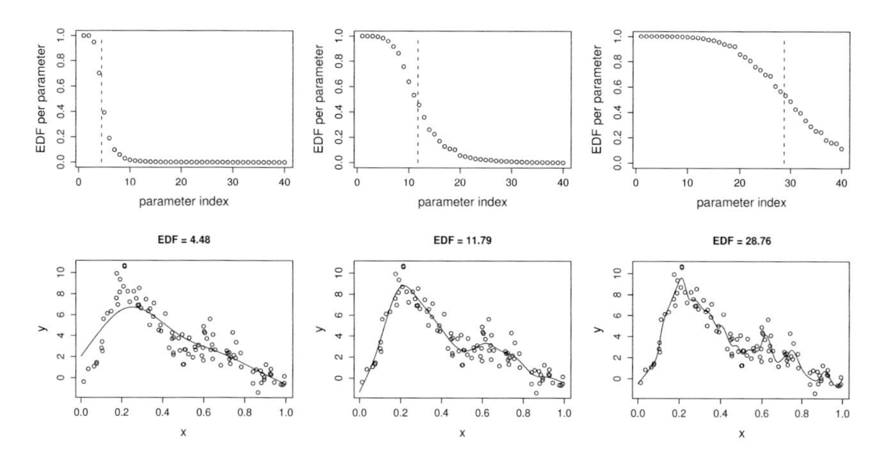

Figure 5.9 *How parameters are shrunk to zero by the penalty in the natural parameterization. All plots relate to a rank 40 cubic regression spline fit to the data shown in the bottom row plots. The top row plots the degrees of freedom associated with each parameter against parameter index using the natural parameterization, for 3 levels of penalization. On each plot the effective degrees of freedom (i.e., the effective number of free parameters) is shown by a vertical dashed line: note that this is simply the sum of the plotted degrees of freedom per parameter. The lower row shows the smooth fit corresponding to each of the top row plots. For this smooth the first two 'natural' parameters are never penalized. Notice how some potentially penalized parameters are effectively un-penalized, some are effectively suppressed completely, while some suffer intermediate penalization. Clearly, it is only approximately true that a penalized fit is equivalent to an un-penalized fit with the number of parameters given by the penalized fit effective degrees of freedom.*

Figure 5.9 illustrates how parameters get shrunk by the penalty, using the natural parameterization. The figures illustrate that at most levels of penalization some subspace of the model space is almost completely suppressed, while some other subspace is left almost unpenalized. This lends some support to the idea that a penalized fit is somehow equivalent to an unpenalized fit with degrees of freedom close to the effective degrees of freedom of the penalized model. However, the fact that many parameters have intermediate penalization means that this support is only limited.

The natural parameterization also makes the penalized smoothers behave more like full spline models than is otherwise immediately apparent. For example, for a full spline smoother the EDF matrix is simply the influence matrix, $\mathbf{A}$, which also defines the Bayesian posterior covariance matrix $\mathbf{A}\sigma^2$ and the equivalent frequentist matrix $\mathbf{A}^2\sigma^2$. In other words, the relationship between these matrices that holds for smoothing splines also holds for general penalized regression smoothers with the natural parameterization.

The penalty's action in effectively suppressing some dimensions of the model space is also readily apparent in the natural parameterization. For most smoothers the penalty assesses the 'wiggliness' of the fitted model, in such a way that smoother functions are generally not simply closer to the zero function. In this circumstance, if increased penalization is to lead to continuously smoother models then, in the natural parameterization, it is clear that the elements of $\mathbf{D}$ must have a spread of (non-negative) values, with the higher values penalizing coefficients corresponding to more wiggly components of the model function. If this is not clear, consider the converse situation in which all elements on the leading diagonal of $\mathbf{D}$ are the same. In that case increased penalization would amount to simply multiplying each model coefficient by the same constant, thereby shrinking the fitted function towards zero without changing its shape.

5.4.3 *Null space penalties*

One feature of most smoothing penalties is that they treat some *null* space of functions as being 'completely smooth' and hence having zero penalty. For example the usual cubic spline penalty, $\int f''(x)^2dx$, is zero for any straight line, $f(x) = \alpha + \beta x$. This means that as $\lambda \to \infty$ the estimated smoother tends to a function in the penalty null space, rather than to the zero function. Hence, although smoothing parameter selection methods effectively perform a great deal of model selection, in selecting f from a wide class of possibilities of differing complexity, λ selection is not usually able to remove a smooth term from a model altogether.

One possibility is to add an extra penalty to a smooth term, which penalizes only functions in the smoothing penalty null space. Such penalties are easily constructed. Consider the eigen-decompositon of the penalty matrix, $\mathbf{S} = \mathbf{U}\mathbf{\Lambda}\mathbf{U}^{\mathsf{T}}$, and let $\tilde{\mathbf{U}}$ denote the columns of $\mathbf{U}$ (eigenvectors of $\mathbf{S}$), corresponding to zero eigenvalues on the leading diagonal of $\mathbf{\Lambda}$. Then $\lambda'\boldsymbol{\beta}^{\mathsf{T}}\tilde{\mathbf{U}}\tilde{\mathbf{U}}^{\mathsf{T}}\boldsymbol{\beta}$ can be used as an extra penalty on the null space of the original penalty. This construction ensures that if all smoothing parameters for a term $\to \infty$ then the term will tend to the zero function. Note that the new penalty has no effect on the components of the smooth not in the original penalty null space. For terms with multiple penalties the same construction applies, but an eigen-decomposition of $\sum_j \mathbf{S}_j$ is used. Marra and Wood (2011) provide more details. The `select` argument of `gam` can be used to impose such penalties in `mgcv`.

A variant on this idea is to replace the zero eigenvalues in the $\mathbf{\Lambda}$ with constants somewhat smaller than the smallest positive entry in $\mathbf{\Lambda}$. Call the result $\tilde{\mathbf{\Lambda}}$. Then the original penalty is replaced by $\lambda\boldsymbol{\beta}^{\mathsf{T}}\mathbf{U}\tilde{\mathbf{\Lambda}}\mathbf{U}^{\mathsf{T}}\boldsymbol{\beta}$, and $\lambda \to \infty$ again causes the smooth term to tend to the zero function. This idea does not work so well with multiple penalties, and results are somewhat dependent on how we choose the 'somewhat smaller' constant. In `mgcv` the 'shrinkage' bases, `"cs"` and `"ts"`, implement this idea.

5.5 Isotropic smoothing

This section considers smooths of one or more variables, in particular smooths that, in the multiple covariate case, produce identical predictions of the response variable under any rotation or reflection of the covariates.¶

¶In the literature 'isotropic' is sometimes used for smooths that are not rotationally invariant, but simply smooth 'equally' with respect to each covariate. This is a slight abuse of language.

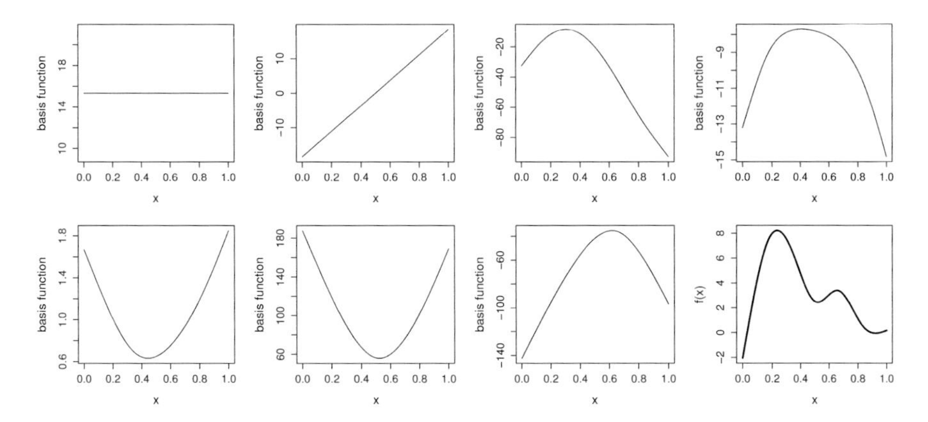

Figure 5.10 *Illustration of a thin plate spline basis for representing a smooth function of one variable fitted to 7 data with penalty order* $m = 2$. *The first 7 panels (starting at top left) show the basis functions, multiplied by coefficients, that are summed to give the smooth curve in the lower right panel. The first two basis functions span the space of functions that are completely smooth, according to the wiggliness measure. The remaining basis functions represent the wiggly component of the smooth curve: these latter functions are shown after absorption of the thin plate spline constraints* $\mathbf{T}^\mathsf{T}\boldsymbol{\delta} = \mathbf{0}$ *into the basis.*

5.5.1 *Thin plate regression splines*

The bases covered so far are each useful in practice, but are open to some criticisms.

1. It is necessary to choose knot locations, in order to use each basis: this introduces an extra degree of subjectivity into the model fitting process.
2. The bases are only useful for representing smooths of one predictor variable.
3. It is not clear to what extent the reduced rank bases are better or worse than any other reduced rank basis that might be used.

In this section, an approach is developed which goes some way to addressing these issues, by producing knot free bases for smooths of any number of predictors, that are in a certain limited sense 'optimal': the thin plate regression splines.

Thin plate splines

Thin plate splines (Duchon, 1977) are a very elegant and general solution to the problem of estimating a smooth function of multiple predictor variables, from noisy observations of the function at particular values of those predictors. Consider the problem of estimating the smooth function $g(\mathbf{x})$, from n observations $(y_i, \mathbf{x}_i)$ such that

$$y_i = g(\mathbf{x}_i) + \epsilon_i$$

where ϵ_i is a random error term and where $\mathbf{x}$ is a d-vector. Thin-plate spline smoothing estimates g by finding the function $\hat{f}$ minimizing

$$\|\mathbf{y} - \mathbf{f}\|^2 + \lambda J_{md}(f), \tag{5.5}$$

where $\mathbf{y}$ is the vector of y_i data and $\mathbf{f} = [f(\mathbf{x}_1), f(\mathbf{x}_2), \ldots, f(\mathbf{x}_n)]^\mathsf{T}$. $J_{md}(f)$ is a penalty functional measuring the 'wiggliness' of f, and λ is a smoothing parameter, controlling the tradeoff between data fitting and smoothness of f. The wiggliness penalty is defined as

$$J_{md} = \int_{\Re^d} \sum_{\nu_1+\ldots+\nu_d=m} \frac{m!}{\nu_1!\ldots\nu_d!} \left(\frac{\partial^m f}{\partial x_1^{\nu_1}\ldots\partial x_d^{\nu_d}} \right)^2 dx_1 \ldots dx_d.^{\|} \tag{5.6}$$

Further progress is only possible if m is chosen so that $2m > d$, and in fact for 'visually smooth' results it is preferable that $2m > d+1$. Subject to the first of these restrictions, it can be shown that the function minimizing (5.5) has the form

$$\hat{f}(\mathbf{x}) = \sum_{i=1}^{n} \delta_i \eta_{md}(\|\mathbf{x} - \mathbf{x}_i\|) + \sum_{j=1}^{M} \alpha_j \phi_j(\mathbf{x}), \tag{5.7}$$

where $\boldsymbol{\delta}$ and $\boldsymbol{\alpha}$ are vectors of coefficients to be estimated, $\boldsymbol{\delta}$ being subject to the linear constraints that $\mathbf{T}^\mathsf{T}\boldsymbol{\delta} = \mathbf{0}$ where $T_{ij} = \phi_j(\mathbf{x}_i)$. The $M = \binom{m+d-1}{d}$ functions ϕ_i are linearly independent polynomials spanning the space of polynomials in $\Re^d$ of degree less than m. The ϕ_i span the space of functions for which J_{md} is zero, i.e., the 'null space' of J_{md}: those functions that are considered 'completely smooth'. For example, for $m = d = 2$ these functions are $\phi_1(\mathbf{x}) = 1$, $\phi_2(\mathbf{x}) = x_1$ and $\phi_3(\mathbf{x}) = x_2$. The remaining basis functions used in (5.7) are defined as

$$\eta_{md}(r) = \begin{cases} \frac{(-1)^{m+1+d/2}}{2^{2m-1}\pi^{d/2}(m-1)!(m-d/2)!} r^{2m-d} \log(r) & d \text{ even} \\ \frac{\Gamma(d/2-m)}{2^{2m}\pi^{d/2}(m-1)!} r^{2m-d} & d \text{ odd.} \end{cases}$$

Now defining matrix $\mathbf{E}$ by $E_{ij} \equiv \eta_{md}(\|\mathbf{x}_i - \mathbf{x}_j\|)$, the thin plate spline fitting problem becomes,

$$\text{minimize } \|\mathbf{y} - \mathbf{E}\boldsymbol{\delta} - \mathbf{T}\boldsymbol{\alpha}\|^2 + \lambda \boldsymbol{\delta}^\mathsf{T}\mathbf{E}\boldsymbol{\delta} \text{ subject to } \mathbf{T}^\mathsf{T}\boldsymbol{\delta} = \mathbf{0}, \tag{5.8}$$

with respect to $\boldsymbol{\delta}$ and $\boldsymbol{\alpha}$. Wahba (1990) or Green and Silverman (1994) provide further information about thin plate splines, and figure 5.10 illustrates a thin plate spline basis in one dimension.

The thin plate spline, $\hat{f}$, is something of an ideal smoother: it has been constructed by defining exactly what is meant by smoothness, exactly how much weight to give to the conflicting goals of matching the data and making $\hat{f}$ smooth, and finding the *function* that best satisfies the resulting smoothing objective. Notice that in

---

‖The general form of the penalty is somewhat intimidating, so an example is useful. In the case of a smooth of two predictors with wiggliness measured using second derivatives, we have

$$J_{22} = \int\int \left(\frac{\partial^2 f}{\partial x_1^2}\right)^2 + 2\left(\frac{\partial^2 f}{\partial x_1 \partial x_2}\right)^2 + \left(\frac{\partial^2 f}{\partial x_2^2}\right)^2 dx_1 dx_2.$$

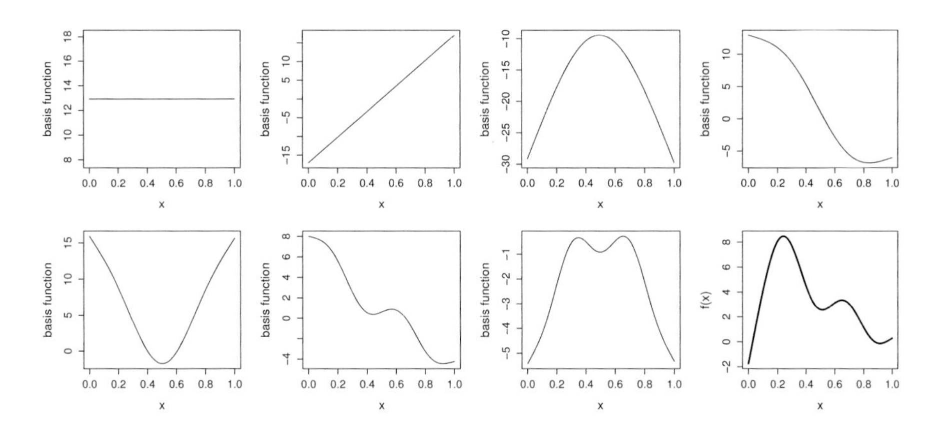

Figure 5.11 *Illustration of a rank 7 thin plate regression spline basis for representing a smooth function of one variable, with penalty order $m = 2$. The first 7 panels (starting at top left) show the basis functions, multiplied by coefficients, that are summed to give the smooth curve in the lower right panel. The first two basis functions span the space of functions that are completely smooth, according to the wiggliness measure. The remaining basis functions represent the wiggly component of the smooth curve: notice how these functions become successively more wiggly while generally tending to contribute less and less to the overall fit.*

doing this we did not have to choose knot positions or select basis functions: both of these emerged naturally from the mathematical statement of the smoothing problem. In addition, thin plate splines can deal with any number of predictor variables, and allow the user some flexibility to select the order of derivative used in the measure of function wiggliness. So, at first sight it might seem that the problems listed at the start of this section are all solved, and thin plate spline bases and penalties should be used to represent all the smooth terms in the model.

The difficulty with thin plate splines is computational cost: these smoothers have as many unknown parameters as there are data (strictly, number of unique predictor combinations), and, except in the single predictor case, the computational cost of model estimation is proportional to the cube of the number of parameters. This is a very high price to pay for using such smooths. Given the discussion in section 5.2, and the fact that the effective degrees of freedom estimated for a model term is usually a small proportion of n, it seems wasteful to use so many parameters to represent the term. This begs the question of whether a low rank approximation could be produced which is as close as possible to the thin plate spline smooth, without incurring prohibitive computational cost.

Thin plate regression splines

Thin plate regression splines are based on the idea of truncating the space of the wiggly components of the thin plate spline (the components with parameters $\boldsymbol{\delta}$), while leaving the components of 'zero wiggliness' unchanged (the $\boldsymbol{\alpha}$ components). Let $\mathbf{E} = \mathbf{U}\mathbf{D}\mathbf{U}^\mathsf{T}$ be the eigen-decomposition of $\mathbf{E}$, where $\mathbf{D}$ is a diagonal matrix of eigenvalues of $\mathbf{E}$ arranged so that $|D_{i,i}| \geq |D_{i-1,i-1}|$ and the columns of $\mathbf{U}$ are the corresponding eigenvectors. Now let $\mathbf{U}_k$ denote the matrix consisting of the first k columns of $\mathbf{U}$ and let $\mathbf{D}_k$ denote the top left $k \times k$ submatrix of $\mathbf{D}$. Restricting $\boldsymbol{\delta}$ to the column space of $\mathbf{U}_k$, by writing $\boldsymbol{\delta} = \mathbf{U}_k \boldsymbol{\delta}_k$, means that (5.8) becomes

$$\text{minimise } \|\mathbf{y} - \mathbf{U}_k \mathbf{D}_k \boldsymbol{\delta}_k - \mathbf{T}\boldsymbol{\alpha}\|^2 + \lambda \boldsymbol{\delta}_k^{\mathsf{T}} \mathbf{D}_k \boldsymbol{\delta}_k \text{ subject to } \mathbf{T}^{\mathsf{T}} \mathbf{U}_k \boldsymbol{\delta}_k = \mathbf{0}$$

w.r.t. $\boldsymbol{\delta}_k$ and $\boldsymbol{\alpha}$. The constraints can be absorbed in the usual manner, described in section 1.8.1 (p. 47). We first find any orthogonal column basis, $\mathbf{Z}_k$, such that $\mathbf{T}^{\mathsf{T}} \mathbf{U}_k \mathbf{Z}_k = \mathbf{0}$. One way to do this is to form the QR decomposition of $\mathbf{U}_k^{\mathsf{T}} \mathbf{T}$: the final $n - M$ columns of the orthogonal factor give a $\mathbf{Z}_k$ (see sections 1.8.1 and B.6). Restricting $\boldsymbol{\delta}_k$ to this space, by writing $\boldsymbol{\delta}_k = \mathbf{Z}_k \tilde{\boldsymbol{\delta}}$, yields the unconstrained problem that must be solved to fit the rank k approximation to the smoothing spline:

$$\text{minimise } \|\mathbf{y} - \mathbf{U}_k \mathbf{D}_k \mathbf{Z}_k \tilde{\boldsymbol{\delta}} - \mathbf{T}\boldsymbol{\alpha}\|^2 + \lambda \tilde{\boldsymbol{\delta}}^{\mathsf{T}} \mathbf{Z}_k^{\mathsf{T}} \mathbf{D}_k \mathbf{Z}_k \tilde{\boldsymbol{\delta}}$$

with respect to $\tilde{\boldsymbol{\delta}}$ and $\boldsymbol{\alpha}$. This has a computational cost of $O(k^3)$. Having fitted the model, evaluation of the spline at any point is easy: simply evaluate $\boldsymbol{\delta} = \mathbf{U}_k \mathbf{Z}_k \tilde{\boldsymbol{\delta}}$ and use (5.7).

Now, the main problem is how to find $\mathbf{U}_k$ and $\mathbf{D}_k$ sufficiently cheaply. A full eigen-decomposition of $\mathbf{E}$ requires $O(n^3)$ operations, which would somewhat limit the utility of the TPRS approach. Fortunately the method of Lanczos iteration can be employed to find $\mathbf{U}_k$ and $\mathbf{D}_k$ at the substantially lower cost of $O(n^2 k)$ operations. See Appendix B, section B.11, for a suitable Lanczos algorithm. If n is large then even $O(n^2 k)$ becomes prohibitive. In that case we can randomly select n_r covariate values from the full set of data and then form the thin plate regression spline basis from this random selection, at $O(n_r^2 k)$ computational cost. The idea is to choose $n \gg n_r \gg k$, to gain efficiency without losing much in the way of practical performance.

Properties of thin plate regression splines

It is clear that thin plate regression splines avoid the problem of knot placement, are relatively cheap to compute, and can be constructed for smooths of any number of predictor variables, but what of their optimality properties? The thin plate splines are optimal in the sense that no smooth function will better minimize (5.5), but to what extent is that optimality inherited by the TPRS approximation? To answer this it helps to think about what would make a good approximation. An ideal approximation would probably result in the minimum possible perturbation of the fitted values of the spline, at the same time as making the minimum possible change to the 'shape' of the fitted spline. It is difficult to see how both these aims could be achieved, for all possible response data, without first fitting the full thin plate spline. But if the criteria are loosened somewhat to minimizing the worst possible changes in shape and fitted value then progress can be made, as follows.

The basis change and truncation can be thought of as replacing $\mathbf{E}$, in the norm in (5.8), by the matrix $\hat{\mathbf{E}} = \mathbf{E}\mathbf{U}_k \mathbf{U}_k^{\mathsf{T}}$, while replacing $\mathbf{E}$, in the penalty term of (5.8), by $\tilde{\mathbf{E}} = \mathbf{U}_k^{\mathsf{T}} \mathbf{U}_k \mathbf{E} \mathbf{U}_k \mathbf{U}_k^{\mathsf{T}}$. Now since the fitted values of the spline are given by $\mathbf{E}\hat{\boldsymbol{\delta}} + \mathbf{T}\boldsymbol{\alpha}$, and $\mathbf{T}$ remains unchanged, the worst possible change in fitted values could be measured by:

$$\hat{e}_k = \max_{\boldsymbol{\delta} \neq \mathbf{0}} \frac{\|(\mathbf{E} - \hat{\mathbf{E}}_k)\boldsymbol{\delta}\|}{\|\boldsymbol{\delta}\|}.$$

(Dividing by $\|\boldsymbol{\delta}\|$ is necessary, since the upper norm otherwise has a maximum at infinity.) The 'shape' of the spline is measured by the penalty term in (5.8), so a suitable measure of the worst possible change in the shape of the spline, caused by the truncation, might be:

$$\tilde{e}_k = \max_{\boldsymbol{\delta} \neq \mathbf{0}} \frac{\boldsymbol{\delta}^\mathsf{T}(\mathbf{E} - \tilde{\mathbf{E}}_k)\boldsymbol{\delta}}{\|\boldsymbol{\delta}\|^2}.$$

It turns out to be quite easy to show that $\hat{e}_k$ and $\tilde{e}_k$ are simultaneously minimized by the choice of $\mathbf{U}_k$ as the truncated basis for $\boldsymbol{\delta}$. That is, there is no matrix of the same dimension as $\mathbf{U}_k$ which would lead to lower $\hat{e}_k$ or $\tilde{e}_k$, if used in place of $\mathbf{U}_k$ (see Wood, 2003).

Note that $\hat{e}_k$ and $\tilde{e}_k$ are really formulated in too large a space. Ideally we would impose the constraints $\mathbf{T}^\mathsf{T}\boldsymbol{\delta} = \mathbf{0}$ on both, but in that case different bases minimize the two criteria. This in turn leads to the question of whether it would not be better to concentrate on just one of the criteria, but this is unsatisfactory, as it leads to results that depend on how the original thin plate spline problem is parameterized. Furthermore, these results can be extremely poor for some parameterizations. For example, if the thin plate spline is parameterized in terms of the fitted values, then the $\hat{e}_k$ optimal approximation is not smooth. Similarly, very poor fitted values result from an $\tilde{e}_k$ optimal approximation to a thin plate spline, if that thin plate spline is parameterized so that the penalty matrix is an identity matrix, with some leading diagonal entries zeroed.

To sum up: thin plate regression splines are probably the best that can be hoped for in terms of approximating the behaviour of a thin plate spline using a basis of any given low rank. They have the nice property of avoiding having to choose 'knot locations', and are reasonably computationally efficient if Lanczos iteration is used to find the truncated eigen-decomposition of $\mathbf{E}$. They also retain the rotational invariance (isotropy) of full thin plate spline. Figures 5.11 and 5.12 provide examples of the basis functions that result from adopting a TPRS approach.

Knot-based approximation

If one is prepared to forgo optimality and choose knot locations, then a simpler approximation is available, which avoids the truncated eigen-decomposition. If knot locations $\{\mathbf{x}_i^* : i = 1 \ldots k\}$ are chosen, then the thin plate spline can be approximated by

$$\hat{f}(\mathbf{x}) = \sum_{i=1}^{k} \delta_i \eta_{md}(\|\mathbf{x} - \mathbf{x}_i^*\|) + \sum_{j=1}^{M} \alpha_j \phi_j(\mathbf{x}) \tag{5.9}$$

where $\boldsymbol{\delta}$ and $\boldsymbol{\alpha}$ are estimated by minimizing

$$\|\mathbf{y} - \mathbf{X}\boldsymbol{\beta}\|^2 + \lambda\boldsymbol{\beta}^\mathsf{T}\mathbf{S}\boldsymbol{\beta} \text{ subject to } \mathbf{C}\boldsymbol{\beta} = \mathbf{0}$$

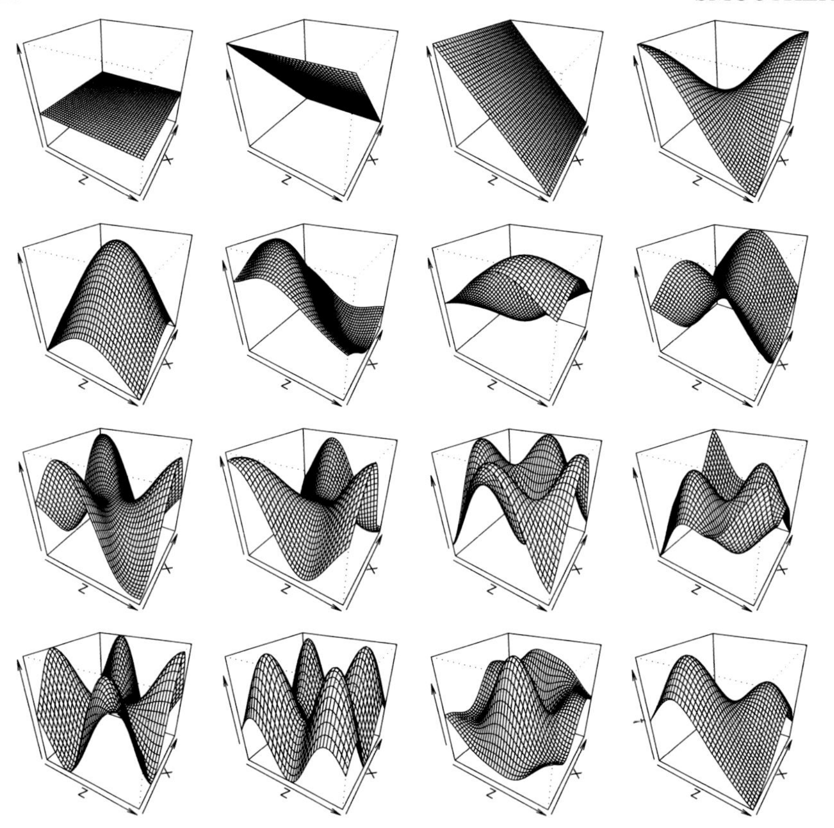

Figure 5.12 *Illustration of a rank 15 thin plate regression spline basis for representing a smooth function of two variables, with penalty order $m = 2$. The first 15 panels (starting at top left) show the basis functions, multiplied by coefficients, that are summed to give the smooth surface in the lower right panel. The first three basis functions span the space of functions that are completely smooth, according to the wiggliness measure, J_{22}. The remaining basis functions represent the wiggly component of the smooth curve: notice how these functions become successively more wiggly.*

w.r.t. $\boldsymbol{\beta}^\mathsf{T} = (\boldsymbol{\delta}^\mathsf{T}, \boldsymbol{\alpha}^\mathsf{T})$. $\mathbf{X}$ is an $n \times (k + M)$ matrix such that

$$X_{ij} = \begin{cases} \eta_{md}(\|\mathbf{x}_i - \mathbf{x}_j^*\|) & j = 1, \ldots, k \\ \phi_{j-k}(x_i) & j = k+1, \ldots, k+M. \end{cases}$$

$\mathbf{S}$ is a $(k + M) \times (k + M)$ matrix with zeroes everywhere except in its upper left $k \times k$ block where $S_{ij} = \eta_{md}(\|\mathbf{x}_i^* - \mathbf{x}_j^*\|)$. Finally, $\mathbf{C}$ is an $M \times (k + M)$ matrix such that

$$C_{ij} = \begin{cases} \phi_i(\mathbf{x}_j^*) & j = 1, \ldots, k \\ 0 & j = k+1, \ldots, k+M. \end{cases}$$

This approximation goes back at least to Wahba (1980). Some care is required to choose the knot locations carefully. In one dimension it is usual to choose quantiles of the empirical distribution of the predictor, or even spacing, but in more dimensions matters are often more difficult. One possibility is to take a random sample of the observed predictor variable combinations; another is to take a 'spatially stratified' sample of the predictor variable combinations. Even spacing is sometimes appropriate, or more sophisticated space filling schemes can be used: Ruppert et al. (2003) provide a useful discussion of the alternatives.

5.5.2 *Duchon splines*

Thin plate splines have been widely adopted in the statistics literature, but are actually only one of a broader class of splines proposed in Duchon (1977). The larger class is particularly useful if we have reason to base the spline penalty on an order of derivative, m, that does not meet the condition $m > d/2(+1/2)$ for the spline to exist (and be smooth). To see why this might be important, the following table shows how the penalty null space dimension, M, has to grow with the number of covariates of the smooth d, if the spline is to have no first derivative discontinuities.

d	1	2	3	4	5	6	7	8	9	10
m	2	2	3	3	4	4	5	5	6	6
M	2	3	10	15	56	84	330	495	2002	3003

Clearly in the rare cases that it is appropriate to smooth isotropically in high dimensions, the null space dimension makes the thin plate spline rather impractical.

Let $\mathcal{F}$ denote Fourier transform, and $\boldsymbol{\tau}$ frequency. By Plancherel's theorem the thin plate spline penalty (5.6) can be expressed

$$J_{md} = \int_{\Re^d} \sum_{\nu_1+\ldots+\nu_d=m} \frac{m!}{\nu_1!\ldots\nu_d!}\left(\mathcal{F}\frac{\partial^m f}{\partial x_1^{\nu_1}\ldots\partial x_d^{\nu_d}}(\boldsymbol{\tau})\right)^2 d\boldsymbol{\tau}$$

and Duchon (1977) actually considers the more general penalty

$$J_{mds} = \int_{\Re^d} \|\boldsymbol{\tau}\|^{2s} \sum_{\nu_1+\ldots+\nu_d=m} \frac{m!}{\nu_1!\ldots\nu_d!}\left(\mathcal{F}\frac{\partial^m f}{\partial x_1^{\nu_1}\ldots\partial x_d^{\nu_d}}(\boldsymbol{\tau})\right)^2 d\boldsymbol{\tau} \tag{5.10}$$

which is only the familiar thin plate spline penalty when $s = 0$. Under the restrictions $-d/2 < s < d/2$ and $m + s > d/2$ ($+1/2$ if we want first derivative continuity), Duchon (1977) shows that the spline with this penalty can be expressed as

$$\hat{f}(\mathbf{x}) = \sum_{i=1}^{n} \delta_i \eta_{2m-2s-d}(\|\mathbf{x}-\mathbf{x}_i\|) + \sum_{j=1}^{M} \alpha_j \phi_j(\mathbf{x}).$$

As for the thin plate spline, the ϕ_j span the space of polynomials of order less than m, but the η_q are now given by

$$\eta_q(t) = \begin{cases} (-1)^{(q+1)/2}|t|^q & q \text{ odd} \\ (-1)^{q/2}|t|^q \log|t| & q \text{ even.} \end{cases}$$

The Duchon spline fitting problem has exactly the form (5.8), but with $E_{ij} =$

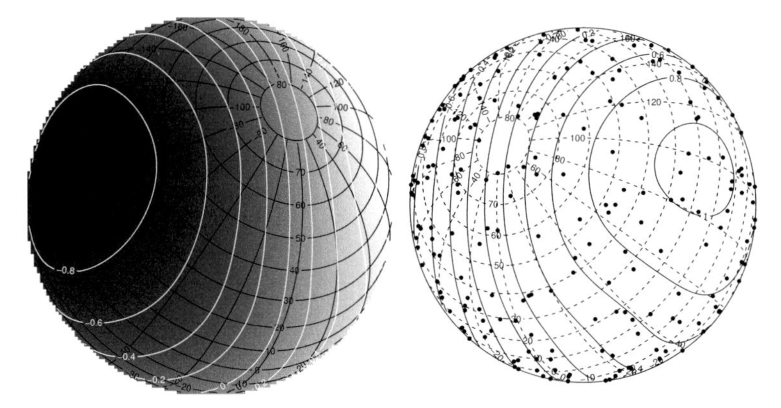

Figure 5.13 *Two visualisations of a spline on the sphere estimated using the* "sos" *basis in* mgcv. *Left shows a 'heat map' of the estimates on the surface of the sphere, with contours overlaid. Right shows a view from a different direction without the heat map but showing the data locations.*

$\eta_{2m-2s-d}(\|\mathbf{x}_j - \mathbf{x}_i\|)$. It follows that exactly the same rank reduction approaches can be taken as in the normal thin plate spline case. `s(x,z,bs="ds",m=c(1,.5))` specifies a Duchon spline in `mgcv` (here with $m = 1$, $s = 1/2$ and $d = 2$). Notice that if we set $s = (d-1)/2$, then $m = 1$ can be used for any dimension d, with the consequence that the null space basis dimension will be $M = d + 1$.

5.5.3 *Splines on the sphere*

Sometimes data arrive on the sphere (e.g., satellite data) and it is appropriate to smooth in a way that recognises this**. Let θ = latitude and ϕ = longitude. The analogue of the second order thin plate spline penalty on the sphere is

$$J(f) = \int_0^{2\pi} \int_0^{\pi} \left(\frac{f_{\phi\phi}}{\sin^2\theta} + \frac{\{\sin(\theta f_\theta)\}_\theta}{\sin\theta} \right)^2 \sin\theta d\theta d\phi \tag{5.11}$$

where subscripts denote differentiation (note the annoying range for θ here). The operator used in the squared bracket is the Laplace-Beltrami operator.

Let $\mathbf{p} = (\phi, \theta)^\mathsf{T}$ and $\gamma(\mathbf{p}, \mathbf{p}') = \arccos\{\sin\theta\sin\theta' + \cos\theta\cos\theta' cos(\phi - \phi')\}$ be the angle between points $\mathbf{p}$ and $\mathbf{p}'$. Wendelberger (1981) shows that the spline with penalty (5.11) has the representation

$$f(\mathbf{p}) = \alpha + \sum_i^n \gamma_i R(\mathbf{p}, \mathbf{p}_i),$$

where if $x = \cos\{\gamma(\mathbf{p}, \mathbf{p}')\}$ then $R(\mathbf{p}, \mathbf{p}') = r(x)$,

$$r(x) = \begin{cases} 1 - \pi^2/6 + \text{Li}_2(1/2 + x/2) & -1 < x \le 0 \\ 1 - \log(1/2 + x/2)\log(1/2 - x/2) - \text{Li}_2(1/2 - x/2) & 1 \ge x > 0, \end{cases}$$

**I am very grateful to Grace Wahba and Jean Duchon for their help with the material in this section.

and $\text{Li}_2 = \sum_{k=1}^{\infty} x^k/k^2$, $|x| \le 1$ (the dilogarithm). Since we only require $\text{Li}_2(x)$ for $|x| \le 1/2$ then in double precision this series converges in 50 terms or fewer.

The estimates of γ and α minimize

$$\|\mathbf{y} - \mathbf{R}\gamma - \mathbf{1}\alpha\|^2 + \lambda\gamma^\mathsf{T}\mathbf{R}\gamma \text{ s.t. } \mathbf{1}^\mathsf{T}\gamma = 0$$

where $\mathbf{R}$ is the matrix with i, j^{th} element $R(\mathbf{p}_i, \mathbf{p}_j)$. Again this is essentially the same structure as the thin plate spline problem, and the same eigen or 'knot' based rank reduction is possible.

Wahba (1981) avoids evaluation of the dilogarithm by modifying the penalty a little so that the corresponding R has the simpler form,

$$R(\mathbf{p}, \mathbf{p}') = \frac{1}{2\pi}\{q_2(1-x)/4 - 1/6\}$$

where $q_2(z) = \log\left(1 + \sqrt{2/z}\right)(3z^2 - 2z) - 6z^{3/2}/\sqrt{2} + 3z + 1$. She also considers different orders of penalty.

Duchon has also proposed smoothing with respect to the 3-dimensional co-ordinates of the points on the sphere's surface, but using a Duchon spline with $m = 2$, $s = 1/2$ and $d = 3$. In limited simulation testing the three approaches have similar performance, with the mean square error for the Duchon approach being slightly better than the Wendelberger (1981) method, which is slightly better than the Wahba (1981) method. In `mgcv` the `"sos"` basis implements all three variants. See figure 5.13.

5.5.4 *Soap film smoothing over finite domains*

Sometimes data are gathered within a domain having a complicated boundary, and it is important not to 'smooth across' boundary features. Figure 5.14 provides an example: the top left figure is a remote sensed image of chlorophyll concentration in the Aral Sea. If we want to smooth the chlorophyll concentrations, it is important not to 'smooth across' the peninsula separating the two arms of the sea. The middle panel in the top row shows an attempt to smooth using a thin plate spline: notice how the densities on the eastern edge of the western arm are too high, as an artefact of smoothing, and the high concentrations at the western edge of the eastern arm. If data coverage is uneven then this effect can become much worse, as the thinned version of the image on the lower row emphasises: the thin plate spline artefacts are then very large. The right panels of the figure show the much better results of smoothing with a 'soap film' smoother (Wood et al., 2008), constructed to avoid such artefacts. This section explains the construction.

Figure 5.15 illustrates the basic idea. Consider a loop of wire (black curve) following the boundary (grey loop) of the region, Ω, of the $x - y$ plane over which we are interested in smoothing. The z coordinate of the loop gives the known function values at the boundary. One definition of a 'completely smooth' function over this domain can be obtained by considering a soap film supported by the boundary wire (in zero gravity): see the surface in the middle panel of figure 5.15. Assuming a

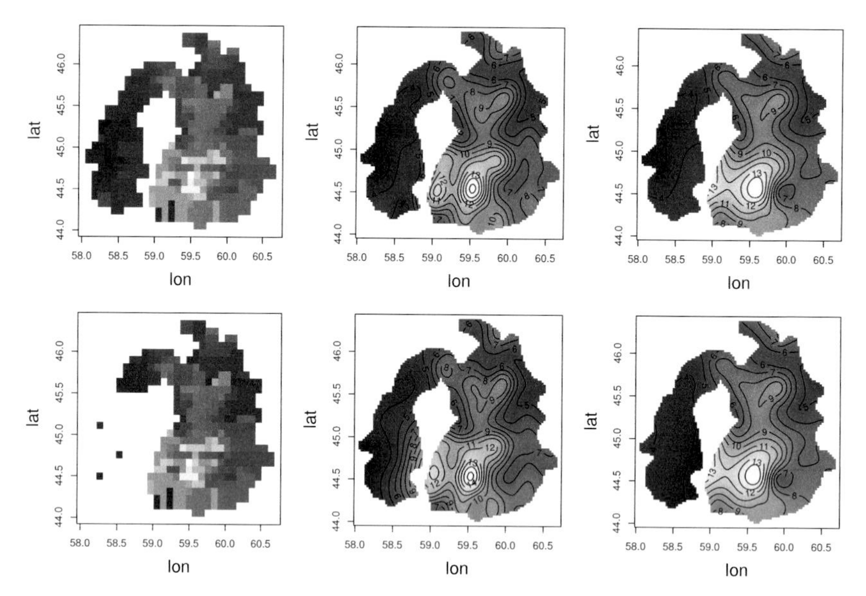

Figure 5.14 *Top row. The left panel is raw remote sensed chlorophyll levels in the Aral Sea. The middle panel is a smoothed version of the left panel data, using a thin plate spline — notice the tendency to 'smooth across' the peninsula. The right panel is a soap film smoother, which avoids smoothing across the peninsula. Bottom row. As top row, but with most of the data in the western arm of the sea dropped (at random), the thin plate spline now shows severe 'leakage' artefacts, while the soap film behaves more reasonably.*

'small' z range for the wire, the height of the soap film inside the boundary is given by the function f satisfying

$$\frac{\partial^2 f}{\partial x^2} + \frac{\partial^2 f}{\partial y^2} = 0$$

and the boundary conditions: i.e., the soap film adopts a minimum surface tension configuration. In order to smooth noisily observed z data over the domain, the soap film should distort smoothly from its minimum energy configuration by moving vertically towards the data (right panel of figure 5.15). An appropriate measure of the total degree of distortion might be:

$$J_\Omega(f) = \int_\Omega \left(\frac{\partial^2 f}{\partial x^2} + \frac{\partial^2 f}{\partial y^2} \right)^2 dx dy.$$

This differs from a thin plate spline penalty functional in three respects: it is only integrated over Ω, rather than the whole x, y plane, there is no mixed second derivative term, and the sum of the second derivative terms is squared, rather than the terms being separately squared. Hence the second derivatives with respect to x and y can be traded off against each other, so that the space of functions for which $J_\Omega(f)$ is zero is infinite dimensional, and functions with zero penalty can be curved enough to meet any smooth boundary condition.

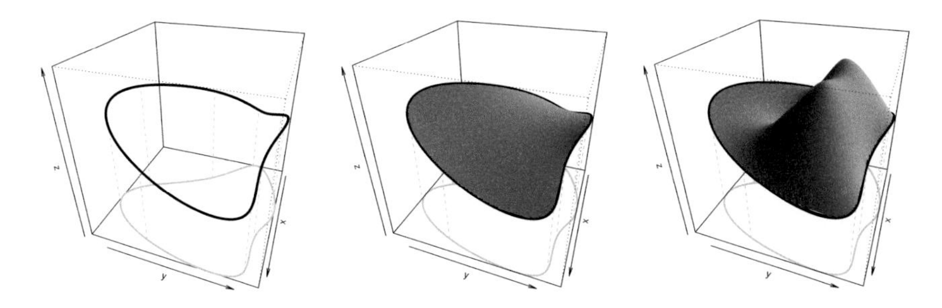

Figure 5.15 *Soap film smoothing idea. Left: the boundary of a finite region in the x, y plane is shown in grey. Suppose the smooth has a known value on the boundary, shown as the thick black curve. Middle: the smoothest surface satisfying the boundary condition is a soap film matching the boundary. Right: the soap film is allowed to distort smoothly towards noisy z data observed over the domain. In reality the boundary condition is usually unknown, so we represent it using a cyclic spline.*

If we have n data points, z_k, which are noisy observations of $h(x_k, y_k)$ where h is a smooth function over the domain, we might seek to estimate h by minimizing (subject to the known boundary conditions)

$$\sum_{i=1}^{n} \{z_i - f(x_i, y_i)\}^2 + \lambda J_\Omega(f) \tag{5.12}$$

w.r.t. f. Here λ is a tunable smoothing parameter. This problem was first considered by Ramsay (2002) (using a different motivation), but the resulting finite element fitting methods do not provide the simple basis-penalty representation that is convenient in the GAM context. To this end the following theorem and lemma from Wood et al. (2008) are helpful

Theorem 2 (Soap film interpolation). *Consider a smooth function $f^*(x, y)$ over the x, y plane. Let B be a collection of closed loops in the x, y plane, such that no two loops intersect and one 'outer' loop encloses all the others. Let Ω be the region made up of all x, y points which are interior to an odd total number of these loops. Suppose that $f^*(x, y)$ is known exactly on B, and that $z_k = f^*(x_k, y_k)$ are observations of f^* at n locations x_k, y_k within Ω. Let $f(x, y)$ be the function which*

1. *interpolates the known f^* values on B and the z_ks at the n points (x_k, y_k);*
2. *satisfies $\partial^2 f/\partial x^2 + \partial^2 f/\partial y^2 = 0$ on B; and*
3. *minimizes*

$$J_\Omega(f) = \int_\Omega \left(\frac{\partial^2 f}{\partial x^2} + \frac{\partial^2 f}{\partial y^2} \right)^2 dx dy.$$

Then f is the function, meeting condition 1, such that

$$\frac{\partial^2 f}{\partial x^2} + \frac{\partial^2 f}{\partial y^2} = \rho, \tag{5.13}$$

where

$$\frac{\partial^2 \rho}{\partial x^2} + \frac{\partial^2 \rho}{\partial y^2} = 0 \tag{5.14}$$

except at the x_k, y_k points, and $\rho = 0$ on B. Equations (5.13) and (5.14) are the Poisson and Laplace equations, respectively.

Proof. See Wood et al. (2008) □

Lemma 1 (Soap film smoothing). *Let the setup be as above, except that the z_k are now measured with error, and $\mathbf{f} = [f(x_1, y_1), f(x_2, y_2), \ldots, f(x_n, y_n)]^\mathsf{T}$. The function, $f(x, y)$, minimizing*

$$\|\mathbf{z} - \mathbf{f}\|^2 + \lambda_f J_\Omega(f) \tag{5.15}$$

subject to the known conditions on B must satisfy (5.13) and (5.14).

Proof. If the minimizing function were $\tilde{f}$, different to f, then we could interpolate the values in $\mathbf{f}$ using a soap film interpolant, thereby leaving $\|\mathbf{z} - \mathbf{f}\|^2$ unchanged but reducing J_Ω, by the minimum J_Ω interpolant property of the soap film interpolant. That is, there is a contradiction unless $\tilde{f} = f$. □

Characterization of the soap film smoother in terms of the solution of (5.13) and (5.14) is the key to evaluation of the basis functions and penalty. Let $\rho_k(x, y)$ denote the function which is zero on B, satisfies (5.14) everywhere in Ω except at the single point x_k, y_k (the kth data location) and satisfies $\int_\Omega \rho_k(x, y) dx dy = 1$. Then any function which satisfies (5.14) everywhere in Ω, except at the set of points $\{x_k, y_k : k = 1, \ldots, n\}$, can be written as

$$\rho(x, y) = \sum_{k=1}^{n} \gamma_k \rho_k(x, y),$$

where the γ_k are coefficients. It is then straightforward to confirm that

$$J_\Omega(f) = \boldsymbol{\gamma}^\mathsf{T} \mathbf{S} \boldsymbol{\gamma},$$

where $\mathbf{S}$ is a matrix of fixed coefficients given by

$$\mathbf{S}_{i,j} = \int_\Omega \rho_i(x, y) \rho_j(x, y) dx dy.$$

The function f can also be written in terms of the γ_k. Let $a(x, y)$ be the solution to (5.13) with $\rho(x, y) = 0 \; \forall \; x, y$, subject to the known boundary condition on B. Now define $g_i(x, y)$ as the solution to (5.13) with $\rho(x, y) = \rho_i(x, y)$ and the boundary condition that $f = 0$ on B; linearity of (5.13) implies that the soap film smoother can be written as

$$f(x, y) = a(x, y) + \sum_{k=1}^{n} \gamma_k g_k(x, y). \tag{5.16}$$

To obtain the basis functions ρ_j, g_j and a, the required partial differential equations can be efficiently solved by multigrid methods (see, e.g., Press et al., 2007). Alternatively, simple finite difference discretization of the derivatives followed by direct solution of the resulting sparse matrix systems using the `Matrix` package in R seems to be even more efficient (Bates and Maechler, 2015; Davis, 2006).

We do not usually want to work with n basis functions for n data, and instead choose a representative set of x, y 'knots' and compute the basis functions only for these. It is also rare to know the value of the smooth on B, and instead we can let the boundary condition be given by a cyclic spline around the boundary. In this case let $a_j(x, y)$ be the solution to (5.13) with $\rho(x, y) = 0 \, \forall \, x, y$, subject to the boundary condition given by setting all the cyclic boundary spline coefficients, β_i, to zero, except for the jth which is set to 1. Then we can substitute $a(x, y) = \sum_j \beta_j a_j(x, y)$ in (5.16). Terms like `s(x,y,bs="so",xt=list(bnd=list(bnd)))` in `mgcv` implement soap film smooths. The list supplied in the `xt` argument defines the boundary. See Wood et al. (2008) and `?soap` for more details. Wang and Ranalli (2007) or Miller and Wood (2014) provide alternative approaches.

5.6 Tensor product smooth interactions

The isotropy of the smooths considered above is often considered to be desirable when modelling something as a smooth function of geographic coordinates,†† but it has some disadvantages. Chief among them is the difficulty of knowing how to scale predictors relative to one another, when both are arguments of the same smooth but they are measured in fundamentally different units. For example, consider a smooth function of a single spatial coordinate and time: the implied relative importance of smoothness in time versus smoothness in space is very different between a situation in which the units are metres and hours, compared to that in which the units are light-years and nanoseconds. One pragmatic approach is to scale all predictors into the unit square, as is often done in LOESS smoothing, but this is essentially arbitrary. A more satisfactory approach uses *tensor product smooths*. Whereas the thin plate spline generalizes from one dimension to multiple dimensions by exchanging a flexible strip for a flexible sheet, the tensor product smooth can be interpreted as exchanging a single flexible strip for a lattice work of strips (as is illustrated in figure 5.17), with the component strips having different stiffness in different directions.

5.6.1 *Tensor product bases*

The basic approach of this section is to start from smooths of single covariates, represented using any basis with an associated quadratic penalty measuring 'wiggliness' of the smooth. From these 'marginal smooths' a 'tensor product' construction is used to build up smooths of several variables. See de Boor (1978) for an important early reference on tensor product spline bases, and Wood (2006a) for this construction.

The methods developed here can be used to construct smooth functions of *any*

††Although it's possible to overstate the case for doing this: in many applications at many locations North-South is not the same as East-West.

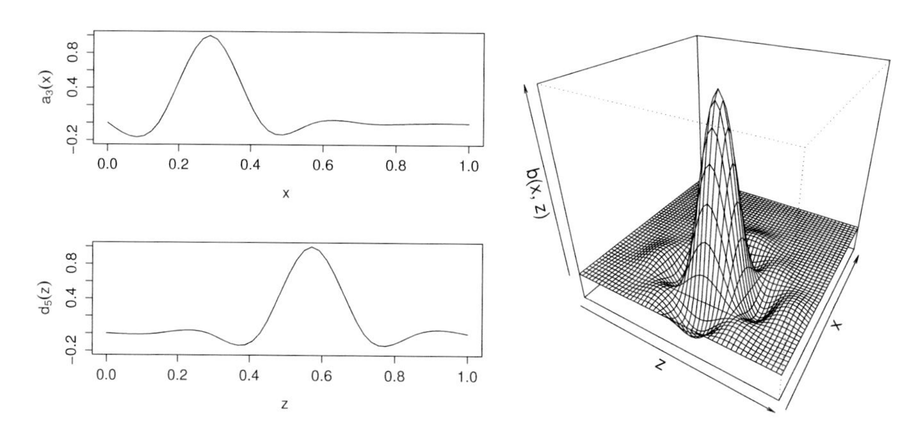

Figure 5.16 *How the product of two marginal basis functions for smooth functions of x and z, separately, results in a basis function for a smooth function of x and z together. The two left panels show the 3rd and 5th basis functions for rank 8 cubic regression spline smooths of x and z, respectively. The right hand plot shows $a_3(x)d_5(z)$, one of 64 similar basis functions of the tensor product smooth derived from these two marginal smooths.*

number of covariates, but the simplest introduction is via the construction of a smooth function of 3 covariates, x, z and v, the generalization then being trivial. The process starts by assuming that we have low rank bases available for representing smooth functions f_x, f_z and f_v of each of the covariates. That is, we can write:

$$f_x(x) = \sum_{i=1}^{I} \alpha_i a_i(x), \quad f_z(z) = \sum_{l=1}^{L} \delta_l d_l(z) \text{ and } f_v(v) = \sum_{k=1}^{K} \beta_k b_k(v),$$

where the α_i, δ_l and β_k are parameters, and the $a_i(x)$, $d_l(z)$ and $b_k(v)$ are known basis functions.

Now consider how f_x, the smooth function of x, could be converted into a smooth function of x and z. What is required is for f_x to vary smoothly with z, and this can be achieved by allowing its parameters, α_i, to vary smoothly with z. Using the basis already available for representing smooth functions of z we could write:

$$\alpha_i(z) = \sum_{l=1}^{L} \delta_{il} d_l(z)$$

which immediately gives

$$f_{xz}(x, z) = \sum_{i=1}^{I} \sum_{l=1}^{L} \delta_{il} d_l(z) a_i(x).$$

Figure 5.16 illustrates this construction. Continuing in the same way, we could now create a smooth function of x, z and v by allowing f_{xz} to vary smoothly with v. Again, the obvious way to do this is to let the parameters of f_{xz} vary smoothly with v, and following the same reasoning as before we get

$$f_{xzv}(x, z, v) = \sum_{i=1}^{I} \sum_{l=1}^{L} \sum_{k=1}^{K} \beta_{ilk} b_k(v) d_l(z) a_i(x).$$

For any particular set of observations of x, z and v, there is a simple relationship between the model matrix, $\mathbf{X}$, evaluating the tensor product smooth at these observations, and the model matrices $\mathbf{X}_x$, $\mathbf{X}_z$ and $\mathbf{X}_v$ that would evaluate the marginal smooths at the same observations. If $\odot$ is the row-wise Kronecker product (see section B.4), then it is easy to show that, given appropriate ordering of the β_{ilk} into a vector $\boldsymbol{\beta}$,

$$\mathbf{X} = \mathbf{X}_x \odot \mathbf{X}_z \odot \mathbf{X}_v.$$

Clearly, (i) this construction can be continued for as many covariates as are required; (ii) the result is independent of the order in which we treat the covariates and (iii) the covariates can themselves be vector covariates. Figure 5.17 attempts to illustrate the tensor product construction for a smooth of two covariates.

5.6.2 *Tensor product penalties*

Having derived a 'tensor product' basis for representing smooth functions, it is also necessary to have some way of measuring function 'wiggliness', if the basis is to be useful for representing smooth functions in a GAM context. Again, it is possible to start from wiggliness measures associated with the marginal smooth functions, and again the three covariate case provides sufficient illustration. Suppose that each marginal smooth has an associated functional that measures function wiggliness and can be expressed as a quadratic form in the marginal parameters. That is

$$J_x(f_x) = \boldsymbol{\alpha}^{\mathsf{T}} \mathbf{S}_x \boldsymbol{\alpha}, \quad J_z(f_z) = \boldsymbol{\delta}^{\mathsf{T}} \mathbf{S}_z \boldsymbol{\delta} \text{ and } J_v(f_v) = \mathcal{B}^{\mathsf{T}} \mathbf{S}_v \mathcal{B}.$$

The $\mathbf{S}_\bullet$ matrices contain known coefficients, and $\boldsymbol{\alpha}$, $\boldsymbol{\delta}$ and $\mathcal{B}$ are vectors of coefficients of the marginal smooths. An example of a penalty functional is the cubic spline penalty, $J_x(f_x) = \int f_x''(x)^2 dx$. Now let $f_{x|zv}(x)$ be $f_{xvz}(x, z, v)$ considered as a function of x only, with z and v held constant, and define $f_{z|xv}(z)$ and $f_{v|xz}(v)$ similarly. A natural way of measuring wiggliness of f_{xzv} is to use:

$$J(f_{xzv}) = \lambda_x \int_{z,v} J_x(f_{x|zv}) dz dv + \lambda_z \int_{x,v} J_z(f_{z|xv}) dx dv + \lambda_v \int_{x,z} J_v(f_{v|xz}) dx dz$$

where the $\lambda_\bullet$ are smoothing parameters controlling the tradeoff between wiggliness in different directions, and allowing the penalty to be invariant to the relative scaling of the covariates. As an example, if cubic spline penalties were used as the marginal penalties, then

$$J(f) = \int_{x,z,v} \lambda_x \left(\frac{\partial^2 f}{\partial x^2}\right)^2 + \lambda_z \left(\frac{\partial^2 f}{\partial z^2}\right)^2 + \lambda_v \left(\frac{\partial^2 f}{\partial v^2}\right)^2 dx dz dv.$$

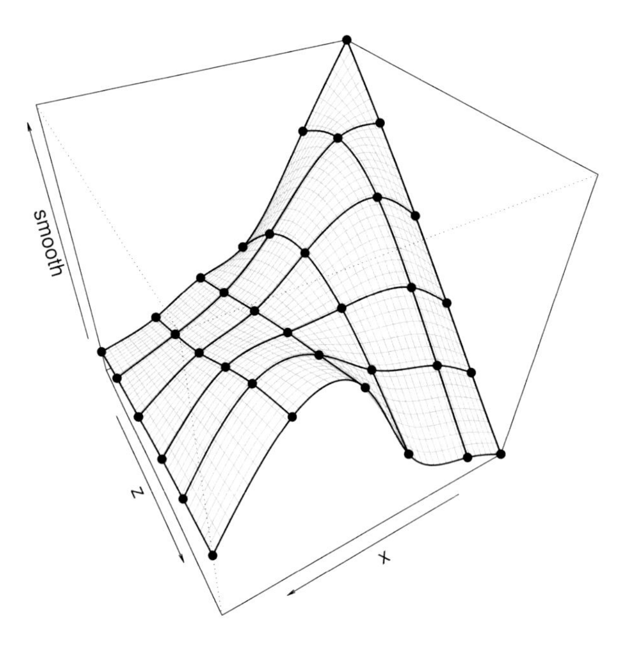

Figure 5.17 *Illustration of a tensor product smooth of two variables x and z, constructed from two rank 6 marginal bases. Following section 5.6.2 a tensor product smooth can always be parameterized in terms of the values of the function at a set of 'knots' spread over the function domain on a regular mesh: i.e. in terms of the heights of the •'s shown. The basis construction can be thought of as follows: start with a smooth of x parameterized in terms of function values at a set of 'knots'; to make the smooth of x vary smoothly with z, simply allow each of its parameters to vary smoothly with z. This can be done by representing each parameter using a smooth of z, also parameterized in terms of function values at a set of 'knots'. Exactly the same smooth arises if we reverse the roles of x and z in this construction. The tensor product smooth penalty in the x direction, advocated in section 5.6.2, is simply the sum of the marginal wiggliness measure for the smooth of x, applied to the thick black curves parallel to the x axes. The z penalty is similarly defined in terms of the marginal penalty of the smooth of z applied to the thick black curves parallel to the z-axis.*

Hence, if the marginal penalties are easily interpretable, in terms of function shape, then so is the induced penalty. Numerical evaluation of the integrals in J is straightforward. As an example consider the penalty in the x direction. The function $f_{x|zv}(x)$ can be written as

$$f_{x|zv}(x) = \sum_{i=1}^{I} \alpha_i(z, v) a_i(x),$$

and it is always possible to find the matrix of coefficients $\mathbf{M}_{z,v}$ such that $\boldsymbol{\alpha}(z, v) = \mathbf{M}_{zv}\boldsymbol{\beta}$ where $\boldsymbol{\beta}$ is the vector of β_{ilk} arranged in some appropriate order. Hence

$$J_x(f_{x|zv}) = \boldsymbol{\alpha}(z, v)^{\mathsf{T}}\mathbf{S}_x\boldsymbol{\alpha}(z, v) = \boldsymbol{\beta}^{\mathsf{T}}\mathbf{M}_{zv}^{\mathsf{T}}\mathbf{S}_x\mathbf{M}_{zv}\boldsymbol{\beta}$$

and so

$$\int_{z,v} J_x(f_{x|zv})dzdv = \boldsymbol{\beta}^\mathsf{T} \int_{z,v} \mathbf{M}_{zv}^\mathsf{T}\mathbf{S}_x\mathbf{M}_{zv}dzdv\boldsymbol{\beta}.$$

The last integral can be performed numerically (in fact Reiss et al. (2014) show how to obtain it directly), and it is clear that the same approach can be applied to all components of the penalty. However, a simple re-parameterization can be used to provide a straightforward approximation to the terms in the penalty, which performs well in practice.

To see how the approach works, consider the marginal smooth f_x. Let $\{x_i^* : i = 1, \ldots, I\}$ be a set of values of x spread evenly through the range of the observed x values. In this case we can always re-parameterize f_x in terms of new parameters

$$\alpha_i' = f_x(x_i^*).$$

Clearly under this re-parameterization $\boldsymbol{\alpha}' = \boldsymbol{\Gamma}\boldsymbol{\alpha}$ where $\Gamma_{ij} = a_i(x_j^*)$. Hence the marginal model matrix becomes $\mathbf{X}_x' = \mathbf{X}_x\boldsymbol{\Gamma}^{-1}$ and the penalty coefficient matrix becomes $\mathbf{S}_x' = \boldsymbol{\Gamma}^{-\mathsf{T}}\mathbf{S}_x\boldsymbol{\Gamma}^{-1}$.

Now suppose that the same sort of re-parameterization is applied to the marginal smooths f_v and f_z. In this case we have that

$$\int_{z,v} J_x(f_{x|zv})dzdv \approx h\sum_{lk} J_x(f_{x|z_l^*v_k^*}),$$

where h is some constant of proportionality related to the spacing of the z_l^* and v_k^*. Similar expressions hold for the other integrals making up J. If $\otimes$ is the Kronecker product (see section B.4), it is straightforward to show that the summation in the above approximation is:

$$J_x^*(f_{xzv}) = \boldsymbol{\beta}^\mathsf{T}\tilde{\mathbf{S}}_x\boldsymbol{\beta} \text{ where } \tilde{\mathbf{S}}_x = \mathbf{S}_x' \otimes \mathbf{I}_L \otimes \mathbf{I}_K$$

and $\mathbf{I}_L$ is the rank L identity matrix. Exactly similar definitions hold for the other components of the penalty so that

$$J_z^*(f_{xzv}) = \boldsymbol{\beta}^\mathsf{T}\tilde{\mathbf{S}}_z\boldsymbol{\beta} \text{ where } \tilde{\mathbf{S}}_z = \mathbf{I}_I \otimes \mathbf{S}_z' \otimes \mathbf{I}_K$$

and

$$J_v^*(f_{xzv}) = \boldsymbol{\beta}^\mathsf{T}\tilde{\mathbf{S}}_v\boldsymbol{\beta} \text{ where } \tilde{\mathbf{S}}_v = \mathbf{I}_I \otimes \mathbf{I}_L \otimes \mathbf{S}_v'.$$

Hence

$$J(f_{xzv}) \approx J^*(f_{xzv}) = \lambda_x J_x^*(f_{xzv}) + \lambda_z J_z^*(f_{xzv}) + \lambda_v J_v^*(f_{xzv}),$$

where any constants, h, have been absorbed into the λ_j. Again, this penalty construction clearly generalizes to any number of covariates. Figure 5.17 attempts to illustrate what the penalties actually measure, for a smooth of two variables.

Given its model matrix and penalties, the coefficients and smoothing parameters of a tensor product smooth can be estimated as GAM components using the methods of Chapter 6. These smooths have the nice property of being invariant to rescaling of the covariates, provided only that the marginal smooths are similarly invariant (which is always the case in practice).

Note that it is possible to omit the re-parameterization of the marginal smooths, in terms of function values, and to work with penalties of the form

$$\boldsymbol{\beta}^{\mathsf{T}}\bar{\mathbf{S}}_z\boldsymbol{\beta} \text{ where } \bar{\mathbf{S}}_z = \mathbf{I}_I \otimes \mathbf{S}_z \otimes \mathbf{I}_K,$$

for example; Eilers and Marx (2003) successfully used this approach to smooth with respect to two variables using tensor products of B-splines. A potential problem is that the penalties no longer have the interpretation in terms of (averaged) function shape that is inherited from the marginal smooths when re-parameterization is used. Another proposal in the literature is to use single penalties of the form:

$$\boldsymbol{\beta}^{\mathsf{T}}\mathbf{S}\boldsymbol{\beta} \text{ where } \mathbf{S} = \mathbf{S}_1 \otimes \mathbf{S}_2 \otimes \cdots \otimes \mathbf{S}_d,$$

but this often leads to severe under-smoothing. The rank of $\mathbf{S}$ is the product of the ranks of the $\mathbf{S}_j$, and in practice this is often far too low for practical work. For example, consider a smooth of 3 predictors constructed as a tensor product smooth of 3 cubic spline bases, each of rank 5. The resulting smooth would have 125 free parameters, but a penalty matrix of rank 27. This means that varying the weight given to the penalty would only result in the effective degrees of freedom for the smooth varying between 98 and 125: not a very useful range. By contrast, for the same marginal bases, the multiple term penalties would have rank 117, leading to a much more useful range of effective degrees of freedom of between 8 and 125.

In `mgcv`, tensor product smooths are specified using `te` terms in model formulae. Any single smoothing parameter smooth that can be specified using an `s` term can be used as a marginal smooth for the term. For example `te(x,z)` would specify a tensor product of two (default) 5-dimensional `"cr"` basis smoothers, resulting in a 25-dimensional basis with two penalties (one for each direction). A more complicated specification is `s(x,z,t,d=c(2,1),bs=c("tp","cr"),k=c(20,6))`:a tensor product smooth constructed from a rank 20 thin plate spline of the 2 variables x and z and a rank 6 cubic regression spline of the single variable t. The result has a basis dimension of 120, and two penalties.

5.6.3 *ANOVA decompositions of smooths*

Sometimes it is useful to include model components such as

$$f_1(x) + f_2(z) + f_3(x, z), \tag{5.17}$$

where f_1 and f_2 are smooth 'main effects' and f_3 is a smooth 'interaction'. Whether nature is really so kind as to arrange matters in a manner so tidily comprehensible is a moot point, but the decomposition can be useful for testing whether additivity ($f_1(x) + f_2(z)$) is appropriate, for example.

In principle, terms like (5.17) can be estimated by using any bases we like for the f_j terms, testing for linear dependencies between them, and imposing constraints to remove any that are found. `mgcv` will do such testing and constraint by default (see below), but such an approach does not leave $f_1 + f_2$ orthogonal to f_3, so that interpretation of the resulting fits is complicated. Ideally we would rather have f_3 constructed to include no components of the form $f_1 + f_2$.

Actually, constructing such interaction terms is very easy if we recognise that the tensor product basis construction is exactly the same as the construction used for any interaction in a linear model, as described in section 1.6.3. It follows that if we subject the marginal smooths of a tensor product to sum-to-zero identifiability constraints *before* constructing the tensor product basis, then the resulting interaction smooths do not include the corresponding main effects (exactly as would occur with any interaction in a linear model). What happens is that the sum-to-zero constraints remove the unit function from the span of the marginals, with the result that the tensor product basis will not include the main effect that results from the product of a marginal basis with the unit functions in the other marginal bases. The sum-to-zero constraints have no effect on the penalty, of course.

Exactly as with other linear model interactions the approach is general: for example the three way interaction terms constructed from three sum-to-zero constrained marginals will exclude not only the bases of the three main effects, but also the bases for the three 2-way interactions (i.e., the terms resulting from the product of the constant function and two other marginal bases). This approach is so simple, obvious and general that it only seems to appear in passing in the literature (in contrast to much more complicated constructions). In `mgcv`, `ti` terms specify such pure interactions, with the syntax being the same as for `te` terms.

Numerical identifiability constraints for nested terms

For completeness this section provides a method for imposing constraints to make model components like (5.17) identifiable, whatever bases are chosen. However, in almost all practical cases use of `ti` type terms is a better choice. The principle is as follows. Working through all smooths, starting from smooths of two variables and working up through smooths of more variables:

1. Identify all smooths of fewer or the same number of variables sharing covariates with the current smooth.
2. Numerically identify any linear dependence of the basis for the current smooth on the other smooths identified as sharing its covariates, and constrain the current smooth to remove this.

Let $\mathbf{X}_1$ be the combined model matrix for all the lower order model terms sharing covariates with the smooth of interest, and $\mathbf{X}_2$ be the model matrix of the smooth of interest. Provided $\mathbf{X}_1$ and $\mathbf{X}_2$ are each of full column rank, we can test for dependence between them by forming their QR decomposition in two steps. We seek

$$[\mathbf{X}_1 : \mathbf{X}_2] = \mathbf{Q}\mathbf{R},$$

where the columns of $\mathbf{Q}$ are columns of an orthogonal matrix, and $\mathbf{R}$ is full rank upper triangular if $[\mathbf{X}_1 : \mathbf{X}_2]$ is of full column rank, and reduced rank if $\mathbf{X}_2$ depends on $\mathbf{X}_1$. First form the QR decomposition

$$\mathbf{X}_1 = \mathbf{Q}_1 \begin{bmatrix} \mathbf{R}_1 \\ \mathbf{0} \end{bmatrix}$$

and let $\mathbf{B}$ be $\mathbf{Q}_1^\mathsf{T}\mathbf{X}_2$ with the first r rows removed, where r is the number of columns of $\mathbf{X}_1$. Now form a second QR decomposition *with pivoting*

$$\mathbf{B} = \mathbf{Q}_2 \begin{bmatrix} \mathbf{R}_2 \\ \mathbf{0} \end{bmatrix}.$$

If some columns of $\mathbf{X}_2$ depend on columns of $\mathbf{X}_1$ then there will be a zero block at the lower right corner of $\mathbf{R}_2$, and columns of $\mathbf{X}_2$ responsible for this block will be identifiable from the record of which columns of $\mathbf{X}_2$ have been pivoted to these final columns. These dependent columns can be removed, to make the model identifiable, or equivalently, the corresponding parameters can be constrained to zero.

If the basis of this algorithm is unclear, note that the implied QR decomposition is

$$[\mathbf{X}_1 : \mathbf{X}_2] = \mathbf{Q}_1 \begin{bmatrix} \mathbf{I} & \mathbf{0} \\ \mathbf{0} & \mathbf{Q}_2 \end{bmatrix} \begin{bmatrix} \mathbf{R}_1 & \bar{\mathbf{B}} \\ \mathbf{0} & \mathbf{R}_2 \\ \mathbf{0} & \mathbf{0} \end{bmatrix},$$

where $\bar{\mathbf{B}}$ is the first r rows of $\mathbf{Q}_1^\mathsf{T}\mathbf{X}_2$.

5.6.4 *Tensor product smooths under shape constraints*

The section 5.3.6 SCOP-spline approach can be combined with the tensor product construction of section 5.6. If we require a shape restriction along a margin of a tensor product smooth then the penalty and model matrix ($\mathbf{XB}$) for that marginal are set up as in section 5.3.6, the other marginals are set up as normal, and the tensor product basis and penalties are set up from the marginals in the usual way (but without the section 5.6 reparameterization). If the jth basis function for the tensor product involves a basis function from the shape constrained margin for which the corresponding coefficient must be positive, then the coefficient, β_j, of the tensor product must also be positive. So we would represent it as $\beta_j = \exp(\gamma_j)$ where γ_j is unrestricted. Sum-to-zero constraints can be imposed by the column centring approach of section 4.3.1 (p. 175).

Again Pya and Wood (2015) provide fuller details, but there is an important restriction on the type of smooths that can be used as unconstrained margins: the basis functions must be non-negative over the region where the marginal monotonicity restriction is to hold. Unusually this restriction is best understood by stating a theorem (not appearing in Pya and Wood, 2015) which establishes the conditions under which tensor product smooths with SCOP-spline monotonic marginals will be monotonic in that margin. The theorem is pretty much proved by stating it.

Theorem 3. Marginal monotonicity. *Let* $t \in \Re$, $\mathbf{x} \in \Re^d$, *for* $d \geq 1$ *and* $f(t, \mathbf{x}) = \sum_j \alpha_j(t) a_j(\mathbf{x})$ *where* $\alpha_j'(t) \geq 0 \; \forall \; j, t$. *Let* $M = \{j : \alpha_j(t) > 0 \, \textit{for some } t\}$. *If* $a_j(\mathbf{x}) \geq 0 \, \forall \, \mathbf{x}, j \in M$ *then*

$$\frac{\partial f(t, \mathbf{x})}{\partial t} \geq 0 \, \forall \, \mathbf{x}.$$

Proof. $\partial f(t, \mathbf{x})/\partial t = \sum_j \alpha_j'(t) a_j(\mathbf{x}) \geq 0$ by the conditions of the theorem. □

Corollary. Now for all $j \in M$ let a_j be basis functions multiplied by non-negative coefficients. If any basis function, a_k, is not strictly positive, then there exist coefficients for which $\partial f(t, \mathbf{x})/\partial t < 0$ for some t.

An obvious example is when all coefficients are zero, apart from the kth. So the positive basis function condition is both necessary and sufficient to guarantee that the tensor product smooth is monotonic with respect to the margin t.

Notice that the same argument that justifies column centring for imposing sum-to-zero constraints in section 4.3.1, appears to justify adding constants to basis functions in order to ensure that they meet the positivity restriction over the region of interest, but care is needed. For example if a thin plate spline marginal in x includes basis functions $b_1(x) = 1$ and the linear function $b_2(x) = x$, then we would inadvertently force an overall positive trend in x over much of the tensor product by adopting this approach. Using a B-spline basis does not suffer from this problem, basically because it looks the same under reversal of the ordering of x.

5.6.5 An alternative tensor product construction

An alternative tensor product construction is due to Fabian Scheipl (`mgcv` 1.7-0, 2010, Wood et al., 2013, online 2012). It is a direct reduced rank analogue of the smoothing spline ANOVA methods of Gu (2013) and Wahba (1990), and has the advantage that the penalty terms are non-overlapping sub-matrices of the identity matrix, making estimation with standard mixed modelling software quite easy (the earlier alternative of Lee and Durbán (2011) does not produce non-overlapping penalties). The smooths are constructed in such a way that the main effects, low order interactions, etc., are explicitly present in the basis and can simply be 'read off' from a fitted term.

We start with any set of (unconstrained) marginal smooths, exactly as for the construction already discussed. The next step is to re-parameterize each marginal so that the penalty becomes the identity matrix, with M leading diagonal entries zeroed, where M is the dimension of the penalty null space. Section 5.4.2 can be used to do this, with the addition of a simple linear rescaling of the parameters in order to equate the positive elements of the penalty matrix to 1 (see also section 5.8). Now let us split up each reparameterized model matrix into columns $\mathbf{X}$ corresponding to the zero elements on the leading diagonal of the penalty and columns $\mathbf{Z}$ corresponding to the unit entries. So

$$\mathbf{f} = \mathbf{X}\boldsymbol{\delta} + \mathbf{Z}\mathbf{b}.$$

For maximum interpretability it helps to have the constant function explicitly present in each marginal basis, but the section 5.4.2 method does not guarantee this. For bases in which the null space basis corresponding to $\mathbf{X}$ is already separated out before re-parameterization it is easy to modify the re-parameterization to ensure that $\mathbf{X}$ has a constant column, but the following simple recipe provides a general automatic re-parameterization method for all cases.

If g is a function in the penalty null space then the additional penalty $\mathcal{P}_N = \sum_i \{g(x_i) - \bar{g}\}^2$ would shrink g towards the space of constant functions. We can write $\mathcal{P}_N = \boldsymbol{\delta}^\mathsf{T}\mathbf{D}^\mathsf{T}\mathbf{D}\boldsymbol{\delta}$ where $\mathbf{D} = \mathbf{X} - \mathbf{1}\mathbf{1}^\mathsf{T}\mathbf{X}/n$ and form the eigen-decomposition

$$\mathbf{D}^\mathsf{T}\mathbf{D} = \mathcal{U}\Omega\mathcal{U}^\mathsf{T}.$$

Reparameterizing so that the null space model matrix is now $\mathbf{X}\mathcal{U}$ ensures that the final column of the new model matrix is constant, provided that the null space of the original penalty includes the constant function in its span.

Having re-parameterized the d marginal smooths so that the jth has unpenalized model matrix $\mathbf{X}^j$ and penalized model matrix $\mathbf{Z}^j$ (with corresponding penalty $\mathbf{I}$), we then initialise a set of matrices $\gamma = \{\mathbf{X}^1, \mathbf{Z}^1\}$, or $\gamma = \{[\mathbf{X}^1], \mathbf{Z}^1\}$ where $[\mathbf{X}^j]$, denotes the set of columns of $\mathbf{X}^j$, each treated as a separate matrix. The following steps are then repeated for i in 2 to d...

1. Form row-wise Kronecker products of $\mathbf{X}^i$ (or of all the columns $[\mathbf{X}^i]$) with all elements of γ.
2. Form row-wise Kronecker products of $\mathbf{Z}^i$ with all elements of γ.
3. Append the matrices resulting from the previous two steps to the set γ.

The model matrix, $\mathbf{X}$, for the tensor product smooth is formed by appending all the elements of γ, columnwise. Each element of γ has an associated identity penalty with an associated smoothing parameter, except for the element(s) which involve(s) no $\mathbf{Z}^j$ term, which is unpenalized. The variant in which $[\mathbf{X}^j]$ replaces $\mathbf{X}^j$ ensures that the smooth is strictly invariant w.r.t. linear rescaling of the covariates, but at the cost of requiring extra penalty terms. In practice the difference between the variants usually seems to be small.

The simplest form of sum-to-zero constraint for such a smooth consists of dropping the constant column from $\mathbf{X}$ while subtracting their means from the remaining columns. Such constraints leave the penalties diagonal and penalizing separate subsets of coefficients. Note that if producing a prediction matrix when using such constraints it is important to subtract the column means of the *original* model matrix used in fitting, and not the column means of the prediction matrix itself. There are also arguments for simply dropping the constant column for fitting, and transforming the coefficients to satisfy the sum to zero constraints after model estimation (see Wood et al., 2013).

What is being penalized?

The construction given above describes the mechanics of setting up the alternative tensor product smoother, without revealing what the corresponding penalties are actually penalizing. As an example, consider the case in which the marginals have penalties based on integrals of squared derivatives. There are two cases: the penalty on the product of the penalty range spaces of two marginals, and the penalty on the product of a range space and a null space.

Consider marginal smooths $h(\mathbf{x})$ and $g(\mathbf{z})$, with penalties

$$\int \sum_i (D_i h)^2 d\mathbf{x} \quad \text{and} \quad \int \sum_j (\Delta_j g)^2 d\mathbf{z}$$

where the D_i and Δ_i are differential operators.

1. Let δ_k denote the kth element of the operator given by $\sum_i D_i \sum_j \Delta_j$. The penalty on the product of the range spaces of the marginal smooths is
$$\int \sum_k (\delta_k f)^2 d\mathbf{x} d\mathbf{z}.$$
2. The product of the null space basis of g's penalty with the basis for h can be written
$$\sum_j \gamma_j(\mathbf{x}) \tilde{c}_j(\mathbf{z})$$
where the $\tilde{c}_j(\mathbf{z})$ are basis functions for the null space of the penalty on g and $\gamma_j(\mathbf{x})$ is a smooth 'coefficient function' represented using the basis for h. The penalty on the product of the null space of g with the range space of h is then
$$\sum_j \int \sum_i (D_i \gamma_j)^2 d\mathbf{x}.$$

Construction of penalties for products of more than two marginals simply applies these rules iteratively (see Wood et al., 2013, for more detail).

For example, when using cubic spline marginals and the $[\mathbf{X}^j]$ variant construction, with penalties $\int h''(x)^2 dx$, then $f(x, z)$ decomposes into the following separately penalized components:

$$f_x(x) + f_z(z) + f_{x1}(x)z + f_{z1}(z)x + f_{xz}(x, z).$$

The penalties on the four univariate terms are the usual cubic spline penalty inherited from the marginal, while the penalty on f_{xz} is $\int \partial^4 f_{xz}/\partial^2 x \partial^2 z dx dz$.

`t2` terms in `mgcv` model formulae are used to specify tensor product smooths using this construction. Use is the same as `te` except that there is an extra logical argument `full` to indicate whether or not to use the $[\mathbf{X}^j]$ variant construction.

5.7 Isotropy versus scale invariance

Thin plate splines and Duchon splines adapt well to irregularly scattered data and have the advantage of only having one smoothing parameter to estimate. They are often used for spatial data as a result. However, the isotropic smoothness assumption is a strong one: isotropic smoothers are sensitive to linear rescaling of single

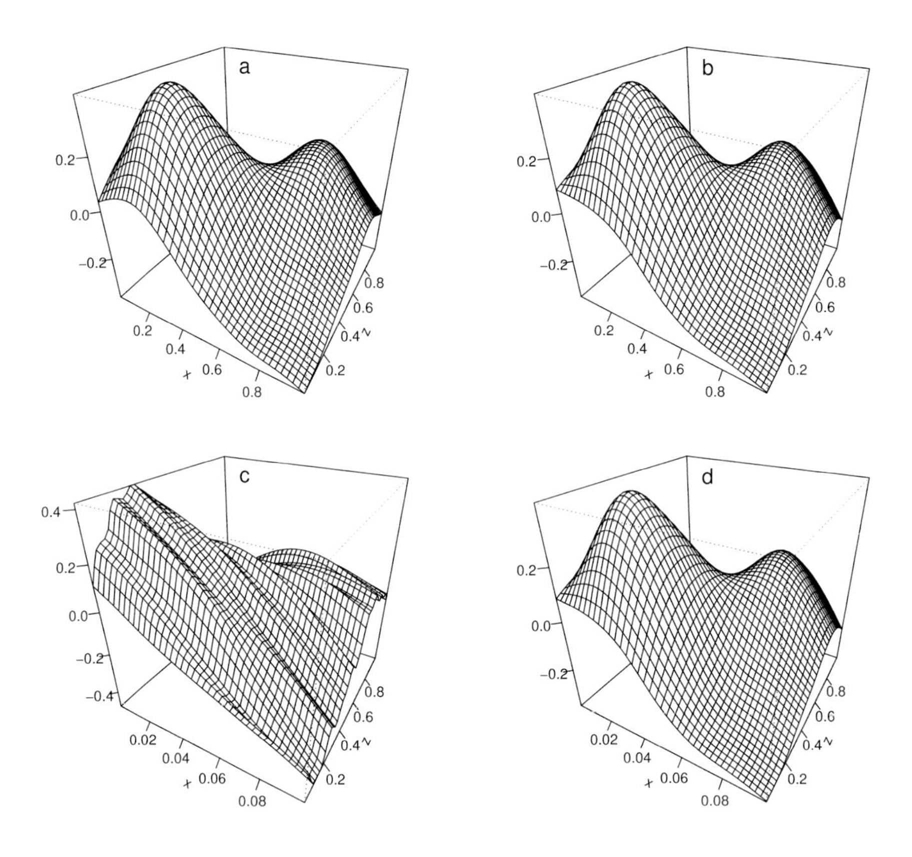

Figure 5.18 *Isotropy versus scale invariance.* (a) *Thin plate spline smooth of noisy observations of a truth looking much like the reconstruction.* (b) *Tensor product smooth of the same data as in a.* (c) *and* (d) *are as a and b, with the single difference that the x values have been divided by 10. Notice how the tensor product smooth (b and d) is invariant to the rescaling, while the isotropic thin plate spline (a and c) is highly sensitive to it.*

covariates (such as occurs under a change of measurement units) in a way that tensor product smooths are not.

Figure 5.18 illustrates this point with four smooths of essentially the same data set. The top row plots show a thin plate spline and a tensor product smooth of the same data, when the smoothing domain is nicely scaled relative to the variability of the true function from which the data were generated. The smooths are very similar (and closely resemble the true function used for data generation). The lower row again shows a thin plate and a tensor product smooth of the same data, but with the x co-ordinates divided by 10. The tensor product smooth is invariant to the change, but the thin plate spline tries to achieve the same smoothness per unit change in x and z, with the result shown in the lower left plot.

5.8 Smooths, random fields and random effects

As we saw in section 4.2.4 (p. 172), one view of the smoothing process is that the smoothing penalty, $\boldsymbol{\beta}^{\mathsf{T}}\mathbf{S}\boldsymbol{\beta}$, is employed during fitting in order to impose the belief that the true function is more likely to be smooth than wiggly. In that case we might as well express the belief in a Bayesian manner, by defining a prior distribution on function wiggliness $f(\boldsymbol{\beta}) \propto \exp(-\lambda\boldsymbol{\beta}^{\mathsf{T}}\mathbf{S}\boldsymbol{\beta}/2)$, which is equivalent to

$$\boldsymbol{\beta} \sim N(\mathbf{0}, \mathbf{S}^{-}/\lambda), \tag{5.18}$$

where $\mathbf{S}^{-}$ is a pseudoinverse of $\mathbf{S}$ (since $\mathbf{S}$ is rank deficient by the dimension of the penalty null space). Given the smoothing basis, smooth model $y_i = f(x_i) + \epsilon_i$, $\epsilon_i \sim N(0, \sigma^2)$ can be expressed as $\mathbf{y} \sim N(\mathbf{X}\boldsymbol{\beta}, \mathbf{I}\sigma^2)$. Exactly as in section 2.4 (p. 78), we can then seek the posterior mean/mode, $\hat{\boldsymbol{\beta}}$, to maximize $f(\mathbf{y}|\boldsymbol{\beta})f(\boldsymbol{\beta})$. Taking logs, multiplying by -1 and discarding irrelevant constants we find that

$$\hat{\boldsymbol{\beta}} = \underset{\boldsymbol{\beta}}{\text{argmin}}\|\mathbf{y} - \mathbf{X}\boldsymbol{\beta}\|^2/\sigma^2 + \lambda\boldsymbol{\beta}^{\mathsf{T}}\mathbf{S}\boldsymbol{\beta}.$$

It is also common to absorb σ^2 into λ so that it does not appear explicitly in the objective function. Notice how the smooth has the mathematical structure of a random effect in a linear mixed model. This is useful computationally, but requires the qualification that the smooth is really best thought of in a Bayesian manner. This is because we do not usually expect $\boldsymbol{\beta}$ to be re-drawn from (5.18) on each replication of the data, as we usually would for a frequentist random effect.

We can also follow section 2.4.1 (p. 79), to obtain the posterior of $\boldsymbol{\beta}$,

$$\boldsymbol{\beta}|\mathbf{y} \sim N(\hat{\boldsymbol{\beta}}, (\mathbf{X}^{\mathsf{T}}\mathbf{X}/\sigma^2 + \lambda\mathbf{S})^{-1}),$$

and, unsurprisingly, the machinery of section 3.4 (p. 147) also allows us to include smooths in generalized linear mixed models in the same way as random effects.

If $\mathbf{f}$ is the vector containing the values of the smooth evaluated at the observed covariate values, then (5.18) implies the prior $\mathbf{f} \sim N(\mathbf{0}, \mathbf{X}\mathbf{S}^{-}\mathbf{X}^{\mathsf{T}})$. This emphasises that we can view the smooth as a Gaussian random field, with posterior mode $\hat{\mathbf{f}}$.

On occasion it is useful to express the smooth more explicitly in mixed model form so that, if $\mathbf{f}$ is the vector such that $f_i = f(x_i)$, then

$$\mathbf{f} = \mathbf{X}'\boldsymbol{\beta}' + \mathbf{Z}\mathbf{b}$$

where $\mathbf{b} \sim N(\mathbf{0}, \mathbf{I}\sigma_b^2)$. Here the columns of $\mathbf{X}'$ form a basis for the null space of the smoothing penalty, while the columns of $\mathbf{Z}$ form a basis for its range space. This is particularly useful if smooths are to be estimated using software designed for estimating linear or generalized linear mixed models. Section 5.4.2 provides one obvious way of computing the required matrices. Following that section's notation, partition $\mathbf{U} = [\mathbf{U}^{+} : \mathbf{U}^{0}]$ where $\mathbf{U}^{+}$ are the eigenvectors corresponding to positive eigenvalues of the smoothing penalty matrix, and $\mathbf{U}^{0}$ are the remaining eigenvectors. Then $\mathbf{X}' = \mathbf{Q}\mathbf{U}^{0}$ and $\mathbf{Z} = \mathbf{Q}\mathbf{U}^{+}\mathbf{D}^{-1/2}$. There are also computationally cheaper ways of

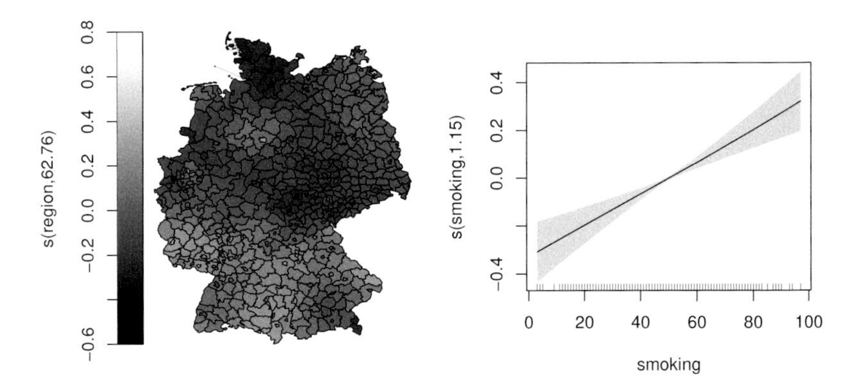

Figure 5.19 *Smooths from a log Poisson additive model of cancer of the larynx recorded by health district in Germany. Left is a GMRF smoother over districts, and right is a cubic spline smooth of smoking level. The near zero width confidence interval in the middle of the smoking variable range is a consequence of the smooth being near linear and the sum to zero identifiability constraint.*

computing the required matrices: for example use a pivoted Cholesky decomposition of the $\mathbf{S}$ matrix, or see section 4.2.4 (p. 172).

This equivalence between smooths, random effects and Gaussian random fields, is discussed in Kimeldorf and Wahba (1970) and Silverman (1985), for example. One consequence is that the computations required to estimate random effects can be used to estimate smooths (e.g., Verbyla et al., 1999; Ruppert et al., 2003), and conversely methods aimed at the smoothing problem can also be used to estimate Gaussian random effects.

5.8.1 *Gaussian Markov random fields*

Suppose that a region is divided into m distinct districts, indexed j, each with a corresponding parameter γ_j, giving the level of some quantity within the district. Let $\text{nei}(j)$ denote the indices of the districts neighbouring district j, and $\overline{\text{nei}}(j)$ denote the elements of $\text{nei}(j)$ for which $i > j$. Now suppose that we would like neighbouring districts to have similar γ_j estimates. One way to achieve this is to impose the penalty

$$J(\gamma) = \sum_{j=1}^{m} \sum_{i \in \overline{\text{nei}}(j)} (\gamma_j - \gamma_i)^2.$$

That is we sum the squared difference between γ_j values for all pairs of neighbouring districts, with each pair entering the summation only once. In that case

$$J(\gamma) = \boldsymbol{\gamma}^{\mathsf{T}}\mathbf{S}\boldsymbol{\gamma},$$

where $S_{ij} = -1$ if $i \in \text{nei}(j)$ and $S_{jj} = n_j$ where n_j is the number of districts neighbouring district j (not including district j itself). As with any of the smoothers covered here, the penalty can be viewed as being induced by an improper Gaussian prior

$$\gamma \sim N(\mathbf{0}, \tau \mathbf{S}^{-})$$

where τ is some precision parameter. So the γ and the neighbourhood structure can be viewed as an (intrinsic) *Gaussian Markov random field* (GMRF), with precision matrix $\mathbf{S}$. It's a *Markov* random field because the precision matrix is sparse.

If we observe n data by district then the GMRF can be viewed as a smoother with model matrix given by an $n \times m$ matrix $\mathbf{X}$ such that $X_{ij} = 1$ if observation i relates to district j and 0 otherwise. The natural parameterization of 5.4.2 can be truncated to produce a reduced rank version.

Terms like `s(loc,bs="mrf",xt=xt)` invoke such smooths in `mgcv`. If `xt` contain a list of polygons defining the district boundaries, then the neighbour relationships can be inferred from these, and they can be used for term plotting, but there are other alternatives for specifying the neighbour relations, and indeed the exact form of the penalty: see `?mrf`. Figure 5.19 illustrates such a smooth, for data on cancer of the larynx: see `?larynx` in the `gamair` package for fitting code. Rue and Held (2005) provide much more on modelling with GMRFs.

5.8.2 *Gaussian process regression smoothers*

Consider again the model $y_i = f(\mathbf{x}_i) + \epsilon_i$, where f is a random field (smooth function). Let $C(\mathbf{x}, \mathbf{x}_i) = c(\|\mathbf{x} - \mathbf{x}_i\|)$ be a non negative function such that $c(0) = 1$ and $c(d) \to 0$ monotonically as $d \to \infty$. Now consider representing f as

$$f(\mathbf{x}) = (1, \mathbf{x}^{\mathsf{T}})\boldsymbol{\beta} + \sum_i b_i C(\mathbf{x}, \mathbf{x}_i), \text{ where } \mathbf{b} \sim N(\mathbf{0}, (\lambda \mathbf{C})^{-1}).$$

$C_{ij} = C(\mathbf{x}_j, \mathbf{x}_i)$ and $\boldsymbol{\beta}$ is a vector of parameters. So if $\mathbf{f}$ is the vector such that $f_i = f(\mathbf{x}_i)$ and $\mathbf{B}$ is the matrix with i^{th} row $(1, \mathbf{x}_i^{\mathsf{T}})$ then $\mathbf{f} = \mathbf{B}^{\mathsf{T}}\boldsymbol{\beta} + \mathbf{C}\mathbf{b}$, and from basic properties of the transformation of covariance matrices the covariance matrix of $\mathbf{f}$ is $\mathbf{C}/\lambda$. Hence C is interpretable as the correlation function of the random field f, and the posterior mode of f is found by finding $\hat{\boldsymbol{\beta}}$, $\hat{\mathbf{b}}$ to minimize

$$\|\mathbf{y} - \mathbf{B}\boldsymbol{\beta} - \mathbf{C}\mathbf{b}\|^2/\sigma^2 + \lambda \mathbf{b}^{\mathsf{T}}\mathbf{C}\mathbf{b}. \tag{5.19}$$

Clearly (5.19) has exactly the structure required to use the rank reduction approaches of section 5.5.1, so these can be applied here too.

This approach is known as Gaussian process regression (or Kriging). Extending Handcock et al. (1994), Kammann and Wand (2003) propose $C(\mathbf{x}, \mathbf{x}_i) = c_0(\|\mathbf{x} - \mathbf{x}_i\|)$ where c_0 is the simplified Matérn correlation function

$$c_0(d) = (1 + d/\rho)\exp(-d/\rho).$$

Kammann and Wand (2003) suggest using $\rho = \max_{ij} \|\mathbf{x}_i - \mathbf{x}_j\|$.[‡‡] In `mgcv` such

[‡‡] Although for efficiency reasons, if using a rank reduced version, one should base this on the 'knots' used in basis creation, rather than the whole dataset.

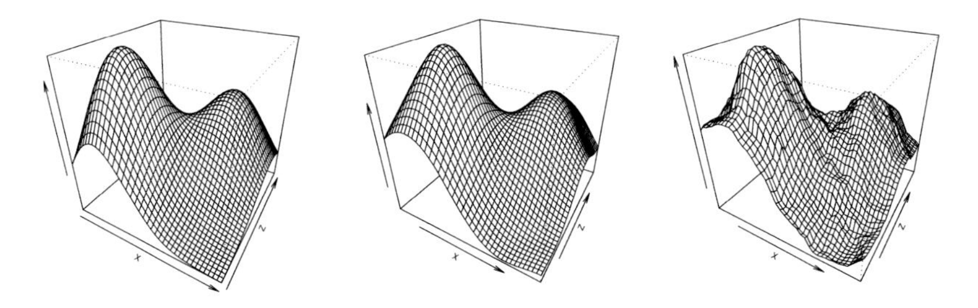

Figure 5.20 *Gaussian process reconstructions of the function shown on the left panel (range 0–0.7), from 300 noisy observations ($\sigma = 0.05$) randomly scattered over the $x - z$ plane. The middle panel uses a Matérn correlation function as suggested by Kammann and Wand (2003). The rougher right panel uses a spherical correlation function with $\rho = .05$.*

'Gaussian process smooths' are specified using terms like `s(x,z,bs="gp")`. The `m` argument can be used to control the form of c. `m[1]` selects from the following

1. The spherical correlation function

$$c(d) = \begin{cases} 1 - 1.5d/\rho + 0.5(d/\rho)^2 & \text{if } d \le \rho \\ 0 & \text{otherwise.} \end{cases}$$

2. $c(d) = \exp\{(-d/\rho)^\kappa\}$ where $0 < \kappa \le 2$. The 'power exponential'.
3. $c(d) = (1 + d/\rho)\exp(-d/\rho)$. Matérn with $\kappa = 1.5$.
4. $c(d) = \{1 + d/\rho + (d/\rho)^2/3\}\exp(-d/\rho)$. Matérn with $\kappa = 2.5$.
5. $c(d) = \{1 + d/\rho + 2(d/\rho)^2/5 + (d/\rho)^3/15\}\exp(-d/\rho)$. Matérn with $\kappa = 3.5$.

See Fahrmeir et al. (2013), section 8.1.3, for more details. If present and positive `m[2]` supplies ρ; otherwise the Kammann and Wand (2003) suggestion is used. `m[3]` supplies κ in the power exponential, if present; otherwise this defaults to 1. If `m` is not supplied then option 3 is used. Figure 5.20 illustrates options 3 and 1: the discontinuity in the spherical correlation function leads to quite rough results.

5.9 Choosing the basis dimension

When using penalized regression splines the modeller chooses the basis dimension as part of the model building process. Typically, this substantially reduces the computational burden of modelling, relative to full spline methods. It also recognizes the fact that, usually, something is seriously wrong if a statistical model really *requires* as many coefficients as there are data. Indeed we saw in section 5.2 that asymptotically the basis dimension need grow only rather slowly with sample size: for example $k = O(n^{1/5})$ for cubic regression splines estimated by REML. But, of course, the asymptotic results tell us nothing about what size of basis to use in any particular application: to know what is optimal we would have to know the 'truth' that we are trying to estimate.

In practice then, choice of basis dimension has to remain a part of model specification. But it is important to note that all the choice is doing is to set an upper limit on the flexibility of a term: the actual flexibility is controlled by the smoothing parameters. Hence, provided that we do not make the basis dimension too restrictively small, its exact value should have only a small effect on the fitted model. However, there is one subtle caveat: a function space with basis dimension 20 will contain a larger space of functions with EDF 5 than will a function space of dimension 10 (the numbers being arbitrary), so that model fit tends to retain some sensitivity to basis dimension, even if the appropriate EDF for a term is well below its basis dimension.

Fortunately informal checking that the basis dimension for a particular term is appropriate is quite easy. Suppose that a smooth term is a function of a covariate $\mathbf{x}$ (which may be vector in general). Let $\text{nei}(i)$ denote the set of indices of the m nearest neighbours of $\mathbf{x}_i$ according to an appropriate measure $\|\mathbf{x}_i - \mathbf{x}_j\|$, and compute the mean absolute or squared difference between the deviance residual ϵ_i and $\{\epsilon_j : j \in \text{nei}(i)\}$. Average this difference over i to get a single measure Δ of the difference between residuals and their neighbours. Now repeat the calculation several hundred times for randomly reshuffled residuals. If there is no residual pattern with respect to covariate $\mathbf{x}$ then Δ should look like an ordinary observation from the distribution of Δ under random re-shuffling, but if the original Δ is unusually small this means that residuals appear positively correlated with their neighbours, suggesting that we may have under-smoothed. In that case if the EDF is also close to the basis dimension then it may be appropriate to increase the basis dimension. Note, however, that un-modelled residual auto-correlation and an incorrectly specified mean variance relationship can both lead to the same result, even when the basis dimension is perfectly reasonable. This residual randomization test is easily automated and computationally efficient (given an efficient nearest neighbour algorithm). It is implemented by default in `gam.check` in `mgcv`.

An alternative to this simple check is to extract the deviance residuals for the full model and try smoothing them w.r.t. the covariates of the term being checked, *but using an increased basis dimension* (e.g., doubled). Smoothing can be done using an identity link, constant variance model. If the original term is really missing variation w.r.t. the covariates, then this approach will pick it up, as the estimated smooth of the residuals will be relatively complicated, rather than very smooth and nearly zero.

In summary, the modeller needs to decide roughly how large a basis dimension is fairly certain to provide just adequate flexibility, in any particular application, and use that. Informal and automatic residual checking can then help to check this assumption.

5.10 Generalized smoothing splines

Much smoothing spline theory is built around some very elegant and general methods based on the theory of reproducing kernel Hilbert spaces. The aim of this section is to give a brief introduction to this theory.

To reduce the level of abstraction, a little, let us revisit the cubic smoothing spline. To this end, first construct a space of functions of x, say, which are 'wiggly' according to the cubic spline penalty. That is, consider a space of functions, $\mathcal{F}$, where the inner product of two functions, f and $g \in \mathcal{F}$, is $\langle g, f \rangle = \int g''(x)f''(x)dx$, and consequently a norm on the space is the cubic spline penalty, $\int f''(x)^2 dx$: except for the zero function, functions for which this norm is zero are currently excluded from $\mathcal{F}$.

There is a rather remarkable theorem, the Riesz representation theorem, which says that there exists a function $R_z \in \mathcal{F}$, such that $f(z) = \langle R_z, f \rangle$, for any $f \in \mathcal{F}$: so if we want to evaluate f at some particular value z, then one way of doing it is to take the inner product of the function, f, with the 'representor of evaluation' function, R_z.

Suppose further that we construct a function of two variables $R(z, x)$, such that for any z, $R(z, x) = R_z(x)$. This function is known as the *reproducing kernel* of the space, since $\langle R(z, \cdot), R(\cdot, t) \rangle = R(z, t)$, i.e.,

$$\int \frac{\partial^2 R(z,x)}{\partial x^2} \frac{\partial^2 R(x,t)}{\partial x^2} dx = R(z,t),$$

by the Riesz representation theorem. Basically, if you take an appropriately defined inner product of R, with itself, you get back R: hence the terminology.

Now, consider the cubic smoothing spline problem of finding the *function* minimizing:

$$\sum_{i=1}^{n} \{y_i - f(x_i)\}^2 + \lambda \int f''(x)^2 dx. \tag{5.20}$$

It turns out that the minimizer is in the space of functions that can be represented as

$$\hat{f}(x) = \beta_0 + \beta_1 x + \sum_{i=1}^{n} \delta_i R(x_i, x),$$

where the β_j and δ_i are coefficients to be estimated. The first two terms on the r.h.s. simply span the space of functions for which $\int f''(x)^2 dx = 0$, the null space of the penalty: clearly β_0 and β_1 can be chosen to minimize the sum of squares term without worrying about the effect on the penalty. The terms in the summation represent the part of $\hat{f}(x)$ that is in $\mathcal{F}$. It is clear that not all functions in $\mathcal{F}$ can be represented as $\sum_{i=1}^{n} \delta_i R(x, x_i)$, so how do we know that the minimizer of 5.20 can?

The answer lies in writing the minimizer $\hat{f}$'s component in $\mathcal{F}$ as $r + \eta$ where $r = \sum_{i=1}^{n} \delta_i R(x_i, x)$, and η is any function in the part of $\mathcal{F}$ which is orthogonal to r. Orthogonality means that each inner product $\langle R(x_i, x), \eta(x) \rangle = 0$, but these inner products evaluate $\eta(x_i)$, so we have that $\eta(x_i) = 0$ for each i: so $\eta(x)$ can have no effect on the sum of squares term in (5.20). In the case of the penalty term, orthogonality of r and η means that:

$$\int \hat{f}''(x)^2 dx = \int r''(x)^2 dx + \int \eta''(x)^2 dx.$$

Obviously, the η which minimizes this is the zero function, $\eta(x) = 0$.

Having demonstrated what form the minimizer has, we must now actually find the minimizing β_j and δ_i. Given that we now have a basis, this is not difficult, although it is necessary to derive two linear constraints ensuring identifiability between the null space components and the reproducing kernel terms in the model, before this can be done.

The above argument is excessively complicated if all that is required is to derive the cubic smoothing spline, but its utility lies in how it generalizes. The cubic spline penalty can be replaced by any other derivative based penalty with associated inner product, and x can be a vector. In fact this *reproducing kernel Hilbert space* approach generalizes in all sorts of wonderful directions, and in contrast to some of the other methods described in this book, the basis for representing component functions of a model emerges naturally from the specification of fitting criteria like (5.20), rather than being a more or less arbitrary choice made by the modeller. See Wahba (1990) or Gu (2013) for a full treatment of this approach and Kim and Gu (2004) for efficient reduced rank methods.

5.11 Exercises

1. It is easy to see that the cubic regression spline defined by equation (5.3) in section 5.3.1 has value β_j and second derivative δ_j at knot x_j, and that value and second derivative are continuous across the knots. Show that the condition that the first derivative of the spline be continuous across x_j (and that the second derivative be zero at x_1 and x_k) leads to equation (5.4).
2. This question refers, again, to the cubic regression spline (5.3) of section 5.3.1.
   (a) Show that the second derivative of the spline can be expressed as
   $$f''(x) = \sum_{i=2}^{k-2} \delta_i d_i(x),$$
   where
   $$d_i(x) = \begin{cases} (x - x_i)/h_{i-1} & x_{i-1} \le x \le x_i \\ (x_{i+1} - x)/h_i & x_i \le x \le x_{i+1} \\ 0 & \text{otherwise.} \end{cases}$$
   (b) Hence show that, in the notation of section 5.3.1,
   $$\int f''(x)^2 dx = \boldsymbol{\delta}^{-\mathsf{T}} \mathbf{B} \boldsymbol{\delta}^{-}.$$
   (It may be helpful to review exercise 6 in Chapter 4, before proceeding with this part.)
   (c) Finally show that
   $$\int f''(x)^2 dx = \boldsymbol{\beta}^\mathsf{T} \mathbf{D}^\mathsf{T} \mathbf{B}^{-1} \mathbf{D} \boldsymbol{\beta}.$$
3. The natural parameterization of section 5.4.2 is particularly useful for understanding the way in which penalization causes bias in estimates, and this question explores this issue.

(a) Find an expression for the bias in a parameter estimator $\hat{\beta}_i''$ in the natural parameterization (bias being defined as $\mathbb{E}\{\hat{\beta}_i'' - \beta_i''\}$). What does this tell you about the bias in components of the model which are un-penalized, or only very weakly penalized, and in components for which the 'true value' of the corresponding parameter is zero or nearly zero?

(b) The mean square error in a parameter estimator (MSE) is defined as $\mathbb{E}\{(\hat{\beta}_i - \beta_i)^2\}$ (dropping the primes for notational convenience). Show that the MSE of the estimator is in fact the estimator variance plus the square of the estimator bias.

(c) Find an expression for the mean square error of the ith parameter of a smooth in the natural parameterization.

(d) Show that the lowest achievable MSE, for any natural parameter, is bounded above by σ^2, implying that penalization always has the potential to reduce the MSE of a parameter *if the right smoothing parameter value is chosen*. Comment on the proportion of the minimum achievable MSE that is contributed by the squared bias term, for different magnitudes of parameter value.

(e) Under the prior from the Bayesian model of section 5.8, but still working in the natural parameterization, show that the expected squared value of the ith parameter is $\sigma^2/(\lambda_i D_{ii})$. Using this result as a guide to the typical size of β_i^2, find an expression for the typical size of the squared bias in a parameter, and find a corresponding upper bound on the squared bias as a proportion of σ^2.

4. This question concerns the P-splines of section 5.3.3, the natural parameterization of section 5.4.2, and a related parameterization in terms of the eigen-basis of the penalty matrix.

(a) Write an R function which will return the model matrix and square root penalty matrix for a P-spline, given a sequence of knots, covariate vector `x`, the required basis dimension `q`, and the orders of B-spline and difference penalty required. (See `?splineDesign` from the `splines` library in R.)

(b) Simulate 100 uniform x data on $(0, 1)$, use your routine to evaluate the basis functions of a rank 9 B-spline basis at the x values, and plot these 9 basis functions. (Use cubic B-splines with a second order difference penalty.)

(c) Using the data and basis from the previous part, re-parameterize in terms of the eigen-basis of the P-spline penalty matrix, and plot the 9 evaluated basis functions corresponding to the re-parameterization. That is, given the eigen-decomposition $\mathbf{S} = \mathbf{U}\mathbf{\Lambda}\mathbf{U}^{\mathsf{T}}$, use the re-parameterization $\boldsymbol{\beta}' = \mathbf{U}^{\mathsf{T}}\boldsymbol{\beta}$.

(d) Re-parameterize the P-spline smoother using its natural parameterization (see section 5.4.2). Again plot the 9 evaluated basis functions corresponding to this parameterization.

5. The following R code simulates some data sampled with noise from a function of two covariates:

```
test1<-function(x,z,sx=0.3,sz=0.4)
{ 1.2*exp(-(x-0.2)^2/sx^2-(z-0.3)^2/sz^2)+
  0.8*exp(-(x-0.7)^2/sx^2-(z-0.8)^2/sz^2)
```

```
}
n <- 200
x <- matrix(runif(2*n),n,2)
f <- test1(x[,1],x[,2])
y <- f + rnorm(n)*.1
```

Write an R function to fit a thin plate spline of two variables (penalty order $m = 2$) to the simulated $\{y_i, \mathbf{x}_i\}$ data, given a smoothing parameter value. Write the function so that it can fit either a full thin plate spline or a 'knot based' thin plate regression spline (see the first and last subsections of section 5.5.1 for details). To deal with the linear constraints on the spline, see section 1.8.1 and `?qr`. The simple augmented linear model approach introduced in section 4.2.2 can be used for actual fitting. Write another function to evaluate the fitted spline at new covariate values. Use your functions to fit thin plate splines to the simulated data and produce contour plots of the results, for several different smoothing parameter values.

6. Read the `mgcv` help page `?smooth.construct`, and then implement a smooth constructor and prediction matrix method function for the piecewise linear smoother from chapter 4. Use it for the first example in `?gam`, with REML smoothing parameter selection.

Chapter 6

GAM Theory

In chapter 4 it was demonstrated how the problem of estimating a generalized additive model becomes the problem of estimating smoothing parameters and model coefficients for a penalized likelihood maximization problem, once a basis for the smooth functions has been chosen, together with associated measures of function wiggliness. In practice the penalized likelihood maximization problem is solved by penalized iteratively re-weighted least squares (PIRLS), while the smoothing parameters can be estimated using a separate criterion such as cross validation or REML. Chapter 5 introduced the wide variety of smoothers that can be represented using a linear basis expansion and quadratic penalty, and therefore fit into this framework.

The purpose of this chapter is to justify and extend the methods introduced in chapter 4. Section 6.1 covers model set-up and estimation given smoothing parameters. Section 6.2 then discusses smoothing parameter estimation criteria, while sections 6.3 to 6.9 introduce the computational methods for efficiently optimizing these criteria. The remainder of the chapter covers the results needed for further inference and model selection. Section 6.10 covers interval estimation and posterior simulation, while sections 6.11 and 6.12 cover model selection using AIC and hypothesis testing.

6.1 Setting up the model

A standard generalized additive model, as considered in chapter 4, has the form

$$g(\mu_i) = \mathbf{A}_i\gamma + \sum_j f_j(x_{ji}),\ y_i \sim \text{EF}(\mu_i, \phi)$$

where $\mathbf{A}_i$ is the ith row of a parametric model matrix, with corresponding parameters γ, f_j is a smooth function of (possibly vector) covariate x_j, and $\text{EF}(\mu_i, \phi)$ denotes an exponential family distribution with mean μ_i and scale parameter ϕ. The y_i are modelled as independent given μ_i.

A slight generalization is

$$g(\mu_i) = \mathbf{A}_i\gamma + \sum_j L_{ij} f_j(x_j),\ y_i \sim \text{EF}(\mu_i, \phi)$$

where L_{ij} is a bounded linear functional of f_j. The simplest example is that L_{ij} is an evaluation functional so that $L_{ij}f_j(x_j) = f_j(x_{ji})$ and we recover the basic GAM. Another example is $L_{ij}f_j(x_j) = z_i f_j(x_{ji})$, where z_i is a covariate: model terms like this are sometimes known as 'varying coefficient' terms (Hastie and Tibshirani, 1993) because $f_j(x_{ji})$ can be viewed as a regression coefficient for z which varies smoothly with x_j ('geographic regression' is another name, used when the arguments of f_j are geographic locations). A more exotic example is the 'signal regression' term $L_{ij}f_j(x_j) = \int f_j(x_j)k_i(x_j)dx_j$ where k_i is an observed function, such as a measured spectrum.

For either form of the model we choose smoothing bases and penalties for each f_j, implying model matrices $\mathbf{X}^{[j]}$ and penalties $\mathbf{S}^{[j]}$ (actually there may be multiple penalty matrices per f_j). If $b_{jk}(x)$ is the kth basis function for f_j then $X_{ik}^{[j]} = b_{jk}(x_{ji})$ for the basic model and $X_{ik}^{[j]} = L_{ij}b_{jk}(x_j)$ in the general case.

An identifiability constraint has to be applied to any smooth which contains $\mathbf{1}$ in the span of its $\mathbf{X}^{[j]}$; otherwise the smooth terms will be confounded with the intercept included in $\mathbf{A}$. The only exception is smooths for which the penalty term has no null space, so that $f_j \to 0$ as the corresponding smoothing parameter(s) tend to infinity (Gaussian random effects are an example of this). In practice we usually have to apply constraints to all smooths in the basic form of the model, but not necessarily all $L_{ij}f_j$ terms (for example varying coefficient terms usually do not require constraint).

Identifiability constraints, of the form $\sum_i f_j(x_{ji}) = 0$, are most usefully absorbed into the basis by reparameterization as described in section 5.4.1 (p. 211). Let $\boldsymbol{\mathcal{X}}^{[j]}$ and* $\boldsymbol{\mathcal{S}}_j$ denote the model matrix and penalty matrix for f_j after this reparameterization. We can then combine $\mathbf{A}$ and the $\boldsymbol{\mathcal{X}}^{[j]}$, column-wise, to create a whole model matrix

$$\mathbf{X} = (\mathbf{A} : \boldsymbol{\mathcal{X}}^{[1]} : \boldsymbol{\mathcal{X}}^{[2]} : \cdots).$$

The corresponding model coefficient vector $\boldsymbol{\beta}$ contains $\boldsymbol{\gamma}$ and the individual smooth term coefficient vectors stacked on end. A total smoothing penalty for the model can then be written as

$$\sum_j \lambda_j \boldsymbol{\beta}^{\mathsf{T}}\mathbf{S}_j\boldsymbol{\beta},$$

where λ_j is a smoothing parameter and $\mathbf{S}_j$ is simply $\boldsymbol{\mathcal{S}}_j$ embedded as a diagonal block in a matrix otherwise containing only zero entries, so that $\lambda_j\boldsymbol{\beta}^{\mathsf{T}}\mathbf{S}_j\boldsymbol{\beta}$ is the penalty for f_j. In practice there may be more than one $\mathbf{S}_j$ for each f_j, for example when f_j is a tensor product or adaptive smoother.

So our model has become the over-parameterized GLM

$$g(\mu_i) = \mathbf{X}_i\boldsymbol{\beta}, \; y_i \sim \text{EF}(\mu_i, \phi),$$

to be estimated by maximisation of

$$l_p(\boldsymbol{\beta}) = l(\boldsymbol{\beta}) - \frac{1}{2\phi}\sum_j \lambda_j \boldsymbol{\beta}^{\mathsf{T}}\mathbf{S}_j\boldsymbol{\beta}. \tag{6.1}$$

*The j subscript is a label here, rather than denoting row j.

The λ_j are smoothing parameters, controlling the trade-off between goodness of fit of the model and model smoothness. Following section 5.8 (p. 239) we can view the penalty as being induced by an improper Gaussian prior $\boldsymbol{\beta} \sim N(\mathbf{0}, \mathbf{S}_\lambda^- \phi)$ where $\mathbf{S}_\lambda = \sum_j \lambda_j \mathbf{S}_j$, with generalized inverse $\mathbf{S}_\lambda^-$. This also gives the estimation problem the structure of the GLMM estimation problem of section 3.4.1 (p. 148).

6.1.1 *Estimating $\boldsymbol{\beta}$ given $\boldsymbol{\lambda}$*

The objective (6.1) is immediately recognisable as the objective in the GLMM optimization problem (3.18) in section 3.4.1, with $\mathbf{X}$ in (6.1) taking the part of $\boldsymbol{\mathcal{X}}$ in section 3.4.1. Hence we can maximize (6.1) via the penalized iteratively re-weighted least squares (PIRLS) iteration:

1. Initialize $\hat{\mu}_i = y_i + \delta_i$ and $\hat{\eta}_i = g(\hat{\mu}_i)$, where δ_i is usually zero, but may be a small constant ensuring that $\hat{\eta}_i$ is finite. Iterate the following two steps to convergence.
2. Compute pseudodata $z_i = g'(\hat{\mu}_i)(y_i - \hat{\mu}_i)/\alpha(\hat{\mu}_i) + \hat{\eta}_i$, and iterative weights $w_i = \alpha(\hat{\mu}_i)/\{g'(\hat{\mu}_i)^2 V(\hat{\mu}_i)\}$.
3. Find $\hat{\boldsymbol{\beta}}$ to minimize the weighted least squares objective

$$\|\mathbf{z} - \mathbf{X}\boldsymbol{\beta}\|_W^2 + \sum_j \lambda_j \boldsymbol{\beta}^\mathsf{T} \mathbf{S}_j \boldsymbol{\beta}$$

   and then update $\hat{\boldsymbol{\eta}} = \mathbf{X}\hat{\boldsymbol{\beta}}$ and $\hat{\mu}_i = g^{-1}(\hat{\eta}_i)$.

Recall that $\|a\|_W^2 = \mathbf{a}^\mathsf{T}\mathbf{W}\mathbf{a}$ and (from section 3.1.2, p. 105) that $V(\mu)$ is the variance function determined by the exponential family distribution (or defining the quasi-likelihood), while $\alpha(\mu_i) = [1 + (y_i - \mu_i)\{V'(\mu_i)/V(\mu_i) + g''(\mu_i)/g'(\mu_i)\}]$. Alternatively we can use a 'Fisher scoring' approach in which the Hessian of the log likelihood is replaced by its expectation, corresponding to setting $\alpha(\mu_i) = 1$.

6.1.2 *Degrees of freedom and scale parameter estimation*

Following on with the link between GAMs and GLMMs and identifying the $\lambda_j \mathbf{S}_j/\phi$ with random effects precision† matrices, then section 3.4.2 (p. 149) shows that an appropriate REML estimator of the scale parameter is

$$\hat{\phi} = \frac{\|\mathbf{z} - \mathbf{X}\hat{\boldsymbol{\beta}}\|_W^2}{n - \tau} \tag{6.2}$$

where

$$\tau = \text{tr}\{(\mathbf{X}^\mathsf{T}\mathbf{W}\mathbf{X} + \mathbf{S}_\lambda)^{-1}\mathbf{X}^\mathsf{T}\mathbf{W}\mathbf{X}\}, \tag{6.3}$$

and $\mathbf{S}_\lambda = \sum_j \lambda_j \mathbf{S}_j$. τ can hence be interpreted as the *effective degrees of freedom* of the model, which also coincides with the notion of degrees of freedom discussed in section 5.4.2 (p. 211), but with $\mathbf{F} = (\mathbf{X}^\mathsf{T}\mathbf{W}\mathbf{X} + \mathbf{S}_\lambda)^{-1}\mathbf{X}^\mathsf{T}\mathbf{W}\mathbf{X}$ accounting for the weights. For a given $\mathbf{z}$ and $\mathbf{W}$, $\mathbf{F}$ again has the interpretation of being the matrix

† (pseudo)inverse covariance matrices.

mapping the un-penalized coefficient estimates to the penalized coefficient estimates (for the working linear model), so that its trace is effectively the average shrinkage undergone by the coefficients, multiplied by the number of coefficients. Effective degrees of freedom for individual smooth terms are obtained by summing the F_{ii} values corresponding to the coefficients β_i of the smooth term (e.g., if the coefficients of a particular smooth are $\beta_5, \ldots, \beta_{11}$ then the smooth's EDF is $\sum_{i=5}^{11} F_{ii}$).

Note that $\|\mathbf{z} - \mathbf{X}\hat{\beta}\|_W^2$ corresponds to the Pearson statistic, so that the REML estimate, $\hat{\phi}$, is the Pearson estimate of the scale parameter and hence susceptible to the problems discussed in section 3.1.5 (p. 110): it is therefore somewhat safer to correct the estimator using (3.11).

An alternative definition of the effective degrees of freedom is sometimes useful. For ease of presentation, consider the Gaussian additive model case where the influence matrix is $\mathbf{A} = \mathbf{X}(\mathbf{X}^\mathsf{T}\mathbf{X} + \mathbf{S}_\lambda)^{-1}\mathbf{X}^\mathsf{T}$ and $\mathbf{F} = (\mathbf{X}^\mathsf{T}\mathbf{X} + \mathbf{S}_\lambda)^{-1}\mathbf{X}^\mathsf{T}\mathbf{X}$. Now consider the expected residual sum of squares for this model

$$\mathbb{E}\left(\|\mathbf{y} - \mathbf{A}\mathbf{y}\|^2\right) = \sigma^2 \left\{n - 2\text{tr}\left(\mathbf{A}\right) + \text{tr}\left(\mathbf{A}\mathbf{A}\right)\right\} + \mathbf{b}^\mathsf{T}\mathbf{b} \tag{6.4}$$

where $\mathbf{b} = \boldsymbol{\mu} - \mathbf{A}\boldsymbol{\mu}$ represents the smoothing bias, which can be estimated as $\hat{\mathbf{b}} = \hat{\boldsymbol{\mu}} - \mathbf{A}\hat{\boldsymbol{\mu}}$. This leads to the alternative variance/scale estimator

$$\hat{\sigma}^2 = \frac{\|\mathbf{y} - \mathbf{A}\mathbf{y}\|^2 - \hat{\mathbf{b}}^\mathsf{T}\hat{\mathbf{b}}}{n - 2\text{tr}\left(\mathbf{A}\right) + \text{tr}\left(\mathbf{A}\mathbf{A}\right)},$$

and to taking $\tau_1 = 2\text{tr}\left(\mathbf{A}\right) - \text{tr}\left(\mathbf{A}\mathbf{A}\right) = 2\text{tr}\left(\mathbf{F}\right) - \text{tr}\left(\mathbf{F}\mathbf{F}\right)$ as the effective degrees of freedom of the model. Term-specific effective degrees of freedom are obtained by summing up the corresponding elements of diag$(2\mathbf{F} - \mathbf{F}\mathbf{F})$.

Another route to τ_1 is to consider the bias corrected fitted values: $\tilde{\boldsymbol{\mu}} = \hat{\boldsymbol{\mu}} + \hat{\mathbf{b}} = 2\hat{\boldsymbol{\mu}} - \mathbf{A}\hat{\boldsymbol{\mu}} = (2\mathbf{A} - \mathbf{A}\mathbf{A})\mathbf{y}$. Since $2\mathbf{A} - \mathbf{A}\mathbf{A}$ is the 'bias corrected' influence matrix, then we might take its trace, τ_1, as the effective degrees of freedom of the bias corrected model. This view of the effective degrees of freedom will prove useful for computing p-values for smooths, in section 6.12.1.

6.1.3 *Stable least squares with negative weights*

This section can safely be omitted at a first reading. If some weights are negative then the obvious approach to solving the working penalized least squares problem in the PIRLS of 6.1.1 is to directly solve[‡]

$$(\mathbf{X}^\mathsf{T}\mathbf{W}\mathbf{X} + \mathbf{S}_\lambda)\hat{\boldsymbol{\beta}} = \mathbf{X}^\mathsf{T}\mathbf{W}\mathbf{z} \tag{6.5}$$

for $\hat{\boldsymbol{\beta}}$, where $\mathbf{W} = \text{diag}(w_i)$, $\mathbf{z}$ is the vector of pseudodata and $\mathbf{S}_\lambda = \sum_j \lambda_j \mathbf{S}_j$. However, if $\mathbf{X}$ is badly scaled (ill-conditioned) then $\mathbf{X}^\mathsf{T}\mathbf{W}\mathbf{X}$ will be even more so, and as in section 1.3.1 (p.12), we would prefer to use orthogonal methods. The difficulty is that negative weights mean that we can not simply QR decompose $\sqrt{\mathbf{W}}\mathbf{X}$.

[‡]That is, form the Cholesky decomposition $\mathbf{R}^\mathsf{T}\mathbf{R} = \mathbf{X}^\mathsf{T}\mathbf{W}\mathbf{X} + \mathbf{S}_\lambda$ and then solve the triangular systems $\mathbf{R}^\mathsf{T}\boldsymbol{\alpha} = \mathbf{X}^\mathsf{T}\mathbf{W}\mathbf{z}$ for $\boldsymbol{\alpha}$ followed by $\mathbf{R}\hat{\boldsymbol{\beta}} = \boldsymbol{\alpha}$ for $\hat{\boldsymbol{\beta}}$.

To make progress, let $\mathbf{W}^-$ denote the diagonal matrix such that $W^-_{ii} = 0$ if $w_i \geq 0$ and $-w_i$ otherwise. Also let $\bar{\mathbf{W}}$ be a diagonal matrix with $\bar{W}_{ii} = |w_i|$. Then

$$\mathbf{X}^\mathsf{T}\mathbf{W}\mathbf{X} = \mathbf{X}^\mathsf{T}\bar{\mathbf{W}}\mathbf{X} - 2\mathbf{X}^\mathsf{T}\mathbf{W}^-\mathbf{X}.$$

So $\mathbf{X}^\mathsf{T}\mathbf{W}\mathbf{X}$ has been split into a component that is straightforward to compute with stably, and a 'correction' term. Starting with the straightforward term, perform a QR decomposition without pivoting,

$$\sqrt{\bar{\mathbf{W}}}\mathbf{X} = \mathcal{Q}\mathcal{R}. \tag{6.6}$$

At this stage a method designed for general use has to test for any lack of identifiability in the model, and deal with it. The test must take account of the penalties, since these can act as regularisers on the problem: for example, many quite reasonable random effects formulations are unidentifiable if treated as fixed effects. First use a pivoted Cholesky decomposition to find $\bar{\mathbf{E}}$, such that

$$\bar{\mathbf{E}}^\mathsf{T}\bar{\mathbf{E}} = \sum_i \mathbf{S}_i/\|\mathbf{S}_i\|_F.$$

The scaling of each component of $\mathbf{S}_\lambda$ by its Frobenius norm,[§] is simply designed to give even scaling of the component matrices, given that the structural identifiability problems of interest are not smoothing parameter dependent. Now, using the factor $\mathcal{R}$, from (6.6), scaled by its Frobenius norm, form a pivoted QR decomposition

$$\begin{bmatrix} \mathcal{R}/\|\mathcal{R}\|_F \\ \bar{\mathbf{E}}/\|\bar{\mathbf{E}}\|_F \end{bmatrix} = \bar{\mathbf{Q}}\bar{\mathbf{R}}$$

and determine the rank, r, of the problem from the pivoted $q \times q$ triangular factor $\bar{\mathbf{R}}$ (see Cline et al., 1979). If $r < q$ then the parameters corresponding to the columns pivoted to the last $q-r$ positions are unidentifiable, and the corresponding columns of $\mathcal{R}$ and rows and columns of the $\mathbf{S}_j$ must be dropped. The unidentifiable parameters themselves can be set to zero.

Having dealt with any identifiability problem, let $\mathbf{E}$ be a matrix such that $\mathbf{E}^\mathsf{T}\mathbf{E} = \mathbf{S}_\lambda$ (any columns of $\mathbf{E}$ corresponding to unidentifiable parameters are dropped), and form a further pivoted QR decomposition

$$\begin{bmatrix} \mathcal{R} \\ \mathbf{E} \end{bmatrix} = \mathbf{Q}\mathbf{R}. \tag{6.7}$$

$\mathbf{R}$ is a matrix square root of $\mathbf{X}^\mathsf{T}\bar{\mathbf{W}}\mathbf{X}+\mathbf{S}$. Now define $n \times r$ matrix $\mathbf{Q}_1 = \mathcal{Q}\mathbf{Q}[1:q,]$, where q is the number of columns of $\mathbf{X}$ and $\mathbf{Q}[1:q,]$ denotes the first q rows of $\mathbf{Q}$. Hence

$$\sqrt{\bar{\mathbf{W}}}\mathbf{X} = \mathbf{Q}_1\mathbf{R}. \tag{6.8}$$

For what follows, the pivoting used in the QR step (6.7) is treated as having been applied to rows and columns of $\mathbf{S}_j$, the columns of $\mathbf{X}$ and to $\boldsymbol{\beta}$.

[§]The square root of the sum of squares of its elements.

Now we need to 'correct' $\mathbf{R}$ to account for the negative weights:

$$\begin{aligned}\mathbf{X}^{\mathsf{T}}\mathbf{W}\mathbf{X}+\mathbf{S}_{\lambda} &= \mathbf{R}^{\mathsf{T}}\mathbf{R}-2\mathbf{X}^{\mathsf{T}}\mathbf{W}^{-}\mathbf{X}\\ &= \mathbf{R}^{\mathsf{T}}(\mathbf{I}-2\mathbf{R}^{-\mathsf{T}}\mathbf{X}^{\mathsf{T}}\mathbf{W}^{-}\mathbf{X}\mathbf{R}^{-1})\mathbf{R}\\ &= \mathbf{R}^{\mathsf{T}}(\mathbf{I}-2\mathbf{R}^{-\mathsf{T}}\mathbf{R}^{\mathsf{T}}\mathbf{Q}_1^{\mathsf{T}}\mathbf{I}^{-}\mathbf{Q}_1\mathbf{R}\mathbf{R}^{-1})\mathbf{R}\\ &= \mathbf{R}^{\mathsf{T}}(\mathbf{I}-2\mathbf{Q}_1^{\mathsf{T}}\mathbf{I}^{-}\mathbf{Q}_1)\mathbf{R},\end{aligned}$$

where $\mathbf{I}^{-}$ denotes the diagonal matrix such that $I^{-}_{ii}=0$ if $w_i>0$ and 1 otherwise, while $\mathbf{W}^{-}=\mathbf{I}^{-}\bar{\mathbf{W}}$. The matrix $\mathbf{I}-2\mathbf{Q}_1^{\mathsf{T}}\mathbf{I}^{-}\mathbf{Q}_1$ is not necessarily positive semi-definite, and so requires careful handling. Letting $\bar{\mathbf{I}}$ denote $\mathbf{I}^{-}$ with the zero rows dropped and forming the singular value decomposition

$$\bar{\mathbf{I}}\mathbf{Q}_1=\mathbf{U}\mathbf{D}\mathbf{V}^{\mathsf{T}} \tag{6.9}$$

we obtain

$$\mathbf{X}^{\mathsf{T}}\mathbf{W}\mathbf{X}+\mathbf{S}_{\lambda}=\mathbf{R}^{\mathsf{T}}(\mathbf{I}-2\mathbf{V}\mathbf{D}^2\mathbf{V}^{\mathsf{T}})\mathbf{R}=\mathbf{R}^{\mathsf{T}}\mathbf{V}(\mathbf{I}-2\mathbf{D}^2)\mathbf{V}^{\mathsf{T}}\mathbf{R} \tag{6.10}$$

(and additionally $\mathbf{X}^{\mathsf{T}}\mathbf{W}\mathbf{X}=\mathcal{R}^{\mathsf{T}}\mathcal{R}-2\mathbf{R}^{\mathsf{T}}\mathbf{V}\mathbf{D}^2\mathbf{V}^{\mathsf{T}}\mathbf{R}$). Now define

$$\mathbf{P}=\mathbf{R}^{-1}\mathbf{V}(\mathbf{I}-2\mathbf{D}^2)^{-1/2}\quad\text{and}\quad\mathbf{K}=\mathbf{Q}_1\mathbf{V}(\mathbf{I}-2\mathbf{D}^2)^{-1/2}. \tag{6.11}$$

If $\bar{\mathbf{z}}$ is the vector such that $\bar{z}_i=z_i$ if $w_i\geq 0$ and $-z_i$ otherwise, then substituting from (6.11), (6.10) and (6.8) into (6.5) and solving gives

$$\hat{\boldsymbol{\beta}}=\mathbf{P}\mathbf{K}^{\mathsf{T}}\sqrt{\bar{\mathbf{W}}}\bar{\mathbf{z}}.$$

This last computation has a condition number close to that of $\mathbf{R}$, which is approximately the square root of the condition number for using $\mathbf{X}^{\mathsf{T}}\mathbf{W}\mathbf{X}+\mathbf{S}_{\lambda}$ directly. For smoothing parameter selection via REML we will also need $|\mathbf{X}^{\mathsf{T}}\mathbf{W}\mathbf{X}+\mathbf{S}_{\lambda}|=|\mathbf{R}|^2|\mathbf{I}-2\mathbf{D}^2|$, and note also that $(\mathbf{X}^{\mathsf{T}}\mathbf{W}\mathbf{X}+\mathbf{S}_{\lambda})^{-1}=\mathbf{P}\mathbf{P}^{\mathsf{T}}$, which is the Bayesian covariance matrix for $\boldsymbol{\beta}$ (see section 6.10).

One important detail remains. At the penalized MLE, $\mathbf{X}^{\mathsf{T}}\mathbf{W}\mathbf{X}+\mathbf{S}_{\lambda}$ will be positive semi-definite, so that $d_i\leq 1/\sqrt{2}$, but at earlier PIRLS steps there is no *guarantee* that the penalized likelihood is positive semi-definite. Hence, if $d_i^2>1/2$, for any i, then a Fisher step should be substituted. That is, set $\alpha_i=1$, so that $w_i\geq 0$ $\forall i$. Then

$$\mathbf{P}=\mathbf{R}^{-1}\quad\text{and}\quad\mathbf{K}=\mathbf{Q}_1$$

and the expression for $\hat{\boldsymbol{\beta}}$, above, simplifies to $\hat{\boldsymbol{\beta}}=\mathbf{P}\mathbf{K}^{\mathsf{T}}\sqrt{\mathbf{W}}\mathbf{z}$, while $|\mathbf{X}^{\mathsf{T}}\mathbf{W}\mathbf{X}+\mathbf{S}_{\lambda}|=|\mathbf{R}|^2$.

At the end of model fitting, $\hat{\boldsymbol{\beta}}$ will need to have the pivoting applied at (6.7) reversed, and the elements of $\hat{\boldsymbol{\beta}}$ that were dropped by the truncation step after (6.6) will have to be re-inserted as zeroes. The covariance matrix $\mathbf{P}\mathbf{P}^{\mathsf{T}}$ will also have to be unpivoted and have zero rows and columns inserted for any unidentifiable coefficients. The leading order cost of the method is the $O(nq^2)$ of the first QR decomposition.

6.2 Smoothness selection criteria

So far we have considered estimation of $\boldsymbol{\beta}$ given the smoothing parameters $\boldsymbol{\lambda}$, but we need to estimate these as well, and this constitutes the most challenging part of model estimation. As introduced in sections 4.2.2 (p. 166) and 4.2.4 (p. 172), there are two classes of method in general use: prediction error methods, such as GCV and AIC, or marginal likelihood methods based on the Bayesian/mixed model view of smoothing. For either class of method there are then two main alternative computational strategies: either the smoothness selection criterion is defined for the model itself, and optimized directly, or the smoothness selection criterion is defined for the working model in the PIRLS fitting iteration, and applied to that working model at each PIRLS step. The latter strategy is the one used by the PQL approach of section 3.4.2: such methods are not guaranteed to converge, but can be very fast, especially when the data and model are large.

6.2.1 *Known scale parameter: UBRE*

Consider estimating smoothing parameters in the simple case of an additive model for data with constant known variance. An appealing approach is to try to ensure that $\hat{\boldsymbol{\mu}}$ is as close as possible to the true $\boldsymbol{\mu} \equiv \mathbb{E}(\mathbf{y})$. An appropriate measure of this proximity might be M, the expected mean square error (MSE) of the model, and in section 1.8.6 (p. 52), the argument¶ leading to (1.13) implies that:

$$M = \mathbb{E}\left(\|\boldsymbol{\mu} - \mathbf{X}\hat{\boldsymbol{\beta}}\|^2/n\right) = \mathbb{E}\left(\|\mathbf{y} - \mathbf{A}\mathbf{y}\|^2\right)/n - \sigma^2 + 2\text{tr}\left(\mathbf{A}\right)\sigma^2/n. \tag{6.12}$$

It seems reasonable to choose smoothing parameters in order to minimize an estimate of M, the un-biased risk estimator (UBRE; Craven and Wahba, 1979),

$$\mathcal{V}_u(\boldsymbol{\lambda}) = \|\mathbf{y} - \mathbf{A}\mathbf{y}\|^2/n - \sigma^2 + 2\text{tr}\left(\mathbf{A}\right)\sigma^2/n, \tag{6.13}$$

which is also Mallows' C_p (Mallows, 1973). Note that the r.h.s. of (6.13) depends on the smoothing parameters through $\mathbf{A}$.

If σ^2 is known then estimating $\boldsymbol{\lambda}$ by minimizing $\mathcal{V}_u$ works well, but problems arise if σ^2 has to be estimated. For example, substituting the approximation

$$\mathbb{E}\left(\|\mathbf{y} - \mathbf{A}\mathbf{y}\|^2\right) = \sigma^2\{n - \text{tr}\left(\mathbf{A}\right)\}, \tag{6.14}$$

implied by (6.2), into (6.12) yields

$$M = \mathbb{E}\left(\|\boldsymbol{\mu} - \mathbf{X}\hat{\boldsymbol{\beta}}\|^2/n\right) = \frac{\text{tr}\left(\mathbf{A}\right)}{n}\sigma^2 \tag{6.15}$$

and the MSE estimator $\tilde{M} = \text{tr}\left(\mathbf{A}\right)\hat{\sigma}^2/n$. Now consider comparison of 1 and 2 parameter unpenalized models using $\tilde{M}$: the 2 parameter model has to reduce $\hat{\sigma}^2$ to less than half the one parameter σ^2 estimate before it would be judged to be an improvement. Clearly, therefore, $\tilde{M}$ is not a suitable basis for model selection.

¶Which makes no assumptions that are invalidated by *penalized* least squares estimation.

6.2.2 Unknown scale parameter: Cross validation

As we have seen, naïvely attempting to minimize the average square error in model predictions of $\mathbb{E}(\mathbf{y})$ will not work well when σ^2 is unknown. An alternative is to base smoothing parameter estimation on mean square *prediction error*: that is, on the average squared error in predicting a new observation y using the fitted model. The expected mean square prediction error is readily shown to be

$$P = \sigma^2 + M.$$

The direct dependence on σ^2 tends to mean that criteria based on P are much more resistant to over-smoothing, which would inflate the σ^2 estimate, than are criteria based on M alone.

The most obvious way to estimate P is to use cross validation (e.g., Stone, 1974). By omitting a datum, y_i, from the model fitting process, it becomes independent of the model fitted to the remaining data. Hence the squared error in predicting y_i is readily estimated, and by omitting all data in turn we arrive at the ordinary cross validation estimate of P, given in section 4.2.3 (p. 169):

$$\mathcal{V}_o = \frac{1}{n}\sum_{i=1}^{n}(y_i - \hat{\mu}_i^{[-i]})^2$$

where $\mu_i^{[-i]}$ denotes the prediction of $\mathbb{E}(y_i)$ obtained from the model fitted to all data except y_i.

Fortunately it is unnecessary to calculate $\mathcal{V}_o$ by performing n model fits, to obtain the n terms $\hat{\mu}_i^{[-i]}$. To see this, first consider the penalized least squares objective which in principle has to be minimized to find the ith term in the OCV score:

$$\sum_{\substack{j=1\\ j\neq i}}^{n}(y_j - \hat{\mu}_j^{[-i]})^2 + \text{ Penalties.}^{\|}$$

Clearly, adding zero to this objective will leave the estimates that minimize it completely unchanged. So we can add the term $(\hat{\mu}_i^{[-i]} - \hat{\mu}_i^{[-i]})^2$ to obtain

$$\sum_{j=1}^{n}(y_j^* - \hat{\mu}_j^{[-i]})^2 + \text{ Penalties,} \tag{6.16}$$

where $\mathbf{y}^* = \mathbf{y} - \bar{\mathbf{y}}^{[i]} + \bar{\boldsymbol{\mu}}^{[i]}$: $\bar{\mathbf{y}}^{[i]}$ and $\bar{\boldsymbol{\mu}}^{[i]}$ are vectors of zeroes except for their i^{th} elements which are y_i and $\hat{\mu}_i^{[-i]}$, respectively.

Fitting by minimizing (6.16) obviously results in i^{th} prediction $\hat{\mu}_i^{[-i]}$, and also in an influence matrix $\mathbf{A}$, which is just the influence matrix for the model fitted to all the data (since (6.16) has the structure of the fitting objective for the model of the whole data). So, considering the ith prediction, we have that,

$$\hat{\mu}_i^{[-i]} = \mathbf{A}_i\mathbf{y}^* = \mathbf{A}_i\mathbf{y} - A_{ii}y_i + A_{ii}\hat{\mu}_i^{[-i]} = \hat{\mu}_i - A_{ii}y_i + A_{ii}\hat{\mu}_i^{[-i]},$$

$^{\|}$The penalties do not depend on which observations are included in the sum of squares term.

where $\hat{\mu}_i$ is from the fit to the full $\mathbf{y}$. Subtraction of y_i from both sides and a little rearrangement then yields

$$y_i - \hat{\mu}_i^{[-i]} = (y_i - \hat{\mu}_i)/(1 - A_{ii}),$$

so that the OCV score becomes

$$\mathcal{V}_o = \frac{1}{n}\sum_{i=1}^{n} \frac{(y_i - \hat{\mu}_i)^2}{(1 - A_{ii})^2}, \tag{6.17}$$

which can clearly be calculated from a single fit of the original model. Stone (1977) demonstrates the asymptotic equivalence of cross validation and AIC.

Leave-several-out cross validation

Leave-several-out cross validation can also be used, and again only a single model fit is needed for the computations. Suppose that we want to build a cross validation objective on the basis of leaving out subsets of data. Let α be the set of indices of one such subset, so that $\hat{\boldsymbol{\mu}}_\alpha^{[-\alpha]}$ contains the predictions of the points indexed by α, when $\mathbf{y}_\alpha$ has been omitted from fitting. Starting by defining $\mathbf{y}^* = \mathbf{y} - \bar{\mathbf{y}}^{[\alpha]} + \bar{\boldsymbol{\mu}}^{[\alpha]}$ the argument follows the same lines as in the leave-one-out case to arrive at the identity

$$\mathbf{y}_\alpha - \hat{\boldsymbol{\mu}}_\alpha^{[-\alpha]} = (\mathbf{I} - \mathbf{A}_{\alpha\alpha})^{-1}(\mathbf{y}_\alpha - \hat{\boldsymbol{\mu}}_\alpha),$$

from which a leave-several-out cross validation score can be constructed. $\mathbf{A}_{\alpha\alpha}$ is the matrix consisting of rows and columns α of $\mathbf{A}$.

Problems with ordinary cross validation

OCV is a reasonable way of estimating smoothing parameters, but suffers from two potential drawbacks. Firstly, it is computationally expensive to minimize in the additive model case, where there may be several smoothing parameters. Secondly, it has a slightly disturbing lack of invariance (see Golub et al., 1979; Wahba, 1990, p. 53).

To appreciate the invariance problem, consider the additive model fitting objective,

$$\|\mathbf{y} - \mathbf{X}\boldsymbol{\beta}\|^2 + \sum_{i=1}^{m} \lambda_i \boldsymbol{\beta}^\mathsf{T}\mathbf{S}_i\boldsymbol{\beta},$$

again. Given smoothing parameters, all inferences about $\boldsymbol{\beta}$ made on the basis of minimizing this objective are identical to the inferences that would be made by using the alternative objective,

$$\|\mathbf{Q}\mathbf{y} - \mathbf{Q}\mathbf{X}\boldsymbol{\beta}\|^2 + \sum_{i=1}^{m} \lambda_i \boldsymbol{\beta}^\mathsf{T}\mathbf{S}_i\boldsymbol{\beta},$$

where $\mathbf{Q}$ is any orthogonal matrix of appropriate dimension. However, the two objectives generally give rise to *different* OCV scores.

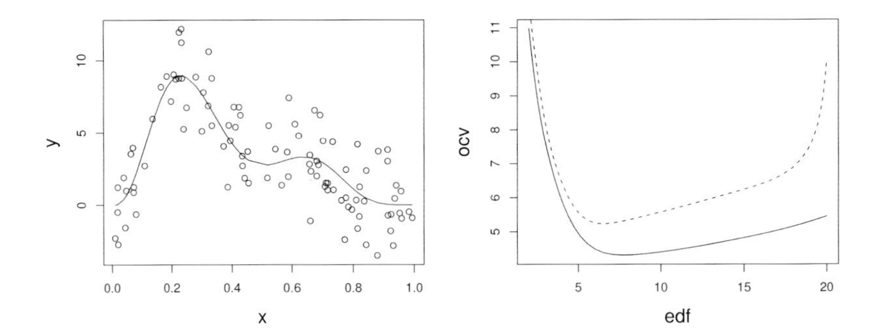

Figure 6.1 *Lack of invariance of ordinary cross validation. The left panel shows a smooth function (continuous line) and some data sampled from it at random x values, with noise added in the y direction. The right panel shows OCV scores against estimated degrees of freedom for a rank 20 penalized regression spline fitted to the data: the continuous curve is the OCV score for the model fitted using the original objective; the dashed line shows the OCV score for the model fitted by an alternative objective. Both objectives result in identical inferences about the model coefficients for any given smoothing parameters, so the difference in OCV score is somewhat unsatisfactory. GCV and UBRE do not suffer from this problem.*

Figure 6.1 illustrates this problem for a smooth of x, y data. The right hand side of figure 6.1 shows $\mathcal{V}_o$ plotted against effective degrees of freedom, for the same rank 20 penalized regression spline fitted to the data in the left panel: the continuous curve is $\mathcal{V}_o$ corresponding to the original objective, while the dashed curve is $\mathcal{V}_o$ for the orthogonally transformed objective. For this example, $\mathbf{Q}$ was obtained from the QR decomposition of $\mathbf{X}$, but similar differences occur for arbitrary orthogonal $\mathbf{Q}$ matrices. Note how the 'optimal' degrees of freedom differ between the two OCV scores, and that this occurs despite the fact that the fitted models at any particular EDF are identical.

6.2.3 Generalized cross validation

The problems with ordinary cross validation arise because, despite parameter estimates, effective degrees of freedom and expected prediction error being invariant to rotation of $\mathbf{y} - \mathbf{X}\boldsymbol{\beta}$ by any orthogonal matrix $\mathbf{Q}$, the elements A_{ii}, on the leading diagonal of the influence matrix, are not invariant and neither are the individual terms in the summation in (6.17). This sensitivity to an essentially arbitrary choice about how fitting is done is unsatisfactory, but what can be done to improve matters?

One approach is to consider what might make for 'good' or 'bad' rotations of $\mathbf{y} - \mathbf{X}\boldsymbol{\beta}$ and to decide to perform cross validation on a particularly nicely rotated problem. One thing that would appear to be undesirable is to base cross validation on data in which a few points have very high leverage relative to the others. That is, highly uneven A_{ii} values are undesirable, as they will tend to cause the cross vali-

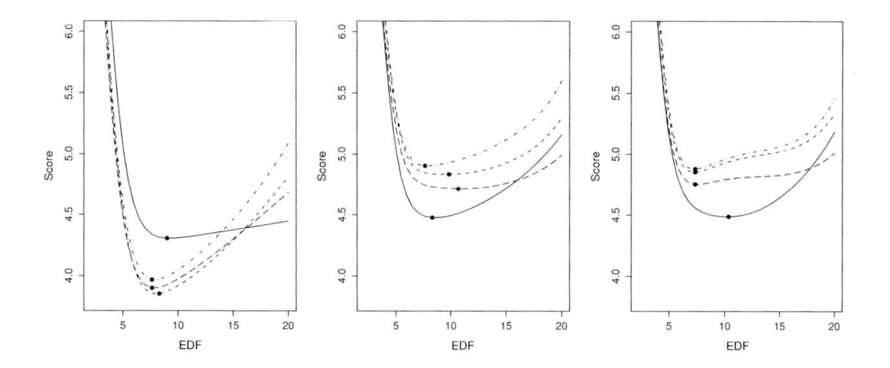

Figure 6.2 *Comparison of different smoothing parameter estimation criteria for rank 20 penalized regression splines fitted to three replicates of the simulated data shown in figure 6.1. The curves are as follows (in ascending order at the 20 EDF end of the left hand panel): the continuous curve is observed mean square error + known σ^2; the long dashed curve is $\mathcal{V}_u + \sigma^2$; the short dashed curve is $\mathcal{V}_g$, the GCV criterion; the alternating dashes and dots show $\mathcal{V}_o$, the OCV criterion. The dot on each curve shows the location of its minimum.*

dation score (6.17) to be dominated by a small proportion of the data. This suggests choosing the rotation $\mathbf{Q}$ in order to make the A_{ii} as even as possible.

In fact it is possible to choose $\mathbf{Q}$ in order to make all the A_{ii} equal. To see this note that if $\mathbf{A}$ is the influence matrix for the original problem, then the influence matrix for the rotated problem is

$$\mathbf{A}_Q = \mathbf{Q}\mathbf{A}\mathbf{Q}^{\mathsf{T}}$$

but if $\mathbf{B}$ is any matrix such that $\mathbf{B}\mathbf{B}^{\mathsf{T}} = \mathbf{A}$ then the influence matrix can be written:

$$\mathbf{A}_Q = \mathbf{Q}\mathbf{B}\mathbf{B}^{\mathsf{T}}\mathbf{Q}^{\mathsf{T}}.$$

Now if the orthogonal matrix $\mathbf{Q}$ is such that each row of $\mathbf{QB}$ has the same Euclidean length, then it is clear that all the elements on the leading diagonal of the influence matrix, $\mathbf{A}_Q$, have the same value, which must be $\text{tr}(\mathbf{A})/n$, since $\text{tr}(\mathbf{A}_Q) = \text{tr}\left(\mathbf{Q}\mathbf{A}\mathbf{Q}^{\mathsf{T}}\right) = \text{tr}\left(\mathbf{A}\mathbf{Q}^{\mathsf{T}}\mathbf{Q}\right) = \text{tr}(\mathbf{A})$.

Does a $\mathbf{Q}$ with this neat row-length-equalizing property actually exist? It is easy to see that it does, by construction. Firstly note that it is always possible to produce an orthogonal matrix to be applied from the left, whose action is to perform a rotation which affects only two rows of a target matrix, such as $\mathbf{B}$: a matrix with this property is known as a Givens rotation. As the angle of rotation, θ, increases smoothly from zero, the Euclidean lengths of the two rows vary smoothly, although the sum of their squared lengths remains constant, as befits a rotation. Once θ reaches 90 degrees, the row lengths are interchanged, since the magnitudes of the elements of the rows have been interchanged. Hence there must exist an intermediate θ at which the row lengths are exactly equal. Let the Givens rotation with this θ be termed the 'averaging rotation' for the two rows. If we now apply an averaging rotation to the pair of rows of $\mathbf{B}$ having smallest and largest Euclidean lengths, then the range of row lengths in the modified $\mathbf{B}$ will automatically be reduced. Iterating this process must eventually lead to all row lengths being equal, and the product of the averaging rotations employed in the iteration yields $\mathbf{Q}$.

With this best rotation of the fitting problem, the ordinary cross validation score (6.17) can be written

$$\mathcal{V}_g = \frac{n\|\mathbf{y} - \hat{\boldsymbol{\mu}}\|^2}{[n - \text{tr}\,(\mathbf{A})]^2}, \tag{6.18}$$

which is known as the *generalized cross validation* score (GCV, Craven and Wahba, 1979; Golub et al., 1979). Notice that we do not have to actually perform the rotation in order to use GCV. Also, since the expected prediction error is unaffected by the rotation, and GCV is just OCV on the rotated problem, GCV must be as valid an estimate of prediction error as OCV, but GCV has the nice property of being invariant.

Figure 6.2 compares GCV, OCV and UBRE scores for some simulated data. Unsurprisingly the criteria tend to be in quite close agreement, and are all in closer agreement with each other than with the observed MSE + σ^2.

6.2.4 Double cross validation

The main practical problem with GCV or OCV is that they are not sensitive enough to overfit. In consequence we tend to do less well than we could at actual prediction. One obvious way around this would be to choose smoothing parameters in order to obtain the closest match between predicted values for the same point when it is excluded and when it is included in the fit. That is, if $\hat{\mu}_i^{[-i]}$ is the prediction of y_i when y_i was dropped from the fitting data, then we could look at

$$c_1 = \frac{1}{n}\sum_{i=1}^{n}\left(\hat{\mu}_i - \hat{\mu}_i^{[-i]}\right)^2.$$

Since $\hat{\mu}_i - \hat{\mu}_i^{[-i]} = y_i - \hat{\mu}_i^{[-i]} - (y_i - \hat{\mu}_i)$ and $y_i - \hat{\mu}_i^{[-i]} = (y_i - \hat{\mu}_i)/(1 - A_{ii})$ it is easy to show that

$$c_1 = \frac{1}{n}\sum_{i=1}^{n}\left\{(y_i - \hat{\mu}_i)A_{ii}/(1 - A_{ii})\right\}^2.$$

Approximating the A_{ii} by their mean, as in GCV, we get

$$c_2 = \frac{1}{n}\sum_{i=1}^{n}(y_i - \hat{\mu}_i)^2\text{tr}(\mathbf{A})^2/\{n - \text{tr}(\mathbf{A})\}^2.$$

On its own this is not very useful, as there is nothing anchoring c_1 to the data, so it makes sense to combine it with the GCV score,

$$c_3 = n\sum_{i=1}^{n}(y_i - \hat{\mu}_i)^2/\{n - \text{tr}(\mathbf{A})\}^2.$$

Exactly how to combine c_2 and c_3 is not completely obvious. A simple approach is

$$c_4 = c_3 + \sqrt{c_3 c_2} + c_2 = \frac{n \sum_{i=1}^n (y_i - \hat{\mu}_i)^2}{\{n - \text{tr}(\mathbf{A})\}^2} \{1 + \text{tr}(\mathbf{A})/n + \text{tr}(\mathbf{A})^2/n^2\},$$

which is equivalent to using $\{n - \text{tr}(\mathbf{A})\}/\sqrt{(1 + \text{tr}(\mathbf{A})/n + \text{tr}(\mathbf{A})^2/n^2}$ as the residual effective degrees of freedom in a GCV score. Assuming $\text{tr}(\mathbf{A})/n$ is small then using the approximation $1/\sqrt{1 + \text{tr}(\mathbf{A})/n + \text{tr}(\mathbf{A})^2/n^2} \approx 1 - 0.5\text{tr}(\mathbf{A})/n$ the residual degrees of freedom becomes $n - 0.5\text{tr}(\mathbf{A}) - \text{tr}(\mathbf{A})\{1 - 0.5\text{tr}(\mathbf{A})/n\}$. Neglecting the term $0.5\text{tr}(\mathbf{A})^2/n$ then yields

$$\mathcal{V}_d = \frac{n\|\mathbf{y} - \hat{\boldsymbol{\mu}}\|^2}{\{n - 1.5\text{tr}(\mathbf{A})\}^2}. \tag{6.19}$$

This score is derived in several ways in the literature, e.g., in Kim and Gu (2004). In `mgcv` it can be obtained by setting the `gamma` parameter of `gam` to 1.5 (provided that GCV smoothing parameter selection is chosen).

6.2.5 *Prediction error criteria for the generalized case*

There are a number of ways of producing smoothing parameter selection criteria for the generalized case, which essentially substitute the model deviance or the Pearson statistic for the residual sum of squares in the UBRE score (6.13) or the GCV score (6.18). In practice the Pearson statistic version tends to under-smooth heavily for binary data, so the deviance based versions are generally preferred. UBRE becomes

$$\mathcal{V}_a(\boldsymbol{\lambda}) = D(\hat{\boldsymbol{\beta}}) + 2\gamma\phi\tau$$

in the case where ϕ is known, and otherwise the GCV score becomes

$$\mathcal{V}_g(\boldsymbol{\lambda}) = nD(\hat{\boldsymbol{\beta}})/(n - \gamma\tau)^2 \tag{6.20}$$

where τ is the model effective degrees of freedom and γ is usually 1, but might be increased to force smoother models, via (6.19) for example.

These criteria are discussed in Hastie and Tibshirani (1990), but this section follows the Wood (2008) extension of Gu and Xiang (2001) to derive them. First write the model deviance as $D(\hat{\boldsymbol{\eta}}) = \sum D_i(\hat{\eta}_i)$, where D_i is the contribution to the deviance associated with the ith datum. Now the mean predictive deviance of the model can be estimated by

$$D_{\text{cv}} = \sum_{i=1}^n D_i(\hat{\eta}_i^{[-i]}),$$

where $\hat{\boldsymbol{\eta}}^{[-i]}$ is the linear predictor obtained by estimating the model from all the data except the ith datum. Minimization of D_{cv} is an attractive way of choosing smoothing parameters as it seeks to minimize the Kullback-Leibler (KL) distance between the estimated model and the truth (since the KL distance depends on the model only via the predictive deviance); however, it is impractically expensive to minimize it directly.

To progress, follow Gu and Xiang (2001) and let $\hat{\boldsymbol{\eta}}^{[-i]}$ be the linear predictor which results if z_i is omitted from the working linear fit at the final stage of the PIRLS. As in section 6.2.4, this can be shown to imply that $\hat{\eta}_i - \hat{\eta}_i^{[-i]} = (z_i - \hat{\eta}_i)A_{ii}/(1-A_{ii})$. But $z_i - \hat{\eta}_i = g'(\hat{\mu}_i)(y_i - \hat{\mu}_i)$, so $\hat{\eta}_i - \hat{\eta}_i^{[-i]} = g'(\hat{\mu}_i)(y_i - \hat{\mu}_i)A_{ii}/(1-A_{ii})$. Now take a first order Taylor expansion

$$D_i(\hat{\eta}_i^{[-i]}) \simeq D_i(\hat{\eta}_i) + \frac{\partial D_i}{\partial \hat{\eta}_i}(\hat{\eta}_i^{[-i]} - \hat{\eta}_i) = D_i(\hat{\eta}_i) - \frac{\partial D_i}{\partial \hat{\eta}_i}\frac{A_{ii}}{1-A_{ii}}g'(\hat{\mu}_i)(y_i - \hat{\mu}_i).$$

Noting that

$$\frac{\partial D_i}{\partial \hat{\eta}_i} = -2\omega_i \frac{y_i - \hat{\mu}_i}{V(\hat{\mu}_i)g'(\hat{\mu}_i)}, \text{ we have } D_i(\hat{\eta}_i^{[-i]}) \simeq D_i(\hat{\eta}_i) + 2\frac{A_{ii}}{1-A_{ii}}\omega_i\frac{(y_i - \hat{\mu}_i)^2}{V(\hat{\mu}_i)}.$$

As in GCV the individual A_{ii} terms are replaced by their average, $\text{tr}(\mathbf{A})/n$, to yield

$$D_i(\hat{\eta}_i^{[-i]}) \simeq D_i(\hat{\eta}_i) + 2\frac{\text{tr}(\mathbf{A})}{n - \text{tr}(\mathbf{A})}\omega_i\frac{(y_i - \hat{\mu}_i)^2}{V(\hat{\mu}_i)},$$

and averaging over the data gives a generalized approximate cross validation score

$$\mathcal{V}_g^* = D(\hat{\boldsymbol{\eta}})/n + \frac{2}{n}\frac{\text{tr}(\mathbf{A})}{n - \text{tr}(\mathbf{A})}\sum_{i=1}^n \omega_i\frac{(y_i - \hat{\mu}_i)^2}{V(\hat{\mu}_i)} = D(\hat{\boldsymbol{\eta}})/n + \frac{2}{n}\frac{\tau}{n-\tau}P(\hat{\boldsymbol{\eta}}),$$

where $P = \sum_i \omega_i(y_i - \hat{\mu}_i)^2/V(\hat{\mu}_i)$ is a Pearson statistic (and the final term on the r.h.s. could also be multiplied by $\gamma \geq 1$). This differs a little from the Gu and Xiang (2001) result because of not assuming canonical link functions.

Notice how $\mathcal{V}_g^*$ is just a linear transformation of (generalized) AIC/UBRE, with the Pearson estimate $\hat{\phi} = P(\hat{\boldsymbol{\eta}})/\{n - \text{tr}(\mathbf{A})\}$ in place of the MLE of ϕ, and $\text{tr}(\mathbf{A})$ as the model degrees of freedom: so if ϕ is known we might as well use $\mathcal{V}_a$. Viewed from this perspective we could also use $D(\hat{\boldsymbol{\eta}})/\{n - \text{tr}(\mathbf{A})\}$ for $\hat{\phi}$, in which case, for $\text{tr}(\mathbf{A}) \ll n$, the resulting criterion would be approximately $\mathcal{V}_g$. These connections are unsurprising: see Stone (1977).

If we set $\gamma = 0.5 \log n$ in $\mathcal{V}_a$ then we get a rescaled generalized version of the Bayesian Information Criterion (BIC; Schwarz, 1978), which could also be approximated by $\mathcal{V}_g$ with the same γ. See, e.g., Wood (2015, §2.5.2) for a derivation. BIC gives much smoother results than the prediction error criteria, but the reader might want to look carefully at its derivation before deciding whether or not to use it.

6.2.6 *Marginal likelihood and REML*

If we take the Bayesian view of smoothing in which smoothing penalties correspond to a Gaussian prior on the model coefficients, then an alternative approach is to

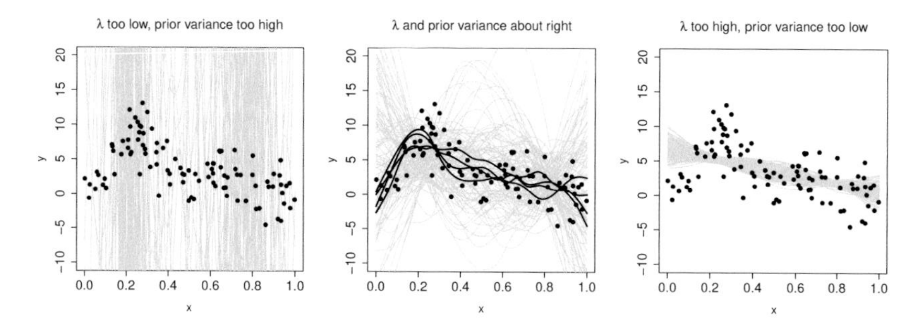

Figure 6.3 *REML can be thought of as randomly drawing from the prior implied by the smoothness penalty, and assessing the average likelihood of the result. If the smoothing parameter is too low (left) then the draws are over variable and almost all fail to pass close to the data. Conversely if the smoothing parameter is too high (right) the curves drawn are all too smooth and consequently also fail to pass close enough to the data to have high likelihood. When the smoothing parameter is about right then the curves have about the right level of smoothness to be able to pass close to the data and have high likelihood, and hence some of them do (shown in black). For plotting purposes all the curves have been centred around the least squares straight line fits to the data.*

choose the smoothing parameters that maximize the Bayesian log *marginal likelihood*: that is, the log of the joint density of the data and coefficients $\boldsymbol{\beta}$, with the coefficients integrated out,

$$\mathcal{V}_r(\boldsymbol{\lambda}) = \log \int f(\mathbf{y}|\boldsymbol{\beta}) f(\boldsymbol{\beta}) d\boldsymbol{\beta}.$$

The integral can be interpreted as the average likelihood of random draws from the prior, as illustrated in figure 6.3, by way of motivation for maximizing it. The approach of estimating parameters by maximizing the marginal likelihood is known as *empirical Bayes* (in contrast to a fully Bayesian approach in which a prior is placed on $\boldsymbol{\lambda}$ and the corresponding posterior density is then obtained). Notice that this Bayesian marginal likelihood has exactly the same form as the REML criterion for a generalized linear mixed model covered in section 3.4 (p. 147), since in that case the random coefficients also have Gaussian distributions. The Laplace approximation (3.17) of section 3.4 can again be used to evaluate the integral, resulting in

$$\mathcal{V}_r(\boldsymbol{\lambda}) \simeq l(\hat{\boldsymbol{\beta}}) - \frac{\hat{\boldsymbol{\beta}}^{\mathsf{T}} \mathbf{S}_\lambda \hat{\boldsymbol{\beta}}}{2\phi} - \frac{\log |\mathbf{S}_\lambda/\phi|_+}{2} - \frac{\log \left|\mathbf{X}^{\mathsf{T}}\mathbf{W}\mathbf{X}/\phi + \mathbf{S}_\lambda/\phi\right|}{2} + \frac{M}{2}\log(2\pi), \tag{6.21}$$

where l is the log likelihood, $\mathbf{W}$ is the diagonal matrix of full Newton weights at convergence of the PIRLS iteration, M is the dimension of the null space of $\mathbf{S}_\lambda$ and $|\mathbf{A}|_+$ denotes the product of the non-zero eigenvalues of $\mathbf{A}$ (a generalized determinant). In the Gaussian additive model case (6.21) is exact, since there are then no neglected terms in the Taylor expansion on which the Laplace approximation is based.

We could also use the frequentist marginal likelihood (3.16) of section 3.4 in which the fixed effects/unpenalized components of the model are not integrated out of the likelihood.** This will tend to lead to smoother models, since the frequentist marginal likelihood tends to underestimate variance components, and smoothing parameters can be thought of as precision (inverse variance) parameters.

The use of marginal likelihood/REML for smoothing parameter selection goes back to Anderssen and Bloomfield (1974), while the Wahba (1985) 'GML' criterion is also REML.

6.2.7 *The problem with* $\log|\mathbf{S}_\lambda|_+$

The main problem in optimizing $\mathcal{V}_r$ is the term $\log|\mathbf{S}_\lambda|_+$ (where ϕ has been absorbed into $\mathbf{S}_\lambda$ to avoid clutter). A simple example nicely illustrates the problem. Consider the real 5×5 matrix $\mathbf{C}$ with QR decomposition $\mathbf{C} = \mathbf{QR}$ so that $|\mathbf{C}| = |\mathbf{R}| = \prod_i R_{ii}$. Suppose that $\mathbf{C} = \mathbf{A} + \mathbf{B}$ where $\mathbf{A}$ is rank 2 with non-zero elements of size $O(a)$, $\mathbf{B}$ is rank 3 with non-zero elements of size $O(b)$ and $a \gg b$. Let the schematic non-zero structure of $\mathbf{C} = \mathbf{A} + \mathbf{B}$ be

$$\begin{bmatrix} \bullet & \bullet & \bullet & & \\ \bullet & \bullet & \bullet & & \\ \bullet & \bullet & \bullet & \cdot & \cdot \\ & & \cdot & \cdot & \cdot \\ & & \cdot & \cdot & \cdot \end{bmatrix} = \begin{bmatrix} \bullet & \bullet & \bullet & & \\ \bullet & \bullet & \bullet & & \\ \bullet & \bullet & \bullet & & \\ & & & & \\ & & & & \end{bmatrix} + \begin{bmatrix} & & & & \\ & & & & \\ & & \cdot & \cdot & \cdot \\ & & \cdot & \cdot & \cdot \\ & & \cdot & \cdot & \cdot \end{bmatrix},$$

where $\bullet$ shows the $O(a)$ elements and $\cdot$ those of $O(b)$. Now QR decomposition (see Golub and Van Loan, 2013) operates by applying successive householder reflections to $\mathbf{C}$, each in turn zeroing the subdiagonal elements of successive columns of $\mathbf{C}$. Let the product of the first 2 reflections be $\mathbf{Q}_2^\mathsf{T}$ and consider the state of the QR decomposition after 2 steps. Schematically $\mathbf{Q}_2^\mathsf{T}\mathbf{C} = \mathbf{Q}_2^\mathsf{T}\mathbf{A} + \mathbf{Q}_2^\mathsf{T}\mathbf{B}$ is

$$\begin{bmatrix} \bullet & \bullet & \bullet & \cdot & \cdot \\ & \bullet & \bullet & \cdot & \cdot \\ & & d_1 & \cdot & \cdot \\ & & d_2 & \cdot & \cdot \\ & & d_3 & \cdot & \cdot \end{bmatrix} = \begin{bmatrix} \bullet & \bullet & \bullet & & \\ & \bullet & \bullet & & \\ & & d_1' & & \\ & & d_2' & & \\ & & d_3' & & \end{bmatrix} + \begin{bmatrix} & & \cdot & \cdot & \cdot \\ & & \cdot & \cdot & \cdot \\ & & d_1'' & \cdot & \cdot \\ & & d_2'' & \cdot & \cdot \\ & & d_3'' & \cdot & \cdot \end{bmatrix}.$$

Because $\mathbf{A}$ is rank 2, d_j' should be 0, and d_j should be d_j'', but computationally $d_j' = O(\epsilon a)$ where ϵ is the machine precision. Hence if ϵa approaches b, we suffer catastrophic loss of precision in $\mathbf{d}$, which will be inherited by R_{33} and the computed value of $|\mathbf{C}|$. Matrices such as $\sum_j \lambda_j \mathbf{S}_j$ can suffer from exactly this problem, since some λ_j can legitimately tend to infinity while others remain small, and the $\mathbf{S}_j$ are often of lower rank than the dimension of their non-zero sub-block. Note that the problem does not lie with the use of the QR decomposition to compute the determinant: the same problem arises with an eigen-decomposition/SVD approach.

**The fact that Bayesians and frequentists have different names for the same quantity is irritatingly typical, but not completely inconsistent: the frequentist marginal likelihood is obtained by integrating out all the random coefficients in the model, but for the Bayesian all the coefficients are random coefficients.

One solution is based on similarity transform. In the case of our simple example, consider the similarity transform $\mathbf{UCU}^\mathsf{T} = \mathbf{UAU}^\mathsf{T} + \mathbf{UBU}^\mathsf{T}$, constructed to produce the following schematic

$$\begin{bmatrix} \bullet & \bullet & \cdot & \cdot & \cdot \\ \bullet & \bullet & \cdot & \cdot & \cdot \\ \cdot & \cdot & \cdot & \cdot & \cdot \\ \cdot & \cdot & \cdot & \cdot & \cdot \\ \cdot & \cdot & \cdot & \cdot & \cdot \end{bmatrix} = \begin{bmatrix} \bullet & \bullet & & & \\ \bullet & \bullet & & & \\ & & & & \\ & & & & \\ & & & & \end{bmatrix} + \begin{bmatrix} \cdot & \cdot & \cdot & \cdot & \cdot \\ \cdot & \cdot & \cdot & \cdot & \cdot \\ \cdot & \cdot & \cdot & \cdot & \cdot \\ \cdot & \cdot & \cdot & \cdot & \cdot \\ \cdot & \cdot & \cdot & \cdot & \cdot \end{bmatrix}.$$

The computation of $\mathbf{UCU}^\mathsf{T}$ is by adding $\mathbf{UBU}^\mathsf{T}$ to $\mathbf{UAU}^\mathsf{T}$ with the theoretically zero elements of the latter set to exact zeroes. $|\mathbf{UCU}^\mathsf{T}| = |\mathbf{C}|$, but computation based on the similarity transformed version no longer suffers from the precision loss problem, no-matter how disparate a and b are in magnitude.

In general we can stabilise the computation of $\log|\mathbf{S}_\lambda|_+$ as follows. Suppose that $\lambda_k\mathbf{S}_k$ has the largest value of $\|\lambda_j\mathbf{S}_j\|$ among the terms in $\mathbf{S}_\lambda = \sum_j \lambda_j\mathbf{S}_j$, and suppose that $\mathbf{S}_k$ is of rank r. In this section let $n = p - r$, where p is the dimension of the $\mathbf{S}$ matrices (so r and n are the dimensions of the range and null spaces of $\mathbf{S}_k$). Now consider the eigen-decomposition $\lambda_k\mathbf{S}_k = \mathbf{UDU}^\mathsf{T}$, and let $\mathbf{S}_{\bar{k}} = \sum_{j\neq k}\lambda_j\mathbf{S}_j$. If $\mathbf{S}^*_\lambda = \mathbf{U}^\mathsf{T}\mathbf{S}_\lambda\mathbf{U}$ then $|\mathbf{S}^*_\lambda|_+ = |\mathbf{S}_\lambda|_+$, and we can partition as follows

$$\mathbf{S}^*_\lambda = \begin{bmatrix} \mathbf{D}_r + \mathbf{U}_r^\mathsf{T}\mathbf{S}_{\bar{k}}\mathbf{U}_r & \mathbf{U}_r^\mathsf{T}\mathbf{S}_{\bar{k}}\mathbf{U}_n \\ \mathbf{U}_n^\mathsf{T}\mathbf{S}_{\bar{k}}\mathbf{U}_r & \mathbf{U}_n^\mathsf{T}\mathbf{S}_{\bar{k}}\mathbf{U}_n \end{bmatrix}$$

where $\mathbf{D}_r$ is the diagonal matrix of the r non-zero eigenvalues of $\mathbf{S}_k$, while the columns of $\mathbf{U}_r$ are the corresponding eigenvectors, and the columns of $\mathbf{U}_n$ are the eigenvectors corresponding to the zero eigenvalues. Computing with the right hand side expression ensures that $\lambda_k\mathbf{S}_k$ does not corrupt parts of the computation involving much smaller magnitude terms.

Now $\mathbf{U}_n^\mathsf{T}\mathbf{S}_{\bar{k}}\mathbf{U}_n = \sum_{j\neq k}\lambda_j\mathbf{U}_n^\mathsf{T}\mathbf{S}_j\mathbf{U}_n$, which has the same form as the original expression for $\mathbf{S}_\lambda$. Hence we can apply the same algorithm to this block – that is identify the largest norm $\lambda_j\mathbf{U}_n^\mathsf{T}\mathbf{S}_j\mathbf{U}_n$, eigen-decompose it and partition the similarity transformed $\mathbf{U}_n^\mathsf{T}\mathbf{S}_{\bar{k}}\mathbf{U}_n$. Continuing in this way we eventually arrive at a transformed $\mathbf{S}_\lambda$ matrix in which the final M rows and columns are 0. Let the non-zero part of this final matrix be the $p - M \times p - M$ matrix $\mathbf{S}'_\lambda$. Then $|\mathbf{S}_\lambda|_+ = |\mathbf{S}'_\lambda|$, and the latter can be computed without large magnitude elements from some of its components corrupting computations that should involve only its smaller element components.

Furthermore we can write $\mathbf{S}'_\lambda = \sum_j \lambda_j\mathbf{S}'_j$, where the $\mathbf{S}'_j$ are the original $\mathbf{S}_j$ matrices after successive application of all the similarity transformations of the recursion, and the dropping of the final M zero valued rows and columns. If $\rho_j = \log\lambda_j$, it then follows that

$$\frac{\partial \log|\mathbf{S}_\lambda|_+}{\partial\rho_j} = \lambda_j \text{tr}(\mathbf{S}'^{-1}_\lambda\mathbf{S}'_j)$$

and

$$\frac{\partial^2 \log|\mathbf{S}_\lambda|_+}{\partial\rho_j\partial\rho_k} = -\lambda_j\lambda_k\text{tr}(\mathbf{S}'^{-1}_\lambda\mathbf{S}'_k\mathbf{S}'^{-1}_\lambda\mathbf{S}'_j) + \delta_j^k\lambda_j\text{tr}(\mathbf{S}'^{-1}_\lambda\mathbf{S}'_j),$$

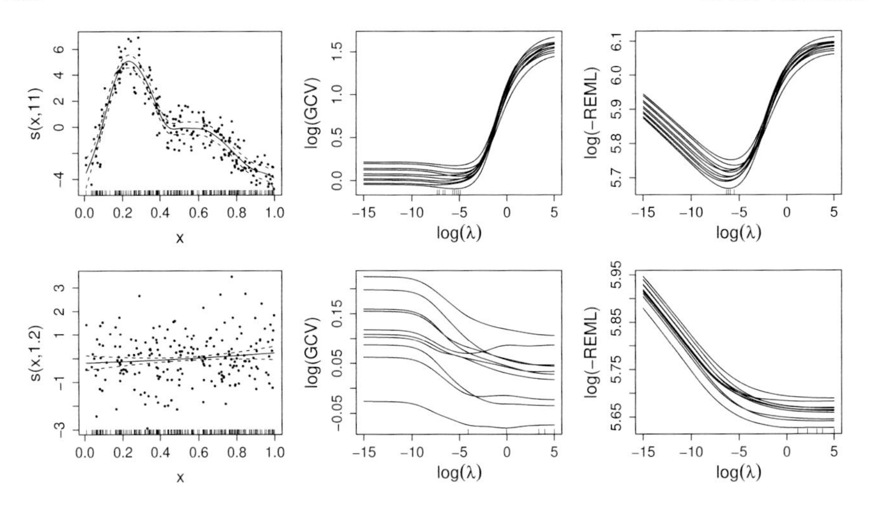

Figure 6.4 *Comparison of GCV and REML objective functions for a high signal to noise situation (top row) and a low signal to noise situation (bottom row). Typical replicate data sets and smooths of those data are shown in the first column. In the middle column are GCV functions for 10 replicates of the data, while the right column shows the REML objective function for the same 10 replicates. Notice that on the over-fitting side the GCV minimum is shallower than the replicate-to-replicate variability in the GCV score. This is not the case for the much more pronounced REML optimum. Notice also the GCV tendency to develop phantom minima when the signal to noise ratio is low.*

where $\delta_j^k = 1$ if $j = k$ and 0 otherwise. The result of all the orthogonal transforms applied during the recursion can be accumulated as a single orthogonal transform as the recursion progresses. Using this to re-parameterize the whole model can only improve the conditioning of terms such as $\log\left|\mathbf{X}^\mathsf{T}\mathbf{W}\mathbf{X}/\phi + \mathbf{S}_\lambda/\phi\right|$ as well.

An objection to the preceding method is that it involves more eigen-decompositions than are strictly necessary to deal with the scaling issue – we really only need to separate the $\lambda_j \mathbf{S}_j$ into *groups* of terms with disparate magnitudes, rather than treating the $\lambda_j \mathbf{S}_j$ singly. The resulting scheme is more efficient, but rather more complicated, and is given in appendix B of Wood (2011).

6.2.8 *Prediction error criteria versus marginal likelihood*

Wahba (1985) showed that asymptotically REML undersmooths relative to GCV, resulting in higher asymptotic mean square error. It is perhaps not surprising that a prediction error criteria should have better asymptotic prediction error performance than a likelihood based criteria, but this is not the whole story. Härdle et al. (1988) show that the rate at which GCV selected smoothing parameters approach their optimal values is very slow, while Reiss and Ogden (2009) show that at finite sample sizes GCV is more likely to develop multiple minima and under-smooth, relative to REML. The tendency of GCV to badly under-smooth on occasion is well known to practitioners, and is the motivation for development of modifications, such as that in section 6.2.4.

Figure 6.4 shows a comparison of GCV and REML, which illustrates why REML tends to be more resistant to occasional severe over-fitting: its optimum tends to be more pronounced relative to sampling variability, and it has a lesser tendency to develop phantom minimum when there is no real signal in the data.

So there is a trade-off between reliability at finite sample sizes and asymptotic mean square error rates. For example in the case of a cubic spline type smoother the best achievable mean square error rate is $O(n^{-8/9})$ (see section 5.2), which we can get close to with GCV smoothing parameter selection. In contrast, with REML smoothness selection we can not do better than $O(n^{-4/5})$, if we want the smoothing penalties to have any effect in the large sample limit.

Another consideration is that likelihood methods provide an easy means to quantify the uncertainty of the smoothing parameters, which is more difficult if prediction error methods are used.

Unpenalized coefficient bias

Since Rice (1986) it has been recognised that the bias affecting penalized terms can also cause bias in unpenalized terms, if the penalized and unpenalized terms involve (non-linearly) correlated covariates. Indeed when such correlation is present, Rice (1986) shows that asymptotically the bias in an unpenalized estimator may sometimes tend to zero no faster than the standard deviation of the unpenalized estimator. It is not clear exactly how the results extend to the case of automatic smoothing parameter selection, but Rice conjectures that models estimated by cross validation could be adversely affected, and that confidence intervals, in particular, may not be reliable. Rice shows that the squared bias in the unpenalized coefficients can be of the same order as the integrated squared bias of the penalized terms. This suggests undersmoothing in order to reduce the bias, which could be viewed as an argument in favour of REML, which smooths less than cross validation in the large sample limit.

In simulations with automatic smoothing parameter selection it is possible to generate detectable bias, if the non-linear dependence of covariates is strong enough, although interval estimates based on section 6.10 seem to be fairly reliable, presumably as a result of including bias as well as sampling variability components (see section 6.10.1). Figure 6.5 shows some illustrative simulation results for a range of sample sizes for the simple model

$$y_i = \alpha z_i + f(x_i) + \epsilon_i, \quad \epsilon_i \sim N(0,1),$$

where $\alpha = 1$ and $f(x) = 0.2x^{11}\{10(1-x)\}^6 + 10(10x)^3(1-x)^{10}$. The x_i are independent $U(0,1)$ while $z_i = 10x^2(x-1)^3 + \varepsilon_i$ and the ε_i are independent Gaussian with standard deviation 0.25 or 0.1. For the high dependency setting the bias is quite substantial until large sample sizes, although consistently modest in relation to the sampling variability. For lower dependency the bias is rather small. As predicted the bias is lower with REML smoothing parameter estimation than with GCV. Confidence interval coverage is close to nominal despite the bias.

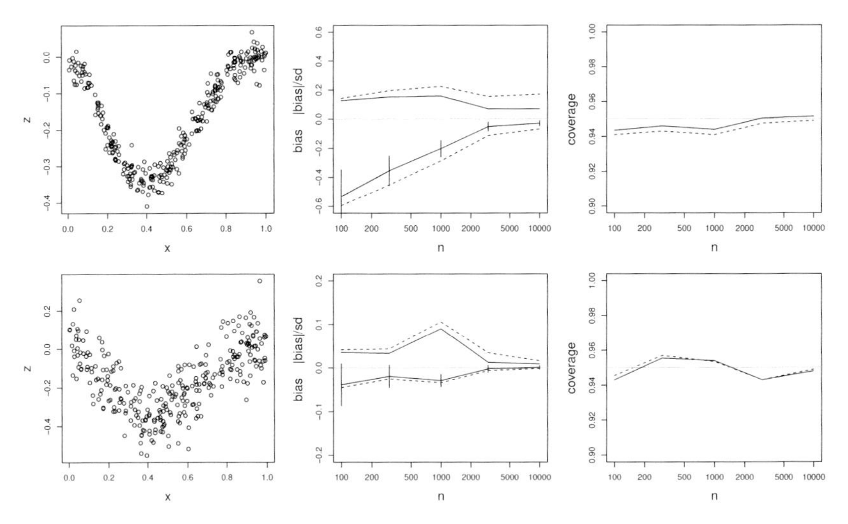

Figure 6.5 *Simulation results illustrating how penalization bias can lead to bias in unpenalized terms, when there is non-linear correlation between covariates. The model is* $y = \alpha z + f(x) + \epsilon$, *with true value of* $\alpha = 1$. *The top row is for high dependence between covariates* x *and* z *while the lower row is for lower dependence. The first column plots the covariates against each other. The next column plots the bias in* $\hat{\alpha}$ *(mostly negative) against sample size and the ratio of the magnitude of the bias to the standard deviation of* $\hat{\alpha}$ *(positive). The standard errors of the bias are shown as vertical bars for the REML case. The right column shows the realized coverage probabilities of nominally 95% confidence intervals for* α. *Solid lines relate to REML based estimates and dashed to GCV.*

6.2.9 *The 'one standard error rule' and smoother models*

Since smoothing parameter estimates are uncertain, we might sometimes prefer to use estimates at the 'smooth model' end of this uncertainty, for reasons of parsimony (and despite the previous section). The 'one standard error rule' (e.g., Hastie et al., 2009) is one way to do this. For a single smoothing parameter, we choose the largest smoothing parameter within one standard error of the minimum of the smoothing selection criterion.

Of the multiple smoothing parameter selection criteria considered here, only REML/ML provide us with easy access to measures of uncertainty, but for these it is relatively straightforward to produce an appropriate multiple smoothing parameter variant of the rule. Asymptotically we have

$$\hat{\boldsymbol{\rho}} \sim N(\boldsymbol{\rho}, \mathbf{V}_\rho)$$

where $\mathbf{V}_\rho$ is the inverse negative Hessian of the marginal or restricted likelihood. There is no point in increasing the smoothing parameters of terms that are already 'completely smooth' (having effectively infinite smoothing parameters, and effective degrees of freedom equal to their penalty null space dimension), so these are dropped from $\hat{\boldsymbol{\rho}}$, and $\mathbf{V}_\rho$ is computed only for the remainder.

The idea is to increase the log smoothing parameters, ρ_i, in proportion to their standard deviations, $s_i = \sqrt{V_\rho(i,i)}$. The constant of proportionality is then computed in order to increase the REML/ML by one standard deviation according to its asymptotic distribution. Specifically, we set the log smoothing parameters to $\hat{\boldsymbol{\rho}} + \mathbf{d}$ where $\mathbf{d} = \alpha \mathbf{s}$ and $\alpha = \sqrt{2p}/\mathbf{s}^\mathsf{T}\mathbf{V}_\rho^{-1}\mathbf{s}$. p is the dimension of $\hat{\boldsymbol{\rho}}$ and $\sqrt{2p}$ the standard deviation of a χ^2_p r.v. `?one.se.rule` in `mgcv` provides example code.

6.3 Computing the smoothing parameter estimates

Finding the smoothing parameter estimates is the most computationally challenging part of GAM estimation, since simultaneously maintaining efficiency and stability is not easy. There are three strategies.

1. Directly optimize one of the general model criteria of sections 6.2.5 or 6.2.6. This optimization will require an 'outer iteration' to optimize the smoothing parameters. Each set of smoothing parameters tried by the outer iteration will require an inner PIRLS iteration to find the model coefficient estimates corresponding to the trial. So we have a nested optimisation procedure. For maximum reliability the outer iteration will use Newton's method, based on computationally exact first and second derivatives. Obtaining these requires that we find the derivatives of the model coefficient estimates with respect to the smoothing parameters.
2. Apply the Gaussian additive model version of the chosen smoothing parameter selection method to the working penalized linear model fitted at each iteration of the PIRLS of section 6.1.1. The prediction error criteria need no special justification to be applied in this context, while REML is justified by the central limit theorem argument of section 3.4.2 (p. 149). This approach is essentially the PQL method of section 3.4.2 and was first proposed (as 'performance oriented iteration') by Gu (1992).
3. Use a simple update formula, the generalized Fellner-Schall method, to update the smoothing parameters at each step of the PIRLS iteration. This approximately optimizes the Laplace approximate REML of the model.

Note that 2 and 3 have the advantage of comparative simplicity and efficiency (especially in big data settings), but they are not guaranteed to converge, and the value of the optimized smoothness selection criterion is not really comparable between model fits. Strategies 1 and 3 generalize beyond models estimable by PIRLS.

6.4 The generalized Fellner-Schall method

The simplest method to implement is the generalized Fellner-Schall method, in which simple explicit formulae are obtained for updating the smoothing parameters in order to increase the Laplace approximate REML score for the model. The original method appears in Fellner (1986) and Schall (1991), and was developed for updating variance components of simple i.i.d. Gaussian random effects in generalized linear mixed models. The method's simplicity led to variations of it being applied in a GAM context (e.g. Rigby and Stasinopoulos, 2014), but the restriction that each random effect had just one variance parameter made it inapplicable for adaptive and tensor product smooths in which the penalty is not separable in this way. Use of the method with smooth interaction terms was hence only possible using the tensor product smooths of section 5.6.5 (p. 235) or an even more special case covered in Rodríguez-Álvarez et al. (2015). These restrictions are removed by the following generalization of the method (Wood and Fasiolo, 2017).

First consider the case of a Gaussian additive model. The log restricted marginal likelihood is

$$l_r(\boldsymbol{\lambda}) = -\frac{\|\mathbf{y} - \mathbf{X}\hat{\boldsymbol{\beta}}_\lambda\|^2 + \hat{\boldsymbol{\beta}}_\lambda^\mathsf{T}\mathbf{S}_\lambda\hat{\boldsymbol{\beta}}_\lambda}{2\sigma^2} + \frac{\log|\mathbf{S}_\lambda/\sigma^2|_+}{2} - \frac{\log|\mathbf{X}^\mathsf{T}\mathbf{X}/\sigma^2 + \mathbf{S}_\lambda/\sigma^2|}{2} + c$$

where $\hat{\boldsymbol{\beta}}_\lambda = \text{argmax}_\beta f_\lambda(\mathbf{y}, \boldsymbol{\beta})$ for a given $\boldsymbol{\lambda}$. Given that $\partial(\|\mathbf{y} - \mathbf{X}\boldsymbol{\beta}\|^2 + \boldsymbol{\beta}^\mathsf{T}\mathbf{S}_\lambda\boldsymbol{\beta})/\partial\boldsymbol{\beta}|_{\hat{\beta}_\lambda} = \mathbf{0}$, by definition of $\hat{\boldsymbol{\beta}}_\lambda$, we have

$$\frac{\partial l_r}{\partial \lambda_j} = \text{tr}(\mathbf{S}_\lambda^-\mathbf{S}_j)/2 - \text{tr}\{(\mathbf{X}^\mathsf{T}\mathbf{X} + \mathbf{S}_\lambda)^{-1}\mathbf{S}_j\}/2 - \hat{\boldsymbol{\beta}}_\lambda^\mathsf{T}\mathbf{S}_j\hat{\boldsymbol{\beta}}_\lambda/(2\sigma^2).$$

By Theorem 4, below, $\text{tr}(\mathbf{S}_\lambda^-\mathbf{S}_j) - \text{tr}\{(\mathbf{X}^\mathsf{T}\mathbf{X} + \mathbf{S}_\lambda)^{-1}\mathbf{S}_j\}$ is non-negative, while $\hat{\boldsymbol{\beta}}_\lambda^\mathsf{T}\mathbf{S}_j\hat{\boldsymbol{\beta}}_\lambda$ is non-negative by the positive semi-definiteness of $\mathbf{S}_j$. $\partial l_r/\partial\lambda_j$ will be negative if

$$\text{tr}(\mathbf{S}_\lambda^-\mathbf{S}_j) - \text{tr}\{(\mathbf{X}^\mathsf{T}\mathbf{X} + \mathbf{S}_\lambda)^{-1}\mathbf{S}_j\} < \hat{\boldsymbol{\beta}}_\lambda^\mathsf{T}\mathbf{S}_j\hat{\boldsymbol{\beta}}_\lambda/\sigma^2,$$

indicating that λ_j should be decreased. If the inequality is reversed then $\partial l_r/\partial\lambda_j$ is positive, indicating that λ_j should be increased. If the inequality becomes an equality then $\partial l_r/\partial\lambda_j = 0$ and λ_j should not be changed. λ_j should remain positive. A simple update meeting these four requirements is

$$\lambda_j^* = \sigma^2\frac{\text{tr}(\mathbf{S}_\lambda^-\mathbf{S}_j) - \text{tr}\{(\mathbf{X}^\mathsf{T}\mathbf{X} + \mathbf{S}_\lambda)^{-1}\mathbf{S}_j\}}{\hat{\boldsymbol{\beta}}_\lambda^\mathsf{T}\mathbf{S}_j\hat{\boldsymbol{\beta}}_\lambda}\lambda_j, \tag{6.22}$$

with λ_j^* set to some predefined upper limit if $\hat{\boldsymbol{\beta}}_\lambda^\mathsf{T}\mathbf{S}_j\hat{\boldsymbol{\beta}}_\lambda$ is so close to zero that the limit would otherwise be exceeded. Formally $\boldsymbol{\Delta} = \boldsymbol{\lambda}^* - \boldsymbol{\lambda}$ is an ascent direction for l_r, by Taylor's theorem and the fact that $\boldsymbol{\Delta}^\mathsf{T}\partial l_r/\partial\boldsymbol{\lambda} > 0$, unless $\boldsymbol{\lambda}$ is already a turning point of l_r. To formally guarantee that the update increases l_r requires step length control; for example, we use the update $\boldsymbol{\delta} = \boldsymbol{\Delta}/2^k$, where k is the smallest integer ≥ 0 such that $l_r(\boldsymbol{\lambda} + \boldsymbol{\delta}) > l_r(\boldsymbol{\lambda})$. In practice such control rarely seems to be needed, although it can be advantageous to maintain an acceleration factor α, and use a step $\boldsymbol{\Delta}\alpha$, where α can be adaptively modified to take longer steps when this further improves l_r.

The update relies on the following, which is the key to the generalization beyond singly penalized smooth terms.

Theorem 4. *Let* $\mathbf{B}$ *be a positive definite matrix and* $\mathbf{S}_\lambda$ *be a positive semi-definite matrix parameterized by* $\boldsymbol{\lambda}$, *and with a null space that is independent of the value of* $\boldsymbol{\lambda}$. *Let positive semi-definite matrix* $\mathbf{S}_j$ *denote the derivative of* $\mathbf{S}_\lambda$ *with respect to* λ_j. *Then tr*$(\mathbf{S}_\lambda^-\mathbf{S}_j)$ − *tr*$\{(\mathbf{B} + \mathbf{S}_\lambda)^{-1}\mathbf{S}_j\} > 0$.

Proof. See Wood and Fasiolo (2017). □

A similar result applies when $\mathbf{B}$ is positive semi-definite. σ^2 also has to be estimated, but as in section 2.4.6 (p. 83), by setting the derivative of l_r with respect to σ^2 to zero and solving we obtain

$$\hat{\sigma}^2 = \|\mathbf{y} - \mathbf{X}\hat{\boldsymbol{\beta}}_\lambda\|^2 / [n - \text{tr}\{(\mathbf{X}^\mathsf{T}\mathbf{X} + \mathbf{S}_\lambda)^{-1}\mathbf{X}^\mathsf{T}\mathbf{X}\}].$$

The method alternates estimation of $\boldsymbol{\beta}$ given $\boldsymbol{\lambda}$, with update of $\boldsymbol{\lambda}$ given $\hat{\boldsymbol{\beta}}$, using (6.22). Wood and Fasiolo (2017) show that the update (6.22) takes longer steps than the EM algorithm, or an obvious acceleration of the EM algorithm, but in the large sample limit the step is smaller than or equal to the equivalent Newton method step.

6.4.1 *General regular likelihoods*

The method can also be applied beyond the Gaussian case, and even beyond GLM likelihoods. For a general regular likelihood the log Laplace approximate marginal likelihood is

$$l_r = l(\hat{\boldsymbol{\beta}}_\lambda) - \hat{\boldsymbol{\beta}}_\lambda^\mathsf{T}\mathbf{S}_\lambda\hat{\boldsymbol{\beta}}_\lambda/2 + \log|\mathbf{S}_\lambda|_+/2 - \log|\mathcal{H}_\lambda|/2 + c,$$

where $\mathcal{H}_\lambda = -\partial^2 l/\partial\boldsymbol{\beta}\partial\boldsymbol{\beta}^\mathsf{T} + \mathbf{S}_\lambda$. Defining $\mathbf{H} = -\partial^2 l/\partial\boldsymbol{\beta}\partial\boldsymbol{\beta}^\mathsf{T}$ and $\mathbf{V}_\lambda^{-1} = \mathcal{H}_\lambda$ or $\mathbb{E}\mathcal{H}_\lambda$, we have

$$\frac{\partial l_r}{\partial \lambda_j} = -\hat{\boldsymbol{\beta}}_\lambda^\mathsf{T}\mathbf{S}_j\hat{\boldsymbol{\beta}}_\lambda/2 + \text{tr}(\mathbf{S}_\lambda^-\mathbf{S}_j)/2 - \text{tr}\{\mathbf{V}_\lambda\mathbf{S}_j\}/2 - \text{tr}\{\mathbf{V}_\lambda\partial\mathbf{H}/\partial\lambda_j\}/2.$$

The dependence of $\partial^2 l/\partial\boldsymbol{\beta}\partial\boldsymbol{\beta}^\mathsf{T}$ on λ_j is inconvenient. However, the performance oriented iteration (PQL) method for $\boldsymbol{\lambda}$ estimation of section 6.5 neglects the dependence of $\partial^2 l/\partial\boldsymbol{\beta}\partial\boldsymbol{\beta}^\mathsf{T}$ on $\boldsymbol{\lambda}$ (Breslow and Clayton, 1993; Gu, 1992). If we follow this precedent then the development follows the Gaussian case and the update is

$$\lambda_j^* = \frac{\text{tr}(\mathbf{S}_\lambda^-\mathbf{S}_j) - \text{tr}\{\mathbf{V}_{\lambda'}\mathbf{S}_j\}}{\boldsymbol{\beta}_\lambda^\mathsf{T}\mathbf{S}_j\boldsymbol{\beta}_\lambda}\lambda_j. \tag{6.23}$$

If $\partial^2 l/\partial\boldsymbol{\beta}\partial\boldsymbol{\beta}^\mathsf{T}$ is independent of $\boldsymbol{\lambda}$ at finite sample size, as is the case for some distribution – link function combinations in a generalized linear model setting, then the update is guaranteed to increase l_r under step size control, but otherwise this is not the case, and in practice the $\boldsymbol{\lambda}$ estimate no longer exactly maximizes l_r.

So again we have a simple way of updating smoothing parameters given model coefficients which can be iteratively alternated with the update of model coefficients given smoothing parameters.

Theorem 4, required to guarantee that $\lambda_j^* > 0$, will hold if V_λ is based on the expected Hessian of the negative log likelihood, but if it is based on the observed Hessian, then this must be positive definite for the Theorem to hold. If the observed Hessian is not positive definite then the expected Hessian, or a suitable nearest positive definite matrix to the observed Hessian, should be substituted.

In the case of a penalized GLM, the general update (6.23) becomes

$$\lambda_j^* = \phi \frac{\text{tr}(\mathbf{S}_\lambda^- \mathbf{S}_j) - \text{tr}\{(\mathbf{X}^\mathsf{T}\mathbf{W}\mathbf{X} + \mathbf{S}_\lambda)^{-1}\mathbf{S}_j\}}{\hat{\boldsymbol{\beta}}_\lambda^\mathsf{T} \mathbf{S}_j \hat{\boldsymbol{\beta}}_\lambda} \lambda_j,$$

where $\mathbf{W}$ is the diagonal matrix of weights at convergence of the usual penalized iteratively re-weighted least squares iteration used to find $\hat{\boldsymbol{\beta}}_\lambda$, and ϕ is the scale parameter, which can be estimated using the obvious equivalent of $\hat{\sigma}^2$.

See exercise 6 for equivalent updates based on AIC and GCV.

6.5 Direct Gaussian case and performance iteration (PQL)

Moving up a level of method complexity, consider the 'performance oriented iteration' of Gu (1992), which is essentially the same approach as the PQL method of Breslow and Clayton (1993). As in section 3.4.2 (p.149), the key idea is to take the working weighted penalized linear model at each step of the Fisher weighted version of the PIRLS of section 6.1.1, and to estimate its smoothing parameters as if it were a simple penalized Gaussian regression model. If we use REML or marginal likelihood for the smoothing parameter estimation then the justification is exactly that given in section 3.4.2 (p. 149). Use of weighted versions of GCV or UBRE requires no special justification: the criteria do not assume Gaussian data, and hence there is nothing about the weighted PIRLS pseudodata that would invalidate their use. If structured correctly the smoothing parameter updates at each step of the method have a lower leading order cost than the computations of $\hat{\boldsymbol{\beta}}$ given smoothing parameters, and it is often the case that the number of PIRLS steps is about the same as would be required with fixed smoothing parameters. Like the generalized Fellner-Schall method, the disadvantage is that beyond the additive Gaussian case the method is not guaranteed to converge. Figure 6.6 shows a simple example iteration.

It remains to develop suitable computational methods for smoothing parameter optimization in the penalized weighted least squares case. First consider the GCV score appropriate for the working model at a PIRLS step

$$\mathcal{V}_g = \frac{n\|\sqrt{\mathbf{W}}(\mathbf{z} - \hat{\boldsymbol{\eta}})\|^2}{\{n - \gamma \text{tr}(\mathbf{A})\}^2}$$

where $\mathbf{A} = \sqrt{\mathbf{W}}\mathbf{X}(\mathbf{X}^\mathsf{T}\mathbf{W}\mathbf{X} + \mathbf{S}_\lambda)^{-1}\mathbf{X}^\mathsf{T}\sqrt{\mathbf{W}}$ and $\sqrt{\mathbf{W}}\hat{\boldsymbol{\eta}} = \mathbf{A}\sqrt{\mathbf{W}}\mathbf{z}$. $\gamma = 1$ gives GCV, whereas $\gamma = 1.5$ gives the DGCV of section 6.2.4. Now if we set $\mathbf{y} = \sqrt{\mathbf{W}}\mathbf{z}$ and $\mathbf{X}' = \sqrt{\mathbf{W}}\mathbf{X}$, then

$$\mathcal{V}_g = \frac{n\|\mathbf{y} - \mathbf{A}\mathbf{y}\|^2}{\{n - \gamma \text{tr}(\mathbf{A})\}^2}$$

where $\mathbf{A} = \mathbf{X}'(\mathbf{X}'^\mathsf{T}\mathbf{X}' + \mathbf{S}_\lambda)^{-1}\mathbf{X}'^\mathsf{T}$, but this latter is exactly the same form as the un-weighted GCV score, and hence it suffices to consider only the un-weighted case, with a similar argument applying to the UBRE score. In what follows the primes are dropped and $\mathbf{X}$ is used to refer to either the weighted or unweighted model matrix, as appropriate.

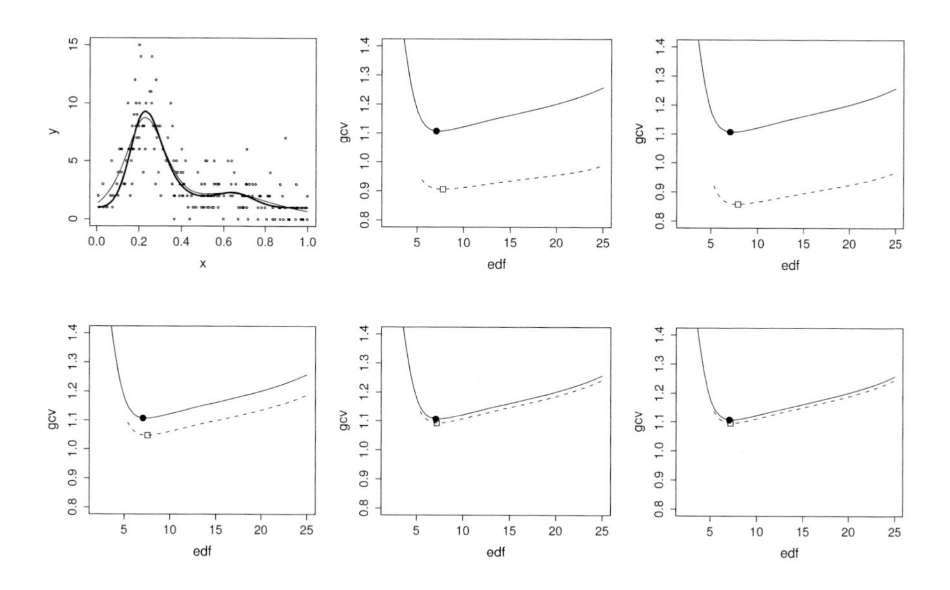

Figure 6.6 *Performance iteration. The top left panel shows 200 simulated x, y data: the x co-ordinates are uniform random deviates and the y co-ordinates are Poisson random variables with x dependent mean given by the thick curve. The data are modelled as $\log(\mu_i) = f(x_i)$, $y_i \sim \text{Poi}(\mu_i)$ with f represented using a rank 25 penalized regression spline. The effective degrees of freedom for f were chosen by performance iteration. Working from middle top to bottom right, the remaining panels illustrate the progress of the performance iteration. The continuous curve is the same in each panel and is the (deviance based) GCV score for the model plotted against EDF. The dashed curves show the GCV score for the working linear model for each PIRLS iteration, against EDF. In each panel the minimum of each curve is marked. Notice how the two curves become increasingly close in the vicinity of the finally selected model, as iteration progresses. In fact the correspondence is unusually close in the replicate shown: often the dashed curve is further below the continuous curve, with the performance iteration therefore suggesting a slightly more complex model. The thin continuous curve, in the upper left panel, shows the model fits corresponding to the minimum of the model GCV score and to the performance iteration estimate: they are indistinguishable.*

6.5.1 *Newton optimization of the GCV score*

Following Wood (2004), $\mathcal{V}$ can be minimized by Newton's method (see section 2.2.1, p.76), which requires that we compute the gradient vector and Hessian matrix of $\mathcal{V}$ w.r.t. the log smoothing parameters $\rho_i = \log(\lambda_i)$. Working on the log scale ensures that the λ_i remain positive. The expensive parts of the enterprise relate to the trace of the influence matrix, $\mathbf{A}$, so it is this influence matrix that must be considered first:

$$\mathbf{A} = \mathbf{X}\left(\mathbf{X}^{\mathsf{T}}\mathbf{X} + \mathbf{H} + \sum_{j=1}^{m}\lambda_j\mathbf{S}_j\right)^{-1}\mathbf{X}^{\mathsf{T}},$$

where $\mathbf{H}$ is a known fixed positive semi-definite matrix, often simply $\mathbf{0}$, but sometimes used to impose minimum values on smoothing parameters, for example. The process of getting $\mathbf{A}$ into a more useful form starts with a QR decomposition of $\mathbf{X}$,

$$\mathbf{X} = \mathbf{QR},$$

where the columns of $\mathbf{Q}$ are columns of an orthogonal matrix, and $\mathbf{R}$ is an upper triangular matrix (see B.6, p. 422). For maximum stability a pivoted QR decomposition should actually be used here (Golub and Van Loan, 2013): this has the consequence that the parameter vector and rows and columns of the $\mathbf{S}_i$ matrices have to be permuted before proceeding, with the inverse permutation applied to the parameter vector and covariance matrix at the end of the estimation procedure.

The next step is to define $\mathbf{S}_\lambda = \mathbf{H} + \sum_{j=1}^{m}\lambda_j\mathbf{S}_j$ and $\mathbf{B}$ as any matrix square root of $\mathbf{S}_\lambda$ such that $\mathbf{B}^{\mathsf{T}}\mathbf{B} = \mathbf{S}_\lambda$. $\mathbf{B}$ can be obtained efficiently by pivoted Cholesky decomposition (see `?chol` in R, for example) or by eigen-decomposition of the symmetric matrix $\mathbf{S}_\lambda$ (see, e.g., Golub and Van Loan, 2013). Augmenting $\mathbf{R}$ with $\mathbf{B}$, a singular value decomposition (Golub and Van Loan, 2013, LAPACK contains a suitable version) is then obtained:

$$\begin{bmatrix} \mathbf{R} \\ \mathbf{B} \end{bmatrix} = \mathbf{UDV}^{\mathsf{T}}.$$

The columns of $\mathbf{U}$ are columns of an orthogonal matrix, and $\mathbf{V}$ is an orthogonal matrix. $\mathbf{D}$ is the diagonal matrix of singular values: examination of these is the most reliable way of detecting numerical rank deficiency of the fitting problem (Golub and Van Loan, 2013; Watkins, 1991). Rank deficiency of the fitting problem is dealt with at this stage by removing from $\mathbf{D}$ the rows and columns containing any singular values that are 'too small', along with the corresponding columns of $\mathbf{U}$ and $\mathbf{V}$. This has the effect of recasting the original fitting problem into a reduced space in which the model parameters are identifiable. 'Too small' is judged with reference to the largest singular value: for example, singular values less than the largest singular value multiplied by the square root of the machine precision might be deleted.

Now let $\mathbf{U}_1$ be the sub-matrix of $\mathbf{U}$ such that $\mathbf{R} = \mathbf{U}_1\mathbf{DV}^{\mathsf{T}}$. This implies that $\mathbf{X} = \mathbf{QU}_1\mathbf{DV}^{\mathsf{T}}$, while $\mathbf{X}^{\mathsf{T}}\mathbf{X} + \mathbf{S}_\lambda = \mathbf{VD}^2\mathbf{V}^{\mathsf{T}}$ and consequently

$$\begin{aligned} \mathbf{A} &= \mathbf{QU}_1\mathbf{DV}^{\mathsf{T}}\mathbf{VD}^{-2}\mathbf{V}^{\mathsf{T}}\mathbf{VDU}_1^{\mathsf{T}}\mathbf{Q}^{\mathsf{T}} \\ &= \mathbf{QU}_1\mathbf{U}_1^{\mathsf{T}}\mathbf{Q}^{\mathsf{T}}. \end{aligned}$$

Hence, $\text{tr}(\mathbf{A}) = \text{tr}\left(\mathbf{U}_1\mathbf{U}_1^{\mathsf{T}}\mathbf{Q}^{\mathsf{T}}\mathbf{Q}\right) = \text{tr}\left(\mathbf{U}_1\mathbf{U}_1^{\mathsf{T}}\right)$. Notice that the main computational cost is the QR decomposition, but thereafter evaluation of $\text{tr}(\mathbf{A})$ is relatively cheap for new trial values of $\boldsymbol{\lambda}$.

For efficient minimization of the smoothness selection criteria, we also need to find the derivatives of the criteria w.r.t. the log smoothing parameters. To this end, it helps to write the influence matrix as $\mathbf{A} = \mathbf{X}\mathbf{G}^{-1}\mathbf{X}^\mathsf{T}$ where $\mathbf{G} = \mathbf{X}^\mathsf{T}\mathbf{X} + \mathbf{S}_\lambda = \mathbf{V}\mathbf{D}^2\mathbf{V}^\mathsf{T}$ and hence $\mathbf{G}^{-1} = \mathbf{V}\mathbf{D}^{-2}\mathbf{V}^\mathsf{T}$. So

$$\frac{\partial \mathbf{G}^{-1}}{\partial \rho_j} = -\mathbf{G}^{-1}\frac{\partial \mathbf{G}}{\partial \rho_j}\mathbf{G}^{-1} = -\lambda_j \mathbf{V}\mathbf{D}^{-2}\mathbf{V}^\mathsf{T}\mathbf{S}_j\mathbf{V}\mathbf{D}^{-2}\mathbf{V}^\mathsf{T}$$

and

$$\frac{\partial \mathbf{A}}{\partial \rho_j} = \mathbf{X}\frac{\partial \mathbf{G}^{-1}}{\partial \rho_j}\mathbf{X}^\mathsf{T} = -\lambda_j \mathbf{Q}\mathbf{U}_1\mathbf{D}^{-1}\mathbf{V}^\mathsf{T}\mathbf{S}_j\mathbf{V}\mathbf{D}^{-1}\mathbf{U}_1^\mathsf{T}\mathbf{Q}^\mathsf{T}.$$

Turning to the second derivatives,

$$\frac{\partial^2 \mathbf{G}^{-1}}{\partial \rho_j \partial \rho_k} = \mathbf{G}^{-1}\frac{\partial \mathbf{G}}{\partial \rho_k}\mathbf{G}^{-1}\frac{\partial \mathbf{G}}{\partial \rho_j}\mathbf{G}^{-1} - \mathbf{G}^{-1}\frac{\partial^2 \mathbf{G}}{\partial \rho_j \partial \rho_k}\mathbf{G}^{-1} + \mathbf{G}^{-1}\frac{\partial \mathbf{G}}{\partial \rho_j}\mathbf{G}^{-1}\frac{\partial \mathbf{G}}{\partial \rho_k}\mathbf{G}^{-1}$$

and hence,

$$\frac{\partial^2 \mathbf{A}}{\partial \rho_j \partial \rho_k} = \mathbf{X}\frac{\partial^2 \mathbf{G}^{-1}}{\partial \rho_j \partial \rho_k}\mathbf{X}^\mathsf{T}.$$

This becomes

$$\frac{\partial^2 \mathbf{A}}{\partial \rho_j \partial \rho_k} = \lambda_j \lambda_k \mathbf{Q}\mathbf{U}_1\mathbf{D}^{-1}\mathbf{V}^\mathsf{T}[\mathbf{S}_k\mathbf{V}\mathbf{D}^{-2}\mathbf{V}^\mathsf{T}\mathbf{S}_j]^\ddagger\mathbf{V}\mathbf{D}^{-1}\mathbf{U}_1^\mathsf{T}\mathbf{Q}^\mathsf{T} + \delta_k^j\frac{\partial \mathbf{A}}{\partial \rho_j}$$

where $\mathbf{B}^\ddagger \equiv \mathbf{B} + \mathbf{B}^\mathsf{T}$ and $\delta_k^j = 1$, if $j = k$, and zero otherwise.

Writing $\alpha = \|\mathbf{y} - \mathbf{A}\mathbf{y}\|^2$, we can now find expressions for the component derivatives needed in order to find the derivatives of the GCV or UBRE/AIC scores. First define: (i) $\mathbf{y}_1 = \mathbf{U}_1^\mathsf{T}\mathbf{Q}^\mathsf{T}\mathbf{y}$; (ii) $\mathbf{M}_j = \mathbf{D}^{-1}\mathbf{V}^\mathsf{T}\mathbf{S}_j\mathbf{V}\mathbf{D}^{-1}$ and (iii) $\mathbf{K}_j = \mathbf{M}_j\mathbf{U}_1^\mathsf{T}\mathbf{U}_1$. Some tedious manipulation then shows that:

$$\mathrm{tr}\left(\frac{\partial \mathbf{A}}{\partial \rho_j}\right) = -\lambda_j \mathrm{tr}\left(\mathbf{K}_j\right), \quad \mathrm{tr}\left(\frac{\partial^2 \mathbf{A}}{\partial \rho_j \partial \rho_k}\right) = 2\lambda_j\lambda_k \mathrm{tr}\left(\mathbf{M}_k\mathbf{K}_j\right) - \delta_k^j \lambda_j \mathrm{tr}\left(\mathbf{K}_j\right),$$

$$\frac{\partial \alpha}{\partial \rho_j} = 2\lambda_j \left[\mathbf{y}_1^\mathsf{T}\mathbf{M}_j\mathbf{y}_1 - \mathbf{y}_1^\mathsf{T}\mathbf{K}_j\mathbf{y}_1\right]$$

and

$$\frac{\partial^2 \alpha}{\partial \rho_j \partial \rho_k} = 2\lambda_j\lambda_k \mathbf{y}_1^\mathsf{T}\left[\mathbf{M}_j\mathbf{K}_k + \mathbf{M}_k\mathbf{K}_j - \mathbf{M}_j\mathbf{M}_k - \mathbf{M}_k\mathbf{M}_j + \mathbf{K}_j\mathbf{M}_k\right]\mathbf{y}_1 + \delta_k^j\frac{\partial \alpha}{\partial \rho_j}.$$

The above derivatives can be used to find the derivatives of $\mathcal{V}_g$ (6.18) or $\mathcal{V}_u$ (6.13) w.r.t. the ρ_i. Defining $\delta = n - \gamma \mathrm{tr}(\mathbf{A})$, so that

$$\mathcal{V}_g = \frac{n\alpha}{\delta^2} \quad \text{and} \quad \mathcal{V}_u = \frac{1}{n}\alpha - \frac{2}{n}\delta\sigma^2 + \sigma^2,$$

then

$$\frac{\partial \mathcal{V}_g}{\partial \rho_j} = \frac{n}{\delta^2}\frac{\partial \alpha}{\partial \rho_j} - \frac{2n\alpha}{\delta^3}\frac{\partial \delta}{\partial \rho_j}$$

and

$$\frac{\partial^2 \mathcal{V}_g}{\partial\rho_j\partial\rho_k} = -\frac{2n}{\delta^3}\frac{\partial\delta}{\partial\rho_k}\frac{\partial\alpha}{\partial\rho_j}+\frac{n}{\delta^2}\frac{\partial^2\alpha}{\partial\rho_j\partial\rho_k}-\frac{2n}{\delta^3}\frac{\partial\alpha}{\partial\rho_k}\frac{\partial\delta}{\partial\rho_j}+\frac{6n\alpha}{\delta^4}\frac{\partial\delta}{\partial\rho_k}\frac{\partial\delta}{\partial\rho_j}-\frac{2n\alpha}{\delta^3}\frac{\partial^2\delta}{\partial\rho_j\partial\rho_k}.$$

Similarly $\frac{\partial\mathcal{V}_u}{\partial\rho_j} = \frac{1}{n}\frac{\partial\alpha}{\partial\rho_j} - 2\frac{\partial\delta}{\partial\rho_j}\frac{\sigma^2}{n}$ and $\frac{\partial^2\mathcal{V}_u}{\partial\rho_j\partial\rho_k} = \frac{1}{n}\frac{\partial^2\alpha}{\partial\rho_j\partial\rho_k} - 2\frac{\partial^2\delta}{\partial\rho_j\partial\rho_k}\frac{\sigma^2}{n}$.

For each trial $\boldsymbol{\lambda}$, these derivatives can be obtained at $O(np^2)$ computational cost, so that Newton's method can be used to find the optimum $\boldsymbol{\lambda}$ efficiently. Given the estimated $\hat{\boldsymbol{\lambda}}$, the best fit $\boldsymbol{\beta}$ vector is simply

$$\hat{\boldsymbol{\beta}} = \mathbf{V}\mathbf{D}^{-1}\mathbf{y}_1,$$

while the Bayesian covariance matrix for the parameters (section 6.10 or 4.2.4) is

$$\mathbf{V}_\beta = \mathbf{V}\mathbf{D}^{-2}\mathbf{V}^\mathsf{T}.$$

Notice that the dominant cost is the $O(np^2)$ cost of the QR decomposition, which is anyway needed to find $\hat{\boldsymbol{\beta}}$. Given this QR decomposition, the $O(p^3)$ cost of the derivatives is usually comparatively modest, making the overall cost per iteration comparable with the generalized Fellner-Schall method.

6.5.2 REML

REML optimization for smooth additive models is also required, and as in the case of GCV/UBRE, the weighted problem transforms to the un-weighted case by pre-multiplying $\mathbf{z}$ and $\mathbf{X}$ by $\sqrt{\mathbf{W}}$. Neglecting uninteresting constants, the REML criterion has the form

$$\mathcal{V}_r(\boldsymbol{\lambda}) = \frac{\|\mathbf{y}-\mathbf{X}\hat{\boldsymbol{\beta}}_\lambda\|^2 + \hat{\boldsymbol{\beta}}_\lambda^\mathsf{T}\mathbf{S}_\lambda\hat{\boldsymbol{\beta}}_\lambda + r}{2\phi} + \frac{\log|\mathbf{X}^\mathsf{T}\mathbf{X}/\phi + \mathbf{S}_\lambda/\phi| - \log|\mathbf{S}_\lambda/\phi|_+}{2} \tag{6.24}$$

where r is a constant, currently equal to zero, and $\hat{\boldsymbol{\beta}}_\lambda$ is the minimiser of

$$\|\mathbf{y}-\mathbf{X}\boldsymbol{\beta}\|^2 + \sum_j \lambda_j\boldsymbol{\beta}^\mathsf{T}\mathbf{S}_j\boldsymbol{\beta}.$$

The efficiency of the REML optimization can be improved by an upfront QR decomposition $\mathbf{X} = \mathbb{Q}\mathbb{R}$ where $n \times p$ matrix $\mathbb{Q}$ has orthogonal columns and $\mathbb{R}$ is upper triangular. Then if $\mathbb{Y} = \mathbb{Q}^\mathsf{T}\mathbf{y}$ and $r = \|\mathbf{y}\|^2 - \|\mathbb{Y}\|^2$, rather than 0, it can readily be checked that $\hat{\boldsymbol{\beta}}_\lambda = \text{argmin}_\beta\|\mathbb{Y} - \mathbb{R}\boldsymbol{\beta}\|^2 + \boldsymbol{\beta}^\mathsf{T}\mathbf{S}_\lambda\boldsymbol{\beta}$ and

$$\mathcal{V}_r(\boldsymbol{\lambda}) = \frac{\|\mathbb{Y}-\mathbb{R}\hat{\boldsymbol{\beta}}_\lambda\|^2 + \hat{\boldsymbol{\beta}}_\lambda^\mathsf{T}\mathbf{S}_\lambda\hat{\boldsymbol{\beta}}_\lambda + r}{2\phi} + \frac{\log|\mathbb{R}^\mathsf{T}\mathbb{R}/\phi + \mathbf{S}_\lambda/\phi| - \log|\mathbf{S}_\lambda/\phi|_+}{2}.$$

The point being that the latter expression has exactly the same structure as (6.24), but after the one off $O(np^2)$ cost of the QR decomposition, the second expression costs only $O(p^3)$ to evaluate, rather than $O(np^2)$. This is convenient when we want to evaluate $\mathcal{V}_r$ several times with different smoothing parameter values. Given that the problem has the same structure with or without this upfront QR reduction, what follows will refer to (6.24), on the understanding that it will apply equally well to the QR reduced form.

Before any fitting it is prudent to check numerically that the model is identifiable. Let $\mathbf{B}$ be any matrix such that $\mathbf{B}^\mathsf{T}\mathbf{B} = \sum_j \mathbf{S}_j/\|\mathbf{S}_j\|$, where $\|\cdot\|$ denotes a matrix norm, and form the pivoted QR decomposition

$$\mathcal{QR} = \begin{bmatrix} \mathbf{X} \\ \mathbf{B} \end{bmatrix}.$$

If the model is not identifiable then $\mathcal{R}$ will not have full rank, with a trailing zero block as a result of the rank deficiency. In such a case the parameters corresponding to the zero block should be dropped from the fit, along with the corresponding columns of $\mathbf{X}$ and rows and columns of each $\mathbf{S}_j$. Zeroes can be inserted in place of the estimates of the dropped parameters after fitting is complete. See function `Rrank` in `mgcv` for reliable estimation of rank from a pivoted triangular factor.

To apply Newton's method we require first and second derivatives of $\mathcal{V}_r$ with respect to the log smoothing parameters $\rho_j = \log(\lambda_j)$.

$$\begin{aligned}\frac{\partial \mathcal{V}_r}{\partial \rho_j} &= \lambda_j \frac{\hat{\boldsymbol{\beta}}^\mathsf{T}\mathbf{S}_j\hat{\boldsymbol{\beta}}}{2\phi} + \frac{1}{2}\frac{\partial \log|\mathbf{X}^\mathsf{T}\mathbf{X}/\phi + \mathbf{S}_\lambda/\phi|}{\partial \rho_j} - \frac{1}{2}\frac{\partial \log|\mathbf{S}_\lambda/\phi|_+}{\partial \rho_j} \\ &\quad -\frac{1}{\phi}\frac{\mathrm{d}\hat{\boldsymbol{\beta}}}{\mathrm{d}\rho_j}\left[(\mathbf{X}^\mathsf{T}\mathbf{X}+\mathbf{S}_\lambda)\hat{\boldsymbol{\beta}} - \mathbf{X}^\mathsf{T}\mathbf{y}\right] \\ &= \lambda_j \frac{\hat{\boldsymbol{\beta}}^\mathsf{T}\mathbf{S}_j\hat{\boldsymbol{\beta}}}{2\phi} + \frac{\lambda_j}{2}\mathrm{tr}\left\{(\mathbf{X}^\mathsf{T}\mathbf{X}+\mathbf{S}_\lambda)^{-1}\mathbf{S}_j\right\} - \frac{1}{2}\frac{\partial \log|\mathbf{S}_\lambda/\phi|_+}{\partial \rho_j},\end{aligned}$$

since the term in square brackets evaluates to 0 by definition of $\hat{\boldsymbol{\beta}}$. When differentiating again, recall that the term in square brackets does not have zero derivative merely because it evaluates to zero at the point $\hat{\boldsymbol{\beta}}$. So we have

$$\begin{aligned}\frac{\partial^2 \mathcal{V}_r}{\partial \rho_j \partial \rho_k} &= \frac{\delta_j^k \lambda_j}{2\phi}\hat{\boldsymbol{\beta}}^\mathsf{T}\mathbf{S}_j\hat{\boldsymbol{\beta}} + \frac{\lambda_j}{2\phi}\left(\frac{\mathrm{d}\hat{\boldsymbol{\beta}}^\mathsf{T}}{\mathrm{d}\rho_k}\mathbf{S}_j\hat{\boldsymbol{\beta}} + \hat{\boldsymbol{\beta}}^\mathsf{T}\mathbf{S}_j\frac{\mathrm{d}\hat{\boldsymbol{\beta}}^\mathsf{T}}{\mathrm{d}\rho_k}\right) - \frac{1}{2}\frac{\partial^2 \log|\mathbf{S}_\lambda/\phi|_+}{\partial \rho_j \partial \rho_k} \\ &\quad - \frac{\lambda_j\lambda_k}{2}\mathrm{tr}\left\{(\mathbf{X}^\mathsf{T}\mathbf{X}+\mathbf{S}_\lambda)^{-1}\mathbf{S}_k(\mathbf{X}^\mathsf{T}\mathbf{X}+\mathbf{S}_\lambda)^{-1}\mathbf{S}_j\right\} \\ &\quad + \frac{\delta_j^k\lambda_j}{2}\mathrm{tr}\left\{(\mathbf{X}^\mathsf{T}\mathbf{X}+\mathbf{S}_\lambda)^{-1}\mathbf{S}_j\right\} - \frac{1}{\phi}\frac{\mathrm{d}\hat{\boldsymbol{\beta}}}{\mathrm{d}\rho_j}(\mathbf{X}^\mathsf{T}\mathbf{X}+\mathbf{S}_\lambda)\frac{\mathrm{d}\hat{\boldsymbol{\beta}}}{\mathrm{d}\rho_k}.\end{aligned}$$

The following subsections discuss computation of the component terms in these derivative expressions.

$\log|\mathbf{S}_\lambda|_+$ *and its derivatives*

As discussed in section 6.2.7, $\log|\mathbf{S}_\lambda/\phi|_+$ is the most numerically taxing term in REML optimization. Since its computation will influence the whole structure of the optimization, consider it first. For maximal computational efficiency it is worth using the fact that $\mathbf{S}_\lambda$ usually has a block diagonal structure, for example

$$\mathbf{S}_\lambda = \begin{bmatrix} \lambda_1 \boldsymbol{\mathcal{S}}_1 & . & . & . & . \\ . & \lambda_2 \boldsymbol{\mathcal{S}}_2 & . & . & . \\ . & . & \sum_j \lambda_j \boldsymbol{\mathcal{S}}_j & . & . \\ . & . & . & . & . \\ . & . & . & . & . \end{bmatrix}.$$

That is, there are some blocks with single smoothing parameters, and others with an additive structure. There are usually also some zero blocks on the diagonal. The block structure means that the generalized determinant, its derivatives with respect to $\rho_k = \log \lambda_k$ and the matrix square root of $\mathbf{S}_\lambda$ can all be computed blockwise. For the above example,

$$\log |\mathbf{S}_\lambda/\phi|_+ = \text{rank}(\boldsymbol{\mathcal{S}}_1) \log(\lambda_1/\phi) + \log |\boldsymbol{\mathcal{S}}_1|_+ + \text{rank}(\boldsymbol{\mathcal{S}}_2) \log(\lambda_2/\phi) \\ + \log |\boldsymbol{\mathcal{S}}_2|_+ + \log \Big| \sum_j \lambda_j \boldsymbol{\mathcal{S}}_j/\phi \Big|_+ + \cdots$$

For any ρ_k relating to a single parameter block we have

$$\frac{\partial \log |\mathbf{S}_\lambda/\phi|_+}{\partial \rho_k} = \text{rank}(\boldsymbol{\mathcal{S}}_k)$$

and zero second derivatives. For multi-λ blocks there will generally be first and second derivatives to compute. There are no second derivatives ‘between-blocks’.

In general, there are three block types, each requiring different preprocessing.

1. Single parameter diagonal blocks. A reparameterization can be used so that all non-zero elements are one, and the rank precomputed.
2. Single parameter dense blocks. An orthogonal reparameterization, based on the eigenvectors of the symmetric eigen-decomposition of the block, can be used to make these blocks look like the previous type (by similarity transform). Again the rank is computed.
3. Multi-λ blocks will require the reparameterization and derivative computation method of section 6.2.7 to be applied for each new $\boldsymbol{\rho}$ proposal, since the numerical problem that the re-parameterization avoids is $\boldsymbol{\rho}$ dependent.

The reparameterizations from each block type are applied to $\mathbf{X}$. The reparameterization information must be stored so that we can return to the original parameterization at the end.

The remaining derivative components

Having applied the initial reparameterizations, and, for each trial $\boldsymbol{\rho}$, any reparameterizations implied by the processing of type 3 blocks (see above), we can now compute the remaining components required by the derivative expressions. Let $\mathbf{E}$ be any matrix such that $\mathbf{E}^\mathsf{T}\mathbf{E} = \mathbf{S}_\lambda$, and form pivoted QR decomposition

$$\mathbf{QR} = \begin{bmatrix} \mathbf{X} \\ \mathbf{E} \end{bmatrix}.$$

If $\mathbf{Q}_1$ denotes the first n rows of $\mathbf{Q}$ then $\log|\mathbf{X}^\mathsf{T}\mathbf{X}/\phi + \mathbf{S}_\lambda/\phi| = 2\sum_j \log(R_{jj}) - p\log\phi$ and

$$\hat{\boldsymbol{\beta}} = \mathbf{R}^{-1}\mathbf{Q}_1^\mathsf{T}\mathbf{y}.$$

$\hat{\boldsymbol{\beta}}$ is in pivoted order here. The derivatives of $\hat{\boldsymbol{\beta}}$ with respect to the log smoothing parameters are obtained as follows. By definition of $\hat{\boldsymbol{\beta}}$ we have that

$$(\mathbf{X}^\mathsf{T}\mathbf{X} + \mathbf{S}_\lambda)\hat{\boldsymbol{\beta}} = \mathbf{X}^\mathsf{T}\mathbf{y}.$$

Differentiating both sides of this equation w.r.t. ρ_j yields

$$\lambda_j\mathbf{S}_j\hat{\boldsymbol{\beta}} + (\mathbf{X}^\mathsf{T}\mathbf{X} + \mathbf{S}_\lambda)\frac{\mathrm{d}\hat{\boldsymbol{\beta}}}{\mathrm{d}\rho_j} = 0$$

and hence

$$\frac{\mathrm{d}\hat{\boldsymbol{\beta}}}{\mathrm{d}\rho_j} = -\lambda_j(\mathbf{X}^\mathsf{T}\mathbf{X} + \mathbf{S}_\lambda)^{-1}\mathbf{S}_j\hat{\boldsymbol{\beta}}.$$

In pivoted order this can be computed as

$$\frac{\mathrm{d}\hat{\boldsymbol{\beta}}}{\mathrm{d}\rho_j} = -\lambda_j\mathbf{R}^{-1}\mathbf{R}^{-\mathsf{T}}\mathbf{S}_j\hat{\boldsymbol{\beta}},$$

where for maximal efficiency it is best to unpivot $\hat{\boldsymbol{\beta}}$, form the vector $\mathbf{S}_j\hat{\boldsymbol{\beta}}$ and then re-pivot it for use in the above expression. Applying the pivoting re-ordering to $\mathbf{S}_j$ would destroy the block structure, making efficient computation less easy.

The final components are the trace terms resulting from the derivatives of $\log|\mathbf{X}^\mathsf{T}\mathbf{X} + \mathbf{S}_\lambda|$. Optimal efficiency is obtained with the following approach

1. Explicitly form $\mathbf{P} = \mathbf{R}^{-1}$, and unpivot the rows of $\mathbf{P}$. Then form $\mathbf{P}\mathbf{P}^\mathsf{T}$.
2. Form the matrices containing the non-zero rows of $\mathbf{S}_k\mathbf{P}\mathbf{P}^\mathsf{T}$ for all k. This step is cheap for all but type 3 blocks.
3. Compute the required derivatives using

$$\mathrm{tr}\left\{(\mathbf{X}^\mathsf{T}\mathbf{X} + \mathbf{S}_\lambda)^{-1}\mathbf{S}_j\right\} = \mathrm{tr}(\mathbf{S}_j\mathbf{P}\mathbf{P}^\mathsf{T})$$

and

$$\mathrm{tr}\left\{(\mathbf{X}^\mathsf{T}\mathbf{X} + \mathbf{S}_\lambda)^{-1}\mathbf{S}_k(\mathbf{X}^\mathsf{T}\mathbf{X} + \mathbf{S}_\lambda)^{-1}\mathbf{S}_j\right\} = \mathrm{tr}(\mathbf{S}_k\mathbf{P}\mathbf{P}^\mathsf{T}\mathbf{S}_j\mathbf{P}\mathbf{P}^\mathsf{T}).$$

The trace terms are very cheap if computed correctly. In general $\mathrm{tr}(\mathbf{AB}) = \sum_{kj} A_{kj}B_{jk}$. Now let $\mathbf{A}$ have non-zero rows only between k_1 and k_2, while $\mathbf{B}$ has non-zero rows only between j_1 and j_2.

$$\mathrm{tr}(\mathbf{A}) = \sum_{k=k_1}^{k_2} A_{kk} \text{ and } \mathrm{tr}(\mathbf{AB}) = \sum_{k=k_1}^{k_2}\sum_{j=j_1}^{j_2} A_{kj}B_{jk}.$$

Normally the initial zero rows would not be stored in which case we have

$$\mathrm{tr}(\mathbf{A}) = \sum_{k=k_1}^{k_2} A_{k-k_1,k} \text{ and } \mathrm{tr}(\mathbf{AB}) = \sum_{k=k_1}^{k_2}\sum_{j=j_1}^{j_2} A_{k-k_1,j}B_{j-j_1,k}.$$

6.5.3 *Some Newton method details*

The Newton steps to optimize the smoothing parameters need to be conducted with the usual precaution of perturbing the (negative) Hessian to ensure that it is positive definite, if it is not already; otherwise a step in the Newton direction need not improve the smoothness selection criterion. Also the Newton step should be repeatedly halved until an improvement in the criterion is obtained, if the initial step does not improve the criterion and the initial point is not already an optimum.

In addition a less standard modification is useful. It is legitimate for some $\lambda_j \to \infty$, but allowing the computed λ_j to become too large can cause overflow or other numerical stability problems. Fortunately, if a λ_j has reached a value that we could reasonably consider to be 'working infinity' then $\partial\mathcal{V}/\partial\rho_j = \partial^2\mathcal{V}/\partial\rho_j^2 = 0$. Hence while any smoothing parameter approximately meets this condition, we can drop it from the optimization. That is, we leave its value unchanged and drop the corresponding row and column of the Hessian and element of the gradient vector, when computing the Newton step. Note that smoothing parameters can always re-enter the optimization if the condition ceases to hold.

6.6 Direct nested iteration methods

The most reliable, but computationally complex, method aims to directly optimize the smoothing criteria given in sections 6.2.5 and 6.2.6 applied directly to the model of interest, rather than a linearisation of it. Barring numerical catastrophes and assuming properly defined likelihoods, this method will converge because it optimizes a single properly defined function of the smoothing parameters (rather than a succession of different functions based on successive linearised models, or on dropping a slightly different term from the REML criteria at each iteration). We can re-use many ideas covered already.

In this case consider a general regular likelihood first, where the penalized log likelihood is $l(\boldsymbol{\beta}) - \boldsymbol{\beta}^\mathsf{T}\mathbf{S}_\lambda\boldsymbol{\beta}/2$ (any scale parameter being treated as already scaling $\boldsymbol{\lambda}$) and the Laplace approximate REML is

$$\mathcal{V}_r = l(\hat{\boldsymbol{\beta}}) - \frac{1}{2}\hat{\boldsymbol{\beta}}^\mathsf{T}\mathbf{S}_\lambda\hat{\boldsymbol{\beta}} + \frac{1}{2}\log|\mathbf{S}_\lambda|_+ - \frac{1}{2}\log|\mathbf{H} + \mathbf{S}_\lambda| + \frac{M}{2}\log(2\pi),$$

where $\mathbf{H}$ is the Hessian matrix of the negative log likelihood ($H_{ij} = -\partial^2 l/\partial\beta_i\partial\beta_j$). At $\hat{\boldsymbol{\beta}}$ the derivative vector of the penalized log likelihood w.r.t. $\boldsymbol{\beta}$ is $\mathbf{0}$, so

$$\frac{\partial\mathcal{V}_r}{\partial\rho_j} = -\frac{\lambda_j}{2}\hat{\boldsymbol{\beta}}^\mathsf{T}\mathbf{S}_j\hat{\boldsymbol{\beta}} + \frac{1}{2}\frac{\partial\log|\mathbf{S}_\lambda|_+}{\partial\rho_j} - \frac{1}{2}\mathrm{tr}\left\{(\mathbf{H} + \mathbf{S}_\lambda)^{-1}\left(\frac{\partial\mathbf{H}}{\partial\rho_j} + \lambda_j\mathbf{S}_j\right)\right\},$$

using the standard result $\partial \log |\mathbf{B}|/\partial x = \text{tr}(\mathbf{B}^{-1}\partial \mathbf{B}/\partial x)$. Continuing,

$$\frac{\partial^2 \mathcal{V}_r}{\partial \rho_j \partial \rho_k} = \frac{\mathrm{d}\hat{\boldsymbol{\beta}}^{\mathsf{T}}}{\mathrm{d}\rho_j}(\mathbf{H}+\mathbf{S}_\lambda)\frac{\mathrm{d}\hat{\boldsymbol{\beta}}}{\mathrm{d}\rho_k} - \delta_k^j \frac{\lambda_j}{2}\hat{\boldsymbol{\beta}}^{\mathsf{T}}\mathbf{S}_j\hat{\boldsymbol{\beta}} + \frac{1}{2}\frac{\partial^2 \log|\mathbf{S}_\lambda|_+}{\partial \rho_j \partial \rho_k}$$
$$+\frac{1}{2}\text{tr}\left\{(\mathbf{H}+\mathbf{S}_\lambda)^{-1}\left(\frac{\partial \mathbf{H}}{\partial \rho_k} + \lambda_k \mathbf{S}_k\right)(\mathbf{H}+\mathbf{S}_\lambda)^{-1}\left(\frac{\partial \mathbf{H}}{\partial \rho_j} + \lambda_j \mathbf{S}_j\right)\right\}$$
$$-\frac{1}{2}\text{tr}\left\{(\mathbf{H}+\mathbf{S}_\lambda)^{-1}\left(\frac{\partial^2 \mathbf{H}}{\partial \rho_j \partial \rho_k} + \delta_k^j \lambda_j \mathbf{S}_j\right)\right\}.$$

The derivatives of $\log |\mathbf{S}_\lambda|_+$ can be computed exactly as in section 6.5.2. The derivatives of $\mathbf{H}$ w.r.t. the log smoothing parameters can be computed via the chain rule if we can compute the derivatives of $\hat{\boldsymbol{\beta}}$ w.r.t. $\boldsymbol{\rho}$. Note that the computation of terms involving the derivatives of $\mathbf{H}$ has to be structured carefully to keep the computational cost down. The required derivatives of $\hat{\boldsymbol{\beta}}$ are obtained by implicit differentiation. By definition of $\hat{\boldsymbol{\beta}}$ we have

$$\left.\frac{\partial l}{\partial \boldsymbol{\beta}}\right|_{\hat{\boldsymbol{\beta}}} - \mathbf{S}_\lambda \hat{\boldsymbol{\beta}} = \mathbf{0},$$

and, hence, differentiating w.r.t. ρ_j,

$$\left.\frac{\partial^2 l}{\partial \boldsymbol{\beta}\partial \boldsymbol{\beta}^{\mathsf{T}}}\right|_{\hat{\boldsymbol{\beta}}}\frac{\mathrm{d}\hat{\boldsymbol{\beta}}}{\mathrm{d}\rho_j} - \lambda_j \mathbf{S}_j \hat{\boldsymbol{\beta}} - \mathbf{S}_\lambda \frac{\mathrm{d}\hat{\boldsymbol{\beta}}}{\mathrm{d}\rho_j} = \mathbf{0} \Rightarrow \frac{\mathrm{d}\hat{\boldsymbol{\beta}}}{\mathrm{d}\rho_j} = -\lambda_j(\mathbf{H}+\mathbf{S}_\lambda)^{-1}\mathbf{S}_j\hat{\boldsymbol{\beta}}.$$

Differentiating again and substituting from the above we get

$$\frac{\partial^2 \hat{\boldsymbol{\beta}}}{\partial \rho_j \partial \rho_k} = (\mathbf{H}+\mathbf{S}_\lambda)^{-1}\left(\frac{\mathrm{d}\mathbf{H}}{\mathrm{d}\rho_k}\frac{\mathrm{d}\hat{\boldsymbol{\beta}}}{\mathrm{d}\rho_j} - \lambda_k \mathbf{S}_k \frac{\mathrm{d}\hat{\boldsymbol{\beta}}}{\mathrm{d}\rho_j} - \lambda_j \mathbf{S}_j \frac{\mathrm{d}\hat{\boldsymbol{\beta}}}{\mathrm{d}\rho_k} - \delta_j^k \lambda_j \mathbf{S}_j \hat{\boldsymbol{\beta}}\right).$$

To apply Newton's method to $\mathcal{V}_r$ we need to be able to compute $\mathcal{V}_r$ and its first two derivatives w.r.t. $\boldsymbol{\rho}$ for each trial value of $\boldsymbol{\rho}$. Typically any initial re-parameterization required by the method of section 6.5.2 will be applied before any optimization starts. Then for each trial value of $\boldsymbol{\rho}$ we have to do the following.

1. Compute $\log |\mathbf{S}_\lambda|_+$ and its derivatives using the method of section 6.5.2, and apply any re-parameterization that this requires.
2. Use Newton's method to maximize the penalized likelihood of the model.
3. Compute $\mathrm{d}\hat{\boldsymbol{\beta}}/\mathrm{d}\boldsymbol{\rho}$.
4. Compute derivative of $\mathcal{V}_r$ terms requiring $\mathrm{d}\hat{\boldsymbol{\beta}}/\mathrm{d}\boldsymbol{\rho}$.
5. Compute $\partial^2\hat{\boldsymbol{\beta}}/\partial\boldsymbol{\rho}\partial\boldsymbol{\rho}^{\mathsf{T}}$.
6. Compute any derivative of $\mathcal{V}_r$ terms requiring $\partial^2\hat{\boldsymbol{\beta}}/\partial\boldsymbol{\rho}\partial\boldsymbol{\rho}^{\mathsf{T}}$.
7. Reverse any re-parameterization from step 1.

Having performed these steps, we have the gradient and Hessian required to compute a Newton step proposing a new value for $\boldsymbol{\rho}$ (which in turn requires re-application of the steps). As usual both the outer and inner Newton iterations should be performed with step halving and, if needed, perturbation of the negative Hessian to positive definiteness (see section 2.2.1, p. 76). The outer Newton iteration can also drop ρ_j terms when they reach 'working infinity' as described in section 6.5.3.

$\mathbf{H}$ and its derivatives are model specific. For any GLM type likelihood the Newton iteration can be performed via PIRLS using the method of section 6.1.3 and we have $\mathbf{H} = \mathbf{X}^\mathsf{T}\mathbf{W}\mathbf{X}/\phi$. The derivatives of $\mathbf{W}$ w.r.t. $\boldsymbol{\rho}$ can be computed from the link and distribution specific derivatives of w_i w.r.t. η_i, which with $\mathbf{X}$ yield the derivatives of w_i w.r.t. $\boldsymbol{\beta}$, and then via $\partial\hat{\boldsymbol{\beta}}/\partial\rho$, the derivatives of $\mathbf{W}$ w.r.t. $\boldsymbol{\rho}$. If needed, ϕ can be treated as an extra smoothing parameter and estimated to maximize $\mathcal{V}_r$ (notice that $\hat{\boldsymbol{\beta}}$ does not depend on ϕ), although for further inference it is still preferable to use the Fletcher estimator (3.11), p. 111. Full details of the computational steps for the GLM case are given in Wood (2011), including how to best structure the computations of the trace terms in the second derivative of $\mathcal{V}_r$ to keep the computational cost to $O(np^2)$ per smoothing parameter.

The GLM case can be generalized to any likelihood where $\hat{\boldsymbol{\beta}}$ can be obtained by PIRLS. Such likelihoods may have extra parameters such as θ of the negative binomial distribution or p of the Tweedie distribution (see section 3.1.9, p. 115), which may be treated in a similar way to smoothing parameters and estimated to maximize $\mathcal{V}_r$. The main additional detail over GLMs is that the dependence of $\hat{\boldsymbol{\beta}}$ and $\mathbf{H}$ on these extra parameters has to be dealt with. Wood et al. (2016) has details and several examples, including models for ordered categorical data.

For likelihoods which can not be optimized via simple PIRLS, then the inner Newton iteration should also be performed with a step that controls for non-identifiable elements of $\boldsymbol{\beta}$. This can be done by testing the rank, r, of a pivoted Cholesky factor of the penalized Hessian at convergence, dropping the parameters pivoted to positions beyond the rth, and then running further Newton steps to allow other elements of $\hat{\boldsymbol{\beta}}$ to adjust, if necessary. In this case any extra parameters are treated as elements of $\boldsymbol{\beta}$. Again Wood et al. (2016) provide the full details, with several examples, including Cox proportional hazard survival models, multivariate additive models and location-scale models.

Note that we can use the $\hat{\boldsymbol{\beta}}$ estimates from the previous outer iteration step to initialise the inner Newton iteration at the current step, with the result that the inner iteration typically requires very few steps to converge, especially after the first few outer steps are completed.

6.6.1 *Prediction error criteria*

The direct nested iteration method was presented above for the Laplace approximate REML score, $\mathcal{V}_r$, but the prediction error criteria of section 6.2.5 can also be used, as can criteria such as AIC or BIC based on the effective degrees of freedom $\tau = \text{tr}\{(\mathbf{H}+\mathbf{S}_\lambda)^{-1}\mathbf{H}\}$. The derivatives of these alternative criteria are based on the derivatives of the deviance or log likelihood w.r.t. the smoothing parameters, which can be readily (and cheaply) computed using the chain rule and the derivatives of $\hat{\boldsymbol{\beta}}$ w.r.t. $\boldsymbol{\rho}$. The derivatives of τ require more care, if the floating point operation cost is to be kept at $O(np^2)$ per smoothing parameter. An additional implementational nuisance is that τ is usually based on the expected version of $\mathbf{H}$ to ensure that τ is always properly bounded between the penalty null space dimension and the number of model coefficients. Implicit differentiation still requires the observed version.

Writing $\mathbf{H}_\rho^k$ for the derivative of $\mathbf{H}$ w.r.t. ρ_k and defining $\mathcal{H}_\lambda = \mathbf{H} + \mathbf{S}_\lambda$ we have

$$\frac{\partial \tau}{\partial \rho_j} = -\text{tr}\left\{\mathcal{H}_\lambda^{-1}\left(\mathbf{H}_\rho^j + \lambda_j \mathbf{S}_j\right)\mathcal{H}_\lambda^{-1}\mathbf{H}\right\} + \text{tr}\left\{\mathcal{H}_\lambda^{-1}\mathbf{H}_\rho^j\right\}$$

and rather tediously

$$\begin{aligned}\frac{\partial^2 \tau}{\partial \rho_j \partial \rho_k} &= \text{tr}\left\{\mathcal{H}_\lambda^{-1}(\mathbf{H}_\rho^k + \lambda_k \mathbf{S}_k)\mathcal{H}_\lambda^{-1}(\mathbf{H}_\rho^j + \lambda_j \mathbf{S}_j)\mathcal{H}_\lambda^{-1}\mathbf{H}\right\} \\ &+ \text{tr}\left\{\mathcal{H}_\lambda^{-1}(\mathbf{H}_\rho^j + \lambda_j \mathbf{S}_j)\mathcal{H}_\lambda^{-1}(\mathbf{H}_\rho^k + \lambda_k \mathbf{S}_k)\mathcal{H}_\lambda^{-1}\mathbf{H}\right\} \\ &- \text{tr}\left\{\mathcal{H}_\lambda^{-1}(\mathbf{H}_\rho^k + \lambda_k \mathbf{S}_k)\mathcal{H}_\lambda^{-1}\mathbf{H}_\rho^j\right\} - \text{tr}\left\{\mathcal{H}_\lambda^{-1}(\mathbf{H}_\rho^j + \lambda_j \mathbf{S}_j)\mathcal{H}_\lambda^{-1}\mathbf{H}_\rho^k\right\} \\ &- \text{tr}\left\{\mathcal{H}_\lambda^{-1}(\mathbf{H}_{\rho\rho}^{jk} + \delta_k^j \lambda_j \mathbf{S}_j)\mathcal{H}_\lambda^{-1}\mathbf{H}\right\} + \text{tr}\left\{\mathcal{H}_\lambda^{-1}\mathbf{H}_{\rho\rho}^{jk}\right\}.\end{aligned}$$

Wood (2011) gives the details for computing these terms efficiently for the case of a GLM likelihood where $\mathbf{H} = \mathbf{X}^\mathsf{T}\mathbf{W}\mathbf{X}/\phi$. Note that if we are only interested in prediction error criteria then it is possible to use a Fisher weights version of PIRLS to estimate $\hat{\boldsymbol{\beta}}$, and to differentiate this PIRLS to obtain an iteration for the derivatives of $\hat{\boldsymbol{\beta}}$ w.r.t. $\boldsymbol{\rho}$. Then we only need the expected version of $\mathbf{H}$, and not the observed version (see Wood, 2008). While somewhat simpler, Fisher-based PIRLS often requires more steps for non-canonical link functions than full Newton, and a Laplace approximate REML based on the expected Hessian provides poorer smoothing parameter estimates in simulations.

6.6.2 *Example: Cox proportional hazards model*

This section may safely be omitted at first reading. Wood et al. (2016) provide details of how the method of section 6.6 can be applied to multivariate additive models, location-scale models and Tweedie, negative binomial, ordered categorical, scaled t and beta models. By way of illustration, this section presents just one example of a model beyond the usual GLM structure, the Cox proportional hazards model (Cox, 1972).[††]. With some care in the structuring of the computations, the computational cost can be kept to $O(Mnp^2)$

The model concerns event time data (also known as survival data). The idea is that for each subject we have observations of the time at which an event occurred (often death, in a medical context), along with some predictor variables that we think predict the risk of failure at any time. For some subjects we may only know that no event occurred up until a particular time. This is known as 'censoring': if the event occurred then it occurred after we last observed the subject. Technically we define the 'hazard', $h(t)$, as the probability density that the event time $T = t$, given that

[††]Actually as shown in section 3.1.10 the model can be estimated as a Poisson regression model, but only by creating pseudodata that raise the computational cost of fitting from $O(np^2)$ to $O(n^2p^2)$

$T \geq t$. If the event is death, the hazard is the instantaneous per capita death rate. The Cox model supposes that there is a baseline hazard function $h_0(t)$, and that a subject's individual hazard function is given by

$$h(t_i) = h_0(t_i) \exp(\mathbf{X}_i \boldsymbol{\beta})$$

where the linear predictor $\mathbf{X}_i \boldsymbol{\beta}$ contains no intercept term. So the role of the linear predictor is to modify the baseline hazard. The baseline hazard itself can be left unspecified by employing a *partial likelihood* for $\boldsymbol{\beta}$ which does not depend on $h_0(t)$ (see section 3.1.10, p. 116, for motivation). If $\mathbf{X}$ represents a smooth additive predictor (with intercept removed) then this partial likelihood can be penalized in the same way as any other likelihood.

Let the n data be of the form $(\tilde{t}_i, \mathbf{X}_i, \delta_i)$, i.e., an event time, model matrix row and an indicator of death (1) or censoring (0). Assume that the data are ordered so that the $\tilde{t}_i$ are non-increasing with i. The time data can conveniently be replaced by a vector $\mathbf{t}$ of n_t unique decreasing event times, and an n vector of indices, r, such that $t_{r_i} = \tilde{t}_i$.

The log partial likelihood, as in Hastie and Tibshirani (1990), is

$$l(\boldsymbol{\beta}) = \sum_{j=1}^{n_t} \left[\sum_{\{i:r_i=j\}} \delta_i \mathbf{X}_i \boldsymbol{\beta} - d_j \log \left\{ \sum_{\{i:r_i \leq j\}} \exp(\mathbf{X}_i \boldsymbol{\beta}) \right\} \right].$$

Now let $\eta_i \equiv \mathbf{X}_i \boldsymbol{\beta}$, $\gamma_i \equiv \exp(\eta_i)$ and $d_j = \sum_{\{i:r_i=j\}} \delta_i$ (i.e., the count of deaths at this event time). Then

$$l(\boldsymbol{\beta}) = \sum_{j=1}^{n_t} \left[\sum_{\{i:r_i=j\}} \delta_i \eta_i - d_j \log \left\{ \sum_{\{i:r_i \leq j\}} \gamma_i \right\} \right].$$

Further define $\gamma_j^+ = \sum_{\{i:r_i \leq j\}} \gamma_i$, so that we have the recursion

$$\gamma_j^+ = \gamma_{j-1}^+ + \sum_{\{i:r_i=j\}} \gamma_i$$

where $\gamma_0^+ = 0$. Then

$$l(\boldsymbol{\beta}) = \sum_{i=1}^{n} \delta_i \eta_i - \sum_{j=1}^{n_t} d_j \log(\gamma_j^+).$$

Turning to the gradient $g_k = \partial l / \partial \beta_k$, we have

$$\mathbf{g} = \sum_{i=1}^{n} \delta_i \mathbf{X}_i - \sum_{j=1}^{n_t} d_j \mathbf{b}_j^+ / \gamma_j^+$$

where $\mathbf{b}_j^+ = \mathbf{b}_{j-1}^+ + \sum_{\{i:r_i=j\}} \mathbf{b}_i$, $\mathbf{b}_i = \gamma_i \mathbf{X}_i$. and $\mathbf{b}_0^+ = \mathbf{0}$. The Hessian $H_{km} =$

$\partial^2 l/\partial\beta_k\partial\beta_m$ is given by

$$\mathbf{H} = \sum_{j=1}^{n_t} d_j \mathbf{b}_j^+ \mathbf{b}_j^{+\mathsf{T}}/\gamma_j^{+2} - d_j \mathbf{A}_j^+/\gamma_j^+$$

where $\mathbf{A}_j^+ = \mathbf{A}_{j-1}^+ + \sum_{\{i:r_i=j\}} \mathbf{A}_i$, $\mathbf{A}_i = \gamma_i \mathbf{X}_i \mathbf{X}_i^\mathsf{T}$ and $\mathbf{A}_0^+ = \mathbf{0}$.

Derivatives with respect to smoothing parameters

To obtain derivatives it will be necessary to obtain expressions for the derivatives of l and $\mathbf{H}$ with respect to $\rho_k = \log(\lambda_k)$. Firstly we have

$$\frac{\partial \eta_i}{\partial \rho_k} = \mathbf{X}_i \frac{\partial \hat{\boldsymbol{\beta}}}{\partial \rho_k}, \quad \frac{\partial \gamma_i}{\partial \rho_k} = \gamma_i \frac{\partial \eta_i}{\partial \rho_k}, \quad \frac{\partial \mathbf{b}_i}{\partial \rho_k} = \frac{\partial \gamma_i}{\partial \rho_k} \mathbf{X}_i \text{ and } \frac{\partial \mathbf{A}_i}{\partial \rho_k} = \frac{\partial \gamma_i}{\partial \rho_k} \mathbf{X}_i \mathbf{X}_i^\mathsf{T}.$$

Similarly $\frac{\partial^2 \eta_i}{\partial \rho_k \partial \rho_m} = \mathbf{X}_i \frac{\partial^2 \hat{\boldsymbol{\beta}}}{\partial \rho_k \partial \rho_m}$, $\frac{\partial^2 \gamma_i}{\partial \rho_k \partial \rho_m} = \gamma_i \frac{\partial \eta_i}{\partial \rho_k}\frac{\partial \eta_i}{\partial \rho_m} + \gamma_i \frac{\partial^2 \eta_i}{\partial \rho_k \partial \rho_m}$ and $\frac{\partial^2 \mathbf{b}_i}{\partial \rho_k \partial \rho_m} = \frac{\partial^2 \gamma_i}{\partial \rho_k \partial \rho_m} \mathbf{X}_i$.

Derivatives sum in the same way as the terms they relate to:

$$\frac{\partial l}{\partial \rho_k} = \sum_{i=1}^{n} \delta_i \frac{\partial \eta_i}{\partial \rho_k} - \sum_{j=1}^{n_t} \frac{d_j}{\gamma_j^+} \frac{\partial \gamma_j^+}{\partial \rho_k},$$

and

$$\frac{\partial^2 l}{\partial \rho_k \partial \rho_m} = \sum_{i=1}^{n} \delta_i \frac{\partial^2 \eta_i}{\partial \rho_k \partial \rho_m} + \sum_{j=1}^{n_t} \left(\frac{d_j}{\gamma_j^{+2}} \frac{\partial \gamma_j^+}{\partial \rho_m} \frac{\partial \gamma_j^+}{\partial \rho_k} - \frac{d_j}{\gamma_j^+} \frac{\partial^2 \gamma_j^+}{\partial \rho_k \partial \rho_m} \right),$$

while

$$\frac{\partial \mathbf{H}}{\partial \rho_k} = \sum_{j=1}^{n_t} \frac{d_j}{\gamma_j^{+2}} \left\{ \mathbf{A}_j^+ \frac{\partial \gamma_j^+}{\partial \rho_k} + \frac{\partial \mathbf{b}^+}{\partial \rho_k} \mathbf{b}^{+\mathsf{T}} + \mathbf{b}^+ \frac{\partial \mathbf{b}^{+\mathsf{T}}}{\partial \rho_k} \right\} - \frac{d_j}{\gamma_j^+} \frac{\partial \mathbf{A}_j^+}{\partial \rho_k} - \frac{2 d_j}{\gamma_j^{+3}} \mathbf{b}^+ \mathbf{b}^{+\mathsf{T}} \frac{\partial \gamma_j^+}{\partial \rho_k}$$

and

$$
\begin{aligned}
\frac{\partial^2 \mathbf{H}}{\partial \rho_k \partial \rho_m} &= \sum_{j=1}^{n_t} \frac{-2d_j}{\gamma_j^{+3}} \frac{\partial \gamma_j^+}{\partial \rho_m} \left\{ \mathbf{A}_j^+ \frac{\partial \gamma_j^+}{\partial \rho_k} + \frac{\partial \mathbf{b}^+}{\partial \rho_k} \mathbf{b}^{+\mathsf{T}} + \mathbf{b}^+ \frac{\partial \mathbf{b}^{+\mathsf{T}}}{\partial \rho_k} \right\} \\
&\quad + \frac{d_j}{\gamma_j^{+2}} \left\{ \frac{\partial \mathbf{A}_j^+}{\partial \rho_m} \frac{\partial \gamma_j^+}{\partial \rho_k} + \mathbf{A}_j^+ \frac{\partial^2 \gamma_j^+}{\partial \rho_k \partial \rho_m} + \frac{\partial^2 \mathbf{b}^+}{\partial \rho_k \partial \rho_m} \mathbf{b}^{+\mathsf{T}} \right. \\
&\quad \left. + \frac{\partial \mathbf{b}^+}{\partial \rho_k} \frac{\partial \mathbf{b}^{+\mathsf{T}}}{\partial \rho_m} + \frac{\partial \mathbf{b}^+}{\partial \rho_m} \frac{\partial \mathbf{b}^{+\mathsf{T}}}{\partial \rho_k} + \mathbf{b}^+ \frac{\partial^2 \mathbf{b}^{+\mathsf{T}}}{\partial \rho_k \partial \rho_m} \right\} \\
&\quad + \frac{d_j}{\gamma_j^{+2}} \frac{\partial \gamma_j^+}{\partial \rho_m} \frac{\partial \mathbf{A}_j^+}{\partial \rho_k} - \frac{d_j}{\gamma_j^+} \frac{\partial^2 \mathbf{A}_j^+}{\partial \rho_k \partial \rho_m} + \frac{6d_j}{\gamma_j^{+4}} \frac{\partial \gamma_j^+}{\partial \rho_m} \mathbf{b}^+ \mathbf{b}^{+\mathsf{T}} \frac{\partial \gamma_j^+}{\partial \rho_k} \\
&\quad - \frac{2d_j}{\gamma_j^{+3}} \left\{ \frac{\partial \mathbf{b}^+}{\partial \rho_m} \mathbf{b}^{+\mathsf{T}} \frac{\partial \gamma_j^+}{\partial \rho_k} + \mathbf{b}^+ \frac{\partial \mathbf{b}^{+\mathsf{T}}}{\partial \rho_m} \frac{\partial \gamma_j^+}{\partial \rho_k} + \mathbf{b}^+ \mathbf{b}^{+\mathsf{T}} \frac{\partial^2 \gamma_j^+}{\partial \rho_k \partial \rho_m} \right\}.
\end{aligned}
$$

In fact with suitable reparameterization it will only be necessary to obtain the second derivatives of the leading diagonal of $\mathbf{H}$, although the full first derivative of $\mathbf{H}$ matrices will be required. All that is actually needed is $\text{tr}\left(\boldsymbol{\mathcal{H}}^{-1} \partial^2 \mathbf{H} / \partial \rho_k \partial \rho_m\right)$. Consider the eigen-decomposition $\boldsymbol{\mathcal{H}}^{-1} = \mathbf{V} \boldsymbol{\Lambda} \mathbf{V}^\mathsf{T}$. We have

$$
\text{tr}\left(\boldsymbol{\mathcal{H}}^{-1} \frac{\partial \mathbf{H}}{\partial \theta}\right) = \text{tr}\left(\boldsymbol{\Lambda} \frac{\partial \mathbf{V}^\mathsf{T} \mathbf{H} \mathbf{V}}{\partial \theta}\right), \quad \text{tr}\left(\boldsymbol{\mathcal{H}}^{-1} \frac{\partial^2 \mathbf{H}}{\partial \theta_k \partial \theta_m}\right) = \text{tr}\left(\boldsymbol{\Lambda} \frac{\partial^2 \mathbf{V}^\mathsf{T} \mathbf{H} \mathbf{V}}{\partial \theta_k \partial \theta_m}\right).
$$

Since $\boldsymbol{\Lambda}$ is diagonal only the leading diagonal of the derivative of the reparameterized Hessian $\mathbf{V}^\mathsf{T} \mathbf{H} \mathbf{V}$ is required, and this can be efficiently computed by simply using the reparameterized model matrix $\mathbf{X} \mathbf{V}$. So the total cost of all derivatives is kept to $O(Mnp^2)$.

Prediction and the baseline hazard

For some purposes it is also important to estimate the baseline hazard. Klein and Moeschberger (2003, pages 283, 359, 381) give the details, which are restated here in a form suitable for efficient computation using the notation and assumptions of the previous sections.

1. The estimated cumulative baseline hazard ($H_0(t) = \int_0^t h_0(x) dx$) is

$$
H_0(t) = \begin{cases} h_j & t_j \le t < t_{j-1} \\ 0 & t < t_{n_t} \\ h_1 & t \ge t_1 \end{cases}
$$

where the following back recursion defines h_j,

$$
h_j = h_{j+1} + \frac{d_j}{\gamma_j^+}, \quad h_{n_t} = \frac{d_{n_t}}{\gamma_{n_t}^+}.
$$

2. The variance of the estimated cumulative hazard is given by the back recursion

$$q_j = q_{j+1} + \frac{d_j}{\gamma_j^{+2}}, \quad q_{n_t} = \frac{d_{n_t}}{\gamma_{n_t}^{+2}}.$$

3. The estimated survival function for time t, covariate vector $\mathbf{x}$, is

$$\hat{S}(t, \mathbf{x}) = \exp\{-H_0(t)\}^{\exp(\mathbf{x}^\mathsf{T}\boldsymbol{\beta})}$$

and consequently $\log \hat{S}(t, \mathbf{x}) = -H_0(t)\exp(\mathbf{x}^\mathsf{T}\boldsymbol{\beta})$. Let $\hat{S}_i$ denote the estimated version for the ith subject, at their event time.

4. The estimated standard deviation of $\hat{S}(t, \mathbf{x})$ is‡‡

$$\exp(\mathbf{x}^\mathsf{T}\hat{\boldsymbol{\beta}})\hat{S}(t, \mathbf{x})(q_i + \mathbf{v}_i^\mathsf{T}\mathbf{V}_\beta\mathbf{v}_i)^{1/2}, \quad \text{if} \quad t_i \le t < t_{i-1} \tag{6.25}$$

where $\mathbf{v}_i = \mathbf{a}_i - \mathbf{x}h_i$, and the vector $\mathbf{a}_i$ is defined by the back recursion

$$\mathbf{a}_i = \mathbf{a}_{i+1} + \mathbf{b}_i^+\frac{d_i}{\gamma_i^{+2}}, \quad \mathbf{a}_{n_t} = \mathbf{b}_{n_t}^+\frac{d_{n_t}}{\gamma_{n_t}^{+2}}.$$

For efficient prediction with standard errors, there seems to be no choice but to compute the n_t, $\boldsymbol{a}_i$ vectors at the end of fitting and store them.

5. Martingale residuals are defined as

$$\hat{M}_j = \delta_j + \log \hat{S}_j,$$

and deviance residuals as

$$\hat{D}_j = \text{sign}(\hat{M}_j)[-2\{\hat{M}_j + \delta_j \log(-\log \hat{S}_j)\}]^{1/2},$$

the latter also being useful for computing a deviance.

6.7 Initial smoothing parameter guesses

The smoothing parameter optimization methods given so far are numerically robust and not dependent on having very good starting values for the smoothing parameters. However, they do require that we start out from smoothing parameter values that are neither effectively zero, nor infinity. Then the effective degrees of freedom of each smooth will be between its minimum and maximum possible values and the model coefficients will be sensitive to small changes in the log smoothing parameters, so that local derivative information is useful for deciding how to update the parameters.

A cheap way of achieving this is to insist that the leading diagonals of $\mathbf{X}^\mathsf{T}\mathbf{W}\mathbf{X}$ and $\mathbf{S}_\lambda$ should be roughly balanced, at least for elements for which the penalty is non-zero. `mgcv` adopts the following simple heuristic. If s denotes the non zero elements of $\text{diag}(\mathbf{S}_j)$ and $\mathbf{d}$ denotes the corresponding elements of $\text{diag}(\mathbf{X}^\mathsf{T}\mathbf{W}\mathbf{X})$, then choose λ_j so that $\text{mean}(\mathbf{d}/(\mathbf{d} + \lambda_j\mathbf{s})) \approx 0.4$. For models estimable by PIRLS, the initial estimate of $\mathbf{W}$ is used. For more general models an initial estimate of the diagonal of the Hessian of the negative log likelihood is substituted for $\text{diag}(\mathbf{X}^\mathsf{T}\mathbf{W}\mathbf{X})$.

‡‡There is a typo in (8.8.5) of Klein and Moeschberger (2003). Expression (10) in Andersen et al. (1996) gives the correct result used here.

6.8 Methods for generalized additive mixed models

As we have seen in sections 4.2.4 and 5.8 (pages 172 and 239) there is a duality between smooths and random effects, that is further emphasised by the use of marginal likelihood methods for smoothing parameter estimation. In consequence simple Gaussian random effects can also be estimated as if they were smooth terms in the model, using the estimation methods already covered in this chapter (or those covered subsequently): the methods covered so far already deal with many generalized additive mixed models.

An alternative is to estimate the smooths in a GAMM as if they were random effects in a (generalized) linear mixed model, using general purpose mixed model methods and software for the purpose. This is the way that `mgcv`'s `gamm` function operates, using `lme` from the `nlme` package for estimation, either directly or within a PQL loop. `gamm4` from the `gamm4` package does the same, but using `lmer` or `glmer` from the `lme4` package as the fitting engine.

The main difficulty with this way of working is that the smooths have to be set up in a form that corresponds to a random effect structure that the software recognises. For any smooth with a single smoothing parameter this is easy. The recipe given in section 5.8 is followed to represent each such smooth explicitly in the form $\mathbf{X}\boldsymbol{\beta}+\mathbf{Z}\mathbf{b}$ where $\boldsymbol{\beta}$ is treated as a vector of (unpenalized) fixed effects and $\mathbf{b} \sim N(\mathbf{0}, \mathbf{I}\sigma_b^2)$ is treated as a vector of i.i.d. random effects. Any sum-to-zero constraints on the smooth terms are usually absorbed into the bases before this re-parameterization (although Wood et al., 2013, discuss several alternatives for constraining terms).

Tensor product smooths are slightly more problematic. The construction of section 5.6.5 (p. 235) is designed so that a simple partitioning of the columns of the smooth's model matrix and its parameter vector, allows the smooth to be represented in the form $\mathbf{X}\boldsymbol{\beta}+\mathbf{Z}_1\mathbf{b}_1+\mathbf{Z}_2\mathbf{b}_2+\ldots$, where $\boldsymbol{\beta}$ is again a vector of fixed parameters, and the $\mathbf{b}_j$ are independent vectors of random effects $\mathbf{b}_j \sim N(\mathbf{0}, \mathbf{I}\sigma_{b_j}^2)$. Such terms are as readily estimated as smooths with single penalties.

When representing tensor product smooths that do not have separable penalties (see section 5.6, p. 227), the positive semi-definite pseudoinverse of the covariance matrix for the smooth coefficients, $\tilde{\boldsymbol{\beta}}$ say, is now of the form $\sum_{i=1}^d \lambda_i \tilde{\mathbf{S}}_i$, where $\tilde{\mathbf{S}}_i$ is defined in section 5.6.2 (p. 229). The degree of rank deficiency of this matrix, M_T, is readily shown to be given by the product of the dimensions of the null spaces of the marginal penalty matrices, $\mathbf{S}_i$ (provided that $\lambda_i > 0 \; \forall \; i$). Again re-parameterization is needed, this time by forming

$$\sum_{i=1}^d \tilde{\mathbf{S}}_i = \mathbf{U}\mathbf{D}\mathbf{U}^\mathsf{T}$$

where $\mathbf{U}$ is an orthogonal matrix of eigenvectors, and $\mathbf{D}$ is a diagonal matrix of eigenvalues, with M_T zero elements at the end of the leading diagonal. Notice that there are no λ_i parameters in the sum that is decomposed: this is reasonable since the null space of the penalty does not depend on these parameters (however given finite precision arithmetic it might be necessary to scale the $\tilde{\mathbf{S}}_i$ matrices in some cases).

It is not now possible to achieve the sort of simple representation of a term that was obtained with a single penalty, so the re-parameterization is actually simpler. Suppose that $\tilde{\mathbf{X}}$ is the original model matrix for the tensor product smooth. Partitioning the eigenvector matrix so that $\mathbf{U} \equiv [\mathbf{U}_R : \mathbf{U}_F]$ where $\mathbf{U}_F$ has M_T columns, it is necessary to define $\mathbf{X} \equiv \tilde{\mathbf{X}}\mathbf{U}_F$, $\mathbf{Z} \equiv \tilde{\mathbf{X}}\mathbf{U}_R$ and $\boldsymbol{\mathcal{S}}_i = \mathbf{U}_R^{\mathsf{T}}\tilde{\mathbf{S}}_i\mathbf{U}_R$. A mixed model representation of the tensor product term (i.e. the linear predictor and random effects distribution) is now

$$\mathbf{X}\boldsymbol{\beta} + \mathbf{Z}\mathbf{b} \text{ where } \mathbf{b} \sim N\left(\mathbf{0}, \left(\sum \lambda_i \boldsymbol{\mathcal{S}}_i\right)^{-1}\right),$$

and the λ_i and $\boldsymbol{\beta}_F$ parameters have to be estimated. This covariance matrix structure is not a form that is available in standard software, but it turns out to be possible to implement, at least in the `nlme` software of Pinheiro and Bates (2000): R package `mgcv` provides an appropriate '`pdMat`' class for `lme` called '`pdTens`'.[†] Given such a class, incorporation of one or more tensor product terms into a (generalized) linear mixed model is straightforward.

6.8.1 *GAMM inference with mixed model estimation*

When estimating GAMMs as mixed models, we also need to compute confidence/credible intervals, as for a GAM. In particular, credible regions for the smooth components are required. To do this, let $\boldsymbol{\beta}$ now contain all the fixed effects *and* the random effects for the smooth terms (only), and let $\bar{\mathbf{X}}$ be the corresponding model matrix. Let $\bar{\mathbf{Z}}$ be the random effects model matrix *excluding* the columns relating to smooths, and let $\bar{\boldsymbol{\psi}}_\theta$ be the corresponding random effects covariance matrix. Now define a covariance matrix

$$\mathbf{V} = \bar{\mathbf{Z}}\bar{\boldsymbol{\psi}}_\theta\bar{\mathbf{Z}}^{\mathsf{T}} + \boldsymbol{\Lambda}\sigma^2,$$

where $\boldsymbol{\Lambda}$ is the estimated residual covariance matrix of the linear mixed model (or working linear mixed model in the generalized case). Essentially the Bayesian approach of sections 4.2.4, 5.8 and 6.10 implies that

$$\boldsymbol{\beta} \sim N(\hat{\boldsymbol{\beta}}, (\bar{\mathbf{X}}^{\mathsf{T}}\mathbf{V}^{-1}\bar{\mathbf{X}} + \mathbf{S})^{-1})$$

where $\mathbf{S} = \sum \lambda_i/\sigma^2\mathbf{S}_i$. Similarly the leading diagonal of

$$\mathbf{F} = (\bar{\mathbf{X}}^{\mathsf{T}}\mathbf{V}^{-1}\bar{\mathbf{X}} + \mathbf{S})^{-1}(\bar{\mathbf{X}}^{\mathsf{T}}\mathbf{V}^{-1}\bar{\mathbf{X}})$$

gives the effective degrees of freedom for each element of $\boldsymbol{\beta}$. Note that it is generally a good idea to exploit the special structure of $\mathbf{V}$ when computing $\bar{\mathbf{X}}^{\mathsf{T}}\mathbf{V}^{-1}\bar{\mathbf{X}}$, otherwise it has the potential to be rather expensive.

[†]Note that the EM optimizer for `lme` is not designed to work with user defined `pdMat` classes, such as `pdTens`, so only Newton optimization is possible when using this class.

6.9 Bigger data methods

We saw in section 6.5.2 that in the context of performance iteration (or a Gaussian-identity link model) the model coefficients and REML criteria can both be computed using only the upper triangular factor $\mathbb{R}$ from the QR decomposition $\sqrt{\mathbf{W}}\mathbf{X} = \mathbb{QR}$, along with $\mathbb{Y} = \mathbb{Q}^\mathsf{T}\sqrt{\mathbf{W}}\mathbf{z}$, $r = \|\sqrt{\mathbf{W}}\mathbf{z}\|^2 - \|\mathbb{Y}\|^2$ and the penalty matrix $\mathbf{S}_\lambda$. So basically, the $n \times p$ (weighted) model matrix can be replaced by its $p \times p$ triangular factor. This means that if we can iteratively compute the $\mathbb{R}$ and $\mathbb{Y}$ from successive blocks of $\sqrt{\mathbf{W}}\mathbf{X}$, then we need never form $\sqrt{\mathbf{W}}\mathbf{X}$ as a whole, thereby allowing much larger sample sizes, n, than could otherwise be accommodated by available computer memory. To see the potential importance of this, consider a model matrix with 10^6 rows and 1000 columns: its memory footprint would be around 8 Gb before we did anything with it, while the corresponding $\mathbb{R}$ requires about 1000 times less storage. Prediction error criteria are also susceptible to this sort of upfront reduction in problem size. For example, the GCV score is

$$\mathcal{V}_g = \frac{n\|\sqrt{\mathbf{W}}(\mathbf{z} - \mathbf{X}\boldsymbol{\beta})\|^2}{[n - \text{tr}\{(\mathbf{X}^\mathsf{T}\mathbf{W}\mathbf{X} + \mathbf{S}_\lambda)^{-1}\mathbf{X}^\mathsf{T}\mathbf{W}\mathbf{X}\}]^2} = \frac{n\|\mathbb{Y} - \mathbb{R}\boldsymbol{\beta}\|^2 + r}{[n - \text{tr}\{(\mathbb{R}^\mathsf{T}\mathbb{R} + \mathbf{S}_\lambda)^{-1}\mathbb{R}^\mathsf{T}\mathbb{R}\}]^2}.$$

To understand how a QR decomposition can be obtained iteratively, first consider the simple case in which $\mathbf{X}$ is partitioned into just two blocks so that $\mathbf{X} = \begin{bmatrix} \mathbf{X}_0 \\ \mathbf{X}_1 \end{bmatrix}$, and similarly $\mathbf{z} = \begin{bmatrix} \mathbf{z}_0 \\ \mathbf{z}_1 \end{bmatrix}$. $\mathbf{X}_0$ and $\mathbf{z}_0$ both have n_0 rows, while $\mathbf{X}_1$ and $\mathbf{z}_1$ both have n_1 rows. $n_0 + n_1 = n$. Now form QR decompositions $\mathbf{X}_0 = \mathbf{Q}_0\mathbf{R}_0$ and $\begin{bmatrix} \mathbf{R}_0 \\ \mathbf{X}_1 \end{bmatrix} = \mathbf{Q}_1\mathbb{R}$. It is routine to check that $\mathbf{X} = \mathbb{QR}$ where $\mathbb{Q} = \begin{bmatrix} \mathbf{Q}_0 & \mathbf{0} \\ \mathbf{0} & \mathbf{I} \end{bmatrix}\mathbf{Q}_1$ ($\mathbf{I}$ is $n_1 \times n_1$ here) and $\mathbb{Q}^\mathsf{T}\mathbf{z} = \mathbf{Q}_1^\mathsf{T}\begin{bmatrix} \mathbf{Q}_0^\mathsf{T}\mathbf{z}_0 \\ \mathbf{z}_1 \end{bmatrix}$. The same applies if $\mathbf{X}$ and $\mathbf{z}$ are both pre-multiplied by $\sqrt{\mathbf{W}}$, and this basic idea can be applied repeatedly if $\mathbf{X}$ is split into more than two blocks, so that we can indeed form $\mathbb{R}$ and $\mathbb{Y}$ while only forming $\mathbf{X}$ block-wise. In fact there are many ways of doing this, including updating $\mathbb{R}$ using one row of $\mathbf{X}$ at a time via Givens rotations. Notice, however, that the leading order operations cost of the decomposition is still $O(np^2)$ when using these schemes — it is only the memory footprint that is reduced. A small memory footprint performance oriented iteration scheme then looks like the following.

Initialization: Let $\mathbf{x}_i$ denote the vector of covariates associated with response variable y_i where $i = 1, \ldots, n$, and divide the integers 1 to n into M non-overlapping subsets $\gamma_1, \ldots, \gamma_M$, of approximately equal size (so that $\cup_i\gamma_i = \{1, \ldots, n\}$ while $\gamma_j \cap \gamma_i = \emptyset$ for all $i \neq j$). M is chosen to avoid running out of computer memory. Let $\bar{\eta}_i = g(y_i + \delta_i)$ (with δ_i as defined in section 6.1.1). Set the PIRLS iteration index $q = 0$ and $D = 0$. Perform any initialization steps necessary to set up the bases for the smooth terms.

Iteration:

1. Set $D_{\text{old}} = D$, $\mathbb{R}$ to be a $0 \times p$ matrix, $\mathbb{Y}$ a 0-vector, $D = 0$ and $r = 0$.
2. Repeat the following steps (a)–(f) for $k = 1, \ldots, M$,
   (a) Set $\mathbf{Y}_0 = \mathbb{Y}$ and $\mathbf{R}_0 = \mathbb{R}$.
   (b) Form the model sub-matrix $\mathbf{X}_k$ for the covariate set $\{\mathbf{x}_i : i \in \gamma_k\}$.
   (c) If $q > 0$ form $\hat{\boldsymbol{\eta}} = \mathbf{X}_k \hat{\boldsymbol{\beta}}_\lambda$, otherwise $\hat{\boldsymbol{\eta}} = \bar{\boldsymbol{\eta}}_{\gamma_k}$.
   (d) Form $\hat{\mu}_i = g^{-1}(\hat{\eta}_i)$, $z_i = g'(\hat{\mu}_i)(y_i - \hat{\mu}_i) + \hat{\eta}_i$, and $w_i = V(\hat{\mu}_i)g'(\hat{\mu}_i)^{-2}$ $\forall i \in \gamma_k$. Let $\mathbf{z}$ be the vector containing these z_i values and $\mathbf{W}$ be the diagonal matrix of corresponding w_i values.
   (e) Set $r \leftarrow r + \|\sqrt{\mathbf{W}}\mathbf{z}\|^2$, calculate the deviance residuals for the current subset of data and add the sum of squares of these to D.
   (f) Form $\mathbf{Q}\mathbb{R} = \begin{bmatrix} \mathbf{R}_0 \\ \sqrt{\mathbf{W}}\mathbf{X}_k \end{bmatrix}$ and $\mathbb{Y} = \mathbf{Q}^\mathsf{T} \begin{bmatrix} \mathbf{Y}_0 \\ \sqrt{\mathbf{W}}\mathbf{z} \end{bmatrix}$ and discard $\mathbf{Q}$.
3. Set $\|\mathbf{r}\|^2 = r - \|\mathbb{Y}\|^2$.
4. If $q > 0$ test for convergence by comparing the current deviance D to the previous deviance D_{old}. Stop if convergence has been reached (or q has exceeded some predetermined limit suggesting failure).
5. Estimate $\boldsymbol{\lambda}$ by optimizing the GCV or REML score based on $\mathbb{R}$, $\mathbb{Y}$ and r.This also yields $\hat{\boldsymbol{\beta}}_\lambda$.
6. $q \leftarrow q + 1$.

At convergence the final $\hat{\boldsymbol{\beta}}_\lambda$ and $\boldsymbol{\lambda}$ are the coefficient and smoothing parameter estimates. The large sample posterior approximation becomes

$$\boldsymbol{\beta}|\mathbf{y} \sim N(\hat{\boldsymbol{\beta}}_\lambda, (\mathbb{R}^\mathsf{T}\mathbb{R} + \mathbf{S}_\lambda)^{-1}\phi),$$

where any ϕ estimate required is obtained as part of REML optimization or as

$$\hat{\phi} = \frac{\|\mathbb{Y} - \mathbb{R}\hat{\boldsymbol{\beta}}_\lambda\|^2 + \|\mathbf{r}\|^2}{n - \text{tr}\{(\mathbb{R}^\mathsf{T}\mathbb{R} + \mathbf{S}_\lambda)^{-1}\mathbb{R}^\mathsf{T}\mathbb{R}\}},$$

although again it might be preferable to substitute the Fletcher (2012) estimate covered in section 3.1.5 (p. 110). See Wood et al. (2015) for more detail.

6.9.1 *Bigger still*

Wood et al. (2017) developed methods that were applied to about 10^7 data and 10^4 model coefficients based on three ideas:

1. Pivoted Cholesky decomposition can be made to scale relatively well to multi-core computing, unlike QR and eigen-decompositions.
2. We can produce a very efficient performance oriented iteration for REML optimization using only Cholesky decomposition and basic matrix products, if we can avoid having to actually evaluate the REML score itself.

3. Often covariates take only a modest number of discrete values (e.g., temperature recorded to the nearest tenth of a degree), or they can be rounded so that this is true. In that case enormous savings in storing and computing with the model matrix are possible, using a set of algorithms developed in Wood et al. (2017).

The basic issue with point 1 is that if we want to do matrix computations in a way that efficiently uses multiple computational cores, then we have to use algorithms which are structured so that the bulk of the work is in the form of matrix-matrix operations on sub-blocks of the matrices concerned. This is because matrix-matrix multiplication has the property that the elements of the matrices are re-visited repeatedly, whereas a matrix-vector multiplication visits each element of the matrix only once. Repeated revisiting of matrix elements is the key to fast memory access, since then the elements concerned can be stored in fast cache memory for almost instant access by the computational cores, provided the (sub) matrices are small enough. If the required elements can not be cached and re-used repeatedly then the computation becomes 'memory bandwidth limited' as the core has to wait for the slow process of loading each matrix element from memory. This bandwidth limitation becomes worse and worse as the number of cores requiring simultaneous memory access increases. Lucas (2004) provides a pivoted Cholesky algorithm in which most of the work is in the form of matrix-matrix operations on sub-blocks of the matrix. In contrast, the Quintana-Ortí et al. (1998) pivoted QR algorithm only has around half of the work performed by these memory efficient matrix-matrix operations, and hence scales poorly to multiple core computing. Eigen-routines are even more difficult.

The idea with point 2 is that we really only need the QR and eigen decompositions to deal with the instability in computing the log determinant terms in the REML score. But the derivatives of the REML criterion are much better behaved, and do not require such careful handling. Now during optimization the Newton step can be computed from the first and second derivatives of the REML, and the REML criterion itself is only needed to check that the step has improved the REML criterion. An alternative is to check that the derivative of the REML at the end of the Newton step is not decreasing (having first checked that we have not converged). This computation only requires derivatives, and for a unimodal REML will guarantee convergence just as well as the original check (with multimodality it could theoretically be defeated by a pathological enough function, and never converge). The Wood et al. (2017) method uses this approach to come up with an iteration that only requires Cholesky decomposition, and interleaves single Newton steps, to update smoothing parameters, with updates of the model coefficients. So efficiency is also gained by not fully optimizing the working REML criterion at each step, since it is usually wasteful to be carefully optimizing a criterion which will anyway be immediately discarded.

The idea with 3 is best seen by example. Suppose that the covariate for the jth model term is discretized into m_j discrete values. Then the model matrix columns for that term can be written as

$$\mathbf{X}_j(i,l) = \bar{\mathbf{X}}_j(k_j(i),l)$$

where $\bar{\mathbf{X}}_j$ has only m_j rows and $\mathbf{k}_j$ is an index vector. Storing $\bar{\mathbf{X}}_j$ and $\mathbf{k}_j$ uses much less memory than storing $\mathbf{X}_j$ directly. There are also savings in computing products with the model matrix. For example

$$\mathbf{X}_j^\mathsf{T}\mathbf{y} = \bar{\mathbf{X}}_j^\mathsf{T}\bar{\mathbf{y}} \text{ where } \bar{y}_l = \sum_{k_j(i)=l} y_i,$$

which costs $O(n) + O(m_j p_j)$ floating point operations, where n is the number of data and p_j the number of columns of $\mathbf{X}_j$. For large n this reduces the floating point operation count by a factor of about p_j, relative to forming $\mathbf{X}_j^\mathsf{T}\mathbf{y}$ directly. Lang et al. (2014) used this idea for single smooth terms. Wood et al. (2017) provide the extensions for working with tensor product smooths and multiple smooth terms where cross products such as $\mathbf{X}_j^\mathsf{T}\mathbf{W}\mathbf{X}_k$ are required for $k \neq j$.

See Wood et al. (2017) for full details and an application to air pollution modelling. The `bam` function in `mgcv` uses the method if the option `discrete=TRUE` is used (in which case argument `nthreads` controls the number of cores to use for computation).

6.10 Posterior distribution and confidence intervals

As covered in sections 4.2.4 (p. 172) and 5.8 (p. 239), if we take the Bayesian view of the smoothing process, in which $\boldsymbol{\beta}$ has a zero mean improper Gaussian prior distribution with precision matrix proportional to $\mathbf{S}_\lambda$, then, following sections 2.4.2 and 3.4.3 (pages 80 and 151), the posterior distribution of $\boldsymbol{\beta}$ is

$$\boldsymbol{\beta}|\mathbf{y}, \boldsymbol{\lambda} \sim N(\hat{\boldsymbol{\beta}}, \mathbf{V}_\beta). \tag{6.26}$$

In the Gaussian identity link case $\mathbf{V}_\beta = (\mathbf{X}^\mathsf{T}\mathbf{X} + \mathbf{S}_\lambda)^{-1}\sigma^2$, following directly from (2.17) in section 2.4.2. In the exponential family case (or for any model estimable by PIRLS), then $\mathbf{V}_\beta = (\mathbf{X}^\mathsf{T}\mathbf{W}\mathbf{X} + \mathbf{S}_\lambda)^{-1}\phi$, from (3.21) in section 3.4.3. In the case of a general regular likelihood this becomes $\mathbf{V}_\beta = (\hat{\boldsymbol{\mathcal{I}}} + \mathbf{S}_\lambda)^{-1}$, where $\hat{\boldsymbol{\mathcal{I}}}$ is the Hessian of the negative log likelihood at $\hat{\boldsymbol{\beta}}$, or its expectation.

These results can be used for computing Bayesian credible intervals for any quantity, α, predicted by the model. The obvious approach is to simulate replicate coefficient vectors from (6.26), and to compute α for each such replicate, to produce a sample from the posterior distribution of $\alpha|\mathbf{y}$. In `mgcv` the `type="lpmatrix"` argument to `predict.gam` facilitates this, by enabling calculation of the matrix $\tilde{\mathbf{X}}$ mapping the coefficients to the model linear predictor for any supplied values of the covariates.

For any quantity that is linear in the model parameters there is no need to simulate and matters are more straightforward. For example, consider a component function of a model, $f(x)$, say. Suppose we want to produce credible intervals for this at a series of values of x. If $\hat{\mathbf{f}}$ is the vector containing $f(x)$ at the evaluation points then we can write $\hat{\mathbf{f}} = \tilde{\mathbf{X}}\boldsymbol{\beta}$, where the matrix $\tilde{\mathbf{X}}$ will have zeros in the columns corresponding to coefficients having nothing to do with $f(x)$, while its other columns contain the basis functions for f evaluated at each of the values of x. Let $\mathbf{v} = \text{diag}(\tilde{\mathbf{X}}\mathbf{V}_\beta\tilde{\mathbf{X}}^\mathsf{T})$. Then $\hat{f}_i \pm z_{\alpha/2}\sqrt{v_i}$ is an approximate $(1-\alpha)100\%$ credible interval for f_i. Note the practical detail that $\mathbf{v}$ would never be computed by forming $\tilde{\mathbf{X}}\mathbf{V}_\beta\tilde{\mathbf{X}}^\mathsf{T}$ and then extracting its leading diagonal; rather, we would only form the leading diagonal elements. For example, in R, `v <- rowSums((X%*%V)*X)` produces the same result as `v <- diag(X%*%V%*%t(X))`, but much more efficiently.

It turns out that the Bayesian credible intervals for component functions of the model have surprisingly good frequentist coverage properties. In simulations from a known truth, a 95% Bayesian credible interval for a component smooth function shows close to 95% observed coverage of the truth, *when averaged across the domain of the function*. So taken as a whole the intervals have close to nominal coverage, but they may do this by trading off over-coverage in some part of the function domain for under-coverage elsewhere. The next subsection considers why (and when) this close to nominal coverage occurs.

6.10.1 Nychka's coverage probability argument

The problem with constructing confidence intervals for smooth terms in a model is the smoothing bias, which has to be corrected or accounted for in order to obtain confidence intervals with good properties. It turns out that the Bayesian confidence intervals can be viewed as including a component accounting for bias and that this explains their coverage properties. The main ideas can be understood by considering the Gaussian identity link case. First consider the decomposition

$$\begin{aligned}\mathbb{E}\|\mathbf{B}(\hat{\boldsymbol\beta}-\boldsymbol\beta)\|^2 &= \mathbb{E}\|\mathbf{B}(\hat{\boldsymbol\beta}-\mathbb{E}\hat{\boldsymbol\beta})\|^2 + \|\mathbf{B}(\mathbb{E}\hat{\boldsymbol\beta}-\boldsymbol\beta)\|^2\\ &= \mathrm{tr}(\mathbf{B}\mathbf{V}_{\hat{\boldsymbol\beta}}\mathbf{B}^\mathsf{T}) + \|\mathbf{B}(\mathbf{F}\boldsymbol\beta-\boldsymbol\beta\|^2\end{aligned}$$

where $\mathbf{B}$ is some general matrix of coefficients, $\mathbf{F} = (\mathbf{X}^\mathsf{T}\mathbf{X}+\mathbf{S}_\lambda)^{-1}\mathbf{X}^\mathsf{T}\mathbf{X}$, $\mathbf{V}_{\hat{\boldsymbol\beta}} = (\mathbf{X}^\mathsf{T}\mathbf{X}+\mathbf{S}_\lambda)^{-1}\mathbf{X}^\mathsf{T}\mathbf{X}(\mathbf{X}^\mathsf{T}\mathbf{X}+\mathbf{S}_\lambda)^{-1}$ and the expected sum of squared differences between $\mathbf{B}\boldsymbol\beta$ and its estimate has been decomposed into a variance and a squared bias component, using the fact that $\mathbb{E}\hat{\boldsymbol\beta} = \mathbf{F}\boldsymbol\beta$.

One way to estimate the squared bias is with $\|\mathbf{B}(\mathbb{E}\hat{\boldsymbol\beta}-\boldsymbol\beta)\|^2 \approx \|\mathbf{B}(\mathbf{F}\hat{\boldsymbol\beta}-\hat{\boldsymbol\beta})\|^2$. Another approach is to use the expectation of $\|\mathbf{B}(\mathbf{F}\boldsymbol\beta-\boldsymbol\beta)\|^2$ according to the prior mean and variance of $\boldsymbol\beta$; i.e., to use the average value of the squared bias term expected if the Bayesian model is correct. The derivation is complicated by $\mathbf{S}_\lambda$'s lack of full rank, which can be handled using a reparameterization like that of section 5.4.2. However, the most transparent derivation simply turns $\mathbf{S}_\lambda$ into a formally full rank matrix using a null space penalty as in section 5.4.3 (p. 214), but with the associated smoothing parameter set to such a small value that $\hat{\boldsymbol\beta}$ is identical to the parameter estimate without the penalty, to machine precision.

Using $\mathbb{E}_\pi$ to denote expectation w.r.t. the prior density, and noting that $\mathbf{F}-\mathbf{I} = -(\mathbf{X}^\mathsf{T}\mathbf{X}+\mathbf{S}_\lambda)^{-1}\mathbf{S}_\lambda$, we then have

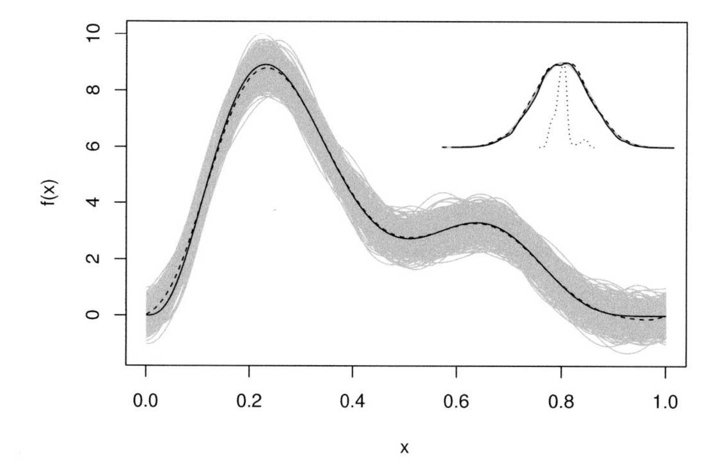

Figure 6.7 *The Nychka (1988) idea. The main black curve shows a true function $f(x)$, while the grey curves show 500 replicate spline estimates $\hat{f}(x)$. The dashed curve is $\mathbb{E}\hat{f}(x)$. Inset at top right are scaled kernel smooth estimates of the distributions of the sampling error, $\hat{f} - \mathbb{E}\hat{f}$ (continuous black); the bias, $\mathbb{E}\hat{f} - f$, evaluated at a random x (dotted) and $\hat{f} - f$ evaluated at a random x (dashed). In grey is the normal approximation to the dashed curve. Evaluation at a random x turns the bias into a random variable, which has substantially lower variance than the approximately normal $\hat{f} - \mathbb{E}\hat{f}$. Hence the sum of the randomized bias and sampling error is approximately normally distributed. The variance of this sum turns out to be well approximated by the Bayesian posterior covariance for f.*

$$
\begin{aligned}
\mathbb{E}\|\mathbf{B}(\hat{\boldsymbol{\beta}} - \boldsymbol{\beta})\|^2 &\simeq \text{tr}(\mathbf{B}\mathbf{V}_{\hat{\boldsymbol{\beta}}}\mathbf{B}^\mathsf{T}) + \mathbb{E}_\pi\|\mathbf{B}(\mathbf{F}\boldsymbol{\beta} - \boldsymbol{\beta})\|^2 \\
&= \text{tr}(\mathbf{B}\mathbf{V}_{\hat{\boldsymbol{\beta}}}\mathbf{B}^\mathsf{T}) + \mathbb{E}_\pi \text{tr}\{\mathbf{B}(\mathbf{F} - \mathbf{I})\boldsymbol{\beta}\boldsymbol{\beta}^\mathsf{T}(\mathbf{F} - \mathbf{I})^\mathsf{T}\mathbf{B}^\mathsf{T}\} \\
&= \text{tr}(\mathbf{B}\mathbf{V}_{\hat{\boldsymbol{\beta}}}\mathbf{B}^\mathsf{T}) + \text{tr}\{\mathbf{B}(\mathbf{F} - \mathbf{I})\mathbf{S}_\lambda^{-1}(\mathbf{F} - \mathbf{I})^\mathsf{T}\mathbf{B}^\mathsf{T}\}\sigma^2 \\
&= \sigma^2 \text{tr}[\mathbf{B}\{(\mathbf{X}^\mathsf{T}\mathbf{X} + \mathbf{S}_\lambda)^{-1}\mathbf{X}^\mathsf{T}\mathbf{X}(\mathbf{X}^\mathsf{T}\mathbf{X} + \mathbf{S}_\lambda)^{-1} \\
&\qquad + (\mathbf{X}^\mathsf{T}\mathbf{X} + \mathbf{S}_\lambda)^{-1}\mathbf{S}_\lambda(\mathbf{X}^\mathsf{T}\mathbf{X} + \mathbf{S}_\lambda)^{-1}\}\mathbf{B}^\mathsf{T}] \\
&= \sigma^2 \text{tr}\{\mathbf{B}(\mathbf{X}^\mathsf{T}\mathbf{X} + \mathbf{S}_\lambda)^{-1}\mathbf{B}^\mathsf{T}\} = \text{tr}(\mathbf{B}^\mathsf{T}\mathbf{B}\mathbf{V}_\beta). \qquad (6.27)
\end{aligned}
$$

The construction of a confidence interval now follows Nychka (1988), and is illustrated in figure 6.7. Suppose that we have observations of the covariate of the smooth of interest, $x_1, x_2, \ldots, x_n$ and let I be a discrete random variable with uniform distribution on $1, 2, \ldots, n$. We seek constants d and c_i such that

$$
\text{ACP} = \Pr\{|\hat{f}(x_I) - f(x_I)| \le dc_I\} = 1 - \alpha, \qquad (6.28)
$$

i.e., such that interval $\hat{f}(x_i) \pm dc_i$ has a probability $1 - \alpha$ of including $f(x_i)$, *on average across the function*. Now define random variables $V = \{\hat{f}(x_I) - \mathbb{E}\hat{f}(x_I)\}/c_I$

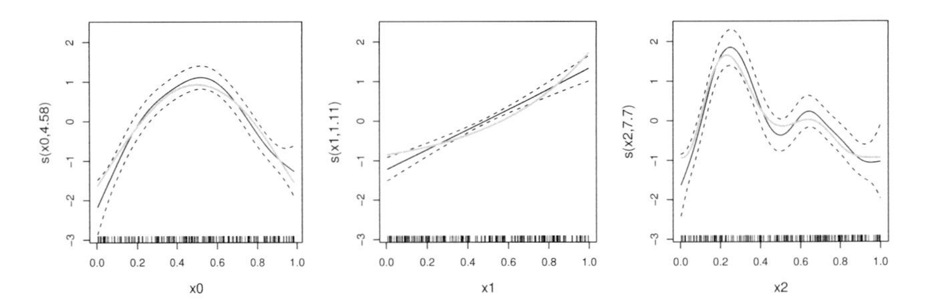

Figure 6.8 *Functions used for the interval coverage simulations of section 6.10.1 in grey, overlaid on some estimates of those functions from a sample of size 200 binomial observations with binomial $n = 5$. Notice that the middle function is quite close to a straight line, and can easily be estimated as such from noisy data, causing failure of the Nychka assumption of sampling error exceeding bias.*

and $B = \{\mathbb{E}\hat{f}(x_I) - f(x_I)\}/c_I$, so that $\{\hat{f}(x_I) - f(x_I)\}/c_I = V + B$. Since $\hat{f}(x_i) - \mathbb{E}\hat{f}(x_i)$ is a zero mean normal random variable, V has a mixture of normals density, or just a normal density if c_i is chosen to give constant variance to $\{\hat{f}(x_i) - \mathbb{E}\hat{f}(x_i)\}/c_i$. By evaluating at a random x_i, Nychka's construction also ensures that the bias term B is a random variable. His argument then proceeds as follows: $\mathbb{E}(B) \approx 0$ while the variance of B is much less than the variance of V. In consequence, if V has a normal density, then $B + V$ will have an approximately normal density.

Now let $\mathbf{C} = \text{diag}(c_i)$, and to obtain approximately constant variance of $B + V$ set $\mathbf{C} = \text{diag}\{\tilde{\mathbf{X}}(\mathbf{X}^\mathsf{T}\mathbf{X} + \mathbf{S}_\lambda)^{-1}\tilde{\mathbf{X}}^\mathsf{T}\}^{1/2}$, where $\tilde{\mathbf{X}}$ is the matrix such that $[\hat{f}(x_1), \hat{f}(x_2), \ldots, \hat{f}(x_n)]^\mathsf{T} = \tilde{\mathbf{X}}\hat{\boldsymbol{\beta}}$. The variance of $B + V$ is then

$$\mathbb{E}\|\mathbf{C}^{-1}(\hat{\mathbf{f}} - \mathbf{f})\|^2/n = \mathbb{E}\|\mathbf{C}^{-1}\tilde{\mathbf{X}}(\hat{\boldsymbol{\beta}} - \boldsymbol{\beta})\|^2/n,$$

which can be estimated, using (6.27), as

$$\text{tr}\{\tilde{\mathbf{X}}^\mathsf{T}\mathbf{C}^{-2}\tilde{\mathbf{X}}(\mathbf{X}^\mathsf{T}\mathbf{X} + \mathbf{S}_\lambda)^{-1}\}\sigma^2/n = \text{tr}\{\mathbf{C}^{-2}\tilde{\mathbf{X}}(\mathbf{X}^\mathsf{T}\mathbf{X} + \mathbf{S}_\lambda)^{-1}\tilde{\mathbf{X}}^\mathsf{T}\}\sigma^2/n = \sigma^2.$$

Hence to satisfy (6.28) using the given c_i we require $d = -z_{\alpha/2}\sigma^2$, where $z_{\alpha/2}$ denotes the $\alpha/2$ critical point of standard normal distribution. The same construction is readily extended to the generalized case: see Marra and Wood (2012), who also construct intervals based on the alternative direct estimate of the bias term, and provide extensive simulation evidence of the intervals' effectiveness.

Interval limitations and simulations

The Nychka (1988) argument rests on the variance being substantially larger than the squared bias, but this assumption can fail for smooth terms that are subject to sum to zero constraints and are estimated to be straight lines (or close to straight lines), when the truth is not quite a straight line. The problem is that such terms have zero (or very small) variance where the estimated line passes through zero, although the bias at such a point is not zero. Over an interval around the zero variance point the variance will not be substantially larger than the squared bias. Figure 6.8 illustrates such a function in its middle panel. Marra and Wood (2012) show that this is not merely a theoretical concern — realised interval coverage probabilities are really reduced in simulations. The suggested fix is to change the target of inference to the smooth term of interest *plus the overall model intercept*. This is equivalent to assuming that identifiability constraints have been applied to all the other model components, but not to the smooth of interest. Adopting this device eliminates the problem of vanishing variance for near straight line terms, and restores close to nominal coverage (of the modified target).

As an illustration of the issues, data were simulated using a true additive linear predictor made up from the three functions shown in grey in figure 6.8. In each case 4 i.i.d. $U(0, 1)$ covariates were simulated. Sample sizes of 200 and 500 were used. Simulations were conducted using Gaussian, Poisson and binomial distributions. In the Gaussian case the linear predictor was scaled to the range $(0, 1)$ and noise levels of $\sigma = 0.2$, 0.05 or 0.02 were employed. In the Poisson case the linear predictor was scaled so that the Poisson mean ranged from 1 to $p_{\text{max}} = 5$, 10 or 15. In the binomial case the linear predictor was scaled so that the probabilities were in the range $(0.02, 0.98)$ and binomial denominators of 1, 3 or 5 were used. Across the function coverage probabilities were assessed for 500 replicates at each combination of distribution, sample size and noise level. Figure 6.9 shows the results. Coverage is reasonably close to nominal, especially at the larger sample size when signal-to-noise ratio is high, except for the function f_1 (the middle panel in figure 6.8) for which coverage is too low, as expected.

The Marra and Wood (2012) proposal fixes this problem (for example by using the `seWithMean=TRUE` argument to `plot.gam`). An alternative, which avoids modifying the interval's target, is to use a variant of the 'one standard error rule' discussed in section 6.2.9. If a term has very low estimated degrees of freedom (for example within 0.5 of the EDF under full penalization), then its smoothing parameter could be *decreased* until the REML or ML score increases by $\sqrt{2}$ (or possibly until its EDF has increased by 1 if this is sooner), and the corresponding confidence interval used. Figure 6.10 shows the results of doing this.

The previous edition of this book (and Wood, 2006b) suggested that coverage probabilities of individual smooths were poor, and attributed this to the neglect of smoothing parameter uncertainty. This turned out to be wrong — the real cause of the poor performance was that the early versions of the smoothing parameter selection method, used in the simulations, tended to estimate smoothing parameters as effectively infinite too often (this happened because the original Wood, 2000, method alternated Newton steps with a global search for an overall smoothing parameter: the latter could cause the optimization to get stuck on flat regions of the GCV score). Substitution of better smoothing parameter estimation methods fixed the problem.

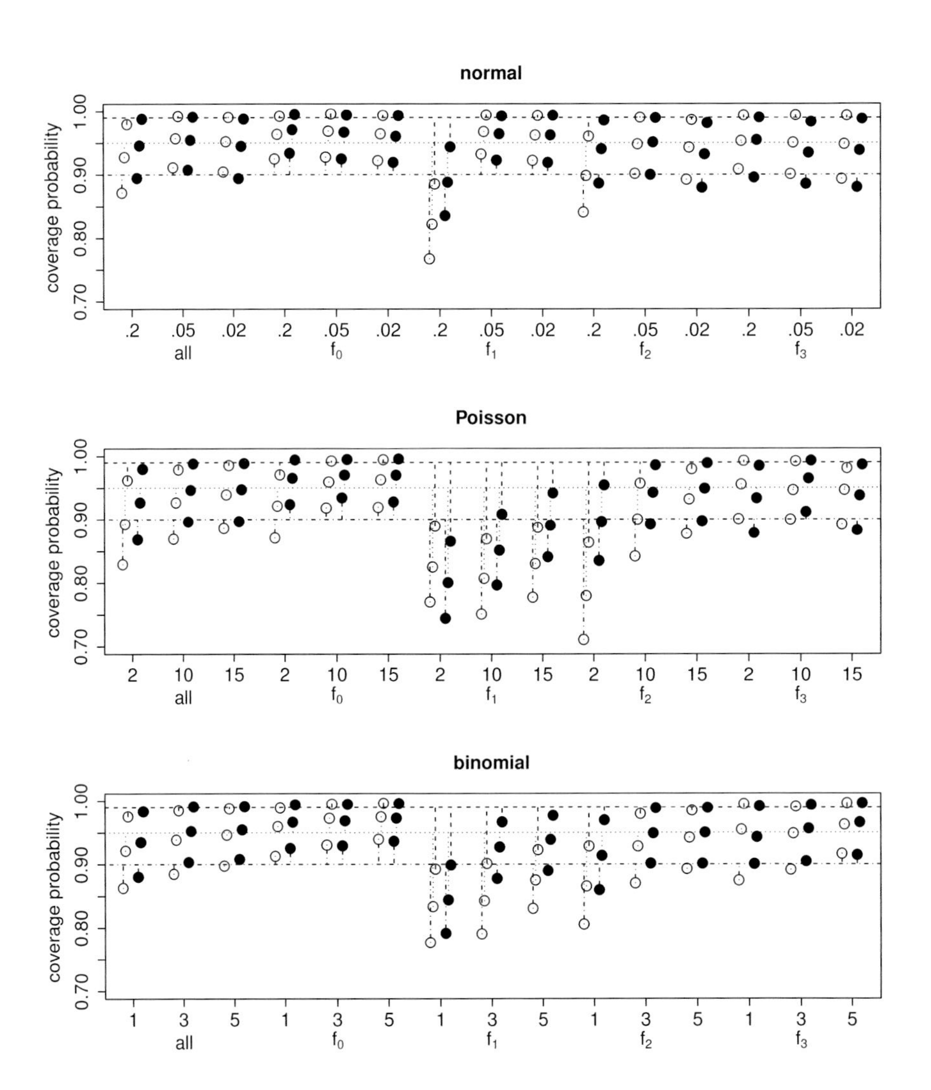

Figure 6.9 *Realised confidence interval coverage probabilities for whole linear predictor, and each component function, at each sample size, for confidence levels 0.9, 0.95 and 0.99. Simulation details are given in section 6.10.1. Open circles are for sample size 200 and filled circles are for sample size 500. The values of the noise-to-signal controlling parameters are shown on the horizontal axis. Notice the poor coverage of f_1, the 'near-straight-line' function.*

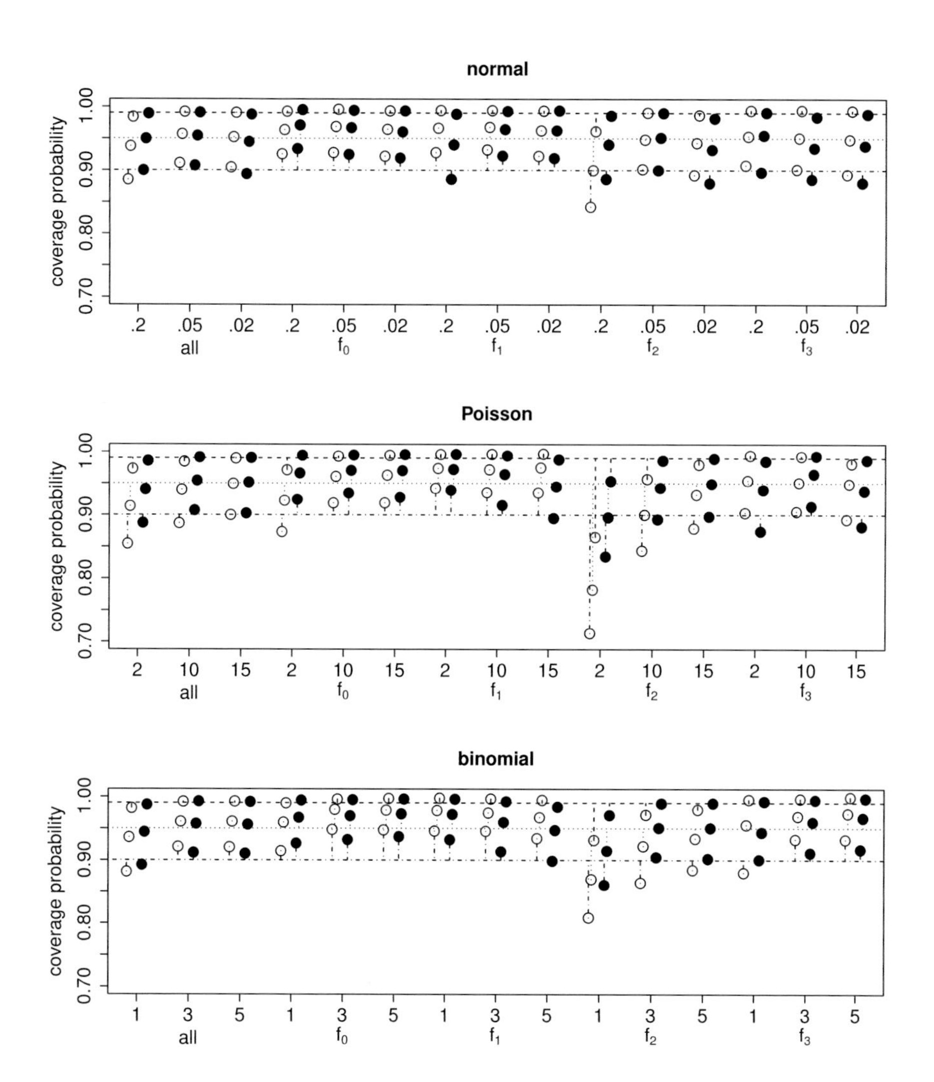

Figure 6.10 *Equivalent to figure 6.9 when the 'one-standard-error-rule' is used to decrease the smoothing parameter for f_1 by an 'insignificant' amount, if its EDF is less than 1.5. A plot using the Marra and Wood (2012) proposal would look similar.*

6.10.2 Whole function intervals

The above intervals have a specified probability *across the function* of including the true function. It is sometimes interesting to have an interval that will include the whole of the true function with a specified probability. That is, the interval is deemed not to cover the truth if it fails to include the truth anywhere in the range of the covariate. It is easy to produce such an interval by posterior simulation. We can rapidly simulate several thousand curves using (6.26), and then search for the factor $\alpha > 1$ by which the standard Bayesian interval has to be expanded in order to completely include a specified proportion of such curves.

6.10.3 Posterior simulation in general

Given the posterior distribution of the parameters (6.26) in section 6.10, possibly with correction (6.31) from section 6.11.1, we have already seen (at the beginning of section 6.10) how to simulate approximately from the posterior density of any function of the coefficients, $\boldsymbol{\beta}$. An alternative (e.g., Borchers et al., 1997; Augustin et al., 1998) might be to use bootstrapping to quantify uncertainty, but this is typically much more expensive, requiring a model fit per bootstrap replicate. In addition bootstrapping is somewhat problematic when there are penalties present. Non-parametric bootstrapping results in some data appearing twice in the bootstrap sample, which leads to under-smoothing (consider leave-one-out cross validation to see why this will happen). Parametric bootstrapping is made awkward by the presence of smoothing bias in the parametric model from which the samples are generated.

If the expense of bootstrapping is tolerable, then a better approach to more complete uncertainty quantification is to take a full Bayesian approach, put priors on the smoothing parameters, and simulate directly from the posterior of the model coefficients. This is not a huge step conceptually, since the methods presented in this book can anyway be viewed as 'empirical Bayes' procedures, tuned to obtain good frequentist properties. It is not within the scope of this book to cover Bayesian simulation approaches in detail. However, the way in which smooths can be re-written as random effects, described in section 5.8 (p. 239), means that it is also straightforward to set up the models described in this book as fully Bayesian models for simulation using JAGS (Plummer, 2003). Wood (2016a) provides the details, and function `jagam` in `mgcv` facilitates the approach. See 7.7.3 (p. 374) for more.

The Gibbs sampling approach of `jagam` is a long way from being optimally efficient, and it pays to design samplers specifically for the case of smooth model terms. The BayesX project does this (e.g. Belitz et al., 2015; Fahrmeir and Lang, 2001; Fahrmeir et al., 2004; Lang and Brezger, 2004; Kneib and Fahrmeir, 2006; Belitz and Lang, 2008; Lang et al., 2014; Klein et al., 2015), generating very efficient software for posterior simulation with GAMs. Another fully Bayesian approach achieves efficiency by avoiding simulation altogether, and using higher order Laplace approximation coupled with sparse matrix methods, to compute posterior densities for quantities of interest from a smooth additive model. This is the iterated nested Laplace approximation (INLA, Rue et al., 2009; Lindgren et al., 2011).

6.11 AIC and smoothing parameter uncertainty

A large part of what is traditionally considered to be model selection is performed by the estimation of smoothing parameters. However, smoothing parameter selection usually stops short of removing terms from a model altogether (although see section 5.4.3, p. 214), and in any case it is often interesting to compare models that are not nested. Akaike's information criterion is one popular method for model selection often used with regression models, but its use requires some care in the context of models containing random effects and smoothers. There are two approaches:

1. *Marginal* AIC is based on the (frequentist) marginal likelihood of the model: that is on the likelihood obtained by treating all penalized coefficients as random effects and integrating them out of the joint density of response data and random effects. The number of coefficients to use for the AIC penalty is then just the number of fixed effects plus the number of variance and smoothing parameters.
2. *Conditional AIC* is based on the likelihood of all the coefficients at their maximum penalized likelihood (MAP) estimates. The number of coefficients in the penalty then has to be based on some estimate of the *effective* number of parameters, in order to account for the fact that the coefficient estimates are penalized.

Philosophically the approaches differ in how they treat smooths. For the marginal approach the smooth is really being treated as a frequentist random effect, re-drawn from its marginal/prior distribution on each replication of the data. But modellers usually view smoothers as essentially fixed effects (which would stay constant under replication of the data). The conditional approach is aligned with this latter view.

The practical problem with the marginal AIC is that frequentist marginal likelihood underestimates variance components (in the context of a smoother, making it too smooth). So the estimation procedure is biased towards simpler models, and AIC will reflect this bias, making it overly conservative. The obvious alternative would be to use the restricted likelihood, REML, in the AIC computation, but the restricted likelihood is only comparable between models with the same fixed effects structure, so that is can not be used to compare models with and without a smooth (unless that smooth has no unpenalized component). Whether it is possible to construct a well founded AIC based on the marginal likelihood evaluated at the REML 'estimates' seems to be an open question.

The traditional approach to conditional AIC uses the model degrees of freedom τ from section 6.1.2 in the AIC penalty term (e.g. Hastie and Tibshirani, 1990), but Greven and Kneib (2010) show that this leads to an AIC that is highly anti-conservative, especially in the context of random effects: this conditional AIC is far too likely to select a model which includes a random effect that is not present in the true model. They also show that the problem lies with the neglect of smoothing parameter uncertainty in τ. A number of fixes have been proposed for this problem (Greven and Kneib, 2010; Yu and Yau, 2012; Säfken et al., 2014), but here we follow the relatively simple method of Wood et al. (2016), which has the advantage of being general enough to apply to all the models covered in this book.

The idea is to compute a first order correction to the posterior distribution for the model coefficients, accounting for smoothing parameter uncertainty. Then the penalty term in the AIC is expressed in terms of the Bayesian covariance matrix of the coefficients. Substituting the corrected covariance matrix into the AIC penalty yields the corrected version of τ to use in the AIC penalty term.

6.11.1 Smoothing parameter uncertainty

Conventionally in a GAM context smoothing parameters have been treated as fixed, ignoring the uncertainty in estimating them. Kass and Steffey (1989) proposed a simple first order correction for this sort of uncertainty in the context of i.i.d. Gaussian random effects in a one way ANOVA type design, and their general approach can be extended. Let $\rho_i = \log \lambda_i$ and $\mathbf{S}_\lambda = \sum_j \lambda_j \mathbf{S}_j$. Writing $\hat{\boldsymbol{\beta}}_\rho$ for $\hat{\boldsymbol{\beta}}$, to emphasise the dependence of $\hat{\boldsymbol{\beta}}$ on the smoothing parameters, we use the Bayesian large sample approximation

$$\boldsymbol{\beta}|\mathbf{y}, \boldsymbol{\rho} \sim N(\hat{\boldsymbol{\beta}}_\rho, \mathbf{V}_\beta) \text{ where } \mathbf{V}_\beta = (\hat{\boldsymbol{\mathcal{I}}} + \mathbf{S}_\lambda)^{-1} \tag{6.29}$$

which is exact in the Gaussian case, along with the large sample approximation

$$\boldsymbol{\rho}|\mathbf{y} \sim N(\hat{\boldsymbol{\rho}}, \mathbf{V}_\rho) \tag{6.30}$$

where $\mathbf{V}_\rho$ is the inverse of the Hessian of the negative log marginal likelihood with respect to $\boldsymbol{\rho}$.[‡] To improve on using (6.29) with $\boldsymbol{\rho}$ fixed at its estimate, note that if (6.29) and (6.30) are correct, while $\mathbf{z} \sim \mathbf{N}(\mathbf{0}, \mathbf{I})$ and independently $\boldsymbol{\rho}^* \sim N(\hat{\boldsymbol{\rho}}, \mathbf{V}_\rho)$, then (using '$\overset{d}{=}$' to denote 'equal in distribution') $\boldsymbol{\beta}|\mathbf{y} \overset{d}{=} \hat{\boldsymbol{\beta}}_{\rho^*} + \mathbf{R}_{\rho^*}^\mathsf{T} \boldsymbol{z}$ where $\mathbf{R}_{\rho^*}^\mathsf{T}\mathbf{R}_{\rho^*} = \mathbf{V}_\beta$ (and $\mathbf{V}_\beta$ depends on $\boldsymbol{\rho}^*$). This provides a way of simulating from $\boldsymbol{\beta}|\mathbf{y}$, but it is computationally expensive as $\hat{\boldsymbol{\beta}}_{\rho^*}$ and $\mathbf{R}_{\rho^*}$ must be recomputed for each sample. As an alternative consider the first order Taylor expansion

$$\boldsymbol{\beta}|\mathbf{y} \overset{d}{=} \hat{\boldsymbol{\beta}}_{\hat{\rho}} + \mathbf{J}(\boldsymbol{\rho} - \hat{\boldsymbol{\rho}}) + \mathbf{R}_{\hat{\rho}}^\mathsf{T}\boldsymbol{z} + \sum_k \left.\frac{\partial \mathbf{R}_\rho^\mathsf{T}\boldsymbol{z}}{\partial \rho_k}\right|_{\hat{\rho}} (\rho_k - \hat{\rho}_k) + r$$

where r is a lower order remainder term and $\mathbf{J} = \mathrm{d}\hat{\boldsymbol{\beta}}/\mathrm{d}\boldsymbol{\rho}|_{\hat{\rho}}$. Dropping r, the expectation of the right hand side is $\hat{\boldsymbol{\beta}}_{\hat{\rho}}$. Denote the elements of $\mathbf{R}_\rho$ by R_{ij}. Routine calculation shows that the three remaining random terms are uncorrelated with covariance matrix

$$\mathbf{V}_\beta' = \mathbf{V}_\beta + \mathbf{V}' + \mathbf{V}'', \mathbf{V}' = \mathbf{J}\mathbf{V}_\rho\mathbf{J}^\mathsf{T} \text{ and } V_{jm}'' = \sum_i^p \sum_l^M \sum_k^M \frac{\partial R_{ij}}{\partial \rho_k} V_{\rho,kl} \frac{\partial R_{im}}{\partial \rho_l} \tag{6.31}$$

(which is computable at $O(Mp^3)$ cost). Appendix B.7 (p. 422) covers computation of the derivative of the Cholesky factor. Dropping $\mathbf{V}''$ we have the Kass and Steffey

[‡] The approximation (6.30) applies in the interior of the parameter space, but if smoothing parameters are effectively infinite then a pseudoinverse of the Hessian or other regularization of the inversion will be required.

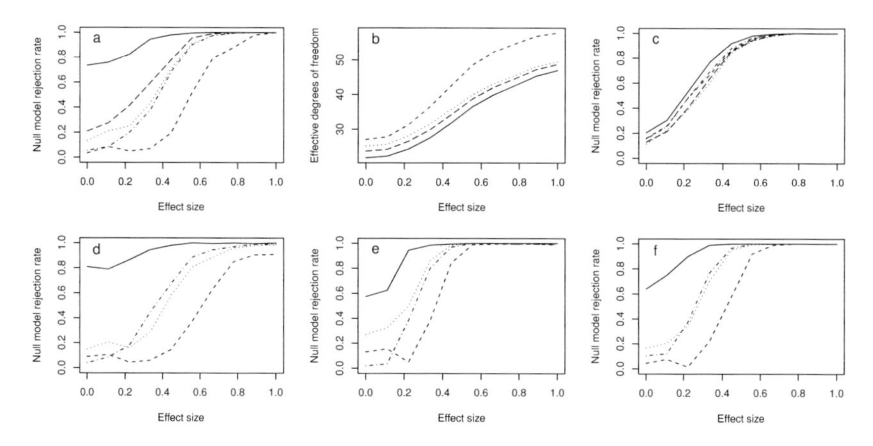

Figure 6.11 *Simulation based illustration (from Wood et al., 2016) of the problems with previous AIC type model selection criteria and the relatively good performance of (6.32). In all panels: (i) the solid curves are for conventional conditional AIC using τ from section 6.1.2, (ii) the dotted curves are for (6.32), (iii) the middle length dashed curves are for AIC using τ_1 from section 6.1.2, (iv) the dashed dot curves are for the marginal likelihood based AIC and (v) the long dashed curves are for the Greven and Kneib (2010) corrected AIC (top row only). (a) Observed probability of selecting the larger model as the effect strength of the differing term is increased from zero, for a 40 level random effect and Gaussian likelihood. (b) Whole model effective degrees of freedom used in the alternative conditional AIC scores for the left hand panel as effect size increases. (c) Same as (a), but where the term differing between the two models was a smooth curve. (d) As (a) but for a Bernoulli likelihood. (e) As (a) for a beta likelihood. (f) As (a) for a Cox proportional hazards partial likelihood.*

(1989) like approximation $\boldsymbol{\beta}|\mathbf{y} \sim N(\hat{\boldsymbol{\beta}}_{\hat{\rho}}, \mathbf{V}^*_{\beta})$ where $\mathbf{V}^*_{\beta} = \mathbf{V}_{\beta} + \mathbf{J}\mathbf{V}_{\rho}\mathbf{J}^{\mathsf{T}}$. (A first order Taylor expansion of $\hat{\boldsymbol{\beta}}$ about $\boldsymbol{\rho}$ yields a similar correction for the frequentist covariance matrix of $\hat{\beta}$: $\mathbf{V}^*_{\hat{\beta}} = (\hat{\boldsymbol{\mathcal{I}}} + \mathbf{S}_{\lambda})^{-1}\hat{\boldsymbol{\mathcal{I}}}(\hat{\boldsymbol{\mathcal{I}}} + \mathbf{S}_{\lambda})^{-1} + \mathbf{J}\mathbf{V}_{\rho}\mathbf{J}^{\mathsf{T}}$, where $\hat{\boldsymbol{\mathcal{I}}}$ is the negative Hessian of the log likelihood.)

Wood et al. (2016) show that the $\mathbf{J}(\boldsymbol{\rho} - \hat{\boldsymbol{\rho}})$ correction is dominant for smooth components that are strongly non-zero. This offers some justification for using the Kass and Steffey (1989) type approximation, but not in a model selection context, where near zero model components are those of most interest: hence in what follows we use (6.31) without dropping $\mathbf{V}''$.

6.11.2 A corrected AIC

Following Appendix A.7 (or Davison, 2003, §4.7), the derivation of AIC with the MLE replaced by the penalized MLE is identical up to the point at which the AIC can be represented as

$$\begin{aligned}\text{AIC} &= -2l(\hat{\boldsymbol{\beta}}) + 2\mathbb{E}\left\{(\hat{\boldsymbol{\beta}} - \boldsymbol{\beta}_d)^\mathsf{T}\boldsymbol{\mathcal{I}}_d(\hat{\boldsymbol{\beta}} - \boldsymbol{\beta}_d)\right\} \\ &= -2l(\hat{\boldsymbol{\beta}}) + 2\text{tr}\left[\mathbb{E}\{(\hat{\boldsymbol{\beta}} - \boldsymbol{\beta}_d)(\hat{\boldsymbol{\beta}} - \boldsymbol{\beta}_d)^\mathsf{T}\}\boldsymbol{\mathcal{I}}_d\right]\end{aligned}$$

where $\boldsymbol{\beta}_d$ is the coefficient vector minimizing the K-L divergence and $\boldsymbol{\mathcal{I}}_d$ is the corresponding expected negative Hessian of the log likelihood. In an unpenalized setting $\mathbb{E}\{(\hat{\boldsymbol{\beta}} - \boldsymbol{\beta}_d)(\hat{\boldsymbol{\beta}} - \boldsymbol{\beta}_d)^\mathsf{T}\}$ is estimated as the observed inverse information matrix $\hat{\boldsymbol{\mathcal{I}}}^{-1}$ and $\tau' = \text{tr}\left\{\mathbb{E}(\hat{\boldsymbol{\beta}} - \boldsymbol{\beta}_d)(\hat{\boldsymbol{\beta}} - \boldsymbol{\beta}_d)^\mathsf{T}\boldsymbol{\mathcal{I}}_d\right\}$ is estimated as $\text{tr}(\hat{\boldsymbol{\mathcal{I}}}^{-1}\hat{\boldsymbol{\mathcal{I}}}) = k$. Penalization means that the expected inverse covariance matrix of $\hat{\boldsymbol{\beta}}$ is no longer well approximated by $\hat{\boldsymbol{\mathcal{I}}}$, but approximating $\boldsymbol{\mathcal{I}}_d$ by $\hat{\boldsymbol{\mathcal{I}}}$ and using (6.27), in section 6.10.1, yields $\tau' = \text{tr}(\mathbf{V}_\beta\hat{\boldsymbol{\mathcal{I}}})$, which can be corrected for smoothing parameter uncertainty using (6.31) to obtain $\tau_2 = \text{tr}(\mathbf{V}'_\beta\hat{\boldsymbol{\mathcal{I}}})$, and

$$\text{AIC} = -2l(\hat{\boldsymbol{\beta}}) + 2\tau_2. \tag{6.32}$$

Figure 6.11 illustrates the relatively good performance of (6.32) compared to alternatives. Most of the panels show 'power curves': the probability of correctly accepting a larger model as the amount of difference between smaller and larger model is increased from zero. When there is no difference between the models then AIC should typically select the larger model in about 16% of cases, by analogy with the fully parametric case in which the probability of AIC choosing a model with one extra spurious parameter is $\Pr(\chi^2_1 > 2)$ (given (A.5), p. 411).

6.12 Hypothesis testing and p-values

Hypothesis testing is an alternative for model selection, especially when we have reason to favour simpler models over more complicated ones. P-values for the parametric model effects can be computed exactly as they would be for an unpenalized model. Suppose that we want to test $\text{H}_0 : \boldsymbol{\beta}_j = \mathbf{0}$, where $\boldsymbol{\beta}_j$ is a subvector of $\boldsymbol{\beta}$ containing only unpenalized (fixed effect) coefficients. If we treat smooths as random effects for this purpose, then we can again use the fact that the frequentist covariance matrix for $\hat{\boldsymbol{\beta}}_j$ can be read from the Bayesian covariance matrix for $\boldsymbol{\beta}$ (see sections 3.4.3 and 2.4.2, pages 151 and 80). So let $\mathbf{V}_{\beta_j}$ denote the block of $\mathbf{V}_\beta$ corresponding to $\boldsymbol{\beta}_j$. Following section 1.3.4, page 14 (approximately for the generalized case),

$$\hat{\boldsymbol{\beta}}_j^\mathsf{T}\mathbf{V}_{\beta_j}^{-1}\hat{\boldsymbol{\beta}}_j/p_j \sim F_{p_j, n-p}$$

if there is a scale parameter estimate involved, or if not then

$$\hat{\boldsymbol{\beta}}_j^\mathsf{T}\mathbf{V}_{\beta_j}^{-1}\hat{\boldsymbol{\beta}}_j \sim \chi^2_{p_j}.$$

p and p_j are the dimensions of $\boldsymbol{\beta}$ and $\boldsymbol{\beta}_j$, respectively. We can also test more general linear hypotheses, $\text{H}_0 : \mathbf{C}\boldsymbol{\beta}_j = \mathbf{d}$ by replacing $\hat{\boldsymbol{\beta}}_j^\mathsf{T}\mathbf{V}_{\beta_j}^{-1}\hat{\boldsymbol{\beta}}_j$ by $(\mathbf{C}\hat{\boldsymbol{\beta}}_j -$

$\mathbf{d})^\mathsf{T}(\mathbf{C}\mathbf{V}_{\beta_j}\mathbf{C}^\mathsf{T})^{-1}(\mathbf{C}\hat{\boldsymbol{\beta}}_j - \mathbf{d})$ in the above distributional results, as in section 1.3.4 (p. 14). For single parameter tests these results can also be re-written as exactly equivalent tests using t_{n-p} or $N(0,1)$ as the reference distributions.

For the penalized terms in a model the p-value computation is less straightforward, because of the effect of penalization on the coefficient estimates. Naïvely we can try to form test statistics like those used above for the parametric terms, but this has two problems. One is that penalization may mean that the $\mathbf{V}_{\beta_j}$ does not have full rank – this can be addressed by using a generalized inverse in the test statistic and modifying the reference distribution accordingly, but even if we do that there are problems. When we invert $\mathbf{V}_{\beta_j}$ in the test statistic then we up-weight the components of $\boldsymbol{\beta}_j$ that have low variance: but many of those components only have low variance because they are so heavily penalized that they contribute almost nothing to the model. Using a test statistic based on up-weighting the model components for which there is least evidence in the data is a recipe for achieving low testing power. In developing more effective procedures there are two cases to consider separately: smooth terms where the null space of the penalty is finite dimensional, and terms with no penalty null space (such as random effects, or a smooth with a first order penalty subject to a sum-to-zero constraint).

6.12.1 *Approximate p-values for smooth terms*

Consider testing $\mathrm{H}_0 : f_j(x) = 0$, for all x in the range of interest. So the aim is to test whether or not function f_j is actually needed in a model. The key idea here is to exploit the relatively good coverage properties of the Bayesian intervals as a result of the Nychka (1988) argument of section 6.10.1. Since the good interval properties rest on considering coverage across the whole interval, it seems sensible to create a test statistic based on the whole interval.

Let $\mathbf{f}_j$ denote the vector of $f_j(x)$ evaluated at the observed covariate values. Further, let $\tilde{\mathbf{X}}$ be such that $\mathbf{f}_j = \tilde{\mathbf{X}}\boldsymbol{\beta}$. Section 6.10.1 shows that we will get well-calibrated confidence intervals for $\mathbf{f}_j$ if we start from the approximate result $\hat{\boldsymbol{\beta}} \sim N(\boldsymbol{\beta}, \mathbf{V}_\beta)$, where $\mathbf{V}_\beta$ is the Bayesian covariance matrix for $\boldsymbol{\beta}$, so that approximately $\hat{\mathbf{f}}_j \sim N(\mathbf{f}_j, \mathbf{V}_{f_j})$ where $\mathbf{V}_{f_j} = \tilde{\mathbf{X}}\mathbf{V}_\beta\tilde{\mathbf{X}}^\mathsf{T}$. The fact that the good calibration rests on the behaviour of the smoothing bias when averaged across-the-function, suggests basing a test statistic for f_j also on across-the-function evaluation of f_j. For example

$$T_r = \hat{\mathbf{f}}_j^\mathsf{T}\mathbf{V}_{f_j}^{r-}\hat{\mathbf{f}}_j$$

where $\mathbf{V}_{f_j}^{r-}$ is a rank r pseudo-inverse of $\mathbf{V}_{f_j}$. The question is what value r should take? The maximum rank of $\mathbf{V}_{f_j}$ is p_j, the number of coefficients of f_j. r can not exceed p_j, but if we simply use $r = p_j$ then we typically obtain poor power, because the very components of f_j that are most heavily penalized to zero are those most heavily up-weighted in the statistic, despite being the components for which the data contain least information. One approach to choosing r is to consider the number of coefficients required to optimally approximate the penalized estimate $\hat{f}$ using an unpenalized estimate. Since the unpenalized estimate has no smoothing bias it is most straightforward to consider approximating the penalized estimate with a first order correction for the smoothing bias applied, and in that case we conclude that the approximately optimal r is the τ_1 version of the term-specific EDF from section 6.1.2 (see Wood, 2013a, §2.2).

Now although this choice of r should ensure that we exclude the most heavily penalized components of f_j from the test statistic, it has the problem of not being integer. We might try rounding it to the nearest integer, but this can be problematic when this means rounding down to the dimension of the penalty null space, since we can then remove important information. To see this, imagine a weakly nonlinear $f_j(x)$, with no overall linear trend component, modelled by a cubic regression spline. If the effective degrees of freedom were 1.45 we would round to $r = 1$, but because there is no linear trend, conclude that the function could not be distinguished from zero, no-matter how unlikely this conclusion might look based on the narrowness of the confidence interval for f_j. The obvious 'fix' is to round up the effective degrees of freedom, but then consider the case of a basically linear term with effective degrees of freedom 1.0001. Rounding up to $r = 2$ would result in a test statistic in which a nearly zero component of f_j had an utterly disproportionate influence on the test statistic, and again this can lead to very poor practical performance.

To avoid these rounding problems it is appealing to modify the test statistic so that r can be set to the un-rounded EDF of f_j, while behaving exactly like the original test statistic for integer r, and having similar properties whether or not r is integer. More specifically, in the known scale parameter case, the null distribution of T_r should be χ^2_r if r is integer, while in any case $\mathbb{E}(T_r) = r$ and var$(T_r) = 2r$, under H_0.

Let $k = \lfloor r \rfloor$ denote the integer part of r (i.e., r rounded down), $\nu = r - k$ and $\rho = \{\nu(1-\nu)/2\}^{1/2}$, and suppose that the columns of $\mathbf{U}$ contain the eigenvectors of $\mathbf{V}_{f_j}$ corresponding to its non-zero eigenvalues Λ_i. Then the desired properties of T_r are obtained by setting

$$\mathbf{V}_{f_j}^{r-} = \mathbf{U} \begin{bmatrix} \lambda_1^{-1} & & & \\ & \ddots & & \\ & & \lambda_{k-2}^{-1} & & \\ & & & \mathbf{B} & \\ & & & & \mathbf{0} \end{bmatrix} \mathbf{U}^{\mathsf{T}},$$

$$\text{where } \mathbf{B} = \tilde{\mathbf{\Lambda}}\tilde{\mathbf{B}}\tilde{\mathbf{\Lambda}}^{\mathsf{T}}, \quad \tilde{\mathbf{\Lambda}} = \begin{bmatrix} \lambda_{k-1}^{-1/2} & 0 \\ 0 & \lambda_k^{-1/2} \end{bmatrix} \text{ and } \tilde{\mathbf{B}} = \begin{bmatrix} 1 & \rho \\ \rho & \nu \end{bmatrix}.$$

Then if $\nu_1 = \{\nu + 1 + (1-\nu^2)^{1/2}\}/2$ and $\nu_2 = \nu + 1 - \nu_1$ (the eigenvalues of $\tilde{B}$), we have that

$$T_r \sim \chi^2_{k-2} + \nu_1\chi^2_1 + \nu_2\chi^2_1, \tag{6.33}$$

under H_0, which is simply $T_r \sim \chi^2_r$, if r is integer (Wood, 2013a).

The c.d.f. of the distribution in (6.33) can be computed by the method of Davies (1980), or approximated by using a gamma$(r/2, 2)$ distribution, which can be made less crude in the upper tail by employing the approximation of Liu et al. (2009) (beyond the 0.9 point, for example). When a scale parameter has been estimated then the p-value is computed as $p = \text{pr}(\chi^2_{k-2} + \nu_1\chi^2_1 + \nu_2\chi^2_1 > t_r\chi^2_\kappa/\kappa)$ where κ is the residual degrees of freedom used to compute the scale estimate, and t_r is the observed T_r. When using an approximate distribution for T_r this can be cheaply evaluated by quadrature. Alternatively the method of Davies (1980) can be used to compute it exactly as $p = \text{pr}(\chi^2_{k-2} + \nu_1\chi^2_1 + \nu_2\chi^2_1 - t_r\chi^2_\kappa/\kappa > 0)$.[§]

A final detail is that there is an ambiguity in the definition of T_r since $\mathbf{B}$ is not diagonal, and eigenvectors are only unique up to a change of sign. This means that although the null distributional result holds for either version, their actual numerical values can vary slightly. A simple fix for this is to compute both p-values, and average them (averaging the two alternative test statistics would change their distribution).

Computing T_r

Naïve formation of $\mathbf{V}^{r-}_{f_j}$ would involve forming the eigen-decomposition of $\mathbf{V}_{f_j}$ at $O(n^3)$ computational cost. This can be greatly improved by first obtaining the QR decomposition

$$\mathbf{X}_j = \mathcal{Q}\begin{bmatrix} \mathbf{R} \\ \mathbf{0} \end{bmatrix}$$

(at a cost of at most $O(np_j^2)$).

Consider the eigen-decomposition $\mathbf{U}\boldsymbol{\Lambda}\mathbf{U}^\mathsf{T} = \mathbf{R}\mathbf{V}_{\beta_j}\mathbf{R}^\mathsf{T}$. We then have

$$\tilde{\mathbf{X}}\mathbf{V}_{\beta_j}\tilde{\mathbf{X}}^\mathsf{T} = \mathcal{Q}\begin{bmatrix} \mathbf{R}\mathbf{V}_{\beta_j}\mathbf{R}^\mathsf{T} & \mathbf{0} \\ \mathbf{0} & \mathbf{0} \end{bmatrix}\mathcal{Q}^\mathsf{T} = \mathcal{Q}\begin{bmatrix} \mathbf{U} & \mathbf{0} \\ \mathbf{0} & \mathbf{I} \end{bmatrix}\begin{bmatrix} \boldsymbol{\Lambda} & \mathbf{0} \\ \mathbf{0} & \mathbf{0} \end{bmatrix}\begin{bmatrix} \mathbf{U}^\mathsf{T} & \mathbf{0} \\ \mathbf{0} & \mathbf{I} \end{bmatrix}\mathcal{Q}^\mathsf{T}.$$

Since $\begin{bmatrix} \mathbf{U} & \mathbf{0} \\ \mathbf{0} & \mathbf{I} \end{bmatrix}$ is an orthogonal matrix, and $\begin{bmatrix} \boldsymbol{\Lambda} & \mathbf{0} \\ \mathbf{0} & \mathbf{0} \end{bmatrix}$ is diagonal this is the eigen-decomposition of $\mathbf{V}_{f_j}$. A normal generalized inverse of $\mathbf{V}_{f_j}$ would invert the r largest eigenvalues, Λ_i, and set the rest to zero, which is equivalent to simply forming the generalized inverse of $\mathbf{R}\mathbf{V}_{\beta_j}\mathbf{R}^\mathsf{T}$ in the middle expression above. Hence we have

$$(\tilde{\mathbf{X}}\mathbf{V}_{\beta_j}\tilde{\mathbf{X}}^\mathsf{T})^{r-} = \mathcal{Q}\begin{bmatrix} (\mathbf{R}\mathbf{V}_{\beta_j}\mathbf{R}^\mathsf{T})^{r-} & \mathbf{0} \\ \mathbf{0} & \mathbf{0} \end{bmatrix}\mathcal{Q}^\mathsf{T} = \mathbf{Q}(\mathbf{R}\mathbf{V}_{\beta_j}\mathbf{R}^\mathsf{T})^{r-}\mathbf{Q}^\mathsf{T},$$

where $\mathbf{Q}$ is the first p_j columns of $\mathcal{Q}$ and the generalized inverse can also be the version with non-integer r from above. It then follows that

$$T_r = \hat{\boldsymbol{\beta}}_j^\mathsf{T}\mathbf{R}^\mathsf{T}\mathbf{Q}^\mathsf{T}\mathbf{Q}(\mathbf{R}\mathbf{V}_{\beta_j}\mathbf{R}^\mathsf{T})^{r-}\mathbf{Q}^\mathsf{T}\mathbf{Q}\mathbf{R}\hat{\boldsymbol{\beta}}_j = \hat{\boldsymbol{\beta}}_j^\mathsf{T}\mathbf{R}^\mathsf{T}(\mathbf{R}\mathbf{V}_{\beta_j}\mathbf{R}^\mathsf{T})^{r-}\mathbf{R}\hat{\boldsymbol{\beta}}_j,$$

which has only $O(p_j^3)$ cost, given the QR decomposition.

The $O(np_j^2)$ cost of the QR decomposition at each step is also avoidable, if we already have available the QR decomposition of the whole model matrix, from model fitting. Suppose $\mathbf{X} = \mathbf{Q}'\mathbf{R}'$, and let $\tilde{\mathbf{R}}$ contain the columns of $\mathbf{R}'$ corresponding to $\tilde{\mathbf{X}}$, so that $\tilde{\mathbf{X}} = \mathbf{Q}\tilde{\mathbf{R}}$. Now form a further QR decomposition $\tilde{\mathbf{R}} = \tilde{\mathbf{Q}}\mathbf{R}$ at $O(pp_j^2)$ cost. Then $\tilde{\mathbf{X}} = \mathbf{Q}\mathbf{R}$ where $\mathbf{Q} = \mathbf{Q}'\tilde{\mathbf{Q}}$. So this reduces the $O(np_j^2) + O(p_j^3)$ cost to

[§]Thanks to Thomas Lumley, for pointing this out.

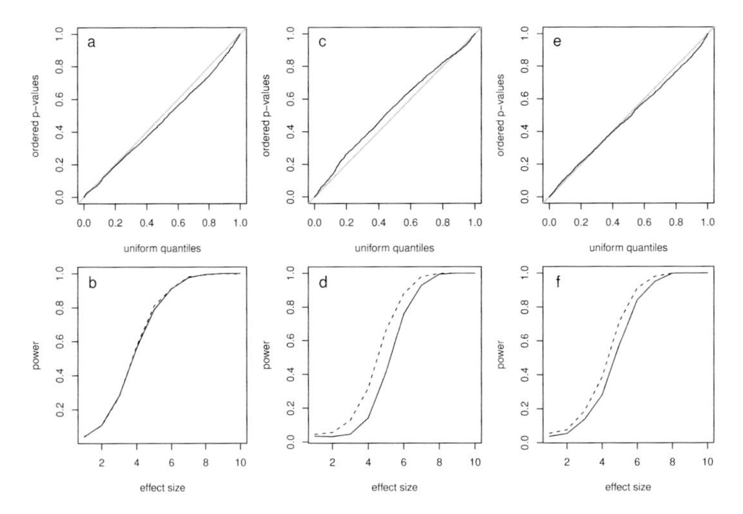

Figure 6.12 *Top row: QQ-plots for null distribution of p-values for smooth terms. The grey line is the ideal reference line. Bottom row: corresponding power at 5% level. The continuous curve is for the test of section 6.12.1 while the dashed curve is based on direct simulation of the null distribution of an F-ratio test statistic. All panels are for testing a smooth embedded in a model with 3 other smooth terms. a,b: a univariate cubic spline; c,d: a bivariate thin plate spline; e,f: a bivariate tensor product smooth.*

$O(pp_j^2) + O(p_j^3)$. Notice that we do not actually need the $\mathbf{Q}$ matrices to compute T_r, and furthermore $\mathbf{R}$ is also the Cholesky factor of $\mathbf{X}^\mathsf{T}\mathbf{X}$.

Actually the readily available $\mathbf{R}$ is often the Cholesky factor of $\mathbf{X}^\mathsf{T}\mathbf{W}\mathbf{X}$, where $\mathbf{W}$ is the diagonal matrix of PIRLS weights at convergence. We can motivate using this $\mathbf{R}$, by modifying T_r to be based on the weighted function values $\sqrt{\mathbf{W}}\hat{\mathbf{f}}$ (at least when the weights are all positive). More generally still, we can use the Cholesky factor of the negative Hessian of the log-likelihood, or its expected value.

Simulation performance

The preceding derivation deals with some practical pathologies and power loss that can occur with more naïve p-value computations (such as the approximate computations used in the first edition of this book), but the distribution of T_r under the null hypothesis rests on the validity of the Nychka argument. We do not need to worry about the bias exceeding the variance when we estimate a close to straight line term under the null, since in that case the straight line is correct. But the distributional result is still approximate. Figure 6.12 shows some simulation results examining the p-value null distribution and power at the 5% level for the test, for 3 different smooth terms embedded in Gaussian additive models with 4 smooth terms. The leftmost panels relate to a rank 10 univariate smooth, while the middle panels are for a rank 50 bivariate thin plate spline and the rightmost are for a rank 49 tensor product smooth. For the power calculations the test was compared to a test based on the simulated

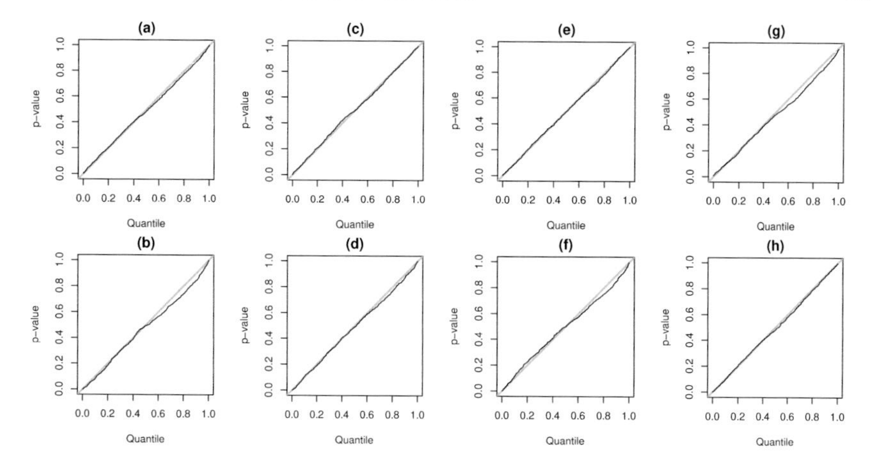

Figure 6.13 *Null distribution of p-values from Wood (2013b), for testing for the presence of a single random effect. QQ-plots for the null distribution of p-values of the section 6.12.2 method overlaid over a grey reference line. Panels (a)–(f) are for a Gaussian responses, following section 3.1.1 of Scheipl et al. (2008). Sample sizes are 30 (a),(b), 100 (c),(d) and 500 (e),(f). Group sizes, from (a) to (f), are 3, 10, 5, 20, 5, 100. Panels (g) and (h) are for binary responses, for a model where there was a smooth term present in addition to the random effect. Sample sizes are 100 for (g) and 400 for (h) with a group size of 20.*

null distribution of an F-ratio statistic (in a simulation study we can simulate exactly from the null distribution of a test statistic). The sample size was 400, 2000 replicates were used for the null distribution simulations and 500 for each power simulation. The test performance is reasonable in each case, although not perfect. For the bivariate smooths the power is a little below the ideal represented by the simulated test. Wood (2013a) contains further simulations.

6.12.2 *Approximate p-values for random effect terms*

The preceding construction fails for simple i.i.d. random effects and smoothers with no null space. One problem is that for a heavily shrunk smooth with no null space the squared smoothing bias can easily become as large as the variance; another is with the notion of the best rank r approximation to a simple i.i.d. random effect. Generally the problem of testing random effects for equality to zero is known to be difficult. The problem is equivalent to testing whether a variance is zero, but zero is at the edge of the feasible parameter space so that a generalized likelihood ratio test can not be used, as the Taylor expansion based arguments underpinning it break down.

An approach that generalizes easily to the generalized multiple smoothing parameter case, is to re-write the test's log likelihood ratio statistic as a quadratic form in those predicted random effects from the larger model fit which would be zeroed by the reduced model embodying the null hypothesis. This quadratic form has a computable distribution, unaffected by 'edge of parameter space' problems. To see how this works, first consider the linear mixed model for n independent response variables, y_i,

$$\boldsymbol{\mu} = \mathbf{X}\boldsymbol{\beta} + \sum_{j=1}^{m} \mathbf{Z}_j \mathbf{b}_j, \;\; \mathbf{b}_j \sim N(\mathbf{0}, \boldsymbol{\psi}_j \sigma^2), \;\; y_i \sim N(\mu_i, \sigma^2),$$

where $\mathbf{Z}_j$ and $\mathbf{X}$ are model matrices, $\boldsymbol{\beta}$ is a vector of fixed parameters and $\boldsymbol{\psi}_j \sigma^2$ is a parameterized covariance matrix for the random effects, $\mathbf{b}_j$. We wish to test $H_0 : \boldsymbol{\psi}_k = \mathbf{0}$ for some k. σ and the $\boldsymbol{\psi}_j$ can be estimated by marginal likelihood or REML maximisation. Given these estimates, the predicted random effects, $\hat{\mathbf{b}}_j$ and the maximum likelihood estimates $\hat{\boldsymbol{\beta}}$ can be found by minimization of

$$\left\| \mathbf{y} - \mathbf{X}\boldsymbol{\beta} - \sum_{j=1}^{m} \mathbf{Z}_j \mathbf{b}_j \right\|^2 + \sum_{j=1}^{m} \mathbf{b}_j^{\mathsf{T}} \hat{\boldsymbol{\psi}}_j^{-} \mathbf{b}_j$$

with respect to the $\mathbf{b}_j$ and $\boldsymbol{\beta}$, where $\hat{\boldsymbol{\psi}}_j^{-}$ is the (pseudo) inverse of $\hat{\boldsymbol{\psi}}_j$. If $\hat{\boldsymbol{\mathcal{B}}} = (\hat{\boldsymbol{\beta}}, \hat{\mathbf{b}}_1, \hat{\mathbf{b}}_2, \ldots)$ then the solution to this minimisation can be written $\hat{\boldsymbol{\mathcal{B}}} = \mathbf{P}\mathbf{y}$. Let $\mathbf{P}_j$ denote the rows of $\mathbf{P}$ such that $\hat{\mathbf{b}}_j = \mathbf{P}_j \mathbf{y}$. Under the null hypothesis $\mathbf{y} \sim N(\mathbf{X}\boldsymbol{\beta}, \boldsymbol{\Sigma}_{-k})$, where $\boldsymbol{\Sigma}_{-k} = (\mathbf{I} + \sum_{j \ne k} \mathbf{Z}_j \boldsymbol{\psi}_j \mathbf{Z}_j^{\mathsf{T}})\sigma^2$, so $\hat{\mathbf{b}}_k \sim N(\mathbf{0}, \mathbf{P}_k \boldsymbol{\Sigma}_{-k} \mathbf{P}_k^{\mathsf{T}})$.

Now suppose that σ and all the $\boldsymbol{\psi}_j$ except $\boldsymbol{\psi}_k$ are fixed at their estimated values, and suppose that we want to test $H_0 : \boldsymbol{\psi}_k = \mathbf{0}$ against $H_1 : \boldsymbol{\psi}_k = \hat{\boldsymbol{\psi}}_k$. Under H_1 we have that the restricted log likelihood is

$$\hat{l}_1' = -\frac{1}{2\hat{\sigma}^2} \left\| y - X\hat{\beta} - \sum_{j=1}^{m} Z_j \hat{b}_j \right\|^2 - \frac{1}{2\hat{\sigma}^2} \sum_{j=1}^{m} \hat{b}_j^{\mathsf{T}} \hat{\psi}_j^{-} \hat{b}_j + c_1 = \frac{\hat{l}_1}{\hat{\sigma}^2} + c_1,$$

while under H_0 it is

$$\hat{l}_0' = -\frac{1}{2\hat{\sigma}^2} \left\| y - X\tilde{\beta} - \sum_{j \ne k}^{m} Z_j \tilde{b}_j \right\|^2 - \frac{1}{2\hat{\sigma}^2} \sum_{j \ne k}^{m} \tilde{b}_j^{\mathsf{T}} \hat{\psi}_j^{-} \tilde{b}_j + c_0 = \frac{\hat{l}_0}{\hat{\sigma}^2} + c_0.$$

The constants c_0 and c_1 are irrelevant, given fixed $\boldsymbol{\psi}_j$ and σ. The vectors $\hat{\boldsymbol{\beta}}$ and $\hat{\mathbf{b}}_j$ are the maximisers of l_1', while $\tilde{\boldsymbol{\beta}}$ and $\tilde{\mathbf{b}}_j$ are the maximizers of l_0'. The test statistic is $\mathcal{W} = 2(\hat{l}_1 - \hat{l}_0)$, which can be expressed as a quadratic form in $\hat{\mathbf{b}}_k$, as shown next.

Without loss of generality, assume that $k = m$, the largest value of j: the $\mathbf{Z}_j$ can always be re-ordered to ensure that this is so. Let $p = \dim(\boldsymbol{\mathcal{B}})$ and $p_k = \dim(\mathbf{b}_k)$. Consider the QR decomposition of the augmented model matrix, i.e.,

$$\tilde{\mathbf{X}} = \begin{bmatrix} \mathbf{X} & \mathbf{Z}_{-k} & \mathbf{Z}_k \\ \mathbf{0} & \mathbf{B}_{-k} & \mathbf{0} \\ \mathbf{0} & \mathbf{0} & \mathbf{B}_k \end{bmatrix} = \mathbf{Q} \begin{bmatrix} \mathbf{R} \\ \mathbf{0} \end{bmatrix}$$

where $\mathbf{Z}_{-k}$ is $\mathbf{Z}$ with the columns of $\mathbf{Z}_k$ omitted, $\mathbf{B}_{-k}$ is a square root of $\sum_{j \ne k} \hat{\boldsymbol{\psi}}_j^{-}$ and $\mathbf{B}_k$ a square root of $\hat{\boldsymbol{\psi}}_k^{-}$. Now $\hat{\boldsymbol{\beta}}$ and the $\hat{\mathbf{b}}_j$ are the minimizers of $\|\mathbf{f} - \mathbf{R}\boldsymbol{\mathcal{B}}\|^2$, where $\mathbf{f}^\mathsf{T}$ is the first p elements of $(\mathbf{y}^\mathsf{T}, \mathbf{0})\mathbf{Q}$, and we define $\mathbf{r}$ to be the remaining n elements. Hence $\hat{\boldsymbol{\mathcal{B}}} = \mathbf{R}^{-1}\mathbf{f}$ and $2\hat{l}_1 = -\|\mathbf{r}\|^2$. Now partition

$$\mathbf{R} = \begin{bmatrix} \mathbf{R}_0 & \mathbf{R}_1 \\ \mathbf{0} & \tilde{\mathbf{R}} \end{bmatrix}$$

where $\mathbf{R}_0$ is the upper left $p - p_k \times p - p_k$ block of $\mathbf{R}$. Letting $\mathbf{f}_0$ denote the first $p - p_k$ rows of $\mathbf{f}$ and $\mathbf{f}_1$ the remaining p_k rows, then $\tilde{\beta}$ and the $\tilde{\mathbf{b}}_j$ are the minimizers of

$$\|\mathbf{f}_0 - \mathbf{R}_0\boldsymbol{\mathcal{B}}_{-k}\|^2$$

where $\boldsymbol{\mathcal{B}}_{-k}$ is $\boldsymbol{\mathcal{B}}$ with $\mathbf{b}_k$ omitted, while $2\hat{l}_0 = -\|\mathbf{r}\|^2 - \|\mathbf{f}_1\|^2$. Hence $2(\hat{l}_1 - \hat{l}_0) = \|\mathbf{f}_1\|^2$. By the upper triangular structure of $\mathbf{R}$, $\mathbf{f}_1 = \tilde{\mathbf{R}}\hat{\mathbf{b}}_k$, so $2(\hat{l}_1 - \hat{l}_0) = \hat{\mathbf{b}}_k^\mathsf{T}\tilde{\mathbf{R}}^\mathsf{T}\tilde{\mathbf{R}}\hat{\mathbf{b}}_k$. Now find a matrix square root, $\mathbf{C}$, of the covariance matrix of $\hat{\mathbf{b}}_k$ under H_0, so that $\mathbf{C}^\mathsf{T}\mathbf{C} = \mathbf{P}_k\boldsymbol{\Sigma}_{-k}\mathbf{P}_k^\mathsf{T}$. If $\mathbf{z}_k$ is a p_k vector of independent $N(0,1)$ random variables, it follows that under H_0,

$$\mathcal{W} = \hat{\mathbf{b}}_k^\mathsf{T}\tilde{\mathbf{R}}^\mathsf{T}\tilde{\mathbf{R}}\hat{\mathbf{b}}_k \sim \mathbf{z}_k^\mathsf{T}\mathbf{C}\tilde{\mathbf{R}}^\mathsf{T}\tilde{\mathbf{R}}\mathbf{C}^\mathsf{T}\mathbf{z}_k = \mathbf{z}_k^\mathsf{T}\mathbf{U}\boldsymbol{\Lambda}\mathbf{U}^\mathsf{T}\mathbf{z}_k \sim \sum_{i=1}^{p_k}\lambda_i\chi_{1i}^2, \tag{6.34}$$

where $\mathbf{U}\boldsymbol{\Lambda}\mathbf{U}^\mathsf{T}$ is the spectral decomposition of $\mathbf{C}\tilde{\mathbf{R}}^\mathsf{T}\tilde{\mathbf{R}}\mathbf{C}^\mathsf{T}$, so that $\mathbf{U}$ is an orthogonal matrix while $\boldsymbol{\Lambda}$ is a diagonal matrix of eigenvalues, λ_i. The cumulative distribution function of such a weighted sum of χ_1^2 random variables can again be computed by the method of Davies (1980), or approximated using Liu et al. (2009). If ψ_j and σ were known, then this test would be exact and is simply a likelihood ratio test.

In the generalized case Laplace approximations to the restricted likelihoods under null and alternative are employed, to arrive at the same results, just with $\sqrt{\mathbf{W}}\mathbf{X}$ replacing $\mathbf{X}$, where $\mathbf{W}$ is the diagonal matrix of Fisher weights from the PIRLS. See Wood (2013b) for details and simulation testing. As with section 6.12.1 test, computation of the p-values can be made more efficient by basing computations on the Cholesky factor of $\mathbf{X}^\mathsf{T}\mathbf{W}\mathbf{X}$ (usually computed by QR decomposition of $\sqrt{\mathbf{W}}\mathbf{X}$).

Crainiceanu and Ruppert (2004) proposed an exact test for the case of a Gaussian linear mixed model with a single variance component and Greven et al. (2008) used this as the basis for an approximate test in the multiple variance component case. Wood (2013b) contains a simulation study (based on Scheipl et al., 2008) comparing the method of this section to the Greven et al. (2008) proposal and to simulation of the likelihood ratio statistic null distribution. Figure 6.13 shows the null distribution of this section's test in the case of testing for a single variance component in Gaussian and binary settings (in the binary setting a smooth term was present in addition to the random effect). Figure 6.14 compares the power of this section's proposal to the exact test of Crainiceanu and Ruppert (2004) in the Gaussian case, and to direct simulation of the null distribution of the test statistic in the binary case. The test of this section performs well (although the Wood, 2013b, simulations still suggest that the section 6.12.1 test is better for smooths).

Notice the substantial limitation that the test can not be used for hypotheses involving correlated random effects. For example, if a model has correlated random

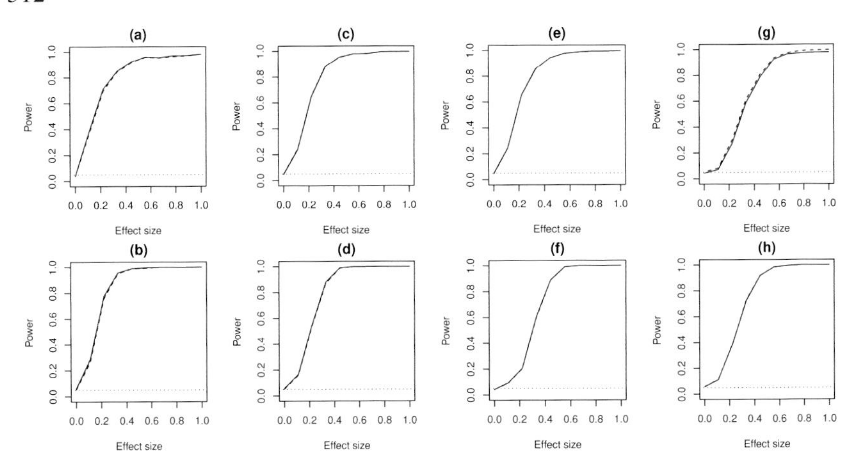

Figure 6.14 *Power comparisons with Crainiceanu and Ruppert (2004) and with direct simulation. Data and models for each panel correspond to those given for the panels in figure 6.13. Power curves, at the 5% level, are plotted as continuous for the new method, dotted for Crainiceanu and Ruppert (2004) and dashed for direct simulation. The curves are often so similar as to be indistinguishable. Effect strength was controlled by increasing the random effect variance from zero, but is shown on a standardised scale. The horizontal dotted line denotes the 5% level. Panels (g) and (h) compare the proposed method with a method using exact simulation of the null distribution of the likelihood ratio statistic for logistic regression with binary data, since Crainiceanu and Ruppert (2004) does not apply in this case.*

intercepts and slopes, we cannot use this approach to directly test whether either the slope or the intercept could be dropped on their own. In practice, however, it is often the case that a test would first be made for correlation between the slopes and intercepts: a normal GLRT is fine for this, since zero correlation is not at the edge of the feasible parameter space. If the correlation is dropped then we could use the test here to test whether either the slope or intercept term is required.

Another caveat is that the test is only approximate because of the fixing of the variance parameters at their estimated values, and the reliance on large sample normality results in the generalized case: we should expect noticeable deterioration of performance when variance components are poorly identified, and when the normality of the $\mathbf{b}_k$ is a questionable approximation.

6.12.3 *Testing a parametric term against a smooth alternative*

Sometimes it is interesting to test whether a smooth is needed, or whether the parametric terms in the null space of the smooth would be sufficient. For instance, do we need a cubic spline term, or would a straight line be good enough? The obvious way to do this is to parameterize the smooth so that the basis for functions in the null space of the penalty is explicitly separated from the basis for the penalty range space. We can then use the results of the previous section to test whether the function component in the penalty range space is zero, meaning that we might as well include the null space terms as simple parametric effects. `mgcv` provides a mechanism for doing this for thin plate regression splines, allowing the smooth to be set up without a null space basis, so that this must be added explicitly. For example

```
summary(gam(y ~ x + s(x,bs="tp",m=c(2,0)))
```

will automatically return the p-value for testing whether the non-linear components of the cubic thin place spline are required. `m=c(2,0)` specifies that a spline with a second order penalty and no null space basis is required.

Such tests were particularly important before efficient methods for smoothing parameter selection were available. They are less useful with automatic smoothing parameter selection, which often estimates a term to be in the penalty null space if that is what the data suggest.

6.12.4 *Approximate generalized likelihood ratio tests*

It is tempting to try to compare GAMs using a generalized likelihood ratio test (appendix A.5, p. 411). One possibility is to use the frequentist marginal likelihood, counting the number of fixed effects plus number of smoothing parameters and variance parameters in order to obtain appropriate degrees of freedom. An alternative is to use the (conditional) likelihood along with effective degrees of freedom. Neither approach works for testing whether a random effect is needed, since in the marginal case the null model is restricting the variance parameters to the edge of the feasible parameter space, and in the conditional case we can not really view effective degrees of freedom as representing the number of unpenalized coefficients needed to approximate the penalized model.

In the case of smooth terms, marginal likelihood tends to be biased towards slightly over-smooth estimates, which can inflate the acceptance rate for null models. Conversely the conditional approach seems to favour the larger alternative model unless we account for smoothing parameter uncertainty. The somewhat informal justification for using a conditional GLRT is that we could approximate a penalized model by an unpenalized model, with a number of coefficients for each smooth given by the effective degrees of freedom of each smooth (see figure 6.15). If this approximation is a good one, then the distribution of the log likelihood of each penalized model and of their difference should be well approximated by the equivalents for the unpenalized approximations. Hence for comparing a simple model 0 with coefficients $\hat{\beta}_0$ to a larger model 1, with coefficients $\hat{\beta}_1$ we can use the approximation, under model 0, that

$$\lambda = 2\{l(\hat{\beta}_1) - l(\hat{\beta}_0)\} \sim \chi^2_{\text{EDF}_1 - \text{EDF}_0}.$$

The approximation of penalized models by unpenalized should be best if we bias correct the penalized model, in which case the appropriate EDF would be the bias corrected version, τ_1, from section 6.1.2. However, under the null model the bias corrections of the null and alternative models should approximately cancel, suggesting that we might not want to bother actually applying the correction when computing the log likelihood ratio. To allow for smoothing parameter uncertainty, we can apply

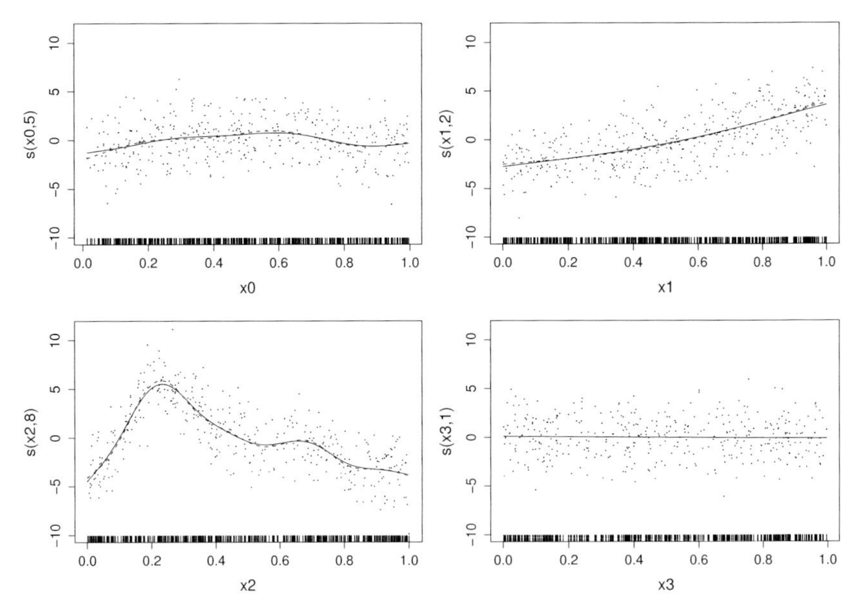

Figure 6.15 *Illustration of the similarity between the estimated terms of a GAM represented using penalized regression splines (solid curves) and pure regression splines (dashed curves — based on thin plate regression splines). Both models are fitted to the same data, and the effective degrees of freedom of each penalized term matches the degrees of freedom of its pure regression equivalent. Partial residuals are shown to give an indication of the scale of variability in the data. In this example the mean square difference between the fitted values from the two models is 1.2% of the residual variance. The correlation between the fitted values from the two models is 0.998.*

the simple EDF correction from section 6.11.2. This latter correction requires that we use REML or ML for smoothing parameter estimation.

Notice that for simple Gaussian random effects we can not generally approximate a penalized term by an unpenalized one with dimension equal to some effective degrees of freedom, and there is no justification for the test.

Figure 6.16 illustrates the results of a small set of simulations in which the truth was a Gaussian additive model $y = f_1(x_1) + f_2(x_2) + f_3(x_3) + \epsilon$, where the f_j were smooth functions and the sample size was 400. For 1000 replicates, the approximate GLRT was used to compare a null model with the structure of the true model to i) a model with an extra spurious term, $f_4(x_4)$; ii) a model with extra spurious terms $f_4(x_4) + f_5(x_3, x_4)$, where the latter was a smooth interaction term; iii) a model with an extra spurious 40 level Gaussian random effect. If the GLRT is functioning correctly then the p-values should be uniformly distributed on the unit interval. Four methods were used to compute the p-values: i) GCV smoothing parameter selection was used with the effective degrees of freedom given by τ from section 6.1.2; ii) was as (i), but using τ_1 from section 6.1.2; iii) REML smoothing parameter estimation and τ used for the EDF; iv) is the method suggested by the previous discussion, that

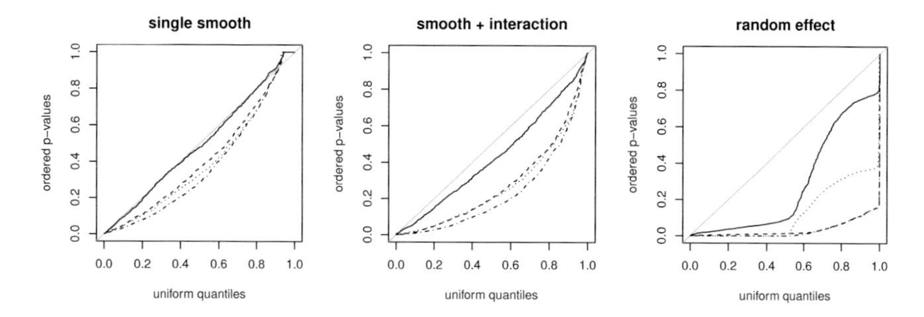

Figure 6.16 *QQ-plots of the realised null distribution of p-values for the approximate GLRT test of section 6.12.4. In each plot the grey line shows the ideal reference line for uniform distribution, the dash-dotted curve is for GCV with EDF given by τ, the dotted curve is GCV with EDF given by τ_1, the dashed curve is using REML and τ, while the solid curve is using REML with τ_1 corrected for smoothing parameter uncertainty. Left is when testing for the presence of a single univariate smooth. The middle plot is for a univariate smooth plus a bivariate interaction. The right hand plot is testing for a single spurious random effect, illustrating that the test is useless in this case.*

is as (iii) but with τ_1 corrected for smoothing parameter uncertainty, as in section 6.11. As expected, the test is clearly useless for comparing models differing in random effect structure. For other comparisons the test seems to provide a reasonable approximation provided the smoothing parameter uncertainty correction is applied, which in practice requires use of REML or ML smoothing parameter selection.

6.13 Other model selection approaches

In GAM estimation the smoothing parameter selection criterion does a great deal of the work of model selection before we apply methods like hypothesis testing and AIC for the final say as to whether a term is required in the model at all. An obvious alternative is to allow the smoothing parameter selection criterion to do all the work of model selection, by associating penalties with the otherwise unpenalized model terms and estimating the associated smoothing parameters. Section 5.4.3 (p. 214) provides the details of how to construct appropriate penalties for smooth terms. In `mgcv` such penalties are imposed using the `select=TRUE` argument to `gam` and the `paraPen` argument also facilitates penalization of strictly parametric terms. Marra and Wood (2011) investigate this approach in more detail.

There is also an extensive literature on GAM term selection in high dimensional contexts where there are large numbers of potential covariates to screen. It is not clear why a smooth additive structure should be a good model structure to use in these settings, but neither is it clear what would be better. The major classes of method are based around boosting (e.g., Tutz and Binder, 2006; Schmid and Hothorn, 2008), and variants of L_1-penalization methods, such as the Lasso and Group Lasso (e.g., Lin et al., 2006; Chouldechova and Hastie, 2015).

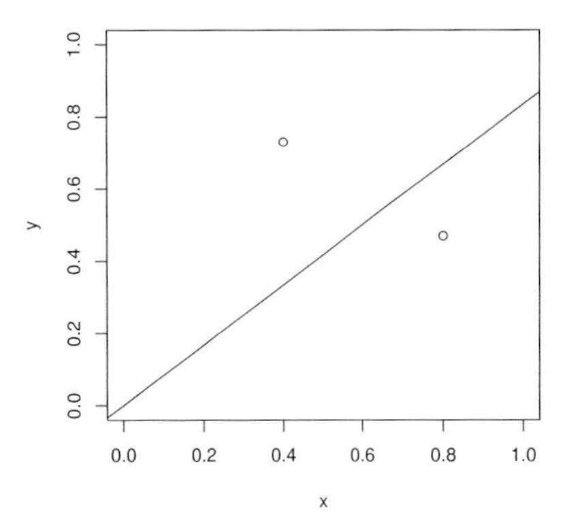

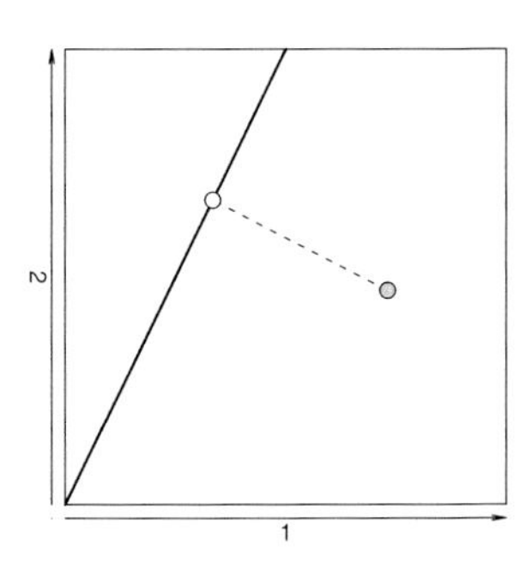

Figure 6.17 *The simple one-parameter linear regression used to illustrate the geometry of penalized regression in section 6.14.1. The left panel shows the least squares fit of a straight line through the origin to two response data. The right panel shows the corresponding unpenalized fitting geometry, in a space in which there is an axis for each response datum, and the response data are therefore represented by a single point (the grey circle). The thick line shows the model subspace: the model fitted values must lie on this line. The open circle is the orthogonal projection of the response data onto the model space, i.e. the fitted values.*

6.14 Further GAM theory

This section considers the geometry of penalization and the original backfitting method for GAM estimation.

6.14.1 The geometry of penalized regression

The geometry of linear and generalized linear model fitting, covered in sections 1.4 (p. 19) and 3.2 (p. 119), becomes more complicated when quadratically penalized estimation is used. In the unpenalized cases a response data vector, $\mathbf{y}$, which is exactly representable by the model (i.e., $g(y_i) = \mathbf{X}_i\boldsymbol{\beta}$ for some $\boldsymbol{\beta}$ and all i), results in model fitted values $\hat{\boldsymbol{\mu}} = \mathbf{y}$. When penalized estimation is used, then this is generally not the case (except for $\mathbf{y}$ that are exactly representable by the model with parameter values for which the penalty is exactly zero). However, there is a geometric interpretation of penalized fitting, in terms of projections in a larger space than the 'data space' that was considered in sections 1.4 and 3.2.

Consider the model

$$\mathbf{y} = \mathbf{X}\boldsymbol{\beta} + \boldsymbol{\epsilon},$$

to be estimated by minimization of the penalized objective function

$$\|\mathbf{y} - \mathbf{X}\boldsymbol{\beta}\|^2 + \lambda\boldsymbol{\beta}^{\mathsf{T}}\mathbf{S}\boldsymbol{\beta}. \tag{6.35}$$

A geometric interpretation of penalized estimation can be obtained by re-writing

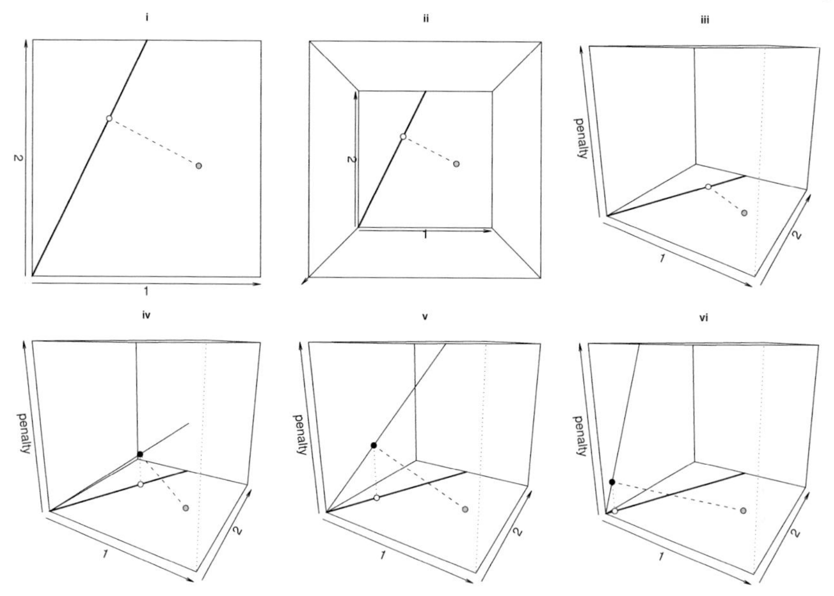

Figure 6.18 *Geometry of penalized linear regression for the model shown in figure 6.17. (i) The unpenalized geometry, exactly as in figure 6.17. (ii) The space shown in panel (i) is augmented by a space corresponding to the penalty term. (iii) Another view of the augmented space from (ii), still corresponding to the unpenalized, $\lambda = 0$, case. (iv) Penalized geometry for $\lambda = 0.1$. The thin line now shows the model subspace spanned by the columns of the augmented model matrix. Fitting amounts to finding the orthogonal projection from the (zero augmented) response data vector (grey circle) onto the augmented model subspace (giving the black circle). The model fitted values for the (un-augmented) response data are obtained by then projecting back onto the original model subspace to get the model fitted values (the open circle). (v) Same as (iv) but for $\lambda = 1$. (vi) Same as (iv) for $\lambda = 10$. Notice how even data lying exactly in the original model subspace would result in fitted values that are different from those data: only zero data are not shrunk in this way.*

(6.35) as

$$\left\| \begin{bmatrix} \mathbf{y} \\ \mathbf{0} \end{bmatrix} - \begin{bmatrix} \mathbf{X} \\ \sqrt{\lambda}\mathbf{S} \end{bmatrix} \boldsymbol{\beta} \right\|^2,$$

i.e., as the ordinary least squares objective for an augmented model, fitted to the response data augmented with zeroes. So penalized regression amounts to orthogonally projecting $\begin{bmatrix} \mathbf{y} \\ \mathbf{0} \end{bmatrix}$ onto the space spanned by the columns of $\begin{bmatrix} \mathbf{X} \\ \sqrt{\lambda}\mathbf{S} \end{bmatrix}$ and then orthogonally projecting back onto the space spanned by $\mathbf{X}$, to get the fitted values corresponding to $\mathbf{y}$.

It is difficult to draw pictures that really illustrate all aspects of this geometry in a satisfactory way, but some insight can be gained by considering the model

$$y_i = x_i\beta + \epsilon_i$$

for two data. Figure 6.17 illustrates the corresponding unpenalized fitting geometry, as in section 1.4. Now consider fitting this model by penalized regression: in fact by simple ridge regression. The model fitting objective is then

$$\sum_{i=1}^{2}(y_i - x_i\beta)^2 + \lambda\beta^2 = \left\| \begin{bmatrix} y_1 \\ y_2 \\ 0 \end{bmatrix} - \begin{bmatrix} x_1 \\ x_2 \\ \sqrt{\lambda} \end{bmatrix} \beta \right\|^2 .$$

The geometry of the model fit is shown in figure 6.18. Notice the way in which increasing λ results in fitted values closer and closer to zero.

For the simple penalty, illustrated in figure 6.18, only $\beta = 0$ results in a zero penalty, and hence only the response data $(0, 0)$ would result in identical fitted values $(0, 0)$. For penalties on multiple parameters, the penalty coefficient matrix, $\mathbf{S}$, is generally d short of full rank, where d is some integer. There is then a rank d subspace of the model space for which the fitting penalty is identically zero, and if the data happen to be in that space (i.e., with no component outside it) then the fitted values will be exactly equal to the response data. For example, data lying exactly on a straight line are left unchanged by a cubic spline smoother.

6.14.2 *Backfitting GAMs*

The name 'generalized additive model' was coined by Hastie and Tibshirani (1986, 1990), who first proposed this class of models, along with methods for their estimation. Their proposed GAM estimation technique of 'backfitting' is elegant and has the advantage that it allows the component functions of an additive model to be represented using almost any smoothing or modelling technique (regression trees for example). The disadvantage is that estimation of the degree of smoothness of a model is hard to integrate into this approach.

The basic idea is to estimate each smooth component of an additive model by iteratively smoothing *partial residuals* from the AM, with respect to the covariate(s) that the smooth relates to. The partial residuals relating to the jth smooth term are the residuals resulting from subtracting all the current model term estimates from the response variable, *except* for the estimate of the jth smooth. Almost any smoothing method (and mixtures of methods) can be employed to estimate the smooths.

Here is a more formal description of the algorithm. Suppose that the object is to estimate the additive model:

$$y_i = \alpha + \sum_{j=1}^{m} f_j(x_{ji}) + \epsilon_i$$

where the f_j are smooth functions, and the covariates, x_j, may sometimes be vector covariates. Let $\hat{\mathbf{f}}_j$ denote the vector whose i^{th} element is the estimate of $f_j(x_{ji})$. The basic backfitting algorithm is as follows.

1. Set $\hat{\alpha} = \bar{y}$ and $\hat{\mathbf{f}}_j = \mathbf{0}$ for $j = 1, \ldots, m$.
2. Repeat steps 3 to 5 until the estimates, $\hat{\mathbf{f}}_j$, stop changing.

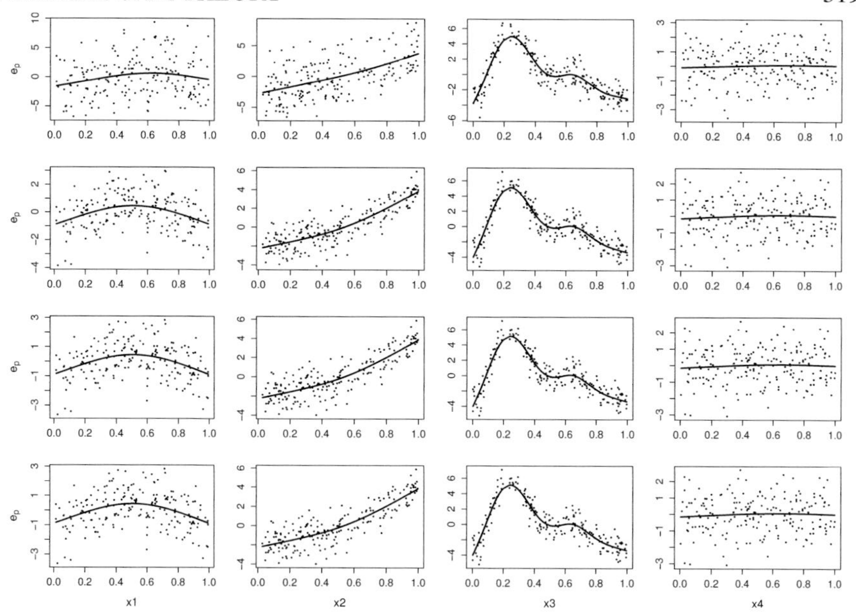

Figure 6.19 *Fitting a 4 term additive model using backfitting. Iteration proceeds across the columns and down the rows (i.e., starts at top left and finishes at bottom right). The jth column relates to the jth smooth and its covariate. The points shown in each plot are the partial residuals for the term being estimated, plotted against the corresponding covariate. By smoothing these residuals, an estimate of the corresponding smooth function (evaluated at the covariate values) is produced: these estimates are shown as thick curves. For example, the third column, second row, shows the partial residuals for f_3, at the second iteration, plotted against x_3: the thick curve is therefore the second iteration estimate of f_3, and results from smoothing the partial residuals with respect to x_3. The estimated functions change moving down the rows, but have stabilised by the last row.*

3. For $j = 1, \ldots, m$ repeat steps 4 and 5.
4. Calculate partial residuals:

$$\mathbf{e}_p^j = \mathbf{y} - \hat{\alpha} - \sum_{k \neq j} \hat{\mathbf{f}}_k.$$

5. Set $\hat{\mathbf{f}}_j$ equal to the result of smoothing e_p^j with respect to x_j.

As an example, here is some R code implementing the basic backfitting algorithm, using R routine `smooth.spline` as the smoothing method. It is assumed that `edf[j]` contains the required degrees of freedom for the jth smooth, that `x` is an `m` column array, with j^{th} column containing the (single) covariate for the jth smooth, and that the response is in `y`.

```
f <- x*0; alpha <- mean(y); ok <- TRUE; rss0 <- 0
while (ok) { # backfitting loop
  for (i in 1:ncol(x)) { # loop through the smooth terms
```

```
    ep <- y - rowSums(f[,-i]) - alpha
    b <- smooth.spline(x[,i],ep,df=edf[i])
    f[,i] <- predict(b,x[,i])$y
  }
  rss <- sum((y-rowSums(f))^2)
  if (abs(rss-rss0)<1e-6*rss) ok <- FALSE
  rss0 <- rss
}
```

Figure 6.19 shows 4 iterations of the backfitting algorithm applied to the estimation of an additive model with 4 smooth components. In this case the data were simulated using the code given in the help file for `gam` in the `mgcv` package. The `edf` array was set to 3, 3, 8 and 3.

To fit GAMs by penalized iteratively re-weighted least squares, a weighted version of the backfitting algorithm is used, and there are some refinements which speed up convergence (which can otherwise be slow if covariates are highly correlated). Notice how the cost of the algorithm depends on the cost of the component-wise smoothers: `smooth.spline` is a very efficient way of smoothing with respect to one variable, for example. Clearly the elegance of backfitting is appealing, as is the flexibility to choose from a very wide variety of smoothing methods for component estimation, but this does come at a price. It is not easy to efficiently integrate smoothness estimation methods with backfitting: the obvious approach of applying GCV to the smoothing of partial residuals is definitely not recommendable, while direct estimation of the influence/hat matrix of the whole model has computational cost of at least $O(mn^2)$, where n is the number of data, so that cross validation or GCV would both be quite expensive.

6.15 Exercises

1. Prove that equation (6.4) in section 6.1.2 is true.
2. Cross validation can fail completely for some problems: this question explores why.
   (a) Consider attempting to smooth some response data y_i by solving the 'ridge regression' problem

$$\text{minimise} \sum_{i=1}^{n}(y_i - \mu_i)^2 + \lambda \sum_{i=1}^{n} \mu_i^2 \text{ w.r.t. } \boldsymbol{\mu},$$

   where λ is a smoothing parameter. Show that for this problem the GCV and OCV scores are identical, and are independent of λ.

   (b) By considering the basic principle underpinning ordinary cross validation, explain what causes the failure of cross validation in part (a).

   (c) Given the explanation of the failure of cross validation for the ridge regression problem in part (a), it might be expected that the following modified approach would work better. Suppose that we have an x_i covariate observed for each y_i

(and for convenience $x_i < x_{i+1} \; \forall \; i$). Define the function $\mu(x)$ to be the piecewise linear interpolant of the points (x_i, μ_i). In this case we could estimate the μ_i by minimizing the following penalized least squares objective

$$\sum_{i=1}^{n} (y_i - \mu_i)^2 + \lambda \int \mu(x)^2 dx$$

w.r.t. the μ_i.

Now consider 3 equally spaced points x_1, x_2, x_3 with corresponding μ values μ_1, μ_2, μ_3. Suppose that $\mu_1 = \mu_3 = \mu^*$, but that we are free to choose μ_2. Show that, in order to minimize $\int_{x_1}^{x_3} \mu(x)^2 dx$, we should set $\mu_2 = -\mu^*/2$. What does this imply about trying to choose λ by cross validation? (Hint: think about what the penalty will do to μ_i if we 'leave out' y_i.)

(d) Would the penalty:

$$\int \mu'(x)^2 dx$$

suffer from the same problem as the penalty in part (c)?

(e) Would you expect to encounter these sorts of problems with penalized regression smoothers? Explain.

3. This question covers PIRLS and smoothing parameter estimation, using the P-spline set up in question 4 of chapter 5. The following simulates some x, y data:

```
f <- function(x) .04*x^11*(10*(1-x))^6+2*(10*x)^3*(1-x)^10
n <- 100;x <- sort(runif(n))
y <- rpois(rep(1,n),exp(f(x)))
```

An appropriate model for the data is $y_i \sim \text{Poi}(\mu_i)$ where $\log(\mu_i) = f(x_i)$: f is a smooth function, representable as a P-spline, and the y_i are independent.

(a) Write an R function to estimate the model by penalized likelihood maximization, given a smoothing parameter. Re-use the function from the first part of the previous question, in order to set up the P-spline for f.

(b) Modify your routine to return the model deviance and model degrees of freedom, so that a GCV score can be calculated for the model.

(c) Write a loop to fit the model and evaluate its (deviance based) GCV score, on a grid of smoothing parameter values. Plot the fitted values from the GCV optimal model, over the original data.

4. The natural parameterization of section 5.4.2 also allows the GCV score for a single smoothing parameter model to be evaluated very efficiently, for different trial values of the smoothing parameter. Consider such a single smooth to be fitted to response data $\mathbf{y}$: this question examines the efficient calculation of the components of the GCV score.

(a) In the notation of section 5.4.2, find an expression for the influence matrix $\mathbf{A}$ in terms of $\mathbf{Q}$, $\mathbf{U}$, $\mathbf{D}$ and λ.

(b) Show that the effective degrees of freedom of the model is simply:

$$\sum_{i=1}^{k} \frac{1}{1 + \lambda D_{ii}}.$$

(c) Again using the natural parameterization, show how calculations can be arranged so that, after some initial setting up, each new evaluation of $\|\mathbf{y} - \mathbf{Ay}\|^2$ costs only $O(k)$ arithmetic operations.

5. In section 6.9.1 methods were discussed for efficient computation with discretised covariates. This question explores these a little more.

(a) Consider an $n \times p$ model matrix $\mathbf{X}$ with m unique rows. If $\mathbf{W}$ is a diagonal matrix, give the (order of the) operations count for forming $\mathbf{X}^\mathsf{T}\mathbf{W}\mathbf{X}$ using $\mathbf{X}$ directly. By adapting the algorithm given in section 6.9.1 find an $O(np) + O(mp^2)$ algorithm to compute $\mathbf{X}^\mathsf{T}\mathbf{W}\mathbf{X}$ from the $m \times p$ matrix of its unique rows, $\bar{\mathbf{X}}$.

(b) If $\mathbf{X}_1$ is $n \times p_1$ with m_1 unique rows, while $\mathbf{X}_2$ is $n \times p_2$ with m_2 unique rows, give the operations costs of finding $\mathbf{X}_1^\mathsf{T}\mathbf{W}\mathbf{X}_2$ from explicitly formed matrices $\mathbf{X}_1$ and $\mathbf{X}_2$, and by applying an appropriate adaptation of your method from part (a).

(c) If $\mathbf{X}_1$ and $\mathbf{X}_2$ have the same number of unique rows, when is it more efficient to compute $\mathbf{X}_1^\mathsf{T}\mathbf{W}\mathbf{X}_2$ rather than $\mathbf{X}_2^\mathsf{T}\mathbf{W}\mathbf{X}_1$ when using the discretised method?

6. The generalized Fellner-Schall method of section 6.4 is based on REML. This question explores alternatives.

(a) Derive equivalent updates for the GCV criterion (6.20), from section 6.2.5, and for the generalized AIC, $-2l + 2\tau$, where τ is the usual EDF.

(b) Using something like `smoothCon(s(times,k=20),data=mcycle)` to construct the model matrices and penalties, fit a smooth to the `mcycle` data from the `MASS` library, estimating λ using your FS update formula for GCV.

(c) Write code to fit a smooth version of the Poisson model of section 3.3.2 using your AIC based FS update formula to estimate smoothing parameters.

7. The deviance based GCV (6.20) in section 6.2.5 includes a tuning parameter γ that can be increased from its default value of 1 to obtain smoother models, increasing the price paid per model effective degree of freedom in the GCV criterion.

(a) An alternative motivation for γ is obtained by considering the GCV score with $\gamma > 1$ as being an estimate of the GCV score that would have been obtained with a smaller sample size. Re-arrange the GCV score to make this interpretation clear.

(b) Using this interpretation, obtain a REML type criterion in which γ plays a similar role to that played in GCV.

8. Section 6.2.7 discusses the stability problems with log determinants that can occur when diagonal blocks making up the matrix overlap, are rank deficient and become badly scaled. The following code gives a simple example of a matrix, `A`, for which determinant calculation becomes numerically unstable as the parameter `lam` tends to zero.

```
set.seed(1); lam <- 1
A1 <- crossprod(diff(diag(3),diff=1))
A2 <- crossprod(matrix(runif(9),3,3))
A <- matrix(0,5,5);A[1:3,1:3] <- A1
A[3:5,3:5] <- A[3:5,3:5] + lam * A2
```

Write code to evaluate the log determinant of `A` at 100 log `lam` values spaced evenly between -40 and -25. Try 4 alternatives: QR, eigen and SVD decomposition of `A` and the similarity transform approach of section 6.2.7 (with QR based computation of the determinant). On a single plot, show each alternative method's answer against log `lam`. The stability problems of the first 3 alternatives should be obvious.

9. This question is about defining informative priors for the degree of smoothness of terms.
   (a) If we want to view smoothing parameters estimated by REML in a Bayesian way, as MAP estimators, what prior is being assumed for the log smoothing parameters?
   (b) How would the estimation of smoothing parameters be modified if we chose to impose a proper prior on the log smoothing parameter?
   (c) It is often much easier to specify a sensible prior for the effective degrees of freedom of model terms than it is to do the same for the smoothing parameters controlling these. Assuming a model containing only singly penalized smooths explain how you could go about incorporating independent Gaussian priors on the termwise effective degrees of freedom into model estimation.
   (d) What is the problem with trying to do this for tensor product smooths?

Chapter 7

GAMs in Practice: mgcv

This chapter covers use of the generalized additive modelling functions provided by R package mgcv: the design of these functions is based largely on Hastie (1993), although to facilitate smoothing parameter estimation, their details have been modified. There are three main modelling functions: gam, already introduced in section 4.6 (p. 182); bam a version of gam designed for larger datasets; and gamm a version of gam for estimating generalized additive mixed models via package nlme, allowing access to a rich range of random effects and correlation structures.* In addition jagam provides an interface to JAGS for Bayesian stochastic simulation. Note that gam can be used with a wide range of distributions beyond the usual GLM exponential family. This generality is not available with bam and gamm. See section 7.12 for a brief overview of the other R packages available for smooth modelling.

When reading this chapter note that R and the mgcv package are subject to continuing efforts to improve them. Sometimes this may involve modifications of numerical optimization behaviour, which may result in noticeable, but (hopefully) statistically unimportant, differences between the output given in this chapter and the corresponding results with more recent versions. Sometimes the exact formatting of output can also change a little.

Familiarity with the mgcv help pages, covered in section 4.6.4 (p. 191), is likely to be helpful when reading this chapter.

7.1 Specifying smooths

For the most part, this chapter covers specification of models by example, but it is worth discussing some general points about the specification of smooths in model formulae up front. There are 4 types of smooth that can be used and mixed.

s() is used for univariate smooths (section 5.3, p. 201), isotropic smooths of several variables (section 5.5, 214) and random effects (section 3.5.2, 154).

te() is used to specify tensor product smooths constructed from any singly penalized marginal smooths usable with s(), according to section 5.6 (p. 227). Examples are provided in sections 7.2.3, 7.4 and 7.7.1, for example.

*gamm4 from the gamm4 package is similar, but using lmer and glmer from package lme4 as the fitting engines

`ti()` is used to specify tensor product interactions with the marginal smooths (and their lower order interactions) excluded, facilitating smooth ANOVA models as discussed in section 5.6.3 (p. 232), and exemplified in section 7.3.

`t2()` is used to specify the alternative tensor product smooth construction discussed in section 5.6.5 (p. 235), which is especially useful for generalized additive mixed modelling with the `gamm4` package described in section 7.7.

The first arguments to all these functions are the covariates of the smooth. Some further arguments control the details of the smoother. The most important are

`bs` is a short character string specifying the type of basis. e.g. `"cr"` for cubic regression spline, `"ds"` for Duchon spline, etc. It may be a vector in the tensor product case, if different types of basis are required for different marginals.

`k` is the basis dimension, or marginal basis dimension (tensor case). It can also be a vector in the tensor case, specifying a dimension for each marginal.

`m` specifies the order of basis and penalty, in a basis specific manner.

`id` labels the smooth. Smooths sharing a label all have the same smoothing parameter (assuming that they are of the same smoother type).

`by` is the name of a variable by which the smooth should be multiplied (metric case), or each level of which should have a separate copy of the smooth (factor case).

The last two items on the list require further explanation. An example formula with a smooth `id` is `y ~ s(x) + s(z1,id="foo") + s(z2,id="foo")`. This forces the smooths of `z1` and `z2` to have the same smoothing parameter: for this to really make sense they are also forced to have the same basis and penalty, provided this is possible.

`by` variables are the means for implementing 'varying coefficient models', such as that used in section 7.5.3. Suppose, for example, that we have metric variables x and z and want to specify a linear predictor term '$f(x_i)z_i$' where f is a smooth function. The model formula entry for this would be `s(x,by=z)`. Only one `by` variable is allowed per smooth, but any smooth with multiple covariates (specified by `s`, `te`, `ti` or `t2`) can also have a `by` variable. Note that, provided the `by` variable takes more than one value, such terms are identifiable without a sum-to-zero constraint, and so they are left unconstrained.

Metric `by` variables combined with a *summation convention* are the means by which linear functionals of smooths can be incorporated into the linear predictor. Examples are provided in sections 7.4.2 and 7.11.1. The idea is that if the covariates of the smooth and the `by` variable are all matrices, then a summation across rows is implied. For example if `X`, `Z` and `L` are all matrices then `s(X,Z,by=L)` specifies the term $\sum_k f(X_{ik}, Z_{ik})L_{ik}$ in the linear predictor. Tensor terms also support the convention.

`by` variables also facilitate 'smooth-factor' interactions, in which we have a separate smooth of one or more covariates for each level of a factor `by` variable. For example, suppose we have metric variables x and z and factor variable g with three levels. Let $g(i)$ denote the level of g corresponding to observation i. Then `te(z,x,by=g)` would contribute the terms '$f_{g(i)}(x_i, z_i)$' to the model linear predictor. That is, which of three separate smooth functions of x and z contributes to the linear predictor depends on which of the three levels of g applies for observation i. Again `s`, `te`, `ti` or `t2` terms all work in the same way regardless of the number of their covariates. To avoid confounding problems the smooths are all subject to sum to zero constraints, which usually means that the main effect of g should also be included in the model specification. For example, `g + te(z,x,by=g)`. Factor by variables can not be mixed with the summation convention.[†] When there are several factor `by` variables then identifiability can get tricky, and it can then be useful to employ *ordered factor* `by` variables. If a factor `by` is an ordered factor then no smooth is generated for its first level.

Often we would like all the smooths generated by a factor `by` to have the same smoothing parameter. The `id` mechanism allows this. For example `te(z,x,by=g,id="a")` causes the smooths for each level of `g` to share the same `id`, and hence all to have the same smoothing parameter.

7.1.1 How smooth specification works

As far as the `mgcv` estimation functions are concerned a smooth (or indeed a random effect) is just a set of model matrix columns and one or more corresponding penalties. This fact is exploited in the code design to make smooth implementation highly modular. Within the code, smooths are implemented via a smooth *constructor* method function. In addition each smooth has a *prediction matrix* method function, which will produce the matrix mapping the coefficients of the smooth to predictions at new covariate values. Finally, smooths can have a *plot* method function (although most use the generic smooth plot method).

The design uses the '`S3`' object orientation built into R. `s()`, `te()`, `ti()` or `t2()` terms in a model formula give rise to a *smooth specification object*, with a class determining the type of smooth to be produced. This object is passed to the smooth constructor for its class (e.g., `smooth.construct.cr.smooth.spec` for the cubic regression spline basis `"cr"`). The constructor returns a *smooth object*, with its own class, containing a model matrix for the smooth, along with quadratic penalty matrices and other information. It also contains any information specific to the smooth that will be needed for prediction or plotting (knot locations, for example). See `?smooth.construct` for information on how to add a smooth class.

The smooth constructor and prediction matrix method functions are usually called via wrapper functions (`smoothCon()` and `PredictMat()`) that handle things like identifiability constraints, matrix arguments and `by` variables automatically for any smooth. Indeed it is quite easy to use `mgcv` only for the purpose of setting up smoothers and penalties. Here is an example setting up a thin plate regression spline basis and using it for unpenalized regression with `lm()`.

```
library(mgcv); library(MASS) ## load for mcycle data.
## set up a smoother...
```

[†] Separate summation convention smooths for each level of a factor are usually obtained by including a smooth for each level of the factor, with factor level specific versions of the `by` matrix set up to zero the term for all levels except the one to which the smooth applies.

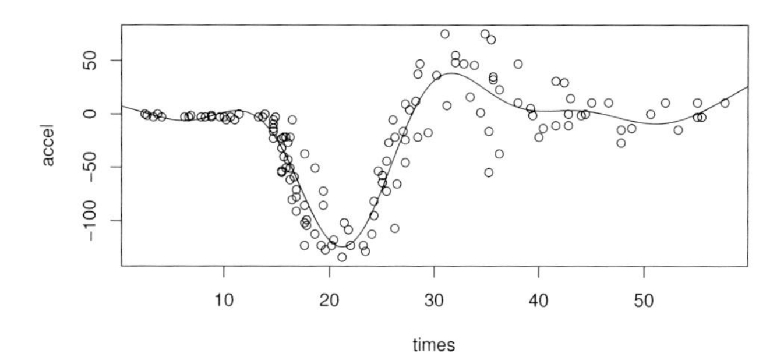

Figure 7.1 *Regression spline fit to the motorcycle data from the* MASS *library, illustrating how the* mgcv smoothCon *and* PredictMat *functions can be used to work with smoothers outside the* mgcv *modelling functions.*

```
sm <- smoothCon(s(times,k=10),data=mcycle,knots=NULL)[[1]]
## use it to fit a regression spline model...
beta <- coef(lm(mcycle$accel~sm$X-1))
with(mcycle,plot(times,accel)) ## plot data
times <- seq(0,60,length=200)  ## create prediction times
## Get matrix mapping beta to spline prediction at 'times'
Xp <- PredictMat(sm,data.frame(times=times))
lines(times,Xp%*%beta) ## add smooth to plot
```

The resulting plot is shown in figure 7.1.

7.2 Brain imaging example

Section 4.6 (p. 182) provided a basic introduction to mgcv. This section examines a more substantial example, in particular covering issues of model selection and checking in more detail. The data are from brain imaging by functional magnetic resonance scanning, and were reported in Landau et al. (2003). The data are available in data frame brain, and are shown in figure 7.2. Each row of the data frame corresponds to one voxel. The columns are: X and Y, giving the locations of each voxel; medFPQ, the brain activity level measurement (median 'Fundamental Power Quotient' over three measurements); region, a code indicating which region the voxel belongs to (0 for 'base' region; 1 for region of experimental interest and 2 for region subjected to direct stimulation) — there are some NA's in this column; meanTheta is the average phase shift at each voxel, which we will not consider further. This section will consider models for medFPQ as a function of X and Y.

Clearly the medFPQ data are quite noisy, and the main purpose of the modelling in this section is simply to partition this very variable signal into a smooth trend component and a 'random variability' component, so that the pattern in the image becomes a bit clearer. For data such as these, where the discretization (into voxels) is

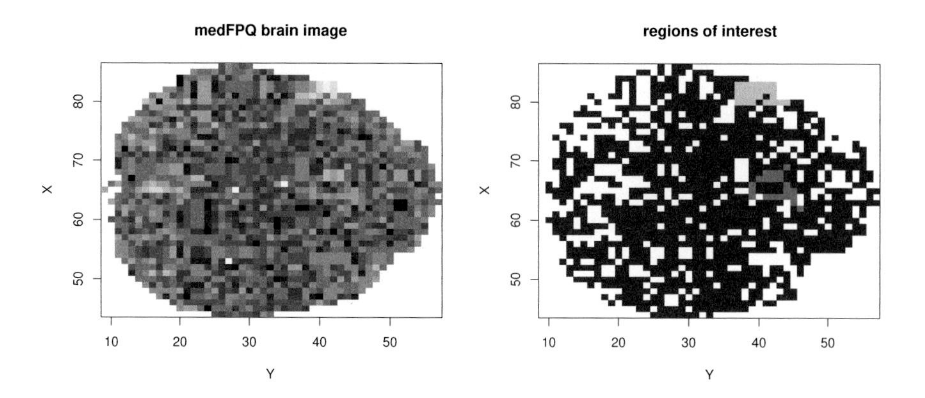

Figure 7.2 *The raw data for the brain imaging data discussed in section 7.2. The left hand plot shows the median of 3 measurements of Fundamental Power Quotient values at each voxel making up this slice of the scan: these are the measurements of brain activity. The right hand panel shows the regions of interest: the 'base region' is the darkest shade, the region directly stimulated experimentally is shown in light grey and the region of experimental interest is shown as dark grey. Unclassified voxels are white.*

essentially arbitrary, there is clearly a case to be made for employing a model which includes local correlation in the 'error' terms. In principle the correlation structures available in `gamm` could be used for this purpose (see, e.g., section 7.7.2, p. 371), but here we will proceed by treating the randomness as being independent between voxels, and let all between voxel correlation be modelled by the trend. This is not simply a matter of convenience, but relates closely to the purpose of the analysis: for these data the main interest lies in cleaning up this particular image: that is, in removing the component of variability that appears to be nothing more than random variation, at the level of the individual voxel. The fact that the underlying mechanism, generating features in the image, may include a component attributable to correlated noise across voxels is only something that need be built into the model if, for example, the objective is to be able to remove this component of the pattern from the image.

7.2.1 Preliminary modelling

An appropriate initial model structure would probably involve modelling `medFPQ` as a smooth function of `X` and `Y`, but before attempting to fit models, it is worth examining the data itself to look for possible problems. When this is done, 2 voxels appear problematic. These voxels have `medFPQ` values recorded as 3×10^{-6} and 4×10^{-7}, while the remaining 1565 voxels have values in the range 0.003 to 20. Residual plots from all attempts to model the data set including these two voxels consistently show them as grotesque outliers. Therefore they were excluded from the following analysis:

```
brain <- brain[brain$medFPQ>5e-3,] # exclude 2 outliers
```

The fairly skewed nature of the response data, medFPQ, and the fact that it is a necessarily positive quantity, suggest that some transformation may be required if a Gaussian error model is to be used. Attempting to use a Gaussian model without transformation confirms this:

```
> m0 <- gam(medFPQ ~ s(Y,X,k=100), data=brain)
> gam.check(m0)
Method: GCV   Optimizer: magic
Smoothing parameter selection converged after 6 iterations.
The RMS GCV score gradient at convergence was 6.236018e-05 .
The Hessian was positive definite.
Model rank =  100 / 100

Basis dim. (k) checking results. Low p-value (k-index<1) may
indicate that k is too low, especially if edf is close to k'.

           k'    edf k-index p-value
s(Y,X) 99.000 86.782   0.863       0
```

gam.check is a routine that produces some basic residual plots, and a little further information about the success or otherwise of the fitting process. Here it reports successful convergence of the GCV optimization and that the model has full rank. An informal basis dimension check is also included based on section 5.9 (p. 242): k' gives the maximum possible EDF for the smooth, to compare with the actual EDF; the k-index is the ratio of the model estimated scale parameter to an estimate based on differencing neighbouring residuals ('neighbouring' according to the arguments of the smooth); the p-value is for the residual randomization test described in section 5.9. Here the results are problematic, but of secondary importance, given the plots shown in figure 7.3. As explained in the figure caption, there are clear problems with the constant variance assumption: the variance is increasing with the mean. From the plots it is not easy to gauge the rate at which the variance of the data is increasing with the mean, but, in the absence of a good physical model of the mechanism underlying the relationship, some guidance can be obtained by a simple informal approach. If we assume that $\text{var}(y_i) \propto \mu_i^\beta$, where $\mu_i = \mathbb{E}(y_i)$ and β is some parameter, then a simple regression estimate of β can be obtained as follows:

```
> e <- residuals(m0); fv <- fitted(m0)
> lm(log(e^2) ~ log(fv))

Call:
lm(formula = log(e^2) ~ log(fv))

Coefficients:
(Intercept)      log(fv)
     -1.961        1.912
```

i.e., $\beta \approx 2$. That is, from the residuals of the simple fit, it appears that the variance of the data increases with the square of the mean. This in turn suggests using the gamma distribution, which has this mean-variance relationship (see table 3.1, p. 104). Alternatively the apparent mean-variance relationship suggests using a log transform of medFPQ to stabilize the variance, but in practice a less extreme 4th root transform

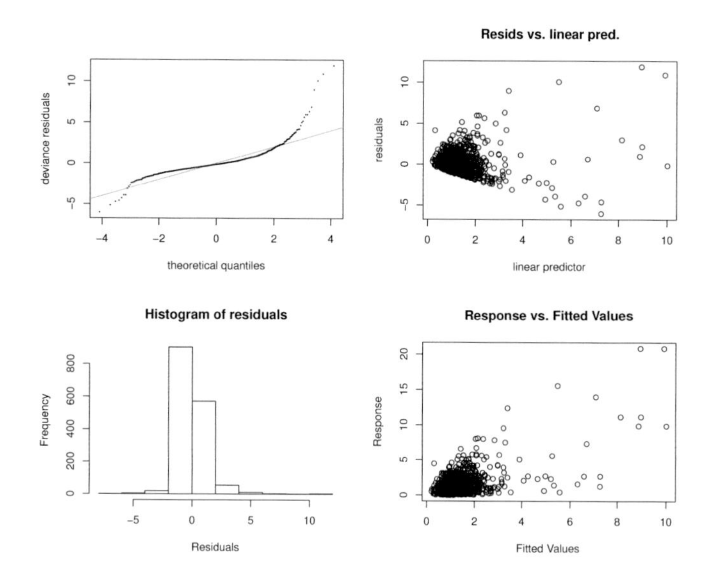

Figure 7.3 *Some basic model checking plots for model* `m0` *fitted to the brain scan data. The upper left normal QQ-plot clearly shows a problem with the Gaussian assumption. This is unsurprising when the upper right plot of residuals versus fitted values (linear predictor) is examined: the constant variance assumption is clearly untenable. The lower left histogram of residuals confirms the pattern evident in the QQ-plot: there are too many residuals in the tails and centre of the distribution relative to its shoulders. In this case the plot of response variable against fitted values, shown in the lower right panel, emphasizes the failure of the constant variance assumption.*

produces better residual plots. With the gamma model it might also be appropriate to use a log link, in order to ensure that all model predicted FPQ values are positive.

The following fits models based first on transforming the data and then on use of the gamma distribution.

```
m1 <- gam(medFPQ^.25 ~ s(Y,X,k=100), data=brain)
gam.check(m1)
m2<-gam(medFPQ~s(Y,X,k=100),data=brain,family=Gamma(link=log))
```

The plots from `gam.check` are shown in figure 7.4, and now show nothing problematic. `gam.check` plots for `m2` are equally good (but not shown). The informal basis dimension test still gives a low p-value, but since the EDF is some way below the basis dimension, and increasing the basis dimension barely changes the fitted values, let's stick with `k=100` here.

The major difference between `m1` and `m2` is in their biasedness on different scales. The model of the transformed data is approximately unbiased on the 4th root of the response scale: this means that it is biased downwards on the response scale itself. The log-gamma model is approximately unbiased on the response scale (only approximately because maximum penalized likelihood estimation is not generally unbiased, but is consistent). This can be seen if we look at the means of the fitted

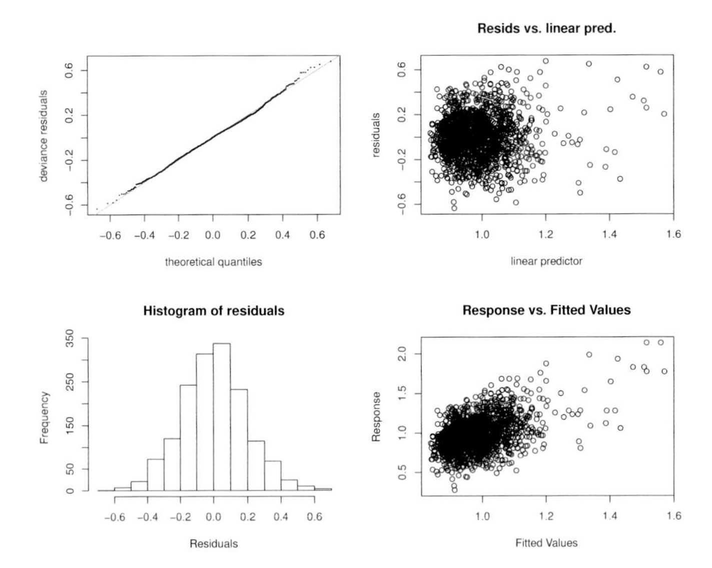

Figure 7.4 *Some basic model checking plots for model* m1 *fitted to the transformed brain scan data. The upper left normal QQ-plot is very close to a straight line, suggesting that the distributional assumption is reasonable. The upper right plot suggests that variance is approximately constant as the mean increases. The histogram of residuals at lower left appears approximately consistent with normality. The lower right plot, of response against fitted values, shows a positive linear relationship with a good deal of scatter: nothing problematic.*

values (response scale) for the two models, and compare these to the mean of the raw data:

```
> mean(fitted(m1)^4);mean(fitted(m2));mean(brain$medFPQ)
[1] 0.985554   # m1 tends to under-estimate
[1] 1.211483   # m2 substantially better
[1] 1.250302
```

Clearly, if the response scale is the scale of prime interest then the gamma model is to be preferred to the model based on normality of the transformed data. So far the best model seems to be the gamma log-link model, m2, which should be examined a little further.

```
> m2

Family: Gamma
Link function: log

Formula:
medFPQ ~ s(Y, X, k = 100)

Estimated degrees of freedom:
60.6  total = 61.61
```

```
GCV score: 0.6216871
> vis.gam(m2,plot.type="contour",too.far=0.03,
+ color="gray",n.grid=60,zlim=c(-1,2))
```

A relatively complex fitted surface has been estimated, with 62 degrees of freedom. The function `vis.gam` provides quite useful facilities for plotting predictions from a `gam` fit against pairs of covariates, either as coloured perspective plots, or as coloured (or grey scale) contour plots. The plot it produces for `m2` is shown in figure 7.5(a). Notice how the activity in the directly stimulated region and the region of interest stand out clearly in this plot.

7.2.2 *Would an additive structure be better?*

Given the large number of degrees of freedom employed by the model `m2`, the question naturally arises of whether a different simpler model structure might achieve a more parsimonious fit. Since this is a book about GAMs, the obvious candidate is the additive model,

$$\log\{\mathbb{E}(\texttt{medFPQ}_i)\} = f_1(\texttt{Y}_i) + f_2(\texttt{X}_i), \quad \texttt{medFPQ}_i \sim \text{gamma}.$$

```
> m3 <- gam(medFPQ ~ s(Y,k=30) + s(X,k=30), data=brain,
+           family=Gamma(link=log))
> m3

Family: Gamma
Link function: log

Formula:
medFPQ ~ s(Y, k = 30) + s(X, k = 30)

Estimated degrees of freedom:
 9.58 20.20  total = 30.77

GCV score: 0.6453502
```

The GCV score is higher for this model, suggesting that it is not an improvement, and a comparison of explained deviances using `summary(m2)` and `summary(m3)` also suggests that the additive model is substantially worse. AIC comparison confirms this:

```
> AIC(m2,m3)
         df      AIC
m2 62.61062 3321.681 ## bivariate smooth
m3 31.77467 3393.738 ## additive model
```

Perhaps the most persuasive argument against the additive structure is provided by the plot of the predicted log activity levels provided in figure 7.5(b): the additive structure produces horizontal and vertical 'stripes' in the plot that have no real support from the data.

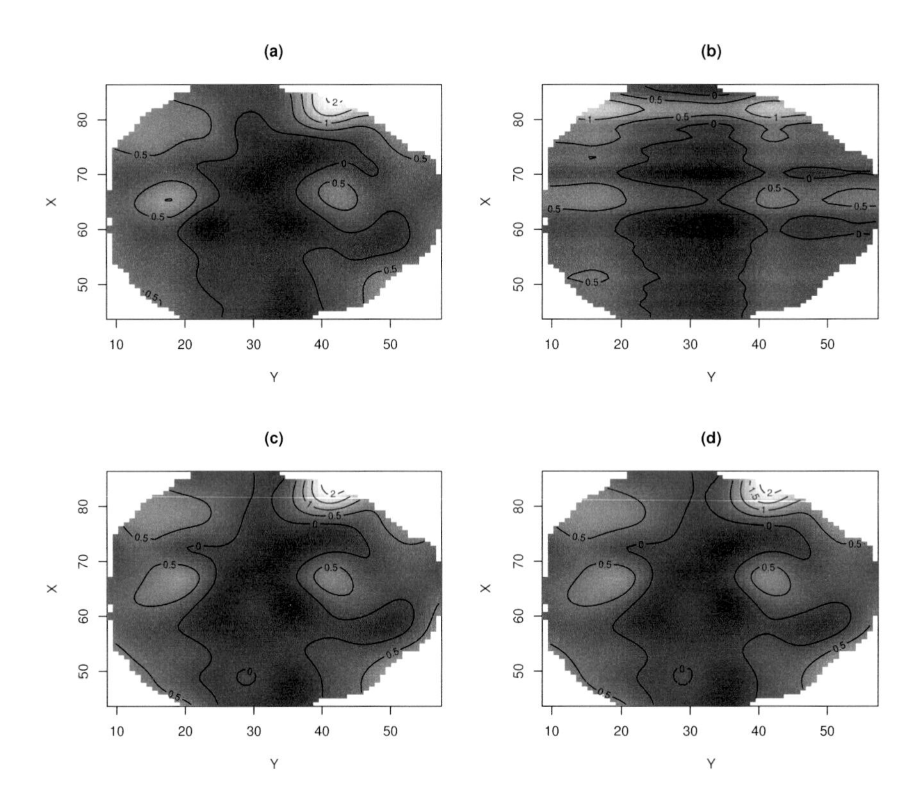

Figure 7.5 *Comparison of 4 models of the brain scan data. All plots show image plots and overlaid contour plots, on the scale of the linear predictor. (a) is model* `m2` *based on an isotropic smooth of* `Y` *and* `X`. *(b) is model* `m3` *based on a sum of smooth functions of* `Y` *and* `X`. *Notice the apparent artefacts in (b), relating to the assumption of an additive structure. (c) plots model* `tm` *and is basically as (a), but using a rank 100 tensor product smooth in place of the isotropic smooth. (d) is for model* `tm1`, *which is as model* `tm` *but with main effects and interaction estimated separately. All plots were produced with something like:* `vis.gam(m2,plot.type="contour",too.far=0.03,color="gray", n.grid=60,zlim=c(-1,2),main="(a)")`

7.2.3 *Isotropic or tensor product smooths?*

As discussed in chapter 5, isotropic smooths, of the sort produced by `s(Y,X)` terms, are usually good choices when the covariates of the smooth are naturally on the same scale, and we expect that the same degree of smoothness is appropriate with respect to both covariate axes. For the brain scan data these conditions probably hold, so the isotropic smooths are probably a good choice.

Nevertheless, it is worth checking what the results look like if we use a scale invariant tensor product smooth of `Y` and `X`, in place of the isotropic smooth (see section 5.6, p. 227). Such smooths are computationally rather efficient (if the marginal bases of a tensor product smooth are cheap to construct, then so is the tensor product basis itself). In addition they provide another way of comparing an additive model to a bivariate smooth model via the `ti` interaction term construction of section 5.6.3 (p. 232). In particular we can fit the model

$$\log\{\mathbb{E}(\texttt{medFPQ}_i)\} = f_1(\texttt{Y}_i) + f_2(\texttt{X}_i) + f_3(\texttt{X}_i, \texttt{Y}_i), \quad \texttt{medFPQ}_i \sim \text{gamma},$$

in which there are main smooth effects and an interaction smooth effect, constructed so that functions of the form $f_1(\texttt{Y}_i) + f_2(\texttt{X}_i)$ are excluded from its basis.

Here is the code, first for the single tensor product smooth and then for the additive plus interaction version:

```
tm<-gam(medFPQ~te(Y,X,k=10),data=brain,family=Gamma(link=log))
tm1 <- gam(medFPQ ~ s(Y,k=10,bs="cr") + s(X,bs="cr",k=10) +
           ti(X,Y,k=10), data=brain, family=Gamma(link=log))
```

The single `s` terms could have been replaced by terms `ti(X,k=10)` resulting in the same model fit. Plots from `tm` and `tm1` are shown in figure 7.5(c) and (d), and residual plots and other checks show that both models are reasonable. However, an AIC comparison suggests that neither tensor product structure is quite as good as the isotropic model for these data:

```
> AIC(m2,tm,tm1)
          df      AIC
m2   62.61062 3321.681 ## isotropic
tm   60.34768 3332.817 ## tensor product
tm1  57.08366 3330.187 ## main + interaction
```

The AIC comparison slightly favours the main effects plus interaction model over the single tensor product smooth. One initially puzzling question is why the AICs are not identical for `tm` and `tm1`, given that they have the same smoothing basis, but simply partitioned differently. The reason is that they do not have the same penalty structure: `tm1` has extra separate penalties on the part of the basis representing the smooth main effects, $f_1(\texttt{Y}_i) + f_2(\texttt{X}_i)$.

Returning to the question of whether an additive model structure or bivariate structure is more appropriate, `tm1` can be examined using `summary` or `anova`.

```
> anova(tm1)

Family: Gamma
Link function: log

Formula:
medFPQ ~ s(Y, k = 10, bs = "cr") + s(X, bs = "cr", k = 10) +
    ti(X, Y, k = 10)

Approximate significance of smooth terms:
           edf Ref.df      F  p-value
s(Y)      8.258  8.750 14.526  < 2e-16
s(X)      7.494  8.314  6.959 2.37e-09
ti(X,Y)  39.332 49.891  2.291 1.15e-06
```

The p-value for ti(X, Y) (computed as in section 6.12.1, p. 305) clearly suggests that the smooth term is significantly non zero: the interaction is needed and an additive structure alone is insufficient.

7.2.4 *Detecting symmetry (with* by *variables)*

It is sometimes of interest to test whether the image underlying some noisy data is symmetric, and this is quite straightforward to accomplish, with the methods covered here. For the brain data we might want to test whether the underlying activity levels are symmetric about the mean X value of 64.5. The symmetry model in this case would be

$$\log\{\mathbb{E}(\texttt{medFPQ}_i)\} = f(\texttt{Y}_i, |\texttt{X}_i - 64.5|), \quad \texttt{medFPQ}_i \sim \text{gamma}$$

and this could be compared with model m2, which does not assume symmetry, although for strict nesting of models we would need to be slightly more sophisticated and use an asymmetry model with a form such as

$$\log\{\mathbb{E}(\texttt{medFPQ}_i)\} = f(\texttt{Y}_i, |\texttt{X}_i - 64.5|) + f_r(\texttt{Y}_i, |\texttt{X}_i - 64.5|).\texttt{right}_i,$$

where f_r is represented using the same basis as f, and $\texttt{right}_i$ is a dummy variable, taking the values 1 or 0, depending on whether $\texttt{medFPQ}_i$ is from the right or left side of the brain.

The following code creates variables required to fit these two models, estimates them, and prints the results.

```
> brain$Xc <- abs(brain$X - 64.5)
> brain$right <- as.numeric(brain$X<64.5)
> m.sy <- gam(medFPQ~s(Y,Xc,k=100),data=brain,
+             family=Gamma(link=log))
> m.as <- gam(medFPQ~s(Y,Xc,k=100)+s(Y,Xc,k=100,by=right),
+             data=brain,family=Gamma(link=log))
> m.sy
[edited]
Estimated degrees of freedom:
51.4  total = 52.44
GCV score: 0.6489799
> m.as
[edited]
Estimated degrees of freedom:
50.5 44.7  total = 96.2
GCV score: 0.6176281
```

The GCV scores suggest that the asymmetric model is better, and the asymmetric model AIC is also 97 lower than the symmetric version. The same question can also be addressed by hypothesis testing:

```
> anova(m.as)
Family: Gamma
Link function: log
```

```
Formula:
medFPQ ~ s(Y, Xc, k = 100) + s(Y, Xc, k = 100, by = right)


Approximate significance of smooth terms:
                 edf Ref.df     F  p-value
s(Y,Xc)        50.48  65.99 4.344  < 2e-16
s(Y,Xc):right 44.72  59.21 2.457 1.06e-08
```

If the symmetric model was adequate then the `s(Y,Xc):right` term should not be significantly different from zero, which is far from being the case here. Again this test uses section 6.12.1 (p. 305). The less well-justified approach of section 6.12.4 (p. 313) could be accessed using `anova(m.sy,m.as)`.

Plots of the different model predictions help to show why symmetry is so clearly rejected (figure 7.6).

```
vis.gam(m.sy,plot.type="contour",view=c("Xc","Y"),too.far=.03,
        color="gray",n.grid=60,zlim=c(-1,2),main="both sides")
vis.gam(m.as,plot.type="contour",view=c("Xc","Y"),
        cond=list(right=0),too.far=.03,color="gray",n.grid=60,
        zlim=c(-1,2),main="left side")
vis.gam(m.as,plot.type="contour",view=c("Xc","Y"),
        cond=list(right=1),too.far=.03,color="gray",n.grid=60,
        zlim=c(-1,2),main="right side")
```

7.2.5 *Comparing two surfaces*

The symmetry testing, considered in the last section, is an example of comparing surfaces — in that case one half of an image with a mirror image of the other half. In some circumstances it is also interesting to compare completely independent surfaces in a similar way. To see how this might work, a second set of brain scan data can be simulated, using the fitted model `m2`, and somewhat perturbed, as follows.

```
brain1 <- brain
mu <- fitted(m2)
n<-length(mu)
ind <- brain1$X<60 & brain1$Y<20
mu[ind] <- mu[ind]/3
set.seed(1)
brain1$medFPQ <- rgamma(rep(1,n),mu/m2$sig2,scale=m2$sig2)
```

Now the data sets can be combined, and dummy variables created to identify which data set each row of the combined data frame relates to.

```
brain2=rbind(brain,brain1)
brain2$sample1 <- c(rep(1,n),rep(0,n))
brain2$sample0 <- 1 - brain2$sample1
```

After which it is straightforward to fit a model with a single combined surface for both data sets, and a second model where the surfaces are allowed to differ. Note that in the latter case a single common surface is estimated, with a difference surface

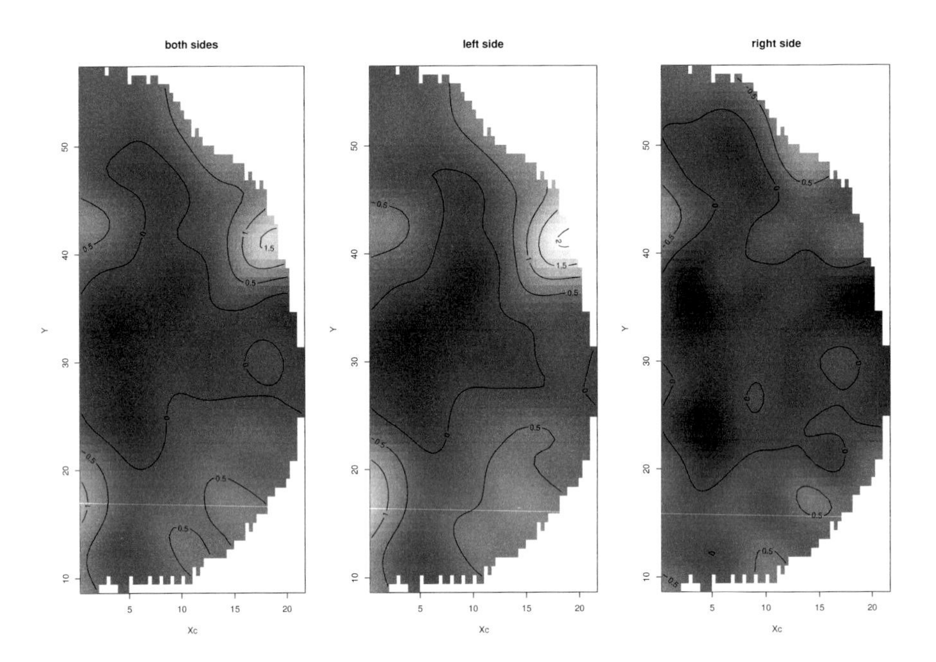

Figure 7.6 *Half brain images for the two models involved in examining possible symmetry in the brain image data. The left panel shows the predictions of log activity for the symmetry model: the left side of the image would be a mirror image of this plot. The middle plot shows a mirror image of the left side brain image under the asymmetry model, while the right plot shows the right side image under the asymmetry model. Clearly the asymmetry model suggests rather different activity levels in the two sides of the image.*

for the second sample. This approach is often preferable to a model with two completely separate surfaces, for reasons of parsimony. If there is no difference between the surfaces, it will still be necessary to estimate two quite complex surfaces, if we adopt separate surfaces for each data set. With the common surface plus difference approach, only one complex surface need be estimated: the difference being close to flat. Similarly, any time that the surfaces are likely to differ in a fairly simple way, it makes sense to use the common surface plus difference model, thereby reducing the number of effective degrees of freedom needed by the model.

```
m.same <- gam(medFPQ~s(Y,X,k=100),data=brain2,
              family=Gamma(link=log))
m.diff <- gam(medFPQ~s(Y,X,k=100)+s(Y,X,by=sample1,k=100),
              data=brain2,family=Gamma(link=log))
```

Examination of the GCV scores for the two models suggests that the second model `m.diff` is preferable to `m.same`. An AIC comparison, or a test of the significance of the difference surface in `mdiff`, confirm this:

```
> AIC(m.same,m.diff)
               df      AIC
m.same 65.09955 6529.403
m.diff 80.84258 6496.320
> anova(m.diff)

Family: Gamma
Link function: log

Formula:
medFPQ ~ s(Y, X, k = 100) + s(Y, X, by = sample1, k = 100)

Approximate significance of smooth terms:
                   edf Ref.df     F  p-value
s(Y,X)           62.46  79.28 6.323  < 2e-16
s(Y,X):sample1 16.38  22.91 2.190 0.000936
```

In this example, the data for the two surfaces were available on exactly the same regular mesh, but the approach is equally applicable for irregular data where the data are not at the same covariate values for the two surfaces (although, of course, the regions of covariate space covered should overlap, for a comparison of this sort to be meaningful).

7.2.6 *Prediction with* `predict.gam`

The `predict` method function, `predict.gam`, enables a `gam` fitted model object to be used for prediction at new values of the model covariates. It is also used to provide estimates of the uncertainty of those predictions, and the user can specify whether predictions should be made on the scale of the response or of the linear predictor. For predictions on the scale of the linear predictor, `predict.gam` also allows predictions to be decomposed into their component terms, and it is also possible to extract the matrix, which, when multiplied by the model parameter vector, yields the vector of predictions at the given set of covariate values.

Usually the covariate values at which predictions are required are supplied as a data frame in argument `newdata`, but if this argument is not supplied then predictions/fitted values are returned for the original covariate values, used for model estimation. Here are some examples. Firstly on the scale of the linear predictor,

```
> predict(m2)[1:5]
        1         2         3         5         6
0.2988590 0.3353680 0.3432145 0.2786153 0.4482840
> pv <- predict(m2,se=TRUE)
> pv$fit[1:5]
        1         2         3         5         6
0.2988590 0.3353680 0.3432145 0.2786153 0.4482840
> pv$se[1:5]
        1         2         3         5         6
0.2582929 0.2124696 0.2120118 0.2193975 0.2234603
```

and then on the response scale,

```
> predict(m2,type="response")[1:5]
       1        2        3        5        6 
1.348320 1.398455 1.409471 1.321299 1.565623 
> pv <- predict(m2,type="response",se=TRUE)
> pv$se[1:5]
        1         2         3         5         6 
0.3482614 0.2971292 0.2988245 0.2898897 0.3498546 
```

For both sets of examples, predictions are produced for all 1564 voxels, but I have only printed the first 5. Note that the standard errors provided on the response scale are approximate, being obtained by the usual Taylor expansion approach.

Usually, predictions are required for new covariate values, rather than simply for the values used in fitting. Suppose, for example, that we are interested in two points in the brain scan image, firstly in the directly stimulated area, $(X = 80.1, Y = 41.8)$, and secondly in the region of experimental interest, $(X = 68.3, Y = 41.8)$. A data frame can be created, containing the covariate values for which predictions are required, and this can be passed to `predict.gam`, as the following couple of examples show.

```
> pd <- data.frame(X=c(80.1,68.3),Y=c(41.8,41.8))
> predict(m2,newdata=pd)
        1         2 
1.2788054 0.5954536 
> predict(m2,newdata=pd,type="response",se=TRUE)
$fit
       1        2 
3.592346 1.813854 

$se.fit
        1         2 
0.5287998 0.2600005 
```

It is also possible to obtain the contributions that each model term, excluding the intercept, makes to the linear predictor, as the following example shows. The additive model `m3` has been used to illustrate this, since it has more than one term.

```
> predict(m3,newdata=pd,type="terms",se=TRUE)
$fit
       s(Y)         s(X)
1 0.2501423  0.518684282
2 0.2501423 -0.003159924


$se.fit
        s(Y)       s(X)
1 0.05856406 0.09826015
2 0.05856406 0.08222815


attr(,"constant")
(Intercept) 
   0.143576 
```

As you can see, named arrays have been returned, the columns of which correspond to each model term. There is one array for the predicted values, and a second for the corresponding standard errors.

Prediction with `lpmatrix`

Because the GAMs discussed in this book have an underlying parametric representation, it is possible to obtain a 'prediction matrix', $\mathbf{X}_p$, which maps the model parameters, $\hat{\boldsymbol{\beta}}$, to the predictions of the linear predictor, $\hat{\boldsymbol{\eta}}_p$. That is, to find $\mathbf{X}_p$ such that

$$\hat{\boldsymbol{\eta}}_p = \mathbf{X}_p\hat{\boldsymbol{\beta}}.$$

`predict.gam` can return $\mathbf{X}_p$, if its `type` argument is set to `"lpmatrix"`.

The following example illustrates this. Since the returned matrix is of dimension 2×100 (with default treatment of identifiability constraints), it has not been printed out, but rather it is demonstrated that it does indeed give the required linear predictor values, when multiplied by the coefficients of the fitted model.

```
> Xp <- predict(m2,newdata=pd,type="lpmatrix")
> fv <- Xp %*% coef(m2)
> fv
       [,1]
1 1.2788054
2 0.5954536
```

Why is $\mathbf{X}_p$ useful? A major use is in the calculation of variances for combinations of linear predictor values. Clearly, if $\hat{\mathbf{V}}_\beta$ is the estimate of the parameter covariance matrix, then from standard probability theory, the estimated covariance matrix of $\boldsymbol{\eta}_p$ must be:

$$\hat{\mathbf{V}}_{\boldsymbol{\eta}_p} = \mathbf{X}_p\hat{\mathbf{V}}_\beta\mathbf{X}_p^{\mathsf{T}}.$$

Now suppose that we are really interested in, for example, the difference, δ, between the linear predictor values at the points in the two regions. This difference could be written as, $\delta = \mathbf{d}^{\mathsf{T}}\boldsymbol{\eta}_p$, where $\mathbf{d}^{\mathsf{T}} = [1, -1]^{\mathsf{T}}$. In that case, standard theory says that:

$$\widehat{\text{var}(\delta)} = \mathbf{d}^{\mathsf{T}}\hat{\mathbf{V}}_{\boldsymbol{\eta}_p}\mathbf{d} = \mathbf{d}^{\mathsf{T}}\mathbf{X}_p\hat{\mathbf{V}}_\beta\mathbf{X}_p^{\mathsf{T}}\mathbf{d}.$$

The following code illustrates this.

```
> d <- t(c(1,-1))
> d %*% fv
          [,1]
[1,] 0.6833517
> d %*% Xp %*% vcov(m2) %*% t(Xp) %*% t(d)
           [,1]
[1,] 0.04184623
```

So, the ability to obtain an explicit 'predictor matrix' makes some variance calculations rather straightforward. Notice that in general some care must be taken with calculations like the preceding one, if `Xp` has a large number of rows. For example, if `Xp` had 10,000 rows and `d` was some corresponding weight vector then it is important to ensure that the calculation does not involve the explicit formation of a $10,000 \times 10,000$ matrix, `Xp%*%vcov(m2)%*%t(Xp)`,

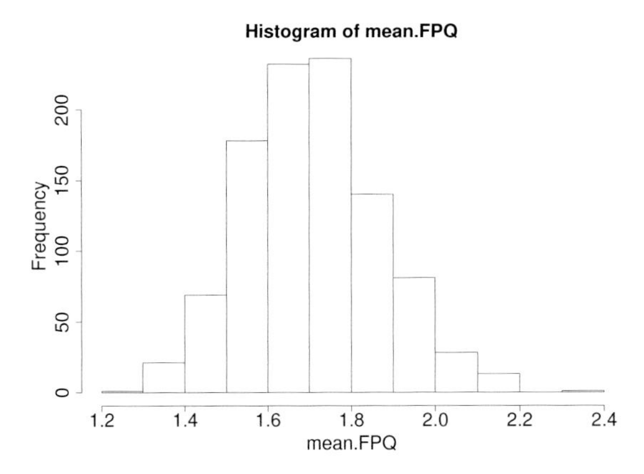

Figure 7.7 *Histogram of 1000 draws from the posterior distribution of average* medFPQ *in the region of experimental interest in the brain scan example.*

which would be costly in both computer time and memory. Fortunately, careful use of brackets will force R to order the operations in a way that avoids this: `(d%*%Xp)%*%vcov(m2)%*%(t(Xp)%*%t(d))` would do the trick.

A further trick along these lines is useful. Quite often we only need the diagonal of some covariance matrix, such as `diag(Xp%*%Vp%*%t(Xp))`. It turns out that this is the same as `rowSums(Xp*(Vp%*%Xp))`. The latter expression is much more efficient than the former as it does not compute the off-diagonal elements that are not of interest.

7.2.7 *Variances of non-linear functions of the fitted model*

The linear theory facilitating the calculations of the previous section is only applicable if we are interested in inference about linear functions of the linear predictor. As soon as the functions of interest become non-linear, as is generally the case when working on the response scale, we need different methods.

In fact, prediction matrices, in conjunction with simulation from the posterior distribution of the parameters, $\boldsymbol{\beta}$, give a simple and general method for obtaining variance estimates (or indeed distribution estimates), for any quantity derived from the fitted model, not simply those quantities that are linear in $\boldsymbol{\beta}$. The idea of this sort of posterior simulation was discussed in section 6.10 (p. 293).

As an example, suppose that we would like to estimate the posterior distribution of the average medFPQ, in the region of experimental interest. Since the quantity of interest is on the response scale, and not the scale of the linear predictor, it is not linear in $\boldsymbol{\beta}$ and the previous section's method is not applicable.

To approach this problem, a prediction matrix is first obtained, which maps the parameters to the values of the linear predictor needed to form the average.

```
ind <- brain$region==1& ! is.na(brain$region)
Xp <- predict(m2,newdata=brain[ind,],type="lpmatrix")
```

Next, a large number of replicate parameter sets are simulated from the posterior distribution of $\boldsymbol{\beta}$, using the `rmvn` function.

```
set.seed(8) ## for repeatability
br <- rmvn(n=1000,coef(m2),vcov(m2)) # simulate from posterior
```

Each column of the matrix `br` is a replicate parameter vector, drawn from the approximate posterior distribution of $\boldsymbol{\beta}$. Given these replicate parameter vectors and the matrix `Xp`, it is a simple matter to obtain the linear predictor implied by each replicate, from which the required averages can easily be obtained:

```
mean.FPQ <- rep(0,1000)
for (i in 1:1000) {
  lp <- Xp %*% br[i,]  # replicate linear predictor
  mean.FPQ[i] <- mean(exp(lp)) # replicate region 1 mean FPQ
}
```

Or, more efficiently, but less readably:

```
mean.FPQ <- colMeans(exp(Xp %*% t(br)))
```

So `mean.FPQ` now contains 1000 replicates from the approximate posterior distribution of the mean FPQ measurement in region 1. The results of `hist(mean.FPQ)` are shown in figure 7.7.

Clearly, this simulation approach is rather general: it is easy to obtain samples from the posterior distribution of any quantity that can be predicted from the fitted model. Notice also that, in comparison to bootstrapping, the approach is very computationally efficient. One disadvantage is that the smoothing parameters are treated as fixed at their estimates, rather than as uncertain. A way around this is to use the smoothing parameter uncertainty correction of section 6.11.1 (p. 302), which is available via `vcov(m,unconditional=TRUE)` provided that `m` is estimated using ML or REML smoothing parameter estimation.

7.3 A smooth ANOVA model for diabetic retinopathy

The `wesdr` data frame (originally from Chong Gu's `gss` package) contains clinical data on whether diabetic patients suffer from diabetic retinopathy or not. Three possible predictors are also available: duration of disease (years), body mass index (mass in kg divided by square of height in metres) and percent glycosylated hemoglobin in the blood. The latter is the percentage of hemoglobin bound to glucose, which reflects long term average blood glucose levels and should be below 6% for non-diabetics. Figure 7.8 shows boxplots of the covariates by disease status. The relationship of risk to the covariates is somewhat unclear.

To try and understand the data, a smooth additive logistic regression model could be employed. Since we might expect to find interactions between the covariates it might make sense to employ the sort of smooth ANOVA model introduced in section

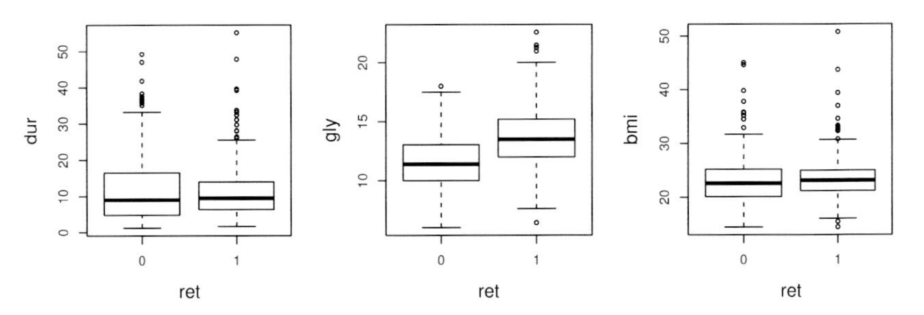

Figure 7.8 *Boxplots of disease duration, percent glycosylated hemoglobin and body mass index, by retinopathy status.*

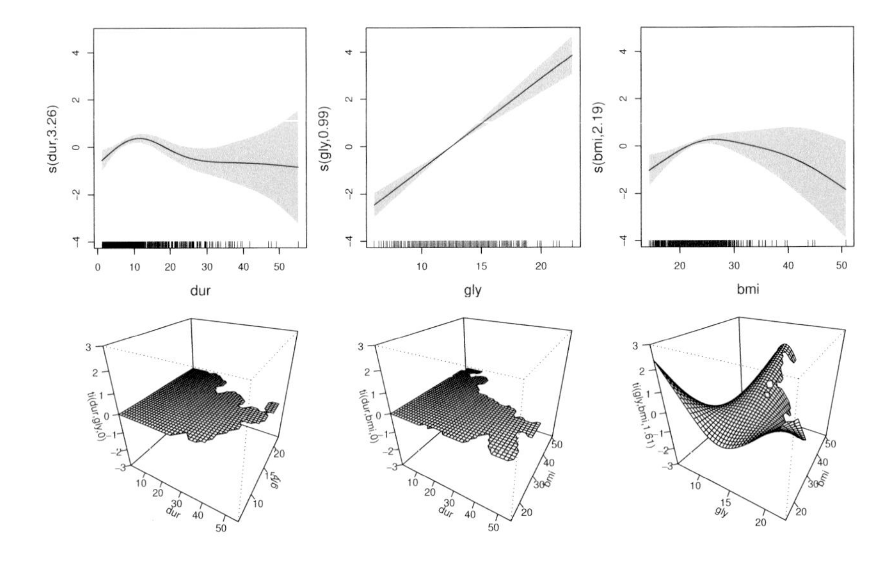

Figure 7.9 *Estimated effects for the smooth ANOVA logistic regression model for retinopathy risk. Notice how only the* `gly-bmi` *interaction is estimated to be non-zero.*

5.6.3 (p. 232). For example $\texttt{ret}_i \sim \text{binom}(1, \mu_i)$,

$$\text{logit}(\mu_i) = f_1(\texttt{dur}_i) + f_2(\texttt{gly}_i) + f_3(\texttt{bmi}_i) + \\ f_4(\texttt{dur}_i, \texttt{gly}_i) + f_5(\texttt{dur}_i, \texttt{bmi}_i) + f_6(\texttt{bmi}_i, \texttt{gly}_i)$$

where f_1 to f_3 are smooth 'main effects', while f_4 to f_6 are smooth 'interactions' constructed to exclude the main effects, as described in section 5.6.3. The model is easily estimated, and for model selection the null space penalties of section 5.4.3 (p. 214) can be used.

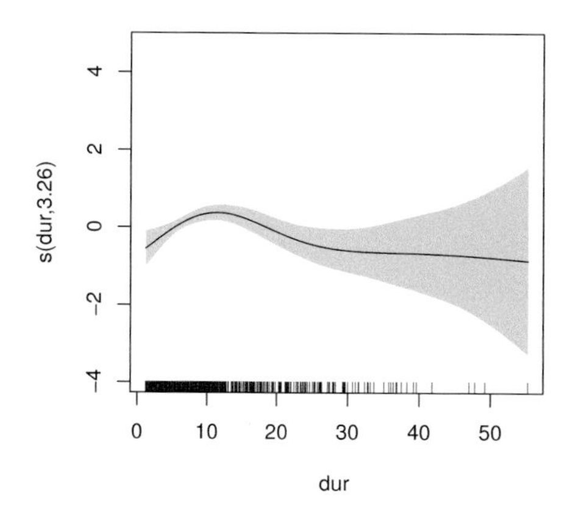

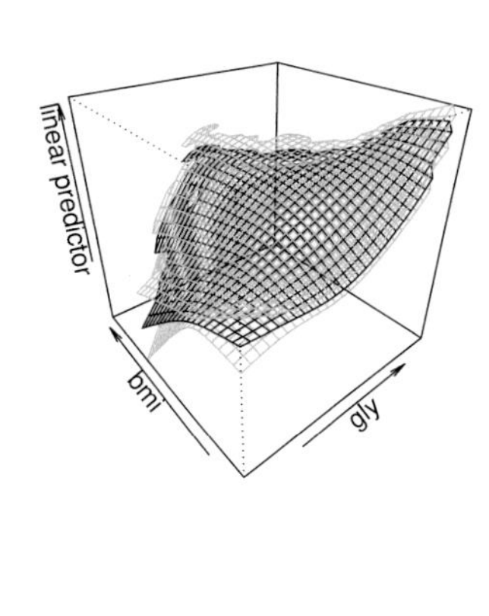

Figure 7.10 *Estimated duration effect and combined effect of body mass index and percent glycosylated hemoglobin from the retinopathy model. On the right, the grey surfaces are at plus or minus one standard error from the estimate. Notice how a higher level of glycosylated hemoglobin seems to be tolerated when body mass index is low.*

```
> b <- gam(ret ~ s(dur,k=k) + s(gly,k=k) + s(bmi,k=k) +
+           ti(dur,gly,k=k) + ti(dur,bmi,k=k) + ti(gly,bmi,k=k),
+        select=TRUE, data=wesdr, family=binomial(), method="ML")
> b
Family: binomial
Link function: logit

Formula:
ret ~ s(dur, k = k) + s(gly, k = k) + s(bmi, k = k) + ti(dur,
    gly, k = k) + ti(dur, bmi, k = k) + ti(gly, bmi, k = k)

Estimated degrees of freedom:
3.2553 0.9892 2.1866 0.0003 0.0001 1.6118  total = 9.04

ML score: 385.7904
```

Notice how f_4 and f_5 are effectively penalized out of the model, as the plots of the effects in figure 7.9 also show. f_6 is estimated to be non-zero and the model `summary` confirms that it is significantly so. Hence we need to interpret the main effects and interaction for `bmi` and `gly` together. One possibility would be to re-estimate the model with the simplified model formula `ret~s(dur,k=k)+te(gly,bmi,k=k)`, but this leads to a small increase in ML score and AIC (since the model is changed slightly by the change in smoothing penalties between the smooth ANOVA model and the simpler model). In figure 7.10 the joint effect of `bmi` and `gly` is therefore visualised using `vis.gam` (with options `se=1` and `colors="bw"`).

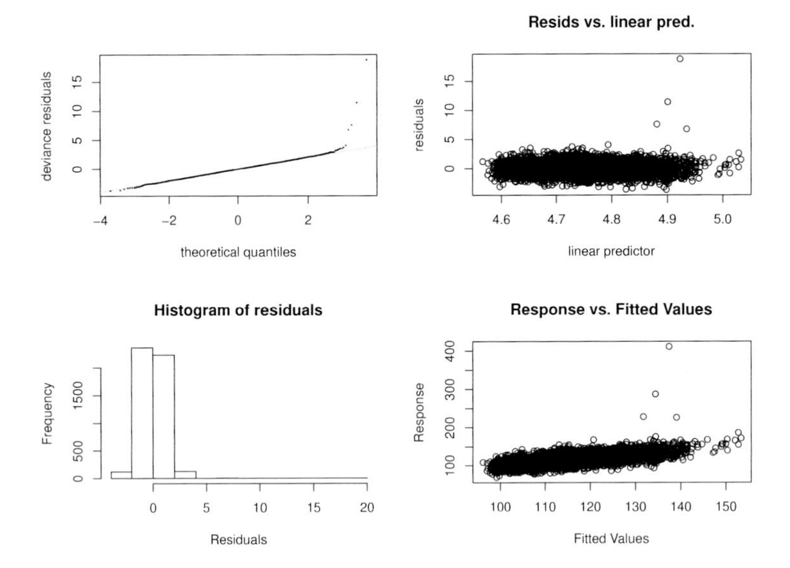

Figure 7.11 *Basic model checking plots for the* `ap0` *air pollution mortality model. For Possion data with moderately high means, the distribution of the standardized residuals should be quite close to normal, so that the QQ-plot is obviously problematic. As all the plots make clear, there are a few gross outliers that are very problematic in this fit.*

It seems that risk increases with increasing glycosylated hemoglobin, as expected, but initially increases less steeply for those with low BMI. Similarly, risk increases initially with disease duration, but there is then some evidence of decline: it could be that the patients only get to have high duration if they tend to have their disease under good control, or it could be that there are just some patients who were not going to get retinopathy, while those less fortunate tended to get it earlier rather than later in their disease progression. Note that not much checking has been done for this example since model checking with binary response data is difficult (see section 7.6 for one possibility), but the flexibility of the assumed model does offer some defence against misspecification problems.

7.4 Air pollution in Chicago

The relationship between air pollution and health is a moderately controversial topic, and there is a great deal of epidemiological work attempting to elucidate the links. In this section a variant of a type of analysis that has become quite prevalent in air-pollution epidemiology is presented. The data are from Peng and Welty (2004) and are contained in a data frame `chicago`. The response of interest is the daily death rate in Chicago, `death`, over a number of years. Possible explanatory variables for the observed death rate are levels of ozone, `o3median`, levels of sulpher dioxide, `so2median`, mean daily temperature, `tmpd`, and levels of particulate matter, pm10median (as generated by diesel exhaust, for example). In addition to these air quality variables, the underlying death rate tends to vary with time (in particular throughout the year), for reasons having little or nothing to do with air quality.

A conventional approach to modelling these data would be to assume that the observed numbers of deaths are Poisson random variables, with an underlying mean that is the product of a basic, time varying, death rate, modified through multiplication by pollution dependent effects. That is, the model would be

$$\log\{\mathbb{E}(\texttt{death}_i)\} = f(\texttt{time}_i) + \beta_1\texttt{pm10median}_i + \beta_2\texttt{so2median}_i + \beta_3\texttt{o3median}_i + \beta_4\texttt{tmpd}_i,$$

where $\texttt{death}_i$ follows a Poisson distribution, and f is a smooth function.

The model is easily fitted and checked.

```
ap0 <- gam(death~s(time,bs="cr",k=200)+pm10median+so2median+
           o3median+tmpd,data=chicago,family=poisson)
gam.check(ap0)
```

A cubic regression spline has been used for f to speed up computation a little. The checking plots are shown in figure 7.11, and show clear problems as a result of some substantial outliers. Plotting the estimated smooth with and without partial residuals emphasises the size of the outliers.

```
par(mfrow=c(2,1))
plot(ap0,n=1000)  # n increased to make plot smooth
plot(ap0,residuals=TRUE,n=1000)
```

The plots are shown in figure 7.12, and four gross outliers, in close proximity to each other, are clearly visible. Examination of the data indicates that the outliers are the four highest daily death rates occurring in the data, and that they occurred on consecutive days,

```
> chicago$death[3111:3125]
[1] 112 97 122 119 116 121 226 411 287 228 159 142 123 102 94
```

Plotting this section of data also indicates that this peak is associated with a period of very high temperatures and high ozone. One immediate possibility is that the model is simply too inflexible, and that some non-linear response of death rate to temperature and ozone is required. This might suggest replacing the linear dependencies on the air quality covariates with smooth functions, so that the model structure becomes:

$$\log\{\mathbb{E}(\texttt{death}_i)\} = f_1(\texttt{time}_i) + f_2(\texttt{pm10median}_i) + f_3(\texttt{so2median}_i) + f_4(\texttt{o3median}_i) + f_5(\texttt{tmpd}_i),$$

where the f_j are smooth functions. This model is easily fitted

```
ap1<-gam(death ~ s(time,bs="cr",k=200)+s(pm10median,bs="cr")+
     s(so2median,bs="cr")+s(o3median,bs="cr")+s(tmpd,bs="cr"),
     data=chicago,family=poisson)
```

but the gam.check plots are almost indistinguishable from those shown in figure 7.11. Figure 7.13 shows the estimated smooths, and indicates a problem with the distribution of pm10median values, in particular, which might be expected to cause

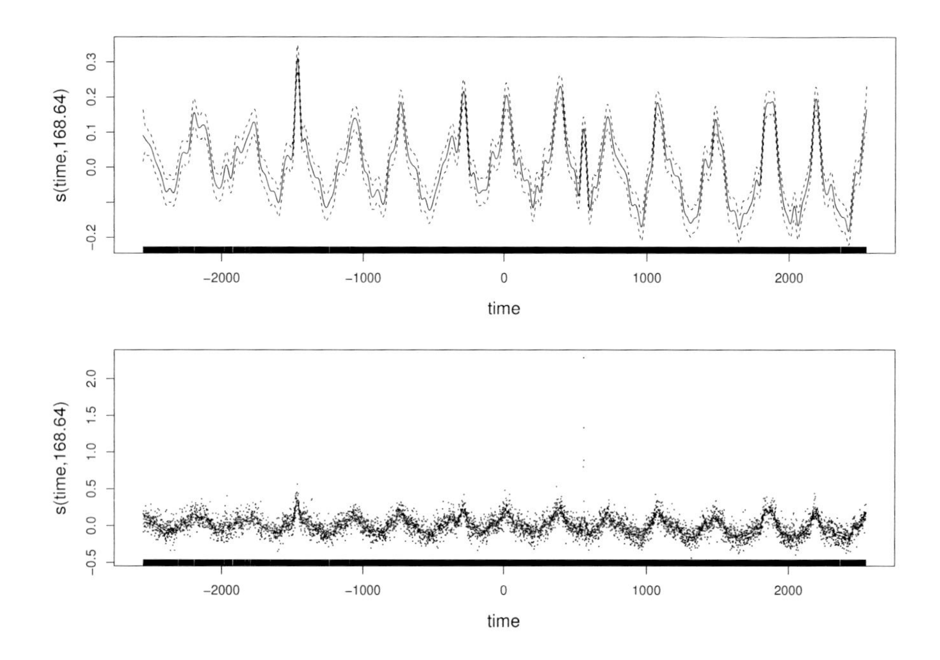

Figure 7.12 *The estimate of the smooth from model* ap0 *shown with and without partial residuals. Note the 4 enormous outliers apparently in close proximity to each other.*

leverage problems. Similar plots, with partial residuals, again indicate a severe failure to fit the four day run of record death rates. Models with smooth interactions of the pollutant variables are no better.

More detailed examination of the data, surrounding the four day mortality surge, shows that the highest temperatures in the temperature record were recorded in the few days *preceding* the high mortalities, when there were also high ozone levels recorded. This suggests that average temperature and pollution levels, over the few days preceding a given mortality rate, might better predict it than the temperature and levels only on the day itself. On reflection, such a model might be more sensible on biological/medical grounds: the pollution levels and temperatures recorded in the data are not high enough to cause immediate acute disease and mortality, and it seems more plausible that any effects would take some time to manifest themselves via, for example, the aggravation of existing medical conditions.

There are then two types of model that might be appropriate. A *single index model* is specified in terms of smooth functions of weighted sums of covariates, where the weights are estimated as part of estimation. In the current case we would be interested in a weighted sum over lagged covariates. An alternative is a *distributed lag model* in which the response depends on a sum of smooth functions of lagged covariates. Usually the smooth functions at different lags vary smoothly with lag: we

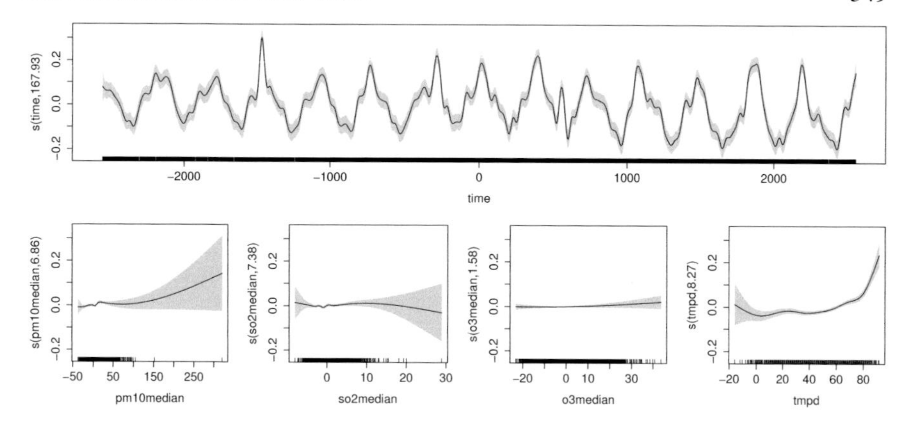

Figure 7.13 *The estimate of the smooth from model* `ap1` *shown without partial residuals. Note the gap in the* `pm10median` *values. Similar plots with partial residuals highlight exactly the same gross outliers as were evident in figure 7.12.*

do not expect the response to yesterday's pollution to have a completely different shape to the response to the day before yesterday's pollution, for example.

7.4.1 A single index model for pollution related deaths

Assuming that the data are arranged in order of increasing time, then an appropriate single index model is

$$\log\{\mathbb{E}(\texttt{death}_i)\} = f_1(\texttt{time}_i) + f_2\left(\sum_{k=0}^{5} \alpha_k \texttt{pm10}_{i-k}\right) + f_3\left(\sum_{k=0}^{5} \alpha_k \texttt{o3}_i, \sum_{k=0}^{5} \alpha_k \texttt{tmp}_i\right)$$

where the α_k are parameters to be estimated. Alternative models with an `so2` effect and/or a 3-way interaction of `pm10`, `o3` and `tmp` do not provide a better fit. Of course it would be possible to argue for different weights, α_k, for the different variables, and more lags could be included: the current structure was chosen to include more lags than seemed likely to have a strong effect while maintaining reasonable computational speed.

As a preliminary for model fitting it helps to prepare a set of 6-column matrices containing the lagged variables in separate columns. For example

```
lagard <- function(x,n.lag=6) {
  n <- length(x); X <- matrix(NA,n,n.lag)
  for (i in 1:n.lag) X[i:n,i] <- x[i:n-i+1]
  X
}
dat <- list(lag=matrix(0:5,nrow(chicago),6,byrow=TRUE))
dat$pm10 <- lagard(chicago$pm10median)
...
```

The easiest way to estimate the single index model is to write a function which accepts the weighting parameter as its first argument, internally re-weights the lagged covariates, and then performs the required GAM fitting call using the re-weighted data. Given such a function we can use R's built in `optim` function to find the weights optimizing the smoothing parameter criterion. Note that without some constraints the weights are not identifiable. They can be made identifiable by insisting that $\alpha_0 > 0$ and $\|\boldsymbol{\alpha}\| = 1$: the function below employs a re-parameterization to enforce this. It is also worth using `bam` (with default REML smoothing parameter estimation) in place of `gam` in the routine, in order to keep the computation time reasonable (using the `discrete=TRUE` option to `bam` would speed up a little more again). A suitable function (adapted from `?single.index`) is:

```
si <- function(theta,dat,opt=TRUE) {
## Return ML if opt==TRUE or fitted gam otherwise.
  alpha <- c(1,theta) ## alpha defined via unconstrained theta
  kk <- sqrt(sum(alpha^2)); alpha <- alpha/kk  ## ||alpha||=1
  o3 <- dat$o3 %*% alpha; tmp <- dat$tmp %*% alpha
  pm10 <- dat$pm10 %*% alpha ## re-weight lagged covariates
  b<- bam(dat$death~s(dat$time,k=200,bs="cr")+s(pm10,bs="cr")+
          te(o3,tmp,k=8),family=poisson) ## fit model
  cat(".") ## give user something to watch
  if (opt) return(b$gcv.ubre) else {
    b$alpha <- alpha  ## add alpha to model object
    b$J <- outer(alpha,-theta/kk^2) ## get dalpha_i/dtheta_j
    for (j in 1:length(theta)) b$J[j+1,j] <- b$J[j+1,j] + 1/kk
    return(b)
  }
} ## si
```

It remains to fit the model and extract the final fitted model.

```
> f1 <- optim(rep(1,5),si,method="BFGS",hessian=TRUE,dat=dat)
> apsi <- si(f1$par,dat,opt=FALSE)
> apsi$alpha
[1]  0.02497  0.66809  0.63504  0.26046  0.26051 -0.11851
```

The fit is fairly slow, taking around half an hour on the laptop computer used to write this, but the results are interesting. Almost no effect is estimated for the temperature on the day of death, the preceding two days are important and the two days before that less so. The fairly small negative coefficient for lag 5 could indicate that increases are somewhat harmful, but could also just be artefact: there might be a case for constraining the α_j to be non-negative here. The `gam.check` plots for `apsi` (not shown) are much improved relative to the previous models. Figure 7.14 shows the estimated effects and a single residual plot. Notice that the rug plot at the bottom of the time effect now has gaps: this is because observations with any NAs in the lagged variables now have to be dropped (the high observations that caused the original bad fits are not dropped this way). The most noticeable effect is the big increase in risk when the weighted sums of ozone and temperature are simultaneously high. The residual plot now looks much improved: there is one large negative outlier, but it is nothing like the magnitude of the previous models' outliers.

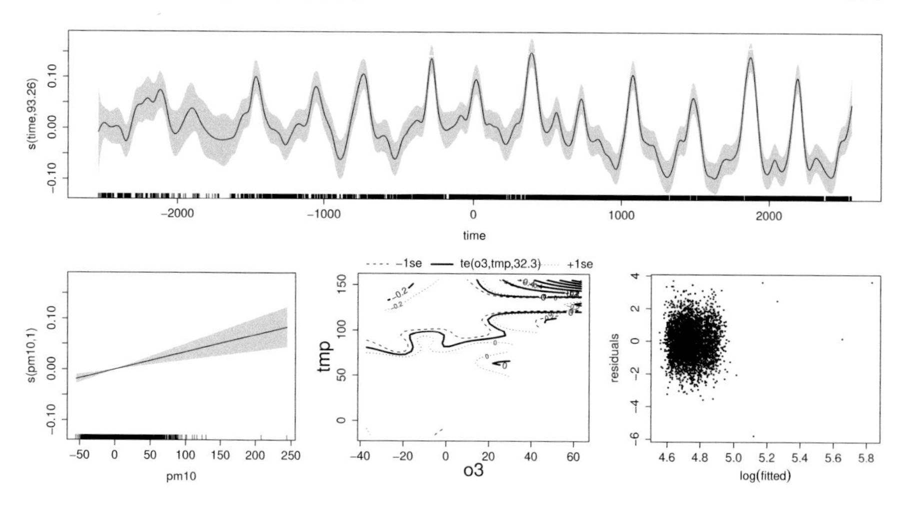

Figure 7.14 *The estimates from the single index model* `apsi`. *Notice the sharp increase in death rate at a combination of high temperature and high ozone over the preceding 4 days. The lower right residual plot for the same model is much improved on the model with only instantaneous daily effects.*

7.4.2 A distributed lag model for pollution related deaths

A distributed lag model represents the effect of a pollutant by a sum of smooth functions of the pollutant at a range of lags. For example the effect of `pm10` on the ith day's deaths might be represented by $\sum_{k=0}^{5} f_k(\texttt{pm10}_{i-k})$, where the f_k are smooth functions to be estimated. This effect is easy to include in a model, via a model formula component something like `s(pm10)+s(pm10.1) ... + s(pm10.5)`, where `pm10.j` contains the `pm.10` observations lagged by j. Of course, we might not be happy that all these effects are estimated completely separately: something like `s(pm10,id=1)+s(pm10.1,id=1) ... + s(pm10.5,id=1)` would force the terms to have the same smoothing parameter. However, the effect estimates could still be wildly different between adjacent lags, which may not be plausible. Another alternative is to insist that the smooth functions themselves vary smoothly with lag. Suppose that f is a bivariate smooth function, while $\texttt{lag}_{ik} = k - 1$ and $\texttt{PM10}_{ik} = \texttt{pm10}_{i-k+1}$. Then an appropriate model for the contribution of `pm10` to the ith day's deaths might be

$$\sum_{k=1}^{6} f(\texttt{PM10}_{ik}, \texttt{lag}_{ik}).$$

Using the 'summation convention' in `mgcv` (see section 7.1), such terms can be specified using a model formula component `te(PM10, lag)`, where `PM10` and `lag` are the six column matrices defined above.

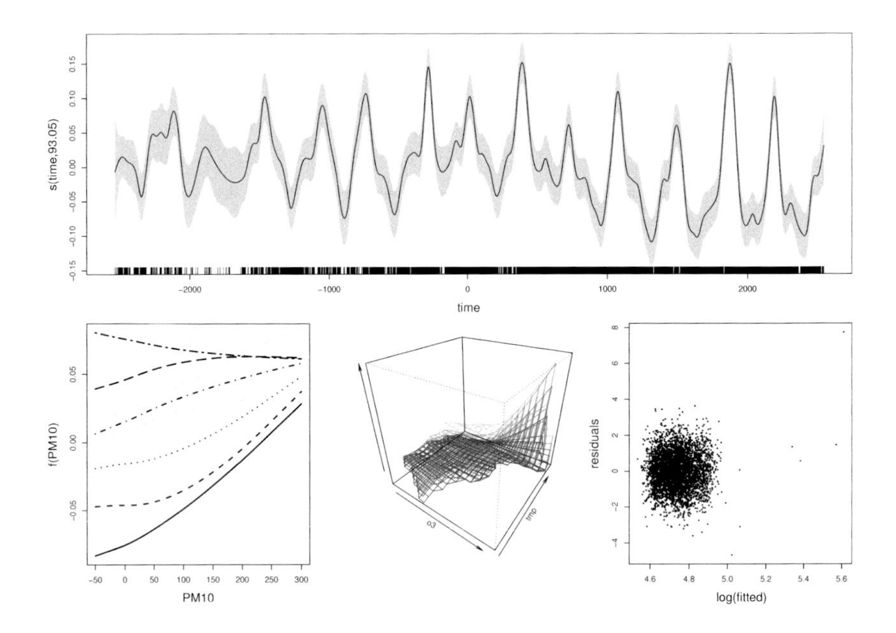

Figure 7.15 *Estimated effects for the distributed lag model of the Chicago air pollution and respiratory deaths data.* Top: *the estimated smooth for the background death rate.* Lower left: *the effect of PM_{10} for each lag from 0 to 5 days. Moving vertically from the bottom left of the plot, the curves are in order of increasing lag. One standard error bands are shown for lags 2 and 3. Notice how there is almost no effect by lags 4 and 5 (curve almost flat), but at shorter lags increased PM_{10} is associated with increased death rate.* Lower middle: *perspective plots of the ozone-temperature interaction by lag (z axis range -0.1 to 0.6). Lighter shades are for longer lags. The dominant feature is high risk at high ozone high temperature combinations, for the middle lags.* Lower right: *the residual plot shows one rather large positive outlier, but is otherwise reasonable.*

A useful model involving the same variables as the single index model is then

$$\log\{\mathbb{E}(\texttt{death}_i)\} = f_1(\texttt{time}_i) + \sum_{k=1}^{6} f_2(\texttt{PM10}_{ik}, \texttt{lag}_{ik}) + \sum_{k=1}^{6} f_3(\texttt{O3}_{ik}, \texttt{TMP}_{ik}, \texttt{lag}_{ik})$$

which can be estimated using

```
apl <- bam(death~s(time,bs="cr",k=200)+te(pm10,lag,k=c(10,5))+
       te(o3,tmp,lag,k=c(8,8,5)),family=poisson,data=dat)
```

A model with an additional distributed lag term for `so2` gives a p-value of 0.28 for the extra term, so there seems little justification for including it. Note that the model with and without `so2` are estimated with different numbers of data, because of missing `so2` data, so comparing them by AIC or GLRT (`anova`) is not valid, unless we re-fit the simpler model only to the subset of data with no missing `so2`.

The distributed lag model fits around 60 times faster than the single index model.

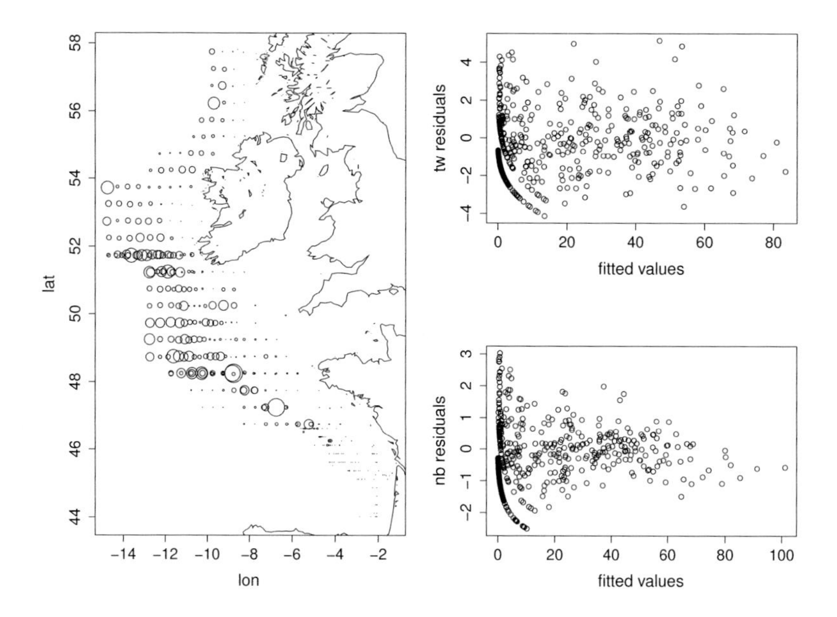

Figure 7.16 *The left hand plot shows the density of stage I mackerel eggs per square metre of sea surface as assessed by net samples in the 1992 mackerel egg survey. Circle areas are proportional to egg density and are centred on the location at which the egg samples were obtained. The right hand plots show residual plots for the Tweedie (upper) and negative binomial (lower) GAMs fitted to these data. The negative binomial deviance residuals appear to be decreasing with the predicted mean, which is problematic.*

It does not do quite as well on the very highest daily death count as the single index model, but does not have the one quite large negative outlier evident from that model. The distributed lag model has the lower AIC of the two. Compact visualization of distributed lag models is not always easy, and figure 7.15 offers a not completely satisfactory attempt (using `predict.gam` to generate the plotting data, and `par(new=TRUE)` in R to overlay perspective plots). The plots are rather consistent with the single index model, with the dominant pollution effect being a big increase in risk associated with high ozone-temperature combinations at intermediate lags.

7.5 Mackerel egg survey example

Worldwide, most commercially exploitable fish stocks are overexploited, with some stocks, such as Newfoundland cod, having famously collapsed. Effective management of stocks rests on sound fish stock assessment, but this is not easy to achieve. The main difficulty is that counting the number of catchable fish of any given species is all but impossible. A standard statistical sampling approach based on trying to catch fish fails, because, for large mobile net-avoiding organisms, there is no way of relating what is caught to what was available to catch. To some extent, surveys with sonar avoid such catchability problems, but suffer from other problems: chiefly that it is often impossible to determine which fish species cause a given sonar signal. Another alternative is to attempt to reconstruct past fish stocks on the basis of records of what fish get landed commercially, an approach known as 'virtual population analysis', but this suffers from the problem that it tells you quite accurately what the stock was several years ago, but is imprecise about its current or recent state.

Egg production methods are a different means of assessing stocks. The idea is to assess the number of eggs produced by a stock, or the rate at which a stock is producing eggs, and then to work out the number (or more often mass) of adult fish required to produce this number or production rate. This works because egg production rates per kg of adult fish *can* be assessed from adults caught in trawls, while eggs can be sampled in an unbiased manner. Egg data are obtained by sending out survey ships to sample eggs at each station of some predefined sampling grid, over the area occupied by a stock. Eggs are usually sampled by hauling a fine meshed net up from the sea bed to the sea surface (or at least from well below the depth at which eggs are expected, to the surface). The number of eggs of the target species in the sample is then counted and the volume of water sampled is known.

7.5.1 Model development

To get the most out of the egg data it is helpful to model the egg distribution, and here GAMs can be useful. The example considered in this section concerns data from a 1992 mackerel egg survey. The data were first modelled using GAMs by Borchers et al. (1997) and are available in `gamair` dataset `mack`. The left hand panel of figure 7.16 shows the location at which samples were taken, and the egg densities found there. As well as longitude and latitude, a number of other possible predictors of egg abundance are available: `salinity`; surface temperature of the ocean, `temp.surf`; water temperature at a depth of 20 m, `temp.20m`; depth of the ocean, `b.depth`; and, finally, distance from the 200 m seabed contour, `c.dist`. The latter predictor reflects the biologists' belief that the fish like to spawn near the edge of the continental shelf, conventionally considered to end at a seabed depth of 200 m. At each sampling location, a net was hauled vertically through the water column from well below the depth at which eggs are found to the surface: the mackerel eggs caught in the net were counted, to give the response variable, `egg.count`.

A reasonable model for the mean egg counts is

$$\mathbb{E}(\texttt{egg.count}_i) = g_i \times (\text{net area})_i,$$

where g_i is the density of eggs, per square metre of sea surface, at the ith sampling location. Taking logs of this equation we get

$$\log\{\mathbb{E}(\texttt{egg.count}_i)\} = d_i + \log\{(\text{net area})_i\},$$

where $d_i = \log(g_i)$ will be modelled as a function of predictor variables, using an additive structure, and $\log\{(\text{net area})_i\}$ will be treated as a model 'offset': that is,

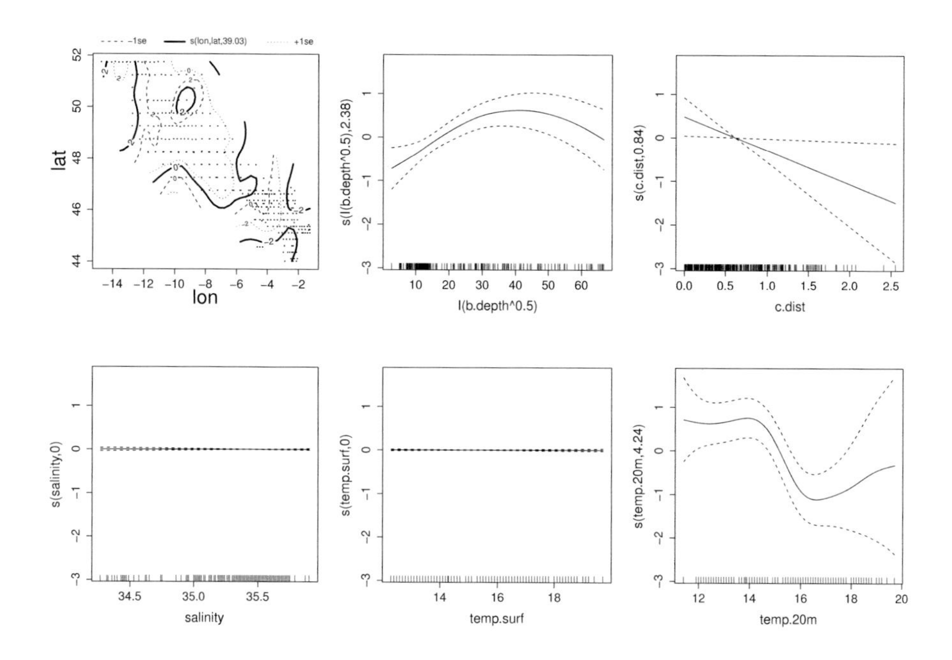

Figure 7.17 *Estimated model terms for the simple additive* `gmtw` *mackerel model.*

as a column of the model matrix with associated parameter fixed at 1. It remains to pick a distribution. Since the response is a count, it is tempting to start with the Poisson distribution, but any attempt to fit the sort of models used here with a Poisson response shows very strong evidence of *over-dispersion*: the residuals are much too large for consistency with a Poisson with unit scale parameter. An alternative is to use the `quasipoisson` family in R, which makes the Poisson-like assumption that $\text{var}(y_i) = \phi\mathbb{E}(y_i)$, but without fixing $\phi = 1$. However, to facilitate things like AIC model selection and REML smoothing parameter estimation[‡] it helps to use full distributions. The obvious alternatives are the Tweedie or negative binomial distributions (see section 3.1.9, p. 115).

For the purposes of this exercise, a simple additive structure will be assumed, with the first model being estimated as follows:

```
mack$log.net.area <- log(mack$net.area)

gmtw <- gam(egg.count ~ s(lon,lat,k=100) + s(I(b.depth^.5))+
        s(c.dist) + s(salinity) + s(temp.surf) + s(temp.20m)+
        offset(log.net.area),data=mack,family=tw,method="REML",
        select=TRUE)
```

[‡]RE/ML smoothing parameter selection has to use the extended quasi-likelihood of McCullagh and Nelder (1989) with `quasi` families, but given Fletcher (2012) it is unclear how reliable this will be in the presence of low expected counts.

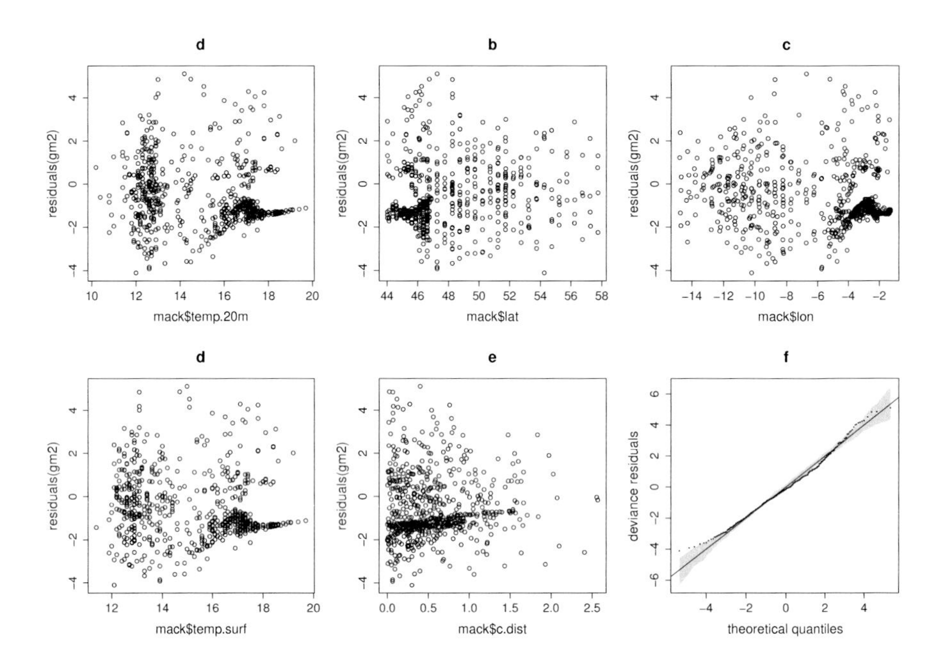

Figure 7.18 *Residual plots for model* gm2. f *is a QQ-plot produced by* qq.gam *showing the ordered deviance residuals against their theoretical quantiles, if the model is correct. The grey band is a reference band.*

Here the model selection penalties discussed in section 5.4.3 (p. 214) are used by setting argument select to TRUE. Tweedie and negative binomial distributions with automatic estimation of their extra parameters are only available with REML or ML smoothing parameter estimation, but these also tend to work better than prediction error criteria with selection penalties, because of reduced tendency to under-smooth. The square root transform of b.depth is to reduce leverage from very high values. The tw family has a log link as default, and will estimate the Tweedie p and scale parameters as part of fitting: the p estimate is 1.33 in this case. A similar call with nb in place of tw will fit a negative binomial based model (θ is estimated as 1.37).

The basic residual plots on the right hand side of figure 7.16 show acceptable behaviour for the deviance residuals of the Tweedie model, but a problematic decline in the variance of the deviance residuals with fitted value for the negative binomial model, indicating that the negative binomial assumes too sharp an increase in the variance with the mean. AIC (using, e.g., AIC(gmtw)) also gives a much smaller value for the Tweedie model. Figure 7.17 shows the effect plots for the Tweedie model, gmtw: the salinity and surf.temp effects have been penalized out of the model altogether. Note however, that there are many missing values for salinity, so that dropping it allows a much larger set of data to be used for estimation. Clearly then, we should not drop surf.temp before refitting without the salinity effect. In fact, doing this still gives a fully penalized `temp.surf` effect, so that is dropped too, to arrive at

```
> gm2 <- gam(egg.count ~ s(lon,lat,k=100) + s(I(b.depth^.5)) +
           s(c.dist) + s(temp.20m) + offset(log.net.area),
           data=mack,family=tw,method="REML")
> gm2

Family: Tweedie(p=1.326)
Link function: log

Formula:
egg.count ~ s(lon, lat, k = 100) + s(I(b.depth^0.5)) +
    s(c.dist) + s(temp.20m) + offset(log.net.area)

Estimated degrees of freedom:
54.48  3.26  1.00  6.21  total = 65.95

REML score: 1566.855
```

Some residual plots for `gm2` are shown in figure 7.18. The plots appear reasonable, with nothing to indicate much wrong with the assumed mean-variance relationship, nor other suspicious patterns. The small dense block of residuals on each plot relates to the densely sampled low abundance block in the south of the survey area. Figure 7.19 plots the non-linear effects for this model. In some respects the high degrees of freedom estimated for the spatial smooth is disappointing: biologically it would be more satisfactory for the model to be based on predictors to which spawning fish might be responding directly. Spatial location can really only be a proxy for something else, or the result of a process in which much of the pattern is driven by spatial correlation.

7.5.2 *Model predictions*

The purpose of this sort of modelling exercise is assessment of the total stock of eggs, which means that predictions are required from the model. In the first instance a simple map of predicted densities is useful. The data frame `mackp` contains the model covariates on a regular grid, over the survey area, as well as an indexing column indicating which grid square the covariates belong to, in an appropriate 2D array. The following code produces the plot on the left hand side of figure 7.20.

```
mackp$log.net.area <- rep(0,nrow(mackp))
lon <- seq(-15,-1,1/4); lat <- seq(44,58,1/4)
zz<-array(NA,57*57); zz[mackp$area.index]<-predict(gm2,mackp)
image(lon,lat,matrix(zz,57,57),col=gray(0:32/32),
      cex.lab=1.5,cex.axis=1.4)
contour(lon,lat,matrix(zz,57,57),add=TRUE)
lines(coast$lon,coast$lat,col=1)
```

Notice the substantial problem that the egg densities remain high at the western boundary of the survey area.

Typically, uncertainty estimates are required for quantities derived from fitted

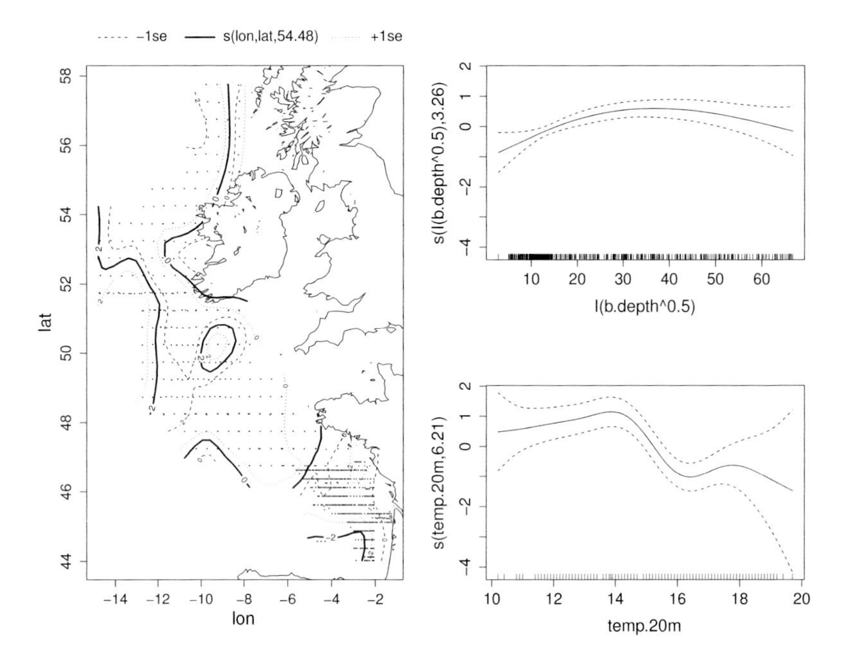

Figure 7.19 *Estimated smooth terms for the mackerel model* `gm2`. *The left panel shows the smooth of location. The upper right panel shows the smooth of seabed depth and the lower right panel shows the smooth of temperature at 20 metres depth. The* `c.dist` *effect is a straight line similar to that shown in figure 7.17.*

model predictions, and as in the brain imaging example, these can be obtained by simulation from the posterior distribution of the model coefficients. For example, the following obtains a sample from the posterior distribution of the mean density of mackerel eggs across the survey area, shown in the upper right panel of figure 7.20.

```
set.seed(4) ## make reproducible
br1 <- rmvn(n=1000,coef(gm2),vcov(gm2))
Xp <- predict(gm2,newdata=mackp,type="lpmatrix")
mean.eggs1 <- colMeans(exp(Xp%*%t(br1)))
hist(mean.eggs1)
```

A disadvantage of such simulations is that they are conditional on the estimated smoothing parameters, $\hat{\boldsymbol{\lambda}}$. That is, we are simulating from the posterior $f(\boldsymbol{\beta}|\mathbf{y}, \hat{\boldsymbol{\lambda}})$, when we would really like to simulate from $f(\boldsymbol{\beta}|\mathbf{y})$. One approach is to use the simple smoothing parameter uncertainty correction of section 6.11.1 (p. 302) to account for this. All we need to do is to substitute `vcov(gm2,unconditional=TRUE)` for `vcov(gm2)` in the above code. Let `mean.eggs` contain the result of this: its histogram is shown at the lower right of figure 7.20. The unconditional distribution is a little wider than the fixed smoothing parameter version.

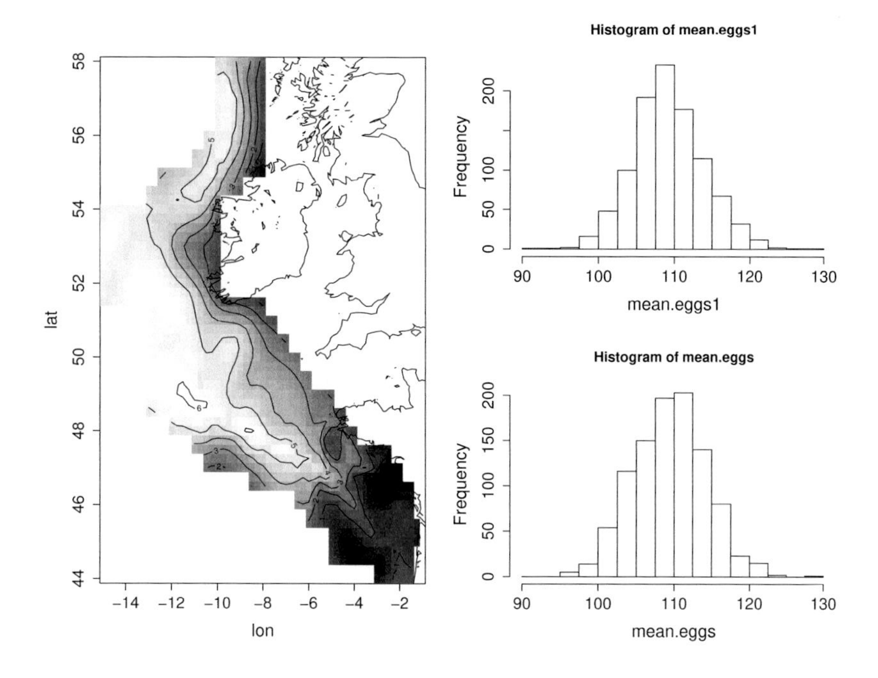

Figure 7.20 *The left hand panel shows predicted log densities of mackerel eggs over the 1992 survey area, according to the model* gm2. *The upper right panel shows the posterior distribution of mean egg densities per square metre sea surface, conditional on the estimated smoothing parameters. The lower left panel is the same, but unconditional.*

7.5.3 *Alternative spatial smooths and geographic regression*

If we were being very fussy, we might object that longitude and latitude are not really an isotropic co-ordinate system, so that a thin plate spline is not appropriate. This view carries an implicit assumption that our model is a much better representation of reality that is likely to be the case, but we could address it anyway, by using a spline on the sphere, as discussed in section 5.5.3 (p. 222). All we need do is to replace the spatial smooth in `gm2` with `s(lon,lat,bs="sos",k=100)`. The AIC gets slightly worse when this is done, but predictions from the model in figure 7.21(a) are almost indistinguishable from the original `gm2` model.

A more useful alternative might be to replace the spatial smooth with a Gaussian process smooth (section 5.8.2, p. 241) in which the autocorrelation function is forced to drop to zero at some point (one degree separation, for example). Replacing the spatial smooth in `gm2` with `s(lon,lat,bs="gp",k=100,m=c(1,1))` achieves this, selecting a spherical correlation model with maximum range 1. A motivation for doing this is that we might avoid the spatial smoother representing long range auto-correlation that would be better represented using the covariates. In fact the AIC only drops by 2 and the estimated remaining effects are almost unchanged.

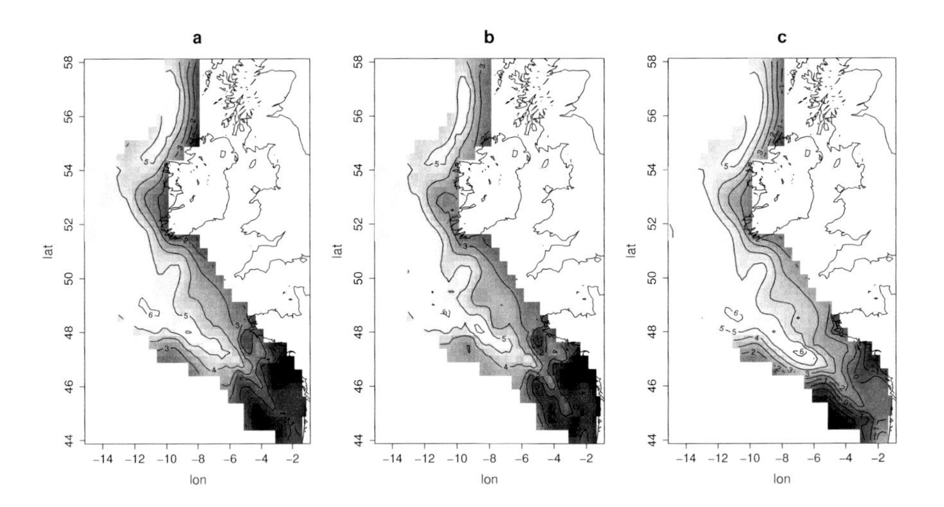

Figure 7.21 (a). *Predictions from the mackerel model with a spline on the spehere used for the spatial effect, to deal 'correctly' with lon-lat locations.* (b). *Predictions from the mackerel model with a Gaussian process smooth of location, using a spherical correlation function.* (c). *Predictions from a geographic regression version of the mackerel model.*

The predictions from the model are shown in figure 7.21(b): they are similar to the original predictions of gm2, but noticeably rougher as a result of the suppression of long-range autocorrelation.

A more radical change in structure would be to use a *geographic regression* model. This is a type of *varying coefficient* model in which each covariate is assumed to have a linear influence on the linear predictor for the response, but the slope parameter of that linear dependence varies smoothly with geographic location. So the log egg density term in the mackerel model might become

$$d_i = f_1(\texttt{lon}_i, \texttt{lat}_i) + f_2(\texttt{lon}_i, \texttt{lat}_i)\texttt{temp.20m}_i + f_3(\texttt{lon}_i, \texttt{lat}_i)\sqrt{\texttt{b.depth}_i}$$

where $f_1 - f_3$ are smooth functions of location (letting the coefficient for location derived variable c.dist vary with location does not seem like a good idea). This model can by estimated using

```
gmgr <- gam(egg.count ~s(lon,lat,k=100)+s(lon,lat,by=temp.20m)
        +s(lon,lat,by=I(b.depth^.5)) +offset(log.net.area),
        data=mack,family=tw,method="REML")
```

It has an AIC value 22 lower than gm2, so is worth serious consideration as an alternative model. Its predictions, shown in figure 7.21c, are noticeably smoother than the previous models, although not radically different in general appearance.

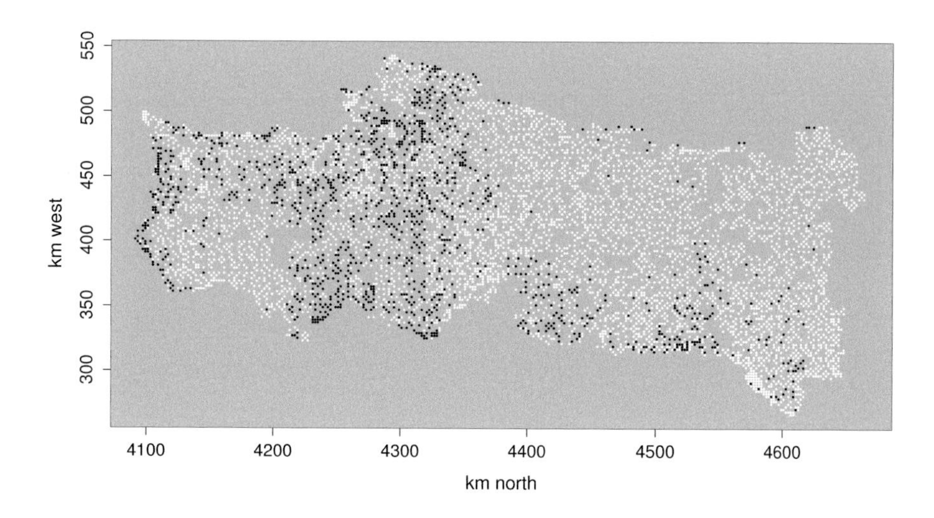

Figure 7.22 *Presence (black) and absence (white) of crested lark in sampled 2 km by 2 km squares in Portugal.*

7.6 Spatial smoothing of Portuguese larks data

Figure 7.22 shows data on the presence or absence of crested lark in each of a set of sampled 2km $\times$ 2km squares, gathered as part of the compilation of the Portuguese Atlas of Breeding Birds. The whole of Portugal was divided into 10km $\times$ 10km squares, each square was further subdivided into 25, 2km $\times$ 2km 'tetrads', and several tetrads (usually 6) were selected from each square. Each selected tetrad was then surveyed, to establish which bird species were breeding within it: crested lark is one of the species surveyed (although it should be noted that there are some problems distinguishing crested lark from Thekla lark).

The compilers of the atlas would like to summarize the information in figure 7.22, as a map, showing how the probability that a tetrad contains breeding crested larks varies over Portugal. An obvious approach is to model the presence – absence data, c_i, as

$$\text{logit}(\mu_i) = f(e_i, n_i)$$

where $\text{logit}(\mu_i) = \log\{\mu_i/(1-\mu_i)\}$, f is a smooth function of location variables e and n, and $c_i \sim \text{binomial}(1, \mu_i)$. The geographic nature of the data suggests using an isotropic smoother to represent f. Given the rather large number of data, it speeds things up to base the TPRS on rather fewer spatial locations than those contained in the entire data set. This is done automatically by the thin plate spline constructor, which will randomly select 2000 of the data locations to use as knots[§] and then build

[§]The same seed is used for exact reproducibility, but without affecting the state of R's random number generator. The number of knots and the random number seed can be changed: see `?tprs`. Alternatively knots can be specified directly via the `knots` argument to `gam`.

the thin plate eigen basis using these. The original data contain spatial locations in metres west and north of some origin. Before fitting let's convert these to kilometres.

```
library(gamair); data(bird)
bird$n <- bird$y/1000;bird$e <- bird$x/1000
m1 <- gam(crestlark ~ s(e,n,k=100),data=bird,family=binomial,
          method="REML")
```

The model is fitted using REML smoothing parameter estimation and a relatively complex model is selected (note that the reported REML score is the negative of the restricted log likelihood, so lower is better).

```
> m1

Family: binomial
Link function: logit

Formula:
crestlark ~ s(e, n, k = 100)

Estimated degrees of freedom:
74.5  total = 75.51

REML score: 2497.513

> plot(m1,scheme=3,contour.col=c1,too.far=0.03,rug=FALSE)
```

produces figure 7.23(a) showing the linear predictor of the model. An equivalent on the response (probability) scale is easily produced using `vis.gam`.

An obvious question with spatial data is whether an alternative spatial smoother might give different, or improved, results. In particular, since Portugal is long and thin, there is quite a big area that is close to the boundary, so we might worry about boundary effects. One way to investigate these is to change the smoothing penalty: for example, a Duchon spline based on first derivative penalisation, with parameter $s = 1/2$ (see section 5.5.2, p. 221), will penalize towards the constant function, which tends to lead to rather restrained boundary behaviour. The model call for this model (`m2`, say) is as the `m1` model call, but with the spatial smoother now set to `s(e,n,bs="ds",m=c(1,.5),k=100)`. Argument `m[1]` specifies a first order derivative penalty, and `m[2]` specifies $s = 1/2$. Figure 7.23(b) shows the estimated linear predictor. There are some differences to `m1`, particularly on the boundary, but the changes are small.

Another alternative would be to use a Gaussian process smoother (see section 5.8.2, p. 241). Judged by REML and AIC the Matérn with $\kappa = 1.5$ seems to fit best in this case. Rather than simply assuming some value for the Matérn range parameter, ρ, it is easy to search for the best fit from a set of plausible alternatives. For example

```
REML <- r <- 1:10*10
for (i in 1:length(r)) {
  mt <- gam(crestlark ~ s(e,n,bs="gp",m=c(3,r[i]),k=100),
            data=bird, family=binomial, method="REML")
  REML[i] <- mt$gcv.ubre
}
```

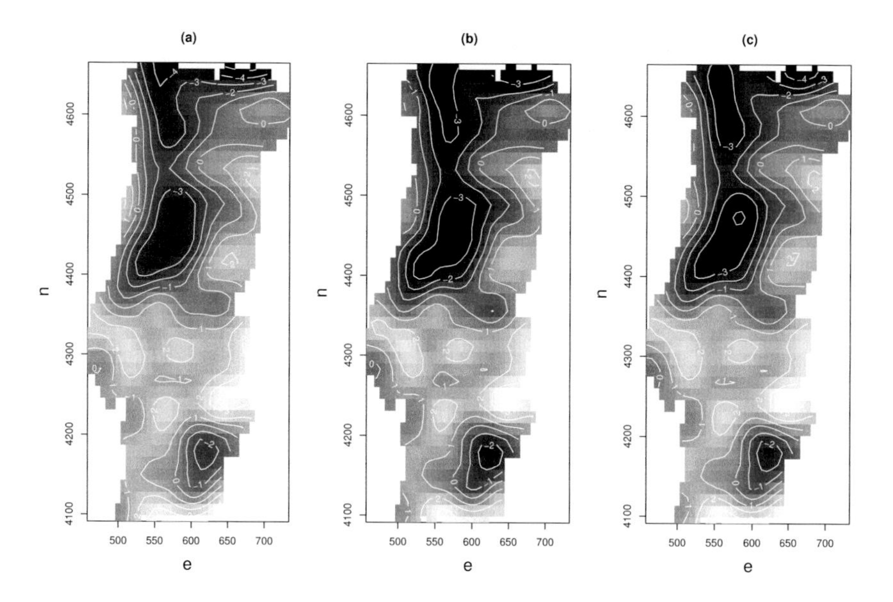

Figure 7.23 (a). *The linear predictor of the crested lark model using a thin plate spline.* (b). *The equivalent to (a) using a Duchon spline with a first derivative penalty.* (c). *The equivalent for a Matérn based Gaussian process smoother.*

shows that there is a minimum at $\rho = 30$. The REML scores at $\rho = 20$ and 40 are both less than 1 different from the value at $\rho = 30$, so there is no need to search on a finer grid. Estimates from the optimum model (`m3`, say) are plotted in figure 7.23c. The plot is rather similar to that from the thin plate spline based model. In fact the thin plate spline model has the lowest REML score of the three alternatives, and AIC also favours this model:

```
> AIC(m1,m2,m3)
         df      AIC
m1 76.07381 4845.400
m2 77.86097 4866.461
m3 75.77660 4863.057
```

Model checking with binary data is somewhat awkward (see exercise 2 in chapter 3), especially for spatial data, but for this application we can simplify matters considerably by choosing to model the data at the 10 km by 10 km square level, in order to obtain a binomial response with easier to interpret residuals. It turns out that the estimated models are almost indistinguishable in terms of predicted probabilities, whether we model raw data or the aggregated data. The `bird` data frame contains a column `QUADRICULA` identifying which 10 km square each tetrad belongs to, so aggregation is easy.

```
bird$tet.n <- bird$N <- rep(1,nrow(bird))
bird$N[is.na(as.vector(bird$crestlark))] <- NA
ba <- aggregate(data.matrix(bird), by=list(bird$QUADRICULA),
                FUN=sum, na.rm=TRUE)
```

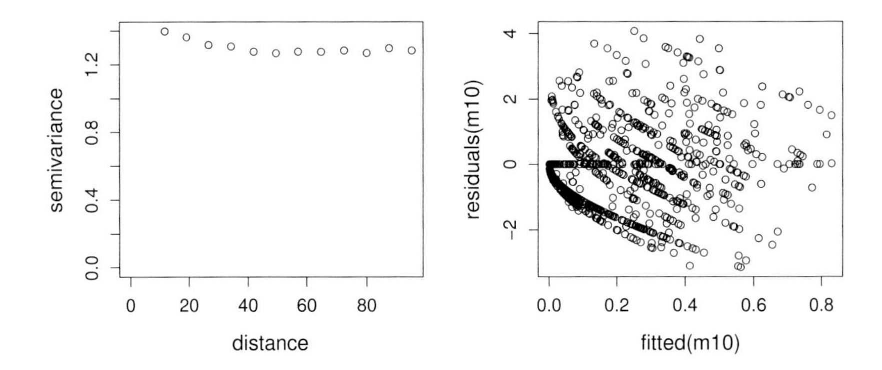

Figure 7.24 *Left: the variogram for the deviance residuals from* m10*, the aggregated crested lark binomial model. It is flat and apparently unproblematic. Right: deviance residuals against fitted probabilities, again showing no problems, except slight over-dispersion.*

```
ba$e <- ba$e/ba$tet.n; ba$n <- ba$n/ba$tet.n
```

The model can now be estimated.

```
m10 <- gam(cbind(crestlark,N-crestlark) ~ s(e,n,k=100),
           data=ba, family=binomial, method="REML")
```

A plot looks very similar to that from m1.

As usual, the (deviance) residuals for m10 should be plotted against fitted values to check model assumptions, but conventional residual plots are unlikely to pick up one potential problem with spatial data: namely spatial auto-correlation in the residuals and consequent violation of the independence assumption. To check this, it is useful to examine the *variogram* of the residuals, and the geoR package has convenient functions for doing this.

```
library(geoR)
coords<-matrix(0,nrow(ba),2);coords[,1]<-ba$e;coords[,2]<-ba$n
gb <- list(data=residuals(m10,type="d"),coords=coords)
plot(variog(gb,max.dist=100))
plot(fitted(m10),residuals(m10))
```

The plotted variogram is shown on the left of figure 7.24. Uncorrelated residuals should give a more or less flat variogram, while un-modelled spatial auto-correlation usually results in a variogram which increases sharply before eventually plateauing. In the current case, the variogram suggests no autocorrelation, or even slight negative autocorrelation, which might suggest very slight overfitting (although see exercise 4).

The residual versus fitted value plot, shown on the right of figure 7.24, is very good for binomial data, although there is a suggestion of overdispersion: that is the data seem slightly more variable than truly binomial data. Refitting m10 with a quasibinomial family tends to confirm this, yielding a scale parameter estimate around 2.

7.7 Generalized additive mixed models with R

Simple random effects can be included in a GAM in `mgcv` using `s(...,bs="re")` as introduced in section 3.5.2 (p. 154), based on the equivalence of smooths and random effects covered in sections 4.2.4 (p. 172) and 5.8 (p. 239). However, if a model requires a more complex random effect structure than can be handled with simple i.i.d. effects, then it makes sense to convert the smooths in the model to a random effects representation that can be estimated with standard mixed modelling functions, as covered in section 6.8 (p. 288). Another ground for doing this is when there are very large numbers of random effects: mixed modelling functions are typically set up to compute efficiently using the sparsity structure of the random effects in a way that `gam` is not.

`mgcv` includes a `gamm` function which fits GAMMs based on linear mixed models as implemented in the `nlme` library. The function performs the re-parameterizations described in section 5.8 (p. 239) and calls `lme` to estimate the re-parameterized model (see section 4.2.4), either directly, or as part of a PQL iteration. It then unscrambles the returned `lme` object so that it looks like a `gam` object, returning a two item list containing the `lme` object for the working mixed model, and the un-scrambled `gam` object.

`gamm` gives full access to the random effect and correlation structures available in `lme`, but it should be noted that it seems to work `lme` quite hard, and it is not difficult to specify models which cause numerical problems in estimation, or failure of the PQL iterations in the generalized case. This is particularly true when explicitly modelling correlation in the data, probably because of the inherent difficulty in separating correlation from trend, when the trend model is itself rather complex. Note also that changes in the underlying optimization methods may lead to slight differences between the results obtained with different `mgcv` and `nlme` versions: these should not be statistically important.

The `gamm4` package provides the equivalent to `gamm` using the `lme4` package. This does not have correlation structures available, but does not rely on PQL for GLMM estimation and is much more efficient than `lme` with ‘crossed’ (as opposed to nested) random effects. The only other limitation is that only the `t2` construction for tensor product smoothing (see 5.6.5, p. 235) can be used with `gamm4`, because of limitations in the way random effects can be specified. Another option is to use Bayesian stochastic simulation: the `jagam` function provides one option for doing this.

Sometimes a random effects model is required with a large number of smooth curves, which really are random effects. For example we might have a random smooth curve per subject in some experiment. These can be set up to be efficiently computed in `gamm` or `gamm4` using the `"fs"` basis. See `?factor.smooth.interaction` and section 7.7.4.

7.7.1 *A space-time GAMM for sole eggs*

To start with, let us revisit the Bristol channel sole data one more time. The GLMs and GLMMs considered in sections 3.3.5 and 3.5 were rather unwieldy, as a result of

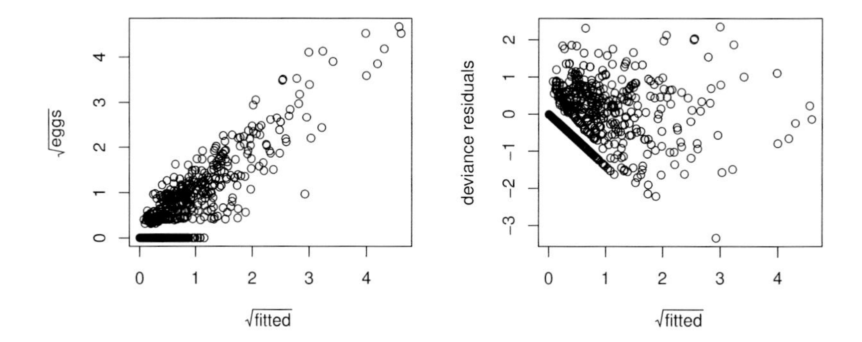

Figure 7.25 *Residual plots for the GAMM of the Bristol channel sole data, fitted by* bam. *The left panel shows the relationship between raw data and fitted values. The right panel shows deviance residuals against fitted values. Both plots appear to be reasonable.*

the large number of polynomial terms involved in specifying the models. If nothing else, a GAMM offers a way of reducing the clumsiness of the model formulation. It is straightforward to write the basic sole model (3.15), plus random effect, as a GAMM,

$$\log(\mu_i) = \log(\Delta_i) + f_1(\texttt{lo}_i, \texttt{la}_i, \texttt{t}_i) - f_2(\texttt{t}_i)\bar{a}_i + b_k,$$

if observation i is from sampling station k. Here f_1 and f_2 are smooth functions, $b_k \sim N(0, \sigma_b^2)$ are the i.i.d. random effects for station and, as before, μ_i is the expected value for $\texttt{egg}_i$ (given $\mathbf{b}$), while Δ_i and $\bar{a}_i$ are the width and average age of the corresponding egg class. lo, la and t are location and time variables. For simplicity the mortality rate term, f_2, is assumed to depend only on time: this assumption could be relaxed, but it is unlikely that the data really contain enough information to say much about spatial variation in mortality rate.

We need to decide what sort of smooths to use to represent f_1 and f_2. For f_1, a tensor product of a thin plate regression spline of lo and la, with a thin plate regression spline (or any other spline) of t, is probably appropriate: isotropy is a reasonable assumption for spatial dependence (although in that case we should really use a more isotropic co-ordinate system than longitude and latitude), but not for the space-time interaction. For f_2 and the default TPRS suffices. The multiplication of f_2 by $\bar{a}_i$ is achieved using the by variable mechanism.

Following on from the previous analyses, here is the call used to fit the model using gamm. Note that, unlike lme, gamm will *only* accept the list form of the random argument. Following the previous analyses a quasipoisson family is used. We first have to create a station factor variable, from the unique sampling station locations.

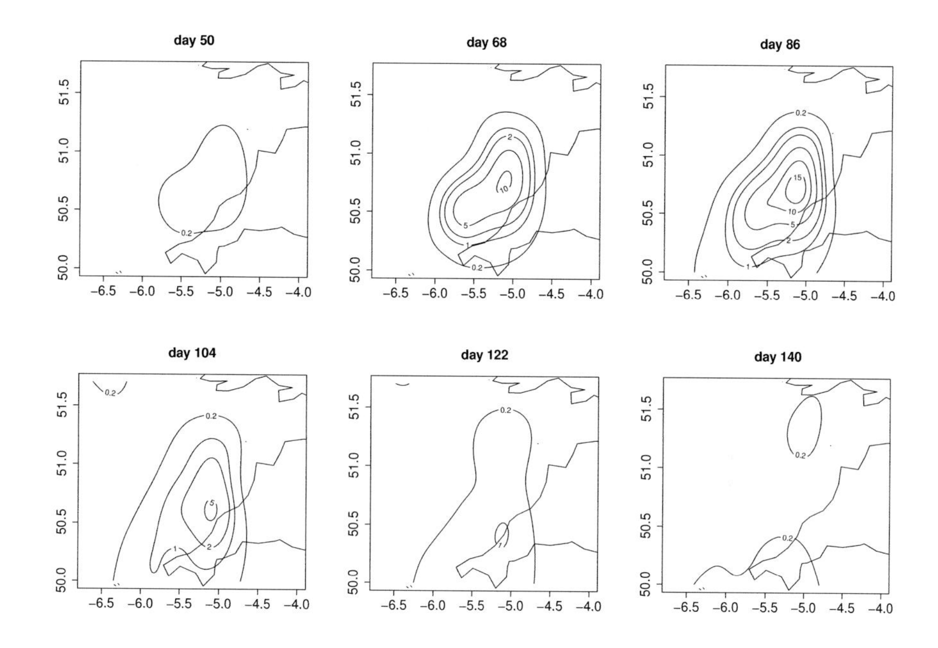

Figure 7.26 *Predicted spawning rates at various times, using the GAMM of Bristol channel sole eggs, presented in section 7.7.1. Note how plots are more peaked and of rather different shape to those from figure 3.19.*

```
solr$station <- factor(with(solr,paste(-la,-lo,-t,sep="")))
som <- gamm(eggs~te(lo,la,t,bs=c("tp","tp"),k=c(25,5),d=c(2,1))
           +s(t,k=5,by=a)+offset(off), family=quasipoisson,
           data=solr,random=list(station=~1))
```

`gamm` returns a list with two components: `lme` is the object returned by `lme`; `gam` is an incomplete object of class `gam`, which can be treated like a `gam` object for prediction, plotting, etc. For example,

```
> som$gam

Family: quasipoisson
Link function: log

Formula:
eggs ~ te(lo, la, t, bs = c("tp", "tp"), k = c(25, 5),
      d = c(2, 1)) + s(t, k = 5, by = a) + offset(off)

Estimated degrees of freedom:
49.77  4.38  total = 55.15
```

A somewhat quicker alternative for fitting, given the modest number of sampling stations, is to use function `bam` as follows

```
som1 <- bam(eggs~te(lo,la,t,bs=c("tp","tp"),k=c(25,5),d=c(2,1))
            + s(t,k=5,by=a)+offset(off)+s(station,bs="re"),
        family=quasipoisson,data=solr)
```

Fitting now takes about one third of the time that gamm takes, and the default is to use REML for the working model estimation. Fitting using gam is also possible: while slower than bam or gamm this would also allow negative binomial or Tweedie fits with estimation of all parameters.

Residual plots for som1 are shown in figure 7.25: these are residuals in which the random effects are at their predicted values. By default, residuals for som$gam from gamm would have the random effects set to zero, while those for som$lme would have the random effects set to their predicted values. This difference also carries over to prediction. Using predict with som$gam will give predictions in which the random effects are set to zero. By contrast predict with som1, fitted by bam (gam would be the same), sets the random effects to their predicted values. However, predict.gam also allows terms to be excluded from predictions, which is equivalent to setting them to zero. For example predict(som1,...,exclude="s(station)") would ensure that the station effect was set to zero in predictions. This feature has been used to produce the predicted spawning rate plots in figure 7.26.

Estimates of the random effect variances are extractable from both gamm and bam/gam fits. For example:

```
> gam.vcomp(som1)

Standard deviations and 0.95 confidence intervals:

               std.dev      lower     upper
...
s(station)   0.9677711 0.82946738 1.1291353
scale        0.5351120 0.51526441 0.5557242
```

or equivalently:

```
> som$lme
...
 Formula: ~1 | station %in% g.0 %in% g
        (Intercept)  Residual
StdDev:   0.9710766 0.5340049
...
```

g.0 and g are dummy variables associated with re-casting the smooths as random effects and can be ignored. The important information is the station random effect standard deviation (0.97) and the residual standard deviation (0.53).

Soap film improvement of boundary behaviour

An obvious objection to the results in figure 7.26 is that the model predicts high fish egg densities on land, and in one or two places near the edge of the survey area predicts low densities where there are almost certainly no eggs at all (but there was no sampling because of the certainty of finding nothing). In practice it could be argued

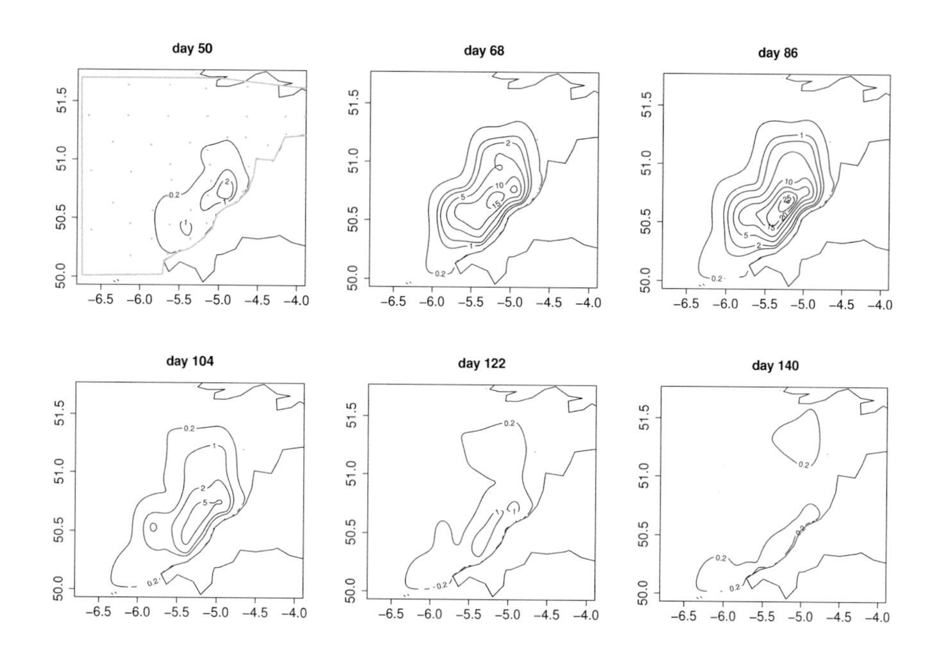

Figure 7.27 *Predicted spawning rates at various times, using the soap film smooth based GAMM of Bristol channel sole eggs, presented in section 7.7.1. Note how the contour plots now descend to near zero at the coast. The top left panel shows the boundary region and knots in grey.*

that this matters little, and that for assessing the total spawning we would anyway only integrate over the sea, and could easily exclude areas where eggs should not be. On the other hand the statistician might have difficulty selling this argument to the scientists who have spent several weeks, often in bad weather, collecting the egg data at considerable expense.† In particular one might worry about a model that predicted high egg densities right up to the beach, when in reality density will decline to near zero there.

One way of achieving a model with better boundary behaviour is to use a finite area soap film smoother (see section 5.5.4, p. 223), in place of the thin plate spline of location, to allow some control of the boundary. In particular we could set things up so that the boundary smooth of the soap film is almost constant over space and time, so that the data will then force the boundary density to be modelled as close to zero. The space time smoother in our model is a tensor product of a spatial smoother and a smooth in time. To use the soap film as the spatial smooth marginal in `mgcv` requires that we split the doubly penalized soap smooth into two singly penalized

†The author's misspent youth included just one week of doing just that, leading to some sympathy with the scientists' position.

components: the boundary interpolating soap film, and the smooth departures from this interpolating film in the interior of the boundary. These can then be made into two separate tensor product smooths with time. mgcv supplies "sf" ('soap film') and "sw" ('soap wiggly') smooths for this purpose. In fact for our current model, where we want the boundary smooth to be constant in space and time, we can drop the tensor product with "sf" altogether (since a constant smooth subject to a sum-to-zero identifiability constraint is simply zero).

Soap film smoothers require that polygons describing the boundary are supplied as the bnd component of the xt argument to the smooth specification. bnd contains a list of lists. There is a list for each boundary defining polygon, and each polygon is a list of two named arrays defining co-ordinates of polygon vertices: the names must match the smoother arguments. For a simple boundary there is only one polygon defined in the list, but if a region also contains islands/ holes then there may be more polygons, defining these. See ?soap for examples. To obtain a suitable boundary polygon for the sole egg data I produced a plot of the sampling station locations and coast, and then used R's locator function to define a suitable simple boundary. Note that this is also a good approach when a boundary is rather complicated, since a simple boundary 'enclosing' the real boundary of interest is usually sufficient, and less likely to produce artefacts than a highly convoluted boundary. The top left panel of figure 7.27 shows the boundary in grey.

We also have to choose the knots of the soap film smoother.[‡] I simply selected a nicely spread out subset of the sampling station locations for one sampling occasion (shown in grey in the top left panel of figure 7.27). The fitting code is as follows.

```
sole$station <- solr$station
som2 <- bam(eggs ~ te(lo,la,t,bs=c("sw","cr"),k=c(40,5),
            d=c(2,1),xt=list(list(bnd=bnd),NULL)) +
            s(t,k=5,by=a) + offset(off) + s(station,bs="re"),
            knots=knots, family=quasipoisson, data=sole)
```

The te statement defining the space-time interaction of the interior soap film and time now requires a list of xt objects to be passed to the marginal smooths. The first item in the list is the important one, containing the soap boundary; the second item can be NULL as the "cr" basis does not require an xt. The rest of the model specification is as before, except that there is now a knots argument supplying the soap film knots.

Figure 7.27 shows the predicted egg density distributions for the new model. The distributions are somewhat more biologically realistic now. The term wise effective degrees of freedom for this model are very similar to the previous model (around 50 for the space time smooth), and both models have an explained deviance of 88%. AIC or REML comparison of the models is not available here: it would be possible to re-fit using a Tweedie distribution to gain access to AIC, but let's move on.

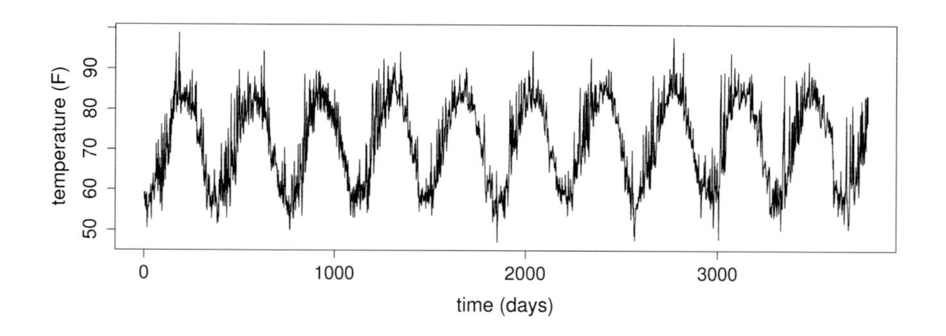

Figure 7.28 *Daily air temperature in Cairo, Egypt from January 1st 1995.*

7.7.2 The temperature in Cairo

Figure 7.28 shows daily temperature in Cairo over nearly a decade. The data are from `http://www.engr.udayton.edu/weather/citylistWorld.htm`. It is clear that the data contain a good deal of noisy short term auto-correlation, and a strong yearly cycle. Much less clear, given these other sources of variation, is whether there is evidence for any increase in average mean temperature over the period: this is the sort of uncertainty that allows climate change sceptics to bamboozle the ill-informed. A reasonable model of these data might therefore be

$$\texttt{temp}_\texttt{i} = f_1(\texttt{time.of.year}_i) + f_2(\texttt{time}_i) + e_i \tag{7.1}$$

where $e_i = \phi e_{i-1} + \epsilon_i$, the ϵ_i being i.i.d. $N(0, \sigma^2)$ random variables. f_1 should be a 'cyclic' function, with value and first 2 derivatives matching at the year ends. The basic idea is that if we model time of year and auto-correlation properly, then we should be in a good position to establish whether there is a significant overall temperature trend.

Model (7.1) is easily fitted:

```
ctamm <- gamm(temp~s(day.of.year,bs="cc",k=20)+s(time,bs="cr"),
         data=cairo,correlation=corAR1(form=~1|year))
```

Note a couple of details: the data cover some leap years, where `day.of.year` runs from 1 to 366. This means that by default the cyclic smooth matches at day 1 and 366, which is correct in non-leap years, but not quite right in the leap years themselves: a very fussy analysis might deal more carefully with this. Secondly, I have nested the AR model for the residuals within year: this vastly speeds up computation, but is somewhat arbitrary.

Examine the `gam` part of the fitted model first:

```
> summary(ctamm$gam)
```

‡The boundary file and knots used here can be found in the `gamair` help file `?sole`.

```
Family: gaussian
Link function: identity

Formula:
temp ~ s(day.of.year, bs = "cc", k = 20) + s(time, bs = "cr")

Parametric coefficients:
            Estimate Std. Error t value Pr(>|t|)
(Intercept)  71.6581     0.1518   472.2   <2e-16 ***
---
Signif. codes:  0 '***' 0.001 '**' 0.01 '*' 0.05 '.' 0.1 ' ' 1

Approximate significance of smooth terms:
                edf Ref.df       F p-value
s(day.of.year) 9.392 18.000 222.803 < 2e-16 ***
s(time)        1.382  1.382   9.633 0.00224 **
---
Signif. codes:  0 '***' 0.001 '**' 0.01 '*' 0.05 '.' 0.1 ' ' 1

R-sq.(adj) =  0.849
  Scale est. = 16.519    n = 3780
```

It seems that there is evidence for a long term trend in temperature, and that the model fits fairly closely. We can also extract things from the `lme` representation of the fitted model, for example 95% confidence intervals for the variance parameters.

```
> intervals(ctamm$lme,which="var-cov")
Approximate 95% confidence intervals

 Random Effects:
  Level: g
                 lower      est.     upper
sd(Xr - 1) 0.06064995 0.1450556 0.3477507
  Level: g.0
                   lower         est.     upper
sd(Xr.0 - 1) 2.817526e-07 0.0003473937 0.9780106

 Correlation structure:
        lower      est.     upper
Phi 0.6598207 0.6840085 0.7067803
attr(,"label")
[1] "Correlation structure:"

 Within-group standard error:
   lower     est.    upper
3.913727 4.064353 4.220775
```

The confidence interval for ϕ is easily picked out, and provides very strong evidence that the AR1 model is preferable to an independence model ($\phi = 0$), while the interval for σ is (3.92,4.23). Note that confidence intervals for the smoothing parameters are also available. Under the `Random Effects` heading the interval for `g.1` re-

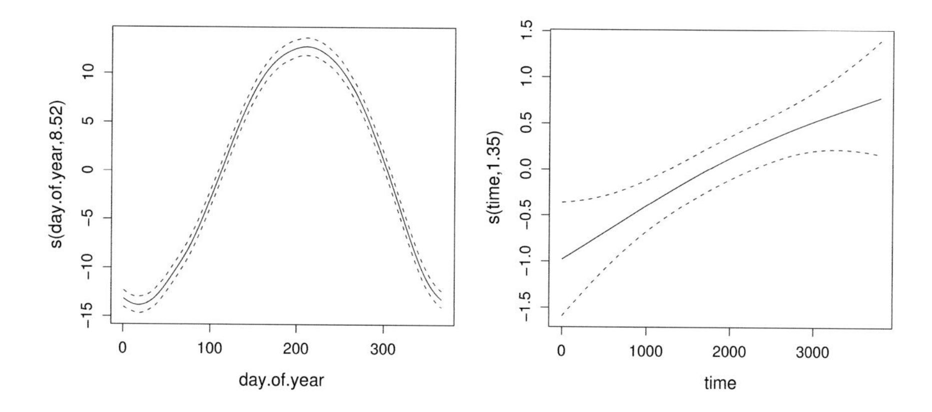

Figure 7.29 *GAMM terms for daily air temperature in Cairo, Egypt from* 1st *January 1995. The left panel is the estimated annual cycle: note that it has a fatter peak and thinner trough than a sinusoid. The right pattern is the estimated long term trend: there appears to have been a rise of around 1.5 F over the period of the data.*

lates to the smoothing parameter for the first smooth, while that for `g.2` relates to the second smooth. The parameterization is not completely obvious: the reported parameters are σ^2/λ_i, as the following confirms

```
ctamm$gam$sig2/ctamm$gam$sp
s(day.of.year)          s(time)
  0.1450556262    0.0003473937
```

We can plot the estimated model terms using

```
plot(ctamm$gam,scale=0)
```

which results in figure 7.29. The temperature increase appears to be quite marked. The simulation techniques covered in section 7.2.7 can be used to simulate from the posterior distribution of the modelled temperature rise, if required.

The `bam` function also allows a simple AR1 model on the residuals (in data frame order). It has the small advantage that we do not need to nest the AR1 correlation structure within year, but the disadvantage that estimation of the correlation parameter ρ is not automatic. However, we can easily search for a REML optimal ρ as follows:

```
REML <- rho <- 0.6+0:20/100
for (i in 1:length(rho)) {
  ctbam <-bam(temp~s(day.of.year,bs="cc",k=20)+s(time,bs="cr"),
                   data=cairo,rho=rho[i])
  REML[i] <- ctbam$gcv.ubre
}
```

The minimum REML is at $\rho = 0.69$, and with this correlation parameter the fitted effects are indistinguishable from those shown in figure 7.29, and the conclusions are identical.

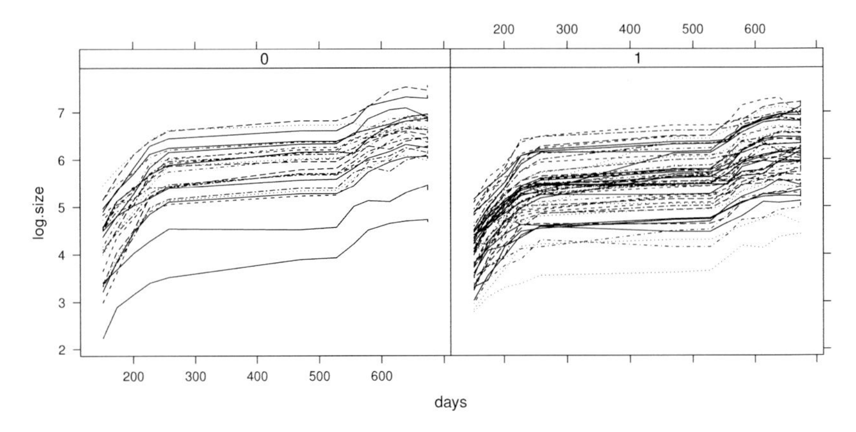

Figure 7.30 *Growth trajectories for 79 Sitka spruce trees, grown in elevated ozone (right) and control (left) conditions.*

7.7.3 *Fully Bayesian stochastic simulation:* `jagam`

The duality between smooths and random effects/random fields (section 5.8, p. 239) means that we can also estimate generalized additive mixed models in a fully Bayesian way. All that is required, in terms of extra model formulation, is to put priors on the smoothing parameters and any variance parameters, and (usually) on any un-penalized model coefficients (such as those representing the linear part of a cubic spline). There are specialist packages for doing this, in particular `R2BayesX` and `INLA`, but in this section we will consider using the general purpose Gibbs sampling package `JAGS` (Plummer, 2003), via its `rjags` R interface.

Figure 7.30 shows 79 growth trajectories of Sitka spruce trees grown in either elevated ozone or control conditions. They are available in the `sitka` data frame from the `SemiPar` package. A simple smooth additive mixed model would be

$$\texttt{log.size}_i = f(\texttt{day}_\texttt{i}) + d_{\texttt{id}_i} + \epsilon_i, \quad d_{\texttt{id}} \sim N(0, \sigma_b^2), \quad \epsilon_i \sim N(0, \sigma^2),$$

where $\texttt{id}_i$ is the identity of tree to which the ith measurement belongs. The model could easily be fitted with `gamm`, `gam` or `bam`, but can also be used to illustrate the use of `jagam`, which opens up the possibility of much more complicated random effect structures.

A `jagam` call is much like a `gam` call, but rather than fitting a model it writes the `JAGS` code to perform Bayesian simulation from the model. Any random effects structure is not specified as part of the `jagam` call, but is instead added into the `JAGS` code template produced by `jagam`. In this way the rich random effects modelling structure available in `JAGS` can be accessed. `rjags` can then be used to simulate from the model. Functions are also provided to interpret the results after simulation. For the `sitka` model

```
jd <- jagam(log.size ~ s(days) + ozone, data=sitka,
            file="sitka0.jags", diagonalize=TRUE)
```

produces a template file of JAGS code, "sitka0.jags", specifying the Sitka model excluding the random effects. It also returns a list, jd, containing the variables referred to in the code. The code can then be edited to add the random effects, yielding the following ("sitka.jags"). All comments except those beginning '!# ADDED' are autogenerated by jagam.

```
model {
  mu0 <- X %*% b ## expected response
  for (i in 1:n) { mu[i] <- mu0[i] + d[id[i]]} #! ADDED r.e.
  for (i in 1:n) { y[i] ~ dnorm(mu[i],tau) } ## response
  scale <- 1/tau ## convert tau to standard GLM scale
  tau ~ dgamma(.05,.005) ## precision parameter prior
  for (i in 1:nd) {d[i] ~ dnorm(0,taud)} #! ADDED r.e. dist
  taud ~ dgamma(.05,.005) #! ADDED r.e. precision prior
  ## Parametric effect priors CHECK tau=1/58^2 is appropriate!
  for (i in 1:2) { b[i] ~ dnorm(0,3e-04) }
  ## prior for s(days)...
  for (i in 3:10) { b[i] ~ dnorm(0, lambda[1]) }
  for (i in 11:11) { b[i] ~ dnorm(0, lambda[2]) }
  ## smoothing parameter priors CHECK...
  for (i in 1:2) {
    lambda[i] ~ dgamma(.05,.005)
    rho[i] <- log(lambda[i])
  }
}
```

So in this case three lines were needed to add the random effect to the model. Notice that JAGS requires all priors to be proper (although they can be very vague). Hence priors have been put on parametric fixed effects, while the improper priors equivalent to smoothing penalties have to be made proper to be usable. The simplest way to achieve this is via the null space penalties of section 5.4.3 (p. 214), and this is the approach taken by jagam. The added random effects code requires variables id and nd, which must be added to jd:

```
jd$jags.data$id <- sitka$id.num
jd$jags.data$nd <- length(unique(sitka$id.num))
```

Now everything is ready to simulate from the model from within R.

```
library(rjags); load.module("glm")
jm <- jags.model("sitka.jags", data=jd$jags.data,
      inits=jd$jags.ini, n.chains=1)
sam <- jags.samples(jm, c("b","rho","scale","mu"),
       n.iter=10000, thin=10)
jam <- sim2jam(sam, jd$pregam)
```

jags.model compiles the model specification in sitka.jags and initializes the model. jags.samples then simulates from the model by Gibbs sampling, returning every 10th simulated value of the model (fixed + spline) coefficients (b), the log smoothing parameter (rho), etc. A total of 10000 iterations is performed. Finally the sampler output is used to create a simple gam object, jam, for further investigation.

Figure 7.31 shows some results from the model, produced by the following code:

```
plot(jam) ## plot the estimated mean growth trajectory
```

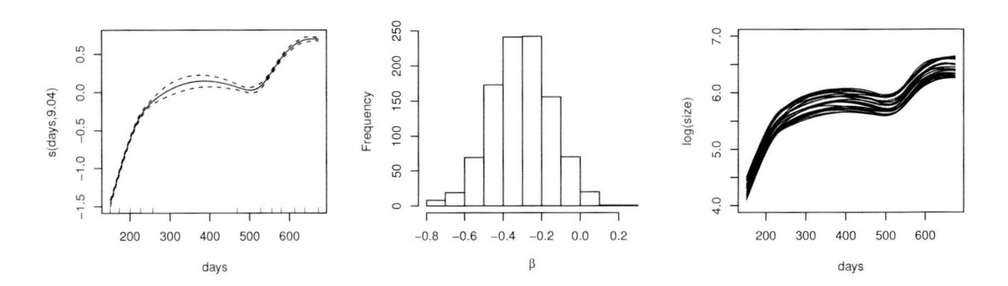

Figure 7.31 *Left: smooth mean Sitka growth currve with 95% credible interval. Middle: histogram of 1000 draws from the posterior of the* `ozone` *effect; there is good evidence for ozone surpressing growth. Right: random growth curves for control trees drawn from the posterior distribution of the model.*

```
hist(sam$b[2,,1]) ## histogram of ozone effect parameter
pd <- data.frame(days=152:674,ozone=days*0)
Xp <- predict(jam,newdata=pd,type="lpmatrix")
ii <- 1:25*20+500 ## draws to select
for (i in ii) { ## draw growth curves
  fv <- Xp%*%sam$b[,i,1] ## growth curve from posterior
  if (i==ii[1]) plot(pd$days,fv,type="l") else lines(pd$days,fv)
}
```

More detail can be found in Wood (2016a). Clearly the real strength of this approach is when a model requires complicated random effects that can be specified in `JAGS` but are difficult to deal with otherwise.

7.7.4 *Random wiggly curves*

An obvious alternative model for the `sitka` data would have a separate random curve for each tree, in addition to the main average growth curve. These random curves would have no un-penalized null space, instead being shrunk towards zero. The `"fs"` basis provides such random curves: one for each level of a factor. The following code would estimate such a model:

```
sitka$id.num <- as.factor(sitka$id.num)
b <- gamm(log.size ~ s(days) + ozone + ozone:days +
          s(days,id.num,bs="fs",k=5), data=sitka)
plot(b$gam,pages=1)
```

An extra slope parameter as well as an intercept has been allowed for the ozone effect. `s(days,id.num,bs="fs",k=5)` specifies a rank 5 cubic spline (default) for each level of `id.num`. The splines all share the same 3 smoothing parameters: one for the usual spline penalty, and one for each term in the penalty null space. `"fs"` smooths are coded somewhat trickily to work especially efficiently with `gamm` (although there are restrictions: see `?factor.smooth.interaction`), but we could also have estimated the model (more slowly) using `gam` with REML or ML

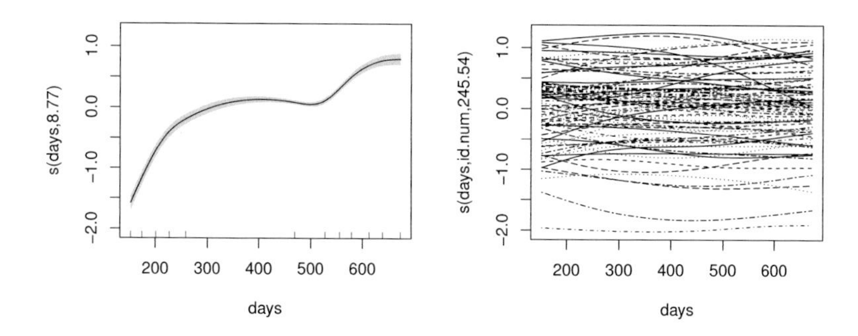

Figure 7.32 *Left: smooth mean Sitka growth currve with 95% credible interval. Right: estimated random departures from mean growth trajectory for each tree.*

smoothing parameter estimation. In either case there is a significant negative slope for enhanced ozone, and the model estimated effects are as shown in figure 7.32.

7.8 Primary biliary cirrhosis survival analysis

Primary biliary cirrhosis is a disease caused by autoimmune damage to the bile ducts of the liver. It can eventually lead to cirrhosis of the liver. The `survival` package in R (Therneau, 2015; Therneau and Grambsch, 2000) contains data frames `pbc` and `pbcseq` from a trial conducted at the Mayo clinic between 1974 and 1984 testing the drug D-penicillamine against a placebo. In the `pbc` dataset a number of baseline measurements are available for each patient, while in the `pbcseq` data frame time varying measurements are available. 6% of the patients exited the trial by receiving a transplant – here these will be treated as censored.[§]

The covariates include patient `id`, `age`, `sex`, treatment (`trt`) and disease `stage`, classified as 1 to 4, where 1 corresponds to subtle bile damage and other early signs and 4 is cirrhosis. The other covariates of interest here are blood measurements of quantities related to liver function. `albumin` is an important blood protein produced in the liver; alkaline phosphotase (`alk.phos`) is an enzyme particularly concentrated in the liver and bile ducts; aspartate aminotransferase (`ast`) is an enzyme commonly used as a biomarker for liver health; `bili`rubin is a product of the breakdown of aged red blood cells processed in the liver and excreted in bile and urine; count of `platelets`, which are the component of blood responsible for clotting, but can be destroyed in a damaged liver; `protime` is a standardised blood clotting time measure related to the previous covariate.

An initial model includes all the preceding covariates, with age and the blood test measurements included as additive smooth effects.

```
pbc$status1 <- as.numeric(pbc$status==2)
```

[§]Actually 3/4 of patients had stage 4 disease by time of transplant, so this sort of ‘censoring’ is really somewhat informative.

```
pbc$stage <- factor(pbc$stage)
b0 <- gam(time ~ trt+sex+stage+s(sqrt(protime))+s(platelet)+
          s(age)+s(bili)+s(albumin)+s(sqrt(ast))+s(alk.phos),
          weights=status1,family=cox.ph,data=pbc)
```

`protime` and `ast` have been square rooted to reduce their skew. The `alk.phos` effect is estimated to be almost flat with wide confidence intervals, and a high p-value. Dropping it also reduces the model AIC. The `ast` effect then continues to look marginal and has a p-value of 0.16: dropping it increases the AIC only marginally, so it might as well be omitted. `stage` then has a p-value of 0.2, its omission increases the AIC slightly, but it is dropped. The `anova` of the final model is then

```
> anova(b)

Family: Cox PH
Link function: identity

Formula:
time ~ trt + sex + s(sqrt(protime)) + s(platelet) + s(age) +
    s(bili) + s(albumin)

Parametric Terms:
    df Chi.sq p-value
trt  1  0.120  0.7294
sex  1  3.439  0.0637


Approximate significance of smooth terms:
                   edf Ref.df Chi.sq  p-value
s(sqrt(protime)) 1.000  1.001 13.337 0.000261
s(platelet)      1.001  1.002  5.789 0.016141
s(age)           6.043  7.172 29.417 0.000145
s(bili)          4.264  5.223 89.545  < 2e-16
s(albumin)       1.000  1.000 31.086 2.47e-08
```

This provides no evidence for a treatment effect. The smooth effects are plotted in figure 7.33 along with a residual plot of the deviance residuals against the 'risk scores' (simply the linear predictor values), using

```
plot(b); plot(b$linear.predictors,residuals(b))
```

The wedge shaped block of residuals at the lower left of the residual plot is caused by the censored observations and is typical of such plots. The couple of high points at the top left are individuals who died despite relatively low risk scores, and may be slight outliers here. Otherwise the plot appears reasonable. There seems to be a slow upward drift in hazard with age, while the other effects are in line with expectations. Hazard increases with blood clotting time and reduced platelet count. There is an initial sharp increase with serum bilirubin which then levels off, consistent with a poorly functioning liver failing to clear bilirubin, while hazard decreases with serum albumin, presumably as ability to produce albumin increases with liver function.

We can easily produce model predicted survival functions for any set of covariate values, using `predict.gam` with `type="response"`. `predict.gam` uses

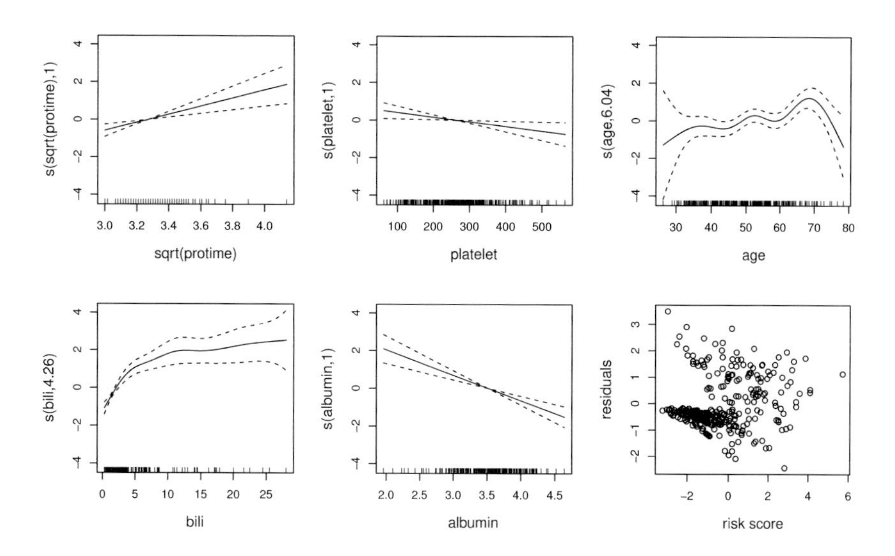

Figure 7.33 *Smooth effects for the primary biliary cirrhosis fixed covariates model along with a corresponding residual plot (lower right).*

(6.25) from section 6.6.2 (p. 287) to compute standard errors on the survivor function scale. These are not ideal for plotting survival function intervals as they can lead to intervals outside $[0, 1]$. However, it is easy to transform to standard errors on the cumulative hazard scale, and then transform cumulative hazard intervals to survival intervals. The following example code plots the estimated survival function and both types of interval (using one standard error) for patient 25's covariates.

```
## create prediction data frame...
np <- 300
newd <- data.frame(matrix(0,np,0))
for (n in names(pbc)) newd[[n]] <- rep(pbc[[n]][25],np)
newd$time <- seq(0,4500,length=np)
## predict and plot the survival function...
fv <- predict(b,newdata=newd,type="response",se=TRUE)
plot(newd$time,fv$fit,type="l",ylim=c(0.,1),xlab="time",
     ylab="survival",lwd=2)
## add crude one s.e. intervals...
lines(newd$time,fv$fit+fv$se.fit,col="grey")
lines(newd$time,fv$fit-fv$se.fit,col="grey")
## and intervals based on cumulative hazard s.e...
se <- fv$se.fit/fv$fit
lines(newd$time,exp(log(fv$fit)+se))
lines(newd$time,exp(log(fv$fit)-se))
```

Figure 7.34 shows such plots for three patients.

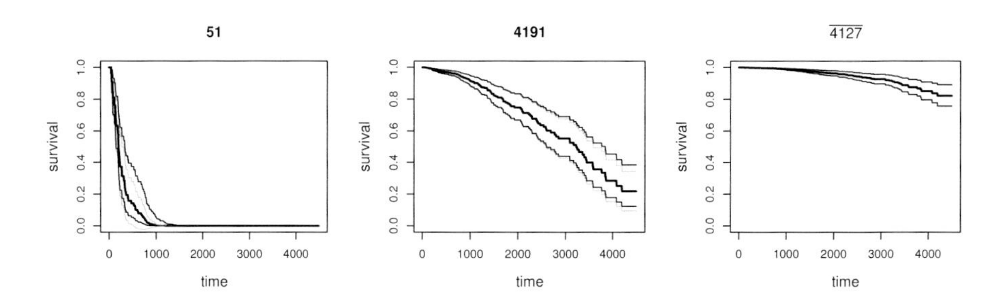

Figure 7.34 *Survival functions and one standard error bands for patients 10, 66 and 25 according to the primary biliary cirrhosis fixed covariates model. The figures above each panel are the patient's actual survival or censoring (barred) time. The thin black lines show the bands produced via transformation from the cumulative hazard scale, whereas the grey lines are approximations directly on the survival scale.*

7.8.1 Time dependent covariates

Now consider the `pbcseq` data, in which the covariates are measured at several times for each patient. The `cox.ph` family is designed only for the fixed covariate case, but the equivalent Poisson model trick of section 3.1.10 (p. 116) can readily deal with this situation, and can just as well be used with penalized smooth models.

A modelling decision must now be made of how to estimate the covariate values at each non-censored event time, as required for the time dependent version of the Cox proportional hazards model. We certainly do not have direct measurements for each patient at each event time. Here let us make the crude assumption that the measurement at an event time is whatever it was the previous time it was measured. This stepwise interpolation is very easy to carry out, but is obviously rather crude: linear or spline interpolation might be preferable.

The first task, then, is to convert the `pbcseq` data frame into a data frame of artificial Poisson data, with the interpolated time varying covariate values. The `?cox.pht` help file example in `mgcv` includes a routine `tdpois` for performing this task, which can be applied as follows (the defaults are for `pbcseq` here!).

```
pbcseq$status1 <- as.numeric(pbcseq$status==2) ## deaths
pb <- tdpois(pbcseq) ## conversion
pb$tf <- factor(pb$futime) ## add factor for event time
```

The resulting `pb` data frame has 28,380 rows — this approach is quite computationally costly. In the fixed covariate case fitting the equivalent Poisson model using `gam` with REML smoothness selection produces the same results as using the `cox.ph` family, but can be quite time consuming, because of the data set size and number of levels of the `tf` factor. Here let's tolerate some approximation and fit using `bam` with the `discrete==TRUE` option (which would still be feasible for millions of data).

Starting with the same dependence on covariates as in the baseline covariates case, backwards model selection drops `alk.phos`, `sex` and `stage` in that order, on the basis of each sequentially being the term with the highest p-value. Each reduc-

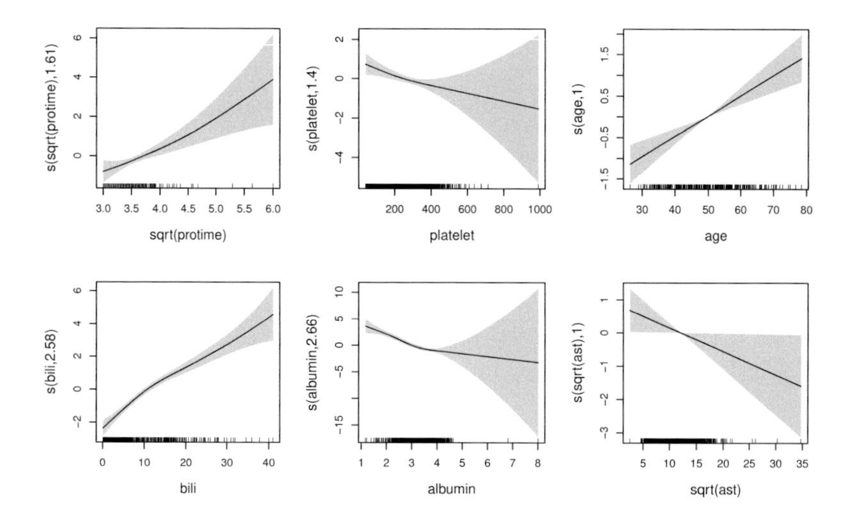

Figure 7.35 *Smooth effects for the primary biliary cirrhosis variable covariates model.*

tion of the model also reduced the AIC. As the trial variable of interest, `trt` was not considered as a candidate for dropping. The fitting call for the selected model was

```
b <- bam(z ~ tf - 1  +  trt + s(sqrt(protime)) + s(platelet) +
         s(age) + s(bili) + s(albumin) + s(sqrt(ast)),
         family=poisson,data=pb,discrete=TRUE,nthreads=2)
```

where as in the `glm` case, '`tf-1`' is the first term in the linear predictor to ensure that we get one coefficient estimated for each event time, rather than an intercept and difference terms. The option `nthreads=2` tells `bam` to use two CPU cores when possible. `anova(b)` or `summary(b)` gives a p-value of about 0.7 for `trt`, so again there is no evidence for a treatment effect.

Residuals (as defined in section 6.6.2, p. 286) are quite easily computed for the time dependent model. The cumulative hazard per subject simply needs to take into account the covariate variability, but this is straightforward.

```
chaz <- tapply(fitted(b),pb$id,sum) ## cum. hazard by subject
d <- tapply(pb$z,pb$id,sum) ## censoring indicator
mrsd <- d - chaz ## Martingale residuals
drsd <- sign(mrsd)*sqrt(-2*(mrsd + d*log(chaz))) ## deviance
```

Figure 7.35 is the result of `plot(b,pages=1,scale=0,scheme=1)`. Compared to the baseline covariates model there is slightly more non-linearity evident in the estimated effects now, although the estimated effect of age is a simple straight line. The direction and broad interpretation of the effects are as before.

Survival function prediction is essentially the same as in section 3.3.3 (p. 136) following the calculations described in section 3.1.10 (p. 116). Here is some code to plot the survivor function for subject 25 again:

```
te <- sort(unique(pb$futime)) ## event times
di <- pbcseq[pbcseq$id==25,] ## data for subject 25
```

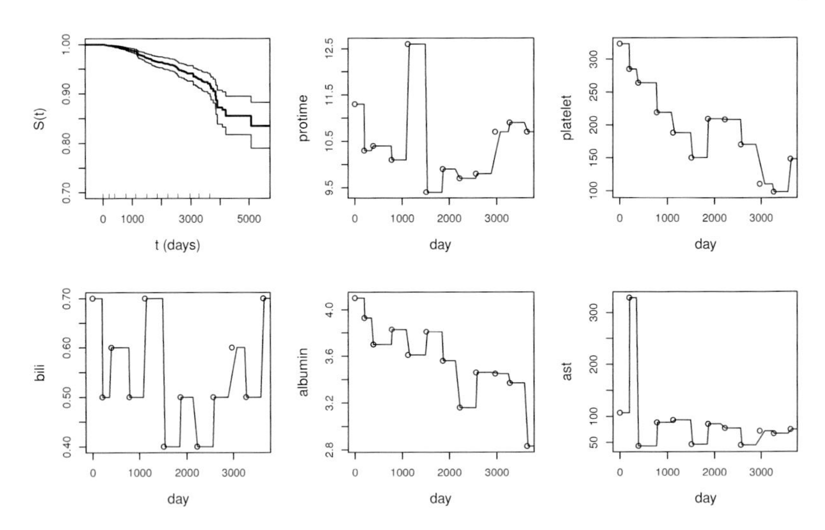

Figure 7.36 *Top left: survival function and one standard error bands for subject 25, according to the primary biliary cirrhosis variable covariates model. Remaining panels: the covariates through time.*

```
## interpolate to te using app from ?cox.pht...
pd <- data.frame(lapply(X=di,FUN=app,t=di$day,to=te))
pd$tf <- factor(te)
X <- predict(b,newdata=pd,type="lpmatrix")
eta <- drop(X%*%coef(b)); H <- cumsum(exp(eta))
J <- apply(exp(eta)*X,2,cumsum)
se <- diag(J%*%vcov(b)%*%t(J))^.5
plot(stepfun(te,c(1,exp(-H))),do.points=FALSE,ylim=c(0.7,1),
     ylab="S(t)",xlab="t (days)",main="",lwd=2)
lines(stepfun(te,c(1,exp(-H+se))),do.points=FALSE)
lines(stepfun(te,c(1,exp(-H-se))),do.points=FALSE)
rug(pbcseq$day[pbcseq$id==25]) ## measurement times
```

This results in the top left panel of figure 7.36. It is also of interest to relate the survivor function to the changes in covariates over time. The following extracts the time varying covariate record and plots it, also overlaying the interpolated version used for fitting, to produce plots like the remaining panels in figure 7.36

```
er <- pbcseq[pbcseq$id==25,]
plot(er$day,er$ast);lines(te,pd$ast)
```

Notice how the reduction in the survival function relative to the constant baseline covariate model is probably driven by deterioration in serum albumin and platelet counts in particular, and some increase in clotting time towards the end of the data.

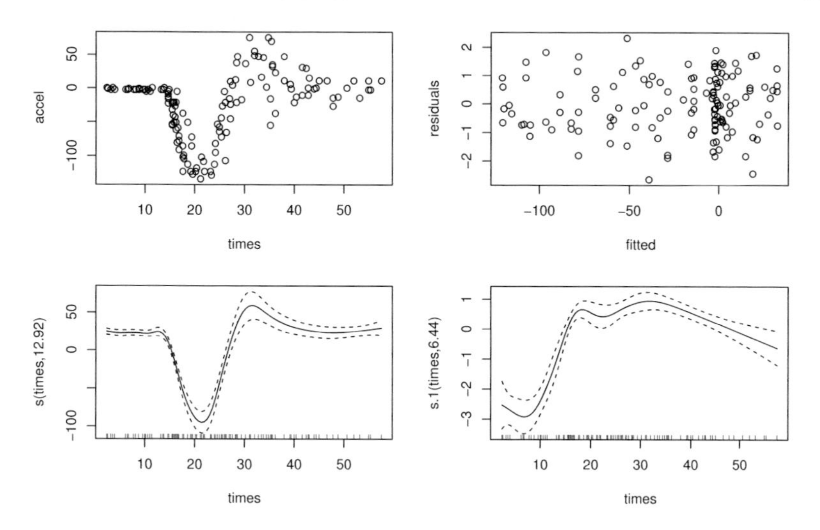

Figure 7.37 *Top left: acceleration of the head of a crash test dummy against time. Top right: residuals of a Gaussian location-scale model against predicted acceleration. Bottom left: estimated smooth for the mean acceleration. Bottom right: estimated smooth for the log (shifted) standard deviation.*

7.9 Location-scale modelling

Sometimes it is useful to allow several parameters of the response distribution to each depend on covariates. When the dependence is via smooth functions, Rigby and Stasinopoulos (2005) call the resulting models 'generalized additive models for location scale and shape' (GAMLSS), while Klein et al. (2015) refer to 'distributional regression'. The methods of section 6.6 (p. 280) handle such models as a special case (see Wood et al., 2016), and some models are available in `mgcv`.¶

A very simple example is a Gaussian location-scale model for the `mcycle` data from the `MASS` library, shown at the top left of figure 7.37. These data are measurements of the acceleration of the head of a crash test dummy in a simulated motorcycle accident, against time. Given the variability in both the mean and the variance of the data then the following model might be appropriate:

$$\texttt{accel}_i \sim N(\mu_i, \sigma_i^2), \quad \mu_i = f_1(\texttt{time}_i), \quad \log(\sigma_i - b) = f_2(\texttt{time}_i),$$

where f_1 and f_2 are smooth functions and b is a small lower bound on σ_i ensuring that we avoid singularities in the likelihood. The adaptive smooths of section 5.3.5 (p. 207) might be useful here, so the following code will fit the model.

```
library(MASS);library(mgcv)
b <- gam(list(accel~s(times,bs="ad"),~s(times,bs="ad")),
         family=gaulss,data=mcycle)
```

¶At time of writing the `gaulss`, `gevlss`, `multinom` and `ziplss` families.

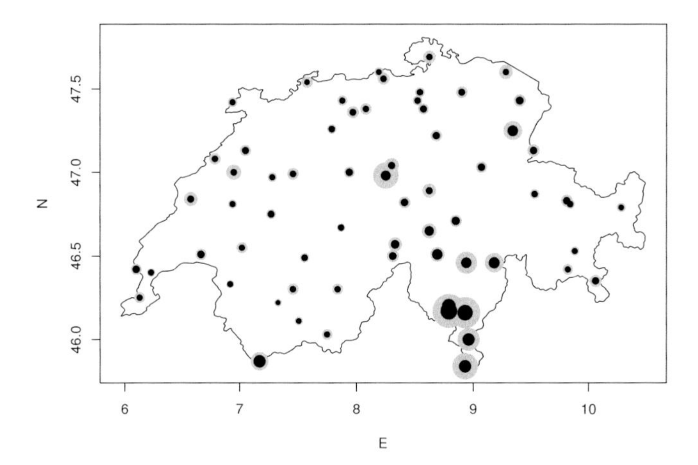

Figure 7.38 *Locations of 65 Swiss weather stations. Grey circle sizes are proportional to the largest 12-hour rainfall seen at the station (1981-2015), while the black circle size shows the average annual maximum.*

Notice how `gam` now requires a list of formulae, with the first specifying the response and the linear predictor for the mean, and the next specifying the linear predictor for the log (shifted) standard deviation. The fitted values for the resulting model will now be a two column matrix. The first column is the predicted mean ($\hat{\mu}_i$), while the second is the square root of the precision (i.e., $1/\hat{\sigma}_i$). `predict` will also return matrix results for `type "response"` and `"link"`. The top right panel of figure 7.37 shows the residuals against fitted mean, while the lower left panel is $\hat{f}_1$ and the lower right is $\hat{f}_2$. See `?gaulss` for more detail. Notice that the variance parameter in models of this class is being estimated by penalized maximum likelihood, and therefore suffers the usual downward bias at modest sample sizes discussed in section 2.4.5 (p. 83).

7.9.1 *Extreme rainfall in Switzerland*

The `swer` data frame contains the highest rainfall recorded in any 12-hour period each year for 65 Swiss weather stations, with covariates, from 1981 to 2015. Figure 7.38 shows where they are (using package `mapdata` for the outline of Switzerland), and gives some impression of the data variability. The generalized extreme value distribution is often used as a model for 'batch extreme' data of this sort. Its p.d.f. is

$$t(y)^{\xi+1}e^{-t(x)}/\sigma \text{ where } t(y) = [1 + \{(y - \mu)/\sigma\}\xi]^{-1/\xi}$$

($\xi \neq 0$) and $t(y) = \exp\{-(y - \mu)/\sigma\}$ ($\xi = 0$). The `mgcv` version uses the simple approximation that $\xi = \pm\epsilon$ for $|\xi| < \epsilon$, where ϵ is a small constant.

The gevlss family in mgcv allows μ, $\log \sigma$ and ξ to depend on smooth linear predictors. Maximum likelihood estimation for the GEV distribution is consistent when $\xi > -1$, and the GEV distribution has finite variance when $\xi < 1/2$, so mgcv uses a modified logit link to keep ξ in the interval $(-1, 1/2)$.

The initial model fit was as follows

```
library(mgcv);library(gamair);data(swer)
b0 <- gam(list(exra ~ s(nao)+ s(elevation)+ climate.region+
                      te(N,E,year,d=c(2,1),k=c(20,5)),
        ~ s(year)+ s(nao)+ s(elevation)+ climate.region+ s(N,E),
        ~ s(elevation)+ climate.region),family=gevlss,data=swer)
```

That is, the annual maximum 12 hourly rainfall, exra, was modelled using a GEV distribution, in which the location parameter, μ, depends on a space-time interaction term, which of 11 climate.regions the station belongs to, station elevation in metres and the annual North Atlantic Oscillation index nao. The nao measures pressure anomalies between the Azores and Iceland, and is somewhat predictive of whether western Europe is dominated by warm wet weather from the Atlantic, or colder continental weather. The scale parameter, $\log \sigma$, was modelled as depending on year and location (N and E) additively, but was otherwise as the location parameter. The shape parameter ξ was modelled as depending on elevation and climate.region. The choice of initial model reflects the fact that the data tend to be more informative about location than scale, and about scale than shape.

Model selection proceeded by first replacing the space time interaction term, te(N,E,year), with an additive structure, s(N,E)+s(year), which decreased the model AIC. After that a backward selection approach was used in which anova was used to obtain p-values for each term, and the term with the highest p-value (over 0.05) was dropped, provided it also decreased the AIC. For this example the results were rather clear cut, ending up with the following.

```
b <- gam(list(exra~ s(nao)+s(elevation)+climate.region+s(N,E),
           ~ s(year)+ s(elevation)+ climate.region+ s(N,E),
         ~ climate.region),family=gevlss,data=swer)
plot(b,scale=0,scheme=c(1,1,3,1,1,3),contour.col="white")
```

The result of the plot call is shown in the top two rows of figure 7.39. Notice the increase in location and scale in the west of the country, not in the rain shadow of high mountains. As expected, elevation generally increases both location and scale, although the complex estimate for the dependence on elevation is quite likely to result from confounding with hidden variables, such as aspect (e.g., west slope versus east slope). The NAO index effect is interesting, with the location parameter apparently having a minimum when the index is around zero: Switzerland is, of course, a long way inland, so this effect is likely to be quite complex. Finally there is a weak year effect evident in the variability.

The bottom row of figure 7.39 plots residuals and raw data against the expected annual maximum 12-hour rainfall. The expected value for the GEV distribution is $\mu + \sigma\{\Gamma(1-\xi) - 1\}/\xi$, so the following code computes it from the three columns of the fitted matrix for the model:

```
mu <- fitted(b)[,1];rho <- fitted(b)[,2]; xi <- fitted(b)[,3]
fv <- mu + exp(rho)*(gamma(1-xi)-1)/xi
```

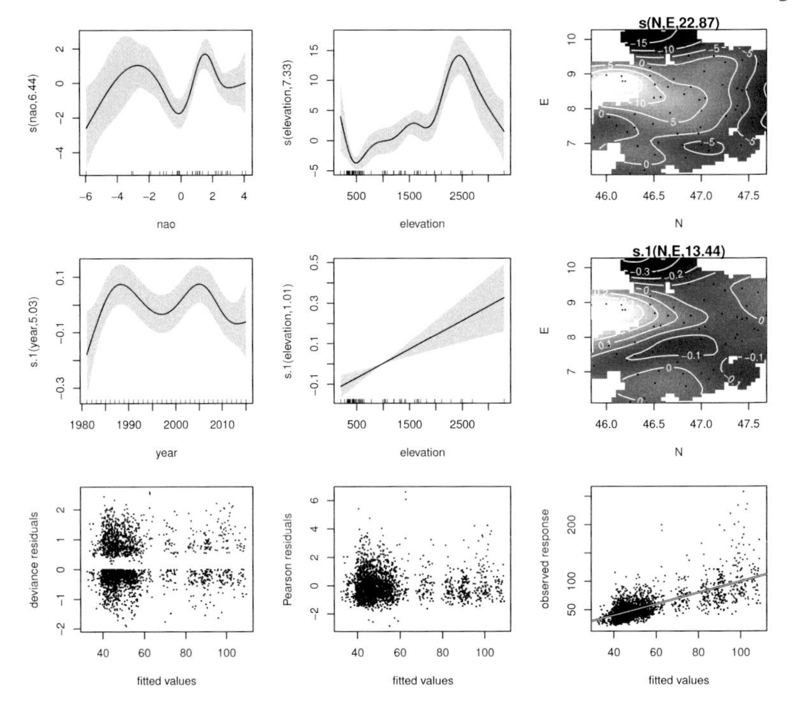

Figure 7.39 *The top two rows show the estimated effects for the Swiss 12-hour extreme rainfall data. The bottom row shows deviance residuals, Pearson residuals and raw data plotted against the model predicted expected annual maximum 12-hour rainfall. The r^2 for the lower right panel is 0.5.*

The `residuals` function computes the deviance and Pearson residuals, as usual.

When modelling extremes some quantile of the fitted distribution is often of interest. For any set of covariates, this can be computed from the quantile function of the GEV given predictions of μ, σ and ξ, but here let's consider a slightly more complicated target requiring some simulation. Specifically, let's target the 98th percentile of the posterior distribution of annual maximum of 12-hour rainfall for each station over its period of operation. This is easily done by simulating from the approximate posterior distribution of the model coefficients, $\boldsymbol{\beta}$, and then simulating replicate data from the resulting GEV. The following code does this for 1000 replicates.

```
Fi.gev <- function(z,mu,sigma,xi) { ## GEV inverse cdf.
  xi[abs(xi)<1e-8] <- 1e-8 ## approximate xi=0, by small xi
  x <- mu + ((-log(z))^-xi-1)*sigma/xi
}
mb <- coef(b);Vb <- vcov(b) ## posterior mean and cov
b1 <- b ## copy fitted model object to modify
n.rep <- 1000; br <- rmvn(n.rep,mb,Vb) ## posterior sim
n <- length(fitted(b))
sim.dat <- cbind(data.frame(rep(0,n*n.rep)),swer$code)
```

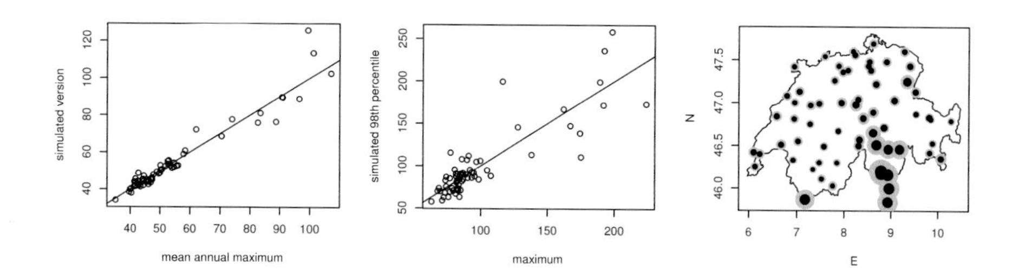

Figure 7.40 *Results of simulating from the posterior annual 12-hour maximum rainfall distribution. The left plot is simulation mean against actual mean, by station. The middle plot is the* 98th *percentile of the annual 12-hour maximum posterior distribution, against maximum observed between 1981 and 2015, by station. The right plot corresponds to figure 7.38, but with grey circle size proportional to the simulated* 98th *percentile and black to the simulated mean.*

```
for (i in 1:n.rep) {
  b1$coefficients <- br[i,] ## copy sim coefs to gam object
  X <- predict(b1,type="response");ii <- 1:n + (i-1)*n
  sim.dat[ii,1] <- Fi.gev(runif(n),X[,1],exp(X[,2]),X[,3])
}
```

So the idea is to simulate 1000 replicate parameter vectors from the posterior for the coefficients, and then for each to predict the corresponding GEV parameters for each year at each station. Given these parameters a value is simulated from the corresponding GEV distribution (by evaluating the quantile function/inverse c.d.f. at a uniform random value). The preceding code is somewhat inefficient for readability purposes. A more efficient version would call `predict` with the `type="lpmatrix"` argument. This produces a single matrix, consisting of a block of columns for each of the three linear predictors. Which matrix column belongs to which linear predictor is given by the `"lpi"` attribute of the returned matrix — a list of three vectors of column indices.

Given `sim.dat` it is now easy to obtain the mean and 98th percentile of the annual maximum for each station.

```
stm <- tapply(sim.dat[,1],sim.dat[,2],mean)
st98 <- tapply(sim.dat[,1],sim.dat[,2],quantile,probs=0.98)
```

Figure 7.40 shows plots of the results of this exercise, indicating that the model is doing a reasonable job in this case.

Chavez-Demoulin and Davison (2005) and Yee and Stephenson (2007) are foundational references for extreme value GAMs (but use backfitting approaches).

7.10 Fuel efficiency of cars: Multivariate additive models

The `mpg` data frame contains data on fuel efficiency of cars along with characteristics of the cars. For each car there is a city driving and highway driving fuel efficiency

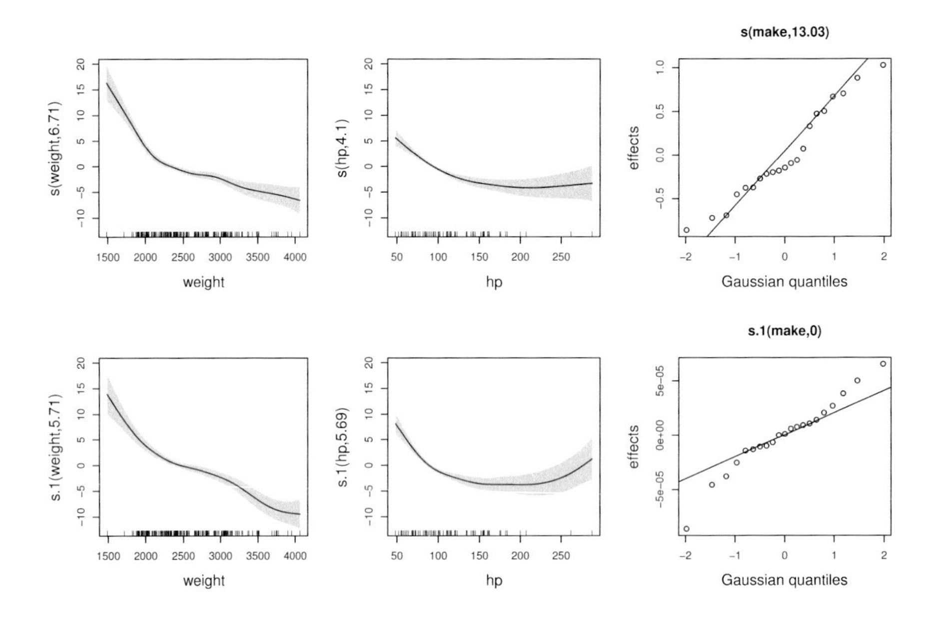

Figure 7.41 *Multivariate additive model for the* mpg *data. Note that terms from the second linear predictor are labelled with an extra '*.1*' on the term name.*

given in miles per gallon (the data are from the US). It makes sense to treat the fuel efficiency measurements as a bivariate response. If we assume bivariate normality then the mvn family in mgcv can be used. This implements a simply multivariate additive model in which the mean of each component of the multivariate response has its own linear predictor, and there is the possibility to share terms between linear predictors. The correlations between the response components are also estimated (but do not depend on covariates).

From physical principles (see, e.g., the excellent MacKay, 2008) we would expect that highway fuel consumption would be dominated by aerodynamic factors, and hence would be best modelled in terms of car dimensions (frontal area in particular), while city fuel consumption should be dominated by weight and other engine characteristics. Unfortunately the physically based models I tried all did worse than a rather boring model involving weight and engine power (hp) and using the same variables for city and highway. Here is the model fitting code

```
library(mgcv); library(gamair); data(mpg)
b <- gam(list(city.mpg ~ fuel +style +drive +s(weight) +s(hp)
                         + s(make,bs="re"),
                hw.mpg ~ fuel +style +drive +s(weight) +s(hp)
                         + s(make,bs="re")),
               family = mvn(d=2) , data = mpg)
```

fuel is diesel or petrol (gasoline), style is a factor for style of car, drive is a

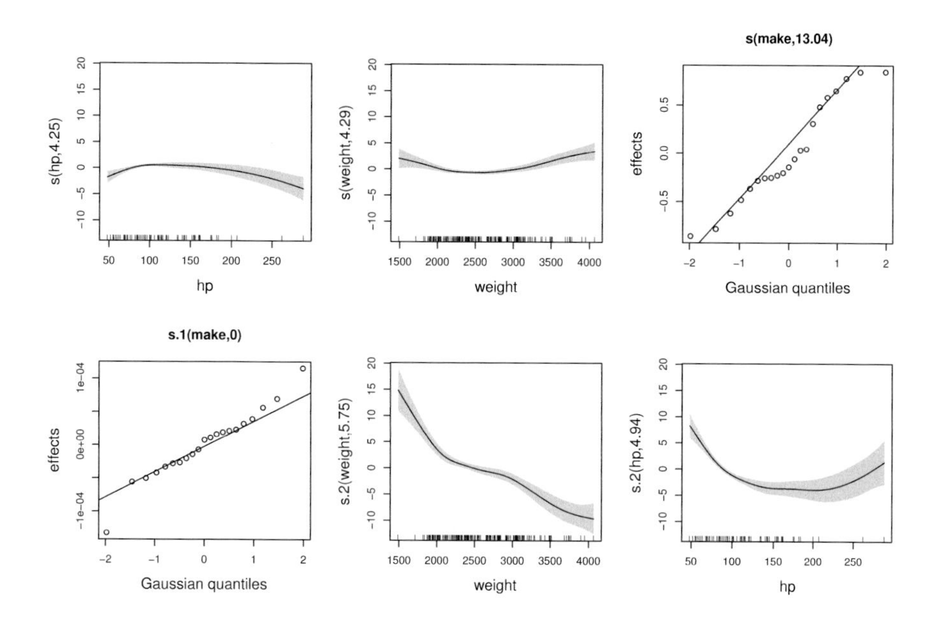

Figure 7.42 *Shared term multivariate additive model for the* `mpg` *data. The effects specified in the third shared model formula are labelled with an extra* '`.2`'.

factor for front, rear or all wheel drive. `make` is a factor for manufacturer — a random effect is assumed for this. There is one formula for each component of the bivariate response. Figure 7.41 shows the results. The random effect of `make` is estimated as effectively zero for highway driving, whereas `make` seems to matter for city driving. The smooths for `weight` and `hp` have strikingly similar shapes for city and highway, and the obvious question arises of whether we could simplify the model, by using the same smooths for both components. To address this question a model can be fitted that uses the same smooths of `weight` and `hp` for both `mpg` measurements, but has an extra smooth of both for the `city.mpg`. If the joint smooths are adequate then the extra smooths should turn out to be unnecessary.

The `gam` call for this model now has three formulae.

```
b1 <- gam(list(city.mpg ~ fuel +style +drive +s(hp) +s(weight)
                          + s(make,bs="re"),
               hw.mpg ~ fuel +style +drive +s(make,bs="re"),
                  1+2 ~ s(weight) +s(hp) -1),
             family = mvn(d=2) , data = mpg)
```

The third formula specifies the shared terms. Its left hand side contains numbers and the '+' symbol, in order to indicate which components of the response the formula relates to: here the same `s(weight)` and `s(hp)` will be used in both the `city.mpg` and `hw.mpg` linear predictors. For models with more components then we could have further shared component formulae of this sort (see `?formula.gam`). This

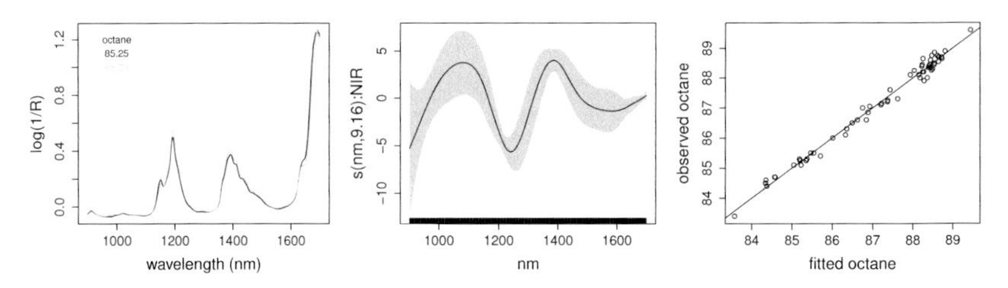

Figure 7.43 *Left: two example near infra-red spectra for gasoline samples with their octane ratings. Middle: estimated coefficient function for the scalar on function regression. Right: observed versus model fitted octane measurements.*

same mechanism can also be used for other multi-formula families. The results of this second model are shown in figure 7.42. Both the figure and a look at the model summary indicate that there are some significant (non constant) differences between the city and highway responses to `weight` and `hp`. The way that city efficiency initially goes up with `hp`, relative to highway, is initially counter intuitive, but probably reflects some relative inefficiency of low powered cars in highway driving. The initial extra effect of `weight` is as expected – extra weight adds disproportionately more consumption in city driving. In fact `b1` has a slightly higher AIC than `b`, so there is not really any good reason to prefer it to the original model.

7.11 Functional data analysis

Functional data analysis (Ramsay and Silverman, 2005) is concerned with models for data where some of the observations involved are best treated as functions. The methods are very closely related to those presented in this book, especially the use of basis expansions and penalties. This section will merely scratch the surface of the topic by presenting a couple of models of this type that can be directly estimated in `mgcv`. See the `refund` package in R (Goldsmith et al., 2016) for much more.

7.11.1 Scalar on function regression

The left hand panel of figure 7.43 shows near infra-red spectra measured for two gasoline/petrol samples (out of 60 available), along with the octane rating of the sample. Octane rating measures the 'knocking resistance' of fuel: the higher the octane rating the higher can be the compression ratio of the engine that burns it. Traditionally, octane rating measurement is somewhat complicated, requiring a variable compression test engine and standardized reference fuel. In comparison obtaining the near infra-red spectrum of fuel is much quicker and cheaper, so it would be good to be able to predict octane rating from a measured spectrum. The `gas` data contain a vector of `octane` measurements, along with a matrix `NIR`, each row of which contains a spectrum for the corresponding `octane` measurement. Matrix `nm` is the corresponding matrix of wavelengths.

A simple model (sometimes known as a 'signal regression model', e.g., Marx and Eilers, 2005) for predicting octane rating from the spectra might be:

$$\texttt{octane}_i = \int f(\nu) k_i(\nu) d\nu + \epsilon_i$$

where $k_i(\nu)$ is a spectrum as a function of frequency, ν, and the ϵ_i are i.i.d. Gaussian errors. f is a smooth function to estimate. Because integration is a linear operation, this model can be estimated by exactly the machinery used for other GAMs. Replacing the integral by a discrete sum ('quadrature') approximation, such a model can be specified and estimated in `mgcv` using the summation convention that applies when a smooth term has matrix arguments. Generically, suppose that a model contains the term `s(X,by=Z)` where `X` and `Z` are matrices of identical dimensions. The contribution to the ith element of the linear predictor from such a term is $\sum_j f(X_{ij}) Z_{ij}$, where f is the smooth function. $\mathbf{Z}$ is assumed to be a matrix of ones, if it is not supplied. The summation convention can be used to include a discrete version of any (bounded) linear functional of a function in the linear predictor of a model, by suitable choice of $\mathbf{X}$ and $\mathbf{Z}$.

The following code fits the model and produces the middle and right hand panels of figure 7.43.

```
b <- gam(octane~s(nm,by=NIR,k=50),data=gas)
plot(b,scheme=1,col=1)
plot(fitted(b),gas$octane)
```

From the right panel it is clear that the model works rather well in this case.

Note that these functional regression terms can be mixed with normal smooth terms in a model, and that the summation convention also allows for the smooth function to vary with other covariates. For example suppose that we would like to use a model

$$y_i = \int f_1(\nu, x_i) k_i(\nu) d\nu + f_2(z_i) + \epsilon_i,$$

In such a case three matrices would be needed: `K` containing $k_i(\nu)$ measurements in each row (possibly multiplied by quadrature weights); the matrix `nu`, each row of which contains the ν values at which $k_i(\nu)$ has been measured (usually each row is identical); and `X`, each row of which simply contains the required x_i value repeated for each column (so the columns of `X` are all identical). Then the model formula would be `y ~ te(nu,X,by=K) + s(z)` (assuming default `k` values). A tensor product smooth usually makes sense in such cases, since ν and x are usually not naturally on the same scale.

Prostate cancer screening

Figure 7.44 shows protein mass spectra for blood samples from 9 patients: 3 with each of a healthy prostate gland, an enlarged gland and prostate cancer. Blood testing being much less invasive than biopsy, it is desirable to be able to predict prostate condition from the spectra. Data described in Adam et al. (2002) are available in the `prostate` dataset which contains 654 prostate classifications (`type`: 1, healthy;

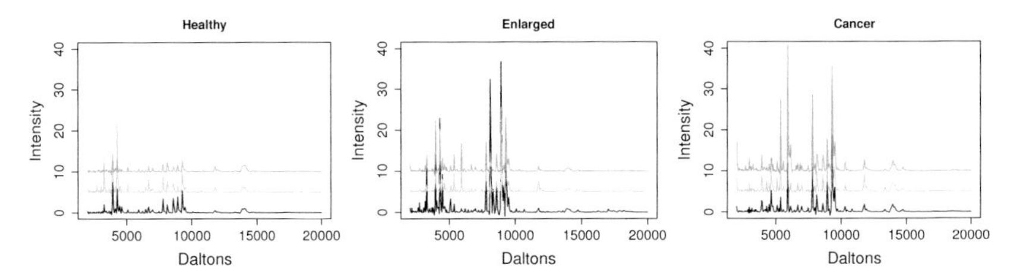

Figure 7.44 *Example protein mass spectra for 3 subjects with healthy, enlarged and cancerous prostate gland. The spectra are vertically shifted for visibility.*

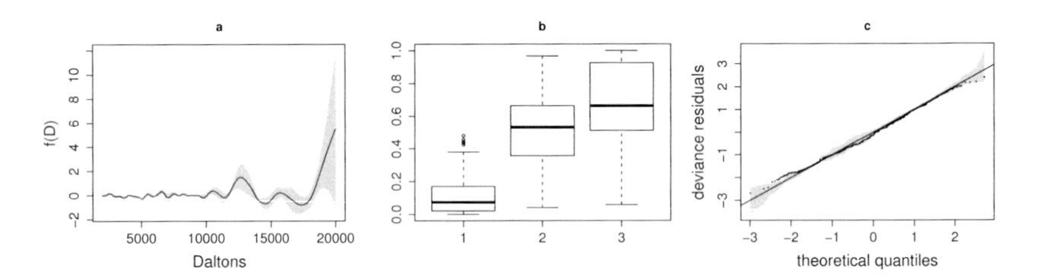

Figure 7.45 (a). *The estimated smooth function and confidence band for the ordered categorical prostate diagnosis model.* (b). *Box plots of the predicted probability of cancer for each subject, by actual status.* (c). *Sorted deviance residuals for the model against simulated theoretical quantiles, with reference band in grey.*

2, enlarged; 3, cancerous) along with the corresponding spectral (`intensity`) for each. `intensity` is a 624×264 matrix, with one spectrum per row. The protein masses at which the spectra were measured are in the matrix `MZ`.

One model to try might be an ordered categorical scalar on function model. The idea is that the linear predictor determines the mean value of a latent random variable, whose value relative to some cut points determines the category observed for a subject. The linear predictor itself is given by the integral of the observed spectrum, multiplied by a coefficient function, to be estimated. So we have that

$$y_i = \begin{cases} 1 & z_i < \theta_1 \\ 2 & \theta_1 \leq z_i < \theta_2 \\ 3 & \theta_2 \leq z_i \end{cases}$$

where z_i is a logistically distributed random variable with mean

$$\mu_i = \int f(m)k_i(m)dm.$$

$k_i(m)$ is the ith spectrum, while $f(m)$ is a smooth function to be estimated. θ_1 is set to -1 for identifiability, but θ_2 is estimated as part of model fitting. In reality the integral is replaced by a discrete sum, of course.

For the current example the fitting code is simply

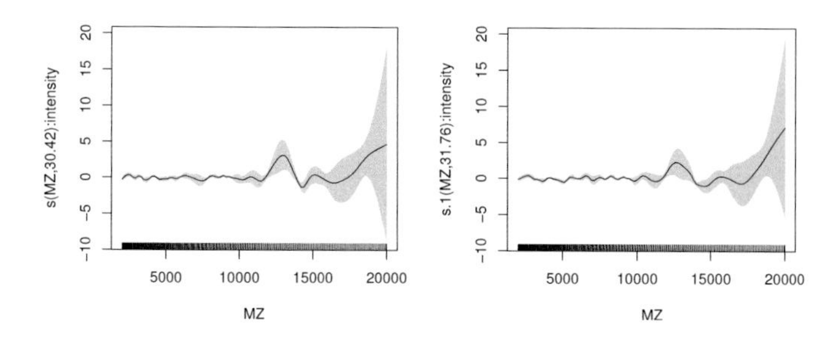

Figure 7.46 *Estimated coefficient functions for the multinomial prostate screening model. Left is for the enlarged class while right is for the cancerous class.*

```
b <- gam(type ~ s(MZ,by=intensity,k=100),family=ocat(R=3),
         data=prostate,method="ML")
```

`ocat` specifies the ordered categorical model with `R=3` classes (see Wood et al., 2016, for full details). The estimated smooth function is quite complex, but also quite robust to the choice of basis and basis dimension. For example, the fit is very similar using an adaptive smoother via `s(MZ,by=intensity,k=100,bs="ad")`. The following code plots the estimated function, shows box plots of the predicted probability of cancer by actual class, and plots a QQ-plot and reference bands for the model's deviance residuals.

```
plot(b,rug=FALSE,scheme=1,xlab="Daltons",ylab="f(D)",
     cex.lab=1.6,cex.axis=1.4)
pb <- predict(b,type="response") ## matrix of class probs
plot(factor(prostate$type),pb[,3])
qq.gam(b,rep=100,lev=.95)
```

The results are shown in figure 7.45, but the middle figure is somewhat disappointing. The model does not distinguish cancer from non-cancer very sharply, and the problem-specific algorithm described in Adam et al. (2002) does better in this case.

A multinomial prostate screening model

Perhaps the problem is the way that the model treats the health status categories as ordered. This approach does not have a strong clinical basis, so we might be better off treating the categories as un-ordered. One model for this is the multinomial logistic regression model implemented in `mgcv` in the `multinom` family.[‖] The idea is that each response observation, y_i, is one of $K+1$ category labels, $0, 1, \ldots, K$. For identifiability reasons category 0 is treated as a reference category, and there is one linear predictor $\eta^{[k]}$ associated with each of the remaining K categories. The linear predictors can contain parametric or smooth terms in the usual way. The probability of y_i taking the value 0 is then $1/\{1+\sum_k \exp(\eta_i^{[k]})\}$, while the probability that it

[‖]Technically this family has the structure of the location scale and shape models of section 7.9 and is implemented in the same way. Of course, the distribution parameters are in a sense all location parameters.

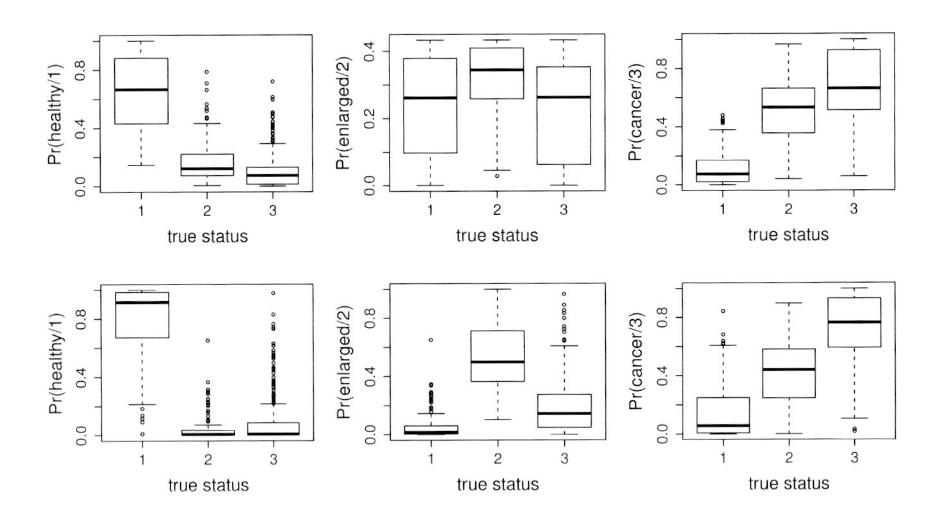

Figure 7.47 *Top row: Box plots of ordered categorical model predicted probabilities for each health category (1-healthy, 2-enlarged, 3-cancer) by actual category. Bottom row: the same for the multinomial model, which appears to be substantially better.*

takes the value $j \neq 0$ is $\exp(\eta^{[j]})/\{1 + \sum_k \exp(\eta_i^{[k]})\}$. In this case we would like both linear predictors to have the same structure used in the ordered categorical case and fitting is therefore as follows.

```
prostate$type1 <- prostate$type - 1 ## recode for multinom
b1 <- gam(list(type1 ~ s(MZ,by=intensity,k=100),
                 ~ s(MZ,by=intensity,k=100)),
          family=multinom(K=2),data=prostate)
```

Figure 7.46 shows the estimated smooths. Notice that while both are somewhat similar to the single coefficient function in the ordered categorical model, there are now some differences between them. `predict(b1,type="response")` will produce the three column matrix of class probabilities for this model. Figure 7.47 gives an impression of the predictive performance of the multinomial model (bottom row) and the ordered categorical model (top row). The multinomial model clearly has much better discriminatory power. For example, if we classify patients according to the category for which their probability is highest, then 89% classified as healthy are really healthy (up from 83%), while 76% of those categorised as having cancer, really do (up from 72%). However, the real gain is in the enlarged category with 76% correctly classified, up from 37%.

Of course, for a real diagnostic procedure we would not classify in this crude way, but would use the probabilities as inputs to a decision rule adjusted to have the best achievable balance of false positive to false negative rate in the cancer classification.

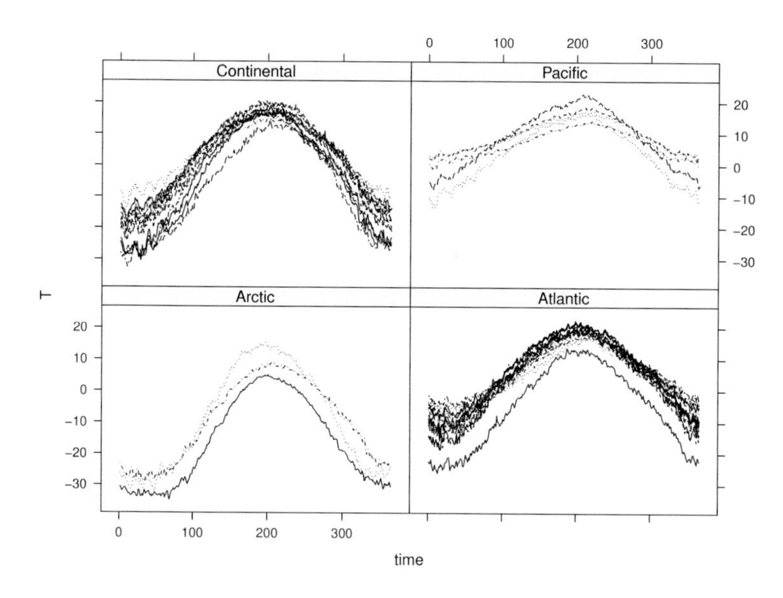

Figure 7.48 *Daily temperature data (degrees centigrade) for one year from 35 locations in Canada, classified by climatic zone.*

7.11.2 Function on scalar regression: Canadian weather

'The Canadian Weather Data' are a much overused example in functional data analysis, but provide a nice illustration of function on scalar regression. They are in the `CanWeather` dataset in `gamair`, and are shown in figure 7.48, which results from

```
require(gamair);require(lattice);data(canWeather)
xyplot(T~time|region,data=CanWeather,type="l",groups=place)
```

Treating each location's temperature profile as a function of time, we would like to model the temperature profiles using the scalar covariates `region` (one of four climate zones) and `latitude`. A possible model is

$$\texttt{T}_i(\texttt{time}) = f_{\texttt{region}(i)}(\texttt{time}) + f(\texttt{time})\texttt{latitude}_i + e_i(\texttt{time})$$

i.e., there are five smooth functions of time to estimate, an intercept function for each climate zone, and a function giving the slope with latitude. It is assumed that the residual function $e_i(\texttt{time})$ is a process that leads to AR1 Gaussian residuals when observed daily.

If we assume independent daily residuals then this model can be estimated using `gam`; however, the AR1 assumption requires either that we use `gamm` or `bam`. `bam` uses the simple computationally efficient AR residual handling of section 1.8.5 (p. 52): an AR1 correlation structure is assumed between consecutive observations in the fitting data, with correlation parameter given by the `rho` argument to `bam`. Breaks between autocorrelated sequences of residuals are indicated by the logical vector argument `AR.start` which is `TRUE` at the start of a new sequence of correlated

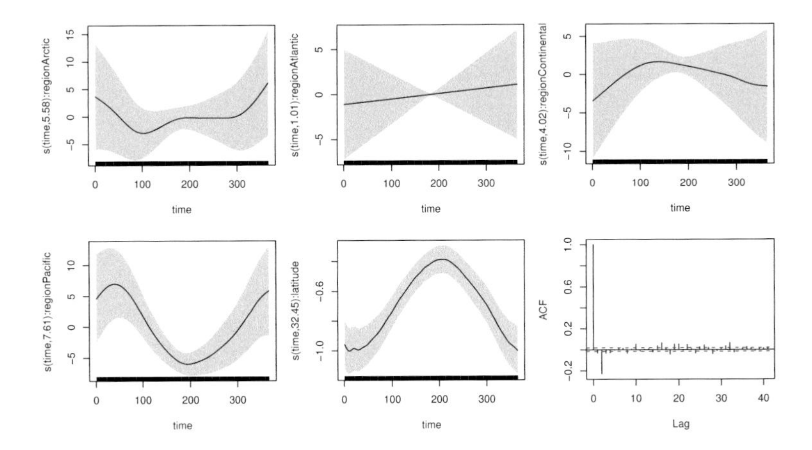

Figure 7.49 *The fitted intercept functions and the* `latitude` *slope function estimates for the function on scalar model of the Canadian Weather data. The lower right panel shows the ACF of the standardized residuals for the fitted model: the negative correlation at lag 2 shows that the AR1 model is certainly not perfect.*

residuals, and `FALSE` otherwise. To choose `rho` a sequence of values can be tried, and the value chosen that minimizes the AIC or negative REML score. Examination of the ACF for a fit with `rho=0` is usually sufficient to find a likely range for `rho`.

The following code fits the model for 10 ρ values evenly spaced between 0.9 and 0.99.** Notice that the intercept function smooths are centred, so the main effect of `region` is also needed.

```
aic <- reml <- rho <- seq(0.9,0.99,by=.01)
for (i in 1:length(rho)) {
  b <- bam(T ~ region + s(time,k=20,bs="cr",by=region) +
           s(time,k=40,bs="cr",by=latitude),
           data=CanWeather,AR.start=time==1,rho=rho[i])
  aic[i] <- AIC(b); reml[i] <- b$gcv.ubre
}
```

Both the AIC and negative REML score have a single minimum at 0.97 here, so this is the value chosen. Figure 7.49 shows the estimated smooth effects. It appears that, as expected, the annual cycle becomes more pronounced with increasing latitude. In practice the Atlantic zone is acting as the reference climate in the model fit, with the Pacific region generally showing warmer winter temperatures. The continental region shows some evidence for colder winters, while the Arctic appears to have slightly warmer winters than the linear latitude model would predict. The lower right plot is the ACF for the standardized residuals returned in `b$std.rsd`: if the AR1 model is perfect then the standardized residuals should appear uncorrelated. The correlation at lag 2 clearly indicates that the model is not perfect, and that something slightly more

**For much larger datasets or models, the option `discrete=TRUE` would be worthwhile – here it is only around twice as quick.

sophisticated would ideally be used. However the AR1 model is certainly much better than an un-correlated model, which gives a very much worse ACF.

7.12 Other packages

There are now far too many packages for smooth modelling in R to hope to present meaningful illustrations of their use without massively lengthening this book, so instead this section will just offer some pointers to some of what is available.

- The original back-fitting approach to GAMs (e.g., Hastie and Tibshirani, 1986, 1990, 1993) is available in Trevor Hastie's `gam` package: a port to R of the original `gam` in S-PLUS. The `SemiPar` package on the other hand focuses on the semi-parametric approach of the book by Ruppert et al. (2003).
- For the full spline smoothing approach (e.g., Wahba, 1990; Wang, 1998b,a; Gu, 2013; Gu and Kim, 2002; Gu and Wahba, 1991) packages `gss` and `assist` are available. `gss` focuses especially on smoothing spline ANOVA models and efficient computation. More recently the `bigspline` package offers yet more efficiency (Helwig and Ma, 2016).
- In a fully Bayesian context `R2BayesX` offers an R interface to the BayesX smooth modelling software (e.g., Fahrmeir et al., 2004; Kneib and Fahrmeir, 2006; Lang et al., 2014; Klein et al., 2015). The `INLA` package is available from `http://www.r-inla.org` and implements the approach of Rue et al. (2009) including the very flexible spatial smoothing methods opened up by Lindgren et al. (2011).
- Smooth distributional regression and more is covered by package `VGAM` (Yee and Wild, 1996) and the `gamlss` packages of Rigby and Stasinopoulos (2005).
- In addition there are many more packages offering related models or some smoothing: for example `scam` (Pya and Wood, 2015) provides shape constrained additive models, `refund` offers functional data analysis, `mboost` provides GAMS via boosting, the `survival` package allows spline effects, and so it goes on.

As one simple example, here is the code for fitting a version of the Chicago air-pollution model using the `gam` package (the covariates were summed over the 4 days preceding death, and degrees of freedom have to be chosen by the modeller).

```
> library(gam)
> bfm <- gam(death~s(time,df=140)+lo(o3,tmp,span=.1),
             family=poisson,control=gam.control(bf.maxit=150))
```

The results are shown in figure 7.50.

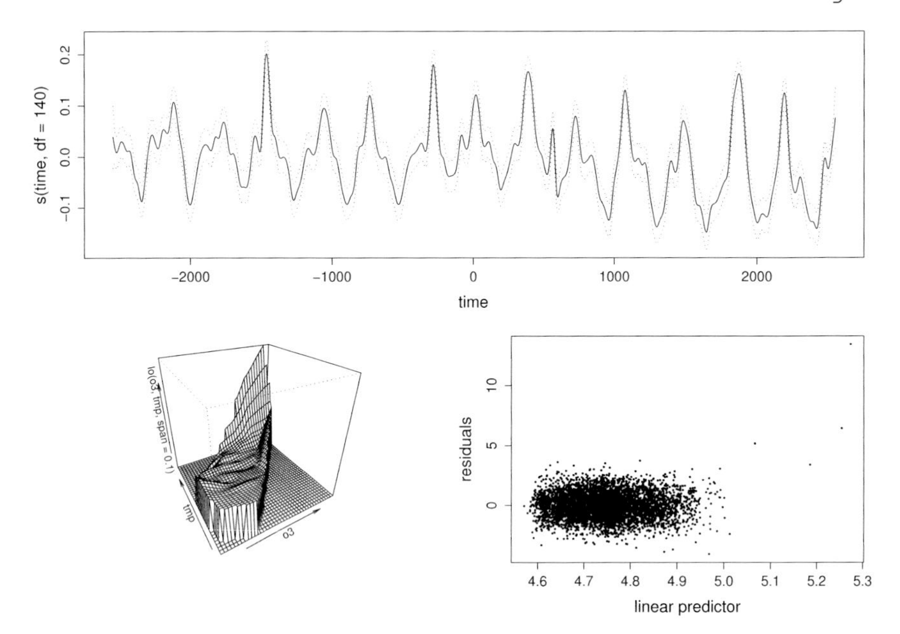

Figure 7.50 *A GAM fitted to the Chicago air pollution data, using* gam *from package* gam. *The upper panel is the estimated smooth of time while the lower left panel is a perspective plot of the estimated ozone – temperature interaction. The lower right plot is a residual plot.*

7.13 Exercises

1. This question re-examines the hubble data from Chapter 1.
   (a) Use gam to fit the model, $\texttt{V}_i = f(\texttt{D}_i) + \epsilon_i$, to the hubble data, where f is a smooth function and the ϵ_i are i.i.d. $N(0, \sigma^2)$. Does a straight line model appear to be most appropriate? How would you interpret the best fit model?
   (b) Examine appropriate residual plots and refit the model with more appropriate distributional assumptions. How are your conclusions from part (a) modified?
2. This question is about using gam for univariate smoothing, the advantages of penalized regression and weighting a smooth model fit. The mcycle data in the MASS package are a classic dataset in univariate smoothing, introduced in Silverman (1985). The data measure the acceleration of the rider's head, against time, in a simulated motorcycle crash.
   (a) Plot the acceleration against time, and use gam to fit a univariate smooth to the data, selecting the smoothing parameter by GCV (k of 30 to 40 is plenty for this example). Plot the resulting smooth, with partial residuals, but without standard errors.
   (b) Use lm and poly to fit a polynomial to the data, with approximately the same degrees of freedom as was estimated by gam. Use termplot to plot

the estimated polynomial and partial residuals. Note the substantially worse fit achieved by the polynomial, relative to the penalized regression spline fit.

(c) It's possible to overstate the importance of penalization in explaining the improvement of the penalized regression spline, relative to the polynomial. Use `gam` to refit an un-penalized thin plate regression spline to the data, with basis dimension the same as that used for the polynomial, and again produce a plot for comparison with the previous two results.

(d) Redo part (c) using an un-penalized cubic regression spline. You should find a fairly clear ordering of the acceptability of the results for the four models tried — what is it?

(e) Now plot the model residuals against time, and comment.

(f) To try and address the problems evident from the residual plot, try giving the first 20 observations the same higher weight, α, while leaving the remaining observations with weight one. Adjust α so that the variance of the first 20 residuals matches that of the remaining residuals. Recheck the residual plots.

(g) Experiment with the order of penalty used in the smooth. Does increasing it affect the model fit?

3. This question uses the `mcycle` data again, in order to more fully explore the influence matrix of a smoother.

(a) Consider a model for response data, $\mathbf{y}$, which has the influence matrix, $\mathbf{A}$, mapping the response data to the fitted values, given a smoothing parameter, λ, i.e., $\hat{\boldsymbol{\mu}} = \mathbf{A}\mathbf{y}$. Show that if we fit the same model, with the same λ, to the response data $\mathbf{I}_j$ (the jth column of the identity matrix), then the resulting model fitted values are the jth column of $\mathbf{A}$.

(b) Using the result from part (a) evaluate the influence matrix, $\mathbf{A}$, for the model fitted in question 2(a).

(c) What value do all the rows of $\mathbf{A}$ sum to? Why?

(d) Any smoothing model that can be represented as $\hat{\boldsymbol{\mu}} = \mathbf{A}\mathbf{y}$, simply replaces each value y_j by a weighted sum of neighbouring y_i values. For example, $\hat{\mu}_j = \sum_i A_{ji} y_i$, where the A_{ji} are the weights in the summation. It is instructive to examine the weights in the summation, so plot the weights used to form $\hat{\mu}_{65}$ against `mcycle$time`. What you have plotted is the 'equivalent kernel' of the fitted spline (at the 65th datum).

(e) Plot all the equivalent kernels, on the same plot. Why do their peak heights vary?

(f) Now vary the smoothing parameter around the GCV selected value. What happens to the equivalent kernel for the 65th datum?

4. This question follows on from questions 2 and 3, and examines the auto-correlation of residuals that results from smoothing.

(a) Based on the best model from question 2, produce a plot of the residual auto-correlation at lag 1 (average correlation between each residual and the previous residual, see `?acf`) against the model effective degrees of freedom. Vary the degrees of freedom between 2 and 40 by varying the `sp` argument to `gam`.

(b) Why do you see positive auto-correlation at very low EDF, and what causes this to reduce as the EDF increases?

(c) The explanation of why auto-correlation becomes negative is not quite so straightforward. Given the insight from question 3, that smoothers operate by forming weighted averages of neighbouring data, the cause of the negative autocorrelation can be understood by examining the simplest weighted average smoother: the k-point running mean. The fitted values from such a smoother are a simple average of the k nearest neighbours of the point in question (including the point itself). k is an odd integer.

i. Write out the form of a typical row, j, of the influence matrix, $\mathbf{A}$, of the simple running mean smoother. Assume that row j is not near the beginning or end of $\mathbf{A}$, and is therefore unaffected by edge effects.

ii. It is easy to show that the residuals are given by $\hat{\boldsymbol{\epsilon}} = (\mathbf{I}-\mathbf{A})\mathbf{y}$ and hence that their covariance matrix is $\mathbf{V}_{\hat{\epsilon}} = (\mathbf{I}-\mathbf{A})(\mathbf{I}-\mathbf{A})\sigma^2$. Find expressions for the elements of $\mathbf{V}_{\hat{\epsilon}}$ on its leading diagonal and on the sub- and super-diagonal, in terms of k. Again, only consider rows/columns that are unaffected by being near the edge of the data.

iii. What do your expressions suggest about residual auto-correlation as the amount of smoothing is reduced?

5. This question is about modelling data with seasonality, and the need to be very careful if trying to extrapolate with GAMs (or any statistical model). The data frame `co2s` contains monthly measurements of CO_2 at the south pole from January 1957 onwards. The columns are `co2`, the month of the year, `month`, and the cumulative number of months since January 1957, `c.month`. There are missing `co2` observations in some months.

(a) Plot the CO_2 observations against cumulative months.

(b) Fit the model $\texttt{co2}_i = f(\texttt{c.month}_i) + \epsilon_i$ where f is a smooth function and the ϵ_i are i.i.d. with constant variance, using the `gam` function. Use the `cr` basis, and a basis dimension of 300.

(c) Obtain the predicted CO_2 for each month of the data, plus 36 months after the end of the data, as well as associated standard errors. Produce a plot of the predictions with twice standard error bands. Are the predictions in the last 36 months credible?

(d) Fit the model $\texttt{co2}_i = f_1(\texttt{c.month}_i) + f_2(\texttt{month}_i) + \epsilon_i$ where f_1 and f_2 are smooth functions, but f_2 is cyclic (you will need to use the `knots` argument of `gam` to ensure that f_2 wraps appropriately: it's important to make sure that January is the same as January, not that December and January are the same!).

(e) Repeat the prediction and plotting in part (c) for the new model. Are the predictions more credible now? Explain the differences between the new results and those from part (c).

6. The data frame `ipo` contains data from Lowry and Schwert (2002) on the number of 'Intitial Public Offerings' (IPOs) per month in the US financial markets between 1960 and 2002. IPOs are the process by which companies go public:

ownership of the company is sold, in the form of shares, as a means of raising capital. Interest focuses on exploring the effect that several variables may have on numbers of IPOs per month, `n.ipo`. Of particular interest are the variables:

`ir` the average initial return from investing in an IPO. This is measured as percentage difference between the offer price of shares, and the share price after the first day of trading in the shares: this reflects by how much the offer price undervalues the company. One might expect companies to pay careful attention to this when deciding on IPO timing.

`dp` is the average percentage difference between the middle of the share price range proposed when the IPO is first filed, and the final offer price. This might be expected to carry information about the direction of changes in market sentiment.

`reg.t` the average time (in days) it takes from filing to offer. Obviously fluctuations in the length of time it takes to register have a direct impact on the number of companies making it through to offering in any given month.

Find a suitable possible model for explaining the number of IPOs in terms of these variables (as well as time, and time of year). Note that it is probably appropriate to look at lagged versions of `ir` and `dp`, since the length of the registration process precludes the number of IPOs in a month being driven by the initial returns in that same month. In the interests of interpretability it is probably worth following the advice of Kim and Gu (2004) and setting `gamma=1.4` in the `gam` call. Look at appropriate model checking plots, and interpret the results of your model fitting.

7. The data frame `wine` contains data on prices and growing characteristics of 25 high quality Bordeaux wines from 1952 to 1998 as reported in: `http://schwert.ssb.rochester.edu/a425/a425.htm`.
`price` gives the average price as a percentage of 1961; `s.temp` is the average temperature (Celsius) over the summer preceding harvest, while `h.temp` is the average temperature at harvest; `w.rain` is the mm of rain in the preceding winter, while `h.rain` give the mm of rain in the harvest month; `year` is obvious. Create a GAM to model `price` in terms of the given predictors. Interpret the effects and use the model to predict the missing prices.

8. Sometimes rather unusual models can be expressed as GAMs. For example the data frame `blowfly` contains `counts` (not samples!) of adults in a laboratory population of blowflies, as reported in the classic ecological papers of Nicholson (1954a,b). One possible model for these data (basically Ellner et al., 1997) is that they are governed by the equation

$$\Delta n_{t+1} = f_b(n_{t-k})n_{t-k} - f_d(n_t)n_t + \epsilon_t,$$

where n_t is the population at time t, $\Delta n_{t+1} = n_{t+1} - n_t$, f_b and f_d are smooth functions and the ϵ_t are i.i.d. errors with constant variance. The idea is that the change in population is given by the difference between the birth and death rates plus a random term. *per capita* birth rates and death rates are smooth functions of populations, and the delayed effect of births on the adult population is because it takes around 12 days to go from egg to adult.

(a) Plot the blowfly data against time.

(b) Fit the proposed model, using gam. You will need to use by variables to do this. Comment on the estimates of f_b and f_d. It is worth making f_b and f_d depend on log populations, rather than the populations themselves, to avoid leverage problems.

(c) Using the beginning of the real data as a starting sequence, iterate your estimated model forward in time, to the end of the data, and plot it. First do this with the noise terms set to zero, and then try again with an error standard deviation of up to 500 (much larger and the population tends to explode). Comment on the results. You will need to artificially restrict the population to the range seen in the real data, to avoid problems caused by extrapolating the model.

(d) Why is the approach used for these data unlikely to be widely applicable?

9. This question is about creating models for calibration of satellite remote sensed data. The data frame chl contains direct ship based measurements of chlorophyll concentrations in the top 5 metres of ocean water, chl, as well as the corresponding satellite estimate chl.sw (actually a multi-year average measurement for the location and time of year), along with ocean depth, bath, day of year, jul.day and location, lon, lat. The data are from the world ocean database (see http://seawifs.gsfc.nasa.gov/SEAWIFS/ for information on SeaWifs). chl and chl.sw do not correlate all that well with each other, probably because the reflective characteristics of water tend to change with time of year and whether the water is near the coast (and hence full of particulate matter) or open ocean. One way of improving the predictive power of the satellite observations might be to model the relationship between directly measured chlorophyll and remote sensed chlorophyll, viewing the relationship as a smooth one that is modified by other factors. Using ocean depth as an indicator for water type (near shore vs. open), a model something like

$$\mathbb{E}(\texttt{chl}_i) = f_1(\texttt{chl.sw}_i) f_2(\texttt{bath}_i) f_3(\texttt{jul.day}_i)$$

might be a reasonable starting point.

(a) Plot the response data and predictors against each other using pairs. Notice that some predictors have very skewed distributions. It is worth trying some simple power transformations in order to get a more even spread of predictors, and reduce the chance of a few points being overly influential: find appropriate transformations.

(b) Using gam, try modelling the data using a model of the sort suggested (but with predictors transformed as in part (a)). Make sure that you use an appropriate family. It will probably help to increase the default basis dimensions used for smoothing, somewhat (especially for jul.day). Use the "cr" basis to avoid excessive computational cost.

(c) In this sort of modelling the aim is to improve on simply using the satellite measurements as predictions of the direct measurements. Given the large number of data, it is easy to end up using rather complex models after quite a lot of examination of the data. It is therefore important to check that the model is not over-fitting. A sensible way of doing this is to randomly select, say, 90% of the data to be fitted by the model, and then to see how well the model predicts the other 10% of data. Do this, using proportion deviance explained as the measure of fit/prediction. Note that the family you used will contain a function `dev.resids`, which you can use to find the deviance of any set of predictions.

10. This question follows on from question 11 Chapter 3, on soybean modelling.
    (a) Using `gamm`, replace the cubic function of time in the GLMM of question 10, with a smooth function of time. You should allow for the possibility that the varieties depend differently on time, and examine appropriate model checking plots.
    (b) Evaluate whether a model with or without a variety effect is more appropriate, and what kind of variety effect is most appropriate.
    (c) Explain why a model with separate smooths for the two different varieties is different to a model with a smooth for one variety and a smooth correction for the other variety.
11. The `med` dataset in `gamair` contains the equivalent of the mackerel survey data from section 7.5, but from 2010. Find a suitable model for these data, and investigate whether the Tweedie or negative binomial distributions offer a better model in this case.

Appendix A

Maximum Likelihood Estimation

This appendix covers the general theory of likelihood and quasi-likelihood used for inference with mixed models, GLMs, GAMs, etc. In particular the results invoked in sections 2.2 and 3.1 are derived here. The emphasis is on explaining the key ideas as simply as possible, so some of the results are proved only for the simple case of a single parameter and i.i.d. data, with the generalizations merely stated. To emphasize that the results given here apply much more widely than GLMs, the parameter that is the object of inference is denoted by $\boldsymbol{\theta}$ in this section: for GLMs this will usually be $\boldsymbol{\beta}$. Good references on the topics covered here are Cox and Hinkley (1974) and Silvey (1970), which are followed quite closely below.

Proofs are only given in outline and two general statistical results are used repeatedly: the 'law of large numbers' (LOLN) and the 'central limit theorem' (CLT). In the i.i.d. context these are as follows. Let $X_1, X_2, \ldots, X_n$ be i.i.d. random variables with mean μ and variance σ^2 (both of which are finite). The LOLN states that as $n \to \infty$, $\bar{X} \to \mu$ (in probability).* The CLT states that as $n \to \infty$ the distribution of $\bar{X}$ tends to $N(\mu, \sigma^2/n)$ whatever the distribution of the X_i. Both results generalize to multivariate and non-i.i.d. settings.

A.1 Invariance

Consider an observation, $\mathbf{y} = [y_1, y_2, \ldots, y_n]^\mathsf{T}$, of a vector of random variables with joint p.m.f. or p.d.f. $f(\mathbf{y}, \theta)$, where θ is a parameter with MLE $\hat{\theta}$. If γ is a parameter such that $\gamma = g(\theta)$, where g is any function, then the maximum likelihood estimate of γ is $\hat{\gamma} = g(\hat{\theta})$, and this property is known as *invariance*.

Invariance holds for any g, but a proof is easiest for the case in which g is a one to one function, so that g^{-1} is well defined. In this case $\theta = g^{-1}(\gamma)$ and maximum likelihood estimation would proceed by maximizing the likelihood

$$L(\gamma) = f(\mathbf{y}, g^{-1}(\gamma))$$

w.r.t. γ. But we know that the maximum of f occurs at $f(\mathbf{y}, \hat{\theta})$, by definition of $\hat{\theta}$, so

*Tending to a limit in probability basically means that the probability of being further than any positive constant ϵ from the limit tends to zero.

it must be the case that L's maximum w.r.t. γ occurs when $\hat{\theta} = g^{-1}(\hat{\gamma})$, i.e.,

$$\hat{\gamma} = g(\hat{\theta})$$

is the MLE of γ. So, when working with maximum likelihood estimation, we can adopt whatever parameterization is most convenient for performing calculations, and simply transform back to the most interpretable parameterization at the end. Note that invariance holds for vector parameters as well.

A.2 Properties of the expected log-likelihood

The key to proving and understanding the large sample properties of maximum likelihood estimators lies in obtaining some results for the expectation of the log-likelihood, and then using the convergence in probability of the log-likelihood to its expected value, which results from the law of large numbers. In this section, some simple properties of the expected log likelihood are derived.

Let $y_1, y_2, \ldots, y_n$ be independent observations from a p.d.f. $f(y, \theta)$, where θ is an unknown parameter with true value θ_0. Treating θ as unknown, the log-likelihood for θ is

$$l(\theta) = \sum_{i=1}^{n} \log\{f(y_i, \theta)\} = \sum_{i=1}^{n} l_i(\theta),$$

where l_i is the log-likelihood, given only the single observation y_i. Treating l as a function of random variables, $Y_1, Y_2, \ldots, Y_n$, means that l is itself a random variable (and the l_i are independent random variables). Hence we can consider expectations of l and its derivatives.

Result 1:

$$\mathbb{E}_0\left(\left.\frac{\partial l}{\partial \theta}\right|_{\theta_0}\right) = 0. \tag{A.1}$$

The subscript on the expectation is to emphasize that the expectation is w.r.t. $f(y, \theta_0)$. The proof is as follows (where it is to be taken that all derivatives are evaluated at θ_0, and there is sufficient regularity that the order of differentiation and integration can be exchanged).

$$\begin{aligned}\mathbb{E}_0\left(\frac{\partial l_i}{\partial \theta}\right) &= \mathbb{E}_0\left[\frac{\partial}{\partial \theta}\log\{f(Y, \theta)\}\right] = \int \frac{1}{f(y, \theta_0)}\frac{\partial f}{\partial \theta} f(y, \theta_0) dy \\ &= \int \frac{\partial f}{\partial \theta} dy = \frac{\partial}{\partial \theta}\int f dy = \frac{\partial 1}{\partial \theta} = 0.\end{aligned}$$

That the same holds for l follows immediately.

Result 1 has the following obvious consequence:

Result 2:

$$\text{var}\left(\left.\frac{\partial l}{\partial \theta}\right|_{\theta_0}\right) = \mathbb{E}_0\left\{\left(\left.\frac{\partial l}{\partial \theta}\right|_{\theta_0}\right)^2\right\}. \tag{A.2}$$

It can further be shown that

Result 3:

$$\mathcal{I} \equiv \mathbb{E}_0 \left\{ \left(\left. \frac{\partial l}{\partial \theta} \right|_{\theta_0} \right)^2 \right\} = -\mathbb{E}_0 \left(\left. \frac{\partial^2 l}{\partial \theta^2} \right|_{\theta_0} \right) \tag{A.3}$$

where $\mathcal{I}$ is referred to as the **information** about θ contained in the data. The terminology refers to the fact that, if the data tie down θ very closely (and accurately), then the log-likelihood will be sharply peaked in the vicinity θ_0 (i.e., high $\mathcal{I}$), whereas data containing little information about θ will lead to an almost flat likelihood, and low $\mathcal{I}$.

The proof of result 3 is simple. For a single observation, result 1 says that

$$\int \frac{\partial \log(f)}{\partial \theta} f dy = 0.$$

Differentiating again w.r.t. θ yields

$$\int \frac{\partial^2 \log(f)}{\partial \theta^2} f + \frac{\partial \log(f)}{\partial \theta} \frac{\partial f}{\partial \theta} dy = 0$$

but

$$\frac{\partial \log(f)}{\partial \theta} = \frac{1}{f} \frac{\partial f}{\partial \theta}$$

and so

$$\int \frac{\partial^2 \log(f)}{\partial \theta^2} f dy = -\int \left(\frac{\partial \log(f)}{\partial \theta} \right)^2 f dy,$$

which is

$$\mathbb{E}_0 \left(\left. \frac{\partial^2 l_i}{\partial \theta^2} \right|_{\theta_0} \right) = -\mathbb{E}_0 \left\{ \left(\left. \frac{\partial l_i}{\partial \theta} \right|_{\theta_0} \right)^2 \right\}.$$

The result follows very easily (given the independence of the l_i).

Note that by result 1 the expected log likelihood has a turning point at θ_0, and since $\mathcal{I}$ is positive, result 3 indicates that this turning point is a maximum. So the expected log likelihood has a maximum at the true parameter value. Unfortunately results 1 and 3 don't establish that this maximum is a global maximum, but a slightly more involved proof shows that this is in fact the case.

Result 4:

$$\mathbb{E}_0\{l(\theta_0)\} \geq \mathbb{E}_0\{l(\theta)\} \;\forall\, \theta. \tag{A.4}$$

The proof is based on Jensen's inequality, which says that if c is a concave function (i.e., has negative second derivative) and Y is a random variable, then

$$\mathbb{E}\{c(Y)\} \leq c\{\mathbb{E}(Y)\}.$$

The inequality is almost a statement of the obvious as figure A.1 illustrates. Now con-

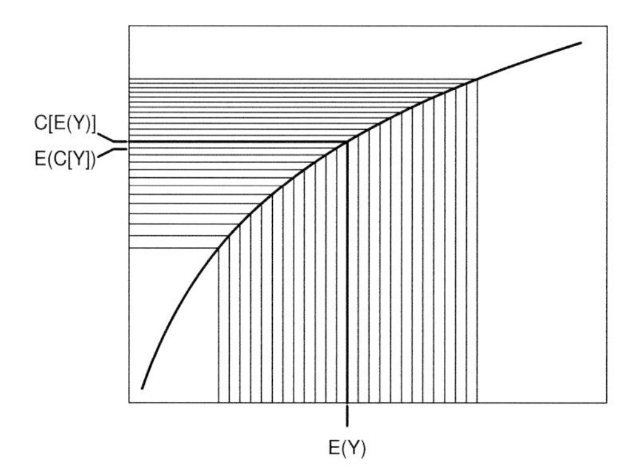

Figure A.1 *Schematic illustration of Jensen's inequality which says that if c is a concave function then $\mathbb{E}\{c(Y)\} \le c\{\mathbb{E}(Y)\}$. The curve shows a concave function, while the lines connect the values of a discrete uniform random variable Y on the horizontal axis to the values of the discrete random variable $c(Y)$ on the vertical axis. It is immediately clear that the way that c spreads out $c(Y)$ values corresponding to low Y values, while bunching together $c(Y)$ values corresponding to high Y values, implies Jensen's inequality. Further reflection suggests (correctly) that Jensen's inequality holds for any distribution.*

sider the concave function, log, and the random variable, $f(Y,\theta)/f(Y,\theta_0)$. Jensen's inequality implies that

$$\mathbb{E}_0\left[\log\left\{\frac{f(Y,\theta)}{f(Y,\theta_0)}\right\}\right] \le \log\left[\mathbb{E}_0\left\{\frac{f(Y,\theta)}{f(Y,\theta_0)}\right\}\right].$$

Consider the right hand side of the inequality.

$$\mathbb{E}_0\left\{\frac{f(Y,\theta)}{f(Y,\theta_0)}\right\} = \int \frac{f(y,\theta)}{f(y,\theta_0)} f(y,\theta_0)dy = \int f(y,\theta)dy = 1.$$

So, since $\log(1) = 0$ the inequality becomes

$$\mathbb{E}_0\left[\log\left\{\frac{f(Y,\theta)}{f(Y,\theta_0)}\right\}\right] \le 0$$

$$\Rightarrow \mathbb{E}_0[\log\{f(Y,\theta)\}] \le \mathbb{E}_0[\log\{f(Y,\theta_0)\}]$$

from which the result follows immediately.

The above results were derived for continuous Y, but also hold for discrete Y: the proofs are almost identical, but with $\sum_{all\ y_i}$ replacing $\int dy$. Note also that, although the results presented here were derived assuming that the data were independent observations from the same distribution, this is in fact much more restrictive than is necessary, and the results hold more generally.

Similarly, the results generalize to vector parameters. Let $\mathbf{u}$ be the vector such that $u_i = \partial l/\partial\theta_i$, and $\mathbf{H}$ be the Hessian matrix of the log-likelihood w.r.t. the parameters so that $H_{ij} = \partial^2 l/\partial\theta_i\partial\theta_j$. Then, in particular,

Result 1 (vector parameter)

$$\mathbb{E}_0(\mathbf{u}) = \mathbf{0} \tag{A.5}$$

and

Result 3 (vector parameter)

$$\mathcal{I} \equiv \mathbb{E}_0(\mathbf{u}\mathbf{u}^\mathsf{T}) = -\mathbb{E}_0(\mathbf{H}). \tag{A.6}$$

A.3 Consistency

Maximum likelihood estimators are often not unbiased, but under quite mild regularity conditions they are *consistent*. This means that as the sample size, on which the estimate is based, tends to infinity, the maximum likelihood estimator tends in probability to the true parameter value. Consistency therefore implies asymptotic† unbiasedness, but it actually implies slightly more than this — for example that the variance of the estimator is decreasing with sample size.

Formally, if θ_0 is the true value of parameter θ, and $\hat{\theta}_n$ is its MLE based on n observations, $y_1, y_2, \ldots, y_n$, then consistency means that

$$\Pr(|\hat{\theta}_n - \theta_0| < \epsilon) \to 1$$

as $n \to \infty$ for any positive ϵ.

To see why MLEs are consistent, consider an outline proof for the case of a single parameter, θ, estimated from independent observations $y_1, y_2, \ldots, y_n$ on a random variable with p.m.f. or p.d.f. $f(y, \theta_0)$. The log-likelihood in this case will be

$$l(\theta) \propto \frac{1}{n} \sum_{i=1}^{n} \log\{f(y_i, \theta)\}$$

where the factor of $1/n$ is introduced purely for later convenience. We need to show that in the large sample limit $l(\theta)$ achieves its maximum at the true parameter value θ_0, but in the previous section it was shown that the expected value of the log likelihood for a single observation attains its maximum at θ_0. The law of large numbers tells us that as $n \to \infty$, $\sum_{i=1}^n \log\{f(Y_i, \theta)\}/n$ tends (in probability) to $\mathbb{E}_0[\log\{f(Y, \theta)\}]$. So in the large sample limit we have that

$$l(\theta_0) \geq l(\theta),$$

i.e., that $\hat{\theta}$ is θ_0.

To show that $\hat{\theta} \to \theta_0$ in some well-ordered manner as $n \to \infty$ requires that we assume some regularity (for example, at minimum, we need to be able to assume that if θ_1 and θ_2 are 'close' then so are $l(\theta_1)$ and $l(\theta_2)$), but in the vast majority of practical situations such conditions hold. Figure A.2 can help illustrate how the

† 'Asymptotic' here meaning 'as sample size tends to infinity'.

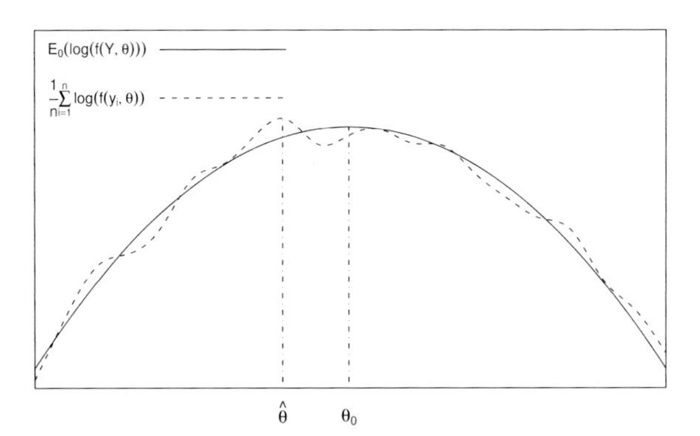

Figure A.2 *Illustration of the idea behind the derivation of consistency of MLEs. The dashed curve is the log-likelihood divided by the sample size, n, while the solid curve is the expectation of the log-likelihood divided by n (equivalently, the expected log-likelihood given one observation). The solid curve has a maximum at the true parameter value, and as sample size tends to infinity the dashed curve tends to the solid curve by the LOLN. Hence, in the large sample limit, the log likelihood has a maximum at the true parameter value, and we expect $\hat{\theta}_n \to \theta_0$ as $n \to \infty$.*

argument works: as the sample size tends to infinity the dashed curve, proportional to the log likelihood, tends in probability to the solid curve, $\mathbb{E}_0[\log\{f(Y, \theta)\}]$, which has its maximum at θ_0, hence $\hat{\theta} \to \theta_0$.

For simplicity of presentation, the above argument dealt only with a single parameter and data that were independent observations of a random variable from one distribution. In fact consistency holds in much more general circumstances: for vector parameters and non-independent data that do not necessarily all come from the same distribution.

A.4 Large sample distribution of $\hat{\theta}$

To obtain the large sample distribution of the MLE, $\hat{\theta}$, we make a Taylor expansion of the derivative of the log likelihood around the true parameter, θ_0, and evaluate this at $\hat{\theta}$.

$$\left.\frac{\partial l}{\partial \theta}\right|_{\hat{\theta}} \simeq \left.\frac{\partial l}{\partial \theta}\right|_{\theta_0} + \left(\hat{\theta} - \theta_0\right) \left.\frac{\partial^2 l}{\partial \theta^2}\right|_{\theta_0}$$

and from the definition of the MLE the left hand side must be zero, so we have that

$$\left(\hat{\theta} - \theta_0\right) \simeq \frac{\partial l/\partial \theta|_{\theta_0}}{-\,\partial^2 l/\partial \theta^2|_{\theta_0}},$$

with equality in the large sample limit (by consistency of $\hat{\theta}$). Now the top of this fraction has expected value zero and variance $\mathcal{I}$ (see results (A.1) and (A.2)), but it is also made up of a sum of i.i.d. random variables, $l_i = \log\{f(Y_i, \theta)\}$, so that by the central limit theorem, as $n \to \infty$, its distribution will tend to $N(0, \mathcal{I})$. By the law of large numbers we also have that, as $n \to \infty$, $- \left.\partial^2 l / \partial\theta^2\right|_{\theta_0} \to \mathcal{I}$ (in probability). So in the large sample limit $(\hat{\theta} - \theta_0)$ is distributed as an $N(0, \mathcal{I})$ r.v. divided by $\mathcal{I}$. So in the limit as $n \to \infty$,

$$\left(\hat{\theta} - \theta_0\right) \sim N(0, \mathcal{I}^{-1}).$$

The result generalizes to vector parameters:

$$\hat{\boldsymbol{\theta}} \sim N(\boldsymbol{\theta}_0, \boldsymbol{\mathcal{I}}^{-1}) \tag{A.7}$$

in the large sample limit. Again the result holds generally and not just for the somewhat restricted form of the likelihood which we have assumed here.

Usually $\boldsymbol{\mathcal{I}}$ will not be known, any more than $\boldsymbol{\theta}$ is, and will have to be estimated by plugging $\hat{\boldsymbol{\theta}}$ into the expression for $\boldsymbol{\mathcal{I}}$, or by the negative of the Hessian ($-\mathbf{H}$) of the log-likelihood evaluated at the MLE For reasonable sample sizes, the *empirical information matrix*, $\hat{\boldsymbol{\mathcal{I}}} = -\mathbf{H}$, is usually an adequate approximation to the information matrix $\boldsymbol{\mathcal{I}}$ itself (this follows from the law of large numbers).

A.5 The generalized likelihood ratio test (GLRT)

Consider an observation, $\mathbf{y}$, on a random vector of dimension n with p.d.f. (or p.m.f.) $f(\mathbf{y}, \boldsymbol{\theta})$, where $\boldsymbol{\theta}$ is a parameter vector. Suppose that we want to test:

$$H_0 : \mathbf{R}(\boldsymbol{\theta}) = \mathbf{0} \quad \text{vs.} \quad H_1 : \mathbf{R}(\boldsymbol{\theta}) \neq \mathbf{0},$$

where $\mathbf{R}$ is a vector valued function of $\boldsymbol{\theta}$, such that H_0 imposes r restrictions on the parameter vector. If H_0 is true then in the limit as $n \to \infty$

$$2\lambda = 2\{l(\hat{\boldsymbol{\theta}}_{H_1}) - l(\hat{\boldsymbol{\theta}}_{H_0})\} \sim \chi^2_r, \tag{A.8}$$

where l is the log-likelihood function and $\hat{\boldsymbol{\theta}}_{H_1}$ is the MLE of $\boldsymbol{\theta}$. $\hat{\boldsymbol{\theta}}_{H_0}$ is the value of $\boldsymbol{\theta}$ satisfying $\mathbf{R}(\boldsymbol{\theta}) = \mathbf{0}$, which maximizes the likelihood (i.e., the restricted MLE). This result is used to calculate approximate p-values for the test.

Tests performed in this way are known as 'generalized likelihood ratio tests', since λ is the log of the ratio of the maximized likelihoods under each hypothesis. In the context of GLMs, the null hypothesis is usually that the correct model is a simplified version of the model under the alternative hypothesis.

A.6 Derivation of $2\lambda \sim \chi^2_r$ under H_0

To simplify matters, first suppose that the parameterization is such that $\boldsymbol{\theta} = \begin{bmatrix} \boldsymbol{\psi} \\ \boldsymbol{\gamma} \end{bmatrix}$, where $\boldsymbol{\psi}$ is r-dimensional, and the null hypothesis can be written $H_0 : \boldsymbol{\psi} = \boldsymbol{\psi}_0$.

In principle it is always possible to re-parameterize a model so that the null has this form.[‡]

Now let the unrestricted MLE be $\begin{bmatrix} \hat{\boldsymbol{\psi}} \\ \hat{\boldsymbol{\gamma}} \end{bmatrix}$, and let $\begin{bmatrix} \boldsymbol{\psi}_0 \\ \hat{\boldsymbol{\gamma}}_0 \end{bmatrix}$ be the MLE under the restrictions defining the null hypothesis. The key to making progress is to be able to express $\hat{\boldsymbol{\gamma}}_0$ in terms of $\hat{\boldsymbol{\psi}}$, $\hat{\boldsymbol{\gamma}}$ and $\boldsymbol{\psi}_0$. In general, this is possible in the large sample limit, provided that the null hypothesis is true, so that $\hat{\boldsymbol{\psi}}$ is close to $\boldsymbol{\psi}_0$. Taking a Taylor expansion of the log likelihood around the unrestricted MLE, $\hat{\boldsymbol{\theta}}$, yields

$$l(\boldsymbol{\theta}) \simeq l(\hat{\boldsymbol{\theta}}) - \frac{1}{2}\left(\boldsymbol{\theta} - \hat{\boldsymbol{\theta}}\right)^{\mathsf{T}} \mathbf{H} \left(\boldsymbol{\theta} - \hat{\boldsymbol{\theta}}\right) \tag{A.9}$$

where $H_{ij} = -\left.\partial^2 l / \partial\theta_i \partial\theta_j\right|_{\hat{\boldsymbol{\theta}}}$ (note the factor of -1 introduced here). Exponentiating this expression, the likelihood can be written

$$L(\boldsymbol{\theta}) \simeq L(\hat{\boldsymbol{\theta}}) \exp\left\{-\left(\boldsymbol{\theta} - \hat{\boldsymbol{\theta}}\right)^{\mathsf{T}} \mathbf{H} \left(\boldsymbol{\theta} - \hat{\boldsymbol{\theta}}\right)/2\right\}.$$

i.e. the likelihood can be approximated by a function proportional to the p.d.f. of an $N(\hat{\boldsymbol{\theta}}, \mathbf{H}^{-1})$ random variable. By standard properties of the multivariate normal p.d.f., if

$$\begin{bmatrix} \boldsymbol{\psi} \\ \boldsymbol{\gamma} \end{bmatrix} \sim N\left(\begin{bmatrix} \hat{\boldsymbol{\psi}} \\ \hat{\boldsymbol{\gamma}} \end{bmatrix}, \begin{bmatrix} \boldsymbol{\Sigma}_{\psi\psi} & \boldsymbol{\Sigma}_{\psi\gamma} \\ \boldsymbol{\Sigma}_{\gamma\psi} & \boldsymbol{\Sigma}_{\gamma\gamma} \end{bmatrix}\right)$$

then

$$\boldsymbol{\gamma}|\boldsymbol{\psi} \sim N(\hat{\boldsymbol{\gamma}} + \boldsymbol{\Sigma}_{\gamma\psi}\boldsymbol{\Sigma}_{\psi\psi}^{-1}(\boldsymbol{\psi} - \hat{\boldsymbol{\psi}}), \boldsymbol{\Sigma}_{\gamma\gamma} - \boldsymbol{\Sigma}_{\gamma\psi}\boldsymbol{\Sigma}_{\psi\psi}^{-1}\boldsymbol{\Sigma}_{\psi\gamma}).$$

Hence if $\boldsymbol{\psi}$ is fixed at $\boldsymbol{\psi}_0$ by hypothesis, then the approximation to the likelihood will be maximized when $\boldsymbol{\gamma}$ takes the value

$$\hat{\boldsymbol{\gamma}}_0 = \hat{\boldsymbol{\gamma}} + \boldsymbol{\Sigma}_{\gamma\psi}\boldsymbol{\Sigma}_{\psi\psi}^{-1}(\boldsymbol{\psi}_0 - \hat{\boldsymbol{\psi}}). \tag{A.10}$$

If the null hypothesis is true, then in the large sample limit $\hat{\boldsymbol{\psi}} \to \boldsymbol{\psi}_0$ (in probability) so that the approximate likelihood tends to the true likelihood, and we can expect (A.10) to hold for the maximizers of the exact likelihood.

Expressing (A.10) in terms of a partitioning of $\boldsymbol{\Sigma} = \mathbf{H}^{-1}$ is not as useful as having the results in terms of the equivalent partitioning of $\mathbf{H}$ itself. Writing $\boldsymbol{\Sigma}\mathbf{H} = \mathbf{I}$ in partitioned form

$$\begin{bmatrix} \boldsymbol{\Sigma}_{\psi\psi} & \boldsymbol{\Sigma}_{\psi\gamma} \\ \boldsymbol{\Sigma}_{\gamma\psi} & \boldsymbol{\Sigma}_{\gamma\gamma} \end{bmatrix} \begin{bmatrix} \mathbf{H}_{\psi\psi} & \mathbf{H}_{\psi\gamma} \\ \mathbf{H}_{\gamma\psi} & \mathbf{H}_{\gamma\gamma} \end{bmatrix} = \begin{bmatrix} \mathbf{I} & \mathbf{0} \\ \mathbf{0} & \mathbf{I} \end{bmatrix},$$

and multiplying out results in four matrix equations, of which two are useful:

$$\boldsymbol{\Sigma}_{\psi\psi}\mathbf{H}_{\psi\psi} + \boldsymbol{\Sigma}_{\psi\gamma}\mathbf{H}_{\gamma\psi} = \mathbf{I}, \tag{A.11}$$

[‡]Of course, to use the result no re-parameterization is necessary — it is only being done here for theoretical convenience when deriving the result. Invariance ensures that re-parameterization is a legitimate thing to do.

$$\boldsymbol{\Sigma}_{\psi\psi}\mathbf{H}_{\psi\gamma} + \boldsymbol{\Sigma}_{\psi\gamma}\mathbf{H}_{\gamma\gamma} = \mathbf{0}. \tag{A.12}$$

Re-arranging (A.12) while noting that, by symmetry, $\mathbf{H}_{\psi\gamma}^{\mathsf{T}} = \mathbf{H}_{\gamma\psi}$ and $\boldsymbol{\Sigma}_{\psi\gamma}^{\mathsf{T}} = \boldsymbol{\Sigma}_{\gamma\psi}$,[§] yields

$$-\mathbf{H}_{\gamma\gamma}^{-1}\mathbf{H}_{\gamma\psi} = \boldsymbol{\Sigma}_{\gamma\psi}\boldsymbol{\Sigma}_{\psi\psi}^{-1}$$

and hence

$$\hat{\gamma}_0 = \hat{\gamma} + \mathbf{H}_{\gamma\gamma}^{-1}\mathbf{H}_{\gamma\psi}(\hat{\psi} - \psi_0). \tag{A.13}$$

For later use it is also worth eliminating $\boldsymbol{\Sigma}_{\psi\gamma}$ from (A.11) and (A.12), which results in

$$\boldsymbol{\Sigma}_{\psi\psi}^{-1} = \mathbf{H}_{\psi\psi} - \mathbf{H}_{\psi\gamma}\mathbf{H}_{\gamma\gamma}^{-1}\mathbf{H}_{\gamma\psi}. \tag{A.14}$$

Now provided that the null hypothesis is true, so that $\hat{\psi}$ is close to ψ_0, we can re-use the expansion (A.9) and write the log-likelihood at the restricted MLE as

$$l(\psi_0, \hat{\gamma}_0) \simeq l(\hat{\psi}, \hat{\gamma}) - \frac{1}{2}\begin{bmatrix} \psi_0 - \hat{\psi} \\ \hat{\gamma}_0 - \hat{\gamma} \end{bmatrix}^{\mathsf{T}} \mathbf{H} \begin{bmatrix} \psi_0 - \hat{\psi} \\ \hat{\gamma}_0 - \hat{\gamma} \end{bmatrix}.$$

Hence

$$2\lambda = 2\{l(\hat{\psi}, \hat{\gamma}) - l(\psi_0, \hat{\gamma}_0)\} \simeq \begin{bmatrix} \psi_0 - \hat{\psi} \\ \hat{\gamma}_0 - \hat{\gamma} \end{bmatrix}^{\mathsf{T}} \mathbf{H} \begin{bmatrix} \psi_0 - \hat{\psi} \\ \hat{\gamma}_0 - \hat{\gamma} \end{bmatrix}.$$

Substituting for $\hat{\gamma}_0$ from (A.13) and writing out $\mathbf{H}$ in partitioned form gives

$$2\lambda \simeq \begin{bmatrix} \psi_0 - \hat{\psi} \\ \mathbf{H}_{\gamma\gamma}^{-1}\mathbf{H}_{\gamma\psi}(\hat{\psi} - \psi_0) \end{bmatrix}^{\mathsf{T}} \begin{bmatrix} \mathbf{H}_{\psi\psi} & \mathbf{H}_{\psi\gamma} \\ \mathbf{H}_{\gamma\psi} & \mathbf{H}_{\gamma\gamma} \end{bmatrix} \begin{bmatrix} \psi_0 - \hat{\psi} \\ \mathbf{H}_{\gamma\gamma}^{-1}\mathbf{H}_{\gamma\psi}(\hat{\psi} - \psi_0) \end{bmatrix}$$

and a short routine slog results in

$$2\lambda \simeq (\hat{\psi} - \psi_0)^{\mathsf{T}}\left[\mathbf{H}_{\psi\psi} - \mathbf{H}_{\psi\gamma}\mathbf{H}_{\gamma\gamma}^{-1}\mathbf{H}_{\gamma\psi}\right](\hat{\psi} - \psi_0).$$

But given (A.14), this means that

$$2\lambda \simeq (\hat{\psi} - \psi_0)^{\mathsf{T}}\boldsymbol{\Sigma}_{\psi\psi}^{-1}(\hat{\psi} - \psi_0). \tag{A.15}$$

Now if H_0 is true, then as $n \to \infty$ this expression will tend towards exactness as $\hat{\psi} \to \psi_0$. Furthermore, by the law of large numbers and (A.6), $\mathbf{H} \to \mathcal{I}$ as $n \to \infty$ (recall that in this section $\mathbf{H}$ is the negative second derivative matrix), which means that $\boldsymbol{\Sigma}$ tends to $\mathcal{I}^{-1}$, and hence $\boldsymbol{\Sigma}_{\psi\psi}$ tends to the covariance matrix of $\hat{\psi}$ (see result A.7). Hence, by the asymptotic normality of the MLE $\hat{\psi}$,

$$2\lambda \sim \chi^2_r$$

under H_0. Having proved the asymptotic distribution of λ under H_0, you might be wondering why it was worth bothering, when we could simply have used the right

[§] And remembering that $(\mathbf{AB})^{\mathsf{T}} = \mathbf{B}^{\mathsf{T}}\mathbf{A}^{\mathsf{T}}$.

hand side of (A.15) directly as the test statistic. This approach is indeed possible, and is known as the Wald test, but it suffers from the disadvantage that at finite sample sizes the magnitude of the test statistic depends on how we choose to parameterize the model. The GLRT, on the other hand, is invariant to the parameterization we choose to use, irrespective of sample size. This invariance seems much more satisfactory — we do not generally want our statistical conclusions to depend on details of how we set up the model, if those details could never be detected by observing data from the model.

A.7 AIC in general

As we have seen, selecting between (nested) models on the basis of which has higher likelihood is generally unsatisfactory, because the model with more parameters always has the higher likelihood. Indeed, the previous section shows that, if we add a redundant parameter to an already correct model, the expected increase in likelihood is $\simeq 1/2$. This problem arises because each additional parameter allows the model to get a little closer to the observed data, by fitting the noise component of the data, as well as the signal. If we were to judge between models on the basis of their fit to new data, not used in estimation, then this problem would not arise. The Akaike Information Criterion (AIC) is an attempt to provide a way of doing this.

The derivation of AIC basically has two parts:

i) The average ability of a maximum likelihood estimated model to predict new data is measured by the expected Kullback-Leibler (K-L) divergence between the estimated model and the true model. This can be shown to be approximately the minimum K-L divergence that the model could possibly achieve, for any parameter values, plus an estimable constant.

ii) The expected negative log likelihood of the model can be approximately expressed as the minimum K-L divergence that could possibly be achieved *minus* the same estimable constant (and another ignorable constant). This immediately suggests an estimator for the expected K-L divergence in terms of the negative log likelihood and the constant.

Superficially similar looking expectations appear to evaluate to rather different quantities in parts (i) and (ii), which can cause confusion. The key difference is that in part (i) we are interested in the estimated model's ability to predict new data, so that the parameter estimators are not functions of the data over whose distribution the expectations are taken. In part (ii), when examining the expected log-likelihood, the parameter estimators are most definitely functions of the data.

Suppose then, that our data were really generated from a density $f_0(\mathbf{y})$ and that our model density is $f_\theta(\mathbf{y})$, where $\boldsymbol{\theta}$ denotes the parameter(s) of f_θ, and $\mathbf{y}$ and $\boldsymbol{\theta}$ will generally be vectors, with $\boldsymbol{\theta}$ having dimension p.

$$K(f_\theta, f_0) = \int [\log\{f_0(\mathbf{y})\} - \log\{f_\theta(\mathbf{y})\}] f_0(\mathbf{y})d\mathbf{y} \tag{A.16}$$

provides a measure of how badly f_θ matches the truth, known as the Kullback-Leibler divergence. So, if $\hat{\boldsymbol{\theta}}$ is the MLE of $\boldsymbol{\theta}$, then $K(f_{\hat{\theta}}, f_0)$ could be used to provide a measure of how well our model is expected to fit a new set of data, not used to estimate $\hat{\boldsymbol{\theta}}$. Note that because we are interested in new data, $\hat{\boldsymbol{\theta}}$ is treated as fixed and not as a function of $\mathbf{y}$ when evaluating (A.16).

Of course, (A.16) can not be used directly, since we are not sure of f_0, but progress can be made by considering a truncated Taylor expansion of $\log(f_\theta)$ about the (unknown) parameters $\boldsymbol{\theta}_K$ which would minimize (A.16).

$$\log\{f_{\hat{\theta}}(\mathbf{y})\} \simeq \log\{f_{\theta_K}(\mathbf{y})\} + (\hat{\boldsymbol{\theta}} - \boldsymbol{\theta}_K)^\mathsf{T}\mathbf{g} + \frac{1}{2}(\hat{\boldsymbol{\theta}} - \boldsymbol{\theta}_K)^\mathsf{T}\mathbf{H}(\hat{\boldsymbol{\theta}} - \boldsymbol{\theta}_K) \quad \text{(A.17)}$$

where $\mathbf{g}$ and $\mathbf{H}$ are the gradient vector and Hessian matrix of first and second derivatives of $\log(f_\theta)$ w.r.t. $\boldsymbol{\theta}$ evaluated at $\boldsymbol{\theta}_K$. It is easy to see that if $\boldsymbol{\theta}_K$ minimizes (A.16) then $\int \mathbf{g} f_0 d\mathbf{y} = \mathbf{0}$, so that substituting (A.17) into (A.16) yields

$$K(f_{\hat{\theta}}, f_0) \simeq K(f_{\theta_K}, f_0) + \frac{1}{2}(\hat{\boldsymbol{\theta}} - \boldsymbol{\theta}_K)^\mathsf{T}\boldsymbol{\mathcal{I}}_K(\hat{\boldsymbol{\theta}} - \boldsymbol{\theta}_K) \quad \text{(A.18)}$$

where $\boldsymbol{\mathcal{I}}_K$ is the information matrix at $\boldsymbol{\theta}_K$. The dependence on $\boldsymbol{\theta}_K$ is not helpful here, but can be removed by taking expectations over the distribution of $\hat{\boldsymbol{\theta}}$, which yields

$$\mathbb{E}\left\{K(f_{\hat{\theta}}, f_0)\right\} \simeq K(f_{\theta_K}, f_0) + p/2, \quad \text{(A.19)}$$

where it has been assumed that the model is close enough to correct that consistency ensures closeness of $\hat{\boldsymbol{\theta}}$ and $\boldsymbol{\theta}_K$ and the large sample distribution of $\hat{\boldsymbol{\theta}}$ implies that twice the last term in (A.18) has a χ^2_p distribution.

Now (A.19) still depends on f_0, and we need to be able to estimate it using what we have available, which does not include knowledge of f_0. A useful estimator of $K(f_{\theta_K}, f_0)$ can be based on $-l(\hat{\boldsymbol{\theta}}) = -\log[f_{\hat{\theta}}(\mathbf{y})]$, but to ensure that it is approximately unbiased, we need to consider $\mathbb{E}[-l(\hat{\boldsymbol{\theta}})]$, where the expectation is now taken allowing for the fact that $\hat{\boldsymbol{\theta}}$ is a function of $\mathbf{y}$.

$$\begin{aligned}
\mathbb{E}\{-l(\hat{\boldsymbol{\theta}})\} &= \mathbb{E}[-l(\boldsymbol{\theta}_K) - \{l(\hat{\boldsymbol{\theta}}) - l(\boldsymbol{\theta}_K)\}] \\
&\simeq -\int \log\{f_{\theta_K}(\mathbf{y})\} f_0(\mathbf{y}) d\mathbf{y} - p/2 \\
&= K(f_{\theta_K}, f_0) - p/2 - \int \log\{f_0(\mathbf{y})\} f_0(\mathbf{y}) d\mathbf{y}
\end{aligned}$$

where again it has been assumed that the model is close enough to correct that we can use the large sample result $2\{l(\hat{\boldsymbol{\theta}}) - l(\boldsymbol{\theta}_K)\} \sim \chi^2_p$. Re-arrangement yields the estimator

$$\widehat{K(f_{\theta_K}, f_0)} = -l(\hat{\boldsymbol{\theta}}) + p/2 + \int \log\{f_0(\mathbf{y})\} f_0(\mathbf{y}) d\mathbf{y},$$

which can be substituted into (A.19) to obtain the estimator

$$\mathbb{E}\left\{\widehat{K(f_{\hat{\theta}}, f_0)}\right\} \simeq -l(\hat{\boldsymbol{\theta}}) + p + \int \log\{f_0(\mathbf{y})\} f_0(\mathbf{y}) d\mathbf{y}.$$

The final term on the r.h.s. depends only on the unknown true model, and will be the same for any set of models to be compared using a given data set. Hence, it can safely be dropped, and $-l(\hat{\boldsymbol{\theta}}) + p$ can be used for model comparison purposes. Twice this quantity is known as the Akaike Information Criterion,

$$AIC = -2l(\hat{\boldsymbol{\theta}}) + 2p. \tag{A.20}$$

Generally we expect estimated models with lower AIC scores to be closer to the true model, in the K-L sense, than models with higher AIC scores. The p term in the AIC score penalizes models with more parameters than necessary, thereby counteracting the tendency of the likelihood to favour ever larger models. Notice how the two $p/2$ terms making up p both come from taking expectations of $(\hat{\boldsymbol{\theta}} - \boldsymbol{\theta}_K)^{\mathsf{T}} \boldsymbol{\mathcal{I}}_K (\hat{\boldsymbol{\theta}} - \boldsymbol{\theta}_K)/2$, the first directly and the second via (A.15). $\mathbb{E}(\hat{\boldsymbol{\theta}} - \boldsymbol{\theta}_K)^{\mathsf{T}} \boldsymbol{\mathcal{I}}_K (\hat{\boldsymbol{\theta}} - \boldsymbol{\theta}_K)$ is the key to finding appropriate effective degrees of freedom for AIC when there is penalization.

The foregoing derivation assumes that the estimated model is 'close' to the true model. If a sequence of nested models includes a model that is acceptably close, then all larger models will also be close. On the other hand, if we start to violate this assumption, by oversimplifying a model, then the resulting decrease in $l(\hat{\boldsymbol{\theta}})$ will typically be much larger than the decrease in p, resulting in a substantial drop in AIC score: basically the likelihood part of the AIC score is well able to discriminate between models that are oversimplified, and those that are not.

A derivation of AIC which deals more carefully with the case in which the model is 'wrong' is given in Davison (2003), from which the above derivation borrows heavily. A more careful accounting makes more precise the nature of the approximation made by using the penalty term, p, when the model is incorrect, but it doesn't make p any less approximate in that case.

A.8 Quasi-likelihood results

The results of sections A.3 to A.6 also apply if the log-likelihood l is replaced by the log quasi-likelihood q. The key to demonstrating this lies in deriving equivalents to results 1 to 4 of section A.2, for the log quasi-likelihood function. Again, for clarity, only a single parameter θ will be considered, but the results generalize.

Consider observations $y_1, y_2, \ldots, y_n$ on independent random variables, each with expectation μ_i. Given a single y_i, let the log quasi-likelihood of μ_i be $q_i(\mu_i)$ (as defined by (3.12) in section 3.1.8), and suppose that all the μ_i depend on a parameter θ, which therefore has log quasi-likelihood function

$$q(\theta) = \sum_{i=1}^{n} q_i(\mu_i).$$

Let θ_0 denote the true parameter value, $\mathbb{E}_0$ denote the expectation operator given that parameter value, and let μ_i^0 denote the true expected value of y_i.

Result 1:

$$\mathbb{E}_0 \left(\left. \frac{\partial q}{\partial \theta} \right|_{\theta_0} \right) = 0. \tag{A.21}$$

The proof is simple:

$$\mathbb{E}_0\left(\left.\frac{\partial q_i}{\partial\theta}\right|_{\theta_0}\right) = \mathbb{E}_0\left(\left.\frac{\partial q_i}{\partial\mu_i}\right|_{\mu_i^0}\right)\left.\frac{\partial\mu_i}{\partial\theta}\right|_{\theta_0} = \frac{\mathbb{E}_0(Y_i)-\mu_i^0}{\phi V(\mu_i^0)}\left.\frac{\partial\mu_i}{\partial\theta}\right|_{\theta_0} = 0,$$

from which the result follows immediately.

An obvious consequence of result 1 is the following.

Result 2:

$$\operatorname{var}\left(\left.\frac{\partial q}{\partial\theta}\right|_{\theta_0}\right) = \mathbb{E}_0\left\{\left(\left.\frac{\partial q}{\partial\theta}\right|_{\theta_0}\right)^2\right\}. \tag{A.22}$$

Furthermore it can be shown that

Result 3:

$$\mathcal{I}_q \equiv \mathbb{E}_0\left\{\left(\left.\frac{\partial q}{\partial\theta}\right|_{\theta_0}\right)^2\right\} = -\mathbb{E}_0\left\{\left.\frac{\partial^2 q}{\partial\theta^2}\right|_{\theta_0}\right\}, \tag{A.23}$$

where $\mathcal{I}_q$ might be termed the **quasi-information** about θ. The proof is straightforward.

$$\begin{aligned}
\mathbb{E}_0\left\{\left(\left.\frac{\partial q_i}{\partial\theta}\right|_{\theta_0}\right)^2\right\} &= \mathbb{E}_0\left\{\left(\left.\frac{\partial q_i}{\partial\mu_i}\right|_{\mu_i^0}\right)^2\right\}\left(\left.\frac{\partial\mu_i}{\partial\theta}\right|_{\theta_0}\right)^2 \\
&= \mathbb{E}_0\left\{\frac{(Y_i-\mu_i^0)^2}{\phi^2 V(\mu_i^0)^2}\right\}\left(\left.\frac{\partial\mu_i}{\partial\theta}\right|_{\theta_0}\right)^2 \\
&= \frac{1}{\phi V(\mu_i^0)}\left(\left.\frac{\partial\mu_i}{\partial\theta}\right|_{\theta_0}\right)^2,
\end{aligned}$$

which can be shown to be equal to $-\mathbb{E}_0\left(\partial^2 q_i/\partial\theta^2\big|_{\theta_0}\right)$ as follows:

$$\begin{aligned}
\mathbb{E}_0\left(\left.\frac{\partial^2 q_i}{\partial\theta^2}\right|_{\theta_0}\right) &= \mathbb{E}_0\left(\left.\frac{\partial^2 q_i}{\partial\mu_i^2}\right|_{\mu_i^0}\right)\left(\left.\frac{\partial\mu_i}{\partial\theta}\right|_{\theta_0}\right)^2 + \mathbb{E}_0\left(\left.\frac{\partial q_i}{\partial\mu_i}\right|_{\mu_i^0}\right)\left.\frac{\partial^2\mu_i}{\partial\theta^2}\right|_{\theta_0} \\
&= \mathbb{E}_0\left(\left.\frac{\partial^2 q_i}{\partial\mu_i^2}\right|_{\mu_i^0}\right)\left(\left.\frac{\partial\mu_i}{\partial\theta}\right|_{\theta_0}\right)^2 \\
&= \mathbb{E}_0\left(\frac{-1}{\phi V(\mu_i^0)} - \frac{Y_i-\mu_i^0}{\phi^2 V(\mu_i^0)^2}\left.\frac{\partial V}{\partial\mu}\right|_{\mu_i^0}\right)\left(\left.\frac{\partial\mu_i}{\partial\theta}\right|_{\theta_0}\right)^2 \\
&= \frac{-1}{\phi V(\mu_i^0)}\left(\left.\frac{\partial\mu_i}{\partial\theta}\right|_{\theta_0}\right)^2.
\end{aligned}$$

The result now follows easily given the independence of the Y_i and hence of the q_i.

Finally we have

Result 4:

$$\mathbb{E}_0\{q(\theta_0)\} \geq \mathbb{E}_0\{q(\theta)\} \;\forall\, \theta. \tag{A.24}$$

Proof is considerably easier than was the case for the equivalent likelihood result. By result 1 we know that $\mathbb{E}_0(q)$ has a turning point at θ_0, and result 3 implies that it must be a maximum (if the data contain any information at all about θ). Furthermore

$$\mathbb{E}_0\left(\frac{\partial q}{\partial \theta}\right) = \sum_{i=1}^{n} \mathbb{E}_0\left(\frac{\partial q_i}{\partial \mu_i}\right)\frac{\partial \mu_i}{\partial \theta} = \sum_{i=1}^{n} \frac{\mu_i^0 - \mu_i}{\phi V(\mu_i)}\frac{\partial \mu_i}{\partial \theta}$$

implying that q decreases monotonically away from θ_0 provided that each $\partial\mu_i/\partial\theta$ has the same sign for all θ (different μ_i can have derivatives of different sign, of course). The restriction is always met for a GLM, since the signs of these derivatives are controlled by the signs of the corresponding elements of the model matrix, which are fixed. Hence the maximum at θ_0 is global.

As in the likelihood case, these quasi-likelihood results can be generalized to vector parameters.

The consistency, large sample distribution and GLRT results of sections A.3 to A.6 can now be re-derived for models estimated by quasi-likelihood maximization. The arguments of sections A.3 to A.6 are unchanged except for the replacement of l by q, l_i by q_i, $\mathcal{I}$ by $\mathcal{I}_q$ and results 1 to 4 of section A.2 by results 1 to 4 of the current section.

Appendix B

Some Matrix Algebra

This appendix provides some useful matrix results, and a brief introduction to some relevant numerical linear algebra.

B.1 Basic computational efficiency

Consider the numerical evaluation of the expression

$$\mathbf{ABy}$$

where $\mathbf{A}$ and $\mathbf{B}$ are $n \times n$ matrices and $\mathbf{y}$ is an n-vector. The following code evaluates this expression in two ways (wrong and right).

```
> n <- 1000
> A <- matrix(runif(n*n),n,n)
> B <- matrix(runif(n*n),n,n)
> y <- runif(n)
> system.time((A%*%B)%*%y)     # wrong way
[1] 5.60 0.01 5.61    NA    NA
> system.time(A%*%(B%*%y))     # right way
[1] 0.02 0.00 0.02    NA    NA
```

So, in this case, the 'wrong way' took 5.6 seconds and the 'right way' less than 0.02 seconds. Why? The wrong way first multiplied $\mathbf{A}$ and $\mathbf{B}$ at a cost of $2n^3$ floating point operations, and then multiplied $\mathbf{AB}$ by $\mathbf{y}$ at the cost of $2n^2$ further operations: total cost $2n^3 + 2n^2$ operations. The right way first formed the vector $\mathbf{By}$ at a cost of $2n^2$ operations and then formed $\mathbf{ABy}$ at a further cost of $2n^2$ operations: total cost $4n^2$ operations. So we should have got around a 500-fold speed up by doing things the right way.

This sort of effect is pervasive: the order in which matrix operations are performed can have a major impact on computational speed. An important example is the evaluation of the trace of a matrix (the sum of the leading diagonal). It is easy to show that $\text{tr}(\mathbf{AB}) = \text{tr}(\mathbf{BA})$, and for non-square matrices, one of these is always cheaper to evaluate than the other. Here is an example

```
> m <- 500;n<-5000
> A <- matrix(runif(m*n),n,m)
> B <- matrix(runif(m*n),m,n)
> system.time(sum(diag(A%*%B)))
```

```
[1] 51.61  0.16 51.93    NA    NA
> system.time(sum(diag(B%*%A)))
[1] 4.92 0.00 4.92   NA   NA
```

In this case formation of $\mathbf{AB}$ costs $2n^2m$ operations, while formation of $\mathbf{BA}$ costs only $2m^2n$: so in the example, the second evaluation is 10 times faster than the first. Actually even the second evaluation is wasteful, as the following code is easily demonstrated to evaluate the trace at a cost of only $2mn$ operations

```
> system.time(sum(t(B)*A))
[1] 0.11 0.00 0.11   NA   NA
```

(`rowSums(t(B)*A)` obtains all the elements on the leading diagonal). Simple tricks like these can lead to very large efficiency gains in matrix computations.

B.2 Covariance matrices

If $\mathbf{X}$ is a random vector and $\boldsymbol{\mu}_x = \mathbb{E}(\mathbf{X})$, then the *covariance matrix* of $\mathbf{X}$ is

$$\mathbf{V}_x = \mathbb{E}\left\{(\mathbf{X} - \boldsymbol{\mu}_x)(\mathbf{X} - \boldsymbol{\mu}_x)^\mathsf{T}\right\}.$$

$\mathbf{V}_x$ is symmetric and positive semi-definite (all its eigenvalues are ≥ 0). The ith element on the leading diagonal of $\mathbf{V}_x$ is the variance of X_i, while the $(i,j)^{\text{th}}$ element is the covariance of X_i and X_j.

If $\mathbf{C}$ is a matrix of fixed real coefficients and $\mathbf{Y} = \mathbf{CX}$ then the covariance matrix of $\mathbf{Y}$ is

$$\mathbf{V}_y = \mathbf{C}\mathbf{V}_x\mathbf{C}^\mathsf{T}.$$

This result is sometimes known as the 'law of propagation of errors'. It is easily proven. First note that $\boldsymbol{\mu}_y \equiv \mathbb{E}(\mathbf{Y}) = \mathbf{C}\mathbb{E}(\mathbf{X}) = \mathbf{C}\boldsymbol{\mu}_x$. Then

$$\begin{aligned}
\mathbf{V}_y &= \mathbb{E}\left\{(\mathbf{Y} - \boldsymbol{\mu}_y)(\mathbf{Y} - \boldsymbol{\mu}_y)^\mathsf{T}\right\} \\
&= \mathbb{E}\left\{(\mathbf{CX} - \mathbf{C}\boldsymbol{\mu}_x)(\mathbf{CX} - \mathbf{C}\boldsymbol{\mu}_x)^\mathsf{T}\right\} \\
&= \mathbf{C}\mathbb{E}\left\{(\mathbf{X} - \boldsymbol{\mu}_x)(\mathbf{X} - \boldsymbol{\mu}_x)^\mathsf{T}\right\}\mathbf{C}^\mathsf{T} \\
&= \mathbf{C}\mathbf{V}_x\mathbf{C}^\mathsf{T}.
\end{aligned}$$

B.3 Differentiating a matrix inverse

Consider differentiation of $\mathbf{A}^{-1}$ w.r.t. x. By definition of a matrix inverse we have that $\mathbf{I} = \mathbf{A}^{-1}\mathbf{A}$. Differentiating this expression w.r.t. x gives

$$\mathbf{0} = \frac{\partial \mathbf{A}^{-1}}{\partial x}\mathbf{A} + \mathbf{A}^{-1}\frac{\partial \mathbf{A}}{\partial x}$$

which re-arranges to

$$\frac{\partial \mathbf{A}^{-1}}{\partial x} = -\mathbf{A}^{-1}\frac{\partial \mathbf{A}}{\partial x}\mathbf{A}^{-1}.$$

B.4 Kronecker product

The Kronecker product of $n \times m$ matrix $\mathbf{A}$ and $p \times q$ matrix $\mathbf{B}$ is the $np \times qm$ matrix

$$\mathbf{A} \otimes \mathbf{B} = \begin{bmatrix} A_{11}\mathbf{B} & A_{12}\mathbf{B} & . & . \\ A_{21}\mathbf{B} & A_{22}\mathbf{B} & . & . \\ . & . & . & . \\ . & . & . & . \end{bmatrix}.$$

In R this is implemented by the operator `%x%`: see `?kronecker` for details.

The construction of tensor product smooths involves taking the Kronecker product row-wise, so that if $\mathbf{C}_i$ denotes the ith row of $n \times p$ matrix $\mathbf{C}$, then the $n \times mp$ row Kronecker product is

$$\mathbf{A} \odot \mathbf{C} = \begin{bmatrix} A_{11}\mathbf{C}_1 & A_{12}\mathbf{C}_1 & . & . \\ A_{21}\mathbf{C}_2 & A_{22}\mathbf{C}_2 & . & . \\ . & . & . & . \\ . & . & . & . \end{bmatrix}.$$

B.5 Orthogonal matrices and Householder matrices

Orthogonal matrices are square matrices which rotate/reflect vectors and have a variety of interesting and useful properties. If $\mathbf{Q}$ is an orthogonal matrix then $\mathbf{Q}\mathbf{Q}^{\mathsf{T}} = \mathbf{Q}^{\mathsf{T}}\mathbf{Q} = \mathbf{I}$, i.e. $\mathbf{Q}^{-1} = \mathbf{Q}^{\mathsf{T}}$. If $\mathbf{x}$ is any vector of appropriate dimension then $\|\mathbf{Q}\mathbf{x}\| = \|\mathbf{x}\|$.* That is, $\mathbf{Q}$ changes the elements of $\mathbf{x}$ without changing its length: i.e. it rotates/reflects $\mathbf{x}$. It follows that all the eigenvalues of $\mathbf{Q}$ have magnitude 1.

A particularly important class of orthogonal matrices are the Householder (or 'reflector') matrices. Let $\mathbf{u}$ be a non-zero vector. Then

$$\mathbf{H} = \mathbf{I} - \gamma\mathbf{u}\mathbf{u}^{\mathsf{T}} \text{ where } \gamma = 2/\|\mathbf{u}\|^2$$

is a *Householder matrix* or *reflector matrix*. Note that $\mathbf{H}$ is symmetric.

The multiplication of a vector by a householder matrix is a prime example of the need to think carefully about the order of operations when working with matrices. $\mathbf{H}$ is never stored as a complete matrix, but rather $\mathbf{u}$ and γ are stored. Then $\mathbf{H}\mathbf{x}$ is evaluated as follows (overwriting $\mathbf{x}$ by $\mathbf{H}\mathbf{x}$).

1. $\alpha \leftarrow \mathbf{u}^{\mathsf{T}}\mathbf{x}$
2. $\mathbf{x} \leftarrow \mathbf{x} - \alpha\gamma\mathbf{u}$

Notice how this requires only $O(n)$ operations, where $n = \dim(\mathbf{x})$ — explicit formation and use of $\mathbf{H}$ is $O(n^2)$.

Householder matrices are important, because if $\mathbf{x}$ and $\mathbf{y}$ are two non-zero vectors of the same length (i.e., $\|\mathbf{x}\| = \|\mathbf{y}\|$), then the Householder matrix such that $\mathbf{H}\mathbf{x} = \mathbf{y}$ is obtained by setting $\mathbf{u} = \mathbf{x} - \mathbf{y}$. This property is exploited in the next section, but note that in practical computation it is important to guard against possible overflow or cancelation error when using these matrices: see, e.g., Watkins (1991, section 3.2) for further details.

---

*Recall that $\|\mathbf{x}\|^2 = \mathbf{x}^{\mathsf{T}}\mathbf{x}$.

B.6 QR decomposition

Any real $n \times m$ ($m \le n$, here) matrix $\mathbf{X}$ can be written as

$$\mathbf{X} = \mathbf{Q}\begin{bmatrix} \mathbf{R} \\ \mathbf{0} \end{bmatrix}$$

where $\mathbf{Q}$ is an orthogonal matrix, and $\mathbf{R}$ is an $m \times m$ upper triangular matrix. $\mathbf{Q}$ can be made up of the product of m Householder matrices. Here is how.

First construct the Householder matrix $\mathbf{H}_1$ which reflects/rotates the first column of $\mathbf{x}$ in such a way that all its elements except the first are zeroed. For example,

$$\mathbf{H}_1 \begin{bmatrix} x_{11} & x_{12} & x_{13} & \cdot \\ x_{21} & x_{22} & x_{23} & \cdot \\ x_{31} & x_{32} & x_{33} & \cdot \\ \cdot & \cdot & \cdot & \cdot \end{bmatrix} = \begin{bmatrix} x'_{11} & x'_{12} & x'_{13} & \cdot \\ 0 & x'_{22} & x'_{23} & \cdot \\ 0 & x'_{32} & x'_{33} & \cdot \\ \cdot & \cdot & \cdot & \cdot \end{bmatrix}.$$

x'_{11} will be the length of the first column of $\mathbf{X}$. The next step is to calculate the Householder matrix, $\mathbf{H}_2$, which will transform the second column of $\mathbf{X}$, so that its first element is unchanged, and every element beyond the second is zeroed. e.g.,

$$\mathbf{H}_2 \begin{bmatrix} x'_{11} & x'_{12} & x'_{13} & \cdot \\ 0 & x'_{22} & x'_{23} & \cdot \\ 0 & x'_{32} & x'_{33} & \cdot \\ \cdot & \cdot & \cdot & \cdot \end{bmatrix} = \begin{bmatrix} x'_{11} & x'_{12} & x'_{13} & \cdot \\ 0 & x''_{22} & x''_{23} & \cdot \\ 0 & 0 & x''_{33} & \cdot \\ \cdot & \cdot & \cdot & \cdot \end{bmatrix}.$$

Notice that the fact that $\mathbf{H}_2$ is only acting on elements from the second row down has two implications: (i) the first row is unchanged by $\mathbf{H}_2$ and (ii) the first column is unchanged, since in this case $\mathbf{H}_2$ simply transforms zero to zero.

Continuing in the same way we eventually reach the situation where

$$\mathbf{H}_m\mathbf{H}_{m-1}\cdots\mathbf{H}_1\mathbf{X} = \begin{bmatrix} \mathbf{R} \\ \mathbf{0} \end{bmatrix}.$$

Hence $\mathbf{Q}^\mathsf{T} = \mathbf{H}_m\mathbf{H}_{m-1}\cdots\mathbf{H}_1$ implying that $\mathbf{Q} = \mathbf{H}_1\mathbf{H}_2\cdots\mathbf{H}_m$.

Note that it is quite common to write the QR decomposition as $\mathbf{X} = \mathbf{QR}$ where $\mathbf{Q}$ is the first m columns of the orthogonal matrix just constructed. Whichever convention is used, $\mathbf{Q}$ is always stored as a series of m Householder matrix components. See `?qr` in R for details about using QR decompositions in R.

B.7 Cholesky decomposition

The Cholesky decomposition is a very efficient way of finding the 'square root' of a positive definite matrix. A positive definite matrix, $\mathbf{D}$, is one for which $\mathbf{x}^\mathsf{T}\mathbf{D}\mathbf{x} > 0$ for *any* non-zero vector $\mathbf{x}$ (of appropriate dimension). Positive definite matrices are symmetric and have strictly positive eigenvalues.

The Cholesky decomposition of $\mathbf{D}$ is

$$\mathbf{D} = \mathbf{R}^\mathsf{T}\mathbf{R}$$

where $\mathbf{R}$ is an upper triangular matrix of the same dimension as $\mathbf{D}$. The algorithm for constructing the Cholesky factor, $\mathbf{R}$, is quite simple. Using the convention that $\sum_{k=1}^{0} x_i \equiv 0$, we have

$$R_{ii} = \sqrt{D_{ii} - \sum_{k=1}^{i-1} R_{ki}^2}, \text{ and } R_{ij} = \frac{D_{ij} - \sum_{k=1}^{i-1} R_{ki} R_{kj}}{R_{ii}}, \quad j > i.$$

Working through these equations in row order, from row one, and starting each row from its leading diagonal component, ensures that all right-hand-side quantities are known at each step (see Golub and Van Loan, 2013; Watkins, 1991, for more). Occasionally it is useful to have the derivative of the Cholesky factor, and by applying the chain rule we obtain

$$\frac{\partial R_{ii}}{\partial \rho} = \frac{1}{2} R_{ii}^{-1} B_{ii}, \quad \frac{\partial R_{ij}}{\partial \rho} = R_{ii}^{-1} \left(B_{ij} - R_{ij} \frac{\partial R_{ii}}{\partial \rho} \right), \text{ where}$$

$$B_{ij} = \frac{\partial D_{ij}}{\partial \rho} - \sum_{k=1}^{i-1} \frac{\partial R_{ki}}{\partial \rho} R_{kj} + R_{ki} \frac{\partial R_{kj}}{\partial \rho}.$$

The Cholesky decomposition costs around $n^3/3$ operations for an $n \times n$ matrix, which makes it a rather efficient way of solving linear systems involving positive definite matrices. For example, the $\mathbf{x}$ satisfying $\mathbf{Dx} = \mathbf{y}$, where $\mathbf{D}$ is positive definite, is the $\mathbf{x}$ satisfying

$$\mathbf{R}^{\mathsf{T}}\mathbf{Rx} = \mathbf{y}$$

where $\mathbf{R}$ is the Cholesky factor of $\mathbf{D}$. Because $\mathbf{R}$ is triangular it is cheap to solve for $\mathbf{Rx}$ and then for $\mathbf{x}$. The following code provides an example in R.

```
n <- 1000
D <- matrix(runif(n*n),n,n)
D <- D%*%t(D)   # create a +ve def. matrix
y <- runif(n)   # and a y
## now solve Dx=y, efficiently, using the Cholesky factor of D
R <- chol(D)
x <- backsolve(R,y,transpose=TRUE)
x <- backsolve(R,x)
```

B.8 Pivoting

The stability of computations involving QR and Cholesky decompositions can be improved by *pivoting*, that is by the re-ordering of the columns (and in the Cholesky case also rows) of the original matrix, in such a way that its QR or Cholesky factorisation has favourable properties. A factorisation of the original matrix can always be recovered by 'un-pivoting' the columns of the triangular factor (although the factor then loses its triangular structure, of course).

Pivoting is performed iteratively as a decomposition progresses. In the QR case the column with the largest Euclidean norm (square root of sum of squares) is first swapped into the first column. After the first Householder reflection has been applied then the norms of the remaining columns below the first row are computed, and the (whole) column with the largest such norm is swapped into the second column. The next Householder reflection is then applied, and the norms of the remaining columns, below the second row, are computed, and so on. A consequence of this approach is that any rank deficiency is manifested as a trailing zero block of the triangular factor, giving a way of testing for rank deficiency.

In the Cholesky case pivoting allows the decomposition to be applied to positive semi-definite matrices, since it is possible to avoid trying to divide by zero, until an acceptable factor has already been found. In either case, solves involving the pivoted triangular factors are usually slightly more stable than solves involving unpivoted factors.

Here is an example in R, of using pivoting to find the Cholesky factor of a matrix that is only positive *semi*-definite.

```
> B <- matrix(runif(200*100),200,100)
> B <- B%*%t(B)    ## B is now +ve semi-definite...
> chol(B)          ## ... so un-pivoted Cholesky fails
Error in chol(B) : the leading minor of order 101 is
not positive definite
> R <- chol(B, pivot = TRUE)      ## ok with pivoting
> piv <- order(attr(R, "pivot")) ## extract pivot index
> R <- R[, piv]       ## unscramble pivoting
> range(B-t(R)%*%R)   ## check that result is valid sqrt
[1] -4.867218e-12  4.444445e-12 ## it is!
```

Of course, what is lost here is that $\mathbf{R}$ is no longer upper triangular.

B.9 Eigen-decomposition

Consider an $n \times n$ matrix $\mathbf{A}$. There exist n scalars λ_i and n corresponding vectors $\mathbf{u}_i$ such that

$$\mathbf{A}\mathbf{u}_i = \lambda_i \mathbf{u}_i,$$

i.e., multiplication of $\mathbf{u}_i$ by $\mathbf{A}$ results in a scalar multiple of $\mathbf{u}_i$. The λ_i are known as the *eigenvalues* (own-values) of $\mathbf{A}$, and the corresponding $\mathbf{u}_i$ are the *eigenvectors* (strictly 'right eigenvectors', since 'left eigenvectors' can also be defined).

In this book we only need eigenvalues and eigenvectors of symmetric matrices; so suppose that $\mathbf{A}$ is symmetric. In that case it is possible to write

$$\mathbf{A} = \mathbf{U}\mathbf{\Lambda}\mathbf{U}^\mathsf{T}$$

where the eigenvectors, $\mathbf{u}_i$, are the columns of the $n \times n$ orthogonal matrix $\mathbf{U}$, while the eigenvalues, λ_i, provide the elements of the diagonal matrix $\mathbf{\Lambda}$. Such a decomposition is known as an *eigen-decomposition* or *spectral decomposition*.

The eigen-decomposition of $\mathbf{A}$ provides an interesting geometric interpretation of how any n vector, $\mathbf{x}$, is transformed by multiplication by $\mathbf{A}$. The product $\mathbf{Ax}$ can be written as $\mathbf{U}\mathbf{\Lambda}\mathbf{U}^\mathsf{T}\mathbf{x}$: so $\mathbf{U}^\mathsf{T}$ first rotates $\mathbf{x}$ into the 'eigenspace' of $\mathbf{A}$; the resulting vector then has each of its elements multiplied by an eigenvalue of $\mathbf{A}$, before being rotated back into the original space, by $\mathbf{U}$.

It is easy to see that for any integer m, $\mathbf{A}^m = \mathbf{U}\boldsymbol{\Lambda}^m\mathbf{U}^\mathsf{T}$, which, given the eigen-decomposition, is rather cheap to evaluate, since $\boldsymbol{\Lambda}$ is diagonal. This result also means that any function with a power series expansion has a natural generalization to the matrix setting. For example, the exponential of a matrix would be

$$\exp(\mathbf{A}) = \mathbf{U}\exp(\boldsymbol{\Lambda})\mathbf{U}^\mathsf{T}$$

where $\exp(\boldsymbol{\Lambda})$ is the diagonal matrix with $\exp(\lambda_i)$ as the ith element on its leading diagonal.

Here are some other useful results related to eigenvalues:

- The 'rank' of a matrix is the number of non-zero eigenvalues that it possesses. The rank is the number of independent rows/columns of a matrix.
- A symmetric matrix is positive definite if all its eigenvalues are strictly positive. i.e., $\mathbf{x}^\mathsf{T}\mathbf{A}\mathbf{x} > 0 \; \forall \; \mathbf{x} \neq \mathbf{0}$.
- A symmetric matrix is positive semi-definite if all its eigenvalues are non-negative. i.e., $\mathbf{x}^\mathsf{T}\mathbf{A}\mathbf{x} \geq 0 \; \forall \; \mathbf{x} \neq \mathbf{0}$.
- The trace of a matrix is also the sum of its eigenvalues.
- The determinant of a matrix is the product of its eigenvalues.

Numerical calculation of eigenvalues and eigenvectors is quite a specialized topic. Algorithms usually start by reducing the matrix to tri-diagonal or bi-diagonal form by successive application of Householder rotations from the left and right. By orthogonality of the Householder rotations, the reduced matrix has the same eigenvalues as the original matrix. The eigenvalues themselves are then found by an iterative procedure, usually the QR-algorithm (not to be confused with the QR decomposition!); see Golub and Van Loan (2013) or Watkins (1991) for details. See `?eigen` for details of the eigen-routines available in R. Sometimes only the largest magnitude eigenvalues and corresponding eigenvectors are required, in which case see section B.11.

B.10 Singular value decomposition

The singular value decomposition is rather like the eigen-decomposition, except that it can be calculated for any real matrix (not just square ones), and its components are all real (eigenvalues and vectors can be complex for non-symmetric matrices). If $\mathbf{X}$ is an $n \times m$ real matrix ($m \leq n$, say) then it is possible to write

$$\mathbf{X} = \mathbf{U}\mathbf{D}\mathbf{V}^\mathsf{T}$$

where the columns of the $n \times m$ matrix $\mathbf{U}$ are the first m columns of an orthogonal matrix (so that $\mathbf{U}^\mathsf{T}\mathbf{U} = \mathbf{I}_m$, but $\mathbf{U}\mathbf{U}^\mathsf{T} \neq \mathbf{I}_n$), $\mathbf{V}$ is an $m \times m$ orthogonal matrix and $\mathbf{D}$ is a diagonal matrix, the diagonal elements of which are the *singular values* of $\mathbf{X}$.

The ith singular value of $\mathbf{X}$ is the positive square root of the ith eigenvalue of $\mathbf{X}^\mathsf{T}\mathbf{X}$, so for a positive semi-definite matrix, the eigenvalues and singular values are the same (for a symmetric matrix the magnitudes will be the same, but signs may differ). Singular values are so called, because there is a sense in which they indicate the 'distance' between a matrix and the 'nearest' singular matrix. The ratio of the largest to smallest singular values is known as the 'condition number' of a matrix, and provides a useful guide to the numerical stability of calculations involving the matrix: again, see Golub and Van Loan (2013) or Watkins (1991) for details.

The number of non-zero singular values gives the rank of a matrix, and the singular value decomposition is the most reliable method for numerically estimating the rank of a matrix.

Again, the calculation of the singular value decomposition starts by transforming the matrix to a bi-diagonal form, after which iteration is used to find the singular values themselves (once again, see Golub and Van Loan, 2013; Watkins, 1991).

B.11 Lanczos iteration

Lanczos iteration (see, e.g., Demmel, 1997) is a method which can be used to obtain the rank k truncated eigen-decomposition of a symmetric matrix, $\mathbf{E}$, in $O(kn^2)$ operations, by iteratively building up a tri-diagonal matrix whose eigenvalues converge (in order of decreasing magnitude) to those required, as iteration proceeds.

The algorithm is iterative, and at the ith iteration produces an $(i \times i)$ symmetric tri-diagonal matrix ($\mathbf{K}_i$, say), the eigenvalues of which approximate the i largest magnitude eigenvalues of the original matrix: these eigenvalues converge as the iteration proceeds, with those of largest magnitude converging first. The eigenvalues and vectors of $\mathbf{K}_i$ can be obtained in order i^3 operations, using the usual QR algorithm[†] with accumulation of the eigenvectors. In principle it should be possible to get away with $O(i^2)$ operations by only using the QR algorithm to find the eigenvalues and then inverse iteration to find the eigenvectors, but I experienced some stability problems when using a simple implementation of inverse iteration to do this, particularly for thin plate regression splines of one predictor variable. In any case $i^3 \ll n^2$ in most cases. The eigenvectors of the original matrix are easily obtained from the eigenvectors of $\mathbf{K}_i$.

A complete version of the algorithm, suitable for finding the truncated decomposition of $\mathbf{E}$, is as follows.

1. Let $\mathbf{b}$ be an arbitrary non-zero n vector.[‡]
2. Set $\mathbf{q}_1 \leftarrow \mathbf{b}/\|\mathbf{b}\|$.
3. Repeat steps (4) to (12) for $j = 1, 2, \ldots$ until enough eigenvectors have converged.
4. Form $\mathbf{c} \leftarrow \mathbf{E}\mathbf{q}_j$.
5. Calculate $\gamma_j \leftarrow \mathbf{q}_j^{\mathsf{T}}\mathbf{c}$.

[†]Not to be confused with the QR decomposition: they are completely different things.

[‡]It may be best to initialize this from a simple random number generator, to reduce the risk of starting out orthogonal to some eigenvector (exact repeatability can be ensured by starting from the same random number generator seed).

6. Reorthogonalize $\mathbf{c}$ to ensure numerical stability, by performing the following step **twice**:
$$\mathbf{c} \leftarrow \mathbf{c} - \sum_{i=1}^{j-1} (\mathbf{c}^{\mathsf{T}} \mathbf{q}_i) \mathbf{q}_i.$$
7. Set $\xi_j \leftarrow \|\mathbf{c}\|$.
8. Set $\mathbf{q}_{j+1} \leftarrow \mathbf{c}/\xi_j$.
9. Let $\mathbf{K}_j$ be the $(j \times j)$ tri-diagonal matrix with $\gamma_1, \ldots \gamma_j$ on the leading diagonal, and $\xi_1, \ldots \xi_{j-1}$ on the leading sub- and super-diagonals.
10. If iteration has proceeded far enough to make it worthwhile, find the eigen-decomposition (spectral decomposition) $\mathbf{K}_j = \mathbf{V}\boldsymbol{\Lambda}\mathbf{V}^{\mathsf{T}}$, where the columns of $\mathbf{V}$ are eigenvectors of $\mathbf{K}_j$ and $\boldsymbol{\Lambda}$ is diagonal with eigenvalues on leading diagonal.
11. Compute 'error bounds' for each $\Lambda_{i,i}$: $|\xi_j V_{i,j}|$.
12. Use the error bounds to test for convergence of the k largest magnitude eigenvalues. Terminate the loop if all are converged.
13. The ith eigenvalue of $\mathbf{E}$ is $\Lambda_{i,i}$. The ith eigenvector of $\mathbf{E}$ is $\mathcal{Q}\mathbf{v}_i$, where $\mathcal{Q}$ is the matrix whose columns are the $\mathbf{q}_j$ (for all j calculated) and $\mathbf{v}_i$ is the ith column of $\mathbf{V}$ (again calculated at the final iteration). Hence $\mathbf{D}_k$ and $\mathbf{U}_k$ can easily be formed.

The given algorithm is stabilised by orthogonalisation against all previous vectors $\mathbf{q}_j$: several selective orthogonalisation schemes have been proposed to reduce the computational burden of this step, but I experienced convergence problems when trying to use these schemes with $\mathbf{E}$ from the single covariate TPRS problem. In any case the computational cost of the method is dominated by the $O(n^2)$ step: $\mathbf{c} \leftarrow \mathbf{E}\mathbf{q}_j$, so the efficiency benefits of using a selective method are unlikely to be very great, in the TPRS case (if $\mathbf{E}$ were sparse then selective methods would offer a benefit).

Finally note that $\mathbf{E}$ need not actually be formed and stored as a whole: it is only necessary that its product with a vector can be formed. For a fuller treatment of the Lanczos method see Demmel (1997), from which the algorithm given here has been modified.

Appendix C

Solutions to Exercises

Solution R code can be found in package `gamair` in the examples sections of help files with names like `ch1.solutions`.

C.1 Chapter 1

1. $t_i = \beta d_i + \epsilon_i$ is a reasonable model. So the least squares estimate of β is

$$\hat{\beta} = \frac{\sum_i x_i y_i}{\sum_i x_i^2} = \frac{0.1 + 1.2 + 2 + 3}{1 + 9 + 16 + 25} = \frac{6.3}{51} \simeq .1235$$

   implying a speed of $51/6.3 \simeq 8.1$ kilometres per hour.

2.

$$\frac{\partial}{\partial \beta} \sum_{i=1}^{n} (y_i - \beta)^2 = -2 \sum_{i=1}^{n} (y_i - \beta) = 0 \Rightarrow \hat{\beta} = \bar{y}.$$

3. None of these.

4.(a)

$$\begin{bmatrix} y_{11} \\ y_{12} \\ y_{21} \\ y_{22} \\ y_{31} \\ y_{32} \end{bmatrix} = \begin{bmatrix} 1 & 0 & 0 \\ 1 & 0 & 0 \\ 1 & 1 & 0 \\ 1 & 1 & 0 \\ 1 & 0 & 1 \\ 1 & 0 & 1 \end{bmatrix} \begin{bmatrix} \alpha \\ \beta_2 \\ \beta_3 \end{bmatrix} + \begin{bmatrix} \epsilon_{11} \\ \epsilon_{12} \\ \epsilon_{21} \\ \epsilon_{22} \\ \epsilon_{31} \\ \epsilon_{32} \end{bmatrix}.$$

   The constraint $\beta_1 = 0$ ensures model identifiability (i.e., full column rank of $\mathbf{X}$): any other constraint doing the same is ok.

(b)

$$\begin{bmatrix} y_{11} \\ y_{12} \\ y_{13} \\ y_{14} \\ y_{21} \\ y_{22} \\ y_{23} \\ y_{24} \\ y_{31} \\ y_{32} \\ y_{33} \\ y_{34} \end{bmatrix} = \begin{bmatrix} 1 & 0 & 0 & 0 & 0 & 0 \\ 1 & 0 & 0 & 1 & 0 & 0 \\ 1 & 0 & 0 & 0 & 1 & 0 \\ 1 & 0 & 0 & 0 & 0 & 1 \\ 1 & 1 & 0 & 0 & 0 & 0 \\ 1 & 1 & 0 & 1 & 0 & 0 \\ 1 & 1 & 0 & 0 & 1 & 0 \\ 1 & 1 & 0 & 0 & 0 & 1 \\ 1 & 0 & 1 & 0 & 0 & 0 \\ 1 & 0 & 1 & 1 & 0 & 0 \\ 1 & 0 & 1 & 0 & 1 & 0 \\ 1 & 0 & 1 & 0 & 0 & 1 \end{bmatrix} \begin{bmatrix} \alpha \\ \beta_2 \\ \beta_3 \\ \gamma_2 \\ \gamma_3 \\ \gamma_4 \end{bmatrix} + \begin{bmatrix} \epsilon_{11} \\ \epsilon_{12} \\ \epsilon_{13} \\ \epsilon_{14} \\ \epsilon_{21} \\ \epsilon_{22} \\ \epsilon_{23} \\ \epsilon_{24} \\ \epsilon_{31} \\ \epsilon_{32} \\ \epsilon_{33} \\ \epsilon_{34} \end{bmatrix}.$$

Here $\beta_1 = \gamma_1 = 0$ have been chosen as constraints ensuring identifiability. Of course, there are many other possibilities (depending on what contrasts are of most interest).

(c)

$$\begin{bmatrix} y_1 \\ y_1 \\ y_3 \\ y_4 \\ y_5 \\ y_6 \end{bmatrix} = \begin{bmatrix} 1 & 0 & 0.1 \\ 1 & 0 & 0.4 \\ 1 & 1 & 0.5 \\ 1 & 1 & 0.3 \\ 1 & 1 & 0.4 \\ 1 & 1 & 0.7 \end{bmatrix} \begin{bmatrix} \alpha \\ \beta_2 \\ \gamma \end{bmatrix} + \begin{bmatrix} \epsilon_1 \\ \epsilon_2 \\ \epsilon_3 \\ \epsilon_4 \\ \epsilon_5 \\ \epsilon_6 \end{bmatrix}.$$

5. The situation described in the question translates into a model of the form:

$$\mu_i = \begin{cases} kx_i + \alpha x_i^2 & \text{alloy 1} \\ kx_i + \beta x_i^2 & \text{alloy 2} \\ kx_i + \gamma x_i^2 & \text{alloy 3} \end{cases}$$

where $y_i \sim N(\mu_i, \sigma^2)$. Written out in full matrix-vector form this is

$$\begin{bmatrix} y_1 \\ y_2 \\ y_3 \\ y_4 \\ y_5 \\ y_6 \\ y_7 \\ y_8 \\ y_9 \\ y_{10} \\ y_{11} \\ y_{12} \\ y_{13} \\ y_{14} \\ y_{15} \\ y_{16} \\ y_{17} \\ y_{18} \end{bmatrix} = \begin{bmatrix} x_1 & x_1^2 & 0 & 0 \\ x_2 & x_2^2 & 0 & 0 \\ x_3 & x_3^2 & 0 & 0 \\ x_4 & x_4^2 & 0 & 0 \\ x_5 & x_5^2 & 0 & 0 \\ x_6 & x_6^2 & 0 & 0 \\ x_1 & 0 & x_1^2 & 0 \\ x_2 & 0 & x_2^2 & 0 \\ x_3 & 0 & x_3^2 & 0 \\ x_4 & 0 & x_4^2 & 0 \\ x_5 & 0 & x_5^2 & 0 \\ x_6 & 0 & x_6^2 & 0 \\ x_1 & 0 & 0 & x_1^2 \\ x_2 & 0 & 0 & x_2^2 \\ x_3 & 0 & 0 & x_3^2 \\ x_4 & 0 & 0 & x_4^2 \\ x_5 & 0 & 0 & x_5^2 \\ x_6 & 0 & 0 & x_6^2 \end{bmatrix} \begin{bmatrix} k \\ \alpha \\ \beta \\ \gamma \end{bmatrix} + \begin{bmatrix} \epsilon_1 \\ \epsilon_2 \\ \epsilon_3 \\ \epsilon_4 \\ \epsilon_5 \\ \epsilon_6 \\ \epsilon_7 \\ \epsilon_8 \\ \epsilon_9 \\ \epsilon_{10} \\ \epsilon_{11} \\ \epsilon_{12} \\ \epsilon_{13} \\ \epsilon_{14} \\ \epsilon_{15} \\ \epsilon_{16} \\ \epsilon_{17} \\ \epsilon_{18} \end{bmatrix}.$$

6.
$$\hat{\boldsymbol{\mu}} = \mathbf{X}(\mathbf{X}^\mathsf{T}\mathbf{X})^{-1}\mathbf{X}^\mathsf{T}\mathbf{y} \Rightarrow \mathbf{X}^\mathsf{T}\hat{\boldsymbol{\mu}} = \mathbf{X}^\mathsf{T}\mathbf{X}(\mathbf{X}^\mathsf{T}\mathbf{X})^{-1}\mathbf{X}^\mathsf{T}\mathbf{y} = \mathbf{X}^\mathsf{T}\mathbf{y}.$$

   If the model has an intercept then one column of $\mathbf{X}$ is a column of ones, so that one of the equations in $\mathbf{X}^\mathsf{T}\hat{\boldsymbol{\mu}} = \mathbf{X}^\mathsf{T}\mathbf{y}$ is simply $\sum_i \hat{\mu}_i = \sum_i y_i$. Re-arrangement of this shows that the residuals must sum to zero.

7. Given that $\mathbb{E}(r_i^2) = \sigma^2$,

$$\mathbb{E}(\|\mathbf{r}\|^2) = \sum_{i=0}^{n-p} \mathbb{E}(r_i^2) = (n-p)\sigma^2,$$

   and the result follows.

8. See `ch1.solutions` in package `gamair`.
9. See `ch1.solutions` in package `gamair`.
10. For (a) and (b) see `ch1.solutions` in package `gamair`.
    (c) The usual argument is really nonsense, and seems to result from confusing continuous and factor variables. In the current case it would condemn us to leaving in the constant, and tolerating very high estimator uncertainty, even though hypothesis testing, AIC, and the physical mechanisms underlying the data, all suggest dropping it.
11. See `ch1.solutions` in package `gamair`.
12. See `ch1.solutions` in package `gamair`.
13. See `ch1.solutions` in package `gamair`.

C.2 Chapter 2

1.(a)

$$y_{ij} = \alpha + b_i + \epsilon_{ij}, \quad b_i \sim N(0, \sigma_b^2) \quad \epsilon_{ij} \sim N(0, \sigma^2),$$

where y_{ij} is the weight of the jth piglet in the ith litter and the random variables b_i and ϵ_{ij} are all mutually independent.

(b) $H_0 : \sigma_b^2 = 0$ against $H_1 : \sigma_b^2 > 0$. i.e., testing whether or not there is a maternal component of piglet weight variability.

(c) There is very strong evidence ($p \approx 10^{-6}$) that $\sigma_b^2 > 0$. i.e., very strong evidence for a maternal component to piglet weight variability.

(d)

$$\hat{\sigma}_b^2 = \hat{\sigma}_0^2 - \hat{\sigma}_1^2/5 = 0.592.$$

2.(a)

$$\begin{aligned} \bar{y}_{i\cdot} &= \frac{1}{J}\sum_{j=1}^{J} y_{ij} = \alpha + b_i + \sum_{j=1}^{J} c_j + \frac{1}{J}\sum_{j=1}^{J} \epsilon_{ij} \\ &= a + e_i \end{aligned}$$

where $a = \alpha + \sum_j c_j$ and $e_i = b_i + \sum_j \epsilon_{ij}/J$. The e_i's are obviously independent since the b_i's and ϵ_{ij}'s are mutually independent and no two e_i's share a b_i or ϵ_{ij}. Normality of the e_i's follows immediately from normality of the b_i's and ϵ_{ij}'s. Furthermore,

$$\mathbb{E}(e_i) = \mathbb{E}(b_i) + \frac{1}{J}\sum_{j=1}^{J} \mathbb{E}(\epsilon_{ij}) = 0$$

and (using independence of the terms)

$$\text{var}(e_i) = \text{var}(b_i) + \frac{1}{J^2}\sum_{j=1}^{J} \text{var}(\epsilon_{ij}) = \sigma_b^2 + \sigma^2/J.$$

So $\sigma_b^2 + \sigma^2/J$ can be estimated from the residual sum of squares, RSS_1, of the least squares fit of the aggregated model to the $\bar{y}_{i\cdot}$ data:

$$\hat{\sigma}_b^2 + \hat{\sigma}^2/J = RSS_1/(I-1)$$

$$\Rightarrow \hat{\sigma}_b^2 = RSS_1/(I-1) - \hat{\sigma}^2/J$$

where $\hat{\sigma}^2$ is given in the question.

(b) Calculations tediously similar to part (a) culminate in:

$$\hat{\sigma}_c^2 = RSS_2/(J-1) - \hat{\sigma}^2/I$$

where RSS_2 is the residual sum of squares for the 2nd aggregated model fit to the $\bar{y}_{\cdot j}$ data.

3.(a) $H_0 : \sigma_a^2 = 0$ would be tested by ANOVA comparison of the fit of the given full model, to the fit of the simplified model implied by H_0,

$$y_{ij} = \mu + \beta x_{ij} + \epsilon_{ij}, \quad \epsilon_{ij} \sim N(0, \sigma^2).$$

$H_0 : \beta = 0$ would be tested by comparing the fit of the full model with the fit of the reduced model implied by this null,

$$y_{ij} = \mu + a_i + \epsilon_{ij}, \quad a_i \sim N(0, \sigma_a^2) \text{ and } \epsilon_{ij} \sim N(0, \sigma^2),$$

using ANOVA, in the same manner as if both were fixed effect linear models.

(b) To estimate σ_a^2 and β the data would be averaged at each level of the random effect a_i to yield

$$\begin{aligned} \bar{y}_{i\cdot} &= \mu + a_i + \beta \bar{x}_{i\cdot} + \frac{1}{J}\sum_{j=1}^{J} \epsilon_{ij} \\ &= \mu + \beta \bar{x}_{i\cdot} + e_i \end{aligned}$$

where $e_i \sim N(0, \sigma_a^2 + \sigma^2/J)$. The least squares fit of this model can be used to estimate β (the full model fit can not be used, since $\hat{\beta}$ will not be independent of the a_i). If RSS_1 and RSS_0 are the residual sums of squares for fitting the full and reduced models, respectively, to the y_{ij} and $\bar{y}_{i\cdot}$ data, then the obvious estimator of σ_a^2 is:

$$\hat{\sigma}_a^2 = \frac{RSS_0}{I-2} - \frac{RSS_1}{J(IJ - I - 1)}.$$

4. (In this question the fixed effects model matrices resulting from any other valid identifiability constraints on the fixed effects are fine, of course.)

(a)

$$\begin{bmatrix} y_{11} \\ y_{12} \\ y_{21} \\ y_{22} \\ y_{31} \\ y_{32} \\ y_{41} \\ y_{42} \end{bmatrix} = \begin{bmatrix} 1 & x_{11} \\ 1 & x_{12} \\ 1 & x_{21} \\ 1 & x_{22} \\ 1 & x_{31} \\ 1 & x_{32} \\ 1 & x_{41} \\ 1 & x_{42} \end{bmatrix} \begin{bmatrix} \alpha \\ \beta \end{bmatrix} + \begin{bmatrix} 1 & 0 & 0 & 0 \\ 1 & 0 & 0 & 0 \\ 0 & 1 & 0 & 0 \\ 0 & 1 & 0 & 0 \\ 0 & 0 & 1 & 0 \\ 0 & 0 & 1 & 0 \\ 0 & 0 & 0 & 1 \\ 0 & 0 & 0 & 1 \end{bmatrix} \begin{bmatrix} a_1 \\ a_2 \\ a_3 \\ a_4 \end{bmatrix} + \begin{bmatrix} \epsilon_{11} \\ \epsilon_{12} \\ \epsilon_{21} \\ \epsilon_{22} \\ \epsilon_{31} \\ \epsilon_{32} \\ \epsilon_{41} \\ \epsilon_{42} \end{bmatrix}$$

with

$$\psi = \mathbf{I}_4 \sigma_a^2.$$

(b)

$$\begin{bmatrix} y_{11} \\ y_{12} \\ y_{21} \\ y_{22} \\ y_{31} \\ y_{32} \\ y_{41} \\ y_{42} \\ y_{51} \\ y_{52} \\ y_{61} \\ y_{62} \end{bmatrix} = \begin{bmatrix} 1 & 0 \\ 1 & 0 \\ 1 & 0 \\ 1 & 0 \\ 1 & 0 \\ 1 & 0 \\ 0 & 1 \\ 0 & 1 \\ 0 & 1 \\ 0 & 1 \\ 0 & 1 \\ 0 & 1 \end{bmatrix} \begin{bmatrix} \alpha_1 \\ \alpha_2 \end{bmatrix} + \begin{bmatrix} 1 & 0 & 0 & 0 & 0 & 0 \\ 1 & 0 & 0 & 0 & 0 & 0 \\ 0 & 1 & 0 & 0 & 0 & 0 \\ 0 & 1 & 0 & 0 & 0 & 0 \\ 0 & 0 & 1 & 0 & 0 & 0 \\ 0 & 0 & 1 & 0 & 0 & 0 \\ 0 & 0 & 0 & 1 & 0 & 0 \\ 0 & 0 & 0 & 1 & 0 & 0 \\ 0 & 0 & 0 & 0 & 1 & 0 \\ 0 & 0 & 0 & 0 & 1 & 0 \\ 0 & 0 & 0 & 0 & 0 & 1 \\ 0 & 0 & 0 & 0 & 0 & 1 \end{bmatrix} \begin{bmatrix} b_1 \\ b_2 \\ b_3 \\ b_4 \\ b_5 \\ b_6 \end{bmatrix} + \begin{bmatrix} \epsilon_{11} \\ \epsilon_{12} \\ \epsilon_{21} \\ \epsilon_{22} \\ \epsilon_{31} \\ \epsilon_{32} \\ \epsilon_{41} \\ \epsilon_{42} \\ \epsilon_{51} \\ \epsilon_{52} \\ \epsilon_{61} \\ \epsilon_{62} \end{bmatrix}$$

with

$$\boldsymbol{\psi} = \mathbf{I}_6 \sigma_b^2.$$

(c)

$$\begin{bmatrix} y_{111} \\ y_{112} \\ y_{113} \\ y_{121} \\ y_{122} \\ y_{123} \\ y_{131} \\ y_{132} \\ y_{133} \\ y_{211} \\ y_{212} \\ y_{213} \\ y_{221} \\ y_{222} \\ y_{223} \\ y_{231} \\ y_{232} \\ y_{233} \end{bmatrix} = \begin{bmatrix} 1 & 0 \\ 1 & 0 \\ 1 & 0 \\ 1 & 0 \\ 1 & 0 \\ 1 & 0 \\ 1 & 0 \\ 1 & 0 \\ 1 & 0 \\ 1 & 1 \\ 1 & 1 \\ 1 & 1 \\ 1 & 1 \\ 1 & 1 \\ 1 & 1 \\ 1 & 1 \\ 1 & 1 \\ 1 & 1 \end{bmatrix} \begin{bmatrix} \mu \\ \alpha_2 \end{bmatrix} + \begin{bmatrix} 1 & 0 & 0 & 1 & 0 & 0 & 0 & 0 & 0 \\ 1 & 0 & 0 & 1 & 0 & 0 & 0 & 0 & 0 \\ 1 & 0 & 0 & 1 & 0 & 0 & 0 & 0 & 0 \\ 0 & 1 & 0 & 0 & 1 & 0 & 0 & 0 & 0 \\ 0 & 1 & 0 & 0 & 1 & 0 & 0 & 0 & 0 \\ 0 & 1 & 0 & 0 & 1 & 0 & 0 & 0 & 0 \\ 0 & 0 & 1 & 0 & 0 & 1 & 0 & 0 & 0 \\ 0 & 0 & 1 & 0 & 0 & 1 & 0 & 0 & 0 \\ 0 & 0 & 1 & 0 & 0 & 1 & 0 & 0 & 0 \\ 1 & 0 & 0 & 0 & 0 & 0 & 1 & 0 & 0 \\ 1 & 0 & 0 & 0 & 0 & 0 & 1 & 0 & 0 \\ 1 & 0 & 0 & 0 & 0 & 0 & 1 & 0 & 0 \\ 0 & 1 & 0 & 0 & 0 & 0 & 0 & 1 & 0 \\ 0 & 1 & 0 & 0 & 0 & 0 & 0 & 1 & 0 \\ 0 & 1 & 0 & 0 & 0 & 0 & 0 & 1 & 0 \\ 0 & 0 & 1 & 0 & 0 & 0 & 0 & 0 & 1 \\ 0 & 0 & 1 & 0 & 0 & 0 & 0 & 0 & 1 \\ 0 & 0 & 1 & 0 & 0 & 0 & 0 & 0 & 1 \end{bmatrix} \begin{bmatrix} b_1 \\ b_2 \\ b_3 \\ (\alpha b)_{11} \\ (\alpha b)_{12} \\ (\alpha b)_{13} \\ (\alpha b)_{21} \\ (\alpha b)_{22} \\ (\alpha b)_{23} \end{bmatrix} + \begin{bmatrix} \epsilon_{111} \\ \epsilon_{112} \\ \epsilon_{113} \\ \epsilon_{121} \\ \epsilon_{122} \\ \epsilon_{123} \\ \epsilon_{131} \\ \epsilon_{132} \\ \epsilon_{133} \\ \epsilon_{211} \\ \epsilon_{212} \\ \epsilon_{213} \\ \epsilon_{221} \\ \epsilon_{222} \\ \epsilon_{223} \\ \epsilon_{231} \\ \epsilon_{232} \\ \epsilon_{233} \end{bmatrix}$$

$$\boldsymbol{\psi} = \begin{bmatrix} \mathbf{I}_3 \sigma_b^2 & \mathbf{0} \\ \mathbf{0} & \mathbf{I}_6 \sigma_{\alpha b}^2 \end{bmatrix}.$$

5.(a)

$$\begin{aligned} \boldsymbol{\Sigma}_{x+z} &= \mathbb{E}[(\mathbf{X} + \mathbf{Z} - \boldsymbol{\mu}_x - \boldsymbol{\mu}_z)(\mathbf{X} + \mathbf{Z} - \boldsymbol{\mu}_x - \boldsymbol{\mu}_z)^\mathsf{T}] \\ &= \mathbb{E}[(\mathbf{X} - \boldsymbol{\mu}_x)(\mathbf{X} - \boldsymbol{\mu}_x)^\mathsf{T}] + \mathbb{E}[(\mathbf{Z} - \boldsymbol{\mu}_z)(\mathbf{X} - \boldsymbol{\mu}_x)^\mathsf{T}] \\ &\quad + \mathbb{E}[(\mathbf{X} - \boldsymbol{\mu}_x)(\mathbf{Z} - \boldsymbol{\mu}_z)^\mathsf{T}] + \mathbb{E}[(\mathbf{Z} - \boldsymbol{\mu}_z)(\mathbf{Z} - \boldsymbol{\mu}_z)^\mathsf{T}] \\ &= \mathbb{E}[(\mathbf{X} - \boldsymbol{\mu}_x)(\mathbf{X} - \boldsymbol{\mu}_x)^\mathsf{T}] + \mathbb{E}[(\mathbf{Z} - \boldsymbol{\mu}_z)(\mathbf{Z} - \boldsymbol{\mu}_z)^\mathsf{T}] \quad \text{(by ind.)} \\ &= \boldsymbol{\Sigma}_x + \boldsymbol{\Sigma}_z. \end{aligned}$$

(b) i.

$$\begin{bmatrix} y_{11} \\ y_{12} \\ y_{13} \\ y_{21} \\ y_{22} \\ y_{23} \\ y_{31} \\ y_{32} \\ y_{33} \end{bmatrix} = \begin{bmatrix} 1 & x_{11} \\ 1 & x_{12} \\ 1 & x_{13} \\ 1 & x_{21} \\ 1 & x_{22} \\ 1 & x_{23} \\ 1 & x_{31} \\ 1 & x_{32} \\ 1 & x_{33} \end{bmatrix} \begin{bmatrix} \alpha \\ \beta \end{bmatrix} + \begin{bmatrix} 1 & 0 & 0 \\ 1 & 0 & 0 \\ 1 & 0 & 0 \\ 0 & 1 & 0 \\ 0 & 1 & 0 \\ 0 & 1 & 0 \\ 0 & 0 & 1 \\ 0 & 0 & 1 \\ 0 & 0 & 1 \end{bmatrix} \begin{bmatrix} b_1 \\ b_2 \\ b_3 \end{bmatrix} + \begin{bmatrix} \epsilon_{11} \\ \epsilon_{12} \\ \epsilon_{13} \\ \epsilon_{21} \\ \epsilon_{22} \\ \epsilon_{23} \\ \epsilon_{31} \\ \epsilon_{32} \\ \epsilon_{33} \end{bmatrix}$$

where $\mathbf{b} \sim N(\mathbf{0}, \mathbf{I}\sigma_b^2)$ and $\boldsymbol{\epsilon} \sim N(\mathbf{0}, \mathbf{I}\sigma^2)$.

ii. The covariance matrix of $\mathbf{y}$ is

$$\begin{bmatrix} 1 & 0 & 0 \\ 1 & 0 & 0 \\ 1 & 0 & 0 \\ 0 & 1 & 0 \\ 0 & 1 & 0 \\ 0 & 1 & 0 \\ 0 & 0 & 1 \\ 0 & 0 & 1 \\ 0 & 0 & 1 \end{bmatrix} \begin{bmatrix} 1 & 0 & 0 \\ 0 & 1 & 0 \\ 0 & 0 & 1 \end{bmatrix} \begin{bmatrix} 1 & 1 & 1 & 0 & 0 & 0 & 0 & 0 & 0 \\ 0 & 0 & 0 & 1 & 1 & 1 & 0 & 0 & 0 \\ 0 & 0 & 0 & 0 & 0 & 0 & 1 & 1 & 1 \end{bmatrix} \sigma_b^2 +$$

$$\begin{bmatrix} 1 & 0 & 0 & 0 & 0 & 0 & 0 & 0 & 0 \\ 0 & 1 & 0 & 0 & 0 & 0 & 0 & 0 & 0 \\ 0 & 0 & 1 & 0 & 0 & 0 & 0 & 0 & 0 \\ 0 & 0 & 0 & 1 & 0 & 0 & 0 & 0 & 0 \\ 0 & 0 & 0 & 0 & 1 & 0 & 0 & 0 & 0 \\ 0 & 0 & 0 & 0 & 0 & 1 & 0 & 0 & 0 \\ 0 & 0 & 0 & 0 & 0 & 0 & 1 & 0 & 0 \\ 0 & 0 & 0 & 0 & 0 & 0 & 0 & 1 & 0 \\ 0 & 0 & 0 & 0 & 0 & 0 & 0 & 0 & 1 \end{bmatrix} \sigma^2$$

which is

$$\begin{bmatrix} \sigma^2+\sigma_b^2 & \sigma_b^2 & \sigma_b^2 & 0 & 0 & 0 & 0 & 0 & 0 \\ \sigma_b^2 & \sigma^2+\sigma_b^2 & \sigma_b^2 & 0 & 0 & 0 & 0 & 0 & 0 \\ \sigma_b^2 & \sigma_b^2 & \sigma^2+\sigma_b^2 & 0 & 0 & 0 & 0 & 0 & 0 \\ 0 & 0 & 0 & \sigma^2+\sigma_b^2 & \sigma_b^2 & \sigma_b^2 & 0 & 0 & 0 \\ 0 & 0 & 0 & \sigma_b^2 & \sigma^2+\sigma_b^2 & \sigma_b^2 & 0 & 0 & 0 \\ 0 & 0 & 0 & \sigma_b^2 & \sigma_b^2 & \sigma^2+\sigma_b^2 & 0 & 0 & 0 \\ 0 & 0 & 0 & 0 & 0 & 0 & \sigma^2+\sigma_b^2 & \sigma_b^2 & \sigma_b^2 \\ 0 & 0 & 0 & 0 & 0 & 0 & \sigma_b^2 & \sigma^2+\sigma_b^2 & \sigma_b^2 \\ 0 & 0 & 0 & 0 & 0 & 0 & \sigma_b^2 & \sigma_b^2 & \sigma^2+\sigma_b^2 \end{bmatrix}.$$

6. (a) Site and source should be fixed; Lot and Wafer should be random.

(b) The thickness for wafer l, lot k, site j and source i is

$$y_{ijkl} = \mu + \alpha_i + \beta_j + b_k + c_{(k)l} + \epsilon_{ijkl}$$

where $b_k \sim N(0, \sigma_b^2)$, $c_{(k)l} \sim N(0, \sigma_c^2)$, $\epsilon_{ijkl} \sim N(0, \sigma^2)$ and all these r.v.s are independent. The subscript, $(k)l$, is used to emphasize that wafer is nested within lot to help clarify the practical fitting of the model (I haven't indicated all such nestings this way!).

(c)
```
> options(contrasts=c("contr.treatment",
+                     "contr.treatment"))
> m1 <- lme(Thickness~Site+Source,Oxide,~1|Lot/Wafer)
> plot(m1)                  # check resids vs. fitted vals
> qqnorm(residuals(m1))   # check resids for normality
> abline(0,sd(resid(m1)))# adding a "reference line"
> qqnorm(m1,~ranef(.,level=1)) # check normality of b_k
> qqnorm(m1,~ranef(.,level=2)) # check normality of c_(k)l
> m2 <- lme(Thickness~Site+Source,Oxide,~1|Lot)
> anova(m1,m2)
   Model df    AIC    BIC  logLik   Test L.Ratio p-value
m1     1  7 455.76 471.30 -220.88
m2     2  6 489.41 502.72 -238.70 1 vs 2 35.6444  <.0001
```

The checking plots suggest no problems at all in this case. `m1` is the full model fit and `m2` is a reduced version, with no $c_{(k)l}$ terms. The `anova(m1,m2)` command performs a generalized likelihood ratio test of the hypothesis that the data were generated by `m2` against the alternative that they were generated by `m1`. The p value is so low in this case that, despite the problems with GLRT tests in this context, we can be very confident in rejecting H_0 and concluding that `m1` is necessary. i.e., there is evidence for wafer to wafer variability above what is explicable by lot to lot variability.

```
> anova(m1)
            numDF denDF   F-value p-value
(Intercept)     1    46 240197.01  <.0001
Site            2    46      0.60  0.5508
Source          1     6      1.53  0.2629
```

There appears to be no evidence for site or source effects, and hence no need for follow up multiple comparisons. Note that for unbalanced data you would usually refit without the redundant fixed effects at this point, in order to gain a little precision, but these data are balanced, so doing so would make no difference.

```
> intervals(m1)
Approximate 95% confidence intervals

 Fixed effects:
                  lower          est.       upper
(Intercept) 1983.237346 1994.9166667 2006.595987
Site2         -2.327301   -0.2500000    1.827301
Site3         -1.243967    0.8333333    2.910634
Source2       -9.888949   10.0833333   30.055615
attr(,"label")
[1] "Fixed effects:"
```

```
 Random Effects:
  Level: Lot
                        lower      est.     upper
sd((Intercept)) 5.831891 10.94954 20.55808
  Level: Wafer
                        lower      est.     upper
sd((Intercept)) 4.057118 5.982933 8.822887

 Within-group standard error:
   lower      est.     upper
2.914196 3.574940 4.385495
```

So the lot to lot variation is substantial, with a standard deviation of between 5.8 and 20.6, while the wafer to wafer variation, within a lot, is not much less, with standard deviation between 4.1 and 8.8. The unaccounted for 'within wafer' variability seems to be a little less, having a standard deviation somewhere between 2.9 and 4.4. There is no evidence that site on the wafer or source have any influence on thickness.

7. See `ch2.solutions` in package `gamair`.
8. See `ch2.solutions` in package `gamair`.
9. (a) Physique and method would be modelled as fixed effects: the estimation of these effects is of direct interest and we would expect to obtain similar estimates in a replicate experiment; the effect of these factors should be fixed properties of the population of interest. Team would be modelled as a random effect — team to team variation is to be expected, but the individual team effects are not of direct interest and would be completely different if the experiment were repeated with different gunners.

   (b) If y_{ijk} denotes the number of rounds for team i, with physique j using method k, then a possible model is

$$y_{ijk} = \mu + \alpha_j + \beta_k + b_i + \epsilon_{ijk}$$

   where the b_i are i.i.d. $N(0, \sigma_b^2)$ and the ϵ_{ijk} are i.i.d. $N(0, \sigma^2)$. A more careful notation might make explicit the nesting of team within build by, for example, replacing b_i by $b_{i(j)}$. An interaction between physique and method might also be considered (as might a random two way interaction of method and team).

   (c) For code see `ch2.solutions` in package `gamair`.

   The residual plots appear quite reasonable in this case (although if you check the random effects themselves these are not so good). Since there is no evidence for an effect of physique the only sensible follow up comparison would compare methods: method one appears to lead to an average increase in firing rate of between 7.5 and 9.5 rounds per minute relative to method 2 (with 95% confidence). By contrast the team to team variability and within team variability are rather low with 95% CIs on their standard deviations of (0.5,2.2) and (1.1,1.9), respectively.

C.3 Chapter 3

1.(a) $\mu \equiv \mathbb{E}(Y) = (1-p) \times 0 + p \times 1 = p$.

(b)

$$\begin{aligned} f(y) &= \exp(\log(\mu^y(1-\mu)^{1-y})) \\ &= \exp(y\log(\mu) + (1-y)\log(1-\mu)) \\ &= \exp\left(y\log\left(\frac{\mu}{1-\mu}\right) + \log(1-\mu)\right). \end{aligned}$$

Now let $\theta = \log(\mu/(1-\mu))$ so that $e^\theta = \mu/(1-\mu)$ and therefore

$$1 + e^\theta = \frac{1}{1-\mu} \Rightarrow -\log(1+e^\theta) = \log(1-\mu) \Rightarrow b(\theta) = \log(1+e^\theta).$$

Hence,

$$f(y) = \exp(y\theta - b(\theta)),$$

which is exponential form with $a(\phi) = \phi = 1$ and $c(y, \phi) \equiv 0$. So the Bernoulli distribution is in the exponential family.

(c) A canonical link is one which when applied to the mean, μ, gives the canonical parameter, so for the Bernoulli it is clearly:

$$g(\mu) = \log\left(\frac{\mu}{1-\mu}\right),$$

a special case of the logit link.

2. See `ch3.solutions` in package `gamair`.

3.(a) See `ch3.solutions` in package `gamair`.

(b) See `ch3.solutions` in package `gamair` for code.

A model in which the counts depend on all two way interactions seems appropriate. Note that the least significant interaction is between the defendant's skin 'colour' and death penalty, but that dropping this term does not seem to be justified (AIC also suggests leaving it in).

(c) From the summary table it is clear that the strongest and most significant association is between skin colour of victim and defendant: blacks kill blacks and whites kill whites, for the most part. Next in terms of significance and strength of effect is the association between victims' colour, and death penalty: the death penalty is less likely if the victim is black. Finally, being black is associated with an increased likelihood of being sentenced to death, once these other associations have been taken into account. Most of these effects can be ascertained by careful examination of the original table, but their significance and the relative strength of the effects are striking results from the modelling.

4.

$$\begin{aligned} \int_{y_i}^{\mu_i} \frac{y_i - z}{\phi z} dz &= \frac{1}{\phi}\left[y_i \log(z) - z\right]_{y_i}^{\mu_i} \\ &= \frac{1}{\phi}\left[y_i \log(\mu_i/y_i) - \mu_i + y_i\right]. \end{aligned}$$

Since the quasi-likelihood of the saturated model is zero, the corresponding deviance is simply

$$2y_i \log(y_i/\mu_i) - 2(y_i - \mu_i),$$

which corresponds to the Poisson deviance.

5. $\sum_i w_i(y_i - X_i\boldsymbol{\beta})^2 \equiv \sum_i(\tilde{y}_i - \tilde{X}_i\boldsymbol{\beta})^2$ where $\tilde{y}_i = \sqrt{w_i}y_i$ and $\tilde{X}_{ij} = \sqrt{w_i}X_{ij}$. Hence re-using the results from section 1.3.8 we have $(\tilde{\mathbf{X}}^\mathsf{T}\tilde{\mathbf{X}})^{-1}\tilde{\mathbf{X}}^\mathsf{T}\tilde{\mathbf{y}}$. Re-writing this in terms of $\mathbf{X}$ and $\mathbf{W}$ yields the required result.

6.(a) $\mathbb{E}(z_i) = g'(\mu_i)\{\mathbb{E}(y_i) - \mu_i\} + \eta_i = 0 + \mathbf{X}_i\boldsymbol{\beta}$

(b)

$$\begin{aligned}\text{var}(z_i) &= \text{var}\{g'(\mu_i)y_i\} \\ &= g'(\mu_i)^2\text{var}(y_i) \\ &= g'(\mu_i)^2V(\mu_i)\phi = w_i^{-1}\phi.\end{aligned}$$

Furthermore, since the y_i are independent, then so are the z_i, so the covariance matrix of the z_i's is $\mathbf{W}^{-1}\phi$, as required.

(c) From the previous question we have that $\hat{\boldsymbol{\beta}} = (\mathbf{X}^\mathsf{T}\mathbf{W}\mathbf{X})^{-1}\mathbf{X}^\mathsf{T}\mathbf{W}\mathbf{z}$. So, by the results of B.2 on transformation of covariance matrices, we have that the covariance matrix of $\hat{\boldsymbol{\beta}}$ is

$$(\mathbf{X}^\mathsf{T}\mathbf{W}\mathbf{X})^{-1}\mathbf{X}^\mathsf{T}\mathbf{W}\mathbf{W}^{-1}\mathbf{W}\mathbf{X}(\mathbf{X}^\mathsf{T}\mathbf{W}\mathbf{X})^{-1}\phi = (\mathbf{X}^\mathsf{T}\mathbf{W}\mathbf{X})^{-1}\phi.$$

Similarly,

$$\mathbb{E}(\hat{\boldsymbol{\beta}}) = (\mathbf{X}^\mathsf{T}\mathbf{W}\mathbf{X})^{-1}\mathbf{X}^\mathsf{T}\mathbf{W}\mathbb{E}(\mathbf{z}) = (\mathbf{X}^\mathsf{T}\mathbf{W}\mathbf{X})^{-1}\mathbf{X}^\mathsf{T}\mathbf{W}\mathbf{X}\boldsymbol{\beta} = \boldsymbol{\beta}.$$

(d) If $\mathbf{X}^\mathsf{T}\mathbf{W}\mathbf{z}$ tends to multivariate normality then so must $\hat{\boldsymbol{\beta}}$. Hence, in the large sample limit, $\hat{\boldsymbol{\beta}} \sim N(\boldsymbol{\beta}, (\mathbf{X}^\mathsf{T}\mathbf{W}\mathbf{X})^{-1}\phi)$. Of course in practical application of GLMs we calculate $\mathbf{z}$ and $\mathbf{W}$ using $\hat{\boldsymbol{\beta}}$, rather than $\boldsymbol{\beta}$, but as sample size tends to infinity $\hat{\boldsymbol{\beta}} \to \boldsymbol{\beta}$, so the result still holds.

7. See `ch3.solutions` in package `gamair`.

8.(a)

$$\frac{1}{\mathbb{E}(c_i)} = \frac{1}{ad_i^m} + t.$$

(b–f) See `ch3.solutions` in package `gamair`.

9. See code in `ch3.solutions` in package `gamair`. The model is clearly inadequate: massive over-dispersion and clear pattern in the residuals. Probably residual autocorrelation should be modelled properly here, reflecting the time-series nature of the data.

10. See code in `ch3.solutions` in package `gamair`. From the resulting plot, the 95% CI is (0.43,0.68), slightly wider than the (0.47, 0.65) found earlier, but in this case almost symmetric. One key difference between these intervals and the intervals covered in the text, is that they are not necessarily symmetric. Another is that parameter values within the interval always have higher likelihood than those outside it, and a third is that they are invariant to re-parameterization.

11. See `ch3.solutions` in package `gamair`.

C.4 Chapter 4

1. See `ch4.solutions` in package `gamair` for the code.
   Note the wild behaviour of the order 10 polynomial, compared to the much better behaviour of the spline model of the same rank.
   The difference in behaviour of the two rank 11 models is perhaps unsurprising. The theoretical justification for polynomial approximations of unknown functions would probably be Taylor's theorem: but this is concerned with getting good approximations in the vicinity of some particular point of interest — it is clear from the theorem that the approximation will eventually become very poor as we move away from that point. The theoretical justifications for splines are much more concerned with properties of the function over the whole region of interest.
2. See `ch4.solutions` in package `gamair`.
3.

$$\begin{aligned}\mathcal{S}_p &= \|\mathbf{y} - \mathbf{X}\boldsymbol{\beta}\|^2 + \lambda\boldsymbol{\beta}^\mathsf{T}\mathbf{S}\boldsymbol{\beta} \\ &= (\mathbf{y} - \mathbf{X}\boldsymbol{\beta})^\mathsf{T}(\mathbf{y} - \mathbf{X}\boldsymbol{\beta}) + \lambda\boldsymbol{\beta}^\mathsf{T}\mathbf{S}\boldsymbol{\beta} \\ &= \mathbf{y}^\mathsf{T}\mathbf{y} - 2\boldsymbol{\beta}^\mathsf{T}\mathbf{X}^\mathsf{T}\mathbf{y} + \boldsymbol{\beta}^\mathsf{T}\left(\mathbf{X}^\mathsf{T}\mathbf{X} + \lambda\mathbf{S}\right)\boldsymbol{\beta}.\end{aligned}$$

   Differentiating $\mathcal{S}_p$ w.r.t. $\boldsymbol{\beta}$ and setting to zero results in the system of equations

$$(\mathbf{X}^\mathsf{T}\mathbf{X} + \lambda\mathbf{S})\hat{\boldsymbol{\beta}} = \mathbf{X}^\mathsf{T}\mathbf{y},$$

   which yields the required result.
4. The result follows from the fact that the upper left $n \times n$ submatrix of $\tilde{\mathbf{X}}(\tilde{\mathbf{X}}^\mathsf{T}\tilde{\mathbf{X}})^{-1}\tilde{\mathbf{X}}^\mathsf{T}$ is $\mathbf{X}(\mathbf{X}^\mathsf{T}\mathbf{X} + \mathbf{S})^{-1}\mathbf{X}^\mathsf{T}$, as the following shows:

$$\begin{aligned}\tilde{\mathbf{X}}(\tilde{\mathbf{X}}^\mathsf{T}\tilde{\mathbf{X}})^{-1}\tilde{\mathbf{X}}^\mathsf{T} &= \begin{bmatrix}\mathbf{X} \\ \mathbf{B}\end{bmatrix}(\mathbf{X}^\mathsf{T}\mathbf{X} + \mathbf{S})^{-1}\begin{bmatrix}\mathbf{X}^\mathsf{T} & \mathbf{B}^\mathsf{T}\end{bmatrix} \\ &= \begin{bmatrix}\mathbf{X}(\mathbf{X}^\mathsf{T}\mathbf{X} + \mathbf{S})^{-1}\mathbf{X}^\mathsf{T} & \mathbf{X}(\mathbf{X}^\mathsf{T}\mathbf{X} + \mathbf{S})^{-1}\mathbf{B}^\mathsf{T} \\ \mathbf{B}(\mathbf{X}^\mathsf{T}\mathbf{X} + \mathbf{S})^{-1}\mathbf{X}^\mathsf{T} & \mathbf{B}(\mathbf{X}^\mathsf{T}\mathbf{X} + \mathbf{S})^{-1}\mathbf{B}^\mathsf{T}\end{bmatrix}.\end{aligned}$$

5. Zero. The most complex model will always be chosen, as this will allow the data to be fitted most closely (only for a set of data configurations of probability zero is this not true — for example data lying exactly on a straight line).
6. Differentiating the basis expansion for f, we get $f''(x) = \boldsymbol{\beta}^\mathsf{T}\mathbf{d}(x)$ where $d_j(x) = b_j''(x)$. Using the fact that a scalar is its own transpose we then have that:

$$\int f''(x)^2 dx = \int \boldsymbol{\beta}^\mathsf{T}\mathbf{d}(x)\mathbf{d}(x)^\mathsf{T}\boldsymbol{\beta} dx = \boldsymbol{\beta}^\mathsf{T}\mathbf{S}\boldsymbol{\beta}$$

   where

$$\mathbf{S} = \int \mathbf{d}(x)\mathbf{d}(x)^\mathsf{T} dx.$$

7.

$$\int \left(\frac{\partial^2 f}{\partial x^2}\right)^2 + 2\left(\frac{\partial^2 f}{\partial x \partial z}\right)^2 + \left(\frac{\partial f^2}{\partial z^2}\right)^2 dxdz = \int \left(\frac{\partial^2 f}{\partial x^2}\right)^2 dxdz + 2\int \left(\frac{\partial^2 f}{\partial x \partial z}\right)^2 dxdz + \int \left(\frac{\partial f^2}{\partial z^2}\right)^2 dxdz.$$

Treating each integral on the r.h.s. in a similar way to the integral in the previous question we find that the penalty can be written as $\boldsymbol{\beta}^\mathsf{T}\mathbf{S}\boldsymbol{\beta}$ where

$$\mathbf{S} = \int \mathbf{d}_{xx}(x,z)\mathbf{d}_{xx}(x,z)^\mathsf{T} + 2\mathbf{d}_{xz}(x,z)\mathbf{d}_{xz}(x,z)^\mathsf{T} + \mathbf{d}_{zz}(x,z)\mathbf{d}_{zz}(x,z)^\mathsf{T} dxdy.$$

Here the jth element of $\mathbf{d}_{xz}(s,z)$ is $\partial^2 b_j/\partial x \partial z$, with similar definitions for $\mathbf{d}_{xx}$ and $\mathbf{d}_{zz}$. Obviously this sort of argument can be applied to all sorts of penalties involving integrals of sums of squared derivatives.

8. See `ch4.solutions` in package `gamair`.
9. See `ch4.solutions` in package `gamair`.

C.5 Chapter 5

1. Consider the spline defined by (5.3). At knot position x_{j+1}, we require that the derivative at x_{j+1} of the section of cubic to the left of x_{j+1} matches the derivative at x_{j+1} of the section of cubic to the right of x_{j+1}. Writing this condition out in full yields

$$-\frac{\beta_j}{h_j} + \frac{\beta_{j+1}}{h_j} + \delta_j\frac{h_j}{6} + \delta_{j+1}\frac{3h_j}{6} - \delta_{j+1}\frac{h_j}{6} = -\frac{\beta_{j+1}}{h_{j+1}} + \frac{\beta_{j+2}}{h_{j+1}} - \delta_{j+1}\frac{3h_{j+1}}{6} + \delta_{j+1}\frac{h_{j+1}}{6} - \delta_{j+2}\frac{h_{j+1}}{6}$$

and simple re-arrangement leads to

$$\frac{1}{h_j}\beta_j - \left(\frac{1}{h_j} + \frac{1}{h_{j+1}}\right)\beta_{j+1} + \frac{1}{h_{j+1}}\beta_{j+2} = \frac{h_j}{6}\delta_j + \left(\frac{h_j}{3} + \frac{h_{j+1}}{3}\right)\delta_{j+1} + \frac{h_{j+1}}{6}\delta_{j+2}.$$

With the additional restriction that $\delta_1 = \delta_k = 0$, this latter system repeated for $j = 1, \ldots, k-2$ is (5.4).

2.(a) Differentiating (5.3) twice yields

$$f''(x) = \delta_j(x_{j+1} - x)/h_j + \delta_{j+1}(x - x_j)/h_j, \;\; x_j \le x \le x_{j+1}.$$

By inspection this can be re-written in the required form.

(b) Writing $\mathbf{d}(x)$ as the vector with i^{th} element $d_{i+1}(x)$ (the first and last d_i's have coefficients zero, so are of no interest), it is easy to show (e.g., exercise 7, Chapter 3) that

$$\int f''(x)^2 dx = \boldsymbol{\delta}^{-\mathsf{T}} \int \mathbf{d}(x)\mathbf{d}(x)^{\mathsf{T}} dx \boldsymbol{\delta}^{-}.$$

Since each $d_i(x)$ is non-zero over only 2 intervals, it is clear that $\int \mathbf{d}(x)\mathbf{d}(x)^{\mathsf{T}} dx$ is tri-diagonal, and it is also obviously symmetric, by construction. The $(i-1)^{\text{th}}$ leading diagonal element is given by

$$\int_{x_{i-1}}^{x_{i+1}} d_i(x)^2 dx = \left[\frac{(x - x_{i-1})^3}{3h_{i-1}^2}\right]_{x_{i-1}}^{x_i} - \left[\frac{(x_{i+1} - x)^3}{3h_i^2}\right]_{x_i}^{x_{i+1}} = \frac{h_{i-1}}{3} + \frac{h_i}{3},$$

where i runs from 2 to $k-1$. In the same vein, the off-diagonal elements $(i-1, i)$ and $(i, i-1)$ are given by:

$$\int_{x_{i-1}}^{x_i} d_i(x) d_{i-1}(x) = \int_{x_{i-1}}^{x_i} \frac{x - x_{i-1}}{h_{i-1}} \frac{x_i - x}{h_{i-1}} dx = \frac{h_{i-1}}{6}.$$

In other words $\int \mathbf{d}(x)\mathbf{d}(x)^{\mathsf{T}} dx = \mathbf{B}$, as required.

(c) From (5.4) we have that $\boldsymbol{\delta}^{-} = \mathbf{B}^{-1}\mathbf{D}\boldsymbol{\beta}$, from which the result follows immediately by substitution (having noted the symmetry of $\mathbf{B}$).

3. (a) In the regular parameterization:

$$\begin{aligned} \mathbb{E}(\hat{\boldsymbol{\beta}}) &= (\mathbf{X}^{\mathsf{T}}\mathbf{X} + \lambda\mathbf{S})^{-1}\mathbf{X}^{\mathsf{T}}\mathbb{E}(\mathbf{y}) \\ &= (\mathbf{X}^{\mathsf{T}}\mathbf{X} + \lambda\mathbf{S})^{-1}\mathbf{X}^{\mathsf{T}}\mathbf{X}\boldsymbol{\beta}. \end{aligned}$$

In the natural parameterization this becomes

$$\mathbb{E}(\hat{\boldsymbol{\beta}}'') = (\mathbf{I} + \lambda\mathbf{D})^{-1}\boldsymbol{\beta}''$$

and hence

$$\text{bias}(\hat{\beta}_i'') = -\beta_i''\lambda D_{ii}/(1 + \lambda D_{ii}).$$

So if $\beta_i'' = 0$ then its estimator is unbiased. Equally if the parameter is unpenalized because $\lambda D_{ii} = 0$ then the estimator is unbiased. Clearly the bias will be small for small true parameter value or low penalization: it is only moderate or strongly penalized model components of substantial magnitude that are subject to substantial bias.

(b)

$$\begin{aligned} \mathbb{E}\{(\hat{\beta}_i - \beta_i)^2\} &= \mathbb{E}\{(\hat{\beta}_i - \mathbb{E}(\hat{\beta}_i) + \mathbb{E}(\hat{\beta}_i) - \beta_i)^2\} \\ &= \mathbb{E}\{(\hat{\beta}_i - \mathbb{E}(\hat{\beta}_i))^2\} + \mathbb{E}\{(\mathbb{E}(\hat{\beta}_i) - \beta_i)^2\} \\ &\quad + \mathbb{E}\{(\hat{\beta}_i - \mathbb{E}(\hat{\beta}_i))(\mathbb{E}(\hat{\beta}_i) - \beta_i)\} \\ &= \text{var}(\hat{\beta}_i) + \text{bias}(\hat{\beta}_i)^2 + 0. \end{aligned}$$

(c) So the MSE, M, of β_i (natural parameterization) is

$$M = \frac{\sigma^2}{(1+\lambda D_{ii})^2} + \frac{(\lambda D_{ii})^2\beta_i^2}{(1+\lambda D_{ii})^2} = \frac{\sigma^2 + (\lambda D_{ii}\beta_i)^2}{(1+\lambda D_{ii})^2}.$$

This expression makes it rather clear how increasing λ decreases variance while increasing bias.

(d) Writing

$$\frac{M}{\sigma^2} = \frac{1+k^2 r}{(1+k)^2}$$

where $\lambda D_{ii} = k$ and $\beta_i^2/\sigma^2 = r$, it's clear that M/σ^2 (and hence M) is minimized by choosing λ so that $k = 1/r$, in which case $M = \sigma^2 r/(1+r)$. In the natural parameterization the unpenalized estimator variance (and hence unpenalized MSE) is σ^2, and $\sigma^2 r/(1+r)$ is clearly always less than this. If λ could be chosen to minimize the MSE for a particular parameter, then from the preceding formulae, it is clear that small magnitude β_i's would lead to high penalization and MSE dominated by the bias term, while large magnitude β_i's would be lightly penalized, with the MSE dominated by the variance.

(e) In the natural parameterization the Bayesian prior variance for β_i is $\sigma^2/(\lambda_i D_{ii})$. Since the prior expected value (for penalized terms) is 0, this means that $\mathbb{E}(\beta_i^2) = \sigma^2/(\lambda_i D_{ii})$ according to the prior. If this is representative of the typical size of β_i^2 then the typical size of the $\text{bias}(\hat{\beta}_i)^2$ would be:

$$\frac{\lambda D_{ii}\sigma^2}{(1+\lambda D_{ii})^2},$$

implying that the squared bias should typically be bounded above by something like $\sigma^2/4$.

4. See `ch5.solutions` in package `gamair`.
5. See `ch5.solutions` in package `gamair` for one possible solution.
6. See `ch5.solutions` in package `gamair` for one possible solution.

C.6 Chapter 6

1.

$$\begin{aligned}
\mathbb{E}(\|\mathbf{y} - \mathbf{A}\mathbf{y}\|^2) &= \mathbb{E}(\|\boldsymbol{\mu} + \boldsymbol{\epsilon} - \mathbf{A}\boldsymbol{\mu} - \mathbf{A}\boldsymbol{\epsilon}\|^2) \\
&= (\boldsymbol{\mu} + \boldsymbol{\epsilon} - \mathbf{A}\boldsymbol{\mu} - \mathbf{A}\boldsymbol{\epsilon})^{\mathsf{T}}(\boldsymbol{\mu} + \boldsymbol{\epsilon} - \mathbf{A}\boldsymbol{\mu} - \mathbf{A}\boldsymbol{\epsilon}) \\
&= \boldsymbol{\mu}^{\mathsf{T}}\boldsymbol{\mu} - \boldsymbol{\mu}^{\mathsf{T}}\mathbf{A}\boldsymbol{\mu} + \boldsymbol{\mu}^{\mathsf{T}}\mathbf{A}\mathbf{A}\boldsymbol{\mu} + \\
&\quad \mathbb{E}(\boldsymbol{\epsilon}^{\mathsf{T}}\boldsymbol{\epsilon}) - 2\mathbb{E}(\boldsymbol{\epsilon}^{\mathsf{T}}\mathbf{A}\boldsymbol{\epsilon}) + \mathbb{E}(\boldsymbol{\epsilon}^{\mathsf{T}}\mathbf{A}\mathbf{A}\boldsymbol{\epsilon}) \\
&= \mathbf{b}^{\mathsf{T}}\mathbf{b} + n\sigma^2 - 2\text{tr}(\mathbf{A})\sigma^2 + \text{tr}(\mathbf{A}\mathbf{A})\sigma^2
\end{aligned}$$

where $\mathbb{E}(\boldsymbol{\epsilon}^{\mathsf{T}}\mathbf{A}\boldsymbol{\epsilon}) = \mathbb{E}\{\text{tr}(\boldsymbol{\epsilon}^{\mathsf{T}}\mathbf{A}\boldsymbol{\epsilon})\} = \mathbb{E}\{\text{tr}(\mathbf{A}\boldsymbol{\epsilon}\boldsymbol{\epsilon}^{\mathsf{T}})\} = \text{tr}(\mathbf{A}\sigma^2)$, and similar have been used.

2.(a) The influence matrix for this problem is obviously $\mathbf{A} = (\mathbf{I} + \lambda \mathbf{I})^{-1}$. Hence $\text{tr}(\mathbf{A}) = n/(1+\lambda)$ and $\mu_i = y_i/(1+\lambda)$. Substituting into the expression for the OCV or GCV scores results in $\mathcal{V}_o = \mathcal{V}_g = \sum_i y_i^2/n$, which does not depend on λ.

(b) If we were to drop a y_i from the model sum of squares term, then the only thing influencing the estimate of μ_i would be the penalty term, which would be minimized by setting $\mu_i = 0$, whatever (positive) value λ takes. This complete decoupling, where each μ_i is influenced only by its own y_i, will clearly cause cross validation to fail: if we leave out y_i then the corresponding μ_i is always estimated as zero, since the other data have no influence on it, and this behaviour occurs for any possible value of λ. In a sense this is unsurprising, since there is actually no covariate in this problem, and hence nothing to indicate which y_i or μ_i values should be close in value.

(c) From the symmetry around x_2 it suffices to examine only the integral of the penalty from x_1 to x_2. Without loss of generality we can take x_1 to be 0. If b is the slope of the line segment joining $(x_1 = 0, \mu_1 = \mu^*)$ to (x_2, μ_2) then $\mu_2 = \mu^* + bx_2$, and it is simpler to work in terms of b, initially. So

$$2P = \int_{x_1}^{x_3} \mu_i(x)^2 dx = 2\int_0^{x_2} (\mu^* + bx)^2 dx$$

so that

$$P = \mu^{*2} x_2 + \mu^* b x_2^2 + b^2 x_2^3/3.$$

To find the b minimizing P, set $\partial P/\partial b = 0$, which implies $b = -3\mu^*/(2x_2)$ and hence $\mu_2 = -\mu^*/2$.

So, consider a set of 3 adjacent points with roughly similar y_i and μ_i values: if we omit the middle point from the fitting, then the action of the penalty will send its μ_i estimate to the other side of zero, from its neighbours. This is a rather unfortunate tendency. It means that the missing datum will always be better predicted with a high λ than with a lower one, since the higher λ will tend to shrink the μ_i for the included data towards zero, which will mean that the μ_i for the omitted datum will also be closer to zero, and hence less far from the omitted datum value. Hence cross validation will have the pathological tendency to always select the model $\mu_i = 0 \; \forall \; i$, an effect which can be demonstrated practically!

(d) The first derivative penalty does not suffer from the problems of the other two penalties. In this case the action of the penalty is merely to try and flatten $\mu(x)$ in the vicinity of an omitted datum: increased flattening with increased λ generally pulls $\mu(x)$ away from the omitted datum, in the way that cross validation implicitly assumes will happen.

(e) Generally, penalized regression smoothers can not decouple in the manner of the smoothers considered in this question: because each smoother has far fewer parameters than data, each μ_i is necessarily dependent on several y_i, rather than just one. It is simply not possible for the penalty to do something bizarre to one μ_i while leaving the others unchanged.

3. See `ch6.solutions` in package `gamair`.

4.(a)

$$\mathbf{A} = \mathbf{QU}(\mathbf{I} + \lambda\mathbf{D})^{-1}\mathbf{U}^\mathsf{T}\mathbf{Q}^\mathsf{T}.$$

(b)

$$\begin{aligned}\text{tr}(\mathbf{A}) &= \text{tr}\left(\mathbf{QU}(\mathbf{I} + \lambda\mathbf{D})^{-1}\mathbf{U}^\mathsf{T}\mathbf{Q}^\mathsf{T}\right) = \text{tr}\left((\mathbf{I} + \lambda\mathbf{D})^{-1}\mathbf{U}^\mathsf{T}\mathbf{Q}^\mathsf{T}\mathbf{QU}\right) \\ &= \text{tr}\left((\mathbf{I} + \lambda\mathbf{D})^{-1}\right) = \sum_{i=1}^{k}\frac{1}{1+\lambda D_{ii}}.\end{aligned}$$

(c)

$$\begin{aligned}\|\mathbf{y} - \mathbf{Ay}\|^2 &= \mathbf{y}^\mathsf{T}\mathbf{y} - 2\mathbf{y}^\mathsf{T}\mathbf{Ay} + \mathbf{y}^\mathsf{T}\mathbf{AAy} \\ &= \mathbf{y}^\mathsf{T}\mathbf{y} - 2\tilde{\mathbf{y}}^\mathsf{T}(\mathbf{I} + \lambda\mathbf{D})^{-1}\tilde{\mathbf{y}} + \tilde{\mathbf{y}}^\mathsf{T}(\mathbf{I} + \lambda\mathbf{D})^{-2}\tilde{\mathbf{y}}\end{aligned}$$

where $\tilde{\mathbf{y}} = \mathbf{U}^\mathsf{T}\mathbf{Q}^\mathsf{T}\mathbf{y}$. Once $\mathbf{y}^\mathsf{T}\mathbf{y}$ and $\tilde{\mathbf{y}}$ have been evaluated, it's clear that $\|\mathbf{y} - \mathbf{Ay}\|^2$ can be evaluated in $O(k)$ operations, since $\mathbf{D}$ is diagonal, and $\tilde{\mathbf{y}}$ is of dimension k. So the GCV score can be calculated in $O(k)$ operations, for each trial λ.

5.(a) The direct cost is $O(np^2)$. Adapting the section 6.9.1 algorithm, we can treat each column of $\mathbf{WX}$ as the $\mathbf{y}$ vector is treated there. Let $\mathbf{z}$ denote column j of $\mathbf{WX}$, and let $k(i)$ give the row index in $\bar{\mathbf{X}}$ of row i of $\mathbf{X}$. Then

$$\mathbf{X}^\mathsf{T}\mathbf{z} = \bar{\mathbf{X}}^\mathsf{T}\bar{\mathbf{z}} \text{ where } \bar{z}_l = \sum_{k(i)=l} w_i\bar{X}(l,j),$$

which costs $O(n) + O(mp)$ operations. Repeating this for each column of the p columns of $\mathbf{WX}$ gives the required algorithm.

(b) The direct method cost is $O(np_1p_2)$. Let $\mathbf{z}$ denote column j of $\mathbf{WX_2}$.

$$\mathbf{X}_1^\mathsf{T}\mathbf{z} = \bar{\mathbf{X}}_1^\mathsf{T}\bar{\mathbf{z}} \text{ where } \bar{z}_l = \sum_{k_1(i)=l} w_i\bar{X}(k_2(i),j),$$

which costs $O(n) + O(m_1p_1)$, and has to be applied for each of the columns of $\mathbf{X}_2$ for a total cost $O(np_2) + O(m_1p_1p_2)$.

(c) So if $p_1 > p_2$, $\mathbf{X}_2^\mathsf{T}\mathbf{WX}_1$ is more efficiently computed than its transpose.

6.(a) $\mathcal{V} = nD(\beta)/(n-\tau)^2$, where τ is the effective degrees of freedom of the model. Differentiating w.r.t. λ_j we have

$$\frac{\partial\mathcal{V}}{\partial\lambda_j} = n\frac{\partial D}{\partial\boldsymbol{\beta}}\frac{\mathrm{d}\hat{\boldsymbol{\beta}}}{\mathrm{d}\lambda_j}/(n-\tau)^2 + \frac{nD}{(n-\tau)^3}\frac{\partial\tau}{\partial\lambda_j}.$$

So λ_j should be increased if

$$\frac{\partial D}{\partial\boldsymbol{\beta}}\frac{\mathrm{d}\hat{\boldsymbol{\beta}}}{\mathrm{d}\lambda_j} + \frac{D}{(n-\tau)}\frac{\partial\tau}{\partial\lambda_j} < 0,$$

decreased if the inequality is reversed and left unchanged if it becomes an equality. Re-arranging, while paying careful attention to signs, we arrive at the FS type update

$$\lambda_j' = \frac{-2D}{(n-\tau)}\frac{\partial \tau}{\partial \lambda_j}\left(\frac{\partial D}{\partial \boldsymbol{\beta}}\frac{\mathrm{d}\hat{\boldsymbol{\beta}}}{\mathrm{d}\lambda_j}\right)^{-1}\lambda_j.$$

If $\hat{\mathbf{H}}$ is the observed Hessian of the deviance w.r.t. the model coefficients then by implicit differentiation we have

$$\frac{\mathrm{d}\hat{\boldsymbol{\beta}}}{\mathrm{d}\lambda_j} = -(\hat{\mathbf{H}}+\mathbf{S}_\lambda)^{-1}\mathbf{S}_j\hat{\boldsymbol{\beta}}.$$

Also if $\mathbf{H} = \mathbb{E}\hat{\mathbf{H}}$, then $\tau = \text{tr}\{(\mathbf{H}+\mathbf{S}_\lambda)^{-1}\mathbf{H}\}$. The full derivative of τ w.r.t. λ_j involves third derivatives of the likelihood, thereby removing the main Fellner-Schall method advantage, but if we again adopt the PQL approximation that $\mathbf{H}$ is independent of λ, then $\partial\tau/\partial\lambda_j = -\text{tr}\{(\hat{\mathbf{H}}+\mathbf{S}_\lambda)^{-1}\mathbf{S}_j(\hat{\mathbf{H}}+\mathbf{S}_\lambda)^{-1}\mathbf{H}\}$, which only involves the quantities anyway required for Newton optimization of the penalized deviance.

The equivalent derivation applied to AIC yields

$$\lambda_j' = \frac{\partial \tau}{\partial \lambda_j}\left(\frac{\partial l}{\partial \boldsymbol{\beta}}\frac{\mathrm{d}\hat{\boldsymbol{\beta}}}{\mathrm{d}\lambda_j}\right)^{-1}\lambda_j.$$

(b) See `ch6.solutions` in package `gamair`.

(c) See `ch6.solutions` in package `gamair`.

7.(a) Writing τ for the EDF,

$$\frac{nD}{(n-\gamma\tau)^2} = \frac{1}{\gamma}\frac{nD/\gamma}{(n/\gamma-\tau)^2}$$

but the latter is proportional to an estimate of the GCV score for a sample of size n/γ, where the deviance expected for a sample of that size has been obtained by dividing the deviance based on n data by γ.

(b) Applying the same principle to REML we get

$$l\hat{\boldsymbol{\beta}})/\gamma - \hat{\boldsymbol{\beta}}^{\mathsf{T}}\mathbf{S}_\lambda\hat{\boldsymbol{\beta}} - \log|\mathbf{S}_\lambda|_+ - \log|\mathbf{H}/\gamma+\mathbf{S}_\lambda| + M\log(2\pi).$$

8. See `ch6.solutions` in package `gamair`.

9.(a) An improper uniform prior on the log smoothing parameters.

(b) All that is required is to add the log prior to the REML during optimization, adding its gradient and Hessian w.r.t. the log smoothing parameters to those of the REML as well, to facilitate Newton method optimization. The inverse of the negative Hessian at convergence then becomes the estimate of the posterior covariance matrix for the smoothing parameters.

(c) In this circumstance there is an invertible mapping from smoothing parameters to termwise effective degrees of freedom, so we can view the REML as being parameterized in terms of either. Hence it suffices to add the log of the prior on the termwise effective degrees of freedom to the REML score during optimization. The required derivatives of this prior w.r.t. the log smoothing parameters are obtainable using the results given in section 6.6.1.

(d) For terms with multiple smoothing parameters the mapping from smoothing parameters to effective degrees of freedom is clearly not invertible, so this simple approach could not be used to yield proper priors.

C.7 Chapter 7

1.(a)
```
data(hubble)
h1 <- gam(y~s(x),data=hubble)
plot(h1) ## model is curved
h0 <- gam(y~x,data=hubble)
h1;h0
AIC(h1,h0)
```

The smooth (curved) model has a lower GCV and lower AIC score than the straight line model. On the face of it there is a suggestion that velocities are lower at very high distances than Hubble's law suggests. This would imply an accelerating expansion!

(b)
```
gam.check(h1) # oh dear
h2 <- gam(V~s(D),data=hubble,family=quasi(var=mu))
gam.check(h2) # not great, but better
h2
```

The residual plots for `h1` are problematic: there is a clear relationship between the mean and the variance. Perhaps a quasi-likelihood approach might solve this. `m2` does have somewhat better residual plots, although they are still not perfect. All evidence for departure from Hubble's law has now vanished.

2.(a)
```
library(MASS)
par(mfrow=c(2,2))
mc <- gam(accel~s(times,k=40),data=mcycle)
plot(mc,residuals=TRUE,se=FALSE,pch=1)
```

Note the way the fitted curve dips too early.

(b)
```
mc1 <- lm(accel~poly(times,11),data=mcycle)
termplot(mc1,partial.resid=TRUE)
```

Notice the wild oscillations, unsupported by the data.

(c)
```
mc2 <- gam(accel~s(times,k=11,fx=TRUE),data=mcycle)
plot(mc2,residuals=TRUE,se=FALSE,pch=1)
```

. . . not much worse than the penalized fit.

(d)
```
mc3 <- gam(accel~s(times,k=11,fx=TRUE,bs="cr"),data=mcycle)
plot(mc3,residuals=TRUE,se=FALSE,pch=1)
```

So, `mc` is a bit better than `mc2` which is a bit better than `mc3` which is much better than `mc1`: i.e., the polynomial does much worse than any sort of spline, while regression splines are a bit worse than penalized splines; however, the TPRS is almost as good as a penalized spline.

(e) `par(mfrow=c(1,1))`
`plot(mcycle$times,residuals(mc))`

The first 20 residuals have much lower variance than the remainder. In addition, just after time 9 there is a substantial cluster of negative residuals, followed by a cluster of positive residuals, suggesting that the model is not capturing the mean acceleration correctly in this region. This is also apparent in the default plot of `mc`, where the model dips too early.

(f) The following was run several times with different α values, before settling on 400 as the weight which causes the final ratio to be approximately 1.

```
mcw <- gam(accel~s(times,k=40),data=mcycle,
           weights=c(rep(400,20),rep(1,113)))
plot(mcw,residuals=TRUE,pch=1)
rsd <- residuals(mcw)
plot(mcycle$times,rsd)
var(rsd[21:133])/var(rsd[1:20])
```

The model and residual plots are now very much better. Although this procedure was somewhat ad hoc it can only be better than ignoring the variance problem in this case.

(g) The following uses the integrated squared third derivative as penalty (`m=3`).

```
gam(accel~s(times,k=40,m=3),data=mcycle,
    weights=c(rep(400,20),rep(1,113)))
```

The original model perhaps failed to go quite deep enough at the minimum of the data: the new curve is fine. Further increase of `m` doesn't result in much further change.

3.(a) Fitting the model to $\mathbf{I}_j$ and evaluating the fitted values, $\hat{\boldsymbol{\mu}}^*$, is equivalent to forming $\hat{\boldsymbol{\mu}}^* = \mathbf{A}\mathbf{I}_j$, but this is clearly just the jth column of $\mathbf{A}$.

(b)
```
library(MASS)
n <- nrow(mcycle)
A <- matrix(0,n,n)
for (i in 1:n) {
  mcycle$y<-mcycle$accel*0;mcycle$y[i] <- 1
  A[,i] <- fitted(gam(y~s(times,k=40),data=mcycle,sp=mc$sp))
}
```

(Actually this could be done more efficiently using the `fit=FALSE` option in `gam`.)

(c) `rowSums(A)` shows that all the rows sum to 1. This has to happen, since by construction the model does not penalize the constant (or linear trend) part of the model. Hence if $\mathbf{b}$ is a vector all elements of which have the same value, b, then we require that $\mathbf{b} = \mathbf{A}\mathbf{b}$. This means that $b_i = \sum_{j=1}^n A_{ij} b_j \; \forall \, i$, but this is equivalent to $b = b \sum_{j=1}^n A_{ij} \; \forall \, i$, which will only happen if $\sum_{j=1}^n A_{ij} = 1 \, \forall \, i$. That is, the rows of $\mathbf{A}$ must each sum to 1.

(d) `plot(mcycle$times,A[,65],type="l",ylim=c(-0.05,0.15))`

Notice how the kernel peaks at the time of the datum that it relates to.

(e) `for (i in 1:n) lines(mcycle$times,A[,i])`

The kernels all have rather similar typical widths: the heights vary according to the number of data within that width. If many data are making a substantial contribution to the weighted sum that defines the fitted value, then the weights must be lower than if fewer data contribute.

(f)
```
par(mfrow=c(2,2))
mcycle$y<-mcycle$accel*0;mcycle$y[65] <- 1
for (k in 1:4) plot(mcycle$times,fitted(
     gam(y~s(times,k=40),data=mcycle,sp=mc$sp*10^(k-1.5))
     ),type="l",ylab="A[65,]",ylim=c(-0.01,0.12))
```

Low smoothing parameters lead to narrow, high kernels, while high smoothing parameters result in wide, low kernels.

4.(a)
```
w <- c(rep(400,20),rep(1,113))
m <- 40;par(mfrow=c(1,1))
sp <- seq(-13,12,length=m) ## trial log(sp)'s
AC1 <- EDF <- rep(0,m)
for (i in 1:m) { ## loop through s.p.'s
 b <- gam(accel~s(times,k=40),data=mcycle,weights=w,
           sp=exp(sp[i]))
 EDF[i] <- sum(b$edf)
 AC1[i] <- acf(residuals(b),plot=FALSE)$acf[2]
}
plot(EDF,AC1,type="l");abline(0,0,col=2)
```

So the lag 1 residual auto-correlation starts positive declines to zero around the GCV best fit model, and then becomes increasingly negative.

(b) At low EDF the model oversmooths and fails to capture the mean of the data. This will lead to clear patterns in the mean of the residuals against time: the ACF picks this residual trend up as positive auto-correlation.

(c) i. So the jth row of the $n \times n$ matrix $\mathbf{A}$ is $j - (k+1)/2$ zeroes, followed by k values, $1/k$, followed by $n - j - (k-1)/2$ further zeroes. That is, something like:

$$\begin{bmatrix} 0 & . & . & 0 & 1/k & 1/k & . & . & 1/k & 0 & . & . & 0 \end{bmatrix}.$$

ii. Using $\mathbf{V}_{\hat{\epsilon}}/\sigma^2 = \mathbf{I} - 2\mathbf{A} + \mathbf{AA}$ and slogging through, it turns out that the leading diagonal elements (in the interior) are $\sigma^2(k-1)/k$, while the elements on the sub- and super-diagonals (the lag 1 covariances) are $-\sigma^2(k+1)/k^2$.

iii. The correlation at lag 1 is clearly

$$-\sqrt{\frac{k+1}{k(k-1)}}.$$

So the residual auto-correlation is always negative, with magnitude decreasing as k increases. Once k is small enough, oversmoothing is avoided, so that the observed ACF reflects this residual auto-correlation, rather than the inadequately modelled trend. It is this increasingly negative auto-correlation that was seen in the `mcycle` example in part (a).

It is tempting to view negative autocorrelation in the residuals as an indication of overfitting, but the preceding analysis indicates that some care would be needed to do this, since the true residual auto-correlation (at lag 1) should always be negative.

5.(a)
```
attach(co2s)
plot(c.month,co2,type="l")
```

(b)
```
b<-gam(co2~s(c.month,k=300,bs="cr"))
```

(c)
```
pd <- data.frame(c.month=1:(n+36))
fv <- predict(b,pd,se=TRUE)
plot(pd$c.month,fv$fit,type="l")
lines(pd$c.month,fv$fit+2*fv$se,col=2)
lines(pd$c.month,fv$fit-2*fv$se,col=2)
```

The prediction of smoothly and sharply decreasing CO_2 is completely out of line with the long term pattern in the data, and is driven entirely by the seasonal dip at the end of the data.

(d)
```
b2 <- gam(co2~s(month,bs="cc")+s(c.month,bs="cr",k=300),
          knots=list(month=seq(1,13,length=10)))
```

Notice how the `knots` argument has been used to ensure that the smooth of `month` wraps around correctly: month 1 should be the same as month 13 (if it ever occurred), not month 12.

(e)
```
pd2 <- data.frame(c.month=1:(n+36),
                  month=rep(1:12,length.out=n+36))
fv <- predict(b2,pd2,se=TRUE)
plot(pd$c.month,fv$fit,type="l")
lines(pd$c.month,fv$fit+2*fv$se,col=2)
lines(pd$c.month,fv$fit-2*fv$se,col=2)
```

The predictions now look *much* more credible, since we now extrapolate the long terms trend, and simply repeat the seasonal cycle on top of it. However, it is worth noting that the smooth of cumulative month (the long term trend) is still estimated to have rather high effective degrees of freedom, and it does still wiggle a lot, suggesting that extrapolation is still a fairly dangerous thing to do, and we had better not rely on it very far into the future.

Note that an alternative way of extrapolating is to add knots beyond the range of the observed data, e.g., with the argument

```
knots=list(...,c.month=seq(1,n+36,length=300))
```

Since the function value at these extra knots can only very subtly alter the shape of the function in the range of the data, the resulting curves tend to have very high associated standard errors: this is probably realistic!

6. There is no unique 'right' answer to this, but here is an outline of how to get started. I looked at `ir` and `dp` lagged for 1 to 4 months. Everything suggested dropping the effect of `ir` at lag 4 from the model, but everything else marginally improved the model, so was left in. Here is the code.

```
n<-nrow(ipo)
## create lagged variables ...
ipo$ir1 <- c(NA,ipo$ir[1:(n-1)])
ipo$ir2 <- c(NA,NA,ipo$ir[1:(n-2)])
ipo$ir3 <- c(NA,NA,NA,ipo$ir[1:(n-3)])
ipo$ir4 <- c(NA,NA,NA,NA,ipo$ir[1:(n-4)])
ipo$dp1 <- c(NA,ipo$dp[1:(n-1)])
ipo$dp2 <- c(NA,NA,ipo$dp[1:(n-2)])
ipo$dp3 <- c(NA,NA,NA,ipo$dp[1:(n-3)])
ipo$dp4 <- c(NA,NA,NA,NA,ipo$dp[1:(n-4)])
## fit initial model and look at it ...
b<-gam(n.ipo~s(ir1)+s(ir2)+s(ir3)+s(ir4)+s(log(reg.t))+
    s(dp1)+s(dp2)+s(dp3)+s(dp4)+s(month,bs="cc")+s(t,k=20),
    data=ipo,knots=list(month=seq(1,13,length=10)),
    family=poisson,gamma=1.4)
par(mfrow=c(3,4))
plot(b,scale=0)
summary(b)
## re-fit model dropping ir4 ...
b1 <- gam(n.ipo~s(ir1)+s(ir2)+s(ir3)+s(log(reg.t))+s(dp1)+
          s(dp2)+s(dp3)+s(dp4)+s(month,bs="cc")+s(t,k=20),
          data=ipo,knots=list(month=seq(1,13,length=10)),
          family=poisson,gamma=1.4)
par(mfrow=c(3,4))
plot(b1,scale=0)
summary(b1)
## residual checking ...
gam.check(b1)
acf(residuals(b1))
```

The final model has good residual plots that largely eliminate the auto-correlation. In the above the degrees of freedom for `t` have been kept fairly low, in order to try and force the model to use the other covariates in preference to `t`, and because too much freedom for the smooth of time tends to lead to it being used to the point of overfitting (lag 1 residual auto-correlation becomes very negative, for example). The most noticeable feature of the plotted smooth effects is the way that IR at all significant lags tends to be associated with an increase in IPO volume up to around 20%, after which there is a decline. If this reflects company behaviour, it certainly makes sense. If IR averages are too low then there is a danger of investors not being interested in IPOs, while excessively high IRs suggest that companies are currently being undervalued excessively, so that unreasonably small amounts of capital will be raised by the IPO. The other plots are harder to interpret!

7. The following is the best model I found.

```
wm<-gam(price~s(h.rain)+s(s.temp)+s(h.temp)+s(year),
     data=wine,family=Gamma(link=identity),gamma=1.4)
plot(wm,pages=1,residuals=TRUE,pch=1,scale=0)
acf(residuals(wm))
gam.check(wm)
```

```
predict(wm,wine,se=TRUE)
```

A Gamma family seems to produce acceptable residual plots. `w.temp` appears to be redundant. Two way interactions between the weather variables only make the GCV and AIC scores worse. `gamma=1.4` is a prudent defence against overfitting, given that the sample size is so small. The effects are easy to interpret: (i) more recent vintages are worth less than older ones; (ii) low harvest rainfall and high summer temperatures are both associated with higher prices; (iii) there is some suggestion of an optimum harvest temperature at 17.5C, but the possible increases at very low and very high harvest temperatures make interpretation difficult (however, this harvest effect substantially increases the proportion deviance explained).

8.(a)
```
plot(bf$day,bf$pop,type="l")
```

(b)
```
## prepare differenced and lagged data ...
bf$dn <- c(NA,bf$pop[2:n]-bf$pop[1:(n-1)])
lag <- 6
bf$n.lag <- c(rep(NA,lag),bf$pop[1:(n-lag)])
bf1 <- bf[(lag+1):n,] # strip out NAs, for convenience
## fit model, note no intercept ...
b<-gam(dn~n.lag+pop+s(log(n.lag),by=n.lag)+
       s(log(pop),by=-pop)-1,data=bf1)
plot(b,pages=1,scale=-1,se=FALSE) ## effects
plot(abs(fitted(b)),residuals(b))
acf(residuals(b))
```

So the per capita birth rate declines with increasing population, which is sensible, given likely competition for resources. The mortality rate increases with population, which is also plausible, although the decline at very high densities is surprising, and may not be real. Note that both functions are negative in places: this is biologically nonsensical. The residual plots are not brilliant, and there is residual auto-correlation at lag 1, so the model is clearly not great.

(c)
```
fv <- bf$pop
e <- rnorm(n)*0 ## increase multiplier for noisy version
min.pop <- min(bf$pop);max.pop <- max(bf$pop)
for (i in (lag+1):n) { ## iteration loop
  dn <- predict(b,data.frame(n.lag=fv[i-lag],pop=fv[i-1]))
  fv[i] <- fv[i-1]+dn + e[i];
  fv[i]<-min(max.pop,max(min.pop,fv[i]))
}
plot(bf$day,fv,type="l")
```

The amplitude without noise is rather low, but does improve with noise; however, the feature that cycles tend to be smoother at the trough and noisier at the peak is not re-captured, suggesting that something more complicated than a constant variance model may be needed. Ideally we would fit the model with constraints forcing the functions to be positive, and possibly monotonic: this is possible, but more complicated (see `?pcls`).

(d) Census data like these are rather unusual. In most cases the model has to deal with measurement error as well, which rather invalidates the approach used in this question.

9.(a)
```
pairs(chl,pch=".")
```

Notice, in particular, that there is fairly good coverage of the predictors `bath` and `jul.day`. Plotting histograms of individual predictors indicates very skewed distributions for `chl.sw` and `bath`: raising these to the power of .4 and .25, respectively, largely eliminates this problem.

(b)
```
fam <- quasi(link=log,var=mu^2)
cm <- gam(chl ~ s(I(chl.sw^.4),bs="cr",k=20)+
      s(I(bath^.25),bs="cr",k=60)+s(jul.day,bs="cr",k=20),
      data=chl,family=fam,gamma=1.4)
gam.check(cm)
summary(cm)
```

The given `quasi` family seems to deal nicely with the mean variance relationship, and a log link is required to linearize the model.

(c)
```
## create fit and validation sets ...
set.seed(2)
n<-nrow(chl);nf <- floor(n*.9)
ind <- sample(1:n,nf,replace=FALSE)
chlf <- chl[ind,];chlv <- chl[-ind,]
## fit to the fit set
cmf<-gam(chl ~ s(I(chl.sw^.4),bs="cr",k=20)+
     s(I(bath^.25),bs="cr",k=60)+s(jul.day,bs="cr",k=20),
     data=chlf,family=fam,gamma=1.4)
## evaluate prop. dev. explained for validation set
y <- chlv$chl;w <- y*0+1
mu <- predict(cmf,chlv,type="response")
pred.dev <- sum(fam$dev.resids(y,mu,w))
null.dev <- sum(fam$dev.resids(y,mean(y),w))
1-pred.dev/null.dev # prop dev. explained
```

I got proportion deviance explained of about 46% for the fitted models and for predicting the validation set, suggesting that the model is not overfitting, and does provide a useful improvement on the raw satellite data. More sophisticated models based on tensor product smooths can be used to improve matters further. For practical use of these models, it is always important to look at their predictions over the whole spatial arena of interest, at a number of times through the year, to check for artefacts (for example, from extrapolating outside the observed predictor space).

10.(a)
```
g1<-gamm(weight ~ Variety + s(Time) +
    s(Time,by=ordered(Variety),data=Soybean,
    family=Gamma(link=log), random=list(Plot=~Time))
plot(g1$lme) ## standard mean variance plot
par(mfrow=c(1,3))
plot(g1$gam,residuals=TRUE,all.terms=TRUE,scale=0)
```

The residual plots look fine. It seems that variety P increases its weight a little more slowly than variety F (don't forget that this is on the log scale).

(b)
```
summary(g1$gam) ## evidence for variety dependence
## could also do following ....
g2 <- gamm(weight~s(Time),family=Gamma(link=log),
      data=Soybean,random=list(Plot=~Time))
g3 <- gamm(weight~Variety+s(Time),family=Gamma(link=log),
      data=Soybean,random=list(Plot=~Time))
## following only a rough guide, but also supports g1 ...
AIC(g1$lme,g2$lme,g3$lme)
```

The summary (in combination with the plotted effects) gives strong evidence that variety P gives higher weights, with the difference decreasing slightly with time. Fitting models without a smooth function of time as a correction for P, or without any effect of variety, both give model fits that seem worse than the original model fit according to the AIC of the working model at convergence of the PQL iterations, confirming the implications of the summary.

(c) If varieties F and P have only slightly different trajectories through time, but we use completely separate smooths for each, then both smooths will require a similar relatively large number of degrees of freedom in order to represent the time trajectory of each variety. On the other hand, if we model the trajectory of P as F's trajectory plus a correction, it is possible that this correction may be very smooth, so that a good fit can be achieved without using up as many degrees of freedom as would be needed for the same fit, using the completely separate smooths. This is in fact what happens for the soybean data.

11. See `ch7.solutions` in package `gamair`.

Bibliography

Adam, B.-L., Y. Qu, J. W. Davis, M. D. Ward, M. A. Clements, L. H. Cazares, O. J. Semmes, P. F. Schellhammer, Y. Yasui, Z. Feng, et al. (2002). Serum protein fingerprinting coupled with a pattern-matching algorithm distinguishes prostate cancer from benign prostate hyperplasia and healthy men. *Cancer Research 62*(13), 3609–3614.

Agarwal, G. G. and W. Studden (1980). Asymptotic integrated mean square error using least squares and bias minimizing splines. *The Annals of Statistics 8*(6), 1307–1325.

Agresti, A. (1996). *An Introduction to Categorical Data Analysis*. New York: Wiley.

Akaike, H. (1973). Information theory and an extension of the maximum likelihood principle. In B. Petran and F. Csaaki (Eds.), *International symposium on information theory*, Budapest: Akadeemiai Kiadi, pp. 267–281.

Andersen, P. K., M. W. Bentzon, and J. P. Klein (1996). Estimating the survival function in the proportional hazards regression model: a study of the small sample size properties. *Scandinavian Journal of Statistics 23*(1), 1–12.

Anderssen, R. and P. Bloomfield (1974). A time series approach to numerical differentiation. *Technometrics 16*(1), 69–75.

Augustin, N. H., D. L. Borchers, E. D. Clarke, S. T. Buckland, and M. Walsh (1998). Spatio-temporal modelling of annual egg production of fish stocks using generalized additive models. *Canadian Journal of Fisheries and Aquatic Sciences 55*, 2608–2621.

Baker, R. R. and M. A. Bellis (1993). Human sperm competition: ejaculate adjustment by males and the function of masturbation. *Animal Behaviour 46*, 861–885.

Bates, D. and M. Maechler (2015). *Matrix: Sparse and Dense Matrix Classes and Methods*. R package version 1.2-2.

Belitz, C., A. Brezger, T. Kneib, S. Lang, and N. Umlauf (2015). BayesX: Software for Bayesian inference in structured additive regression models.

Belitz, C. and S. Lang (2008). Simultaneous selection of variables and smoothing parameters in structured additive regression models. *Computational Statistics & Data Analysis 53*(1), 61–81.

Borchers, D. L., S. T. Buckland, I. G. Priede, and S. Ahmadi (1997). Improving the precision of the daily egg production method using generalized additive models. *Canadian Journal of Fisheries and Aquatic Sciences 54*, 2727–2742.

Bravington, M. V. (2013). *debug: MVB's debugger for R.* R package version 1.3.1.

Breslow, N. E. (1972). Contribution to the discussion of the paper by DR Cox. *Journal of the Royal Statistical Society, Series B 34*(2), 216–217.

Breslow, N. E. and D. G. Clayton (1993). Approximate inference in generalized linear mixed models. *Journal of the American Statistical Association 88*, 9–25.

Brosnan, M. and I. Walker (2009). A preliminary investigation into the potential role of waist hip ratio (WHR) preference within the assortative mating hypothesis of autistic spectrum disorders. *Journal of Autism and Developmental Disorders 39*(1), 164–171.

Chambers, J. M. (1993). Linear Models. In J. M. Chambers and T. J. Hastie (Eds.), *Statistical Models in S*, pp. 95–144. Chapman & Hall.

Chavez-Demoulin, V. and A. C. Davison (2005). Generalized additive modelling of sample extremes. *Journal of the Royal Statistical Society: Series C (Applied Statistics) 54*(1), 207–222.

Chouldechova, A. and T. Hastie (2015). Generalized additive model selection. *arXiv preprint arXiv:1506.03850.*

Claeskens, G., T. Krivobokova, and J. D. Opsomer (2009). Asymptotic properties of penalized spline estimators. *Biometrika 96*(3), 529–544.

Cline, A. K., C. B. Moler, G. W. Stewart, and J. H. Wilkinson (1979). An estimate for the condition number of a matrix. *SIAM Journal on Numerical Analysis 16*(2), 368–375.

Collett, D. (2015). *Modelling Survival Data in Medical Research* (3 ed.). CRC Press.

Cox, D. (1972). Regression models and life tables. *Journal of the Royal Statistical Society. Series B (Methodological) 34*(2), 187–220.

Cox, D. D. (1983). Asymptotics for m-type smoothing splines. *The Annals of Statistics 11*(2), 530–551.

Cox, D. R. and D. V. Hinkley (1974). *Theoretical Statistics.* London: Chapman & Hall.

Crainiceanu, C. M. and D. Ruppert (2004). Likelihood ratio tests in linear mixed models with one variance component. *Journal of the Royal Statistical Society: Series B (Statistical Methodology) 66*(1), 165–185.

Craven, P. and G. Wahba (1979). Smoothing noisy data with spline functions. *Numerische Mathematik 31*, 377–403.

Davies, R. B. (1980). Algorithm as 155: The distribution of a linear combination of χ^2 random variables. *J. R. Statist. Soc. C 29*, 323–333.

Davis, T. A. (2006). *Direct Methods for Sparse Linear Systems.* Philadelphia: SIAM.

Davison, A. C. (2003). *Statistical Models.* Cambridge: Cambridge University Press.

de Boor (1978). *A Practical Guide to Splines.* New York: Springer.

de Boor, C. (2001). *A Practical Guide to Splines* (Revised ed.). New York: Springer.

De Hoog, F. and M. Hutchinson (1987). An efficient method for calculating smoothing splines using orthogonal transformations. *Numerische Mathematik 50*(3), 311–319.

Demmel, J. (1997). *Applied Numerical Linear Algebra*. Philadelphia: SIAM.

Demmler, A. and C. Reinsch (1975). Oscillation matrices with spline smoothing. *Numerische Mathematik 24*(5), 375–382.

Dempster, A., N. Laird, and D. Rubin (1977). Maximum likelihood from incomplete data via the EM algorithm (with discussion). *Journal of the Royal Statistical Society, Series B 39*, 1–38.

Dixon, C. E. (2003). *Multi-dimensional modelling of physiologically and temporally structured populations*. Ph. D. thesis, University of St Andrews.

Dobson, A. J. and A. Barnett (2008). *An Introduction to Generalized Linear Models*. CRC Press.

Duchon, J. (1977). Splines minimizing rotation-invariant semi-norms in Solobev spaces. In W. Schemp and K. Zeller (Eds.), *Construction Theory of Functions of Several Variables*, Berlin, pp. 85–100. Springer.

Dunn, P. K. and G. K. Smyth (2005). Series evaluation of Tweedie exponential dispersion model densities. *Statistics and Computing 15*(4), 267–280.

Eilers, P. H. C. and B. D. Marx (1996). Flexible smoothing with B-splines and penalties. *Statistical Science 11*(2), 89–121.

Eilers, P. H. C. and B. D. Marx (2003). Multivariate calibration with temperature interaction using two-dimensional penalized signal regression. *Chemometrics and Intelligent Laboratory Systems 66*, 159–174.

Ellner, S. P., B. E. Kendall, S. N. Wood, E. McCauley, and C. J. Briggs (1997). Inferring mechanism from time-series data: delay differential equations. *Physica D 110*(3-4), 182–194.

Fahrmeir, L., T. Kneib, and S. Lang (2004). Penalized structured additive regression for space-time data: A Bayesian perspective. *Statistica Sinica 14*(3), 731–761.

Fahrmeir, L., T. Kneib, S. Lang, and B. Marx (2013). *Regression Models*. Springer.

Fahrmeir, L. and S. Lang (2001). Bayesian inference for generalized additive mixed models based on markov random field priors. *Applied Statistics 50*, 201–220.

Faraway, J. J. (2014). *Linear Models with R* (2nd ed.). CRC Press.

Fellner, W. H. (1986). Robust estimation of variance components. *Technometrics 28*(1), 51–60.

Fletcher, D. (2012). Estimating overdispersion when fitting a generalized linear model to sparse data. *Biometrika 99*(1), 230–237.

Freedman, W. L., B. F. Madore, B. K. Gibson, L. Ferrarese, D. D. Kelson, S. Sakai, J. R. Mould, R. C. Kennicutt, H. C. Ford, J. A. Graham, J. P. Huchra, S. M. Hughes, G. D. Illingworth, L. M. Macri, and P. B. Stetson (2001). Final results from the Hubble Space Telescope key project to measure the Hubble constant. *The Astrophysical Journal 553*(1), 47–72.

Goldsmith, J., F. Scheipl, L. Huang, J. Wrobel, J. Gellar, J. Harezlak, M. W. McLean, B. Swihart, L. Xiao, C. Crainiceanu, and P. T. Reiss (2016). *refund: Regression with Functional Data*. R package version 0.1-16.

Golub, G. H., M. Heath, and G. Wahba (1979). Generalized cross validation as a method for choosing a good ridge parameter. *Technometrics 21*(2), 215–223.

Golub, G. H. and C. F. Van Loan (2013). *Matrix computations* (4th ed.). Baltimore: Johns Hopkins University Press.

Green, P. J. and B. W. Silverman (1994). *Nonparametric Regression and Generalized Linear Models*. Chapman & Hall.

Greven, S., C. M. Crainiceanu, H. Küchenhoff, and A. Peters (2008). Restricted likelihood ratio testing for zero variance components in linear mixed models. *Journal of Computational and Graphical Statistics 17*(4), 870–891.

Greven, S. and T. Kneib (2010). On the behaviour of marginal and conditional AIC in linear mixed models. *Biometrika 97*(4), 773–789.

Gu, C. (1992). Cross-validating non-gaussian data. *Journal of Computational and Graphical Statistics 1*, 169–179.

Gu, C. (2013). *Smoothing Spline ANOVA Models* (2nd ed.). New York: Springer.

Gu, C. and Y. J. Kim (2002). Penalized likelihood regression: general approximation and efficient approximation. *Canadian Journal of Statistics 34*(4), 619–628.

Gu, C. and G. Wahba (1991). Minimizing GCV/GML scores with multiple smoothing parameters via the Newton method. *SIAM Journal on Scientific and Statistical Computing 12*(2), 383–398.

Gu, C. and D. Xiang (2001). Cross-validating non-gaussian data: Generalized approximate cross-validation revisited. *Journal of Computational and Graphical Statistics 10*(3), 581–591.

Hall, P. and J. D. Opsomer (2005). Theory for penalised spline regression. *Biometrika 92*(1), 105–118.

Hand, D. J., F. Daly, A. D. Lunn, K. J. McConway, and E. Ostrowski (1994). *A Handbook of Small Data Sets*. Chapman & Hall.

Handcock, M. S., K. Meier, and D. Nychka (1994). Comment. *Journal of the American Statistical Association 89*(426), 401–403.

Härdle, W., P. Hall, and J. S. Marron (1988). How far are automatically chosen regression smoothing parameters from their optimum? *Journal of the American Statistical Association 83*(401), 86–95.

Hastie, T. and R. Tibshirani (1986). Generalized additive models (with discussion). *Statistical Science 1*, 297–318.

Hastie, T. and R. Tibshirani (1990). *Generalized Additive Models*. Chapman & Hall.

Hastie, T. and R. Tibshirani (1993). Varying-coefficient models. *Journal of the Royal Statistical Society, Series B 55*(4), 757–796.

Hastie, T., R. Tibshirani, and J. Friedman (2009). *The Elements of Statistical Learn-*

ing. Springer.

Hastie, T. J. (1993). Generalized additive models. In J. Chambers and T. Hastie (Eds.), *Statistical Models in S*. London: Chapman & Hall.

Helwig, N. E. and P. Ma (2016). Smoothing spline ANOVA for super-large samples: Scalable computation via rounding parameters. *arXiv preprint arXiv:1602.05208*.

Horwood, J. (1993). The Bristol channel sole (*solea solea (l.)*): A fisheries case study. *Advances in Marine Biology 29*, 215–367.

Horwood, J. and M. Greer Walker (1990). Determinacy of fecundity in sole (*solea solea*) from the Bristol channel. *Journal of the Marine Biological Association of the United Kingdom 70*, 803–813.

Kammann, E. E. and M. P. Wand (2003). Geoadditive models. *Applied Statistics 52*(1), 1–18.

Kass, R. E. and D. Steffey (1989). Approximate Bayesian inference in conditionally independent hierarchical models (parametric empirical Bayes models). *Journal of the American Statistical Association 84*(407), 717–726.

Kauermann, G., T. Krivobokova, and L. Fahrmeir (2009). Some asymptotic results on generalized penalized spline smoothing. *Journal of the Royal Statistical Society: Series B (Statistical Methodology) 71*(2), 487–503.

Kim, Y. J. and C. Gu (2004). Smoothing spline Gaussian regression: more scalable computation via efficient approximation. *Journal of the Royal Statistical Society, Series B 66*, 337–356.

Kimeldorf, G. S. and G. Wahba (1970). A correspondence between Bayesian estimation on stochastic processes and smoothing by splines. *The Annals of Mathematical Statistics 41*(2), 495–502.

Klein, J. and M. Moeschberger (2003). *Survival Analysis: Techniques for Censored and Truncated Data* (2nd ed.). New York: Springer.

Klein, N., T. Kneib, S. Lang, and A. Sohn (2015). Bayesian structured additive distributional regression with an application to regional income inequality in Germany. *Annals of Applied Statistics 9*, 1024–1052.

Kneib, T. and L. Fahrmeir (2006). Structured additive regression for categorical space–time data: A mixed model approach. *Biometrics 62*(1), 109–118.

Laird, N. M. and J. H. Ware (1982). Random-effects models for longitudinal data. *Biometrics 38*, 963–974.

Lancaster, P. and K. Šalkauskas (1986). *Curve and Surface Fitting: An Introduction*. London: Academic Press.

Landau, S., I. C. Ellison-Wright, and E. T. Bullmore (2003). Tests for a difference in timing of physiological response between two brain regions measured by using functional magnetic resonance imaging. *Applied Statistics 53*(1), 63–82.

Lang, S. and A. Brezger (2004). Bayesian P-splines. *Journal of Computational and Graphical Statistics 13*, 183–212.

Lang, S., N. Umlauf, P. Wechselberger, K. Harttgen, and T. Kneib (2014). Multilevel structured additive regression. *Statistics and Computing 24*(2), 223–238.

Lee, D.-J. and M. Durbán (2011). P-spline anova-type interaction models for spatio-temporal smoothing. *Statistical Modelling 11*(1), 49–69.

Lin, Y., H. H. Zhang, et al. (2006). Component selection and smoothing in multivariate nonparametric regression. *The Annals of Statistics 34*(5), 2272–2297.

Lindgren, F., H. Rue, and J. Lindström (2011). An explicit link between Gaussian fields and Gaussian Markov random fields: the stochastic partial differential equation approach. *Journal of the Royal Statistical Society: Series B (Statistical Methodology) 73*(4), 423–498.

Liu, H., Y. Tang, and H. H. Zhang (2009). A new chi-square approximation to the distribution of non-negative definite quadratic forms in non-central normal variables. *Computational Statistics & Data Analysis 53*(4), 853–856.

Lowry, M. and G. W. Schwert (2002). IPO market cycles: Bubbles or sequential learning? *The Journal of Finance 67*(3), 1171–1198.

Lucas, C. (2004). LAPACK-style codes for level 2 and 3 pivoted Cholesky factorizations. *LAPACK Working Paper*.

MacKay, D. (2008). *Sustainable Energy – without the hot air*. UIT Cambridge.

Mallows, C. L. (1973). Some comments on c_p. *Technometrics 15*, 661–675.

Marra, G. and S. N. Wood (2011). Practical variable selection for generalized additive models. *Computational Statistics & Data Analysis 55*(7), 2372–2387.

Marra, G. and S. N. Wood (2012). Coverage properties of confidence intervals for generalized additive model components. *Scandinavian Journal of Statistics 39*(1), 53–74.

Marx, B. D. and P. H. Eilers (2005). Multidimensional penalized signal regression. *Technometrics 47*(1), 13–22.

McCullagh, P. and J. A. Nelder (1989). *Generalized Linear Models* (2nd ed.). London: Chapman & Hall.

Miller, D. L. and S. N. Wood (2014). Finite area smoothing with generalized distance splines. *Environmental and Ecological Statistics 21*(4), 715–731.

Nelder, J. A. and R. W. M. Wedderburn (1972). Generalized linear models. *Journal of the Royal Statistical Society, Series A 135*, 370–384.

Nicholson, A. J. (1954a). Compensatory reactions of populations to stresses and their evolutionary significance. *Australian Journal of Zoology 2*, 1–8.

Nicholson, A. J. (1954b). An outline of the dynamics of animal populations. *Australian Journal of Zoology 2*, 9–65.

Nychka, D. (1988). Bayesian confidence intervals for smoothing splines. *Journal of the American Statistical Association 83*(404), 1134–1143.

Parker, R. and J. Rice (1985). Discussion of Silverman (1985). *Journal of the Royal Statistical Society, Series B 47*(1), 40–42.

Patterson, H. D. and R. Thompson (1971). Recovery of interblock information when block sizes are unequal. *Biometrika 58*, 545–554.

Peng, R. D. and L. J. Welty (2004). The NMMAPSdata package. *R News 4*(2), 10–14.

Peto, R. (1972). Contribution to the discussion of the paper by DR Cox. *Journal of the Royal Statistical Society, Series B 34*(2), 205–207.

Pinheiro, J. C. and D. M. Bates (2000). *Mixed-effects models in S and S-PLUS*. New York: Springer-Verlag.

Plummer, M. (2003). JAGS: A program for analysis of Bayesian graphical models using Gibbs sampling. In *Proceedings of the 3rd International Workshop on Distributed Statistical Computing (DSC 2003). March*, pp. 20–22.

Press, W., S. Teukolsky, W. Vetterling, and B. Flannery (2007). *Numerical Recipes* (3rd ed.). Cambridge: Cambridge University Press.

Pya, N. and S. N. Wood (2015). Shape constrained additive models. *Statistics and Computing 25*(3), 543–559.

Quintana-Ortí, G., X. Sun, and C. H. Bischof (1998). A BLAS-3 version of the QR factorization with column pivoting. *SIAM Journal on Scientific Computing 19*(5), 1486–1494.

R Core Team (2016). *R: A Language and Environment for Statistical Computing*. Vienna, Austria: R Foundation for Statistical Computing.

Ramsay, J. and B. Silverman (2005). *Functional Data Analysis* (2nd ed.). Springer.

Ramsay, T. (2002). Spline smoothing over difficult regions. *Journal of the Royal Statistical Society: Series B (Statistical Methodology) 64*(2), 307–319.

Reinsch, C. H. (1967). Smoothing by spline functions. *Numerische Mathematik 10*, 177–183.

Reiss, P. T., L. Huang, H. Chen, and S. Colcombe (2014). Varying-smoother models for functional responses. *arXiv preprint arXiv:1412.0778*.

Reiss, P. T. and T. R. Ogden (2009). Smoothing parameter selection for a class of semiparametric linear models. *Journal of the Royal Statistical Society: Series B (Statistical Methodology) 71*(2), 505–523.

Rice, J. (1986). Convergence rates for partially splined models. *Statistics & Probability Letters 4*(4), 203–208.

Rigby, R. and D. M. Stasinopoulos (2005). Generalized additive models for location, scale and shape. *Journal of the Royal Statistical Society: Series C (Applied Statistics) 54*(3), 507–554.

Rigby, R. A. and D. M. Stasinopoulos (2014). Automatic smoothing parameter selection in GAMLSS with an application to centile estimation. *Statistical Methods in Medical Research 23*(4), 318–332.

Rodríguez-Álvarez, M. X., D.-J. Lee, T. Kneib, M. Durbán, and P. Eilers (2015). Fast smoothing parameter separation in multidimensional generalized P-splines: the SAP algorithm. *Statistics and Computing 25*(5), 941–957.

Rue, H. and L. Held (2005). *Gaussian Markov Random Fields: theory and applications*. CRC Press.

Rue, H., S. Martino, and N. Chopin (2009). Approximate Bayesian inference for latent Gaussian models by using integrated nested Laplace approximations. *Journal of the royal statistical society: Series B 71*(2), 319–392.

Ruppert, D., M. P. Wand, and R. J. Carroll (2003). *Semiparametric Regression*. Cambridge University Press.

Säfken, B., T. Kneib, C.-S. van Waveren, and S. Greven (2014). A unifying approach to the estimation of the conditional Akaike information in generalized linear mixed models. *Electronic Journal of Statistics 8*, 201–225.

Schall, R. (1991). Estimation in generalized linear models with random effects. *Biometrika 78*(4), 719–727.

Scheipl, F., S. Greven, and H. Küchenhoff (2008). Size and power of tests for a zero random effect variance or polynomial regression in additive and linear mixed models. *Computational Statistics & Data Analysis 52*(7), 3283–3299.

Schmid, M. and T. Hothorn (2008). Boosting additive models using component-wise P-splines. *Computational Statistics & Data Analysis 53*(2), 298–311.

Schoenberg, I. J. (1964). Spline functions and the problem of graduation. *Proceedings of the National Academy of Sciences 52*, 947–950.

Schwarz, G. (1978). Estimating the dimension of a model. *Annals of Statistics 6*(2), 461–464.

Shun, Z. and P. McCullagh (1995). Laplace approximation of high dimensional integrals. *Journal of the Royal Statistical Society. Series B (Methodological) 57*(4), 749–760.

Silverman, B. W. (1985). Some aspects of the spline smoothing approach to non-parametric regression curve fitting. *Journal of the Royal Statistical Society B 47*(1), 1–53.

Silvey, S. D. (1970). *Statistical Inference*. London: Chapman & Hall.

Smith, A. F. (1967). Diagnostic value of serum-creatinine-kinase in a coronary care unit. *Lancet 2*, 178.

Smout, S., C. Asseburg, J. Matthiopoulos, C. Fernández, S. Redpath, S. Thirgood, and J. Harwood (2010). The functional response of a generalist predator. *PLoS ONE 5*(5), e10761.

Speckman, P. (1985). Spline smoothing and optimal rates of convergence in nonparametric regression models. *The Annals of Statistics 13*(3), 970–983.

Stone, C. J. (1982). Optimal global rates of convergence for nonparametric regression. *The Annals of Statistics 10*(4), 1040–1053.

Stone, M. (1974). Cross-validatory choice and assessment of statistical predictions (with discussion). *Journal of the Royal Statistical Society, Series B 36*, 111–147.

Stone, M. (1977). An asymptotic equivalence of choice of model by cross-validation and Akaike's criterion. *Journal of the Royal Statistical Society, Series B 39*, 44–47.

Therneau, T. (2015). A package for survival analysis in S. r package version 2.38. *URL http://CRAN. R-project. org/package= survival. Box.*

Therneau, T. M. and P. M. Grambsch (2000). *Modeling Survival Data: Extending the Cox Model.* Springer Science & Business Media.

Tutz, G. and H. Binder (2006). Generalized additive modeling with implicit variable selection by likelihood-based boosting. *Biometrics 62*(4), 961–971.

Tweedie, M. (1984). An index which distinguishes between some important exponential families. In *Statistics: Applications and New Directions: Proc. Indian Statistical Institute Golden Jubilee International Conference*, pp. 579–604.

Venables, B. and B. R. Ripley (2002). *Modern Applied Statistics in S* (4th ed.). Springer.

Verbyla, A. P., B. R. Cullis, M. G. Kenward, and S. J. Welham (1999). The analysis of designed experiments and longitudinal data by using smoothing splines. *Journal of the Royal Statistical Society: Series C (Applied Statistics) 48*(3), 269–311.

Wahba, G. (1980). Spline bases, regularization, and generalized cross validation for solving approximation problems with large quantities of noisy data. In E. Cheney (Ed.), *Approximation Theory III.* London: Academic Press.

Wahba, G. (1981). Spline interpolation and smoothing on the sphere. *SIAM Journal on Scientific and Statistical Computing 2*(1), 5–16.

Wahba, G. (1985). A comparison of GCV and GML for choosing the smoothing parameter in the generalized spline smoothing problem. *The Annals of Statistics 13*(4), 1378–1402.

Wahba, G. (1990). *Spline Models for Observational Data.* Philadelphia: SIAM.

Wang, H. and M. G. Ranalli (2007). Low-rank smoothing splines on complicated domains. *Biometrics 63*(1), 209–217.

Wang, Y. (1998a). Mixed effects smoothing spline analysis of variance. *Journal of the Royal Statistical Society, Series B 60*, 159–174.

Wang, Y. (1998b). Smoothing spline models with correlated random errors. *Journal of the American Statistical Association 93*(441), 341–348.

Watkins, D. S. (1991). *Fundamentals of Matrix Computation.* New York: Wiley.

Wendelberger, J. (1981). *Smoothing noisy data with multidimensional splines and generalized cross validation.* Ph. D. thesis, Departement of Statistics, University of Wisconsin.

Whitehead, J. (1980). Fitting Cox's regression model to survival data using GLIM. *Applied Statistics 29*(3), 268–275.

Wood, S. N. (2000). Modelling and smoothing parameter estimation with multiple quadratic penalties. *Journal of the Royal Statistical Society, Series B 62*, 413–428.

Wood, S. N. (2003). Thin plate regression splines. *Journal of the Royal Statistical Society, Series B 65*, 95–114.

Wood, S. N. (2004). Stable and efficient multiple smoothing parameter estimation for generalized additive models. *Journal of the American Statistical Association 99*, 673–686.

Wood, S. N. (2006a). Low-rank scale-invariant tensor product smooths for generalized additive mixed models. *Biometrics 62*(4), 1025–1036.

Wood, S. N. (2006b). On confidence intervals for generalized additive models based on penalized regression splines. *Australian & New Zealand Journal of Statistics 48*(4), 445–464.

Wood, S. N. (2008). Fast stable direct fitting and smoothness selection for generalized additive models. *Journal of the Royal Statistical Society: Series B (Statistical Methodology) 70*(3), 495–518.

Wood, S. N. (2011). Fast stable restricted maximum likelihood and marginal likelihood estimation of semiparametric generalized linear models. *Journal of the Royal Statistical Society: Series B (Statistical Methodology) 73*(1), 3–36.

Wood, S. N. (2013a). On p-values for smooth components of an extended generalized additive model. *Biometrika 100*(1), 221–228.

Wood, S. N. (2013b). A simple test for random effects in regression models. *Biometrika 100*(4), 1005–1010.

Wood, S. N. (2015). *Core Statistics*. Cambridge University Press.

Wood, S. N. (2016a). Just another Gibbs additive modeller: Interfacing JAGS and mgcv. *Journal of Statistical Software 75*(7).

Wood, S. N. (2016b). P-splines with derivative based penalties and tensor product smoothing of unevenly distributed data. *Statistics and Computing*, 1–5.

Wood, S. N., M. V. Bravington, and S. L. Hedley (2008). Soap film smoothing. *Journal of the Royal Statistical Society: Series B (Statistical Methodology) 70*(5), 931–955.

Wood, S. N. and M. Fasiolo (2017). A generalized Fellner-Schall method for smoothing parameter optimization with application to Tweedie location, scale and shape models. *Biometrics* (in press).

Wood, S. N., Y. Goude, and S. Shaw (2015). Generalized additive models for large data sets. *Journal of the Royal Statistical Society: Series C (Applied Statistics) 64*(1), 139–155.

Wood, S. N., Z. Li, G. Shaddick, and N. H. Augustin (2017). Generalized additive models for gigadata: modelling the UK black smoke network daily data. *Journal of the American Statistical Association* (in press).

Wood, S. N., N. Pya, and B. Säfken (2016). Smoothing parameter and model selection for general smooth models (with discussion). *Journal of the American Statistical Association 111*, 1548–1575.

Wood, S. N., F. Scheipl, and J. J. Faraway (2013). Straightforward intermediate rank tensor product smoothing in mixed models. *Statistics and Computing 23*(3), 341–360.

Yee, T. W. and A. G. Stephenson (2007). Vector generalized linear and additive extreme value models. *Extremes 10*(1-2), 1–19.

Yee, T. W. and C. Wild (1996). Vector generalized additive models. *Journal of the Royal Statistical Society. Series B (Methodological) 58*(3), 481–493.

Yu, D. and K. K. Yau (2012). Conditional Akaike information criterion for generalized linear mixed models. *Computational Statistics & Data Analysis 56*(3), 629–644.

Index

χ^2 test, 109
$|\cdot|_+$, 263

additive model, 174–180
  fitting, 177
age of Universe, 4
aggregate, 64
AIC, 76
  and smoothing parameter uncertainty, 301–304
  corrected, 304
  cult of, 110
  generalized, 262
  GLMs, 110, 128
  in general , 414–416
  linear model, 52
  marginal and conditional, 301
  warning, 352
AIC, 190
analysis of deviance, 108, 109, 129, 141
ANOVA, 16, 45, 56, 59, 67
  decomposition of function, 233, 234, 344
anova, 45, 190
  glm, 134
AR model, 52, 396
  with bam, 373
augmented linear model, 169, 177, 247, 317
autocorrelation, 371

B-spline, 204
  basis, 204
  cSplineDes, 205
  splineDesign, 205
backfitting, 318–320, 398
balanced data, 65
bam, 352
  random effects, 367
  underlying methods, 290
  with AR1 model, 373, 396
basis, 201
  B-spline, 204
  choice of dimension, 185, 242
  cubic regression spline, 201
  expansion, 162
  polynomial, 162
  tensor product, 227–229
  thin plate spline, 216
Bayesian covariance matrix, 212, 276, 294
  smoothing uncertainty corrected, 303, 343
Bayesian smoothing model, 172–174, 178, 239–242, 293, 375
  simulation from posterior, 300, 342, 357, 387
  simulation with JAGS, 374
  well calibrated inference, 294
BIC, 262
big data GAM methods, 290–294, *see* bam
binary data, 361
  residual checking, 156
binomial distribution, 362
binomial family, 362
Bonferroni correction, 98
by variable, 326, 336, 338
  and centering constraints, 366
  factor and ordered, 327

canonical parameter, 103, 111
causation, 36
central limit theorem, 405

Cholesky decomposition, 422
co-linearity, 36
compact support, 165
confidence interval
  and transformation, 9
  calculation, 9, 35, 135
  definition, 9
  for smoothing parameter, 373
  GLM, 108, 135
  GLRT inversion, 159
  limitations, 296
  linear model, 8, 14, 35
  Nychka's coverage argument, 294
  performance, 297
  whole function, 300
  Wilk's, 159
  zero width, 211, 240
confounding, 36, 62
consistency, 267
  MLE, 409
  regression spline, 200
contingency tables, 138
`contour`, 58
contrast matrix, 49
convergence rates, 199, 267
Cook's distance, 26
cool stuff
  air pollution, 346–353
  Bayesian smoothing, 172
  Cox model and Poisson regression, 116
  Duchon splines, 221
  GCV derivation, 256–260
  general Fellner-Schall method, 271
  GLM geometry, 119–124
  linear model geometry, 19–22
  linear model theory, 11–16
  multinomial prostate screening, 393
  natural parameterization of smooth, 211–214
  non-linear fitting geometry, 54–56
  Nychka's coverage argument, 294
  penalized fitting geometry, 316–318
  random effects p-values, 309
  smooth ANOVA models, 232
  smooth survival models, 377–382
  soap film smoothing, 223
  Sole egg GAMM, 365–370
  sperm competition, 22–36
  Swiss extreme rainfall, 384
covariance matrix, 420
  computing diagonal, 342
  of linear transformation, 420
Cox proportional hazards model, 116, 136, 377
  `cox.ph` family, 377
  Poisson equivalent, 117, 137, 380
  residuals, 119, 378, 381
  time varying, 116, 380
Cox proportional hazards model., 283
cross validation, 171
  deviance based, 261
  double, 261
  failure of, 320
  generalized, 258
  generalized approximate, 261
  k-fold, 257
  leave-one-out, *see* OCV
  leave-several-out, 257
cubic regression spline, 201
  penalty, 202
cubic spline, 196, 198, 200, 243
  approximation properties, 197
  natural, 196
  penalty, 167, 198
  smoothest interpolant, 196
cyclic spline, 202, 371
  `cSplineDes`, 205
  penalty, 203

daily temperature, 371
Datasets, xix
  `gamair` package, xix, 4
  AIDS in Belgium, 131, 159
  `aral`, 223
  belief in afterlife, 138

bird, 363
blowfly, 401
bone, 136
brain, 328
cairo, 371
chicago, 346
chl, 402
co2s, 400
engine wear, 165
Florida death penalty, 157
fuel efficiency mpg, 388
harrier, 158
heart attacks and CK, 125
hubble, 4, 398
ipo, 400
larynx, 240
mack, 354
MASS
mcycle, 328, 384, 398
Rubber, 57
nlme
Gun, 99
Loblolly, 87
Machines, 72, 91, 98
Oxide, 98
Rail, 68, 87
Soybean, 160
octane, 390
R
InsectSprays, 59
ldeaths, 159
PlantGrowth, 43
trees, 59, 178, 182, 183
warpbreaks, 58
retinopathy, 343
sitka, 374
sole, 142, 151, 366
sperm.comp1, 22
sperm.comp2, 22, 31
stomata, 62
survival
pbc, 377
Swiss rain swer, 384
wesdr, 343
where to obtain, xix
wine, 401
Demmler-Reinsch parameterization, 211
design matrix, 10
determinant
derivative, 278
generalized, 263
stable computation, 264, 278
deviance, 108
approximate distribution, 109
null, 128
proportion explained, 128
residual, 128
scaled, 109
dilogarithm, 223
discretization of covariates, 293
distributed lag model, 352
distributional regression, 383
gaulss family, 383
generalized extreme value distribution, 384
gevlss family, 384
drop1, 45

EDF, 83
alternative definition, 252
and basis dimension, 186, 243
GAM, 252
natural parameterization, 212
smoothness uncertainty corrected, 304
termwise, 252
effective degrees of freedom, *see* EDF
eigen decomposition, 217, 424
eigenvalue, 424
eigenvector, 424
EM algorithm, 84
Euclidean norm, 12
exponential family, 101, 103–105
mean, 103
variance, 105
extrapolation
dangers of, 400, 450, 453

F-ratio statistic, 15, 29, 109
F-test, 15, 45, 67, 71, 109

factor variables, 11, 38, 45
  contrasts, 48
  dummy variables, 11, 38
  identifiability, 39
  interactions, 41
  use in R, 43–45
Fellner-Schall method, 269–272
  general likelihood, 271
Fisher scoring, 77, 107
Fisher weights, 106
fitted values
  distribution of, 17
fixed effects
  distribution, 80, 151
follow up comparisons, 98
formula, 46
  I(), 27, 46
functional data analysis, 390–397
  function on scalar regression, 395
  scalar on function regression, 390

GACV, 262
GAM, 161, 180–181
  backfit, 318–320
  Bayesian model
    non-linear functions, 300
  bias, 246
  coefficient estimators, 276
  constraints
    identifiability, 211, 234, 250
    shape, 208, 397
  hypothesis testing, 337
  influence matrix, 274
  motivation, 147
  proximity of penalized and unpenalized versions, 314
  setup, 249
  termwise p-values, 190, 337
gam, 182, 183, 185, 186, 330, 331, 333, 335, 336, 338, 347, 356, 397
  and random effects, 94
  anova.gam, 189, 339
  arguments, 185
  controlling, 184–187
    fitting, 185
  gamma argument, 186
  knots, 362
  knots argument, 362
  multivariate, 388
  parametric terms, 189
  plot.gam, 183, 186, 188
    all.terms, 189
    contour plot interpretation, 187
    n argument, 347
    too.far, 187
  predict.gam, 339, 340, 343, 358
    lpmatrix, 341, 387
  prediction matrix, 341
  print.gam, 183, 188
  smooth specification, 325
  summary.gam, 190
  summation convention, 326, 392
  tensor product smooths, 188
  TPRS terms, 188
  variance components, 368
gam.check, 330, 331, 347
gam.method, 331
gam.vcomp, 368
GAMLSS, *see* distributional regression
GAMM
  inference, 289
  with R, 365
gamm
  convergence problems, 88
  correlation structures, 371
  random argument, 87, 367
  use of, 365
gamma distribution, 188, 333
Gamma family, 188, 333
Gauss Markov theorem, 18
Gauss-Newton method, 54
Gaussian Markov random field, 240
Gaussian process regression, 241, 359, 363
Gaussian random field, 239
GCV, 171, 258
  compared to REML, 267

derivatives, 274–276
deviance based, 261
failure of, 444
GAM case, 181
single penalty case, 321
generalized additive model, *see* GAM
generalized cross validation, *see* GCV
generalized extreme value distribution, 384
generalized likelihood ratio test, 75, 108, 129, 411–414
assumptions, 90, 91
generalized linear mixed model, *see* GLMM
generalized linear model, *see* GLM
generalized smoothing splines, 243–245
geographic regression, *see* varying coefficient model, 360
geometry
generalized linear model, 119–124
IRLS, 121
IRLS convergence, 122
least squares, 19
linear model, 19
general covariance structure, 51
nested models, 21
non-linear least squares, 55
orthogonal fitting, 20
penalized least squares, 316–318
`geoR` package, 364
GLM, 101
binomial data, 126
estimation, 105
estimator distribution, 107
fitted values
properties, 112
geometry, 119
hypothesis testing, 108
in R, 124–136, 139–141, 143–146
likelihood, 105
Poisson data, 131
quasi-likelihood, 113–115
residuals, 112
deviance, 113
Pearson, 112
working, 113
theory, 102–115
`glm`, 122, 124, 127, 129, 132, 133, 139, 140, 143
`AIC`, 128
`anova`, 141, 145
`summary.glm`, 143
`update`, 145
GLMM, 147, 154
in R, 151
`glmmPQL`, 151
convergence problems, 88

hat matrix, 16
hazard, 116, 284
baseline, 118, 286
cumulative, 118, 138
help pages in R, 191
Householder matrix, 421
Hubble's law, 1
hypothesis testing
GAM, 339
approximate GLRT, 313
smooth terms, 305–309
unpenalized components, 304
GLM, 108, 129, 134, 141
GLRT, 411–414
linear model, 7–8, 13, 14, 45
random effects, 309–312
smooth vs. linear, 313

idempotency, 16
implicit differentiation, 279
influence matrix, 16, 168, 274
properties, 16, 399
information, 407
information matrix, 108, 409
empirical or observed, 411
interaction
factors, 41
smooth, 227, 344

invariance
    and cross validation, 257–260
    GLRT, 414
    of MLEs, 405
    rotational, 187, 219, 227, 361
    scale, 187, 227
IRLS
    convergence, 122
    derivation, 105
    geometry, 121
    penalized, 149, 180, 251, 272, 321
        low memory footprint, 290

Jacobian, 55, 119
`jagam`, 300, 375
Jensen's inequality, 407

knots, 166, 219
Kriging, 241
Kronecker product, 421
Kullback-Leibler divergence, 75, 415

Lanczos iteration, 218, 426
Laplace approximation, 147, 412
law of large numbers, 405
least squares, 12
    estimators, 12
    geometry, 19
    non-independent data, 50
    non-linear, 54
    residual variance estimation, 57
    with negative weights, 252–254
likelihood
    consistency of MLEs, 409
    distribution of estimators, 410
    linear model, 49
    marginal, 79, 264, 345
    partial, 117
    penalized, 148, 180, 250
    profile, 79, 81
    properties of expected log, 406
    restricted, *see* REML
    theory , 405–418
linear constraints, 40, 47, 59, 211, 233
    GAM, 250
linear contrasts, 48
linear functional terms, 250
    summation convention, 326, 351, 391, 392
linear mixed model, 61
    general case, 77–92
        grouped data, 86
        `R` functions, 86–92
    model selection, 85
    simple balanced case, 61–74
        2 way design, 69–73
        aggregated model, 64, 67, 71
        estimation in R, 68, 72
        general principles, 65–66, 74
        hypothesis testing, 72
        interactions, 70
        oneway ANOVA, 66–69
        variance components, 65, 68, 72, 73
    why bother, 61–66
linear model, 2, 9
    ANOVA, 14
    checking, 25
    coefficients, 27
    estimator distribution, 13
    F-ratio, 14
    fitted values, 16
        properties, 57
    `formula`, 5, 46
    likelihood, 49
    model matrix, 10, 24
    model selection, 30–35, 45
        stepwise, 58
    polynomial, 58
    prediction, 36
    residuals, 16
    t-ratio, 13, 28
    theory , 11–18
    traditional results, 17
linear predictor, 101
link function, 101
    canonical, 111
`lm`, 4, 6, 24, 27, 30–35, 44, 45, 166, 169
    associated functions, 23

for mixed models, 63
plot.lm, 25
print.lm, 27
step, 58
summary.lm, 28
use of, 24
lme
convergence problems, 88
correlation argument, 89
form of model, 86
intervals, 372
lmeControl, 88
msMaxIter, 88
niterEM, 88
plot.lme, 89
use of, 87
lme4 package, 86, 93
location-scale models, *see* distributional regression
log-linear models, 138
logistic regression, 125, 344
lpmatrix, 341

Mallows' statistic, 52, 255
MAP estimates, 80, 148
marginal likelihood, 79, 264, 345, 392
matrix
determinant, 425
differentiating, 420
efficient computation, 419
partitioned inverse, 80
positive definite, 425
positive semi-definite, 425
square root, 274
trace, 425
maximum likelihood estimation, 74, 405
linear mixed model, 78
linear model, 49
mean square error, 169, 255
metric variable, 38
mgcv, 94, 182
help pages, 191
mixed effects model
general case, 88
model formula, *see* formula
model matrix, 10, 78
model selection
drop1, 45
examples, see chapter 7, 325
GAM, 214, 255–269, 301–315
GLM, 108–110, 145
GLMM, 153
linear mixed model, 85
stepwise, 30
multinomial distribution
multinom family, 394
multinomial model, 393
multiple comparisons, 99
multivariate additive model
mvn family, 388

natural spline, 196
negative binomial
nb family, 356
negative binomial distribution, 116
nesting, 62
Newton's method, 77, 280
GLM , 106
REML optimization, 277
smoothing parameter optimization, 274
nlme package, 86
null space, 168, 221

OCV, 170, 256–258
not invariant, 257
offset, 143, 355
offset, 159, 356
one s.e. rule, 268, 297
ordered categorical
ocat family, 393
ordered categorical model, 392
ordinary cross validation, *see* OCV
orthogonal matrix, 421
outliers
dealing with, 27, 32, 329, 347
overdispersion, 115, 143
overfitting
checking for, 402

P-spline, 204, 246, 321
  penalty, 206
p-value, 29, 141
  interpretation, 29
partial residuals, *see* residuals, partial
Pearson statistic, 110
penalized regression, 167
  estimator, 168
  fitting, 169
  shrinkage, 212
penalty
  adaptive, 208
  cubic spline, 198
  discrete, 167, 206, 209
    diff, 169
  GAM, 250
  null space, 168, 214, 216, 221, 315
  tensor product, 229, 237
performance iteration, 272
pivoting, 423
poisson family, 356
Poisson regression, 131, 347
polynomial, 88, 162
polynomial basis, 164
polynomial model
  ensuring stability, 88, 143
posterior simulation, 300, 342, 387
PQL, 149
  justification, 150
  smoothing parameter estimation, 272
prediction
  excluding terms, 154, 368
  lpmatrix, 341
  random effects, 80
  response scale, 339
  term-wise, 340
prediction error, 170, 256
PredictMat, 327
propagation of errors, 420

QQ-plot, 26
QR decomposition, 12, 58, 234, 422
  iterative accumulation, 290
quasi family, 152
quasi-information, 417
quasi-likelihood, 114
  penalized, *see* PQL
  theory , 416–418

R model formulae, 46
R^2, 29
  adjusted, 29, 190
random effects, 61, 64
  equivalence with smooths, 239
  gamm, 366
  jagam and JAGS, 374
  mgcv, 94, 155
  predictions, 80, 151
random fields, 240
randomization, 37
rank deficiency, 233–234, 274
reflector matrix, *see* Householder matrix
REML, 83, 150, 264
  compared to GCV, 267
  Fellner-Schall optimization, 270
  Laplace approximation, 147
  smoothing parameter estimation, 277, 362, 396
reproducing kernel, 244
reproducing kernel Hilbert space, 243
residual plots
  against fitted values, 26
  Cook's distance, 26
  GLMM, 153
  GLMs, 128
  linear mixed model, 73
  linear model, 6
  QQ-plot, 26
  scale location, 26
  spatial, 364
  zero line, 146
residual sum of squares, 13
residuals
  autocorrelation, 371, 400
  checking binary, 156, 363
  deviance, 113, 119
    distribution, 113

distribution of, 17
GAM, 330, 347
GLMs, 112
improvement by transformation, 330–332
linear model, 16
Martingale, 119, 137
observed unbiasedness, 112
partial, 183, 319
Pearson, 112
properties, 57
standardized, 17, 26

`s`, 187
`bs` argument, 185
`k` argument, 185
choice of, 242
scale parameter, 103, 109, 320
estimation, 110, 251
Fletcher estimate, 111
Pearson estimate, 111
scope of statistics, 2
semi-parametric models, 184, 189
shrinkage smoother, 214
signal regression, 391
single index model, 349, 350
singular value decomposition, 425
singularity, 274
`smoothCon`, 327
smoothing bias, 212, 294, 442
and un-penalized terms, 267
smoothing parameter, 168, 198
computation, 269–293
confidence interval, 373
estimation, 169–174, 179, 255–293, 321
efficiently for single smooth, 321
Fellner-Schall method, 269–272
force equality, 326
initialization, 287
nested optimization, 280–287
performance iteration/PQL, 272–280
uncertainty, 303, 343
smooths, *see also* spline
ANOVA decomposition, 233, 344
as mixed model terms, 289
equivalence with random effects, 239
Gaussian process, 363
isotropic, 214–227
isotropic versus tensor, 238
of several variables, 187–188, 214–242
prediction matrix, 327
running mean, 400
smooth constructors, 327
soap film, 223–227, 370
specification in `mgcv`, 325
tensor product, *see* tensor product smooth
sole eggs, 142
spectral decomposition, *see* eigen decomposition
spline, *see* aso cubic spline195
adaptive, 207
B-spline, 204
basis size choice, 242
comparison with polynomial, 440
cubic, 196, 198, 200, 245
cubic regression, 201
cubic smoothing, 198, 243
cyclic, 202
Duchon, 221, 223, 362
equivalent kernel, 399
natural parameterization, 211–214, 245, 246
on the sphere, 222, 359
P-spline, 204, 246
continuous penalty, 207
cyclic, 206
discrete penalty, 206
piecewise linear, 164
random subject specific, 376
reproducing kernel approach, 243–245
shape constrained, 208, 234
smoothing, 195
tensor product, 227

theoretical properties, 195–198, 440
thin plate, 215
thin plate regression (TPRS), 215
`step`, 33
`summary`
`gam`, 190
`glm`, 134, 135
`lm`, 28
summation convention, 351
survival analysis, *see* Cx proportional hazards model377
survival function, 118, 379, 381
confidence interval, 119

Taylor expansion, 76, 147
`te`, 187, 335
`bs` argument, 367
`k` argument, 188
`te()`, 352
tensor product smooth, 227–237
alternative penalties, 232
as mixed model term, 288–289
basis , 227–229
comparison with TPRS, 238, 334–336
marginal bases, 228
marginal penalties, 229
penalty, 229–232, 237
penalty re-parameterization, 231
pure interaction, 233, 344
shape constrained, 234
tent function
basis, 165
model matrix, 165
tent functions, 165
thin plate regression splines, *see* TPRS
thin plate spline, 215, 247
basis, 216
penalty, 216
`ti()`, 344
TPRS, 215–221
basis, 217, 220
comparison with tensor product, 238, 334–336
construction, 217
efficient computation, 218
knot based, 219, 247
properties, 218, 399, 448
speeding up, 362
transformation
to stabilize variance, 26, 330–332
Tweedie distribution, 115
`tw` family, 355

UBRE, 255
derivatives, 274–276
deviance based, 261
`update`, 145

validation set, 402
variance estimation, 13, 65, 68, 71, 252, 303, 368
variance function, 105
variogram, 364
varying coefficient model, 326, 336
mixed effects version, 366
`vis.gam`, 333, 334, 337
with uncertainty, 345

Wald test, 414
wiggliness penalty, 198